A Heat Transfer Textbook

FOURTH EDITION

John H. Lienhard IV
Department of Mechanical Engineering
University of Houston

John H. Lienhard V
Department of Mechanical Engineering
Massachusetts Institute of Technology

Dover Publications, Inc
Mineola, New York

Bibliographical Note

This Dover edition, first published in 2011, is a corrected, revised, and updated fourth edition of the work originally published in 1981 by Prentice-Hall, Inc., Englewood Cliffs, New Jersey.

Library of Congress Cataloging-in-Publication Data

Lienhard, John H., 1930–
A heat transfer textbook / John H. Lienhard IV and John H. Lienhard V.
p. cm.
"Corrected, revised, and updated fourth edition of the work originally published in 1981."
Includes bibliographical references and index.
ISBN-13: 978-0-486-47931-6
ISBN-10: 0-486-47931-5
1. Heat—Transmission. I. Lienhard, John H., 1961– II. Title.

TJ260.L445 2011
621.402'2—dc22

2010032504

Manufactured in the United States by Courier Corporation
47931503 2014
www.doverpublications.com

Preface

From the first edition of *A Heat Transfer Textbook* in 1981, this book was meant to serve students as they set out to understand heat transfer in its many aspects. Whether the reader studies independently or in a classroom is beside the point, since learning (in either case) means formulating and surmounting one's own questions. Where the book succeeds, it will be because students encounter a series of "Oh, now I see!" moments.

The original edition went through several printings and was followed in 1987 by more printings of a second edition—this time with a new chapter on Mass Transfer by John H. Lienhard V who had played only a minor role in John H. Lienhard IV's first edition. After that, we each became involved in other pursuits and were unable to prepare the needed third edition. So the book went out of print.

In the late 1990s, we developed a third edition which we decided to distribute free of charge on the Internet. We obtained funding from the Dell Star Program, did major updating, and posted it in 2000 as a part of MIT's then new OpenCourseWare initiative. In that form, the book underwent many subsequent revisions and changes as we moved to keep up-to-date with rapidly-changing technology. Indeed, the early versions of the third edition are substantially different from the last versions. By 2010, these Internet versions had a quarter million downloads by people on all seven continents and in essentially every country in the world. We also published a small number of paperback versions of the third edition through Phlogiston Press.

Now Dover Publications, Inc., has published this affordable fourth edition. In it, we've made many additional interstitial changes to the last version of the third edition (specifically, Version 31), including corrections, updates, and various text edits. We are calling it a fourth edition to reflect ten years of accumulated revision of the third edition.

With this edition we continue offering engineering juniors, seniors, and first-year graduate students their grounding in heat transfer - in conduction, convection, radiation, phase-change, and an introduction to the kindred subject of mass transfer. We have designed the book in such a way as to accommodate differing levels according to the instructor's use of it (or the student's independent selections). Accordingly, each element of the subject begins simply and may be carried through to the more sophisticated material as the instructor (or the reader) chooses.

We have strived to combine clarity with unusual care to get things right and complete. We take care to deal with the implications, limitations, and meaning of the many aspects of the subject. We have also worked to connect the subject to the real world that it serves, and to develop insight into the many phenomena connected with the subject.

In the interest of grounding students in real-world issues, we begin the book with a three-chapter introduction that takes them through the essential modes of heat transfer and gives them an understanding of the function and design of heat exchangers. With that background, students find the later and more complex aspects of the subject much more meaningful.

Thus, the first three chapters provide background that is needed throughout the chapters that follow. But only the earlier parts of subsequent chapters accumulate in this way. For example, we freely use the dimensional analysis that appears early in Chapter 4. However, while the material on fin design at the end of that chapter is an important application, much used in many industries, it is not material that one has to know to continue through subsequent chapters.

These goals are much the same as we've pursued over the past 30 years. However, we've taken the occasion of this new edition as an opportunity to refocus and renew our purposes. The Internet version will still be available to people who want it, say, available on their laptops (http://ahtt.mit.edu). But Dover Publications now offers this companion hard copy to those of us who want the convenience of a physical book.

We owe thanks to many people. These include our colleagues and students at MIT, the University of Houston, and elsewhere who have provided suggestions, advice, and corrections; and most especially the many thousands of people worldwide who have emailed us with thanks and encouragement to continue this project.

JHL IV, University of Houston
JHL V, Massachusetts Institute of Technology
December 2010

Contents

Part I

The General Problem of Heat Exchange

1. Introduction

The radiation of the sun in which the planet is incessantly plunged, penetrates the air, the earth, and the waters; its elements are divided, change direction in every way, and, penetrating the mass of the globe, would raise its temperature more and more, if the heat acquired were not exactly balanced by that which escapes in rays from all points of the surface and expands through the sky. ***The Analytical Theory of Heat,*** **J. Fourier**

1.1 Heat transfer

People have always understood that something flows from hot objects to cold ones. We call that flow *heat.* In the eighteenth and early nineteenth centuries, scientists imagined that all bodies contained an invisible fluid which they called *caloric.* Caloric was assigned a variety of properties, some of which proved to be inconsistent with nature (e.g., it had weight and it could not be created nor destroyed). But its most important feature was that it flowed from hot bodies into cold ones. It was a very useful way to think about heat. Later we shall explain the flow of heat in terms more satisfactory to the modern ear; however, it will seldom be wrong to imagine caloric flowing from a hot body to a cold one.

The flow of heat is all-pervasive. It is active to some degree or another in everything. Heat flows constantly from your bloodstream to the air around you. The warmed air buoys off your body to warm the room you are in. If you leave the room, some small buoyancy-driven (or *convective*) motion of the air will continue because the walls can never be perfectly isothermal. Such processes go on in all plant and animal life and in the air around us. They occur throughout the earth, which is hot at its core and cooled around its surface. The only conceivable domain free from heat flow would have to be isothermal and totally isolated from any other region. It would be "dead" in the fullest sense of the word — devoid of any process of any kind.

The overall driving force for these heat flow processes is the cooling (or leveling) of the thermal gradients within our universe. The heat flows that result from the cooling of the sun are the primary processes that we experience naturally. The conductive cooling of Earth's center and the radiative cooling of the other stars are processes of secondary importance in our lives.

The life forms on our planet have necessarily evolved to match the magnitude of these energy flows. But while most animals are in balance with these heat flows, we humans have used our minds, our backs, and our wills to harness to harness and control energy flows that are far more intense than those we experience naturally[1]. To emphasize this point we suggest that the reader make an experiment.

Experiment 1.1

Generate as much power as you can, in some way that permits you to measure your own work output. You might lift a weight, or run your own weight up a stairwell, against a stopwatch. Express the result in watts (W). Perhaps you might collect the results in your class. They should generally be less than 1 kW or even 1 horsepower (746 W). How much less might be surprising.

Thus, when we do so small a thing as turning on a 150 W light bulb, we are manipulating a quantity of energy substantially greater than a human being could produce in sustained effort. The power consumed by an oven, toaster, or hot water heater is an order of magnitude beyond our capacity. The power consumed by an automobile can easily be three orders of magnitude greater. If all the people in the United States worked continuously like galley slaves, they could barely equal the output of even a single city power plant.

Our voracious appetite for energy has steadily driven the intensity of actual heat transfer processes upward until they are far greater than those normally involved with life forms on earth. Until the middle of the thirteenth century, the energy we use was drawn indirectly from the sun

[1]Some anthropologists think that the term *Homo technologicus* (those who use technology) serves to define human beings, as apart from animals, better than the older term *Homo sapiens* (those who are wise). We may not be as much wiser than the animals as we think we are, but only we do serious sustained tool making.

using comparatively gentle processes — animal power, wind and water power, and the combustion of wood. Then population growth and deforestation drove the English to using coal. By the end of the seventeenth century, England had almost completely converted to coal in place of wood. At the turn of the eighteenth century, the first commercial steam engines were developed, and that set the stage for enormously increased consumption of coal. Europe and America followed England in these developments.

The development of fossil energy sources has been a bit like Jules Verne's description in *Around the World in Eighty Days* in which, to win a race, a crew burns the inside of a ship to power the steam engine. The combustion of nonrenewable fossil energy sources (and, more recently, the fission of uranium) has led to remarkably intense energy releases in power-generating equipment. The energy transferred as heat in a nuclear reactor is on the order of *one million watts per square meter.*

A complex system of heat and work transfer processes is invariably needed to bring these concentrations of energy back down to human proportions. We must understand and control the processes that divide and diffuse intense heat flows down to the level on which we can interact with them. To see how this works, consider a specific situation. Suppose we live in a town where coal is processed into fuel-gas and coke. Such power supplies used to be common, and they may return if natural gas supplies ever dwindle. Let us list a few of the process heat transfer problems that must be solved before we can drink a glass of iced tea.

- A variety of high-intensity heat transfer processes are involved with combustion and chemical reaction in the gasifier unit itself.

- The gas goes through various cleanup and pipe-delivery processes to get to our stoves. The heat transfer processes involved in these stages are generally less intense.

- The gas is burned in the stove. Heat is transferred from the flame to the bottom of the teakettle. While this process is small, it is intense because boiling is a very efficient way to remove heat.

- The coke is burned in a steam power plant. The heat transfer rates from the combustion chamber to the boiler, and from the wall of the boiler to the water inside, are very intense.

- The steam passes through a turbine where it is involved with many heat transfer processes, including some condensation in the last stages. The spent steam is then condensed in any of a variety of heat transfer devices.

- Cooling must be provided in each stage of the electrical supply system: the winding and bearings of the generator, the transformers, the switches, the power lines, and the wiring in our houses.

- The ice cubes for our tea are made in an electrical refrigerator. It involves three major heat exchange processes and several lesser ones. The major ones are the condensation of refrigerant at room temperature to reject heat, the absorption of heat from within the refrigerator by evaporating the refrigerant, and the balancing heat leakage from the room to the inside.

- Let's drink our iced tea quickly because heat transfer from the room to the water and from the water to the ice will first dilute, and then warm, our tea if we linger.

A society based on power technology teems with heat transfer problems. Our aim is to learn the principles of heat transfer so we can solve these problems and design the equipment needed to transfer thermal energy from one substance to another. In a broad sense, all these problems resolve themselves into collecting and focusing large quantities of energy for the use of people, and then distributing and interfacing this energy with people in such a way that they can use it on their own puny level.

We begin our study by recollecting how heat transfer was treated in the study of thermodynamics and by seeing why thermodynamics is not adequate to the task of solving heat transfer problems.

1.2 Relation of heat transfer to thermodynamics

The First Law with work equal to zero

The subject of thermodynamics, as taught in engineering programs, makes constant reference to the heat transfer between systems. The First Law of Thermodynamics for a closed system takes the following form on a

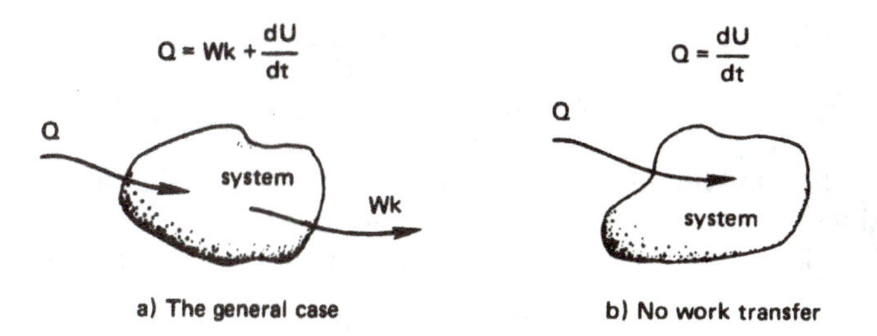

Figure 1.1 The First Law of Thermodynamics for a closed system.

rate basis:

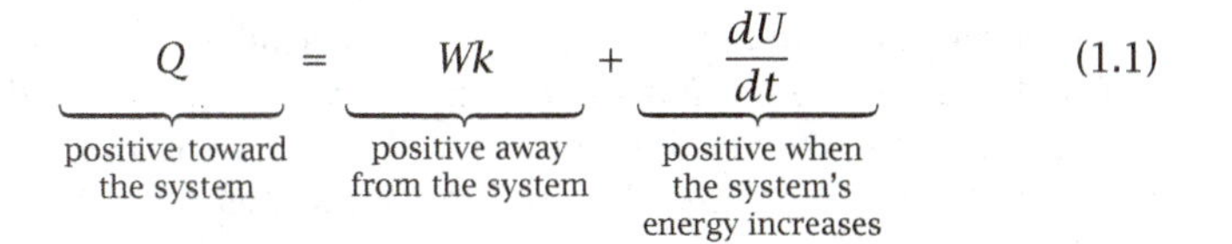

$$\underbrace{Q}_{\substack{\text{positive toward}\\ \text{the system}}} = \underbrace{Wk}_{\substack{\text{positive away}\\ \text{from the system}}} + \underbrace{\frac{dU}{dt}}_{\substack{\text{positive when}\\ \text{the system's}\\ \text{energy increases}}} \tag{1.1}$$

where Q is the heat transfer rate and Wk is the work transfer rate. They may be expressed in joules per second (J/s) or watts (W). The derivative dU/dt is the rate of change of internal thermal energy, U, with time, t. This interaction is sketched schematically in Fig. 1.1a.

The analysis of heat transfer processes can generally be done without reference to any work processes, although heat transfer might subsequently be combined with work in the analysis of real systems. If $p\,dV$ work is the only work that occurs, then eqn. (1.1) is

$$Q = p\,\frac{dV}{dt} + \frac{dU}{dt} \tag{1.2a}$$

This equation has two well-known special cases:

$$\text{Constant volume process:} \qquad Q = \frac{dU}{dt} = mc_v\,\frac{dT}{dt} \tag{1.2b}$$

$$\text{Constant pressure process:} \qquad Q = \frac{dH}{dt} = mc_p\,\frac{dT}{dt} \tag{1.2c}$$

where $H \equiv U + pV$ is the enthalpy, and c_v and c_p are the specific heat capacities at constant volume and constant pressure, respectively.

When the substance undergoing the process is incompressible (so that V is constant for any pressure variation), the two specific heats are equal:

$c_v = c_p \equiv c$. The proper form of eqn. (1.2a) is then

$$\boxed{Q = \frac{dU}{dt} = mc\,\frac{dT}{dt}} \tag{1.3}$$

Since solids and liquids can frequently be approximated as being incompressible, we shall often make use of eqn. (1.3).

If the heat transfer were reversible, then eqn. (1.2a) would become[2]

$$\underbrace{T\frac{dS}{dt}}_{Q_{\text{rev}}} = \underbrace{p\frac{dV}{dt}}_{Wk_{\text{rev}}} + \frac{dU}{dt} \tag{1.4}$$

That might seem to suggest that Q can be evaluated independently for inclusion in either eqn. (1.1) or (1.3). However, it cannot be evaluated using $T\,dS$, because real heat transfer processes are all irreversible and S is not defined as a function of T in an irreversible process. The reader will recall that engineering thermodynamics might better be named thermo*statics*, because it only describes the equilibrium states on either side of irreversible processes.

Since the rate of heat transfer cannot be predicted using $T\,dS$, how can it be determined? If $U(t)$ were known, then (when $Wk = 0$) eqn. (1.3) would give Q, but $U(t)$ is seldom known *a priori.*

The answer is that a new set of physical principles must be introduced to predict Q. The principles are *transport laws*, which are not a part of the subject of thermodynamics. They include Fourier's law, Newton's law of cooling, and the Stefan-Boltzmann law. We introduce these laws later in the chapter. The important thing to remember is that a description of heat transfer requires that additional principles be combined with the First Law of Thermodynamics.

Reversible heat transfer as the temperature gradient vanishes

Consider a wall connecting two thermal reservoirs as shown in Fig. 1.2. As long as $T_1 > T_2$, heat will flow *spontaneously* and *irreversibly* from 1 to 2. In accordance with our understanding of the Second Law of Thermodynamics, we expect the entropy of the universe to increase as a consequence of this process. If $T_2 \longrightarrow T_1$, the process will approach being quasistatic and reversible. But the rate of heat transfer will also approach

[2] T = absolute temperature, S = entropy, V = volume, p = pressure, and "rev" denotes a reversible process.

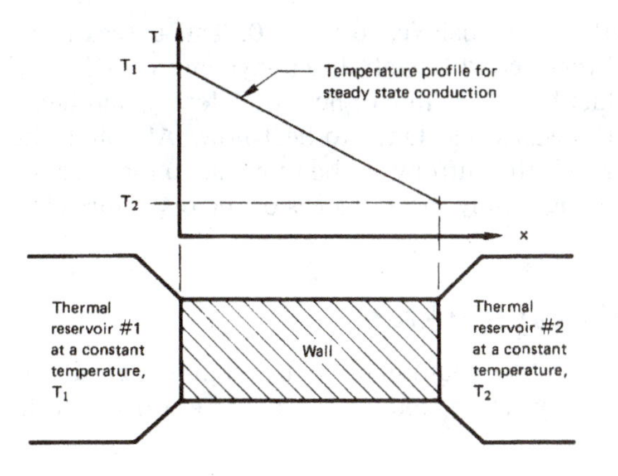

Figure 1.2 Irreversible heat flow between two thermal reservoirs through an intervening wall.

zero if there is no temperature difference to drive it. Thus all real heat transfer processes generate entropy.

Now we come to a dilemma: If the irreversible process occurs at steady state, the properties of the wall do not vary with time. We know that the entropy of the wall depends on its state and must therefore be constant. How, then, does the entropy of the universe increase? We turn to this question next.

Entropy production

The entropy increase of the universe as the result of a process is the sum of the entropy changes of *all* elements that are involved in that process. The *rate of entropy production* of the universe, $\dot{S}_{\text{Un}}$, resulting from the preceding heat transfer process through a wall is

$$\dot{S}_{\text{Un}} = \dot{S}_{\text{res 1}} + \underbrace{\dot{S}_{\text{wall}}}_{\substack{= 0, \text{ since } S_{\text{wall}} \\ \text{must be constant}}} + \dot{S}_{\text{res 2}} \tag{1.5}$$

where the dots denote time derivatives (i.e., $\dot{x} \equiv dx/dt$). Since the reservoir temperatures are constant,

$$\dot{S}_{\text{res}} = \frac{Q}{T_{\text{res}}}. \tag{1.6}$$

Now $Q_{\text{res 1}}$ is negative and equal in magnitude to $Q_{\text{res 2}}$, so eqn. (1.5) becomes

$$\dot{S}_{\text{Un}} = \left| Q_{\text{res 1}} \right| \left(\frac{1}{T_2} - \frac{1}{T_1} \right). \tag{1.7}$$

The term in parentheses is positive, so $\dot{S}_{\text{Un}} > 0$. This agrees with Clausius's statement of the Second Law of Thermodynamics.

Notice an odd fact here: The rate of heat transfer, Q, and hence $\dot{S}_{\text{Un}}$, is determined by the wall's resistance to heat flow. Although the wall is the agent that causes the entropy of the universe to increase, its own entropy does not change. Only the entropies of the reservoirs change.

1.3 Modes of heat transfer

Figure 1.3 shows an analogy that might be useful in fixing the concepts of heat conduction, convection, and radiation as we proceed to look at each in some detail.

Heat conduction

Fourier's law. Joseph Fourier (see Fig. 1.4) published his remarkable book *Théorie Analytique de la Chaleur* in 1822. In it he formulated a very complete exposition of the theory of heat conduction.

He began his treatise by stating the empirical law that bears his name: *the heat flux,*[3] q (W/m^2), *resulting from thermal conduction is proportional to the magnitude of the temperature gradient and opposite to it in sign.* If we call the constant of proportionality, k, then

$$q = -k \frac{dT}{dx} \tag{1.8}$$

The constant, k, is called the *thermal conductivity*. It obviously must have the dimensions W/m·K, or J/m·s·K, or Btu/h·ft·°F if eqn. (1.8) is to be dimensionally correct.

The heat flux is a vector quantity. Equation (1.8) tells us that if temperature decreases with x, q will be positive—it will flow in the x-direction. If T increases with x, q will be negative—it will flow opposite the x-direction. In either case, q will flow from higher temperatures to lower temperatures. Equation (1.8) is the one-dimensional form of Fourier's law. We develop its three-dimensional form in Chapter 2, namely:

$$\vec{q} = -k \nabla T$$

[3]The heat flux, q, is a heat rate per unit area and can be expressed as Q/A, where A is an appropriate area.

Help! The barn is on fire.

Let the *water* be analogous to *heat,* and let the *people* be analogous to the *heat transfer medium.* Then:

Case 1 **The hose directs water from (W) to (B) independently of the medium. This is analogous to *thermal radiation* in a vacuum or in most gases.**

Case 2 **In the bucket brigade, water goes from (W) to (B) through the medium. This is analogous to *conduction.***

Case 3 **A single runner, representing the medium, carries water from (W) to (B). This is analogous to *convection.***

Figure 1.3 An analogy for the three modes of heat transfer.

Figure 1.4 Baron Jean Baptiste Joseph Fourier (1768–1830). Joseph Fourier lived a remarkable double life. He served as a high government official in Napoleonic France and he was also an applied mathematician of great importance. He was with Napoleon in Egypt between 1798 and 1801, and he was subsequently prefect of the administrative area (or "Department") of Isère in France until Napoleon's first fall in 1814. During the latter period he worked on the theory of heat flow and in 1807 submitted a 234-page monograph on the subject. It was given to such luminaries as Lagrange and Laplace for review. They found fault with his adaptation of a series expansion suggested by Daniel Bernoulli in the eighteenth century. Fourier's theory of heat flow, his governing differential equation, and the now-famous "Fourier series" solution of that equation did not emerge in print from the ensuing controversy until 1822. (Etching from *Portraits et Histoire des Hommes Utiles, Collection de Cinquante Portraits,* Société Montyon et Franklin 1839-1840).

Example 1.1

The front of a slab of lead (k = 35 W/m·K) is kept at 110°C and the back is kept at 50°C. If the area of the slab is 0.4 m^2 and it is 0.03 m thick, compute the heat flux, q, and the heat transfer rate, Q.

Solution. For the moment, we presume that dT/dx is a constant equal to $(T_{\text{back}} - T_{\text{front}})/(x_{\text{back}} - x_{\text{front}})$; we verify this in Chapter 2. Thus, eqn. (1.8) becomes

$$q = -35\left(\frac{50 - 110}{0.03}\right) = +70{,}000 \text{ W/m}^2 = 70 \text{ kW/m}^2$$

and

$$Q = qA = 70(0.4) = 28 \text{ kW}$$ ■

In one-dimensional heat conduction problems, there is never any real problem in deciding which way the heat should flow. It is therefore sometimes convenient to write Fourier's law in simple scalar form:

$$q = k\,\frac{\Delta T}{L} \tag{1.9}$$

where L is the thickness in the direction of heat flow and q and ΔT are both written as positive quantities. When we use eqn. (1.9), we must remember that q always flows from high to low temperatures.

Thermal conductivity values. It will help if we first consider how conduction occurs in, for example, a gas. We know that the molecular velocity depends on temperature. Consider conduction from a hot wall to a cold one in a situation in which gravity can be ignored, as shown in Fig. 1.5. The molecules near the hot wall collide with it and are agitated by the molecules of the wall. They leave with generally higher speed and collide with their neighbors to the right, increasing the speed of those neighbors. This process continues until the molecules on the right pass their kinetic energy to those in the cool wall. Within solids, comparable processes occur as the molecules vibrate within their lattice structure and as the lattice vibrates as a whole. This sort of process also occurs, to some extent, in the electron "gas" that moves through the solid. The processes are more efficient in solids than they are in gases. Notice that

$$-\frac{dT}{dx} = \underbrace{\frac{q}{k} \propto \frac{1}{k}}_{\substack{\text{since, in steady} \\ \text{conduction, } q \text{ is} \\ \text{constant}}} \tag{1.10}$$

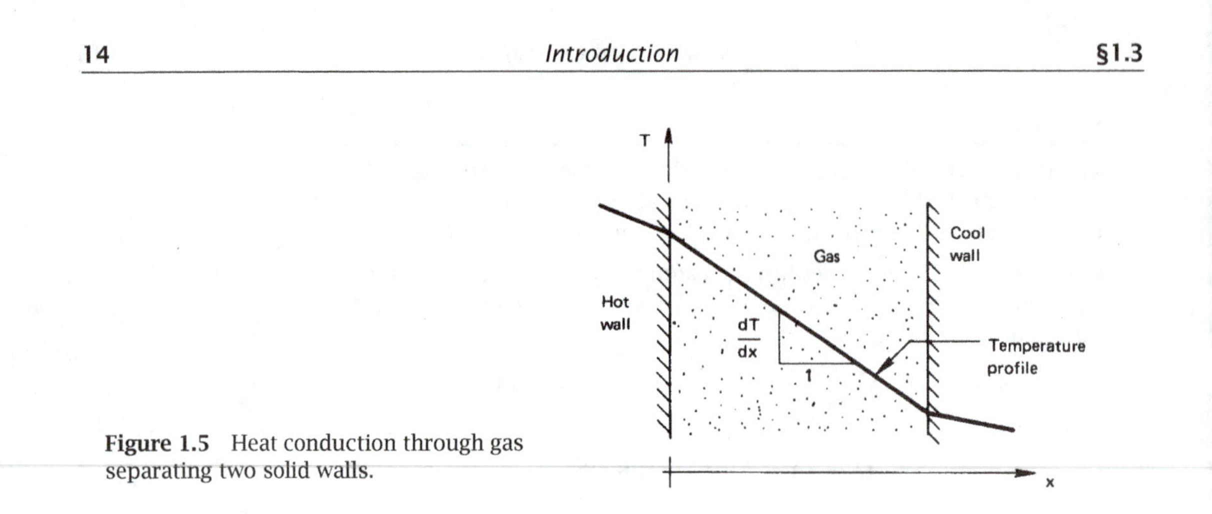

Figure 1.5 Heat conduction through gas separating two solid walls.

Thus solids, with generally higher thermal conductivities than gases, yield smaller temperature gradients for a given heat flux. In a gas, by the way, k is proportional to molecular speed and molar specific heat, and inversely proportional to the cross-sectional area of molecules.

This book deals almost exclusively with S.I. units, or *Système International d'Unités.* Since much reference material will continue to be available in English units, we should have at hand a conversion factor for thermal conductivity:

$$1 = \frac{\text{J}}{0.0009478\ \text{Btu}} \cdot \frac{\text{h}}{3600\ \text{s}} \cdot \frac{\text{ft}}{0.3048\ \text{m}} \cdot \frac{1.8°\text{F}}{\text{K}}$$

Thus the conversion factor from W/m·K to its English equivalent, Btu/h·ft·°F, is

$$\boxed{1 = 1.731\,\frac{\text{W/m·K}}{\text{Btu/h·ft·°F}}} \tag{1.11}$$

Consider, for example, copper—the common substance with the highest conductivity at ordinary temperature:

$$k_{\text{Cu at room temp}} = (383\ \text{W/m·K}) \Big/ 1.731\,\frac{\text{W/m·K}}{\text{Btu/h·ft·°F}} = 221\ \text{Btu/h·ft·°F}$$

The range of thermal conductivities is enormous. As we see from Fig. 1.6, k varies by a factor of about 10^5 between gases and diamond at room temperature. This variation can be increased to about 10^7 if we include the effective conductivity of various cryogenic "superinsulations." (These involve powders, fibers, or multilayered materials that have been

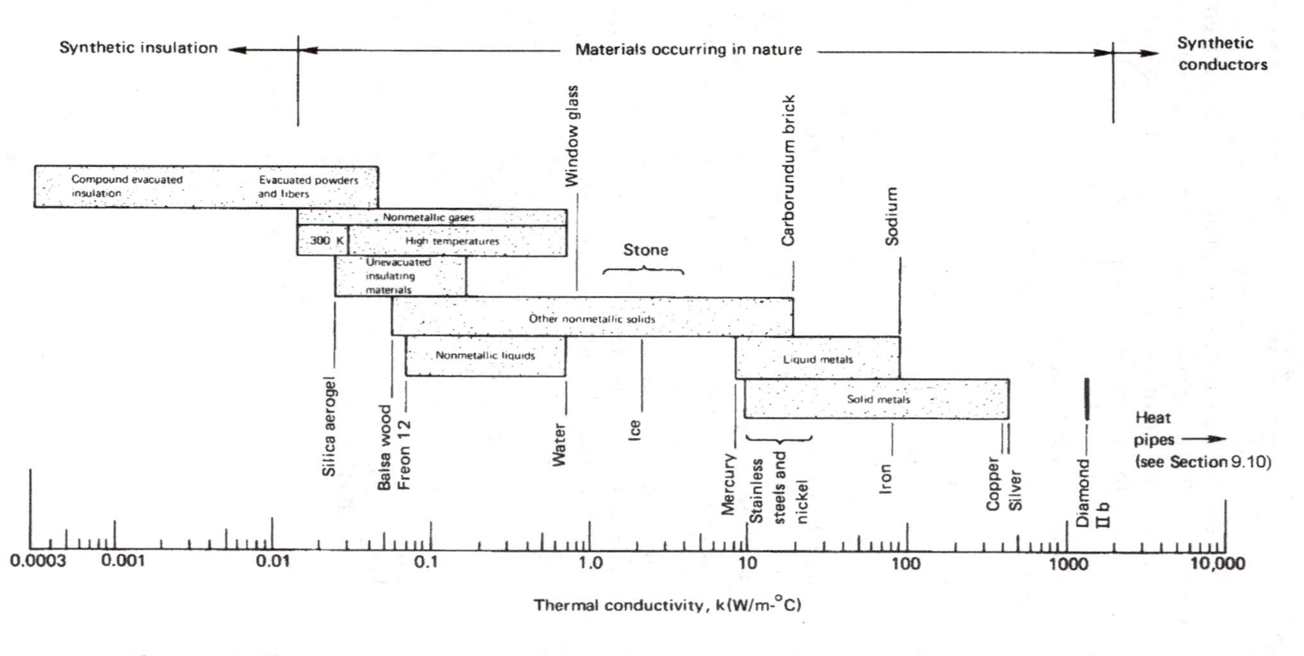

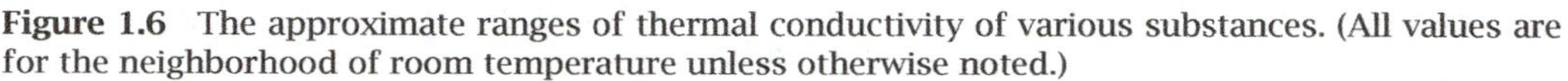

Figure 1.6 The approximate ranges of thermal conductivity of various substances. (All values are for the neighborhood of room temperature unless otherwise noted.)

evacuated of all air.) The reader should study and remember the order of magnitude of the thermal conductivities of different types of materials. This will be a help in avoiding mistakes in future computations, and it will be a help in making assumptions during problem solving. Actual numerical values of the thermal conductivity are given in Appendix A (which is a broad listing of many of the physical properties you might need in this course) and in Figs. 2.2 and 2.3.

Example 1.2

A copper slab (k = 372 W/m·K) is 3 mm thick. It is protected from corrosion on each side by a 2-mm-thick layer of stainless steel (k = 17 W/m·K). The temperature is 400°C on one side of this composite wall and 100°C on the other. Find the temperature distribution in the copper slab and the heat conducted through the wall (see Fig. 1.7).

SOLUTION. If we recall Fig. 1.5 and eqn. (1.10), it should be clear that the temperature drop will take place almost entirely in the stainless steel, where k is less than 1/20 of k in the copper. Thus, the copper will be virtually isothermal at the average temperature of $(400 + 100)/2 = 250$°C. Furthermore, the heat conduction can be estimated in a 4 mm slab of stainless steel as though the copper were not even there. With the help of Fourier's law in the form of eqn. (1.8), we get

$$q = -k\frac{dT}{dx} \simeq 17\ \text{W/m·K} \cdot \left(\frac{400 - 100}{0.004}\right)\ \text{K/m} = 1275\ \text{kW/m}^2$$

The accuracy of this rough calculation can be improved by considering the copper. To do this we first solve for $\Delta T_{\text{s.s.}}$ and ΔT_{Cu} (see Fig. 1.7). Conservation of energy requires that the steady heat flux through all three slabs must be the same. Therefore,

$$q = \left(k\frac{\Delta T}{L}\right)_{\text{s.s.}} = \left(k\frac{\Delta T}{L}\right)_{\text{Cu}}$$

but

$$\begin{aligned} (400 - 100)^\circ\text{C} &\equiv \Delta T_{\text{Cu}} + 2\Delta T_{\text{s.s.}} \\ &= \Delta T_{\text{Cu}}\left[1 + 2\frac{(k/L)_{\text{Cu}}}{(k/L)_{\text{s.s.}}}\right] \\ &= (30.18)\Delta T_{\text{Cu}} \end{aligned}$$

Solving this, we obtain $\Delta T_{\text{Cu}} = 9.94$ K. So $\Delta T_{\text{s.s.}} = (300 - 9.94)/2 = 145$ K. It follows that $T_{\text{Cu, left}} = 255$°C and $T_{\text{Cu, right}} = 245$°C.

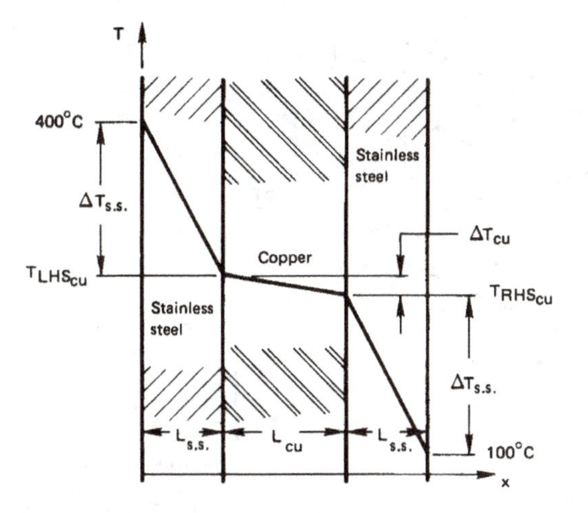

Figure 1.7 Temperature drop through a copper wall protected by stainless steel (Example 1.2).

The heat flux can be obtained by applying Fourier's law to any of the three layers. We consider either stainless steel layer and get

$$q = 17\frac{\text{W}}{\text{m}\cdot\text{K}}\frac{145\text{ K}}{0.002\text{ m}} = 1233\text{ kW/m}^2$$

Thus our initial approximation was accurate within a few percent. ■

One-dimensional heat diffusion equation. In Example 1.2 we had to deal with a major problem that arises in heat conduction problems. The problem is that Fourier's law involves two dependent variables, T and q. To eliminate q and first solve for T, we introduced the First Law of Thermodynamics implicitly: Conservation of energy required that q was the same in each metallic slab.

The elimination of q from Fourier's law must now be done in a more general way. Consider a one-dimensional element, as shown in Fig. 1.8. From Fourier's law applied at each side of the element, as shown, the net heat conduction out of the element during general unsteady heat flow is

$$q_{\text{net}}A = Q_{\text{net}} = -kA\frac{\partial^2 T}{\partial x^2}\delta x \tag{1.12}$$

To eliminate the heat loss Q_{net} in favor of T, we use the general First Law statement for closed, nonworking systems, eqn. (1.3):

$$-Q_{\text{net}} = \frac{dU}{dt} = \rho c A\frac{d(T - T_{\text{ref}})}{dt}\delta x = \rho c A\frac{dT}{dt}\delta x \tag{1.13}$$

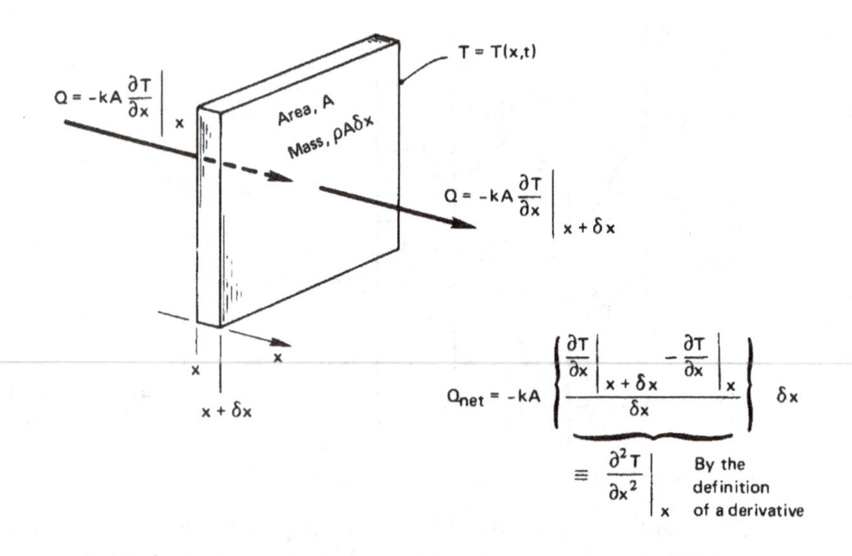

Figure 1.8 One-dimensional heat conduction through a differential element.

where ρ is the density of the slab and c is its specific heat capacity.[4] Equations (1.12) and (1.13) can be combined to give

$$\frac{\partial^2 T}{\partial x^2} = \frac{\rho c}{k}\frac{\partial T}{\partial t} \equiv \frac{1}{\alpha}\frac{\partial T}{\partial t} \tag{1.14}$$

This is the *one-dimensional heat diffusion equation.* Its importance is this: By combining the First Law with Fourier's law, we have eliminated the unknown Q and obtained a differential equation that can be solved for the temperature distribution, $T(x,t)$. It is the primary equation upon which all of heat conduction theory is based.

The heat diffusion equation includes a new property which is as important to transient heat conduction as k is to steady-state conduction.

[4]The reader might wonder if c should be c_p or c_v. This is a strictly incompressible equation so $c_p = c_v = c$. The compressible equation involves additional terms, and this particular term emerges with c_p in it in the conventional rearrangements of terms.

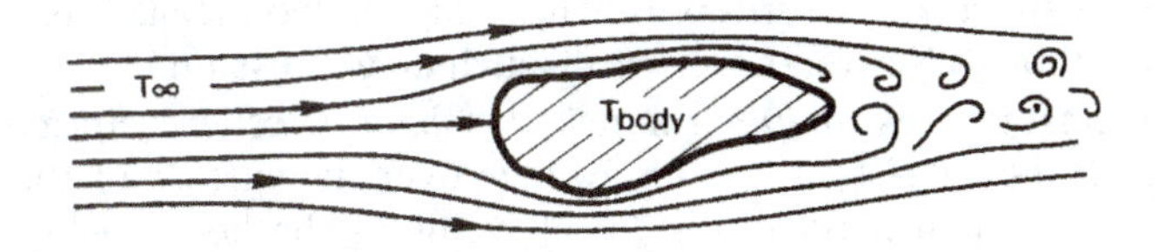

Figure 1.9 The convective cooling of a heated body.

This is the thermal diffusivity, α:

$$\alpha \equiv \frac{k}{\rho c} \frac{\text{J}}{\text{m·s·K}} \frac{\text{m}^3}{\text{kg}} \frac{\text{kg·K}}{\text{J}} = \alpha \;\; \text{m}^2/\text{s} \;\; (\text{or ft}^2/\text{hr}).$$

The thermal diffusivity is a measure of how quickly a material can carry heat away from a hot source. Since material does not just transmit heat but must be warmed by it as well, α involves both the conductivity, k, and the volumetric heat capacity, ρc.

Heat Convection

The physical process. Consider a typical convective cooling situation. Cool gas flows past a warm body, as shown in Fig. 1.9. The fluid immediately adjacent to the body forms a thin slowed-down region called a *boundary layer.* Heat is conducted into this layer, which sweeps it away and, farther downstream, mixes it into the stream. We call such processes of carrying heat away by a moving fluid *convection.*

In 1701, Isaac Newton considered the convective process and suggested that the cooling would be such that

$$\frac{dT_{\text{body}}}{dt} \propto T_{\text{body}} - T_\infty \tag{1.15}$$

where T_∞ is the temperature of the oncoming fluid. This statement suggests that energy is flowing from the body. But if the energy of the body is constantly replenished, the body temperature need not change. Then with the help of eqn. (1.3) we get, from eqn. (1.15) (see Problem 1.2),

$$Q \propto T_{\text{body}} - T_\infty \tag{1.16}$$

This equation can be rephrased in terms of $q = Q/A$ as

$$\boxed{q = \overline{h}\left(T_{\text{body}} - T_\infty\right)} \tag{1.17}$$

This is the steady-state form of Newton's law of cooling, as it is usually quoted, although Newton never wrote such an expression.

The constant h is the *film coefficient* or *heat transfer coefficient.* The bar over h indicates that it is an average over the surface of the body. Without the bar, h denotes the "local" value of the heat transfer coefficient at a point on the surface. The units of h and $\overline{h}$ are W/m^{2}K or J/s·m^2·K. The conversion factor for English units is:

$$1 = \frac{0.0009478\ \text{Btu}}{\text{J}} \cdot \frac{\text{K}}{1.8°\text{F}} \cdot \frac{3600\ \text{s}}{\text{h}} \cdot \frac{(0.3048\ \text{m})^2}{\text{ft}^2}$$

or

$$\boxed{1 = 0.1761 \frac{\text{Btu/h}\cdot\text{ft}^2\cdot°\text{F}}{\text{W/m}^2\text{K}}} \tag{1.18}$$

It turns out that Newton oversimplified the process of convection when he made his conjecture. Heat convection is complicated and $\overline{h}$ can depend on the temperature difference $T_{\text{body}} - T_\infty \equiv \Delta T$. In Chapter 6 we find that h really *is* independent of ΔT in situations in which fluid is forced past a body and ΔT is not too large. This is called *forced convection.*

When fluid buoys up from a hot body or down from a cold one, h varies as some weak power of ΔT—typically as $\Delta T^{1/4}$ or $\Delta T^{1/3}$. This is called *free* or *natural convection.* If the body is hot enough to boil a liquid surrounding it, h will typically vary as ΔT^2.

For the moment, we restrict consideration to situations in which Newton's law is either true or at least a reasonable approximation to real behavior.

We should have some idea of how large h might be in a given situation. Table 1.1 provides some illustrative values of h that have been observed or calculated for different situations. They are only illustrative and should not be used in calculations because the situations for which they apply have not been fully described. Most of the values in the table could be changed a great deal by varying quantities (such as surface roughness or geometry) that have not been specified. The determination of h or $\overline{h}$ is a fairly complicated task and one that will receive a great deal of our attention. Notice, too, that $\overline{h}$ can change dramatically from one situation to the next. Reasonable values of h range over about six orders of magnitude.

Table 1.1 Some illustrative values of convective heat transfer coefficients

Situation	$\overline{h}$, W/m²K
Natural convection in gases	
• 0.3 m vertical wall in air, $\Delta T = 30°$C	4.33
Natural convection in liquids	
• 40 mm O.D. horizontal pipe in water, $\Delta T = 30°$C	570
• 0.25 mm diameter wire in methanol, $\Delta T = 50°$C	4,000
Forced convection of gases	
• Air at 30 m/s over a 1 m flat plate, $\Delta T = 70°$C	80
Forced convection of liquids	
• Water at 2 m/s over a 60 mm plate, $\Delta T = 15°$C	590
• Aniline-alcohol mixture at 3 m/s in a 25 mm I.D. tube, $\Delta T = 80°$C	2,600
• Liquid sodium at 5 m/s in a 13 mm I.D. tube at 370°C	75,000
Boiling water	
• During film boiling at 1 atm	300
• In a tea kettle	4,000
• At a peak pool-boiling heat flux, 1 atm	40,000
• At a peak flow-boiling heat flux, 1 atm	100,000
• At approximate maximum convective-boiling heat flux, under optimal conditions	10^6
Condensation	
• In a typical horizontal cold-water-tube steam condenser	15,000
• Same, but condensing benzene	1,700
• Dropwise condensation of water at 1 atm	160,000

Example 1.3

The heat flux, q, is 6000 W/m² at the surface of an electrical heater. The heater temperature is 120°C when it is cooled by air at 70°C. What is the average convective heat transfer coefficient, $\overline{h}$? What will the heater temperature be if the power is reduced so that q is 2000 W/m²?

Solution.

$$\overline{h} = \frac{q}{\Delta T} = \frac{6000}{120 - 70} = 120 \text{ W/m}^2\text{K}$$

If the heat flux is reduced, $\overline{h}$ should remain unchanged during forced convection. Thus

$$\Delta T = T_{\text{heater}} - 70°\text{C} = \frac{q}{\overline{h}} = \frac{2000\ \text{W/m}^2}{120\ \text{W/m}^2\text{K}} = 16.67\ \text{K}$$

so $T_{\text{heater}} = 70 + 16.67 = 86.67°\text{C}$ ■

Lumped-capacity solution. We now wish to deal with a very simple but extremely important, kind of convective heat transfer problem. The problem is that of predicting the transient cooling of a convectively cooled object, such as we showed in Fig. 1.9. With reference to Fig. 1.10, we apply our now-familiar First law statement, eqn. (1.3), to such a body:

$$\underbrace{Q}_{-\overline{h}A(T-T_\infty)} = \underbrace{\frac{dU}{dt}}_{\frac{d}{dt}[\rho c V(T-T_{\text{ref}})]} \tag{1.19}$$

where A and V are the surface area and volume of the body, T is the temperature of the body, $T = T(t)$, and T_{ref} is the arbitrary temperature at which U is defined equal to zero. Thus[5]

$$\frac{d(T-T_\infty)}{dt} = -\frac{\overline{h}A}{\rho c V}(T - T_\infty) \tag{1.20}$$

The general solution to this equation is

$$\ln(T - T_\infty) = -\frac{t}{(\rho c V/\overline{h}A)} + C \tag{1.21}$$

The group $\rho c V/\overline{h}A$ is the *time constant*, $\boldsymbol{T}$. If the initial temperature is $T(t = 0) \equiv T_i$, then $C = \ln(T_i - T_\infty)$, and the cooling of the body is given by

$$\boxed{\frac{T - T_\infty}{T_i - T_\infty} = e^{-t/\boldsymbol{T}}} \tag{1.22}$$

All of the physical parameters in the problem have now been "lumped" into the time constant. It represents the time required for a body to cool

[5] Is it clear why $(T - T_{\text{ref}})$ has been changed to $(T - T_\infty)$ under the derivative? Remember that the derivative of a constant (like T_{ref} or T_∞) is zero. We can therefore introduce $(T - T_\infty)$ without invalidating the equation, and get the same dependent variable on both sides of the equation.

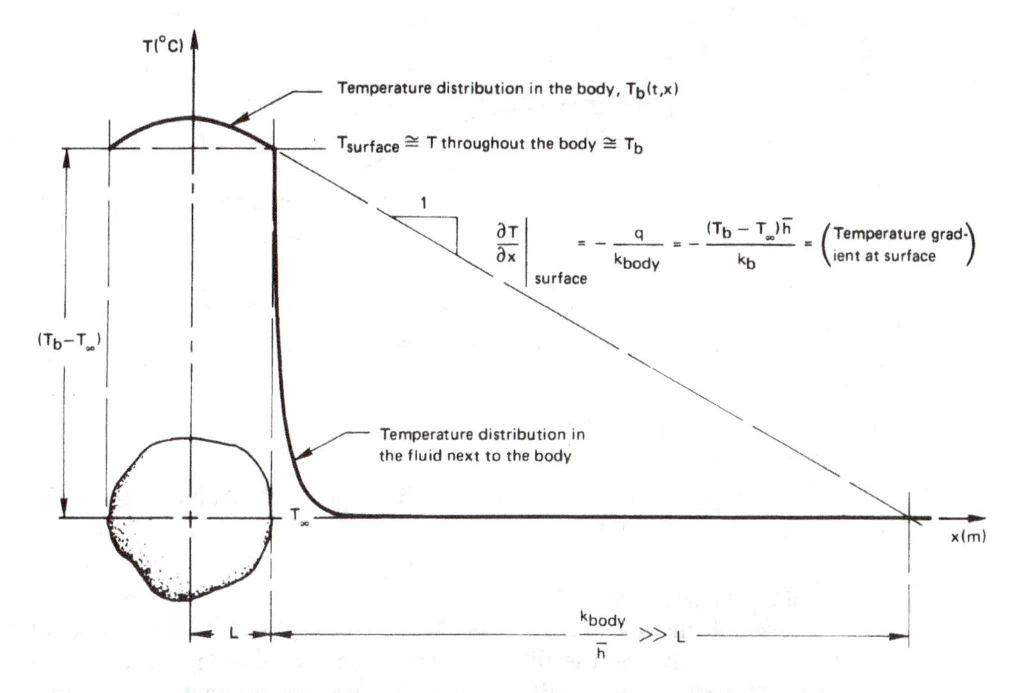

Figure 1.10 The cooling of a body for which the Biot number, $\overline{h}L/k_b$, is small.

to $1/e$, or 37% of its initial temperature difference above (or below) T_∞. The ratio t/T can also be interpreted as

$$\frac{t}{T} = \frac{\overline{h}At \text{ (J/°C)}}{\rho cV \text{ (J/°C)}} = \frac{\text{capacity for convection from surface}}{\text{heat capacity of the body}} \tag{1.23}$$

Notice that the thermal conductivity is missing from eqns. (1.22) and (1.23). The reason is that we have assumed that the temperature of the body is nearly uniform, and this means that internal conduction is not important. We see in Fig. 1.10 that, if $L/(k_b/\overline{h}) \ll 1$, the temperature of the body, T_b, is almost constant within the body at any time. Thus

$$\frac{\overline{h}L}{k_b} \ll 1 \text{ implies that } T_b(x,t) \simeq T(t) \simeq T_{\text{surface}}$$

and the thermal conductivity, k_b, becomes irrelevant to the cooling process. This condition must be satisfied or the lumped-capacity solution

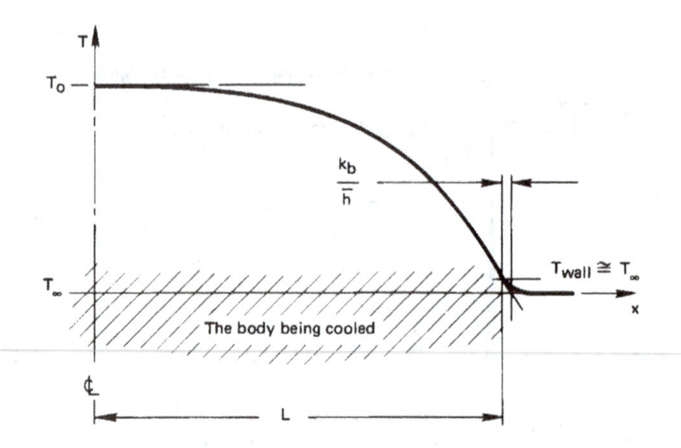

Figure 1.11 The cooling of a body for which the Biot number, $\overline{h}L/k_b$, is large.

will not be accurate.

We call the group $\overline{h}L/k_b$ the *Biot number*[6], Bi. If Bi were large, of course, the situation would be reversed, as shown in Fig. 1.11. In this case $\text{Bi} = \overline{h}L/k_b \gg 1$ and the convection process offers little resistance to heat transfer. We could solve the heat diffusion equation

$$\frac{\partial^2 T}{\partial x^2} = \frac{1}{\alpha}\frac{\partial T}{\partial t}$$

subject to the simple boundary condition $T(x,t) = T_\infty$ when $x = L$, to determine the temperature in the body and its rate of cooling in this case. The Biot number will therefore be the basis for determining what sort of problem we have to solve.

To calculate the rate of entropy production in a lumped-capacity system, we note that the entropy change of the universe is the sum of the entropy decrease of the body and the more rapid entropy increase of

[6]Pronounced **Bee**-oh. J.B. Biot, although younger than Fourier, worked on the analysis of heat conduction even earlier—in 1802 or 1803. He grappled with the problem of including external convection in heat conduction analyses in 1804 but could not see how to do it. Fourier read Biot's work and by 1807 had determined how to analyze the problem. (Later we encounter a similar dimensionless group called the Nusselt number, $\text{Nu} = \overline{h}L/k_{\text{fluid}}$. The latter relates only to the boundary layer and not to the body being cooled. We deal with it extensively in the study of convection.)

the surroundings. The source of irreversibility is heat flow through the boundary layer. Accordingly, we write the time rate of change of entropy of the universe, $dS_{\text{Un}}/dt \equiv \dot{S}_{\text{Un}}$, as

$$\dot{S}_{\text{Un}} = \dot{S}_b + \dot{S}_{\text{surroundings}} = \frac{-Q_{\text{rev}}}{T_b} + \frac{Q_{\text{rev}}}{T_\infty}$$

or

$$\dot{S}_{\text{Un}} = -\rho c V \frac{dT_b}{dt}\left(\frac{1}{T_\infty} - \frac{1}{T_b}\right).$$

We can multiply both sides of this equation by dt and integrate the right-hand side from $T_b(t = 0) \equiv T_{b0}$ to T_b at the time of interest:

$$\Delta S = -\rho c V \int_{T_{b0}}^{T_b} \left(\frac{1}{T_\infty} - \frac{1}{T_b}\right) dT_b. \tag{1.24}$$

Equation 1.24 will give a positive ΔS whether $T_b > T_\infty$ or $T_b < T_\infty$ because the sign of dT_b will always opposed the sign of the integrand.

Example 1.4

A thermocouple bead is largely solder, 1 mm in diameter. It is initially at room temperature and is suddenly placed in a 200°C gas flow. The heat transfer coefficient $\overline{h}$ is 250 W/m^{2}K, and the effective values of k, ρ, and c are 45 W/m·K, 9300 kg/m^3, and c = 0.18 kJ/kg·K, respectively. Evaluate the response of the thermocouple.

Solution. The time constant, $\boldsymbol{T}$, is

$$\begin{aligned} T &= \frac{\rho c V}{\overline{h} A} = \frac{\rho c}{\overline{h}} \frac{\pi D^3/6}{\pi D^2} = \frac{\rho c D}{6\overline{h}} \\ &= \frac{(9300)(0.18)(0.001)}{6(250)} \frac{\text{kg}}{\text{m}^3} \frac{\text{kJ}}{\text{kg·K}} \text{m} \frac{\text{m}^2\text{·K}}{\text{W}} \frac{1000\ \text{W}}{\text{kJ/s}} \\ &= 1.116\ \text{s} \end{aligned}$$

Therefore, eqn. (1.22) becomes

$$\frac{T - 200°\text{C}}{(20 - 200)°\text{C}} = e^{-t/1.116} \quad \text{or} \quad T = 200 - 180\, e^{-t/1.116}\ °\text{C}$$

This result is plotted in Fig. 1.12, where we see that, for all practical purposes, this thermocouple catches up with the gas stream in less than 5 s. Indeed, it should be apparent that any such system will

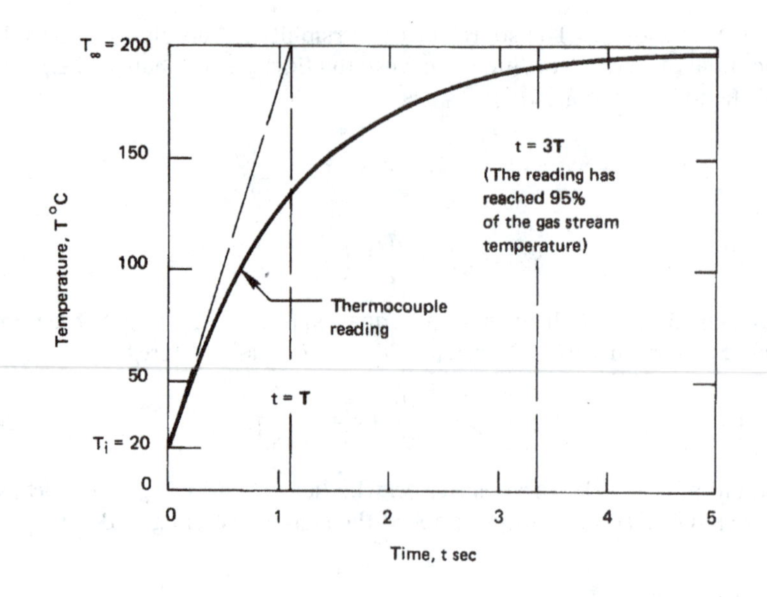

Figure 1.12 Thermocouple response to a hot gas flow.

come within 95% of the signal in three time constants. Notice, too, that if the response could continue at its initial rate, the thermocouple would reach the signal temperature in one time constant.

This calculation is based entirely on the assumption that $\text{Bi} \ll 1$ for the thermocouple. We must check that assumption:

$$\text{Bi} \equiv \frac{\overline{h}L}{k} = \frac{(250\ \text{W/m}^2\text{K})(0.001\ \text{m})/2}{45\ \text{W/m·K}} = 0.00278$$

This is very small indeed, so the assumption is valid. ■

Experiment 1.2

Invent and carry out a simple procedure for evaluating the time constant of a fever thermometer in your mouth.

Radiation

Heat transfer by thermal radiation. All bodies constantly emit energy by a process of electromagnetic radiation. The intensity of such energy flux depends upon the temperature of the body and the nature of its surface. Most of the heat that reaches you when you sit in front of a fire is radiant energy. Radiant energy browns your toast in an electric toaster and it warms you when you walk in the sun.

Objects that are cooler than the fire, the toaster, or the sun emit much less energy because the energy emission varies as the fourth power of absolute temperature. Very often, the emission of energy, or *radiant heat transfer*, from cooler bodies can be neglected in comparison with convection and conduction. But heat transfer processes that occur at high temperature, or with conduction or convection suppressed by evacuated insulations, usually involve a significant fraction of radiation.

Experiment 1.3

Open the freezer door to your refrigerator. Put your face near it, but stay far enough away to avoid the downwash of cooled air. This way you cannot be cooled by convection and, because the air between you and the freezer is a fine insulator, you cannot be cooled by conduction. Still your face will feel cooler. The reason is that you radiate heat directly into the cold region and it radiates very little heat to you. Consequently, your face cools perceptibly.

The electromagnetic spectrum. Thermal radiation occurs in a range of the electromagnetic spectrum of energy emission. Accordingly, it exhibits the same wavelike properties as light or radio waves. Each quantum of radiant energy has a wavelength, λ, and a frequency, ν, associated with it.

The full electromagnetic spectrum includes an enormous range of energy-bearing waves, of which heat is only a small part. Table 1.2 lists the various forms over a range of wavelengths that spans 17 orders of magnitude. Only the tiniest "window" exists in this spectrum through which we can *see* the world around us. Heat radiation, whose main component is usually the spectrum of infrared radiation, passes through the much larger window—about three orders of magnitude in λ or ν.

Table 1.2 Forms of the electromagnetic wave spectrum

Characterization	*Wavelength, λ*	
Cosmic rays	< 0.3 pm	
Gamma rays	0.3–100 pm	
X rays	0.01–30 nm	
Ultraviolet light	3–400 nm	***Thermal Radiation 0.1–1000 μm***
Visible light	0.4–0.7 μm	
Near infrared radiation	0.7–30 μm	
Far infrared radiation	30–1000 μm	
Millimeter waves	1–10 mm	
Microwaves	10–300 mm	
Shortwave radio & TV	300 mm–100 m	
Longwave radio	100 m–30 km	

Black bodies. The model for the perfect thermal radiator is a so-called *black body.* This is a body which absorbs all energy that reaches it and reflects nothing. The term can be a little confusing, since such bodies *emit* energy. Thus, if we possessed infrared vision, a black body would glow with "color" appropriate to its temperature. of course, perfect radiators *are* "black" in the sense that they absorb all visible light (and all other radiation) that reaches them.

It is necessary to have an experimental method for making a perfectly black body. The conventional device for approaching this ideal is called by the German term *hohlraum,* which literally means "hollow space". Figure 1.13 shows how a hohlraum is arranged. It is simply a device that traps all the energy that reaches the aperture.

What are the important features of a thermally black body? First consider a distinction between heat and infrared radiation. *Infrared radiation* refers to a particular range of wavelengths, while *heat* refers to the whole range of radiant energy flowing from one body to another. Suppose that a radiant heat flux, q, falls upon a translucent plate that is not black, as shown in Fig. 1.14. A fraction, α, of the total incident energy, called the *absorptance,* is absorbed in the body; a fraction, ρ, called the *reflectance,* is reflected from it; and a fraction, τ, called the

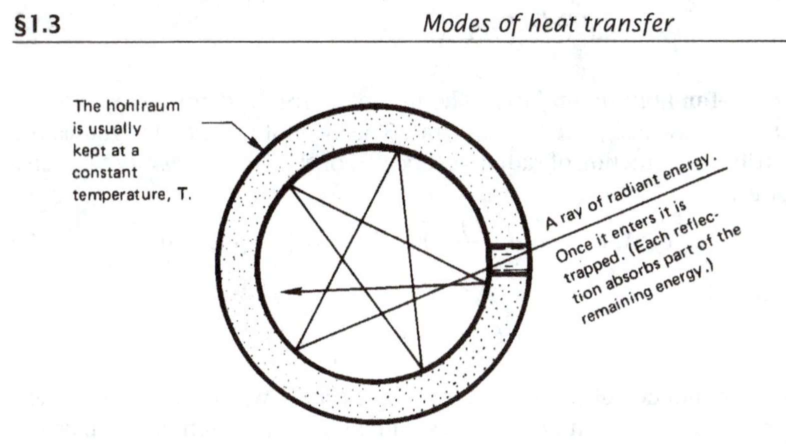

Figure 1.13 Cross section of a spherical hohlraum. The hole has the attributes of a nearly perfect thermal black body.

transmittance, passes through. Thus

$$1 = \alpha + \rho + \tau \tag{1.25}$$

This relation can also be written for the energy carried by each wavelength in the distribution of wavelengths that makes up *heat* from a source at any temperature:

$$1 = \alpha_\lambda + \rho_\lambda + \tau_\lambda \tag{1.26}$$

All radiant energy incident on a black body is absorbed, so that α_b or $\alpha_{\lambda_b} = 1$ and $\rho_b = \tau_b = 0$. Furthermore, the energy emitted from a black body reaches a theoretical maximum, which is given by the Stefan-Boltzmann law. We look at this next.

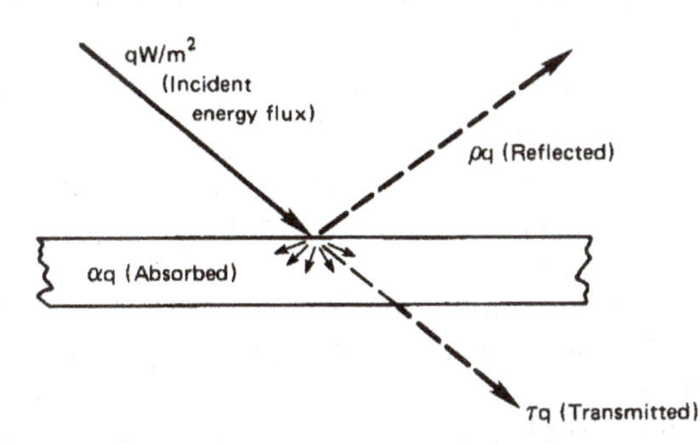

Figure 1.14 The distribution of energy incident on a translucent slab.

The Stefan-Boltzmann law. The flux of energy radiating from a body is commonly designated $e(T)$ W/m^2. The symbol $e_\lambda(\lambda, T)$ designates the distribution function of radiative flux in λ, or the *monochromatic emissive power:*

$$e_\lambda(\lambda, T) = \frac{de(\lambda, T)}{d\lambda} \quad \text{or} \quad e(\lambda, T) = \int_0^\lambda e_\lambda(\lambda, T)\, d\lambda \qquad (1.27)$$

Thus

$$e(T) \equiv E(\infty, T) = \int_0^\infty e_\lambda(\lambda, T)\, d\lambda$$

The dependence of $e(T)$ on T for a black body was established experimentally by Stefan in 1879 and explained by Boltzmann on the basis of thermodynamics arguments in 1884. The Stefan-Boltzmann law is

$$\boxed{e_b(T) = \sigma T^4} \qquad (1.28)$$

where the Stefan-Boltzmann constant, σ, is 5.670400×10^{-8} W/m$^2\cdot$K^4 or 1.714×10^{-9} Btu/hr$\cdot$ft$^2\cdot{}^\circ$R^4, and T is the absolute temperature.

e_λ vs. λ. Nature requires that, at a given temperature, a body will emit a unique distribution of energy in wavelength. Thus, when you heat a poker in the fire, it first glows a dull red—emitting most of its energy at long wavelengths and just a little bit in the visible regime. When it is white-hot, the energy distribution has been both greatly increased and shifted toward the shorter-wavelength visible range. At each temperature, a black body yields the highest value of e_λ that a body can attain.

The very accurate measurements of the black-body energy spectrum by Lummer and Pringsheim (1899) are shown in Fig. 1.15. The locus of maxima of the curves is also plotted. It obeys a relation called *Wien's law:*

$$(\lambda T)_{e_\lambda = \max} = 2898\ \mu\text{m}\cdot\text{K} \qquad (1.29)$$

About three-fourths of the radiant energy of a black body lies to the right of this line in Fig. 1.15. Notice that, while the locus of maxima leans toward the visible range at higher temperatures, only a small fraction of the radiation is visible even at the highest temperature.

Predicting how the monochromatic emissive power of a black body depends on λ was an increasingly serious problem at the close of the nineteenth century. The prediction was a keystone of the most profound scientific revolution the world has seen. In 1901, Max Planck made the

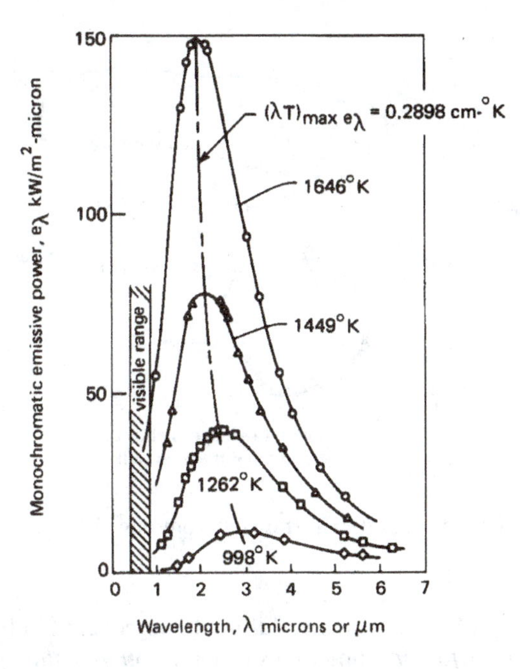

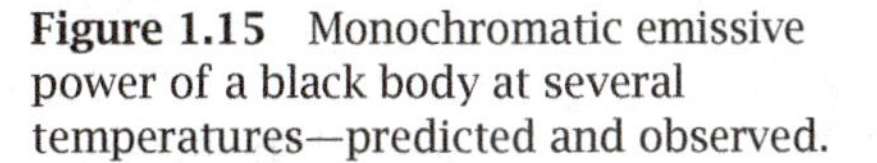
Figure 1.15 Monochromatic emissive power of a black body at several temperatures—predicted and observed.

prediction, and his work included the initial formulation of quantum mechanics. He found that

$$e_{\lambda_b} = \frac{2\pi h c_o^2}{\lambda^5 \left[\exp(hc_o/k_B T\lambda) - 1\right]} \tag{1.30}$$

where c_o is the speed of light, 2.99792458×10^8 m/s; h is Planck's constant, 6.62606876×10^{-34} J·s; and k_B is Boltzmann's constant, 1.3806503×10^{-23} J/K.

Radiant heat exchange. Suppose that a heated object (1 in Fig. 1.16a) radiates only to some other object (2) and that both objects are thermally black. All heat leaving object 1 arrives at object 2, and all heat arriving at object 1 comes from object 2. Thus, the net heat transferred from object 1 to object 2, Q_{net}, is the difference between $Q_{1\text{ to }2} = A_1 e_b(T_1)$ and $Q_{2\text{ to }1} = A_1 e_b(T_2)$

$$Q_{\text{net}} = A_1 e_b(T_1) - A_1 e_b(T_2) = A_1 \sigma\left(T_1^4 - T_2^4\right) \tag{1.31}$$

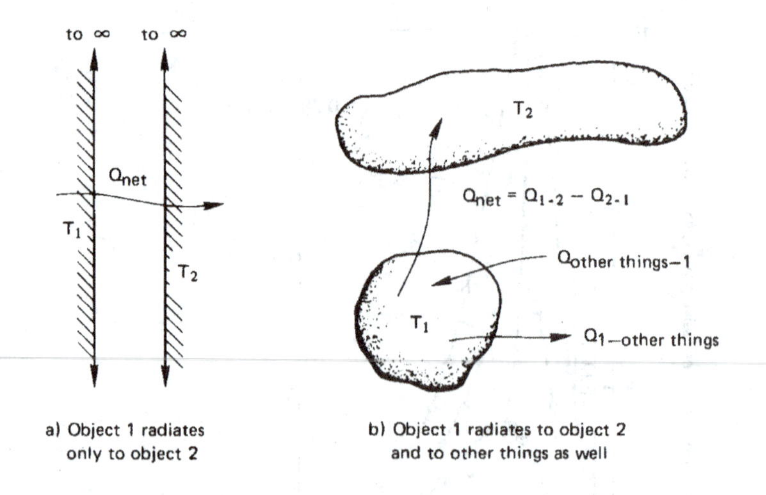

Figure 1.16 The net radiant heat transfer from one object to another.

If the first object "sees" other objects in addition to object 2, as indicated in Fig. 1.16b, then a *view factor* (sometimes called a *configuration factor* or a *shape factor*), $F_{1\text{-}2}$, must be included in eqn. (1.31):

$$\boxed{Q_{\text{net}} = A_1 F_{1\text{-}2}\, \sigma\left(T_1^4 - T_2^4\right)} \tag{1.32}$$

We may regard $F_{1\text{-}2}$ as the fraction of energy leaving object 1 that is intercepted by object 2.

Example 1.5

A black thermocouple measures the temperature in a chamber with black walls. If the air around the thermocouple is at 20°C, the walls are at 100°C, and the heat transfer coefficient between the thermocouple and the air is 75 W/m^{2}K, what temperature will the thermocouple read?

Solution. The heat convected away from the thermocouple by the air must exactly balance that radiated to it by the hot walls if the system is in steady state. Furthermore, $F_{1\text{-}2} = 1$ since the thermocouple

(1) radiates all its energy to the walls (2):

$$\overline{h}A_{tc}\,(T_{tc} - T_{\text{air}}) = -Q_{\text{net}} = -A_{tc}\sigma\left(T_{tc}^4 - T_{\text{wall}}^4\right)$$

or, with T_{tc} in °C,

$$75(T_{tc} - 20)\ \text{W/m}^2 = 5.6704 \times 10^{-8}\left[(100 + 273)^4 - (T_{tc} + 273)^4\right]\ \text{W/m}^2$$

since T for radiation must be in kelvin. Trial-and-error solution of this equation yields $T_{tc} = 28.4$°C. ■

We have seen that non-black bodies absorb less radiation than black bodies, which are perfect absorbers. Likewise, non-black bodies emit less radiation than black bodies, which also happen to be perfect emitters. We can characterize the emissive power of a non-black body using a property called *emittance*, ε:

$$e_{\text{non-black}} = \varepsilon e_b = \varepsilon\sigma T^4 \tag{1.33}$$

where $0 < \varepsilon \le 1$. When radiation is exchanged between two bodies that are not black, we have

$$Q_{\text{net}} = A_1\mathcal{F}_{1\text{-}2}\,\sigma\left(T_1^4 - T_2^4\right) \tag{1.34}$$

where the *transfer factor*, $\mathcal{F}_{1\text{-}2}$, depends on the emittances of both bodies as well as the geometrical "view".

The expression for $\mathcal{F}_{1\text{-}2}$ is particularly simple in the important special case of a small object, 1, in a much larger isothermal environment, 2:

$$\mathcal{F}_{1\text{-}2} = \varepsilon_1 \quad \text{for} \quad A_1 \ll A_2 \tag{1.35}$$

Example 1.6

Suppose that the thermocouple in Example 1.5 was not black and had an emittance of $\varepsilon = 0.4$. Further suppose that the walls were not black and had a much larger surface area than the thermocouple. What temperature would the thermocouple read?

Solution. Q_{net} is now given by eqn. (1.34) and $\mathcal{F}_{1\text{-}2}$ can be found with eqn. (1.35):

$$\overline{h}A_{tc}\,(T_{tc} - T_{\text{air}}) = -A_{tc}\varepsilon_{tc}\sigma\left(T_{tc}^4 - T_{\text{wall}}^4\right)$$

or

$$75(T_{tc} - 20)\ \text{W/m}^2 = (0.4)(5.6704 \times 10^{-8})\left[(100 + 273)^4 - (T_{tc} + 273)^4\right]\ \text{W/m}^2$$

Trial-and-error yields $T_{tc} = 23.5°\text{C}$. ■

Radiation shielding. The preceding examples point out an important practical problem than can be solved with radiation shielding. The idea is as follows: If we want to measure the true air temperature, we can place a thin foil casing, or shield, around the thermocouple. The casing is shaped to obstruct the thermocouple's "view" of the room but to permit the free flow of the air around the thermocouple. Then the shield, like the thermocouple in the two examples, will be cooler than the walls, and the thermocouple it surrounds will be influenced by this much cooler radiator. If the shield is highly reflecting on the outside, it will assume a temperature still closer to that of the air and the error will be still less. Multiple layers of shielding can further reduce the error.

Radiation shielding can take many forms and serve many purposes. It is an important element in superinsulations. A glass firescreen in a fireplace serves as a radiation shield because it is largely opaque to radiation. It absorbs heat radiated by the fire and reradiates that energy (ineffectively) at a temperature much lower than that of the fire.

Experiment 1.4

Find a small open flame that produces a fair amount of soot. A candle, kerosene lamp, or a cutting torch with a fuel-rich mixture should work well. A clean blue flame will not work well because such gases do not radiate much heat. First, place your finger in a position about 1 to 2 cm to one side of the flame, where it becomes uncomfortably hot. Now take a piece of fine mesh screen and dip it in some soapy water, which will fill up the holes. Put it between your finger and the flame. You will see that your finger is protected from the heating until the water evaporates.

Water is relatively transparent to light. What does this experiment show you about the transmittance of water to infrared wavelengths?

1.4 A look ahead

What we have done up to this point has been no more than to reveal the tip of the iceberg. The basic mechanisms of heat transfer have been explained and some quantitative relations have been presented. However, this information will barely get you started when you are faced with a real heat transfer problem. Three tasks, in particular, must be completed to solve actual problems:

- The heat diffusion equation must be solved subject to appropriate boundary conditions if the problem involves heat conduction of any complexity.
- The convective heat transfer coefficient, h, must be determined if convection is important in a problem.
- The factor $F_{1\text{-}2}$ or $\mathcal{F}_{1\text{-}2}$ must be determined to calculate radiative heat transfer.

Any of these determinations can involve a great deal of complication, and most of the chapters that lie ahead are devoted to these three basic problems.

Before becoming engrossed in these three questions, we shall first look at the archetypical applied problem of heat transfer–namely, the design of a heat exchanger. Chapter 2 sets up the elementary analytical apparatus that is needed for this, and Chapter 3 shows how to do such design if $\overline{h}$ is already known. This will make it easier to see the importance of undertaking the three basic problems in subsequent parts of the book.

1.5 Problems

We have noted that this book is set down almost exclusively in S.I. units. The student who has problems with dimensional conversion will find Appendix B helpful. The only use of English units appears in some of the problems at the end of each chapter. A few such problems are included to provide experience in converting back into English units, since such units will undoubtedly persist in the U.S.A. for many more years.

Another matter often leads to some discussion between students and teachers in heat transfer courses. That is the question of whether a problem is "theoretical" or "practical". Quite often the student is inclined to view as "theoretical" a problem that does not involve numbers or that requires the development of algebraic results.

The problems assigned in this book are all intended to be useful in that they do one or more of five things:

1. They involve a calculation of a type that actually arises in practice (e.g., Problems 1.1, 1.3, 1.8 to 1.18, and 1.21 through 1.25).

2. They illustrate a physical principle (e.g., Problems 1.2, 1.4 to 1.7, 1.9, 1.20, 1.32, and 1.39). These are probably closest to having a "theoretical" objective.

3. They ask you to use methods developed in the text to develop other results that would be needed in certain applied problems (e.g., Problems 1.10, 1.16, 1.17, and 1.21). Such problems are usually the most difficult and the most valuable to you.

4. They anticipate development that will appear in subsequent chapters (e.g., Problems 1.16, 1.20, 1.40, and 1.41).

5. They require that you develop your ability to handle numerical and algebraic computation effectively. (This is the case with most of the problems in Chapter 1, but it is especially true of Problems 1.6 to 1.9, 1.15, and 1.17).

Partial numerical answers to some of the problems follow them in brackets. Tables of physical property data useful in solving the problems are given in Appendix A.

Actually, we wish to look at the *theory*, *analysis*, and *practice* of heat transfer—all three—according to Webster's definitions:

Theory: "a systematic statement of principles; a formulation of apparent relationships or underlying principles of certain observed phenomena."

Analysis: "the solving of problems by the means of equations; the breaking up of any whole into its parts so as to find out their nature, function, relationship, etc."

Practice: "the *doing* of something as an application of knowledge."

Problems

1.1 A composite wall consists of alternate layers of fir (5 cm thick), aluminum (1 cm thick), lead (1 cm thick), and corkboard (6 cm thick). The temperature is 60°C on the outside of the for and 10°C on the outside of the corkboard. Plot the temperature gradient through the wall. Does the temperature profile suggest any simplifying assumptions that might be made in subsequent analysis of the wall?

1.2 Verify eqn. 1.16.

1.3 $q = 5000\ \mathrm{W/m^2}$ in a 1 cm slab and $T = 140$°C on the cold side. Tabulate the temperature drop through the slab if it is made of

- Silver
- Aluminum
- Mild steel (0.5 % carbon)
- Ice
- Spruce
- Insulation (85 % magnesia)
- Silica aerogel

Indicate which situations would be unreasonable and why.

1.4 Explain in words why the heat diffusion equation, eqn. (1.13), shows that in transient conduction the temperature depends on the thermal diffusivity, α, but we can solve steady conduction problems using just k (as in Example 1.1).

1.5 A 1 m rod of pure copper 1 cm^2 in cross section connects a 200°C thermal reservoir with a 0°C thermal reservoir. The system has already reached steady state. What are the rates of change of entropy of (a) the first reservoir, (b) the second reservoir, (c) the rod, and (d) the whole universe, as a result of the process? Explain whether or not your answer satisfies the Second Law of Thermodynamics. [(d): +0.0120 W/K.]

1.6 Two thermal energy reservoirs at temperatures of 27°C and −43°C, respectively, are separated by a slab of material 10 cm thick and 930 cm^2 in cross-sectional area. The slab has a thermal conductivity of 0.14 W/m·K. The system is operating at steady-state conditions. What are the rates of change of entropy of (a) the higher temperature reservoir, (b) the lower temperature reservoir, (c) the slab, and (d) the whole universe as a result of this process? (e) Does your answer satisfy the Second Law of Thermodynamics?

1.7 (a) If the thermal energy reservoirs in Problem 1.6 are suddenly replaced with adiabatic walls, determine the final equilibrium temperature of the slab. (b) What is the entropy change for the slab for this process? (c) Does your answer satisfy the Second Law of Thermodynamics in this instance? Explain. The density of the slab is 26 lb/ft^3 and the specific heat is 0.65 Btu/lb·°F. [(b): 30.81 J/K].

1.8 A copper sphere 2.5 cm in diameter has a uniform temperature of 40°C. The sphere is suspended in a slow-moving air stream at 0°C. The air stream produces a convection heat transfer coefficient of 15 W/m^{2}K. Radiation can be neglected. Since copper is highly conductive, temperature gradients in the sphere will smooth out rapidly, and its temperature can be taken as uniform throughout the cooling process (i.e., Bi $\ll$ 1). Write the instantaneous energy balance between the sphere and the surrounding air. Solve this equation and plot the resulting temperatures as a function of time between 40°C and 0°C.

1.9 Determine the total heat transfer in Problem 1.8 as the sphere cools from 40°C to 0°C. Plot the net entropy increase resulting from the cooling process above, ΔS vs. T (K). [Total heat transfer = 1123 J.]

1.10 A truncated cone 30 cm high is constructed of Portland cement. The diameter at the top is 15 cm and at the bottom is 7.5 cm. The lower surface is maintained at 6°C and the top at 40°C. The other surface is insulated. Assume one-dimensional heat transfer and calculate the rate of heat transfer in watts from top to bottom. To do this, note that the heat transfer, Q, must be the same at every cross section. Write Fourier's law locally, and integrate it from top to bottom to get a relation between this unknown Q and the known end temperatures. [$Q = -0.70$ W.]

1.11 A hot water heater contains 100 kg of water at 75°C in a 20°C room. Its surface area is 1.3 m^2. Select an insulating material, and specify its thickness, to keep the water from cooling more than 3°C/h. (Notice that this problem will be greatly simplified if the temperature drop in the steel casing and the temperature drop in the convective boundary layers are negligible. Can you make such assumptions? Explain.)

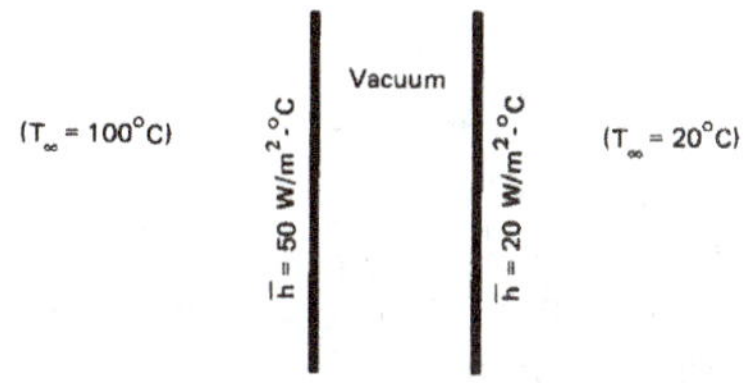

(T∞ = 100°C)
Vacuum
$\bar{h}$ = 50 W/m²·°C
$\bar{h}$ = 20 W/m²·°C
(T∞ = 20°C)

Figure 1.17 Configuration for Problem 1.12

1.12 What is the temperature at the left-hand wall shown in Fig. 1.17. Both walls are thin, very large in extent, highly conducting, and thermally black. [$T_{\text{right}} = 42.5$°C.]

1.13 Develop S.I. to English conversion factors for:

- The thermal diffusivity, α
- The heat flux, q
- The density, ρ
- The Stefan-Boltzmann constant, σ
- The view factor, $F_{1\text{-}2}$
- The molar entropy
- The specific heat per unit mass, c

In each case, begin with basic dimension J, m, kg, s, °C, and check your answers against Appendix B if possible.

1.14 Three infinite, parallel, black, opaque plates transfer heat by radiation, as shown in Fig. 1.18. Find T_2.

1.15 Four infinite, parallel, black, opaque plates transfer heat by radiation, as shown in Fig. 1.19. Find T_2 and T_3. [$T_2 = 75.53$°C.]

1.16 Two large, black, horizontal plates are spaced a distance L from one another. The top one is warm at a controllable tem-

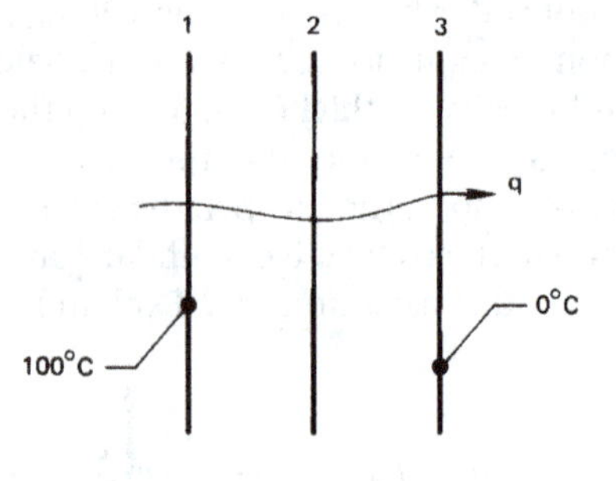

Figure 1.18 Configuration for Problem 1.14

perature, T_h, and the bottom one is cool at a specified temperature, T_c. A gas separates them. The gas is stationary because it is warm on the top and cold on the bottom. Write the equation $q_{\text{rad}}/q_{\text{cond}} = \text{fn}(N, \Theta \equiv T_h/T_c)$, where N is a dimensionless group containing σ, k, L, and T_c. Plot N as a function of Θ for $q_{\text{rad}}/q_{\text{cond}} = 1$, 0.8, and 1.2 (and for other values if you wish).

Now suppose that you have a system in which $L = 10$ cm, $T_c = 100$ K, and the gas is hydrogen with an average k of 0.1 W/m·K. Further suppose that you wish to operate in such a way that the conduction and radiation heat fluxes are identical. Identify the operating point on your curve and report the value of T_h that you must maintain.

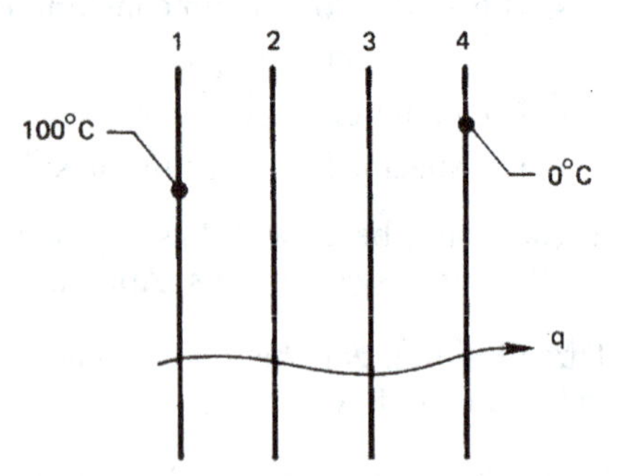

Figure 1.19 Configuration for Problem 1.15

1.17 A blackened copper sphere 2 cm in diameter and uniformly at 200°C is introduced into an evacuated black chamber that is maintained at 20°C.

- Write a differential equation that expresses $T(t)$ for the sphere, assuming lumped thermal capacity.
- Identify a dimensionless group, analogous to the Biot number, than can be used to tell whether or not the lumped-capacity solution is valid.
- Show that the lumped-capacity solution is valid.
- Integrate your differential equation and plot the temperature response for the sphere.

1.18 As part of a space experiment, a small instrumentation package is released from a space vehicle. It can be approximated as a solid aluminum sphere, 4 cm in diameter. The sphere is initially at 30°C and it contains a pressurized hydrogen component that will condense and malfunction at 30 K. If we take the surrounding space to be at 0 K, how long may we expect the implementation package to function properly? Is it legitimate to use the lumped-capacity method in solving the problem? (*Hint:* See the directions for Problem 1.17.) [Time = 5.8 weeks.]

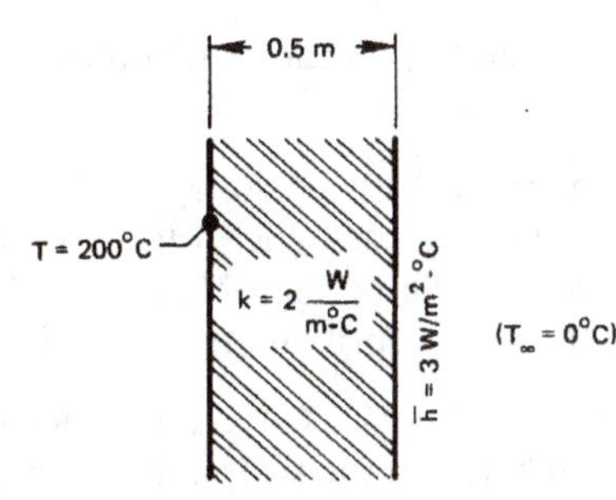

Figure 1.20 Configuration for Problem 1.19

1.19 Consider heat conduction through the wall as shown in Fig. 1.20. Calculate q and the temperature of the right-hand side of the wall.

1.20 Throughout Chapter 1 we have assumed that the steady temperature distribution in a plane uniform wall in linear. To prove this, simplify the heat diffusion equation to the form appropriate for steady flow. Then integrate it twice and eliminate the two constants using the known outside temperatures T_{left} and T_{right} at $x = 0$ and $x =$ wall thickness, L.

1.21 The thermal conductivity in a particular plane wall depends as follows on the wall temperature: $k = A + BT$, where A and B are constants. The temperatures are T_1 and T_2 on either side if the wall, and its thickness is L. Develop an expression for q.

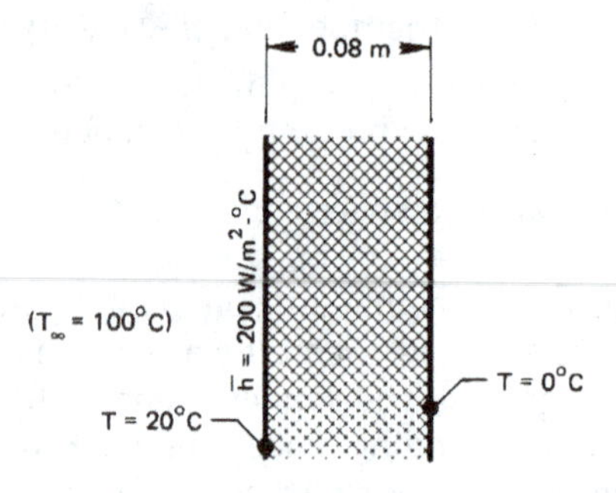

Figure 1.21 Configuration for Problem 1.22

1.22 Find k for the wall shown in Fig. 1.21. Of what might it be made?

1.23 What are T_i, T_j, and T_r in the wall shown in Fig. 1.22? [T_j = 16.44°C.]

1.24 An aluminum can of beer or soda pop is removed from the refrigerator and set on the table. If $\overline{h}$ is 13.5 W/m^{2}K, estimate when the beverage will be at 15°C. Ignore thermal radiation. Verify any important approximation you must make.

1.25 One large, black wall at 27°C faces another whose surface is 127°C. The gap between the two walls is evacuated. If the second wall is 0.1 m thick and has a thermal conductivity of 17.5 W/m·K, what is its temperature on the back side? (Assume steady state.)

1.26 A 1 cm diameter, 1% carbon steel sphere, initially at 200°C, is cooled by natural convection, with air at 20°C. In this case, $\overline{h}$ is not independent of temperature. Instead, $\overline{h} = 3.51(\Delta T°\text{C})^{1/4}$ W/m^{2}K. Plot T_{sphere} as a function of t. Verify the lumped-capacity assumption.

1.27 A 3 cm diameter, black spherical heater is kept at 1100°C. It radiates through an evacuated space to a surrounding spherical

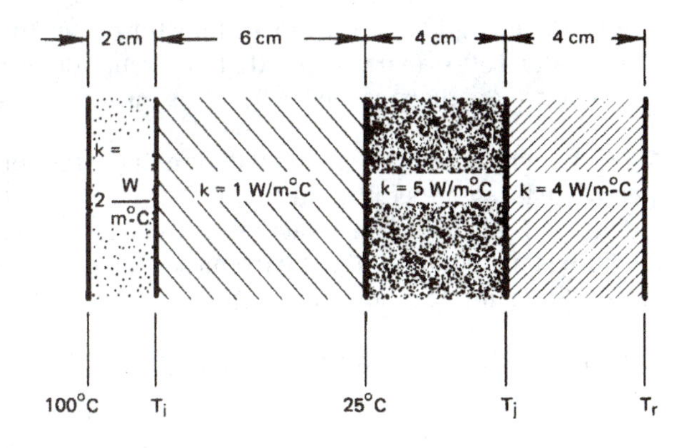

Figure 1.22 Configuration for Problem 1.23

shell of Nichrome V. The shell has a 9 cm inside diameter and is 0.3 cm thick. It is black on the inside and is held at 25°C on the outside. Find (a) the temperature of the inner wall of the shell and (b) the heat transfer, Q. (Treat the shell as a plane wall.)

1.28 The sun radiates 650 W/m^2 on the surface of a particular lake. At what rate (in mm/hr) would the lake evaporate away if all of this energy went to evaporating water? Discuss as many other ways you can think of that this energy can be distributed (h_{fg} for water is 2,257,000 J/kg). Do you suppose much of the 650 W/m^2 goes to evaporation?

1.29 It is proposed to make picnic cups, 0.005 m thick, of a new plastic for which $k = k_o(1 + aT^2)$, where T is expressed in °C, $k_o = 0.15$ W/m·K, and $a = 10^{-4}$ °C^{-2}. We are concerned with thermal behavior in the extreme case in which $T = 100$°C in the cup and 0°C outside. Plot T against position in the cup wall and find the heat loss, q.

1.30 A disc-shaped wafer of diamond 1 lb is the target of a very high intensity laser. The disc is 5 mm in diameter and 1 mm deep. The flat side is pulsed intermittently with 10^{10} W/m^2 of energy for one microsecond. It is then cooled by natural convection from that same side until the next pulse. If $\overline{h} = 10$ W/m^{2}K and

T_∞=30°C, plot T_{disc} as a function of time for pulses that are 50 s apart and 100 s apart. (Note that you must determine the temperature the disc reaches before it is pulsed each time.)

1.31 A 150 W light bulb is roughly a 0.006 m diameter sphere. Its steady surface temperature in room air is 90°C, and $\overline{h}$ on the outside is 7 W/m^{2}K. What fraction of the heat transfer from the bulb is by radiation *directly* from the filament through the glass? (State any additional assumptions.)

1.32 How much entropy does the light bulb in Problem 1.31 produce?

1.33 Air at 20°C flows over one side of a thin metal sheet ($\overline{h}$ = 10.6 W/m^{2}K). Methanol at 87°C flows over the other side ($\overline{h}$ = 141 W/m^{2}K). The metal functions as an electrical resistance heater, releasing 1000 W/m^2. Calculate (a) the heater temperature, (b) the heat transfer from the methanol to the heater, and (c) the heat transfer from the heater to the air.

1.34 A planar black heater is simultaneously cooled by 20°C air ($\overline{h}$ = 14.6 W/m^{2}K) and by radiation to a parallel black wall at 80°C. What is the temperature of the heater if it delivers 9000 W/m^2?

1.35 An 8 oz. can of beer is taken from a 3°C refrigerator and placed in a 25°C room. The 6.3 cm diameter by 9 cm high can is placed on an insulated surface ($\overline{h}$ = 7.3 W/m^{2}K). How long will it take to reach 12°C? Ignore thermal radiation, and discuss your other assumptions.

1.36 A resistance heater in the form of a thin sheet runs parallel with 3 cm slabs of cast iron on either side of an evacuated cavity. The heater, which releases 8000 W/m^2, and the cast iron are very nearly black. The outside surfaces of the cast iron slabs are kept at 10°C. Determine the heater temperature and the inside slab temperatures.

1.37 A black wall at 1200°C radiates to the left side of a parallel slab of type 316 stainless steel, 5 mm thick. The right side of the slab is to be cooled convectively and is not to exceed 0°C. Suggest a convective process that will achieve this.

1.38 A cooler keeps one side of a 2 cm layer of ice at $-10°$C. The other side is exposed to air at 15°C. What is $\overline{h}$ just on the edge of melting? Must $\overline{h}$ be raised or lowered if melting is to progress?

1.39 At what minimum temperature does a black heater deliver its maximum monochromatic emissive power in the visible range? Compare your result with Fig. 10.2.

1.40 The local heat transfer coefficient during the laminar flow of fluid over a flat plate of length L is equal to $F/x^{1/2}$, where F is a function of fluid properties and the flow velocity. How does $\overline{h}$ compare with $h(x = L)$? (x is the distance from the leading edge of the plate.)

1.41 An object is initially at a temperature above that of its surroundings. We have seen that many kinds of convective processes will bring the object into equilibrium with its surroundings. Describe the characteristics of a process that will do so with the least net increase of the entropy of the universe.

1.42 A 250°C cylindrical copper billet, 4 cm in diameter and 8 cm long, is cooled in air at 25°C. The heat transfer coefficient is 5 W/m^{2}K. Can this be treated as lumped-capacity cooling? What is the temperature of the billet after 10 minutes?

1.43 The sun's diameter is 1,392,000 km, and it emits energy as if it were a black body at 5777 K. Determine the rate at which it emits energy. Compare this with a value from the literature. What is the sun's energy output in a year?

Bibliography of Historical and Advanced Texts

We include no specific references for the ideas introduced in Chapter 1 since these are common to introductory thermodynamics or physics books. References 1–7 are some texts which have strongly influenced the field. The rest are relatively advanced texts or handbooks which go beyond the present textbook.

References

[1.1] J. Fourier. *The Analytical Theory of Heat.* Dover Publications, Inc., New York, 1955.

[1.2] L. M. K. Boelter, V. H. Cherry, H. A. Johnson, and R. C. Martinelli. *Heat Transfer Notes.* McGraw-Hill Book Company, New York, 1965. Originally issued as class notes at the University of California at Berkeley between 1932 and 1941.

[1.3] M. Jakob. *Heat Transfer.* John Wiley & Sons, New York, 1949.

[1.4] H. S. Carslaw and J. C. Jaeger. *Conduction of Heat in Solids.* Oxford University Press, New York, 2nd edition, 1959. Very comprehensive.

[1.5] H. Schlichting and K. Gersten. *Boundary-Layer Theory.* Springer-Verlag, Berlin, 8th edition, 2000. Very comprehensive development of boundary layer theory. A classic.

[1.6] R. B. Bird, W. E. Stewart, and E. N. Lightfoot. *Transport Phenomena.* John Wiley & Sons, Inc., New York, 2nd edition, 2002.

[1.7] W. M. Kays and A. L. London. *Compact Heat Exchangers.* McGraw-Hill Book Company, New York, 3rd edition, 1984.

[1.8] V. S. Arpaci. *Conduction Heat Transfer.* Ginn Press/Pearson Custom Publishing, Needham Heights, Mass., 1991. Abridgement of the 1966 edition, omitting numerical analysis.

[1.9] W. M. Kays, M. E. Crawford, and B. Weigand. *Convective Heat and Mass Transfer.* McGraw-Hill Book Company, New York, 4th edition, 2005. Coverage is mainly of boundary layers and internal flows.

[1.10] F.M. White. *Viscous Fluid Flow.* McGraw-Hill, Inc., New York, 2nd edition, 1991. Excellent development of fundamental results for boundary layers and internal flows.

[1.11] A. Bejan. *Convection Heat Transfer.* John Wiley & Sons, New York, 3rd edition, 2004. This book makes good use of scaling arguments.

[1.12] R. Siegel and J. R. Howell. *Thermal Radiation Heat Transfer.* Taylor and Francis-Hemisphere, Washington, D.C., 4th edition, 2001.

[1.13] J. G. Collier and J. R. Thome. *Convective Boiling and Condensation.* Oxford University Press, Oxford, 3rd edition, 1994.

[1.14] G. F. Hewitt, editor. *Heat Exchanger Design Handbook 1998.* Begell House, New York, 1998.

[1.15] A. F. Mills. *Mass Transfer.* Prentice-Hall, Inc., Upper Saddle River, 2001. Mass transfer from a mechanical engineer's perspective with strong coverage of convective mass transfer.

[1.16] D. S. Wilkinson. *Mass Transfer in Solids and Fluids.* Cambridge University Press, Cambridge, 2000. A systematic development of mass transfer with a materials science focus and an emphasis on modeling.

[1.17] D. R. Poirier and G. H. Geiger. *Transport Phenomena in Materials Processing.* The Minerals, Metals & Materials Society, Warrendale, Pennsylvania, 1994. A comprehensive introduction to heat, mass, and momentum transfer from a materials science perspective.

[1.18] G. Chen. *Nanoscale Energy Transport and Conversion.* Oxford University Press, New York, 2005.

[1.19] W. M. Rohsenow, J. P. Hartnett, and Y. I. Cho, editors. *Handbook of Heat Transfer.* McGraw-Hill, New York, 3rd edition, 1998.

2. Heat conduction concepts, thermal resistance, and the overall heat transfer coefficient

It is the fire that warms the cold, the cold that moderates the heat...the general coin that purchases all things...

Don Quixote, **M. de Cervantes, 1615**

2.1 The heat diffusion equation

Objective

We must now develop some ideas that will be needed for the design of heat exchangers. The most important of these is the notion of an overall heat transfer coefficient. This is a measure of the general resistance of a heat exchanger to the flow of heat, and usually it must be built up from analyses of component resistances. Although we shall count radiation among these resistances, this overall heat transfer coefficient is most often dominated by convection and conduction.

We need to know values of $\overline{h}$ to handle convection. Calculating $\overline{h}$ becomes sufficiently complex that we defer it to Chapters 6 and 7. For the moment, we shall take the appropriate value of $\overline{h}$ as known information and concentrate upon its use in the overall heat transfer coefficient.

The heat conduction component also becomes more complex than the planar analyses we did in Chapter 1. But its calculation is within our present scope. Therefore we devote this Chapter to deriving the full heat conduction, or heat diffusion, equation, solving it in some fairly straightforward cases, and using our results in the overall coefficient. We undertake that task next.

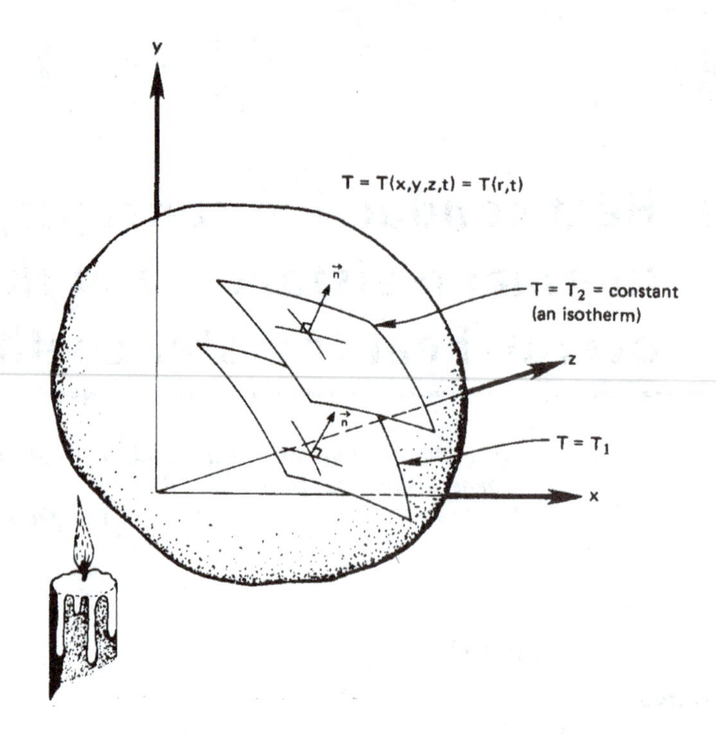

Figure 2.1 A three-dimensional, transient temperature field.

Consider the general temperature distribution in a three-dimensional body as depicted in Fig. 2.1. For some reason, say heating from one side, the temperature of the body varies with time and space. This field $T = T(x, y, z, t)$ or $T(\vec{r}, t)$, defines instantaneous isothermal surfaces, T_1, T_2, and so on.

We next consider a very important vector associated with the scalar, T. The vector that has both the magnitude and direction of the maximum increase of temperature at each point is called the *temperature gradient*, ∇T:

$$\nabla T \equiv \vec{i}\frac{\partial T}{\partial x} + \vec{j}\frac{\partial T}{\partial y} + \vec{k}\frac{\partial T}{\partial z} \qquad (2.1)$$

Fourier's law

"Experience"—that is, physical observation—suggests two things about the heat flow that results from temperature nonuniformities in a body. These are:

$$\frac{\vec{q}}{|\vec{q}|} = -\frac{\nabla T}{|\nabla T|} \quad \begin{cases}\text{This says that } \vec{q} \text{ and } \nabla T \text{ are exactly opposite one} \\ \text{another in direction}\end{cases}$$

and

$$|\vec{q}| \propto |\nabla T| \quad \begin{cases}\text{This says that the magnitude of the heat flux is di-} \\ \text{rectly proportional to the temperature gradient}\end{cases}$$

Notice that the heat flux is now written as a quantity that has a specified direction as well as a specified magnitude. Fourier's law summarizes this physical experience succinctly as

$$\boxed{\vec{q} = -k\nabla T} \tag{2.2}$$

which resolves itself into three components:

$$q_x = -k\frac{\partial T}{\partial x} \qquad q_y = -k\frac{\partial T}{\partial y} \qquad q_z = -k\frac{\partial T}{\partial z}$$

The coefficient k—the thermal conductivity—also depends on position and temperature in the most general case:

$$k = k[\vec{r}, T(\vec{r}, t)] \tag{2.3}$$

Fortunately, most materials (though not all of them) are very nearly homogeneous. Thus we can usually write $k = k(T)$. The assumption that we really want to make is that k is constant. Whether or not that is legitimate must be determined in each case. As is apparent from Fig. 2.2 and Fig. 2.3, k almost always varies with temperature. It always rises with T in gases at low pressures, but it may rise or fall in metals or liquids. The problem is that of assessing whether or not k is approximately constant in the range of interest. We could safely take k to be a constant for iron between 0° and 40°C (see Fig. 2.2), but we would incur error between −100° and 800°C.

It is easy to prove (Problem 2.1) that if k varies linearly with T, and if heat transfer is plane and steady, then $q = k\Delta T/L$, with k evaluated at the average temperature in the plane. If heat transfer is not planar

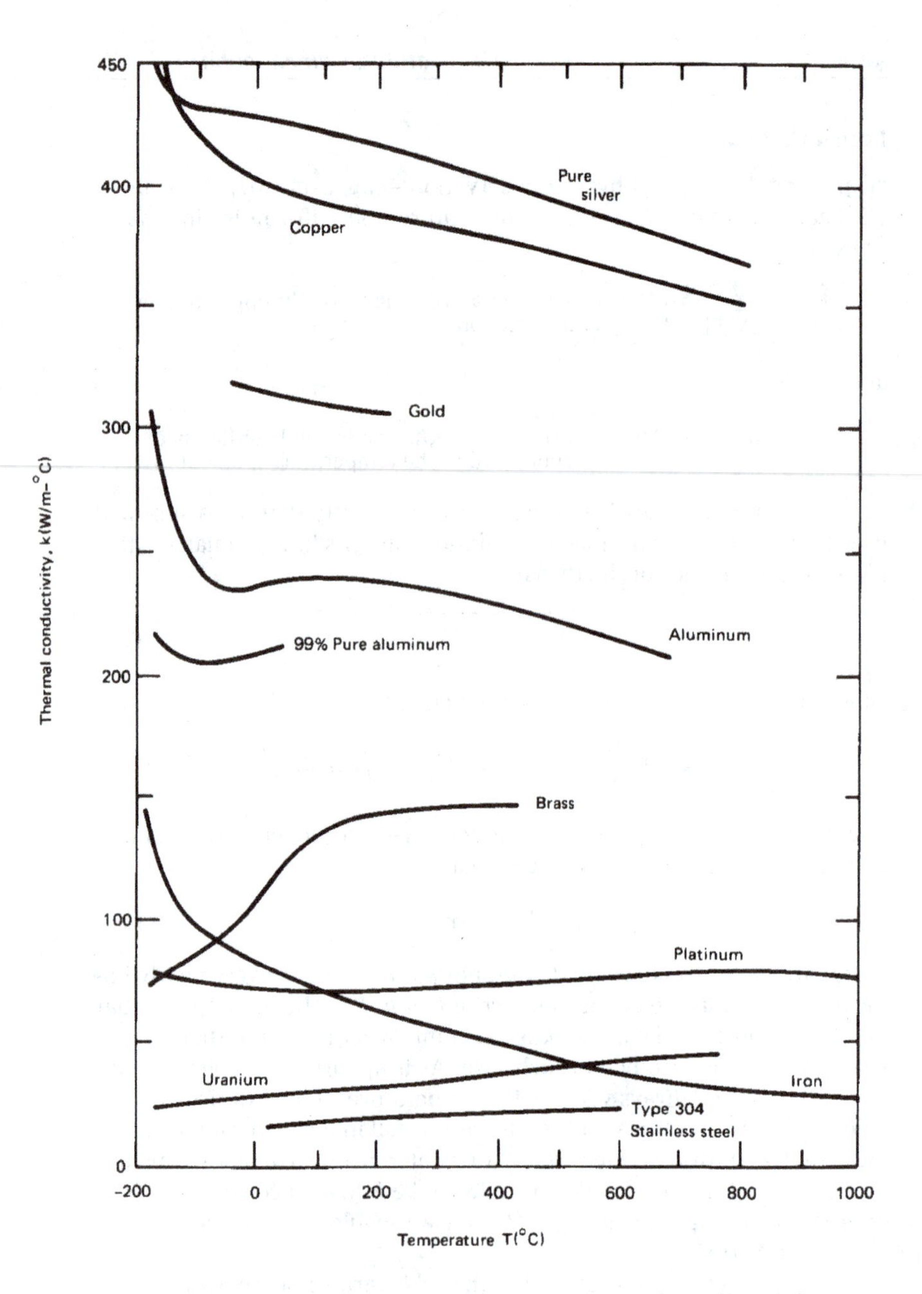

Figure 2.2 Variation of thermal conductivity of metallic solids with temperature

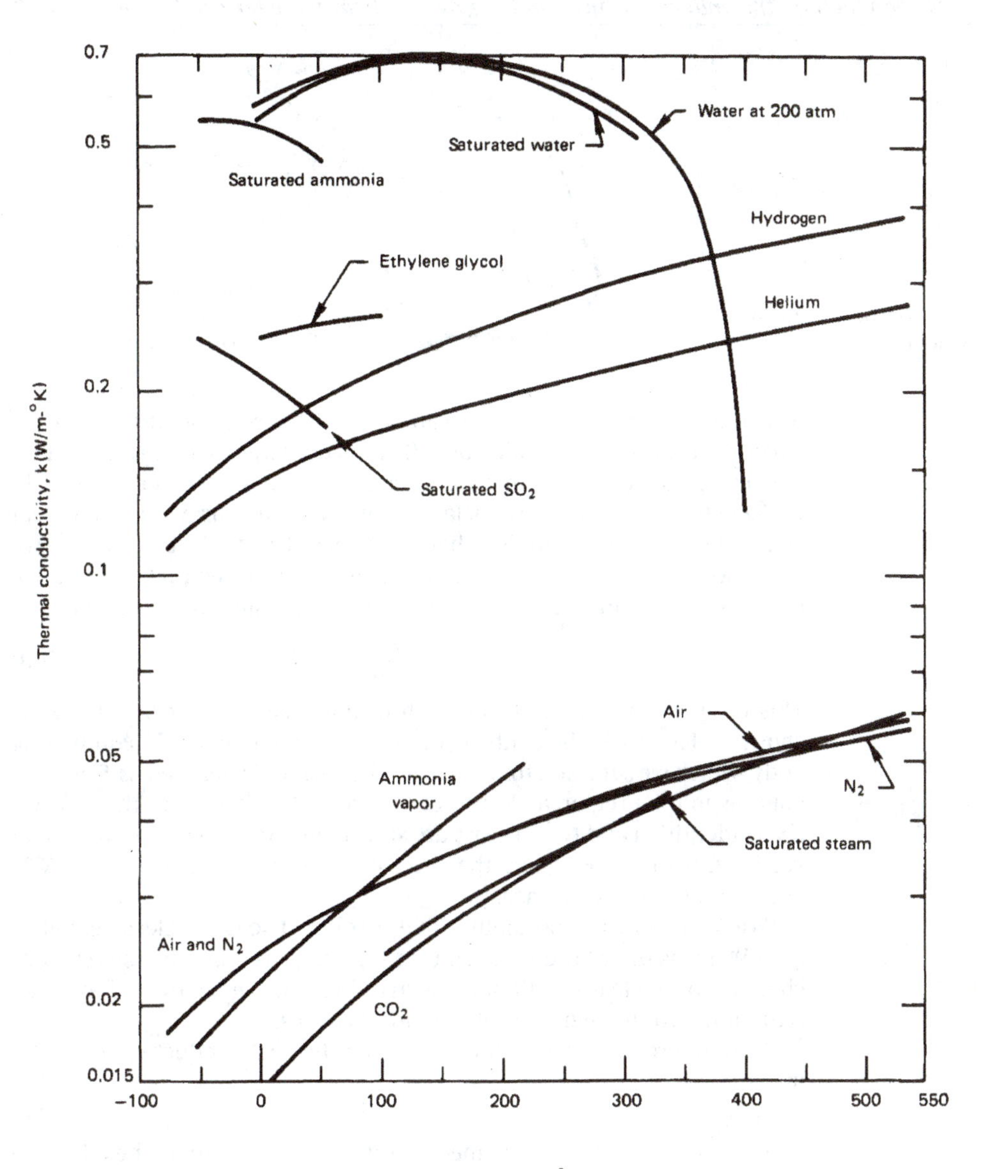

Figure 2.3 The temperature dependence of the thermal conductivity of liquids and gases that are either saturated or at 1 atm pressure.

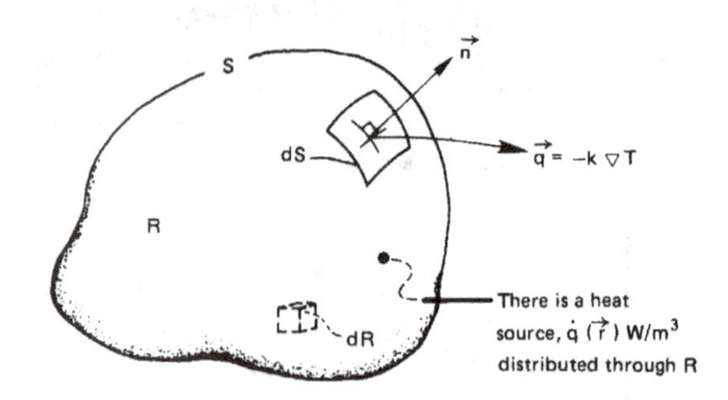

Figure 2.4 Control volume in a heat-flow field.

or if k is not simply $A + BT$, it can be much more difficult to specify a single accurate effective value of k. If ΔT is not large, one can still make a reasonably accurate approximation using a constant average value of k.

Now that we have Fourier's law in three dimensions, we see that heat conduction is more complex than it appeared to be in Chapter 1. We must now write the heat conduction equation in three dimensions. We begin, as we did in Chapter 1, with the First Law statement, eqn. (1.3):

$$Q = \frac{dU}{dt} \tag{1.3}$$

This time we apply eqn. (1.3) to a three-dimensional control volume, as shown in Fig. 2.4.[1] The control volume is a finite region of a conducting body, which we set aside for analysis. The surface is denoted as S and the volume and the region as R; both are at rest. An element of the surface, dS, is identified and two vectors are shown on dS: one is the unit normal vector, $\vec{n}$ (with $|\vec{n}| = 1$), and the other is the heat flux vector, $\vec{q} = -k\nabla T$, at that point on the surface.

We also allow the possibility that a volumetric heat release equal to $\dot{q}(\vec{r})$ W/m^3 is distributed through the region. This might be the result of chemical or nuclear reaction, of electrical resistance heating, of external radiation into the region or of still other causes.

With reference to Fig. 2.4, we can write the heat conducted *out* of dS, in watts, as

$$(-k\nabla T) \cdot (\vec{n} dS) \tag{2.4}$$

The heat generated (or consumed) within the region R must be added to the total heat flow *into* S to get the overall rate of heat addition to R:

$$Q = -\int_S (-k\nabla T) \cdot (\vec{n} dS) + \int_R \dot{q}\, dR \tag{2.5}$$

[1]Figure 2.4 is the three-dimensional version of the control volume shown in Fig. 1.8.

The rate of energy increase of the region R is

$$\frac{dU}{dt} = \int_R \left(\rho c \frac{\partial T}{\partial t}\right) dR \tag{2.6}$$

where the derivative of T is in partial form because T is a function of both $\vec{r}$ and t.

Finally, we combine Q, as given by eqn. (2.5), and dU/dt, as given by eqn. (2.6), into eqn. (1.3). After rearranging the terms, we obtain

$$\int_S k\nabla T \cdot \vec{n} dS = \int_R \left[\rho c \frac{\partial T}{\partial t} - \dot{q}\right] dR \tag{2.7}$$

To get the left-hand side into a convenient form, we introduce Gauss's theorem, which converts a surface integral into a volume integral. Gauss's theorem says that if $\vec{A}$ is any continuous function of position, then

$$\int_S \vec{A} \cdot \vec{n} dS = \int_R \nabla \cdot \vec{A} dR \tag{2.8}$$

Therefore, if we identify $\vec{A}$ with $(k\nabla T)$, eqn. (2.7) reduces to

$$\int_R \left(\nabla \cdot k\nabla T - \rho c \frac{\partial T}{\partial t} + \dot{q}\right) dR = 0 \tag{2.9}$$

Next, since the region R is arbitrary, the integrand must vanish identically.[2] We therefore get the *heat diffusion equation* in three dimensions:

$$\boxed{\nabla \cdot k\nabla T + \dot{q} = \rho c \frac{\partial T}{\partial t}} \tag{2.10}$$

The limitations on this equation are:

- Incompressible medium. (This was implied when no expansion work term was included.)

- No convection. (The medium cannot undergo any relative motion. However, it *can* be a liquid or gas as long as it sits still.)

[2]Consider $\int f(x)\,dx = 0$. If $f(x)$ were, say, $\sin x$, then this could only be true over intervals of $x = 2\pi$ or multiples of it. For eqn. (2.9) to be true for *any* range of integration one might choose, the terms in parentheses must be zero everywhere.

If the variation of k with T is small, k can be factored out of eqn. (2.10) to get

$$\boxed{\nabla^2 T + \frac{\dot{q}}{k} = \frac{1}{\alpha}\frac{\partial T}{\partial t}} \tag{2.11}$$

This is a more complete version of the heat conduction equation [recall eqn. (1.14)] and α is the thermal diffusivity which was discussed after eqn. (1.14). The term $\nabla^2 T \equiv \nabla \cdot \nabla T$ is called the *Laplacian.* It arises thus in a Cartesian coordinate system:

$$\nabla \cdot k\nabla T \simeq k\nabla \cdot \nabla T = k\left(\vec{i}\frac{\partial}{\partial x} + \vec{j}\frac{\partial}{\partial y} + \vec{k}\frac{\partial}{\partial x}\right) \cdot \left(\vec{i}\frac{\partial T}{\partial x} + \vec{j}\frac{\partial T}{\partial y} + \vec{k}\frac{\partial T}{\partial z}\right)$$

or

$$\boxed{\nabla^2 T = \frac{\partial^2 T}{\partial x^2} + \frac{\partial^2 T}{\partial y^2} + \frac{\partial^2 T}{\partial z^2}} \tag{2.12}$$

The Laplacian can also be expressed in cylindrical or spherical coordinates. The results are:

- Cylindrical:

$$\nabla^2 T \equiv \frac{1}{r}\frac{\partial}{\partial r}\left(r\frac{\partial T}{\partial r}\right) + \frac{1}{r^2}\frac{\partial^2 T}{\partial \theta^2} + \frac{\partial^2 T}{\partial z^2} \tag{2.13}$$

- Spherical:

$$\nabla^2 T \equiv \frac{1}{r}\frac{\partial^2 (rT)}{\partial r^2} + \frac{1}{r^2 \sin\theta}\frac{\partial}{\partial \theta}\left(\sin\theta\frac{\partial T}{\partial \theta}\right) + \frac{1}{r^2 \sin^2\theta}\frac{\partial^2 T}{\partial \phi^2} \tag{2.14a}$$

or

$$\equiv \frac{1}{r^2}\frac{\partial}{\partial r}\left(r^2\frac{\partial T}{\partial r}\right) + \frac{1}{r^2 \sin\theta}\frac{\partial}{\partial \theta}\left(\sin\theta\frac{\partial T}{\partial \theta}\right) + \frac{1}{r^2 \sin^2\theta}\frac{\partial^2 T}{\partial \phi^2} \tag{2.14b}$$

where the coordinates are as described in Fig. 2.5.

2.2 Solutions of the heat diffusion equation

We are now in position to calculate the temperature distribution and/or heat flux in bodies with the help of the heat diffusion equation. In every

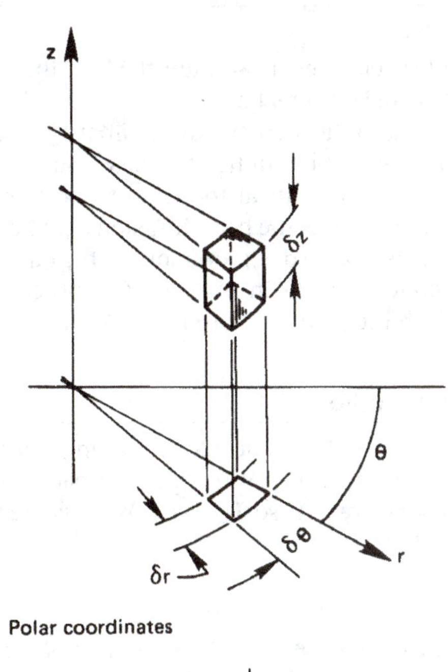

Polar coordinates

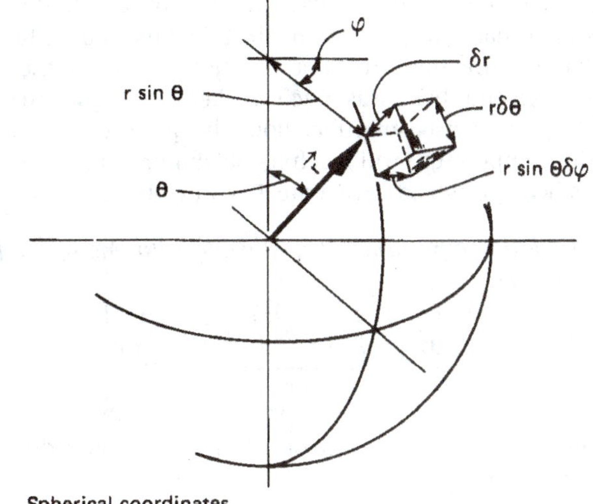

Spherical coordinates

Figure 2.5 Cylindrical and spherical coordinate systems.

case, we first calculate $T(\vec{r}, t)$. Then, if we want the heat flux as well, we differentiate T to get q from Fourier's law.

The heat diffusion equation is a partial differential equation (p.d.e.) and the task of solving it may seem difficult, but we can actually do a lot with fairly elementary mathematical tools. For one thing, in one-dimensional steady-state situations the heat diffusion equation becomes an ordinary differential equation (o.d.e.); for another, the equation is linear and therefore not too formidable, in any case. Our procedure can be laid out, step by step, with the help of the following example.

Example 2.1 Basic Method

A large, thin concrete slab of thickness L is "setting." Setting is an exothermic process that releases $\dot{q}$ W/m^3. The outside surfaces are kept at the ambient temperature, so $T_w = T_\infty$. What is the maximum internal temperature?

Solution.

Step 1. *Pick the coordinate scheme that best fits the problem and identify the independent variables that determine T.* In the example, T will probably vary only along the thin dimension, which we will call the x-direction. (We should want to know that the edges are insulated and that L was much smaller than the width or height. If they are, this assumption should be quite good.) Since the interior temperature will reach its maximum value when the process becomes steady, we write $T = T(x \text{ only})$.

Step 2. *Write the appropriate d.e., starting with one of the forms of eqn. (2.11).*

$$\frac{\partial^2 T}{\partial x^2} + \underbrace{\frac{\partial^2 T}{\partial y^2} + \frac{\partial^2 T}{\partial z^2}}_{\substack{=0, \text{ since} \\ T \neq T(y \text{ or } z)}} + \frac{\dot{q}}{k} = \underbrace{\frac{1}{\alpha}\frac{\partial T}{\partial t}}_{\substack{=0, \text{ since} \\ \text{steady}}}$$

Therefore, since $T = T(x \text{ only})$, the equation reduces to the ordinary d.e.

$$\frac{d^2 T}{dx^2} = -\frac{\dot{q}}{k}$$

Step 3. *Obtain the general solution of the d.e.* (This is usually the easiest step.) We simply integrate the d.e. twice and get

$$T = -\frac{\dot{q}}{2k}x^2 + C_1 x + C_2$$

Step 4. *Write the "side conditions" on the d.e.—the initial and boundary conditions.* This is the trickiest part and the one that most seriously tests our physical or "practical" understanding any heat conduction problem.

Normally, we have to make two specifications of temperature on each position coordinate and one on the time coordinate to get rid of the constants of integration in the general solution. (These matters are discussed at greater length in Chapter 4.)

In this case we know two boundary conditions:

$$T(x = 0) = T_w \quad \text{and} \quad T(x = L) = T_w$$

Very Important Warning: Never, never introduce inaccessible information in a boundary or initial condition. Always stop and ask yourself, "Would I have access to a numerical value of the temperature (or other data) that I specify at a given position or time?" If the answer is no, then your result will be useless.

Step 5. *Substitute the general solution in the boundary and initial conditions and solve for the constants.* This process gets very complicated in the transient and multidimensional cases. Numerical methods are often needed to solve the problem. However, the steady one-dimensional problems are usually easy. In the example, by evaluating at $x = 0$ and $x = L$, we get:

$$T_w = -0 + 0 + C_2 \qquad \text{so} \qquad C_2 = T_w$$

$$T_w = -\frac{\dot{q}L^2}{2k} + C_1 L + \underbrace{C_2}_{=T_w} \qquad \text{so} \qquad C_1 = \frac{\dot{q}L}{2k}$$

Step 6. *Put the calculated constants back in the general solution to get the particular solution to the problem.* In the example problem we obtain:

$$T = -\frac{\dot{q}}{2k}x^2 + \frac{\dot{q}}{2k}Lx + T_w$$

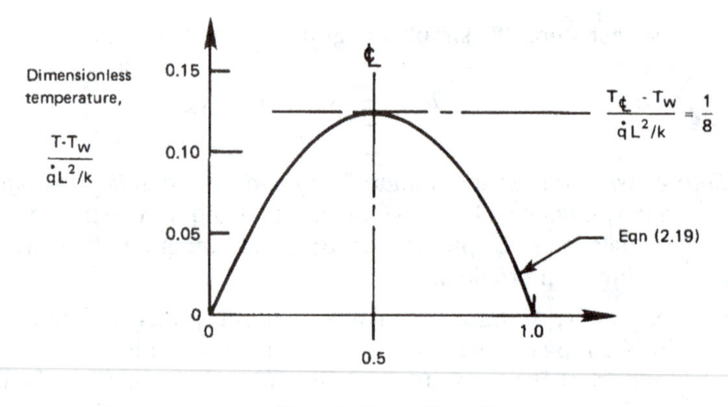

Figure 2.6 Temperature distribution in the setting concrete slab Example 2.1.

When we put this in neat dimensionless form, we can plot the result in Fig. 2.6 without having to know specific values of its parameters:

$$\frac{T - T_w}{\dot{q}L^2/k} = \frac{1}{2}\left[\frac{x}{L} - \left(\frac{x}{L}\right)^2\right] \tag{2.15}$$

Step 7. *Play with the solution—look it over—see what it has to tell you. Make any checks you can think of to be sure it is correct.* In this case, the resulting temperature distribution is parabolic and, as we would expect, symmetrical. It satisfies the boundary conditions at the wall and maximizes in the center. By nondimensionalizing the result, we can represent all situations with a simple curve. That is highly desirable when the calculations are not simple, as they are here. (Even here T actually depends on *five* different things, and its solution is a single curve on a two-coordinate graph.)

Finally, we check to see if the heat flux at the wall is correct:

$$q_{\text{wall}} = -k\frac{\partial T}{\partial x}\bigg|_{x=0} = k\left[\frac{\dot{q}}{k}x - \frac{\dot{q}L}{2k}\right]_{x=0} = -\frac{\dot{q}L}{2}$$

Thus, half of the total energy generated in the slab comes out of the front side, as we would expect. The solution appears to be correct.

Step 8. *If the temperature field is now correctly established, we can, if we wish, calculate the heat flux at any point in the body by substituting $T(\vec{r}, t)$ back into Fourier's law.* We did this already, in Step 7, to check our solution. ■

We offer additional examples in this section and the following one. In the process, we develop some important results for future use.

Example 2.2 The Simple Slab

A slab shown in Fig. 2.7 is at a steady state with dissimilar temperatures on either side and no internal heat generation. We want the temperature distribution and the heat flux through it.

Solution. These can be found quickly by following the steps set down in Example 2.1:

Step 1. $T = T(x)$ for steady x-direction heat flow

Step 2. $\dfrac{d^2T}{dx^2} = 0$, the steady 1-D heat equation with no $\dot{q}$

Step 3. $T = C_1x + C_2$ is the general solution of that equation

Step 4. $T(x = 0) = T_1$ and $T(x = L) = T_2$ are the b.c.s

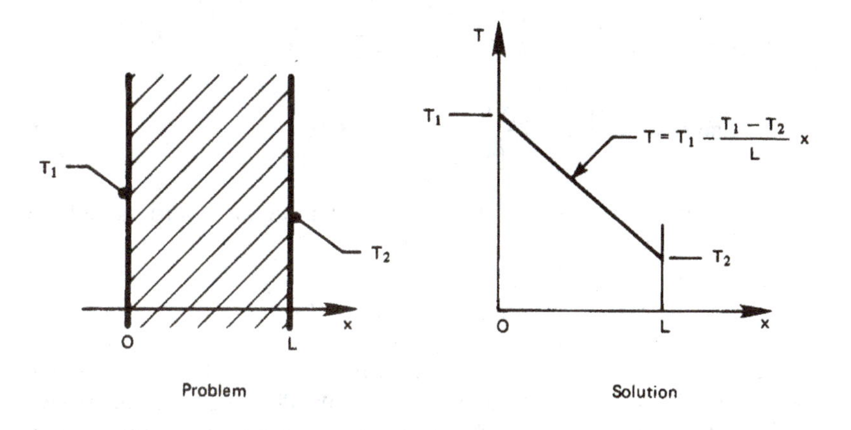

Figure 2.7 Heat conduction in a slab (Example 2.2).

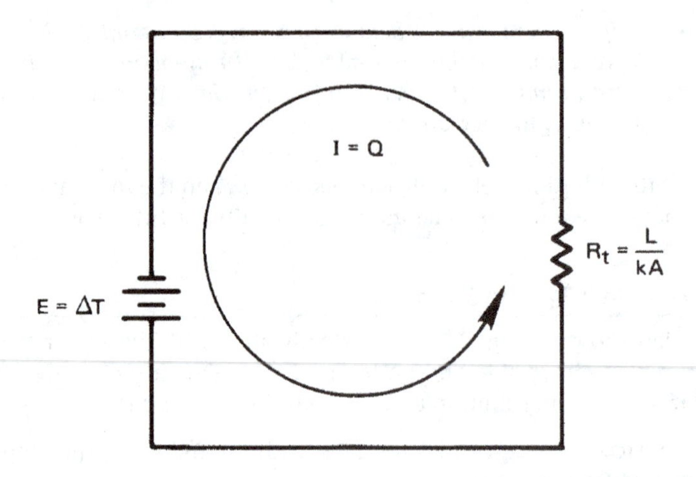

Figure 2.8 Ohm's law analogy to conduction through a slab.

Step 5. $T_1 = 0 + C_2$, so $C_2 = T_1$; and $T_2 = C_1 L + C_2$, so $C_1 = \dfrac{T_2 - T_1}{L}$

Step 6. $T = T_1 + \dfrac{T_2 - T_1}{L} x$; or $\dfrac{T - T_1}{T_2 - T_1} = \dfrac{x}{L}$

Step 7. We note that the solution satisfies the boundary conditions and that the temperature profile is linear.

Step 8. $q = -k \dfrac{dT}{dx} = -k \dfrac{d}{dx}\left(T_1 - \dfrac{T_1 - T_2}{L} x\right)$

so that $\boxed{q = k \dfrac{\Delta T}{L}}$ ■

This result, which is the simplest heat conduction solution, calls to mind Ohm's law. Thus, if we rearrange it:

$$Q = \frac{\Delta T}{L/kA} \quad \text{is like} \quad I = \frac{E}{R}$$

where L/kA assumes the role of a *thermal resistance*, to which we give the symbol R_t. R_t has the dimensions of (K/W). Figure 2.8 shows how we can represent heat flow through the slab with a diagram that is perfectly analogous to an electric circuit.

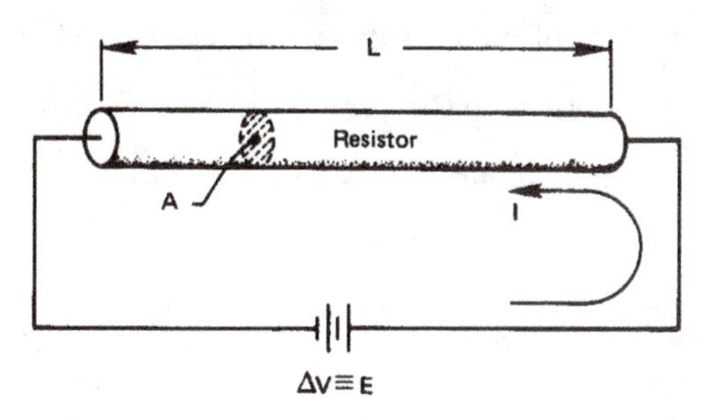

Figure 2.9 The one-dimensional flow of current.

2.3 Thermal resistance and the electrical analogy

Fourier's, Fick's, and Ohm's laws

Fourier's law has several extremely important analogies in other kinds of physical behavior, of which the electrical analogy is only one. These analogous processes provide us with a good deal of guidance in the solution of heat transfer problems. And, conversely, heat conduction analyses can often be adapted to describe those processes.

Let us first consider *Ohm's law* in three dimensions:

$$\text{flux of electrical charge} = \frac{\vec{I}}{A} \equiv \vec{J} = -\gamma\nabla V \tag{2.16}$$

$\vec{I}$ amperes is the vectorial electrical current, A is an area normal to the current vector, $\vec{J}$ is the flux of current or *current density*, γ is the electrical conductivity in cm/ohm$\cdot$cm^2, and V is the voltage.

To apply eqn. (2.16) to a one-dimensional current flow, as pictured in Fig. 2.9, we write eqn. (2.16) as

$$J = -\gamma\frac{dV}{dx} = \gamma\frac{\Delta V}{L}, \tag{2.17}$$

but ΔV is the applied voltage, E, and the resistance of the wire is $R \equiv L/\gamma A$. Then, since $I = J\,A$, eqn. (2.17) becomes

$$I = \frac{E}{R} \tag{2.18}$$

which is the familiar, but restrictive, one-dimensional statement of Ohm's law.

Fick's law is another analogous relation. It states that during mass diffusion, the flux, $\vec{j}_1$, of a dilute component, 1, into a second fluid, 2, is proportional to the gradient of its mass concentration, m_1. Thus

$$\vec{j}_1 = -\rho \mathcal{D}_{12} \nabla m_1 \tag{2.19}$$

where the constant $\mathcal{D}_{12}$ is the binary diffusion coefficient.

Example 2.3

Air fills a thin tube 1 m in length. There is a small water leak at one end where the water vapor concentration builds to a mass fraction of 0.01. A desiccator maintains the concentration at zero on the other side. What is the steady flux of water from one side to the other if $\mathcal{D}_{12}$ is 2.84×10^{-5} m²/s and $\rho = 1.18$ kg/m³?

SOLUTION.

$$\left|\vec{j}_{\text{water vapor}}\right| = 1.18 \frac{\text{kg}}{\text{m}^3}\left(2.84 \times 10^{-5} \frac{\text{m}^2}{\text{s}}\right)\left(\frac{0.01 \text{ kg H}_2\text{O/kg mixture}}{1 \text{ m}}\right)$$

$$= 3.35 \times 10^{-7} \frac{\text{kg}}{\text{m}^2 \cdot \text{s}}$$

■

Contact resistance

The usefulness of the electrical resistance analogy is particularly apparent at the interface of two conducting media. No two solid surfaces ever form perfect thermal contact when they are pressed together. Since some roughness is always present, a typical plane of contact will always include tiny air gaps as shown in Fig. 2.10 (which is drawn with a highly exaggerated vertical scale). Heat transfer follows two paths through such an interface. Conduction through points of solid-to-solid contact is very effective, but conduction through the gas-filled interstices, which have low thermal conductivity, can be very poor. Thermal radiation across the gaps is also inefficient.

We treat the contact surface by placing an interfacial conductance, h_c, in series with the conducting materials on either side. The coefficient h_c is similar to a heat transfer coefficient and has the same units, W/m²K. If ΔT is the temperature difference across an interface of area A, then $Q = Ah_c\Delta T$. It follows that $Q = \Delta T/R_t$ for a contact resistance $R_t = 1/(h_c A)$ in K/W.

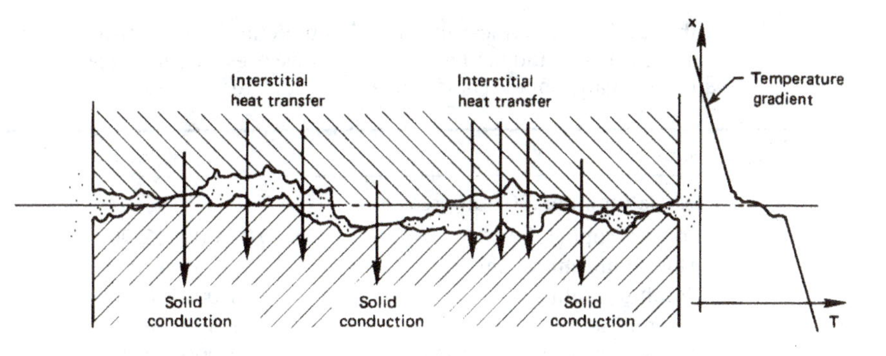

Figure 2.10 Heat transfer through the contact plane between two solid surfaces.

The interfacial conductance, h_c, depends on the following factors:

- The surface finish and cleanliness of the contacting solids.
- The materials that are in contact.
- The pressure with which the surfaces are forced together. This may vary over the surface, for example, in the vicinity of a bolt.
- The substance (or lack of it) in the interstitial spaces. Conductive shims or fillers can raise the interfacial conductance.
- The temperature at the contact plane.

The influence of contact pressure is usually a modest one up to around 10 atm in most metals. Beyond that, increasing plastic deformation of the local contact points causes h_c to increase more dramatically at high pressure. Table 2.1 gives typical values of contact resistances which bear out most of the preceding points. These values have been adapted from [2.1, Chpt. 3] and [2.2]. Theories of contact resistance are discussed in [2.3] and [2.4].

Example 2.4

Heat flows through two stainless steel slabs ($k = 18$ W/m·K) that are pressed together. The slab area is $A = 1$ m^2. How thick must the slabs be for contact resistance to be negligible?

Table 2.1 Some typical interfacial conductances for normal surface finishes and moderate contact pressures (about 1 to 10 atm). Air gaps not evacuated unless so indicated.

Situation	h_c (W/m^{2}K)
Iron/aluminum (70 atm pressure)	45,000
Copper/copper	10,000 – 25,000
Aluminum/aluminum	2,200 – 12,000
Graphite/metals	3,000 – 6,000
Ceramic/metals	1,500 – 8,500
Stainless steel/stainless steel	2,000 – 3,700
Ceramic/ceramic	500 – 3,000
Stainless steel/stainless steel (evacuated interstices)	200 – 1,100
Aluminum/aluminum (low pressure and evacuated interstices)	100 – 400

Solution. With reference to Fig. 2.11, the total or *equivalent* resistance is found by adding these resistances, which are in series:

$$R_{t_{\text{equiv}}} = \frac{L}{kA} + \frac{1}{h_c A} + \frac{L}{kA} = \frac{1}{A}\left(\frac{L}{18} + \frac{1}{h_c} + \frac{L}{18}\right)$$

Since h_c is about 3,000 W/m^{2}K,

$$\frac{2L}{18} \text{ must be } \gg \frac{1}{3000} = 0.00033$$

Thus, L must be large compared to 18(0.00033)/2 = 0.003 m if contact resistance is to be ignored. If L = 3 cm, the error is about 10%. ■

Resistances for cylinders and for convection

As we continue developing our method of solving one-dimensional heat conduction problems, we find that other avenues of heat flow may also be expressed as thermal resistances, and introduced into the solutions that we obtain. We also find that, once the heat conduction equation has been solved, the results themselves may be used as new thermal resistances.

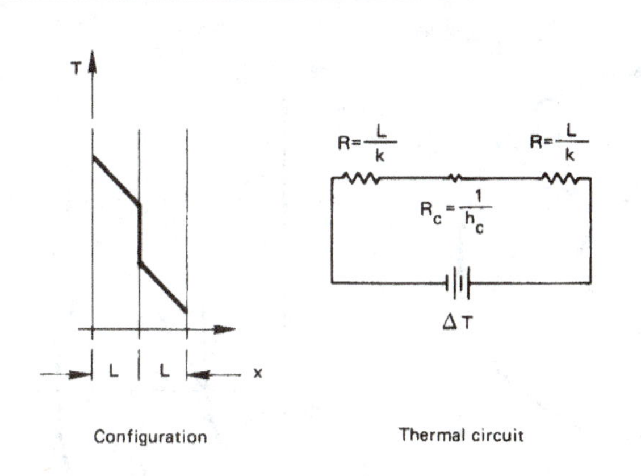

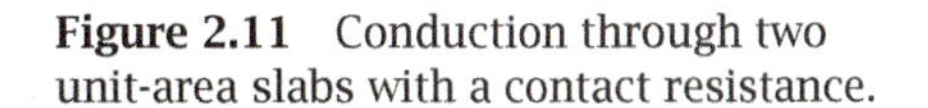
Figure 2.11 Conduction through two unit-area slabs with a contact resistance.

Example 2.5 Radial Heat Conduction in a Tube

Find the temperature distribution and the heat flux for the long hollow cylinder shown in Fig. 2.12.

Solution.

Step 1. $T = T(r)$

Step 2.

$$\frac{1}{r}\frac{\partial}{\partial r}\left(r\frac{\partial T}{\partial r}\right) + \underbrace{\frac{1}{r^2}\frac{\partial^2 T}{\partial \phi^2} + \frac{\partial^2 T}{\partial z^2}}_{=0,\ \text{since}\ T \neq T(\phi, z)} + \underbrace{\frac{\dot{q}}{k}}_{=0} = \underbrace{\frac{1}{\alpha}\frac{\partial T}{\partial t}}_{=0,\ \text{since steady}}$$

Step 3. Integrate once: $r\dfrac{\partial T}{\partial r} = C_1$; integrate again: $T = C_1 \ln r + C_2$

Step 4. $T(r = r_i) = T_i$ and $T(r = r_o) = T_o$

Step 5.

$$\begin{aligned} T_i &= C_1 \ln r_i + C_2 \\ T_o &= C_1 \ln r_o + C_2 \end{aligned} \quad \Longrightarrow \quad \left\{ \begin{aligned} C_1 &= \frac{T_i - T_o}{\ln(r_i/r_o)} = -\frac{\Delta T}{\ln(r_o/r_i)} \\ C_2 &= T_i + \frac{\Delta T}{\ln(r_o/r_i)} \ln r_i \end{aligned} \right.$$

Step 6. $T = T_i - \dfrac{\Delta T}{\ln(r_o/r_i)}(\ln r - \ln r_i)$ or

$$\boxed{\frac{T - T_i}{T_o - T_i} = \frac{\ln(r/r_i)}{\ln(r_o/r_i)}} \tag{2.20}$$

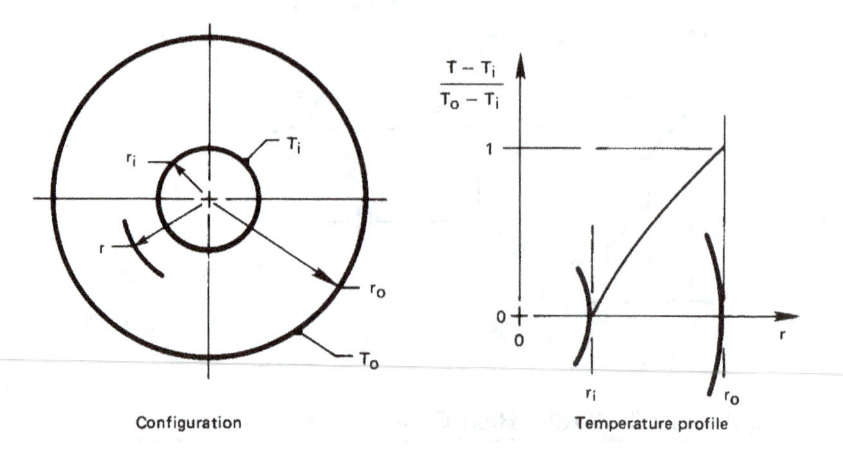

Figure 2.12 Heat transfer through a cylinder with a fixed wall temperature (Example 2.5).

Step 7. The solution is plotted in Fig. 2.12. We see that the temperature profile is logarithmic and that it satisfies both boundary conditions. Furthermore, it is instructive to see what happens when the wall of the cylinder is very thin, or when r_i/r_o is close to 1. In this case:

$$\ln(r/r_i) \simeq \frac{r}{r_i} - 1 = \frac{r - r_i}{r_i}$$

and

$$\ln(r_o/r_i) \simeq \frac{r_o - r_i}{r_i}$$

Thus eqn. (2.20) becomes

$$\frac{T - T_i}{T_o - T_i} = \frac{r - r_i}{r_o - r_i}$$

which is a simple linear profile. This is the same solution that we would get in a plane wall.

Step 8. At any station, r, with $\Delta T = T_i - T_o$:

$$q_{\text{radial}} = -k\frac{\partial T}{\partial r} = +\frac{k\Delta T}{\ln(r_o/r_i)}\frac{1}{r}$$

So the heat *flux* falls off inversely with radius. That is reasonable, since the same heat flow must pass through an increasingly large surface as the radius increases. Let us see if this is the case for a cylinder of length l:

$$Q\text{ (W)} = (2\pi r l)\, q = \frac{2\pi k l \Delta T}{\ln(r_o/r_i)} \neq f(r) \tag{2.21}$$

Finally, we again recognize Ohm's law in this result and write the thermal resistance for a cylinder:

$$R_{t_\text{cyl}} = \frac{\ln(r_o/r_i)}{2\pi l k} \left(\frac{\text{K}}{\text{W}}\right) \tag{2.22}$$

This can be compared with the resistance of a plane wall:

$$R_{t_\text{wall}} = \frac{L}{kA} \left(\frac{\text{K}}{\text{W}}\right)$$

Both resistances are inversely proportional to k, but each reflects a different geometry. ■

In the preceding examples, the boundary conditions were all the same—a temperature specified at an outer edge. Next let us suppose that the temperature is specified in the environment away from a body, with a heat transfer coefficient between the environment and the body.

Example 2.6 A Convective Boundary Condition

A convective heat transfer coefficient around the outside of the cylinder in Example 2.5 provides thermal resistance between the cylinder and an environment at $T = T_\infty$, as shown in Fig. 2.13. Find the temperature distribution and heat flux in this case.

SOLUTION.

Step 1 through 3. These are the same as in Example 2.5.

Step 4. The first boundary condition is $T(r = r_i) = T_i$. The second boundary condition must be expressed as an energy balance at the outer wall (recall Section 1.3).

$$q_\text{convection} = q_{\substack{\text{conduction}\\ \text{at the wall}}}$$

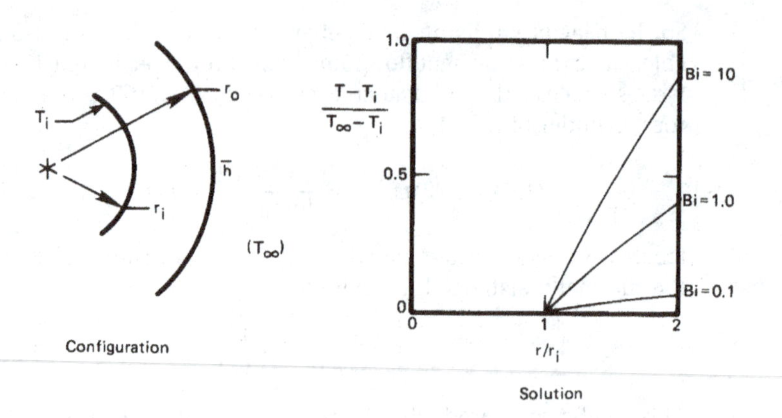

Figure 2.13 Heat transfer through a cylinder with a convective boundary condition (Example 2.6).

or

$$\overline{h}(T - T_\infty)_{r=r_o} = -k \left.\frac{\partial T}{\partial r}\right|_{r=r_o}$$

Step 5. From the first boundary condition we obtain $T_i = C_1 \ln r_i + C_2$. It is easy to make mistakes when we substitute the general solution into the second boundary condition, so we will do it in detail:

$$\overline{h}\Big[(C_1 \ln r + C_2) - T_\infty\Big]_{r=r_o} = -k\left[\frac{\partial}{\partial r}(C_1 \ln r + C_2)\right]_{r=r_o} \qquad (2.23)$$

A common error is to substitute $T = T_o$ on the lefthand side instead of substituting the entire general solution. That will do no good, because T_o is not an accessible piece of information. Equation (2.23) reduces to:

$$\overline{h}(T_\infty - C_1 \ln r_o - C_2) = \frac{kC_1}{r_o}$$

When we combine this with the result of the first boundary con-

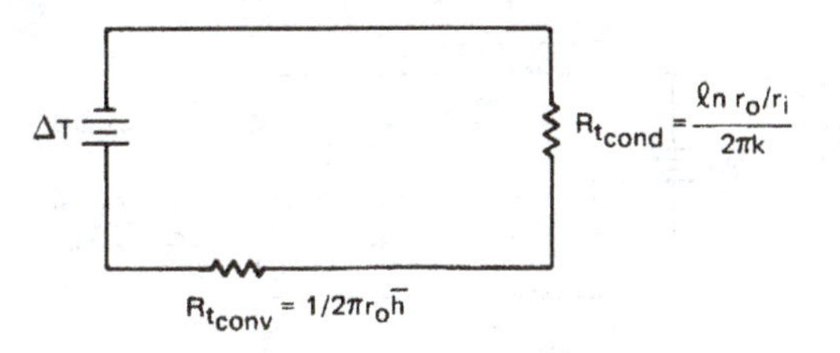

Figure 2.14 Thermal circuit with two resistances.

dition to eliminate C_2:

$$C_1 = -\frac{T_i - T_\infty}{k/(\overline{h}r_o) + \ln(r_o/r_i)} = \frac{T_\infty - T_i}{1/\mathrm{Bi} + \ln(r_o/r_i)}$$

Then

$$C_2 = T_i - \frac{T_\infty - T_i}{1/\mathrm{Bi} + \ln(r_o/r_i)} \ln r_i$$

Step 6.

$$T = \frac{T_\infty - T_i}{1/\mathrm{Bi} + \ln(r_o/r_i)} \ln(r/r_i) + T_i$$

This can be rearranged in fully dimensionless form:

$$\frac{T - T_i}{T_\infty - T_i} = \frac{\ln(r/r_i)}{1/\mathrm{Bi} + \ln(r_o/r_i)} \tag{2.24}$$

Step 7. Let us fix a value of r_o/r_i—say, 2—and plot eqn. (2.24) for several values of the Biot number. The results are included in Fig. 2.13. Some very important things show up in this plot. When $\mathrm{Bi} \gg 1$, the solution reduces to the solution given in Example 2.5. It is as though the convective resistance to heat flow were not there. That is exactly what we anticipated in Section 1.3 for large Bi. When $\mathrm{Bi} \ll 1$, the opposite is true: $(T - T_i)/(T_\infty - T_i)$ remains on the order of Bi, and internal conduction can be neglected. How big is big and how small is small? We do not really have to specify exactly. But in this case $\mathrm{Bi} < 0.1$ signals constancy of temperature inside the cylinder with about ±3%. $\mathrm{Bi} > 20$ means that we can neglect convection with about 5% error.

Step 8. $q_{\mathrm{radial}} = -k\dfrac{\partial T}{\partial r} = k\dfrac{T_i - T_\infty}{1/\mathrm{Bi} + \ln(r_o/r_i)}\dfrac{1}{r}$

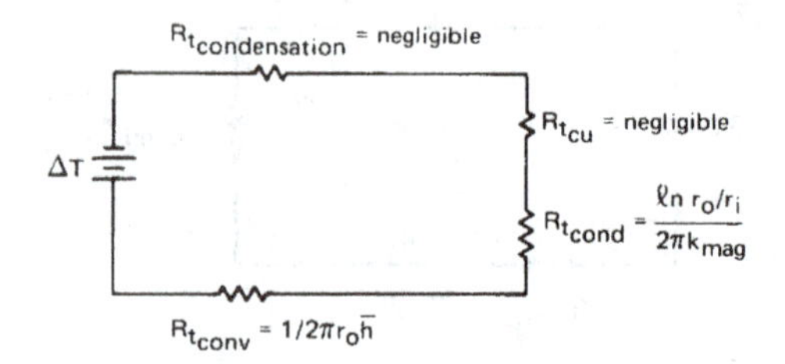

Figure 2.15 Thermal circuit for an insulated tube.

This can be written in terms of Q (W) $= q_{\text{radial}}\,(2\pi r l)$ for a cylinder of length l:

$$Q = \frac{T_i - T_\infty}{\dfrac{1}{\overline{h}\,2\pi r_o l} + \dfrac{\ln(r_o/r_i)}{2\pi k l}} = \frac{T_i - T_\infty}{R_{t_{\text{conv}}} + R_{t_{\text{cond}}}} \tag{2.25}$$

Equation (2.25) is once again analogous to Ohm's law. But this time the denominator is the sum of two thermal resistances, as would be the case in a series circuit. We accordingly present the analogous electrical circuit in Fig. 2.14.

The presence of convection on the outside surface of the cylinder causes a new thermal resistance of the form

$$R_{t_{\text{conv}}} = \frac{1}{\overline{h}A} \tag{2.26}$$

where A is the surface area over which convection occurs. ■

Example 2.7 Critical Radius of Insulation

An interesting consequence of the preceding result can be brought out with a specific example. Suppose that we insulate a 0.5 cm O.D. copper steam line with 85% magnesia to prevent the steam from condensing too rapidly. The steam is under pressure and stays at 150°C. The copper is thin and highly conductive—obviously a tiny resistance in series with the convective and insulation resistances, as we see in Fig. 2.15. The condensation of steam inside the tube also offers very little resistance.[3] But on the outside, a heat transfer coefficient of $\overline{h}$

[3]Condensation heat transfer is discussed in Chapter 8. It turns out that $\overline{h}$ is generally enormous during condensation so that $R_{t_{\text{condensation}}}$ is tiny.

$= 20$ W/m^{2}K offers fairly high resistance. It turns out that insulation can actually *improve* heat transfer in this case.

The two significant resistances, for a cylinder of unit length (l = 1 m), are

$$R_{t_{\text{cond}}} = \frac{\ln(r_o/r_i)}{2\pi k l} = \frac{\ln(r_o/r_i)}{2\pi(0.074)} \quad \text{K/W}$$

$$R_{t_{\text{conv}}} = \frac{1}{2\pi r_o \overline{h}} = \frac{1}{2\pi(20)r_o} \quad \text{K/W}$$

Figure 2.16 is a plot of these resistances and their sum. A very interesting thing occurs here. $R_{t_{\text{conv}}}$ falls off rapidly when r_o is increased, because the outside area is increasing. Accordingly, the total resistance passes through a minimum in this case. Will it always do so? To find out, we differentiate eqn. (2.25), again setting $l = 1$ m:

$$\frac{dQ}{dr_o} = \frac{(T_i - T_\infty)}{\left(\dfrac{1}{2\pi r_o \overline{h}} + \dfrac{\ln(r_o/r_i)}{2\pi k}\right)^2}\left(-\frac{1}{2\pi r_o^2 \overline{h}} + \frac{1}{2\pi k r_o}\right) = 0$$

When we solve this for the value of $r_o = r_{\text{crit}}$ at which Q is maximum and the total resistance is minimum, we obtain

$$\text{Bi} = 1 = \frac{\overline{h} r_{\text{crit}}}{k} \tag{2.27}$$

In the present example, adding insulation will *increase* heat loss instead of reducing it, until $r_{\text{crit}} = k/\overline{h} = 0.0037$ m or $r_{\text{crit}}/r_i = 1.48$. Indeed, insulation will not even start to do any good until $r_o/r_i = 2.32$ or $r_o = 0.0058$ m. We call r_{crit} the *critical radius* of insulation. ■

There is an interesting catch here. For most cylinders, $r_{\text{crit}} < r_i$ and the critical radius idiosyncrasy is of no concern. If our steam line had a 1 cm outside diameter, the critical radius difficulty would not have arisen. When cooling smaller diameter cylinders, such as electrical wiring, the critical radius must be considered, but one need not worry about it in the design of most large process equipment.

Resistance for thermal radiation

We saw in Chapter 1 that the net radiation exchanged by two objects is given by eqn. (1.34):

$$Q_{\text{net}} = A_1 \mathcal{F}_{1\text{-}2}\, \sigma\left(T_1^4 - T_2^4\right) \tag{1.34}$$

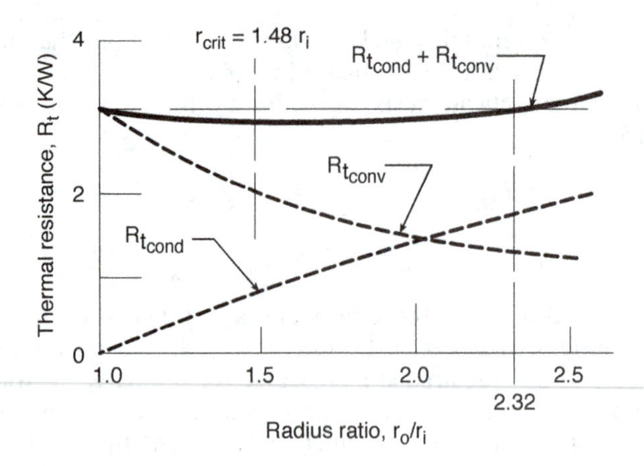

Figure 2.16 The critical radius of insulation (Example 2.7), written for a cylinder of unit length (l = 1 m).

When T_1 and T_2 are close, we can approximate this equation using a *radiation heat transfer coefficient*, h_{rad}. Specifically, suppose that the temperature difference, $\Delta T = T_1 - T_2$, is small compared to the mean temperature, $T_m = (T_1 + T_2)/2$. Then we can make the following expansion and approximation:

$$\begin{aligned} Q_{\text{net}} &= A_1 \mathcal{F}_{1\text{-}2}\, \sigma \left(T_1^4 - T_2^4\right) \\ &= A_1 \mathcal{F}_{1\text{-}2}\, \sigma (T_1^2 + T_2^2)(T_1^2 - T_2^2) \\ &= A_1 \mathcal{F}_{1\text{-}2}\, \sigma \underbrace{(T_1^2 + T_2^2)}_{= 2T_m^2 + (\Delta T)^2/2} \underbrace{(T_1 + T_2)}_{=2T_m} \underbrace{(T_1 - T_2)}_{=\Delta T} \\ &\cong A_1 \underbrace{\left(4\sigma T_m^3 \mathcal{F}_{1\text{-}2}\right)}_{\equiv h_{\text{rad}}} \Delta T \end{aligned} \tag{2.28}$$

where the last step assumes that $(\Delta T)^2/2 \ll 2T_m^2$ or $(\Delta T/T_m)^2/4 \ll 1$. Thus, we have identified the radiation heat transfer coefficient

$$\left.\begin{aligned} Q_{\text{net}} &= A_1 h_{\text{rad}} \Delta T \\ h_{\text{rad}} &= 4\sigma T_m^3 \mathcal{F}_{1\text{-}2} \end{aligned}\right\} \quad \text{for} \quad (\Delta T/T_m)^2/4 \ll 1 \tag{2.29}$$

This leads us immediately to the introduction of a radiation thermal resistance, analogous to that for convection:

$$R_{t_{\text{rad}}} = \frac{1}{A_1 h_{\text{rad}}} \tag{2.30}$$

For the special case of a small object (1) in a much larger environment (2), the transfer factor is given by eqn. (1.35) as $\mathcal{F}_{1\text{-}2} = \varepsilon_1$, so that

$$h_{\text{rad}} = 4\sigma T_m^3 \varepsilon_1 \tag{2.31}$$

If the small object is black, its emittance is $\varepsilon_1 = 1$ and h_{rad} is maximized. For a black object radiating near room temperature, say $T_m = 300$ K,

$$h_{\text{rad}} = 4(5.67 \times 10^{-8})(300)^3 \cong 6 \text{ W/m}^2\text{K}$$

This value is of approximately the same size as $\overline{h}$ for natural convection into a gas at such temperatures. Thus, the heat transfer by thermal radiation and natural convection into gases are similar. Both effects must be taken into account. In forced convection in gases, on the other hand, $\overline{h}$ might well be larger than h_{rad} by an order of magnitude or more, so that thermal radiation can be neglected.

Example 2.8

An electrical resistor dissipating 0.1 W has been mounted well away from other components in an electronical cabinet. It is cylindrical with a 3.6 mm O.D. and a length of 10 mm. If the air in the cabinet is at 35°C and at rest, and the resistor has $\overline{h} = 13$ W/m^{2}K for natural convection and $\varepsilon = 0.9$, what is the resistor's temperature? Assume that the electrical leads are configured so that little heat is conducted into them.

Solution. The resistor may be treated as a small object in a large isothermal environment. To compute h_{rad}, let us estimate the resistor's temperature as 50°C. Then

$$T_m = (35 + 50)/2 \cong 43°\text{C} = 316 \text{ K}$$

so

$$h_{\text{rad}} = 4\sigma T_m^3 \varepsilon = 4(5.67 \times 10^{-8})(316)^3(0.9) = 6.44 \text{ W/m}^2\text{K}$$

Heat is lost by natural convection and thermal radiation acting in parallel. To find the equivalent thermal resistance, we combine the two parallel resistances as follows:

$$\frac{1}{R_{t_{\text{equiv}}}} = \frac{1}{R_{t_{\text{rad}}}} + \frac{1}{R_{t_{\text{conv}}}} = Ah_{\text{rad}} + A\overline{h} = A(h_{\text{rad}} + \overline{h})$$

Thus,

$$R_{t_{\text{equiv}}} = \frac{1}{A(h_{\text{rad}} + \overline{h})}$$

A calculation shows $A = 133\ \text{mm}^2 = 1.33 \times 10^{-4}\ \text{m}^2$ for the resistor surface. Thus, the equivalent thermal resistance is

$$R_{t_{\text{equiv}}} = \frac{1}{(1.33 \times 10^{-4})(13 + 6.44)} = 386.8\ \text{K/W}$$

Since

$$Q = \frac{T_{\text{resistor}} - T_{\text{air}}}{R_{t_{\text{equiv}}}}$$

We find

$$T_{\text{resistor}} = T_{\text{air}} + Q \cdot R_{t_{\text{equiv}}} = 35 + (0.1)(386.8) = 73.68\ ^\circ\text{C}$$

We guessed a resistor temperature of 50°C in finding h_{rad}. Recomputing with this higher temperature, we have $T_m = 327$ K and $h_{\text{rad}} = 7.17\ \text{W/m}^2\text{K}$. If we repeat the rest of the calculation, we get a new value $T_{\text{resistor}} = 72.3$°C. Further iteration is not needed.

Since the use of h_{rad} is an approximation, we should check its applicability:

$$\frac{1}{4}\left(\frac{\Delta T}{T_m}\right)^2 = \frac{1}{4}\left(\frac{72.3 - 35.0}{327}\right)^2 = 0.00325 \ll 1$$

In this case, the approximation is a very good one. ■

Example 2.9

Suppose that power to the resistor in Example 2.8 is turned off. How long does it take to cool? The resistor has $k \cong 10$ W/m·K, $\rho \cong 2000\ \text{kg/m}^3$, and $c_p \cong 700$ J/kg·K.

Solution. The lumped capacity model, eqn. (1.22), may be applicable. To find out, we check the resistor's Biot number, noting that

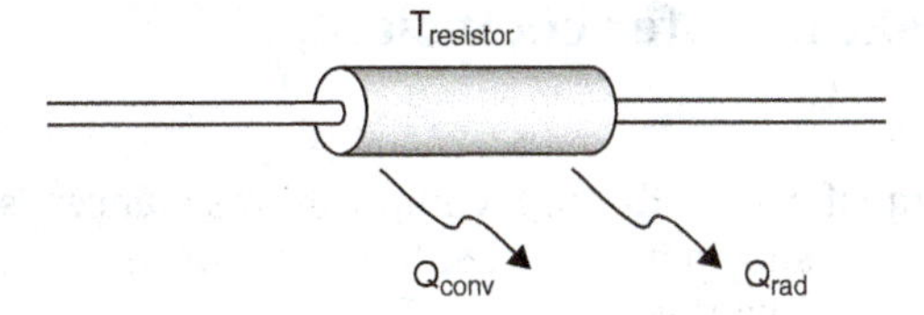

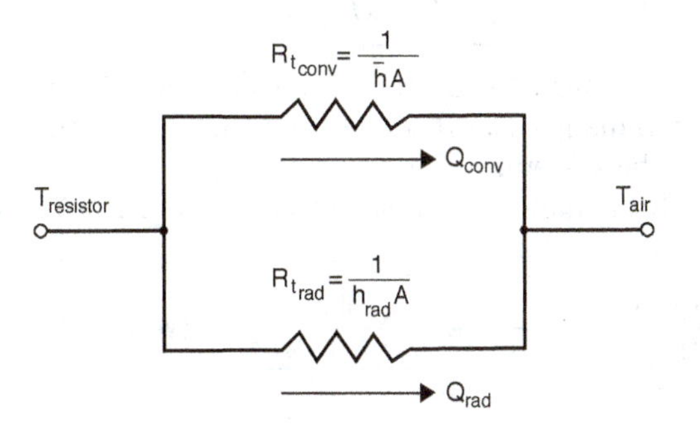

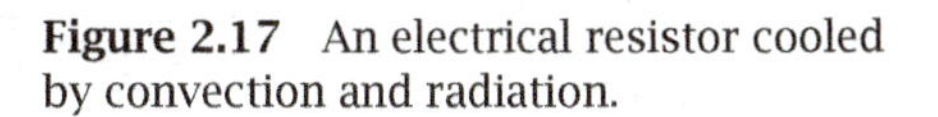

Figure 2.17 An electrical resistor cooled by convection and radiation.

the parallel convection and radiation processes have an *effective* heat transfer coefficient $h_{\text{eff}} = \overline{h} + h_{\text{rad}} = 20.17\ \text{W/m}^2\text{K}$. Then,

$$\text{Bi} = \frac{h_{\text{eff}} r_o}{k} = \frac{(20.17)(0.0036/2)}{10} = 0.0036 \ll 1$$

so eqn. (1.22) can be used to describe the cooling process. The time constant is

$$T = \frac{\rho c_p V}{h_{\text{eff}} A} = \frac{(2000)(700)\pi(0.010)(0.0036)^2/4}{(20.17)(1.33 \times 10^{-4})} = 53.1\ \text{s}$$

From eqn. (1.22) with $T_0 = 72.3°\text{C}$

$$T_{\text{resistor}} = 35.0 + (72.3 - 35.0)e^{-t/53.1}\ °\text{C}$$

Ninety-five percent of the total temperature drop has occured when $t = 3T = 159$ s. ■

2.4 Overall heat transfer coefficient, U

Definition

We often want to transfer heat through composite resistances, such as the series of resistances shown in Fig. 2.18. It is very convenient to have a number, U, that works like this[4]:

$$\boxed{Q = U A \,\Delta T} \tag{2.32}$$

This number, called the *overall heat transfer coefficient*, is defined largely by the system, and in many cases it proves to be insensitive to the operating conditions of the system.

In Example 2.6, for instance, two resistances are in series. We can use the value Q given by eqn. (2.25) to get

$$U = \frac{Q\ (\mathrm{W})}{[2\pi r_o l\ (\mathrm{m^2})]\ \Delta T\ (\mathrm{K})} = \frac{1}{\dfrac{1}{\overline{h}} + \dfrac{r_o \ln(r_o/r_i)}{k}} \quad (\mathrm{W/m^2K}) \tag{2.33}$$

We have based U on the outside area, $A_o = 2\pi r_o l$, in this case. We might instead have based it on inside area, $A_i = 2\pi r_i l$, and obtained

$$U = \frac{1}{\dfrac{r_i}{\overline{h} r_o} + \dfrac{r_i \ln(r_o/r_i)}{k}} \tag{2.34}$$

It is therefore important to remember which area an overall heat transfer coefficient is based on. It is particularly important that A and U be consistent when we write $Q = U A \,\Delta T$.

In general, for any composite resistance, the overall heat transfer coefficient may be obtained from the equivalent resistance. The equivalent resistance is calculated taking account of series and parallel resistors, as in Examples 2.4 and 2.8. Then, because $Q = \Delta T / R_{t_{\mathrm{equiv}}} = U A \,\Delta T$, it follows that $UA = 1/R_{t_{\mathrm{equiv}}}$.

Example 2.10

Estimate the overall heat transfer coefficient for the tea kettle shown in Fig. 2.19. Note that the flame convects heat to the thin aluminum. The heat is then conducted through the aluminum and finally convected by boiling into the water.

[4]This U must not be confused with internal energy. The two terms should always be distinct in context.

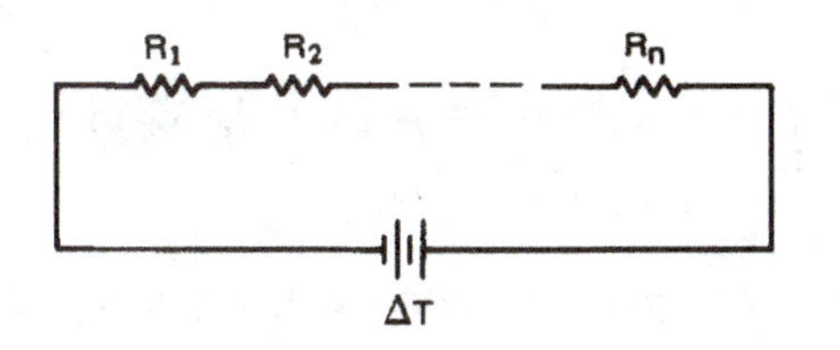

Figure 2.18 A thermal circuit with many resistances in series. The equivalent resistance is $R_{t_{\text{equiv}}} = \sum_i R_i$.

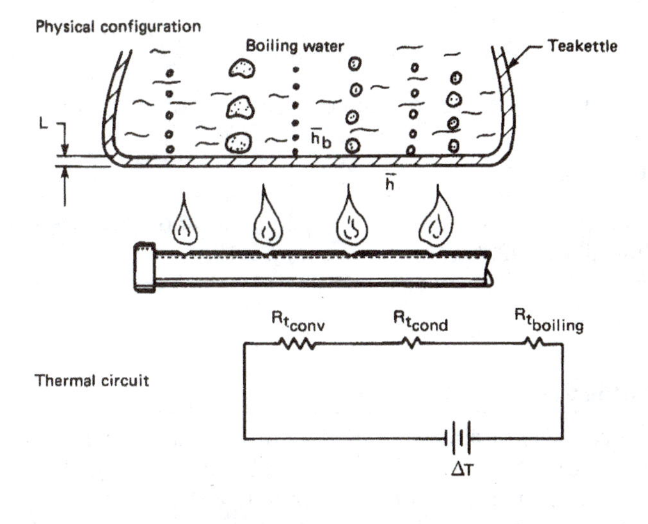

Figure 2.19 Heat transfer through the bottom of a tea kettle.

Solution. We need not worry about deciding which area to base A on, in this case, because the area normal to the heat flux vector does not change. We simply write the heat flow

$$Q = \frac{\Delta T}{\sum R_t} = \frac{T_{\text{flame}} - T_{\text{boiling water}}}{\dfrac{1}{\overline{h}A} + \dfrac{L}{k_{\text{Al}}A} + \dfrac{1}{\overline{h}_b A}}$$

and apply the definition of U

$$U = \frac{Q}{A\Delta T} = \frac{1}{\dfrac{1}{\overline{h}} + \dfrac{L}{k_{\text{Al}}} + \dfrac{1}{\overline{h}_b}}$$

Let us see what typical numbers would look like in this example: $\overline{h}$ might be around 200 W/m^{2}K; L/k_{Al} might be 0.001 m/(160 W/m·K) or 1/160,000 W/m^{2}K; and $\overline{h}_b$ is quite large—perhaps about 5000 W/m^{2}K.

Thus:

$$U \simeq \frac{1}{\dfrac{1}{200} + \dfrac{1}{160{,}000} + \dfrac{1}{5000}} = 192.1 \text{ W/m}^2\text{K}$$ ■

It is clear that the first resistance is dominant, as is shown in Fig. 2.19. Notice that in such cases

$$UA \longrightarrow 1/R_{t_{\text{dominant}}} \tag{2.35}$$

where A is any area (inside or outside) in the thermal circuit.

Experiment 2.1

Boil water in a paper cup over an open flame and explain why you can do so. [Recall eqn. (2.35) and see Problem 2.12.]

Example 2.11

A wall consists of alternating layers of pine and sawdust, as shown in Fig. 2.20). The sheathes on the outside have negligible resistance and $\overline{h}$ is known on the sides. Compute Q and U for the wall.

Solution. So long as the wood and the sawdust do not differ dramatically from one another in thermal conductivity, we can approximate the wall as a parallel resistance circuit, as shown in the figure.[5] The equivalent thermal resistance of the circuit is

$$R_{t_{\text{equiv}}} = R_{t_{\text{conv}}} + \frac{1}{\left(\dfrac{1}{R_{t_{\text{pine}}}} + \dfrac{1}{R_{t_{\text{sawdust}}}}\right)} + R_{t_{\text{conv}}}$$

Thus

$$Q = \frac{\Delta T}{R_{t_{\text{equiv}}}} = \frac{T_{\infty_1} - T_{\infty_r}}{\dfrac{1}{\overline{h}A} + \dfrac{1}{\left(\dfrac{k_p A_p}{L} + \dfrac{k_s A_s}{L}\right)} + \dfrac{1}{\overline{h}A}}$$

[5] For this approximation to be exact, the resistances must be equal. If they differ radically, the problem must be treated as two-dimensional.

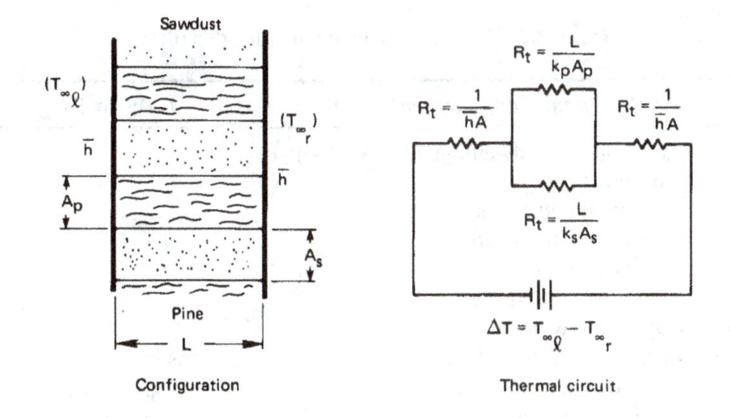

Figure 2.20 Heat transfer through a composite wall.

and

$$U = \frac{Q}{A\Delta T} = \frac{1}{\dfrac{2}{\bar{h}} + \dfrac{1}{\left(\dfrac{k_p}{L}\dfrac{A_p}{A} + \dfrac{k_s}{L}\dfrac{A_s}{A}\right)}}$$ ■

The approach illustrated in this example is very widely used in calculating *U* values for the walls and roofs houses and buildings. The thermal resistances of each structural element — insulation, studs, siding, doors, windows, etc. — are combined to calculate *U* or $R_{t_{\text{equiv}}}$, which is then used together with weather data to estimate heating and cooling loads [2.5].

Typical values of *U*

In a fairly general use of the word, a heat exchanger is anything that lies between two fluid masses at different temperatures. In this sense a heat exchanger might be designed either to impede or to enhance heat exchange. Consider some typical values of *U* shown in Table 2.2, which were assembled from a variety of technical sources. If the exchanger is intended to improve heat exchange, *U* will generally be much greater than 40 W/m^{2}K. If it is intended to impede heat flow, it will be less than 10 W/m^{2}K—anywhere down to almost perfect insulation. You should have some numerical concept of relative values of *U*, so we recommend

Table 2.2 Typical ranges or magnitudes of U

Heat Exchange Configuration	U (W/m^{2}K)
Walls and roofs dwellings with a 24 km/h outdoor wind:	
• Insulated roofs	0.3–2
• Finished masonry walls	0.5–6
• Frame walls	0.3–5
• Uninsulated roofs	1.2–4
Single-pane windows	$\sim 6^\dagger$
Air to heavy tars and oils	As low as 45
Air to low-viscosity liquids	As high as 600
Air to various gases	60–550
Steam or water to oil	60–340
Liquids in coils immersed in liquids	110–2,000
Feedwater heaters	110–8,500
Air condensers	350–780
Steam-jacketed, agitated vessels	500–1,900
Shell-and-tube ammonia condensers	800–1,400
Steam condensers with 25°C water	1,500–5,000
Condensing steam to high-pressure boiling water	1,500–10,000

† Main heat loss is by infiltration.

that you scrutinize the numbers in Table 2.2. Some things worth bearing in mind are:

- The fluids with low thermal conductivities, such as tars, oils, or any of the gases, usually yield low values of $\overline{h}$. When such fluid flows on one side of an exchanger, U will generally be pulled down.
- Condensing and boiling are very effective heat transfer processes. They greatly improve U but they cannot override one very small value of $\overline{h}$ on the other side of the exchange. (Recall Example 2.10.)
- For a high U, *all* resistances in the exchanger must be low.
- The highly conducting liquids, such as water and liquid metals, give high values of $\overline{h}$ and U.

Fouling resistance

Figure 2.21 shows one of the simplest forms of a heat exchanger—a pipe. The inside is new and clean on the left, but on the right it has built up a layer of scale. In conventional freshwater preheaters, for example, this scale is typically $MgSO_4$ (magnesium sulfate) or $CaSO_4$ (calcium sulfate) which precipitates onto the pipe wall after a time. To account for the resistance offered by these buildups, we must include an additional, highly empirical resistance when we calculate U. Thus, for the pipe shown in Fig. 2.21,

$$U\Big|_{\substack{\text{older pipe}\\ \text{based on } A_i}} = \frac{1}{\dfrac{1}{h_i} + \dfrac{r_i \ln(r_o/r_p)}{k_{\text{insul}}} + \dfrac{r_i \ln(r_p/r_i)}{k_{\text{pipe}}} + \dfrac{r_i}{r_o h_o} + R_f}$$

where R_f is a fouling resistance for a unit area of pipe (in $\text{m}^2\text{K/W}$). And clearly

$$\boxed{R_f \equiv \frac{1}{U_{\text{old}}} - \frac{1}{U_{\text{new}}}} \tag{2.36}$$

Some typical values of R_f are given in Table 2.3. These values have been adapted from [2.6] and [2.7]. Notice that fouling has the effect of adding a resistance in series on the order of 10^{-4} $\text{m}^2\text{K/W}$. It is rather like another heat transfer coefficient, $\overline{h}_f$, on the order of 10,000 $\text{W/m}^2\text{K}$ in series with the other resistances in the exchanger.

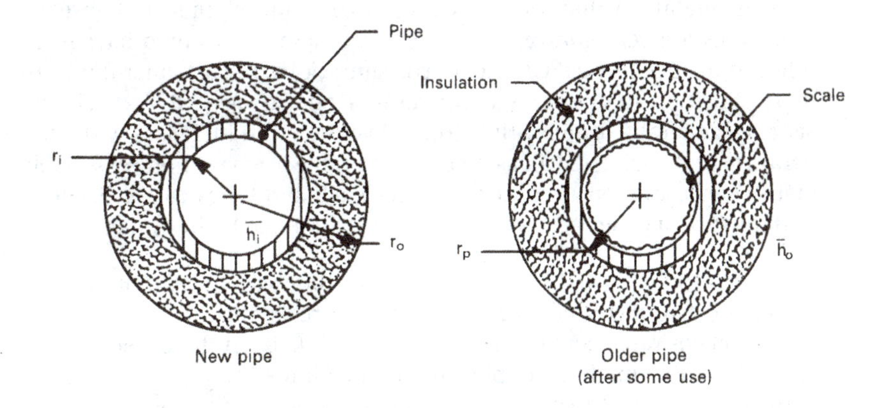

Figure 2.21 The fouling of a pipe.

Table 2.3 Some typical fouling resistances for a unit area.

Fluid and Situation	*Fouling Resistance* R_f (m²K/W)
Distilled water	0.0001
Seawater	0.0001 – 0.0004
Treated boiler feedwater	0.0001 – 0.0002
Clean river or lake water	0.0002 – 0.0006
About the worst waters used in heat exchangers	< 0.0020
No. 6 fuel oil	0.0001
Transformer or lubricating oil	0.0002
Most industrial liquids	0.0002
Most refinery liquids	0.0002 – 0.0009
Steam, non-oil-bearing	0.0001
Steam, oil-bearing (e.g., turbine exhaust)	0.0003
Most stable gases	0.0002 – 0.0004
Flue gases	0.0010 – 0.0020
Refrigerant vapors (oil-bearing)	0.0040

The tabulated values of R_f are given to only one significant figure because they are very approximate. Clearly, exact values would have to be referred to specific heat exchanger configurations, to particular fluids, to fluid velocities, to operating temperatures, and to age [2.8, 2.9]. The resistance generally drops with increased velocity and increases with temperature and age. The values given in the table are based on reasonable maintenance and the use of conventional shell-and-tube heat exchangers. With misuse, a given heat exchanger can yield much higher values of R_f.

Notice too, that if $U \lesssim 1{,}000$ W/m²K, fouling will be unimportant because it will introduce a negligibly small resistance in series. Thus, in a water-to-water heat exchanger, for which U is on the order of 2000 W/m²K, fouling might be important; but in a finned-tube heat exchanger with hot gas in the tubes and cold gas passing across the fins on them, U might be around 200 W/m²K, and fouling will be usually be insignificant.

Example 2.12

You have unpainted aluminum siding on your house and the engineer has based a heat loss calculation on $U = 5$ W/m^{2}K. You discover that air pollution levels are such that R_f is 0.0005 m^{2}K/W on the siding. Should the engineer redesign the siding?

Solution. From eqn. (2.36) we get

$$\frac{1}{U_{\text{corrected}}} = \frac{1}{U_{\text{uncorrected}}} + R_f = 0.2000 + 0.0005 \text{ m}^2\text{K/W}$$

Therefore, fouling is entirely irrelevant to domestic heat loads. ■

Example 2.13

Since the engineer did not fail you in the preceding calculation, you entrust him with the installation of a heat exchanger at your plant. He installs a water-cooled steam condenser with $U = 4000$ W/m^{2}K. You discover that he used water-side fouling resistance for distilled water but that the water flowing in the tubes is not clear at all. How did he do this time?

Solution. Equation (2.36) and Table 2.3 give

$$\begin{aligned}\frac{1}{U_{\text{corrected}}} &= \frac{1}{4000} + (0.0006 \text{ to } 0.0020)\\ &= 0.00085 \text{ to } 0.00225 \text{ m}^2\text{K/W}\end{aligned}$$

Thus, U is reduced from 4,000 to between 444 and 1,176 W/m^{2}K. Fouling is crucial in this case, and the engineer was in serious error.■

2.5 Summary

Four things have been done in this chapter:

- The heat diffusion equation has been established. A method has been established for solving it in simple problems, and some important results have been presented. (We say much more about solving the heat diffusion equation in Part II of this book.)

- We have explored the electric analogy to steady heat flow, paying special attention to the concept of thermal resistance. We exploited the analogy to solve heat transfer problems in the same way we solve electrical circuit problems.

- The overall heat transfer coefficient has been defined, and we have seen how to build it up out of component resistances.

- Some practical problems encountered in the evaluation of overall heat transfer coefficients have been discussed.

Three very important things have *not* been considered in Chapter 2:

- In all evaluations of U that involve values of $\overline{h}$, we have taken these values as given information. In any real situation, we must determine correct values of $\overline{h}$ for the specific situation. Part III deals with such determinations.

- When fluids flow through heat exchangers, they give up or gain energy. Thus, the driving temperature difference varies through the exchanger. (Problem 2.14 asks you to consider this difficulty in its simplest form.) That complicates heat exchanger design, and we learn how to deal with it in Chapter 3.

- The heat transfer coefficients themselves vary with position inside many types of heat exchangers, causing U to be position-dependent.

Problems

2.1 Prove that if k varies linearly with T in a slab, and if heat transfer is one-dimensional and steady, then q may be evaluated precisely using k evaluated at the mean temperature in the slab.

2.2 Invent a numerical method for calculating the steady heat flux through a plane wall when $k(T)$ is an arbitrary function. Use the method to predict q in an iron slab 1 cm thick if the temperature varies from $-100°$C on the left to $400°$C on the right. How far would you have erred if you had taken $k_{\text{average}} = (k_{\text{left}} + k_{\text{right}})/2$?

2.3 The steady heat flux at one side of a slab is a known value q_o. The thermal conductivity varies with temperature in the slab, and the variation can be expressed with a power series as

$$k = \sum_{i=0}^{i=n} A_i T^i$$

(a) Start with eqn. (2.10) and derive an equation that relates T to position in the slab, x. (b) Calculate the heat flux at any position in the wall from this expression using Fourier's law. Is the resulting q a function of x?

2.4 Combine Fick's law with the principle of conservation of mass (of the dilute species) in such a way as to eliminate j_1, and obtain a second-order differential equation in m_1. Discuss the importance and the use of the result.

2.5 Solve for the temperature distribution in a thick-walled pipe if the bulk interior temperature and the exterior air temperature, T_{∞_i}, and T_{∞_o}, are known. The interior and the exterior heat transfer coefficients are $\overline{h}_i$ and $\overline{h}_o$, respectively. Follow the method in Example 2.6 and put your result in the dimensionless form:

$$\frac{T - T_{\infty_i}}{T_{\infty_i} - T_{\infty_o}} = \text{fn}\left(\text{Bi}_i, \text{Bi}_o, r/r_i, r_o/r_i\right)$$

2.6 Put the boundary conditions from Problem 2.5 into dimensionless form so that the Biot numbers appear in them. Let the Biot numbers approach infinity. This should get you back to the boundary conditions for Example 2.5. Therefore, the solution that you obtain in Problem 2.5 should reduce to the solution of Example 2.5 when the Biot numbers approach infinity. Show that this is the case.

2.7 Write an accurate explanation of the idea of *critical radius of insulation* that your kid brother or sister, who is still in grade school, could understand. (If you do not have an available kid, borrow one to see if your explanation really works.)

2.8 The slab shown in Fig. 2.22 is embedded on five sides in insulating materials. The sixth side is exposed to an ambient temperature through a heat transfer coefficient. Heat is generated in the slab at the rate of 1.0 kW/m^3 The thermal conductivity

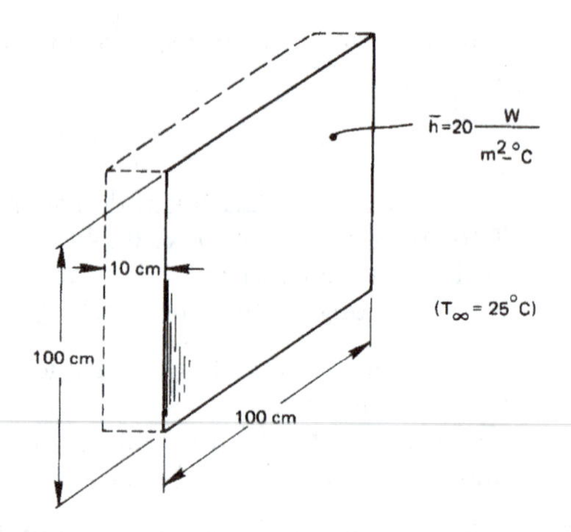

Figure 2.22 Configuration for Problem 2.8.

of the slab is 0.2 W/m·K. (a) Solve for the temperature distribution in the slab, noting any assumptions you must make. Be careful to clearly identify the boundary conditions. (b) Evaluate T at the front and back faces of the slab. (c) Show that your solution gives the expected heat fluxes at the back and front faces.

2.9 Consider the composite wall shown in Fig. 2.23. The concrete and brick sections are of equal thickness. Determine T_1, T_2, q, and the percentage of q that flows through the brick. To do this, approximate the heat flow as one-dimensional. Draw the thermal circuit for the wall and identify all four resistances before you begin.

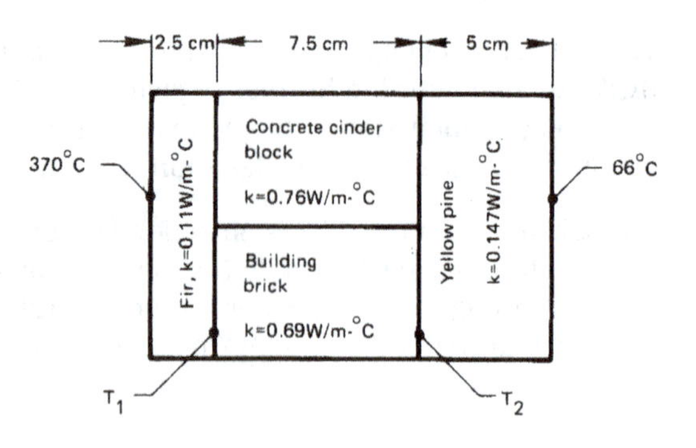

Figure 2.23 Configuration for Problem 2.9.

2.10 Compute Q and U for Example 2.11 if the wall is 0.3 m thick. Five (each) pine and sawdust layers are 5 and 8 cm thick, respectively; and the heat transfer coefficients are 10 on the left and 18 on the right. $T_{\infty_l} = 30°C$ and $T_{\infty_r} = 10°C$.

2.11 Compute U for the slab in Example 1.2.

2.12 Consider the tea kettle in Example 2.10. Suppose that the kettle holds 1 kg of water (about 1 liter) and that the flame impinges on 0.02 m^2 of the bottom. (a) Find out how fast the water temperature is increasing when it reaches its boiling point, and calculate the temperature of the bottom of the kettle immediately below the water if the gases from the flame are at 500°C when they touch the bottom of the kettle. Assume that the heat capacitance of the aluminum kettle is negligible. (b) There is an old parlor trick in which one puts a *paper* cup of water over an open flame and boils the water without burning the paper (see Experiment 2.1). Explain this *using an electrical analogy*. [(a): $dT/dt = 0.37°C/s$.]

2.13 Copper plates 2 mm and 3 mm in thickness are processed rather lightly together. Non-oil-bearing steam condenses under pressure at $T_{sat} = 200°C$ on one side ($\overline{h} = 12{,}000$ W/m^{2}K) and methanol boils under pressure at 130°Con the other ($\overline{h} =$ 9000 W/m^{2}K). Estimate U and q initially and after extended service. List the relevant thermal resistances in order of decreasing importance and suggest whether or not any of them can be ignored.

2.14 0.5 kg/s of air at 20°C moves along a channel that is 1 m from wall to wall. One wall of the channel is a heat exchange surface ($U = 300$ W/m^{2}K) with steam condensing at 120°C on its back. Determine (a) q at the entrance; (b) the rate of increase of temperature of the fluid with x at the entrance; (c) the temperature and heat flux 2 m downstream. [(c): $T_{2m} = 89.7°C$.]

2.15 An isothermal sphere 3 cm in diameter is kept at 80°C in a large clay region. The temperature of the clay far from the sphere is kept at 10°C. How much heat must be supplied to the sphere to maintain its temperature if $k_{clay} = 1.28$ W/m·K? (*Hint:* You must solve the boundary value problem not in the sphere but in the clay surrounding it.) [$Q = 16.9$ W.]

2.16 Is it possible to increase the heat transfer from a convectively cooled isothermal sphere by adding insulation? Explain fully.

2.17 A wall consists of layers of metals and plastic with heat transfer coefficients on either side. U is 255 W/m^{2}K and the overall temperature difference is 200°C. One layer in the wall is stainless steel (k = 18 W/m·K) 3 mm thick. What is ΔT across the stainless steel?

2.18 A 1% carbon-steel sphere 20 cm in diameter is kept at 250°C on the outside. It has an 8 cm diameter cavity containing boiling water ($\overline{h}_{\text{inside}}$ is very high) which is vented to the atmosphere. What is Q through the shell?

2.19 A slab is insulated on one side and exposed to a surrounding temperature, T_∞, through a heat transfer coefficient on the other. There is nonuniform heat generation in the slab such that $\dot{q}$ =[A (W/m^4)][x (m)], where x = 0 at the insulated wall and x = L at the cooled wall. Derive the temperature distribution in the slab.

2.20 800 W/m^3 of heat is generated within a 10 cm diameter nickel-steel sphere for which k = 10 W/m·K. The environment is at 20°C and there is a natural convection heat transfer coefficient of 10 W/m^{2}K around the outside of the sphere. What is its center temperature at the steady state? [21.37°C.]

2.21 An outside pipe is insulated and we measure its temperature with a thermocouple. The pipe serves as an electrical resistance heater, and $\dot{q}$ is known from resistance and current measurements. The inside of the pipe is cooled by the flow of liquid with a known bulk temperature. Evaluate the heat transfer coefficient, $\overline{h}$, in terms of known information. The pipe dimensions and properties are known. (*Hint:* Remember that $\overline{h}$ is not known and we cannot use a boundary condition of the third kind at the inner wall to get $T(r)$.)

2.22 Consider the hot water heater in Problem 1.11. Suppose that it is insulated with 2 cm of a material for which k = 0.12 W/m·K, and suppose that $\overline{h}$ = 16 W/m^{2}K. Find (a) the time constant T for the tank, neglecting the casing and insulation; (b) the initial rate of cooling in °C/h; (c) the time required for the water to cool from its initial temperature of 75°C to 40°C; (d) the percentage of additional heat loss that would result if an outer casing for the insulation were held on by eight steel rods, 1 cm in diameter, between the inner and outer casings.

2.23 A slab of thickness L is subjected to a constant heat flux, q_1, on the left side. The right-hand side if cooled convectively by an environment at T_∞. (a) Develop a dimensionless equation for the temperature of the slab. (b) Present dimensionless equation for the left- and right-hand wall temperatures as well. (c) If the wall is firebrick, 10 cm thick, q_1 is 400 W/m^2, $\overline{h}$ = 20 W/m^{2}K, and T_∞ = 20°C, compute the lefthand and righthand temperatures.

2.24 Heat flows steadily through a stainless steel wall of thickness L_{ss} = 0.06 m, with a variable thermal conductivity of k_{ss} = 1.67 + 0.0143 T(°C). It is partially insulated on the right side with glass wool of thickness L_{gw} = 0.1 m, with a thermal conductivity of k_{gw} = 0.04. The temperature on the left-hand side of the stainless stell is 400°Cand on the right-hand side if the glass wool is 100°C. Evaluate q and T_i.

2.25 Rework Problem 1.29 with a heat transfer coefficient, $\overline{h}_o$ = 40 W/m^{2}K on the outside (i.e., on the cold side).

2.26 A scientist proposes an experiment for the space shuttle in which he provides underwater illumination in a large tank of water at 20°C, using a 3 cm diameter spherical light bulb. What is the maximum wattage of the bulb in zero gravity that will not cause the water to boil on the surface of the bulb?

2.27 A cylindrical shell is made of two layers- an inner one with inner radius = r_i and outer radius = r_c and an outer one with inner radius = r_c and outer radius = r_o. There is a contact resistance, h_c, between the shells. The materials are different, and $T_1(r = r_i) = T_i$ and $T_2(r = r_o) = T_o$. Derive an expression for the inner temperature of the outer shell (T_{2_c}).

2.28 A 1 kW commercial electric heating rod, 8 mm in diameter and 0.3 m long, is to be used in a highly corrosive gaseous environment. Therefore, it has to be provided with a cylindrical sheath of fireclay. The gas flows by at 120°C, and $\overline{h}$ is 230 W/m^{2}K outside the sheath. The surface of the heating rod cannot exceed 800°C. Set the maximum sheath thickness and the outer temperature of the fireclay. (*Hint:* use heat flux and temperature boundary conditions to get the temperature distribution. Then use the additional convective boundary condition to obtain the sheath thickness.)

2.29 A very small diameter, electrically insulated heating wire runs down the center of a 7.5 mm diameter rod of type 304 stainless steel. The outside is cooled by natural convection ($\overline{h} = 6.7$ W/m^{2}K) in room air at 22°C. If the wire releases 12 W/m, plot T_{rod} vs. radial position in the rod and give the outside temperature of the rod. (Stop and consider carefully the boundary conditions for this problem.)

2.30 A contact resistance experiment involves pressing two slabs of different materials together, putting a known heat flux through them, and measuring the outside temperatures of each slab. Write the general expression for h_c in terms of known quantities. Then calculate h_c if the slabs are 2 cm thick copper and 1.5 cm thick aluminum, if q is 30,000 W/m^2, and if the two temperatures are 15°C and 22.1°C.

2.31 A student working heat transfer problems late at night needs a cup of hot cocoa to stay awake. She puts milk in a pan on an electric stove and seeks to heat it as rapidly as she can, without burning the milk, by turning the stove on high and stirring the milk continuously. Explain how this works using an analogous electric circuit. Is it possible to bring the entire bulk of the milk up to the burn temperature without burning part of it?

2.32 A small, spherical hot air balloon, 10 m in diameter, weighs 130 kg with a small gondola and one passenger. How much fuel must be consumed (in kJ/h) if it is to hover at low altitude in still 27°C air? ($\overline{h}_{\text{outside}} = 215$ W/m^{2}K, as the result of natural convection.)

2.33 A slab of mild steel, 4 cm thick, is held at 1,000°C on the back side. The front side is approximately black and radiates to black surroundings at 100°C. What is the temperature of the front side?

2.34 With reference to Fig. 2.3, develop an empirical equation for $k(T)$ for ammonia vapor. Then imagine a hot surface at T_w parallel with a cool horizontal surface at a distance H below it. Develop equations for $T(x)$ and q. Compute q if $T_w = 350°C$, $T_{cool} = -5°C$, and $H = 0.15$ m.

2.35 A type 316 stainless steel pipe has a 6 cm inside diameter and an 8 cm outside diameter with a 2 mm layer of 85% magnesia insulation around it. Liquid at 112°C flows inside, so $\overline{h}_i = 346$ W/m^{2}K. The air around the pipe is at 20°C, and $\overline{h}_0 = 6$ W/m^{2}K. Calculate U based on the inside area. Sketch the equivalent electrical circuit, showing all known temperatures. Discuss the results.

2.36 Two highly reflecting, horizontal plates are spaced 0.0005 m apart. The upper one is kept at 1000°C and the lower one at 200°C. There is air in between. Neglect radiation and compute the heat flux and the midpoint temperature in the air. Use a power-law fit of the form $k = a(T°C)^b$ to represent the air data in Table A.6.

2.37 A 0.1 m thick slab with $k = 3.4$ W/m·K is held at 100°C on the left side. The right side is cooled with air at 20°C through a heat transfer coefficient, and $\overline{h} = (5.1 \text{ W/m}^2(K)^{-5/4})(T_{wall} - T_\infty)^{1/4}$. Find q and T_{wall} on the right.

2.38 Heat is generated at 54,000 W/m^3 in a 0.16 m diameter sphere. The sphere is cooled by natural convection with fluid at 0°C, and $\overline{h} = [2 + 6(T_{surface} - T_\infty)^{1/4}]$ W/m^{2}K, $k_{sphere} = 9$ W/m·K. Find the surface temperature and center temperature of the sphere.

2.39 Layers of equal thickness of spruce and pitch pine are laminated to make an insulating material. How should the laminations be oriented in a temperature gradient to achieve the best effect?

2.40 The resistances of a thick cylindrical layer of insulation must be increased. Will Q be lowered more by a small increase of the outside diameter or by the same decrease in the inside diameter?

2.41 You are in charge of energy conservation at your plant. There is a 300 m run of 6 in. O.D. pipe carrying steam at 250°C. The company requires that any insulation must pay for itself in one year. The thermal resistances are such that the surface of the pipe will stay close to 250°C in air at 25°C when $\overline{h} = 10$ W/m²K. Calculate the annual energy savings in kW·h that will result if a 1 in layer of 85% magnesia insulation is added. If energy is worth 6 cents per kW·h and insulation costs $75 per installed linear meter, will the insulation pay for itself in one year?

2.42 An exterior wall of a wood-frame house is typically composed, from outside to inside, of a layer of wooden siding, a layer glass fiber insulation, and a layer of gypsum wall board. Standard glass fiber insulation has a thickness of 3.5 inch and a conductivity of 0.038 W/m·K. Gypsum wall board is normally 0.50 inch thick with a conductivity of 0.17 W/m·K, and the siding can be assumed to be 1.0 inch thick with a conductivity of 0.10 W/m·K.

a. Find the overall thermal resistance of such a wall (in K/W) if it has an area of 400 ft².

b. Convection and radiation processes on the inside and outside of the wall introduce more thermal resistance. Assuming that the effective outside heat transfer coefficient (accounting for both convection and radiation) is $\overline{h}_o = 20$ W/m²K and that for the inside is $\overline{h}_i = 10$ W/m²K, determine the total thermal resistance for heat loss from the indoors to the outdoors. Also obtain an overall heat transfer coefficient, U, in W/m²K.

c. If the interior temperature is 20°C and the outdoor temperature is −5°C, find the heat loss through the wall in watts and the heat flux in W/m².

d. Which of the five thermal resistances is dominant?

2.43 We found that the thermal resistance of a cylinder was $R_{t_{\text{cyl}}} = (1/2\pi kl)\ln(r_o/r_i)$. If $r_o = r_i + \delta$, show that the thermal resistance of a thin-walled cylinder ($\delta \ll r_i$) can be approximated by that for a slab of thickness δ. Thus, $R_{t_{\text{thin}}} = \delta/(kA_i)$, where $A_i = 2\pi r_i l$ is the inside surface area of the cylinder. How much error is introduced by this approximation if $\delta/r_i = 0.2$? (*Hint:* Use a Taylor series.)

2.44 A Gardon gage measures a radiation heat flux by detecting a temperature difference [2.10]. The gage consists of a circular constantan membrane of radius R, thickness t, and thermal conductivity k_{ct} which is joined to a heavy copper heat sink at its edges. When a radiant heat flux q_{rad} is absorbed by the membrane, heat flows from the interior of the membrane to the copper heat sink at the edge, creating a radial temperature gradient. Copper leads are welded to the center of the membrane and to the copper heat sink, making two copper-constantan thermocouple junctions. These junctions measure the temperature difference ΔT between the center of the membrane, $T(r = 0)$, and the edge of the membrane, $T(r = R)$.

The following approximations can be made:

- The membrane surface has been blackened so that it absorbs all radiation that falls on it
- The radiant heat flux is much larger than the heat lost from the membrane by convection or re-radiation. Thus, all absorbed radiant heat is removed from the membrane by conduction to the copper heat sink, and other loses can be ignored
- The gage operates in steady state
- The membrane is thin enough ($t \ll R$) that the temperature in it varies only with r, i.e., $T = T(r)$ only.

Answer the following questions.

a. For a fixed copper heat sink temperature, $T(r = R)$, sketch the shape of the temperature distribution in the membrane, $T(r)$, for two arbitrary heat radiant fluxes q_{rad1} and q_{rad2}, where $q_{rad1} > q_{rad2}$.

b. Find the relationship between the radiant heat flux, q_{rad}, and the temperature difference obtained from the thermocouples, ΔT. [*Hint:* Treat the absorbed radiant heat flux as if it were a volumetric heat source of magnitude q_{rad}/t (W/m^3)].

2.45 You have a 12 oz. (375 mL) can of soda at room temperature (70°F) that you would like to cool to 45°F before drinking. You rest the can on its side on the plastic rods of the refrigerator shelf. The can is 2.5 inches in diameter and 5 inches long. The can's emittance is $\varepsilon = 0.4$ and the natural convection heat transfer coefficient around it is a function of the temperature difference between the can and the air: $\overline{h} = 2\,\Delta T^{1/4}$ for ΔT in kelvin.

Assume that thermal interactions with the refrigerator shelf are negligible and that buoyancy currents inside the can will keep the soda well mixed.

a. Estimate how long it will take to cool the can in the refrigerator compartment, which is at 40°F.

b. Estimate how long it will take to cool the can in the freezer compartment, which is at 5°F.

c. Are your answers for parts 1 and 2 the same? If not, what is the main reason that they are different?

References

[2.1] W. M. Rohsenow and J. P. Hartnett, editors. *Handbook of Heat Transfer.* McGraw-Hill Book Company, New York, 1973.

[2.2] R. F. Wheeler. Thermal conductance of fuel element materials. USAEC Rep. HW-60343, April 1959.

[2.3] M. M. Yovanovich. Recent developments in thermal contact, gap and joint conductance theories and experiment. In *Proc. Eight Intl. Heat Transfer Conf.*, volume 1, pages 35–45. San Francisco, 1986.

[2.4] C. V. Madhusudana. *Thermal Contact Conductance.* Springer-Verlag, New York, 1996.

[2.5] American Society of Heating, Refrigerating, and Air-Conditioning Engineers, Inc. *2001 ASHRAE Handbook—Fundamentals.* Altanta, 2001.

[2.6] R. K. Shah and D. P. Sekulic. Heat exchangers. In W. M. Rohsenow, J. P. Hartnett, and Y. I. Cho, editors, *Handbook of Heat Transfer*, chapter 17. McGraw-Hill, New York, 3rd edition, 1998.

[2.7] Tubular Exchanger Manufacturer's Association. *Standards of Tubular Exchanger Manufacturer's Association.* New York, 4th and 6th edition, 1959 and 1978.

[2.8] H. Müller-Steinhagen. Cooling-water fouling in heat exchangers. In T. F. Irvine, Jr., J. P. Hartnett, Y. I. Cho, and G. A. Greene, editors, *Advances in Heat Transfer*, volume 33, pages 415–496. Academic Press, Inc., San Diego, 1999.

[2.9] W. J. Marner and J. W. Suitor. Fouling with convective heat transfer. In S. Kakaç, R. K. Shah, and W. Aung, editors, *Handbook of Single-Phase Convective Heat Transfer*, chapter 21. Wiley-Interscience, New York, 1987.

[2.10] R. Gardon. An instrument for the direct measurement of intense thermal radiation. *Rev. Sci. Instr.*, 24(5):366–371, 1953.

Most of the ideas in Chapter 2 are also dealt with at various levels in the general references following Chapter 1.

3. Heat exchanger design

The great object to be effected in the boilers of these engines is, to keep a small quantity of water at an excessive temperature, by means of a small amount of fuel kept in the most active state of combustion...No contrivance can be less adapted for the attainment of this end than one or two large tubes traversing the boiler, as in the earliest locomotive engines.

The Steam Engine Familiarly Explained and Illustrated,
Dionysus Lardner, 1836

3.1 Function and configuration of heat exchangers

The archetypical problem that any heat exchanger solves is that of getting energy from one fluid mass to another, as we see in Fig. 3.1. A simple or composite wall of some kind divides the two flows and provides an element of thermal resistance between them. Direct contact heat exchangers are an exception to this configuration. Figure 3.2 shows one such arrangement in which steam is bubbled into water. The steam condenses and the water is heated at the same time. In other arrangements, immiscible fluids might contact each other or noncondensible gases might be bubbled through liquids.

Our interest here is in heat exchangers with a dividing wall between the two fluids. They come in an enormous variety of configurations, but most commercial exchangers reduce to one of three basic types. Figure 3.3 shows these types in schematic form. They are:

- *The simple parallel or counterflow configuration.* These arrangements are versatile. Figure 3.4 shows how the counterflow arrangement is bent around in a so-called Heliflow compact heat exchanger configuration.

- *The shell-and-tube configuration.* Figure 3.5 shows the U-tubes of a two-tube-pass, one-shell-pass exchanger being installed in the

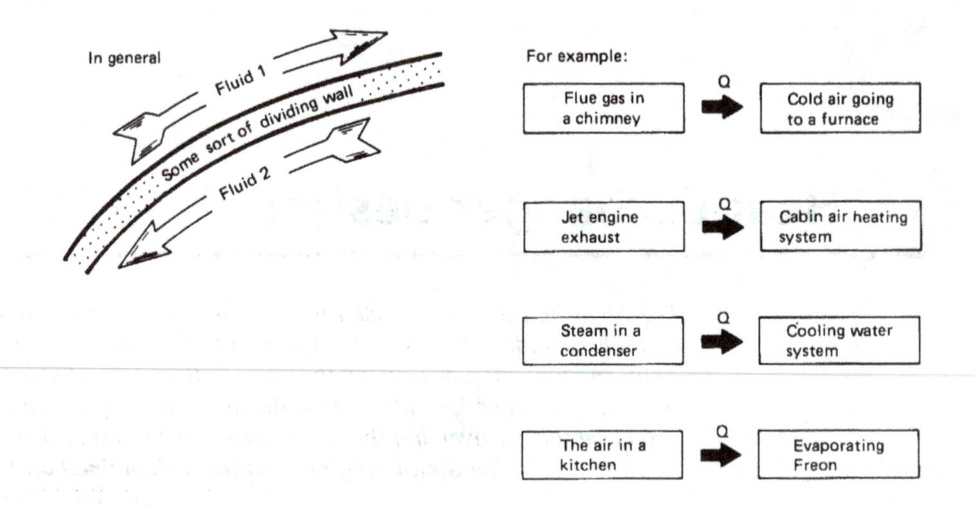

Figure 3.1 Heat exchange.

supporting baffles. The shell is yet to be added. Most of the really large heat exchangers are of the shell-and-tube form.

- *The cross-flow configuration.* Figure 3.6 shows typical cross-flow units. In Fig. 3.6a and c, both flows are *unmixed.* Each flow must stay in a prescribed path through the exchanger and is not allowed to "mix" to the right or left. Figure 3.6b shows a typical plate-fin cross-flow element. Here the flows are also unmixed.

Figure 3.7, taken from the standards of the Tubular Exchanger Manufacturer's Association (TEMA) [3.1], shows four typical single-shell-pass heat exchangers and establishes nomenclature for such units.

These pictures also show some of the complications that arise in translating simple concepts into hardware. Figure 3.7 shows an exchanger with a single tube pass. Although the shell flow is baffled so that it crisscrosses the tubes, it still proceeds from the hot to cold (or cold to hot) end of the shell. Therefore, it is like a simple parallel (or counterflow) unit. The kettle reboiler in Fig. 3.7d involves a divided shell-pass flow configuration over two tube passes (from left to right and back to the "channel header"). In this case, the isothermal shell flow could be flowing in any direction—it makes no difference to the tube flow. Therefore, this exchanger is also equivalent to either the simple parallel or counterflow configuration.

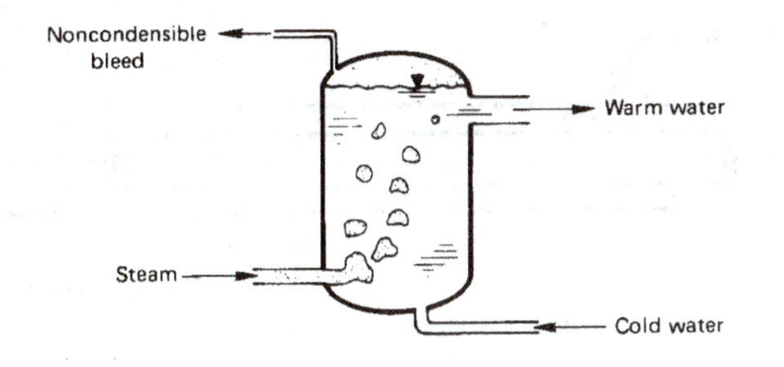

Figure 3.2 A direct-contact heat exchanger.

Notice that a salient feature of shell-and-tube exchangers is the presence of baffles. Baffles serve to direct the flow normal to the tubes. We find in Part III that heat transfer from a tube to a flowing fluid is usually better when the flow moves across the tube than when the flow moves along the tube. This augmentation of heat transfer gives the complicated shell-and-tube exchanger an advantage over the simpler single-pass parallel and counterflow exchangers.

However, baffles bring with them a variety of problems. The flow patterns are very complicated and almost defy analysis. A good deal of the shell-side fluid might unpredictably leak through the baffle holes in the axial direction, or it might bypass the baffles near the wall. In certain shell-flow configurations, unanticipated vibrational modes of the tubes might be excited. Many of the cross-flow configurations also baffle the fluid so as to move it across a tube bundle. The plate-and-fin configuration (Fig. 3.6b) is such a cross-flow heat exchanger.

In all of these heat exchanger arrangements, it becomes clear that a dramatic investment of human ingenuity is directed towards the task of augmenting the heat transfer from one flow to another. The variations are endless, as you will quickly see if you try Experiment 3.1.

Experiment 3.1

Carry a notebook with you for a day and mark down every heat exchanger you encounter in home, university, or automobile. Classify each according to type and note any special augmentation features.

The results of most of what follows in this chapter appear on the Internet in many forms. Information that we derive and present graphically

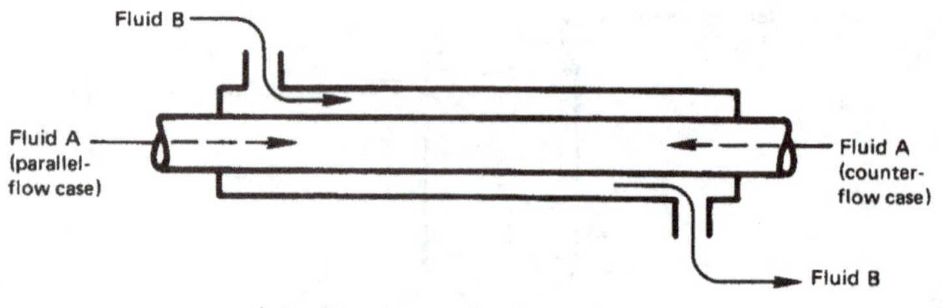

a) Parallel and counterflow heat exchangers

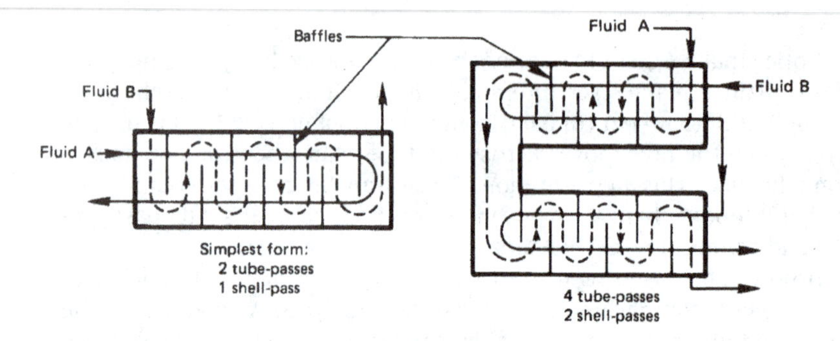

b) Two kinds of shell-and-tube heat exchangers

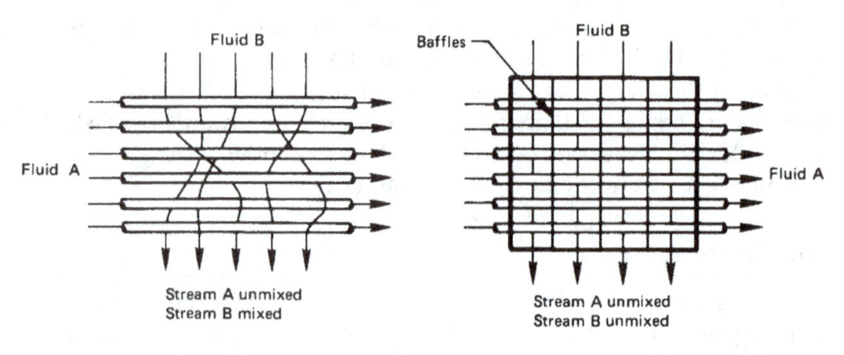

c) Two kinds of cross-flow exchangers

Figure 3.3 The three basic types of heat exchangers.

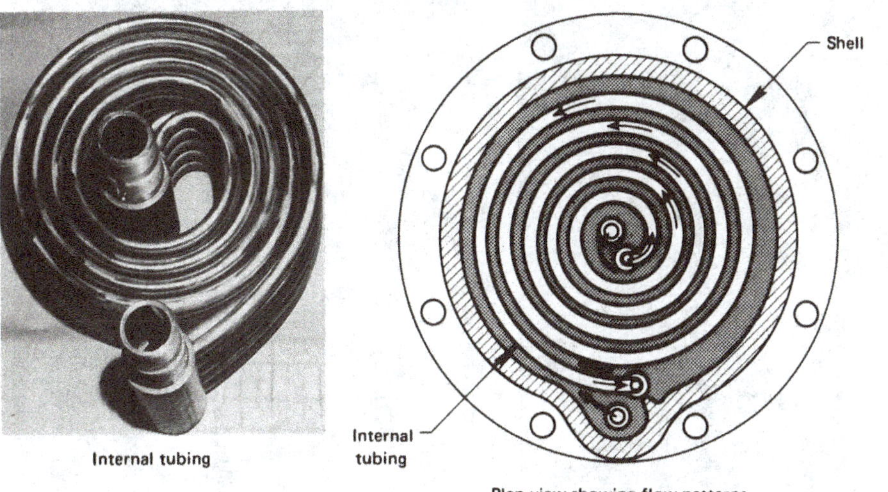

Figure 3.4 Heliflow compact counterflow heat exchanger. (Photograph coutesy of Graham Manufacturing Co., Inc., Batavia, New York.)

is buried in a variety of very effective canned routines for heat exchanger selection. But our job as engineers is not merely to select, it is also to develop new and better systems for exchanging heat. We must be able to look under the hood of those selection programs.

This under the hood analysis of heat exchangers first becomes complicated when we account for the fact that two flow streams change one another's temperature. We turn next, in Section 3.2, to the problem of predicting an appropriate mean temperature difference. Then, in Section 3.3 we develop a strategy for use when this mean cannot be determined initially.

3.2 Evaluation of the mean temperature difference in a heat exchanger

Logarithmic mean temperature difference (LMTD)

To begin with, we take U to be a constant value. This is fairly reasonable in compact single-phase heat exchangers. In larger exchangers, particu-

Above and left: A very large feed-water preheater. Tubes are shown withdrawn from the shell on the left. Inset above shows baffles before tubes are inserted. (Photos courtesy of Southwest Engineering Co., Subsidiary of Cronus Industries, Inc., Los Angeles, Calif.)

Below: Small "Swinglok" exchanger with tube-bundle removed from shell. (Photo courtesy of Graham Manufacturing Co. Inc., Batavia, New York.)

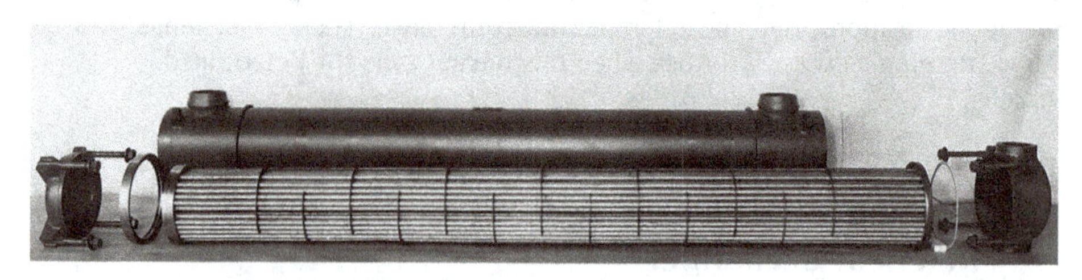

Figure 3.5 Typical commercial one-shell-pass, two-tube-pass heat exchangers.

a. A 1980 Chevette radiator. Cross-flow exchanger with neither flow mixed. Edges of flat vertical tubes can be seen.

b. A section of an automotive air conditioning condenser. The flow through the horizontal wavy fins is allowed to mix with itself while the two-pass flow through the U-tubes remains unmixed.

c. The basic 1 ft. × 1 ft.× 2 ft. module for a waste heat recuperator. It is a plate-fin, gas-to-air cross-flow heat exchanger with neither flow mixed.

Figure 3.6 Several commercial cross-flow heat exchangers. (Photographs courtesy of Harrison Radiator Division, General Motors Corporation.)

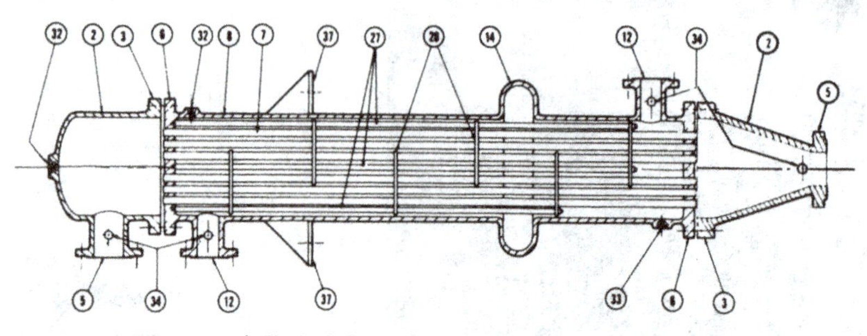

a) Single shell-pass, single tube-pass exchanger

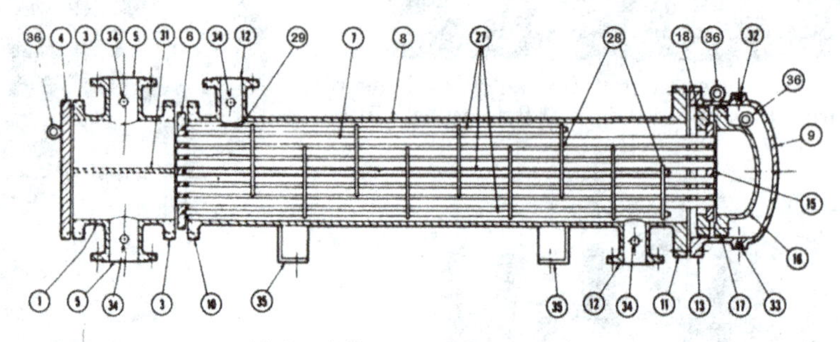

b) One shell-pass, two tube-pass exchanger

1. Stationary head-channel
2. Stationary head-bonnet
3. Stationary head-flange-channel or bonnet
4. Channel cover
5. Stationary head nozzle
6. Stationary tube sheet
7. Tubes
8. Shell
9. Shell cover
10. Shell flange-stationary head end
11. Shell flange-rear head end
12. Shell nozzle
13. Shell cover flange
14. Expansion joint
15. Floating tube sheet
16. Floating head cover
17. Floating head flange
18. Floating head backing device
19. Split shear ring
20. Slip-on backing flange
21. Floating head cover-external
22. Floating tube sheet skirt
23. Packing box
24. Packing
25. Packing gland
26. Lantern ring
27. Tie rods and spacers
28. Transverse baffles or support plates
29. Impingement plate
30. Longitudinal baffle
31. Pass partition
32. Vent connection
33. Drain connection
34. Instrument connection
35. Support saddle
36. Lifting lug
37. Support bracket
38. Weir
39. Liquid level connection

Figure 3.7 Four typical heat exchanger configurations (continued on next page). (Drawings courtesy of the Tubular Exchanger Manufacturers' Association.)

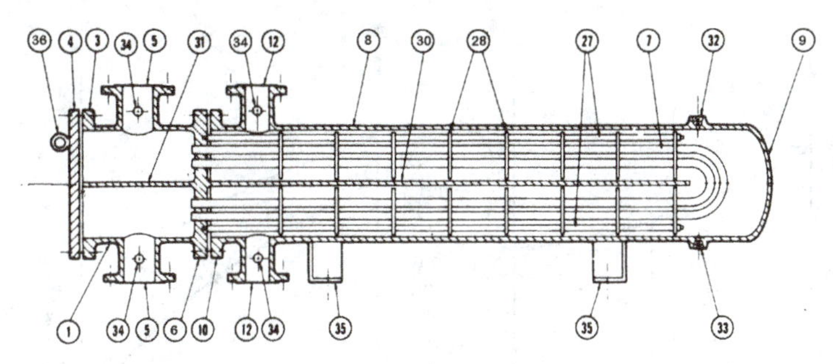

c) Two tube-pass, two shell-pass exchanger

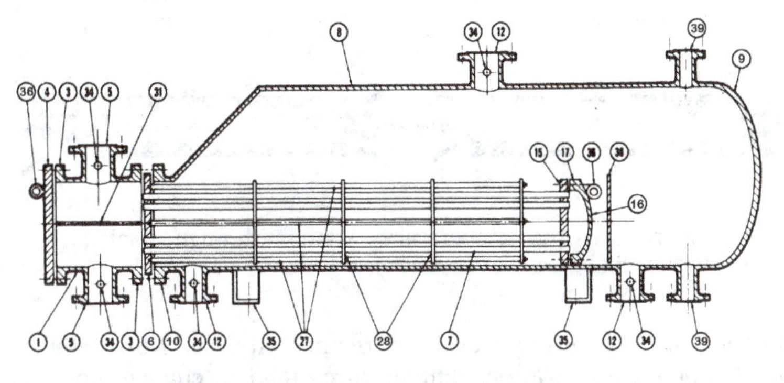

d) One split shell-pass, two tube-pass, kettle type of exchanger

Figure 3.7 Continued

larly in shell-and-tube configurations and large condensers, U is apt to vary with position in the exchanger and/or with local temperature. But in situations in which U is fairly constant, we can deal with the varying temperatures of the fluid streams by writing the overall heat transfer in terms of a mean temperature difference between the two fluid streams:

$$Q = UA\,\Delta T_{\text{mean}} \tag{3.1}$$

Our problem then reduces to finding the appropriate mean temperature difference that will make this equation true. Let us do this for the simple parallel and counterflow configurations, as sketched in Fig. 3.8.

The temperature of both streams is plotted in Fig. 3.8 for both single-pass arrangements—the parallel and counterflow configurations—as a

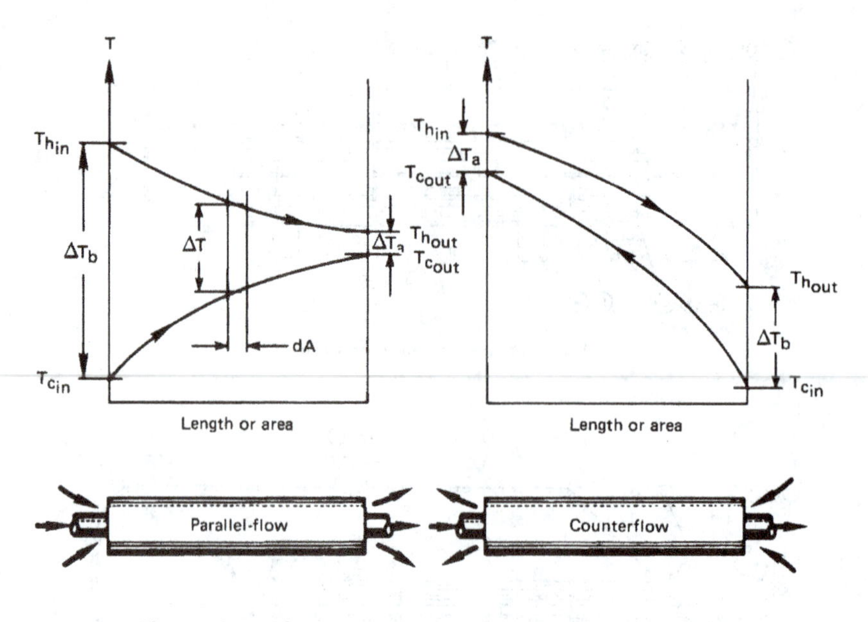

Figure 3.8 The temperature variation through single-pass heat exchangers.

function of the length of travel (or area passed over). Notice that, in the parallel-flow configuration, temperatures tend to change more rapidly with position and less length is required. But the counterflow arrangement achieves generally more complete heat exchange from one flow to the other.

Figure 3.9 shows another variation on the single-pass configuration. This is a condenser in which one stream flows through with its temperature changing, but the other simply condenses at uniform temperature. This arrangement has some special characteristics, which we point out shortly.

The determination of ΔT_{mean} for such arrangements proceeds as follows: the differential heat transfer within either arrangement (see Fig. 3.8) is

$$dQ = U\Delta T\, dA = -(\dot{m}c_p)_h\, dT_h = \pm(\dot{m}c_p)_c\, dT_c \tag{3.2}$$

where the subscripts h and c denote the hot and cold streams, respectively; the upper and lower signs are for the parallel and counterflow cases, respectively; and dT denotes a change from left to right in the

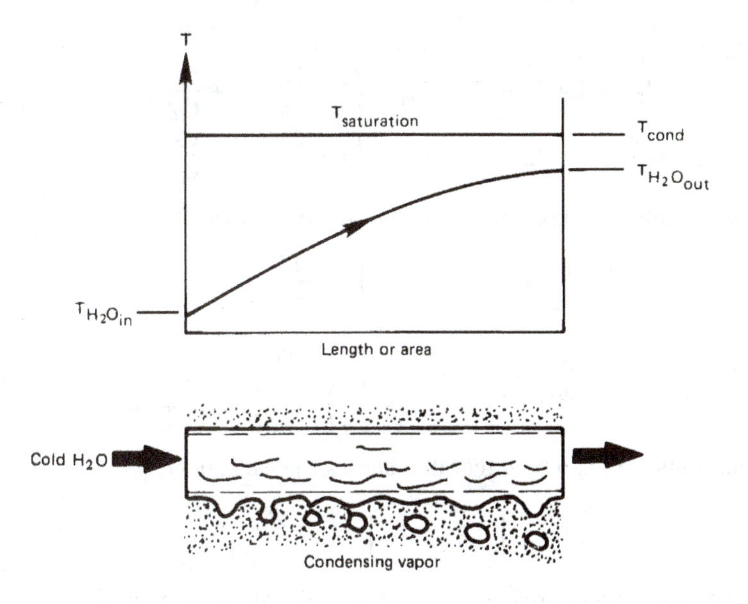

Figure 3.9 The temperature distribution through a condenser.

exchanger. We give symbols to the total heat capacities of the hot and cold streams:

$$C_h \equiv (\dot{m}c_p)_h \text{ W/K} \qquad \text{and} \qquad C_c \equiv (\dot{m}c_p)_c \text{ W/K} \tag{3.3}$$

Thus, for either heat exchanger, $\mp C_h dT_h = C_c dT_c$. This equation can be integrated from the lefthand side, where $T_h = T_{h_\text{in}}$ and $T_c = T_{c_\text{in}}$ for parallel flow or $T_h = T_{h_\text{in}}$ and $T_c = T_{c_\text{out}}$ for counterflow, to some arbitrary point inside the exchanger. The temperatures inside are thus:

$$\text{parallel flow:} \qquad T_h = T_{h_\text{in}} - \frac{C_c}{C_h}(T_c - T_{c_\text{in}}) \qquad = T_{h_\text{in}} - \frac{Q}{C_h} \tag{3.4a}$$

$$\text{counterflow:} \qquad T_h = T_{h_\text{in}} - \frac{C_c}{C_h}(T_{c_\text{out}} - T_c) \qquad = T_{h_\text{in}} - \frac{Q}{C_h} \tag{3.4b}$$

where Q is the total heat transfer from the entrance to the point of interest. Equations (3.4) can be solved for the local temperature differences:

$$\Delta T_{\text{parallel}} = T_h - T_c = T_{h_{\text{in}}} - \left(1 + \frac{C_c}{C_h}\right) T_c + \frac{C_c}{C_h} T_{c_{\text{in}}} \tag{3.5a}$$

$$\Delta T_{\text{counter}} = T_h - T_c = T_{h_{\text{in}}} - \left(1 - \frac{C_c}{C_h}\right) T_c - \frac{C_c}{C_h} T_{c_{\text{out}}} \tag{3.5b}$$

Substitution of these in $dQ = C_c dT_c = U\Delta T\, dA$ yields

$$\left.\frac{U dA}{C_c}\right|_{\text{parallel}} = \frac{dT_c}{\left[-\left(1 + \dfrac{C_c}{C_h}\right) T_c + \dfrac{C_c}{C_h} T_{c_{\text{in}}} + T_{h_{\text{in}}}\right]} \tag{3.6a}$$

$$\left.\frac{U dA}{C_c}\right|_{\text{counter}} = \frac{dT_c}{\left[-\left(1 - \dfrac{C_c}{C_h}\right) T_c - \dfrac{C_c}{C_h} T_{c_{\text{out}}} + T_{h_{\text{in}}}\right]} \tag{3.6b}$$

Equations (3.6) can be integrated across the exchanger:

$$\int_0^A \frac{U}{C_c} dA = \int_{T_{c\text{in}}}^{T_{c\text{out}}} \frac{dT_c}{[---]} \tag{3.7}$$

If U and C_c can be treated as constant, this integration gives

$$\text{parallel:} \quad \ln\left[\frac{-\left(1 + \dfrac{C_c}{C_h}\right) T_{c_{\text{out}}} + \dfrac{C_c}{C_h} T_{c_{\text{in}}} + T_{h_{\text{in}}}}{-\left(1 + \dfrac{C_c}{C_h}\right) T_{c_{\text{in}}} + \dfrac{C_c}{C_h} T_{c_{\text{in}}} + T_{h_{\text{in}}}}\right] = -\frac{UA}{C_c}\left(1 + \frac{C_c}{C_h}\right)$$

$$\text{counter:} \quad \ln\left[\frac{-\left(1 - \dfrac{C_c}{C_h}\right) T_{c_{\text{out}}} - \dfrac{C_c}{C_h} T_{c_{\text{out}}} + T_{h_{\text{in}}}}{-\left(1 - \dfrac{C_c}{C_h}\right) T_{c_{\text{in}}} - \dfrac{C_c}{C_h} T_{c_{\text{out}}} + T_{h_{\text{in}}}}\right] = -\frac{UA}{C_c}\left(1 - \frac{C_c}{C_h}\right) \tag{3.8}$$

If U were variable, the integration leading from eqn. (3.7) to eqns. (3.8) is where its variability would have to be considered. Any such variability of U can complicate eqns. (3.8) terribly. Presuming that eqns. (3.8) are valid, we can simplify them with the help of the definitions of ΔT_a and ΔT_b, given in Fig. 3.8:

$$\text{parallel:} \quad \ln\left[\frac{(1 + C_c/C_h)(T_{c_{\text{in}}} - T_{c_{\text{out}}}) + \Delta T_b}{\Delta T_b}\right] = -UA\left(\frac{1}{C_c} + \frac{1}{C_h}\right)$$

$$\text{counter:} \quad \ln \frac{\Delta T_a}{(-1 + C_c/C_h)(T_{c_{\text{in}}} - T_{c_{\text{out}}}) + \Delta T_a} = -UA\left(\frac{1}{C_c} - \frac{1}{C_h}\right) \tag{3.9}$$

Conservation of energy ($Q_c = Q_h$) requires that

$$\boxed{\frac{C_c}{C_h} = -\frac{T_{h_\text{out}} - T_{h_\text{in}}}{T_{c_\text{out}} - T_{c_\text{in}}}} \tag{3.10}$$

Then eqn. (3.9) and eqn. (3.10) give

$$\text{parallel:} \quad \ln\left[\frac{\overbrace{(T_{c_\text{in}} - T_{c_\text{out}}) + (T_{h_\text{out}} - T_{h_\text{in}})}^{\Delta T_a - \Delta T_b} + \Delta T_b}{\Delta T_b}\right]$$

$$= \ln\left(\frac{\Delta T_a}{\Delta T_b}\right) = -UA\left(\frac{1}{C_c} + \frac{1}{C_h}\right)$$

$$\text{counter:} \quad \ln\left(\frac{\Delta T_a}{\Delta T_b - \Delta T_a + \Delta T_a}\right) = \ln\left(\frac{\Delta T_a}{\Delta T_b}\right) = -UA\left(\frac{1}{C_c} - \frac{1}{C_h}\right) \tag{3.11}$$

Finally, we write $1/C_c = (T_{c_\text{out}} - T_{c_\text{in}})/Q$ and $1/C_h = (T_{h_\text{in}} - T_{h_\text{out}})/Q$ on the right-hand side of either of eqns. (3.11) and get for either parallel or counterflow,

$$\boxed{Q = UA\left(\frac{\Delta T_a - \Delta T_b}{\ln(\Delta T_a/\Delta T_b)}\right)} \tag{3.12}$$

The appropriate ΔT_mean for use in eqn. (3.11) is thus the group on the right, which we call the *logarithmic mean temperature difference* (LMTD):

$$\boxed{\Delta T_\text{mean} = \text{LMTD} \equiv \frac{\Delta T_a - \Delta T_b}{\ln\left(\dfrac{\Delta T_a}{\Delta T_b}\right)}} \tag{3.13}$$

Example 3.1

The idea of a logarithmic mean difference is not new to us. We have already encountered it in Chapter 2. Suppose that we had asked, "What mean radius of pipe would have allowed us to compute the conduction through the wall of a pipe as though it were a slab of thickness $L = r_o - r_i$?" (see Fig. 3.10). To answer this, we write

$$Q = kA\frac{\Delta T}{L} = k(2\pi r_\text{mean} l)\left(\frac{\Delta T}{r_o - r_i}\right)$$

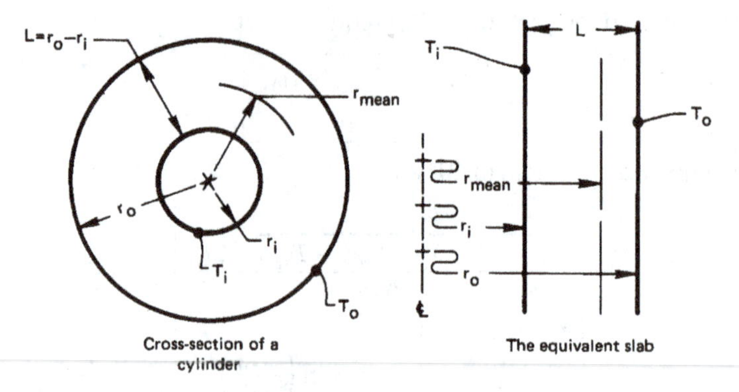

Figure 3.10 Calculation of the mean radius for heat conduction through a pipe.

and then compare it to eqn. (2.21):

$$Q = 2\pi k l \Delta T \frac{1}{\ln(r_o/r_i)}$$

It follows that

$$r_{\text{mean}} = \frac{r_o - r_i}{\ln(r_o/r_i)} = \text{logarithmic mean radius} \qquad \blacksquare$$

Example 3.2 Balanced Counterflow Heat Exchanger

Suppose that the heat capacity rates of a counterflow heat exchanger are equal, $C_h = C_c$. Such an exchanger is said to be *balanced.* From eqn. (3.5b), it follows the local temperature different in the exchanger is constant throughout, $\Delta T_{\text{counter}} = T_{h_{\text{in}}} - T_{h_{\text{out}}} = \Delta T_a = \Delta T_b$. Does the LMTD reduce to this value?

Solution. If we substitute $\Delta T_a = \Delta T_b$ in eqn. (3.13), we get

$$\text{LMTD} = \frac{\Delta T_b - \Delta T_b}{\ln(\Delta T_b/\Delta T_b)} = \frac{0}{0} = \text{indeterminate}$$

Therefore it is necessary to use L'Hospital's rule:

$$\lim_{\Delta T_a \to \Delta T_b} \frac{\Delta T_a - \Delta T_b}{\ln(\Delta T_a/\Delta T_b)} = \frac{\left.\dfrac{\partial}{\partial \Delta T_a}(\Delta T_a - \Delta T_b)\right|_{\Delta T_a = \Delta T_b}}{\left.\dfrac{\partial}{\partial \Delta T_a}\ln\left(\dfrac{\Delta T_a}{\Delta T_b}\right)\right|_{\Delta T_a = \Delta T_b}}$$

$$= \left.\left(\frac{1}{1/\Delta T_a}\right)\right|_{\Delta T_a = \Delta T_b} = \Delta T_a = \Delta T_b$$

So LMTD does indeed reduce to the intuitively obvious result when the capacity rates are balanced. ■

Example 3.3

Water enters the tubes of a small single-pass heat exchanger at 20°C and leaves at 40°C. On the shell side, 25 kg/min of steam condenses at 60°C. Calculate the overall heat transfer coefficient and the required flow rate of water if the area of the exchanger is 12 m^2. (The latent heat, h_{fg}, is 2358.7 kJ/kg at 60°C.)

Solution.

$$Q = \dot{m}_{\text{condensate}} \cdot h_{fg}\big|_{60°\text{C}} = \frac{25(2358.7)}{60} = 983 \text{ kJ/s}$$

and with reference to Fig. 3.9, we can calculate the LMTD without naming the exchanger "parallel" or "counterflow", since the condensate temperature is constant.

$$\text{LMTD} = \frac{(60-20)-(60-40)}{\ln\left(\dfrac{60-20}{60-40}\right)} = 28.85 \text{ K}$$

Then

$$U = \frac{Q}{A(\text{LMTD})} = \frac{983(1000)}{12(28.85)} = 2839 \text{ W/m}^2\text{K}$$

and

$$\dot{m}_{\text{H}_2\text{O}} = \frac{Q}{c_p \Delta T} = \frac{983{,}000}{4174(20)} = 11.78 \text{ kg/s}$$ ■

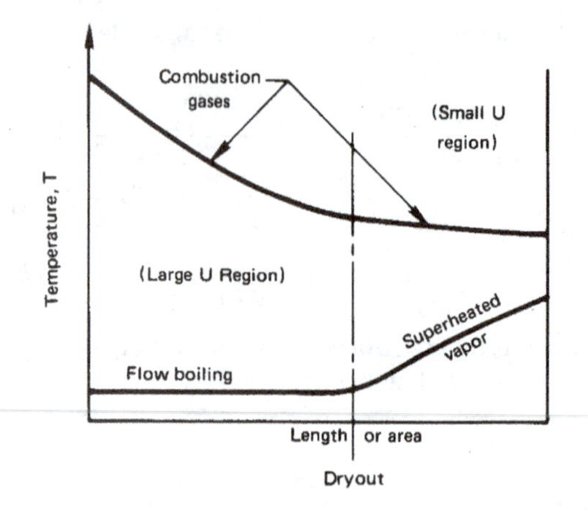

Figure 3.11 A typical case of a heat exchanger in which U varies dramatically.

Extended use of the LMTD

Limitations. The use of an LMTD is limited in two basic ways. The first is that it is restricted to the single-pass parallel and counterflow configurations. This restriction can be overcome by adjusting the LMTD for other configurations—a matter that we take up in the following subsection.

The second limitation—our use of a constant value of U—is harder to deal with. The value of U must be negligibly dependent on T to complete the integration of eqn. (3.7). Even if $U \neq \text{fn}(T)$, the changing flow configuration and the variation of temperature can still give rise to serious variations of U within a given heat exchanger. Figure 3.11 shows a typical situation in which the variation of U within a heat exchanger might be great. In this case, the mechanism of heat exchange on the water side is completely altered when the liquid is finally boiled away. If U were uniform in each portion of the heat exchanger, then we could treat it as two different exchangers in series.

However, the more common difficulty is that of designing heat exchangers in which U varies continuously with position within it. This problem is most severe in large industrial shell-and-tube configurations[1]

[1]Actual heat exchangers can have areas in excess of 10,000 m^2. Large power plant condensers and other large exchangers are often remarkably big pieces of equipment.

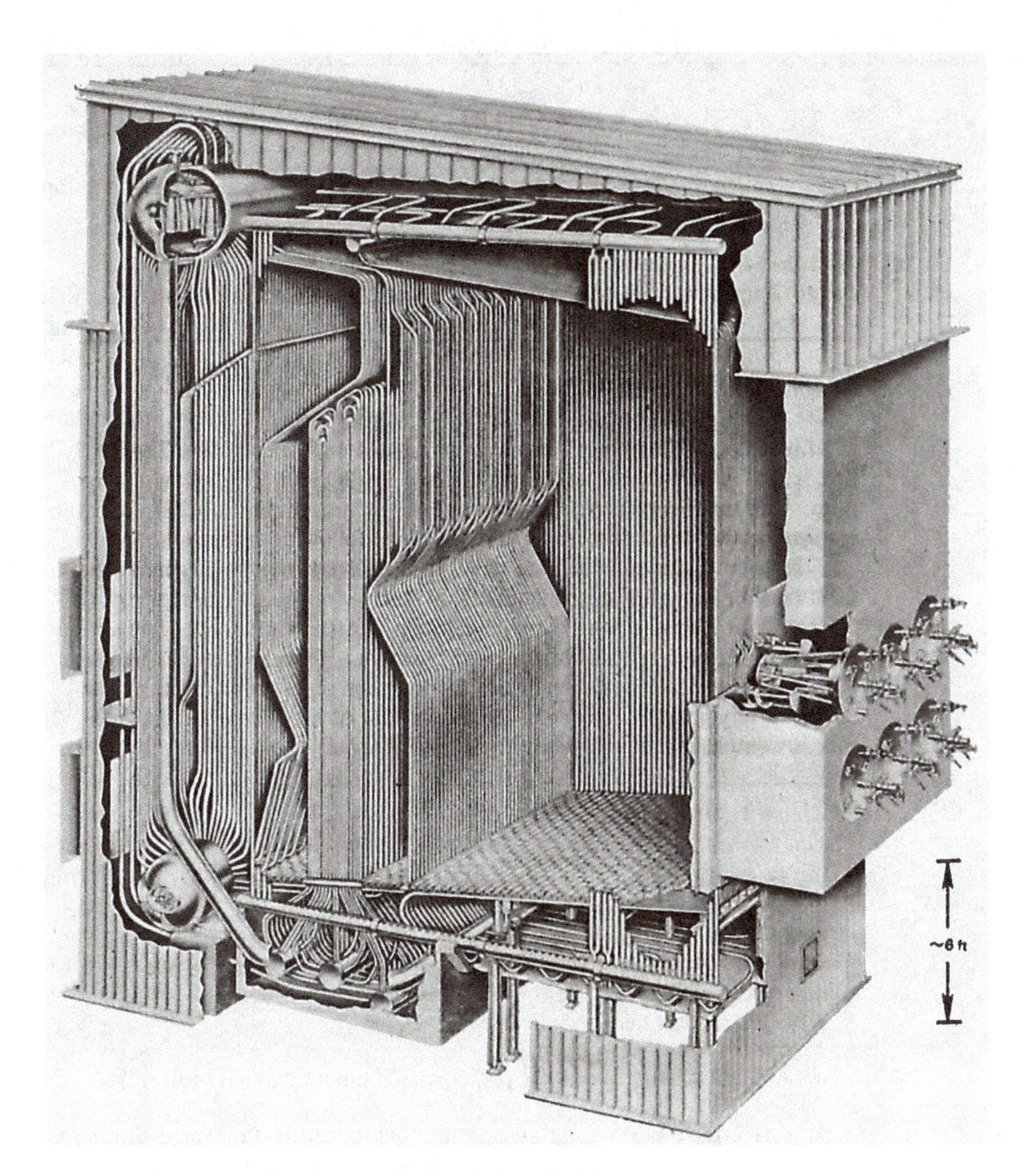

Figure 3.12 The heat exchange surface for a steam generator. This PFT-type integral-furnace boiler, with a surface area of 4560 m^2, is not particularly large. About 88% of the area is in the furnace tubing and 12% is in the boiler (Photograph courtesy of Babcock and Wilcox Co.)

(see, e.g., Fig. 3.5 or Fig. 3.12) and less serious in compact heat exchangers with less surface area. If U depends on the location, analyses such as we have just completed [eqn. (3.1) to eqn. (3.13)] must be done using an average U defined as $\int_0^A U dA/A$.

LMTD correction factor, *F*. Suppose we have a heat exchanger in which U can reasonably be taken constant, but one that involves such configurational complications as multiple passes and/or cross-flow. In such cases we must rederive the appropriate mean temperature difference in the same way as we derived the LMTD. Each configuration must be analyzed separately and the results are generally more complicated than eqn. (3.13).

This task was undertaken on an *ad hoc* basis during the early twentieth century. In 1940, Bowman, Mueller and Nagle [3.2] organized such calculations for the common range of heat exchanger configurations. In each case they wrote

$$Q = UA(\text{LMTD}) \cdot F\left(\underbrace{\frac{T_{t_{\text{out}}} - T_{t_{\text{in}}}}{T_{s_{\text{in}}} - T_{t_{\text{in}}}}}_{P}, \underbrace{\frac{T_{s_{\text{in}}} - T_{s_{\text{out}}}}{T_{t_{\text{out}}} - T_{t_{\text{in}}}}}_{R}\right) \tag{3.14}$$

where T_t and T_s are temperatures of tube and shell flows, respectively. The factor F is an LMTD correction that varies from one to zero, depending on conditions. The dimensionless groups P and R have the following physical significance:

- P is the relative influence of the overall temperature difference $(T_{s_{\text{in}}} - T_{t_{\text{in}}})$ on the tube flow temperature. It must obviously be less than one.

- R, according to eqn. (3.10), equals the heat capacity ratio C_t/C_s.

- If one flow remains at constant temperature (as, for example, in Fig. 3.9), then either P or R will equal zero. In this case the simple LMTD will be the correct ΔT_{mean} and F must go to one.

The factor F is defined in such a way that *the LMTD should always be calculated for the equivalent counterflow single-pass exchanger with the same hot and cold temperatures.* This is explained in Fig. 3.13.

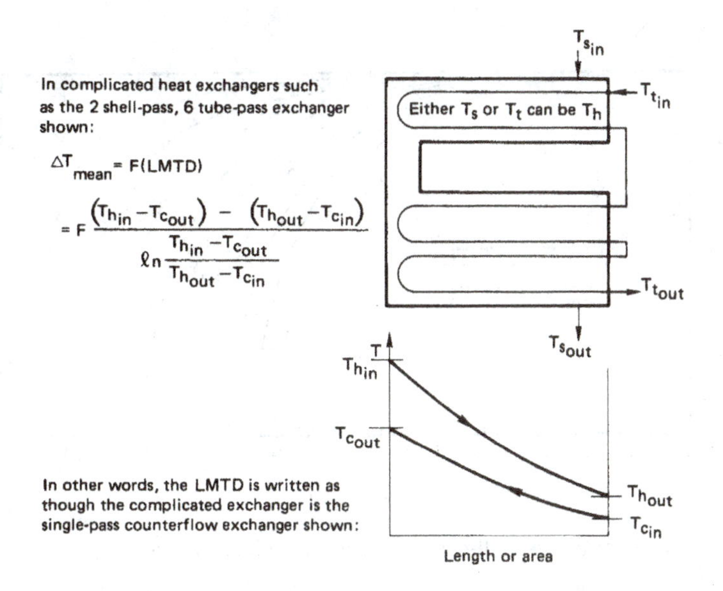

Figure 3.13 The basis of the LMTD in a multipass exchanger, prior to correction.

Bowman *et al.* [3.2] summarized all the equations for F, in various configurations, that had been dervied by 1940. They presented them graphically in not-very-accurate figures that have been widely copied. The TEMA [3.1] version of these curves has been recalculated for shell-and-tube heat exchangers, and it is more accurate. We include two of these curves in Fig. 3.14(a) and Fig. 3.14(b). TEMA presents many additional curves for more complex shell-and-tube configurations. Figures 3.14(c) and 3.14(d) are the Bowman *et al.* curves for the simplest cross-flow configurations. Gardner and Taborek [3.3] redeveloped Fig. 3.14(c) over a different range of parameters. They also showed how Fig. 3.14(a) and Fig. 3.14(b) must be modified if the number of baffles in a tube-in-shell heat exchanger is large enough to make it behave like a series of cross-flow exchangers.

We have simplified Figs. 3.14(a) through 3.14(d) by including curves only for $R \leqslant 1$. Shamsundar [3.4] noted that for $R > 1$, one may obtain F using a simple reciprocal rule. He showed that so long as a heat exchanger has a uniform heat transfer coefficient and the fluid properties are

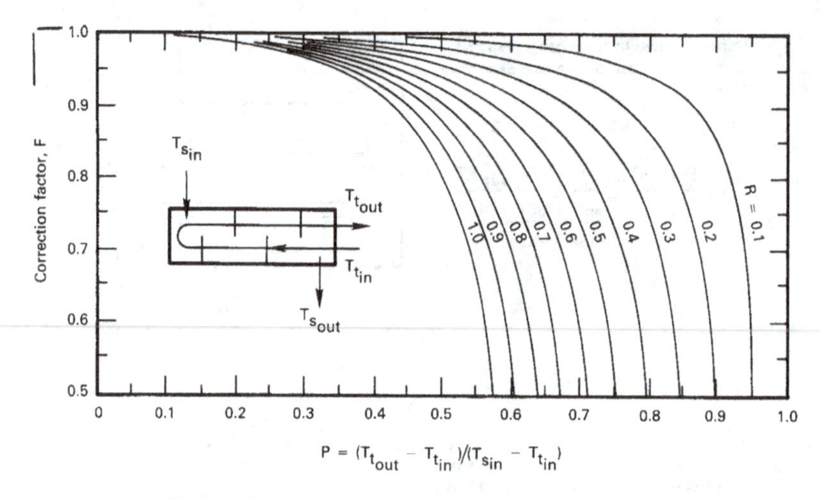

a. *F* for a one-shell-pass, four, six-, . . . tube-pass exchanger.

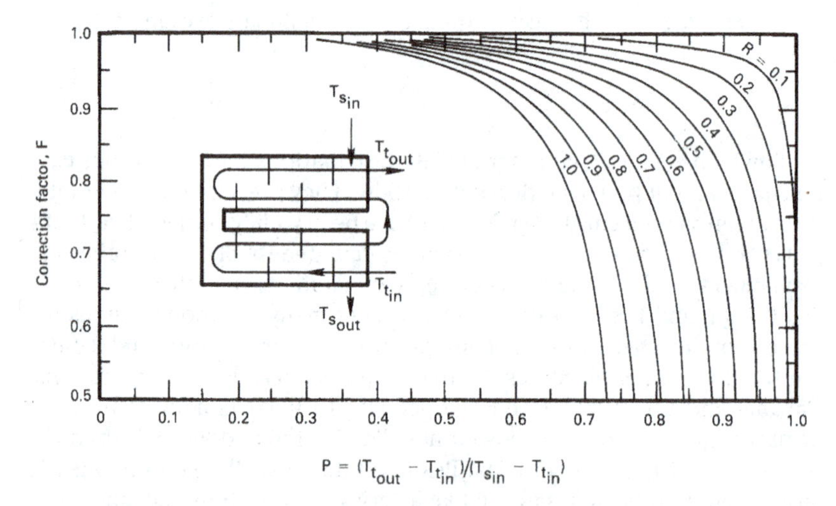

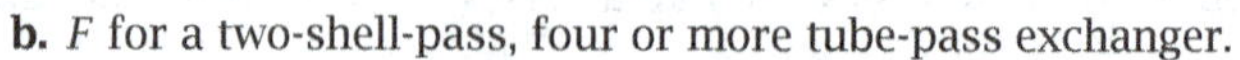

b. *F* for a two-shell-pass, four or more tube-pass exchanger.

Figure 3.14 LMTD correction factors, *F*, for multipass shell-and-tube heat exchangers and one-pass cross-flow exchangers.

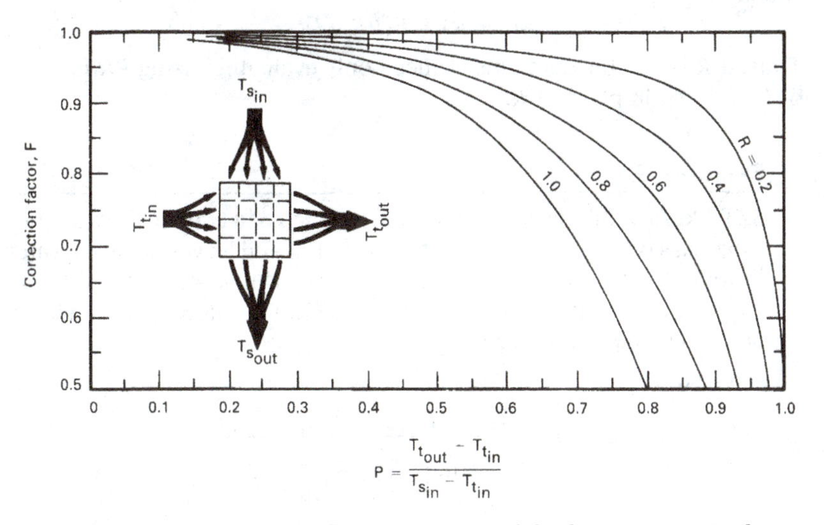

c. *F* for a one-pass cross-flow exchanger with both passes unmixed.

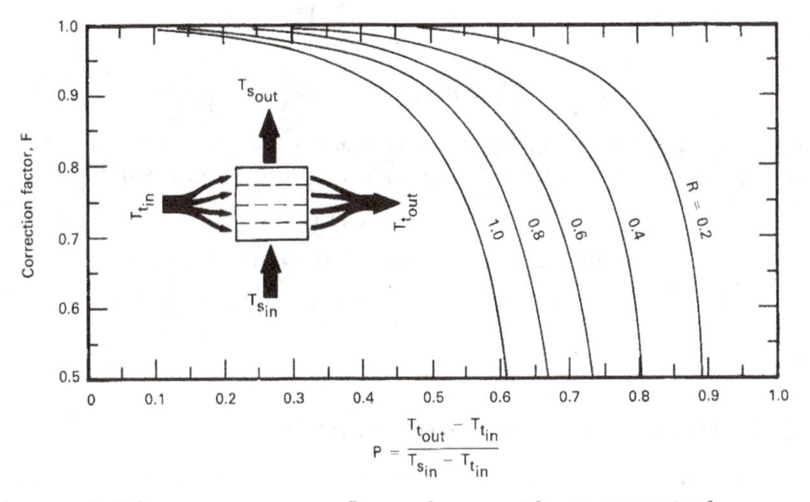

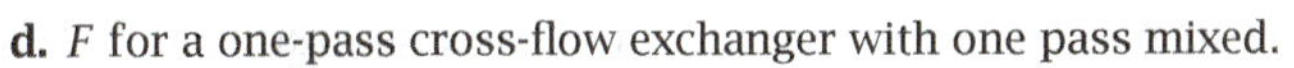

d. *F* for a one-pass cross-flow exchanger with one pass mixed.

Figure 3.14 LMTD correction factors, *F*, for multipass shell-and-tube heat exchangers and one-pass cross-flow exchangers.

constant,

$$F(P,R) = F(PR, 1/R) \tag{3.15}$$

Thus, if R is greater than one, we need only evaluate F using PR in place of P and $1/R$ in place of R.

Example 3.4

5.795 kg/s of oil flows through the shell side of a two-shell pass, four-tube-pass oil cooler. The oil enters at 181°C and leaves at 38°C. Water flows in the tubes, entering at 32°C and leaving at 49°C. In addition, $c_{p_{\text{oil}}} = 2282$ J/kg·K and $U = 416$ W/m²K. Find how much area the heat exchanger must have.

Solution.

$$\text{LMTD} = \frac{(T_{h_\text{in}} - T_{c_\text{out}}) - (T_{h_\text{out}} - T_{c_\text{in}})}{\ln\left(\dfrac{T_{h_\text{in}} - T_{c_\text{out}}}{T_{h_\text{out}} - T_{c_\text{in}}}\right)}$$

$$= \frac{(181 - 49) - (38 - 32)}{\ln\left(\dfrac{181 - 49}{38 - 32}\right)} = 40.76 \text{ K}$$

$$R = \frac{181 - 38}{49 - 32} = 8.412 \qquad P = \frac{49 - 32}{181 - 32} = 0.114$$

Since $R > 1$, we enter Fig. 3.14(b) using $P = 8.412(0.114) = 0.959$ and $R = 1/8.412 = 0.119$ and obtain $F = 0.92$.[2] It follows that:

$$Q = UAF(\text{LMTD})$$

$$5.795(2282)(181 - 38) = 416(A)(0.92)(40.76)$$

$$A = 121.2 \text{ m}^2$$

■

3.3 Heat exchanger effectiveness

We are now in a position to predict the performance of an exchanger once we know its configuration *and* the imposed temperature differences. Unfortunately, we do not often know that much about a system before the design is complete.

[2] Notice that, for a 1 shell-pass exchanger, these R and P lines do not quite intersect [see Fig. 3.14(a)]. Therefore, no single-shell exchanger would give these values.

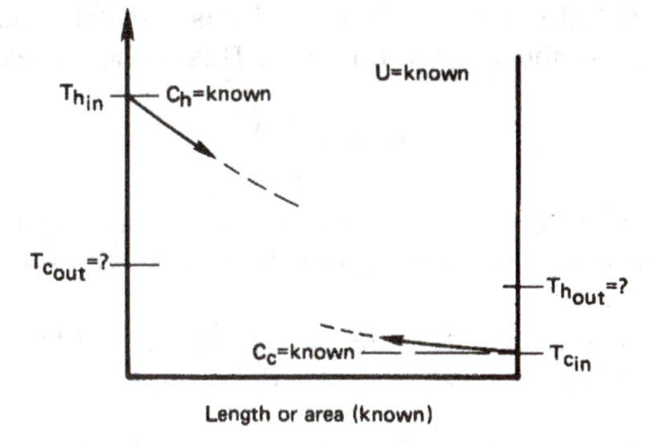

Figure 3.15 A design problem in which the LMTD cannot be calculated a priori.

Often we begin with information such as is shown in Fig. 3.15. If we sought to calculate Q in such a case, we would have to do so by guessing an exit temperature such as to make $Q_h = Q_c = C_h \Delta T_h = C_c \Delta T_c$. Then we could calculate Q from UA(LMTD) or UAF(LMTD) and check it against Q_h. The answers would differ, so we would have to guess new exit temperatures and try again.

Such problems can be greatly simplified with the help of the so-called *effectiveness-NTU method.* This method was first developed in full detail by Kays and London [3.5] in 1955, in a book titled *Compact Heat Exchangers.* We should take particular note of the title. It is with compact heat exchangers that the present method can reasonably be used, since the overall heat transfer coefficient is far more likely to remain fairly uniform.

The heat exchanger effectiveness is defined as

$$\varepsilon \equiv \frac{\text{actual heat transferred}}{\begin{array}{c}\text{maximum heat that could possibly be}\\ \text{transferred from one stream to the other}\end{array}}$$

In mathematical terms, this is

$$\varepsilon = \frac{C_h(T_{h_{\text{in}}} - T_{h_{\text{out}}})}{C_{\text{min}}(T_{h_{\text{in}}} - T_{c_{\text{in}}})} = \frac{C_c(T_{c_{\text{out}}} - T_{c_{\text{in}}})}{C_{\text{min}}(T_{h_{\text{in}}} - T_{c_{\text{in}}})} \tag{3.16}$$

where C_{min} is the smaller of C_c and C_h.

It follows that

$$Q = \varepsilon C_{\text{min}}(T_{h_{\text{in}}} - T_{c_{\text{in}}}) \tag{3.17}$$

A second definition that we will need was originally made by E.K.W. Nusselt, whom we meet again in Part III. This is the *number of transfer units* (NTU):

$$\text{NTU} \equiv \frac{UA}{C_{\text{min}}} \tag{3.18}$$

This dimensionless group can be viewed as a comparison of the heat rate capacity of the heat exchanger, expressed in W/K, with the heat capacity rate of the flow.

We can immediately reduce the parallel-flow result from eqn. (3.9) to the following equation, based on these definitions:

$$-\left(\frac{C_{\text{min}}}{C_c} + \frac{C_{\text{min}}}{C_h}\right)\text{NTU} = \ln\left[-\left(1 + \frac{C_c}{C_h}\right)\varepsilon\frac{C_{\text{min}}}{C_c} + 1\right] \tag{3.19}$$

We solve this for ε and, regardless of whether C_{min} is associated with the hot or cold flow, obtain for the parallel single-pass heat exchanger:

$$\varepsilon \equiv \frac{1 - \exp\left[-(1 + C_{\text{min}}/C_{\text{max}})\text{NTU}\right]}{1 + C_{\text{min}}/C_{\text{max}}} = \text{fn}\left(\frac{C_{\text{min}}}{C_{\text{max}}}, \text{NTU only}\right) \tag{3.20}$$

The corresponding expression for the counterflow case is

$$\varepsilon = \frac{1 - \exp\left[-(1 - C_{\text{min}}/C_{\text{max}})\text{NTU}\right]}{1 - (C_{\text{min}}/C_{\text{max}})\exp\left[-(1 - C_{\text{min}}/C_{\text{max}})\text{NTU}\right]} \tag{3.21}$$

Equations (3.20) and (3.21) are given in graphical form in Fig. 3.16. Similar calculations give the effectiveness for the other heat exchanger configurations (see [3.5] and Problem 3.38), and we include some of the resulting effectiveness plots in Fig. 3.17. The use of effectiveness to rate the performance of an existing heat exchanger and to fix the size of a new one are illustrated in the following two examples.

Example 3.5

Consider the following parallel-flow heat exchanger specification:

$$\begin{aligned} \text{cold flow enters at } 40^\circ\text{C:} \quad & C_c = 20{,}000 \text{ W/K} \\ \text{hot flow enters at } 150^\circ\text{C:} \quad & C_h = 10{,}000 \text{ W/K} \\ A = 30 \text{ m}^2 \qquad & U = 500 \text{ W/m}^2\text{K}. \end{aligned}$$

Determine the heat transfer and the exit temperatures.

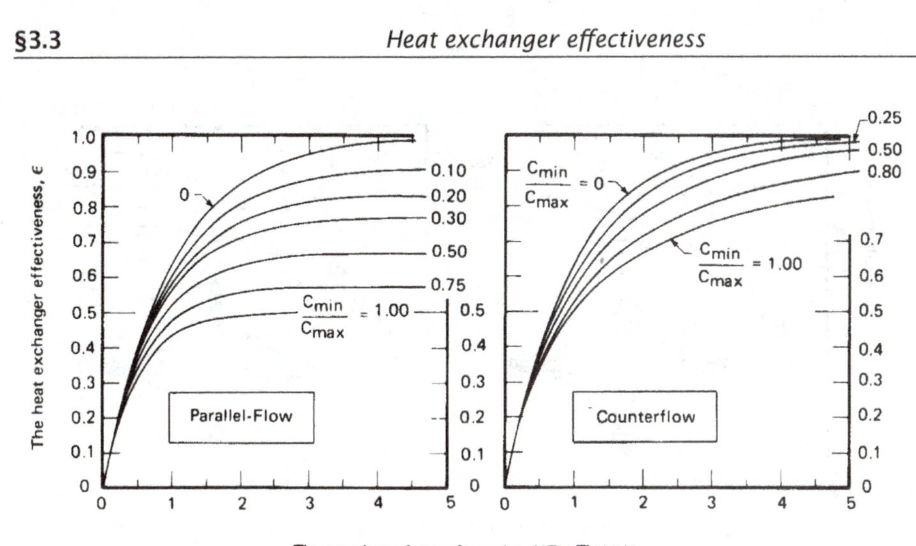

Figure 3.16 The effectiveness of parallel and counterflow heat exchangers. (Data provided by A.D. Kraus.)

Solution. In this case we do not know the exit temperatures, so it is not possible to calculate the LMTD. Instead, we can go either to the parallel-flow effectiveness chart in Fig. 3.16 or to eqn. (3.20), using

$$\text{NTU} = \frac{UA}{C_{\text{min}}} = \frac{500(30)}{10{,}000} = 1.5$$

$$\frac{C_{\text{min}}}{C_{\text{max}}} = 0.5$$

and we obtain $\varepsilon = 0.596$. Now from eqn. (3.17), we find that

$$\begin{aligned} Q = \varepsilon\, C_{\text{min}}(T_{h_{\text{in}}} - T_{c_{\text{in}}}) &= 0.596(10{,}000)(110) \\ &= 655{,}600 \text{ W} = 655.6 \text{ kW} \end{aligned}$$

Finally, from energy balances, such as are expressed in eqn. (3.4), we get

$$T_{h_{\text{out}}} = T_{h_{\text{in}}} - \frac{Q}{C_h} = 150 - \frac{655{,}600}{10{,}000} = 84.44°\text{C}$$

$$T_{c_{\text{out}}} = T_{c_{\text{in}}} + \frac{Q}{C_c} = 40 + \frac{655{,}600}{20{,}000} = 72.78°\text{C} \qquad \blacksquare$$

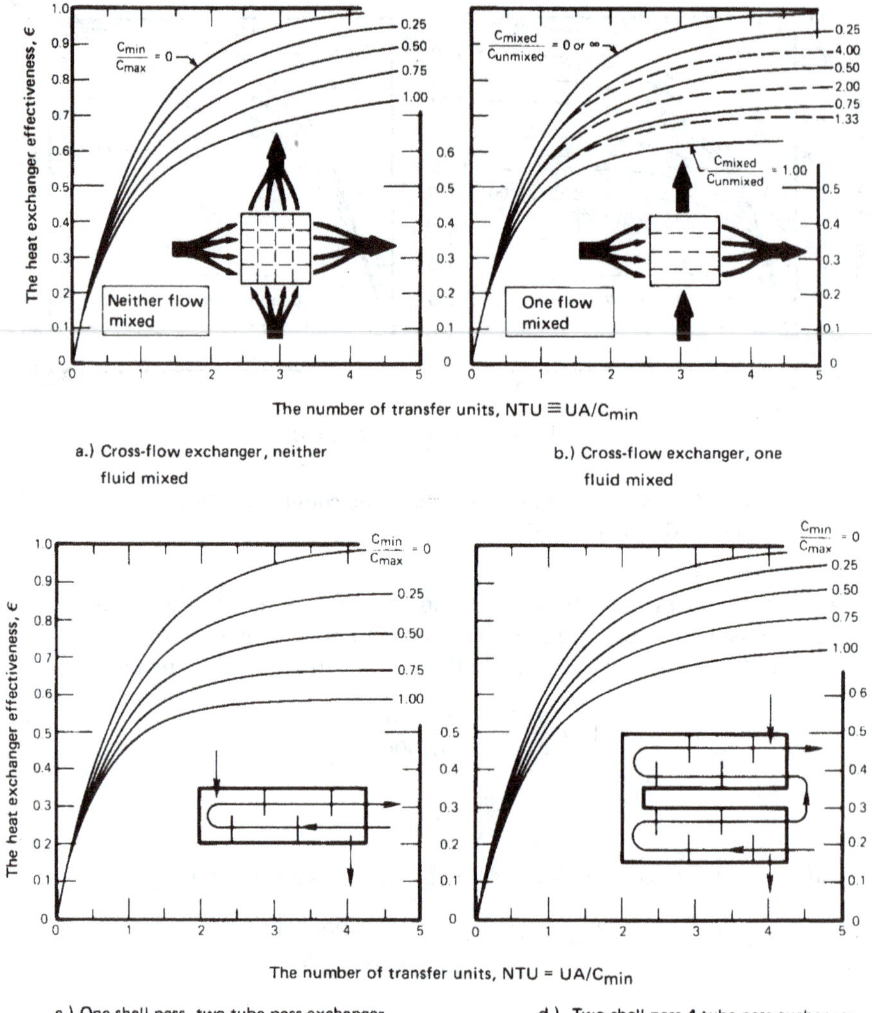

Figure 3.17 The effectiveness of some other heat exchanger configurations. (Data provided by A.D. Kraus.)

Example 3.6

Suppose that we had the same kind of exchanger as we considered in Example 3.5, but that the area remained an unspecified design variable. Calculate the area that would bring the hot flow out at 90°C.

Solution. Once the exit cold fluid temperature is known, the problem can be solved with equal ease by either the LMTD or the effectiveness approach. An energy balance [eqn. (3.4a)] gives

$$T_{c_{\text{out}}} = T_{c_{\text{in}}} + \frac{C_h}{C_c}(T_{h_{\text{in}}} - T_{h_{\text{out}}}) = 40 + \frac{1}{2}(150 - 90) = 70°\text{C}$$

Then, using the effectiveness method,

$$\varepsilon = \frac{C_h(T_{h_{\text{in}}} - T_{h_{\text{out}}})}{C_{\text{min}}(T_{h_{\text{in}}} - T_{c_{\text{in}}})} = \frac{10{,}000(150 - 90)}{10{,}000(150 - 40)} = 0.5455$$

so from Fig. 3.16 we read NTU $\simeq 1.15 = UA/C_{\text{min}}$. Thus

$$A = \frac{10{,}000(1.15)}{500} = 23.00 \text{ m}^2$$

We could also have calculated the LMTD:

$$\text{LMTD} = \frac{(150 - 40) - (90 - 70)}{\ln(110/20)} = 52.79 \text{ K}$$

so from $Q = UA(\text{LMTD})$, we obtain

$$A = \frac{10{,}000(150 - 90)}{500(52.79)} = 22.73 \text{ m}^2$$

The answers differ by 1%, which reflects graph reading inaccuracy. ■

Single stream heat exchangers. When the temperature of either fluid in a heat exchanger is uniform, the problem of analyzing heat transfer is greatly simplified. We have already noted that no F-correction is needed to adjust the LMTD in this case. The reason is that when only one fluid changes in temperature, the configuration of the exchanger becomes irrelevant. Any such exchanger is equivalent to a single fluid stream flowing through an isothermal pipe.[3]

The single stream limit, in which one stream's temperature is constant, occurs when heat capacity rate ratio $C_{\text{min}}/C_{\text{max}}$ goes to zero. The

[3] We make use of this notion in Section 7.4, when we analyze heat convection in pipes and tubes.

heat capacity rate ratio might *approach* zero because the flow rate or specific heat of one stream is very large compared to the other, as when a high flow mass rate of water cools a very low mass flow rate of air. Alternatively, it might *be* infinite because the flow is absorbing or giving up latent heat (as in Fig. 3.9). Since all heat exchangers are equivalent in this case, it follows that the equation for the effectiveness in any configuration must reduce to the same common expression. This limiting expression can be derived directly from energy-balance considerations (see Problem 3.11), but we obtain it here by letting $C_{\text{min}}/C_{\text{max}} \to 0$ in either eqn. (3.20) or eqn. (3.21). The result is

$$\varepsilon_{\text{single stream}} = 1 - e^{-\text{NTU}} \tag{3.22}$$

Eqn. (3.22) defines the curve for $C_{\text{min}}/C_{\text{max}} = 0$ in all six of the effectiveness graphs in Fig. 3.16 and Fig. 3.17.

Balanced counterflow heat exchangers. In Example 3.2, we saw that when the heat capacity rates are balanced in a counterflow heat exchanger, so that $C_h = C_c$ (or $C_{\text{max}} = C_{\text{min}}$), the temperature difference between the hot and cold streams is constant. In this case, the effectiveness equation [eqn. 3.21] limits to

$$\varepsilon = \frac{\text{NTU}}{1 + \text{NTU}} \tag{3.23}$$

(see Problem 3.37). The balanced counterflow arrangement is used for heat recovery in power cycles and ventilation systems; for example, a warm exhaust air stream may be used to preheat an incoming cold air stream.

3.4 Heat exchanger design

The preceding sections provided means for designing heat exchangers that generally work well in the design of smaller exchangers—typically, the kind of compact cross-flow exchanger used in transportation equipment. Larger shell-and-tube exchangers pose two kinds of difficulty in relation to U. The first is the variation of U through the exchanger, which we have already discussed. The second difficulty is that convective heat transfer coefficients are very hard to predict for the complicated flows that move through a baffled shell.

We shall achieve considerable success in using analysis to predict $\overline{h}$'s for various convective flows in Part III. The determination of $\overline{h}$ in a baffled shell remains a problem that cannot be solved analytically. Instead, it is normally computed with the help of empirical correlations or with the aid of large commercial computer programs that include relevant experimental correlations. The problem of predicting $\overline{h}$ when the flow is boiling or condensing is even more complicated and generally requires the use of empirical correlations of $\overline{h}$ data.

Apart from predicting heat transfer, a host of additional considerations must be addressed in designing heat exchangers. The primary ones are minimizing pumping power and fixed costs.

The pumping power calculation, which we do not treat here in any detail, is based on the principles discussed in a first course on fluid mechanics. It generally takes the following form for each stream of fluid through the heat exchanger:

$$\begin{aligned} \text{pumping power} = \left(\dot{m}\,\frac{\text{kg}}{\text{s}}\right)\left(\frac{\Delta p}{\rho}\,\frac{\text{N/m}^2}{\text{kg/m}^3}\right) &= \frac{\dot{m}\Delta p}{\rho}\left(\frac{\text{N}\cdot\text{m}}{\text{s}}\right) \\ &= \frac{\dot{m}\Delta p}{\rho}\;(\text{W}) \end{aligned} \tag{3.24}$$

where $\dot{m}$ is the mass flow rate of the stream, Δp the pressure drop of the stream as it passes through the exchanger, and ρ the fluid density.

Determining the pressure drop can be relatively straightforward in a single-pass pipe-in-tube heat exchanger or extremely difficult in, say, a shell-and-tube exchanger. The pressure drop in a straight run of pipe, for example, is given by

$$\Delta p = f\left(\frac{L}{D_h}\right)\frac{\rho u_{\text{av}}^2}{2} \tag{3.25}$$

where L is the length of pipe, D_h is the hydraulic diameter, u_{av} is the mean velocity of the flow in the pipe, and f is the Darcy-Weisbach friction factor (see Fig. 7.6).

Optimizing the design of an exchanger is not just a matter of making Δp as small as possible. Often, heat exchange can be augmented by employing fins or roughening elements in an exchanger. (We discuss such elements in Chapter 4; see, e.g., Fig. 4.6). Such augmentation will invariably increase the pressure drop, but it can also reduce the fixed cost of an exchanger by increasing U and reducing the required area. Furthermore, it can reduce the required flow rate of, say, coolant, by increasing the effectiveness and thus balance the increase of Δp in eqn. (3.24).

To better understand the course of the design process, faced with such an array of trade-offs of advantages and penalties, we follow Taborek's [3.6] list of design considerations for a large shell-and-tube exchanger:

- Decide which fluid should flow on the shell side and which should flow in the tubes. Normally, this decision will be made to minimize the pumping cost. If, for example, water is being used to cool oil, the more viscous oil would flow in the shell. Corrosion behavior, fouling, and the problems of cleaning fouled tubes also weigh heavily in this decision.
- Early in the process, the designer should assess the cost of the calculation in comparison with:
  (a) The converging accuracy of computation.
  (b) The investment in the exchanger.
  (c) The cost of miscalculation.
- Make a rough estimate of the size of the heat exchanger using, for example, U values from Table 2.2 and/or anything else that might be known from experience. This serves to circumscribe the subsequent trial-and-error calculations; it will help to size flow rates and to anticipate temperature variations; and it will help to avoid subsequent errors.
- Evaluate the heat transfer, pressure drop, and cost of various exchanger configurations that appear reasonable for the application. This is usually done with specialized computer programs that have been developed and are constantly being improved as new research is included in them.

These are the sort of steps incorporated into the many available software packages for heat exchanger design. However, few students of heat transfer will be called upon to use these routines. Instead, most *will* be called upon at one time or another to design smaller exchangers in the range 0.1 to 10 m^2. The heat transfer calculation can usually be done effectively with the methods described in this chapter. Some useful sources of guidance in the pressure drop calculation are the *Heat Exchanger Design Handbook* [3.7], the data in Idelchik's collection [3.8], the TEMA design book [3.1], and some of the other references at the end of this chapter.

In such a calculation, we start off with one fluid to heat and one to cool. Perhaps we know the flow heat capacity rates (C_c and C_h), certain temperatures, and/or the amount of heat that is to be transferred. The problem can be annoyingly wide open, and nothing can be done until it is somehow delimited. The normal starting point is the specification of an exchanger configuration, and to make this choice one needs experience. The descriptions in this chapter provide a kind of first level of experience. References [3.5, 3.7, 3.9, 3.10, 3.11, 3.12, 3.13] provide a second level. Manufacturer's catalogues are an excellent source of more advanced information.

Once the exchanger configuration is set, U will be approximately set and the area becomes the basic design variable. The design can then proceed along the lines of Section 3.2 or 3.3. If it is possible to begin with a complete specification of inlet and outlet temperatures,

$$\underbrace{Q}_{C\Delta T} = \underbrace{U}_{\text{known}} \underbrace{AF(\text{LMTD})}_{\text{calculable}}$$

Then A can be calculated and the design completed. Usually, a reevaluation of U and some iteration of the calculation is needed.

More often, we begin without full knowledge of the outlet temperatures. In such cases, we normally have to invent an appropriate trial-and-error method to get the area and a more complicated sequence of trials if we seek to optimize pressure drop and cost by varying the configuration as well. If the C's are design variables, the U will change significantly, because $\overline{h}$'s are generally velocity-dependent and more iteration will be needed.

We conclude Part I of this book facing a variety of incomplete issues. Most notably, we face a serious need to be able to determine convective heat transfer coefficients. The prediction of $\overline{h}$ depends on a knowledge of heat conduction. We therefore turn, in Part II, to a much more thorough study of heat conduction analysis than was undertaken in Chapter 2. In addition to setting up the methods ultimately needed to predict $\overline{h}$'s, Part II also deals with many other issues that have great practical importance in their own right.

Problems

3.1 Can you have a cross-flow exchanger in which both flows are mixed? Discuss.

3.2 Find the appropriate mean radius, $\overline{r}$, that will make $Q = kA(\overline{r})\Delta T/(r_o - r_i)$, valid for the one-dimensional heat conduction through a thick spherical shell, where $A(\overline{r}) = 4\pi\overline{r}^2$ (*cf.* Example 3.1).

3.3 Rework Problem 2.14, using the methods of Chapter 3.

3.4 2.4 kg/s of a fluid have a specific heat of 0.81 kJ/kg·K enter a counterflow heat exchanger at 0°C and are heated to 400°C by 2 kg/s of a fluid having a specific heat of 0.96 kJ/kg·K entering the unit at 700°C. Show that to heat the cooler fluid to 500°C, all other conditions remaining unchanged, would require the surface area for a heat transfer to be increased by 87.5%.

3.5 A cross-flow heat exchanger with both fluids unmixed is used to heat water ($c_p = 4.18$ kJ/kg·K) from 40°C to 80°C, flowing at the rate of 1.0 kg/s. What is the overall heat transfer coefficient if hot engine oil ($c_p = 1.9$ kJ/kg·K), flowing at the rate of 2.6 kg/s, enters at 100°C? The heat transfer area is 20 m^2. (Note that you can use either an effectiveness or an LMTD method. It would be wise to use both as a check.)

3.6 Saturated non-oil-bearing steam at 1 atm enters the shell pass of a two-tube-pass shell condenser with thirty 20 ft tubes in each tube pass. They are made of schedule 160, ¾ in. steel pipe (nominal diameter). A volume flow rate of 0.01 ft^3/s of water entering at 60°F enters each tube. The condensing heat transfer coefficient is 2000 Btu/h·ft^2·°F, and we calculate $\overline{h}$ = 1380 Btu/h·ft^2·°F for the water in the tubes. Estimate the exit temperature of the water and mass rate of condensate [$\dot{m}_c \simeq$ 8393 lb$_m$/h.]

3.7 Consider a counterflow heat exchanger that must cool 3000 kg/h of mercury from 150°F to 128°F. The coolant is 100 kg/h of water, supplied at 70°F. If U is 300 W/m^{2}K, complete the design by determining reasonable value for the area and the exit-water temperature. [$A = 0.147$ m^2.]

3.8 An automobile air-conditioner gives up 18 kW at 65 km/h if the outside temperature is 35°C. The refrigerant temperature is constant at 65°C under these conditions, and the air rises 6°C in temperature as it flows across the heat exchanger tubes. The heat exchanger is of the finned-tube type shown in Fig. 3.6b, with $U \simeq 200\ \text{W/m}^2\text{K}$. If $U \sim (\text{air velocity})^{0.7}$ and the mass flow rate increases directly with the velocity, plot the percentage reduction of heat transfer in the condenser as a function of air velocity between 15 and 65 km/h.

3.9 Derive eqn. (3.21).

3.10 Derive the infinite NTU limit of the effectiveness of parallel and counterflow heat exchangers at several values of $C_{\text{min}}/C_{\text{max}}$. Use common sense and the First Law of Thermodynamics, and refer to eqn. (3.2) and eqn. (3.21) only to check your results.

3.11 Derive the equation $\varepsilon = (\text{NTU}, C_{\text{min}}/C_{\text{max}})$ for the heat exchanger depicted in Fig. 3.9.

3.12 A single-pass heat exchanger condenses steam at 1 atm on the shell side and heats water from 10°C to 30°C on the tube side with $U = 2500\ \text{W/m}^2\text{K}$. The tubing is thin-walled, 5 cm in diameter, and 2 m in length. (a) Your boss asks whether the exchanger should be counterflow or parallel-flow. How do you advise her? Evaluate: (b) the LMTD; (c) $\dot{m}_{H_2O}$; (d) ε. [$\varepsilon \simeq 0.222$.]

3.13 Air at 2 kg/s and 27°C and a stream of water at 1.5 kg/s and 60°C each enter a heat exchanger. Evaluate the exit temperatures if $A = 12\ \text{m}^2$, $U = 185\ \text{W/m}^2\text{K}$, and:

a. The exchanger is parallel flow;

b. The exchanger is counterflow [$T_{h_{\text{out}}} \simeq 54.0$°C.];

c. The exchanger is cross-flow, one stream mixed;

d. The exchanger is cross-flow, neither stream mixed. [$T_{h_{\text{out}}} = 53.62$°C.]

3.14 Air at 0.25 kg/s and 0°C enters a cross-flow heat exchanger. It is to be warmed to 20°C by 0.14 kg/s of air at 50°C. The streams are unmixed. As a first step in the design process, plot U against A and identify the approximate range of area for the exchanger.

3.15 A particular two shell-pass, four tube-pass heat exchanger uses 20 kg/s of river water at 10°C on the shell side to cool 8 kg/s of processed water from 80°C to 25°C on the tube side. At what temperature will the coolant be returned to the river? If U is 800 W/m^{2}K, how large must the exchanger be?

3.16 A particular cross-flow process heat exchanger operates with the fluid mixed on one side only. When it is new, $U = 2000$ W/m^{2}K, $T_{c_{in}} = 25°C$, $T_{c_{out}} = 80°C$, $T_{h_{in}} = 160°C$, and $T_{h_{out}} = 70°C$. After 6 months of operation, the plant manager reports that the hot fluid is only being cooled to 90°C and that he is suffering a 30% reduction in total heat transfer. What is the fouling resistance after 6 months of use? (Assume no reduction of cold-side flow rate by fouling.)

3.17 Water at 15°C is supplied to a one-shell-pass, two-tube-pass heat exchanger to cool 10 kg/s of liquid ammonia from 120°C to 40°C. You anticipate a U on the order of 1500 W/m^{2}K when the water flows in the tubes. If A is to be 90 m^2, choose the correct flow rate of water.

3.18 Suppose that the heat exchanger in Example 3.5 had been a two shell-pass, four tube-pass exchanger with the hot fluid moving in the tubes. (a) What would be the exit temperature in this case? [$T_{c_{out}} = 75.09°C$.] (b) What would be the area if we wanted the hot fluid to leave at the same temperature that it does in the example?

3.19 Plot the maximum tolerable fouling resistance as a function of U_{new} for a counterflow exchanger, with given inlet temperatures, if a 30% reduction in U is the maximum that can be tolerated.

3.20 Water at 0.8 kg/s enters the tubes of a two-shell-pass, four-tube-pass heat exchanger at 17°C and leaves at 37°C. It cools 0.5 kg/s of air entering the shell at 250°C with $U = 432$ W/m^{2}K. Determine: (a) the exit air temperature; (b) the area of the heat exchanger; and (c) the exit temperature if, after some time, the tubes become fouled with $R_f = 0.0005$ m^{2}K/W. [(c) $T_{air_{out}} = 140.5°C$.]

3.21 You must cool 78 kg/min of a 60%-by-mass mixture of glycerin in water from 108°C to 50°C using cooling water available at 7°C. Design a one-shell-pass, two-tube-pass heat exchanger if

$U = 637$ W/m^{2}K. Explain any design decision you make and report the area, $T_{H_2O_{out}}$, and any other relevant features.

3.22 A mixture of 40%-by-weight glycerin, 60% water, enters a smooth 0.113 m I.D. tube at 30°C. The tube is kept at 50°C, and $\dot{m}_{mixture}$ = 8 kg/s. The heat transfer coefficient inside the pipe is 1600 W/m^{2}K. Plot the liquid temperature as a function of position in the pipe.

3.23 Explain in physical terms why all effectiveness curves Fig. 3.16 and Fig. 3.17 have the same slope as NTU $\to$ 0. Obtain this slope from eqns. (3.20) and (3.21) and give an approximate equation for Q in this limit.

3.24 You want to cool air from 150°C to 60°C but you cannot afford a custom-built heat exchanger. You find a used cross-flow exchanger (both fluids unmixed) in storage. It was previously used to cool 136 kg/min of NH_3 vapor from 200°C to 100°C using 320 kg/min of water at 7°C; U was previously 480 W/m^{2}K. How much air can you cool with this exchanger, using the same water supply, if U is approximately unchanged? (Actually, you would have to modify U using the methods of Chapters 6 and 7 once you had the new air flow rate, but that is beyond our present scope.)

3.25 A one tube-pass, one shell-pass, parallel-flow, process heat exchanger cools 5 kg/s of gaseous ammonia entering the shell side at 250°C and boils 4.8 kg/s of water in the tubes. The water enters subcooled at 27°C and boils when it reaches 100°C. $U = 480$ W/m^{2}K before boiling begins and 964 W/m^{2}K thereafter. The area of the exchanger is 45 m^2, and h_{fg} for water is 2.257×10^6 J/kg. Determine the quality of the water at the exit.

3.26 0.72 kg/s of superheated steam enters a crossflow heat exchanger at 240°C and leaves at 120°C. It heats 0.6 kg/s of water entering at 17°C. $U = 612$ W/m^{2}K. By what percentage will the area differ if a both-fluids-unmixed exchanger is used instead of a one-fluid-unmixed exchanger? [−1.8%]

3.27 Compare values of F from Fig. 3.14(c) and Fig. 3.14(d) for the same conditions of inlet and outlet temperatures. Is the one with the higher F automatically the more desirable exchanger? Discuss.

3.28 Compare values of ε for the same NTU and C_{min}/C_{max} in parallel and counterflow heat exchangers. Is the one with the higher ε automatically the more desirable exchanger? Discuss.

3.29 The *irreversibility rate* of a process is equal to the rate of entropy production times the lowest absolute sink temperature accessible to the process. Calculate the irreversibility (or lost work) for the heat exchanger in Example 3.4. What kind of configuration would reduce the irreversibility, given the same end temperatures.

3.30 Plot T_{oil} and T_{H_2O} as a function of position in a very long counterflow heat exchanger where water enters at 0°C, with C_{H_2O} = 460 W/K, and oil enters at 90°C, with C_{oil} = 920 W/K, U = 742 W/m²K, and A = 10 m². Criticize the design.

3.31 Liquid ammonia at 2 kg/s is cooled from 100°C to 30°C in the shell side of a two shell-pass, four tube-pass heat exchanger by 3 kg/s of water at 10°C. When the exchanger is new, U = 750 W/m²K. Plot the exit ammonia temperature as a function of the increasing tube fouling factor.

3.32 A one shell-pass, two tube-pass heat exchanger cools 0.403 kg/s of methanol from 47°C to 7°C on the shell side. The coolant is 2.2 kg/s of Freon 12, entering the tubes at −33°C, with U = 538 W/m²K. A colleague suggests that this arrangement wastes Freon. She thinks you could do almost as well if you cut the Freon flow rate all the way down to 0.8 kg/s. Calculate the new methanol outlet temperature that would result from this flow rate, and evaluate her suggestion.

3.33 The factors dictating the heat transfer coefficients in a certain two shell-pass, four tube-pass heat exchanger are such that U increases as $(\dot{m}_{shell})^{0.6}$. The exchanger cools 2 kg/s of air from 200°C to 40°C using 4.4 kg/s of water at 7°C, and U = 312 W/m²K under these circumstances. If we double the air flow, what will its temperature be leaving the exchanger? [$T_{air_{out}}$ = 61°C.]

3.34 A flow rate of 1.4 kg/s of water enters the tubes of a two-shell-pass, four-tube-pass heat exchanger at 7°C. A flow rate of 0.6 kg/s of liquid ammonia at 100°C is to be cooled to 30°C on the shell side; $U = 573$ W/m^{2}K. (a) How large must the heat exchanger be? (b) How large must it be if, after some months, a fouling factor of 0.0015 will build up in the tubes, and we still want to deliver ammonia at 30°C? (c) If we make it large enough to accommodate fouling, to what temperature will it cool the ammonia when it is new? (d) At what temperature does water leave the new, enlarged exchanger? [(d) $T_{H_2O} = 49.9$°C.]

3.35 Both C's in a parallel-flow heat exchanger are equal to 156 W/K, $U = 327$ W/m^{2}K and $A = 2$ m^2. The hot fluid enters at 140°C and leaves at 90°C. The cold fluid enters at 40°C. If both C's are halved, what will be the exit temperature of the hot fluid?

3.36 A 1.68 ft^2 cross-flow heat exchanger with one fluid mixed condenses steam at atmospheric pressure ($\overline{h} = 2000$ Btu/h·ft^2·°F) and boils methanol ($T_{sat} = 170$°F and $\overline{h} = 1500$ Btu/h·ft^2·°F) on the other side. Evaluate U (neglecting resistance of the metal), LMTD, F, NTU, ε, and Q.

3.37 Eqn. (3.21) is troublesome when $C_{min}/C_{max} = 1$. Show that ε is given by eqn. 3.23 in this case. Compare it with Fig. 3.16.

3.38 The effectiveness of a cross-flow exchanger with neither fluid mixed can be calculated from the following approximate formula:

$$\varepsilon = 1 - \exp\left[\exp(-\text{NTU}^{0.78} r) - 1](\text{NTU}^{0.22}/r)\right]$$

where $r \equiv C_{min}/C_{max}$. How does this compare with correct values?

3.39 Calculate the area required in a two-tube-pass, one-shell-pass condenser that is to condense 10^6 kg/h of steam at 40°C using water at 17°C. Assume that $U = 4700$ W/m^{2}K, the maximum allowable temperature rise of the water is 10°C, and $h_{fg} = 2406$ kJ/kg.

3.40 An engineer wants to divert 1 gal/min of water at 180°F from his car radiator through a small cross-flow heat exchanger with neither flow mixed, to heat 40°F water to 140°F for shaving when he goes camping. If he produces a pint per minute of hot water, what will be the area of the exchanger and the temperature of the returning radiator coolant if $U = 720$ W/m^{2}K?

3.41 In a process for forming lead shot, molten droplets of lead are showered into the top of a tall tower. The droplets fall through air and solidify before they reach the bottom of the tower. The solid shot is collected at the bottom. To maintain a steady state, cool air is introduced at the bottom of the tower and warm air is withdrawn at the top. For a particular tower, the droplets are 1 mm in diameter and at their melting temperature of 600 K when they are released. The latent heat of solidification is 850 kJ/kg. They fall with a mass flow rate of 200 kg/hr. There are 2430 droplets per cubic meter of air inside the tower. Air enters the bottom at 20°C with a mass flow rate of 1100 kg/hr. The tower has an internal diameter of 1 m with adiabatic walls.

a. Sketch, qualitatively, the temperature distributions of the shot and the air along the height of the tower.
b. If it is desired to remove the shot at a temperature of 60°C, what will be the temperature of the air leaving the top of the tower?
c. Determine the air temperature at the point where the lead has just finished solidifying.
d. Determine the height that the tower must have in order to function as desired. The heat transfer coefficient between the air and the droplets is $\overline{h} = 318$ W/m^{2}K.

3.42 The entropy change per unit mass of a fluid taken from temperature T_i to temperature T_o at constant pressure is $s_o - s_i = c_p \ln(T_o/T_i)$ in J/K·kg. (a) Apply the Second Law of Thermodynamics to a control volume surrounding a counterflow heat exchanger to determine the rate of entropy generation, $\dot{S}_{\text{gen}}$, in W/K. (b) Write $\dot{S}_{\text{gen}}/C_{\min}$ as a function of ε, the heat capacity rate ratio, and $(T_{h,i}/T_{c,i} - 1)$. (c) Show that $\dot{S}_{\text{gen}}/C_{\min}$ is minimized if $C_{\min} = C_{\max}$ for given values of ε and $(T_{h,i}/T_{c,i} - 1)$.

References

[3.1] Tubular Exchanger Manufacturer's Association. *Standards of Tubular Exchanger Manufacturer's Association.* New York, 4th and 6th edition, 1959 and 1978.

[3.2] R. A. Bowman, A. C. Mueller, and W. M. Nagle. Mean temperature difference in design. *Trans. ASME,* 62:283–294, 1940.

[3.3] K. Gardner and J. Taborek. Mean temperature difference: A reappraisal. *AIChE J.,* 23(6):770–786, 1977.

[3.4] N. Shamsundar. A property of the log-mean temperature-difference correction factor. *Mechanical Engineering News,* 19(3): 14–15, 1982.

[3.5] W. M. Kays and A. L. London. *Compact Heat Exchangers.* McGraw-Hill Book Company, New York, 3rd edition, 1984.

[3.6] J. Taborek. Evolution of heat exchanger design techniques. *Heat Transfer Engineering,* 1(1):15–29, 1979.

[3.7] G. F. Hewitt, editor. *Heat Exchanger Design Handbook 1998.* Begell House, New York, 1998.

[3.8] E. Fried and I. E. Idelchik. *Flow Resistance: A Design Guide for Engineers.* Hemisphere Publishing Corp., New York, 1989.

[3.9] R. H. Perry, D. W. Green, and J. Q. Maloney, editors. *Perry's Chemical Engineers' Handbook.* McGraw-Hill Book Company, New York, 7th edition, 1997.

[3.10] D. M. Considine. *Energy Technology Handbook.* McGraw-Hill Book Company, New York, 1975.

[3.11] A. P. Fraas. *Heat Exchanger Design.* John Wiley & Sons, Inc., New York, 2nd edition, 1989.

[3.12] R. K. Shah and D. P. Sekulic. Heat exchangers. In W. M. Rohsenow, J. P. Hartnett, and Y. I. Cho, editors, *Handbook of Heat Transfer,* chapter 17. McGraw-Hill, New York, 3rd edition, 1998.

[3.13] R. K. Shah and D. P. Sekulic. *Fundamentals of Heat Exchanger Design.* John Wiley & Sons, Inc., Hoboken, NJ, 2003.

PART II

ANALYSIS OF HEAT CONDUCTION

4. Analysis of heat conduction and some steady one-dimensional problems

The effects of heat are subject to constant laws which cannot be discovered without the aid of mathematical analysis. The object of the theory which we are about to explain is to demonstrate these laws; it reduces all physical researches on the propagation of heat to problems of the calculus whose elements are given by experiment.

The Analytical Theory of Heat, **J. Fourier, 1822**

4.1 The well-posed problem

The heat diffusion equation was derived in Section 2.1 and some attention was given to its solution. Before we go further with heat conduction problems, we must describe how to state such problems so they can really be solved. This is particularly important in approaching the more complicated problems of transient and multidimensional heat conduction that we have avoided up to now.

A well-posed heat conduction problem is one in which all the relevant information needed to obtain a unique solution is stated. A well-posed and hence solvable heat conduction problem will always read as follows:

Find $T(x, y, z, t)$ such that:

1.
$$\nabla \cdot (k\nabla T) + \dot{q} = \rho c \frac{\partial T}{\partial t}$$

for $0 < t < \mathcal{T}$ (where $\mathcal{T}$ can $\longrightarrow \infty$), and for (x, y, z) belonging to some region, R, which might extend to infinity.[1]

[1] (x, y, z) might be any coordinates describing a position $\vec{r}$: $T(x, y, z, t) = T(\vec{r}, t)$.

2. $T = T_i(x, y, z)$ at $t = 0$

   This is called an *initial condition*, or i.c.

   (a) Condition 1 above is not imposed at $t = 0$.

   (b) Only one i.c. is required. However,

   (c) The i.c. is not needed:

      i. In the steady-state case: $\nabla \cdot (k\nabla T) + \dot{q} = 0$.

      ii. For "periodic" heat transfer, where $\dot{q}$ or the boundary conditions vary periodically with time, and where we ignore the starting transient behavior.

3. T must also satisfy two *boundary conditions*, or b.c.'s, for each coordinate. The b.c.'s are very often of three common types.

   (a) *Dirichlet conditions*, or b.c.'s of the *first kind*:

   T is specified on the boundary of R for $t > 0$. We saw such b.c.'s in Examples 2.1, 2.2, and 2.5.

   (b) *Neumann conditions*, or b.c.'s of the *second kind*:

   The derivative of T normal to the boundary is specified on the boundary of R for $t > 0$. Such a condition arises when the heat flux, $k(\partial T/\partial x)$, is specified on a boundary or when , with the help of insulation, we set $\partial T/\partial x$ equal to zero.[2]

   (c) b.c.'s of the *third kind*:

   A derivative of T in a direction normal to a boundary is proportional to the temperature on that boundary. Such a condition most commonly arises when convection occurs at a boundary, and it is typically expressed as

$$-k \left.\frac{\partial T}{\partial x}\right|_{\text{bndry}} = \overline{h}(T - T_\infty)_{\text{bndry}}$$

   when the body lies to the left of the boundary on the x-coordinate. We have already used such a b.c. in Step 4 of Example 2.6, and we have discussed it in Section 1.3 as well.

This list of b.c.'s is not complete, by any means, but it includes a great number of important cases.

[2]Although we write $\partial T/\partial x$ here, we understand that this might be $\partial T/\partial z$, $\partial T/\partial r$, or any other derivative in a direction locally normal to the surface on which the b.c. is specified.

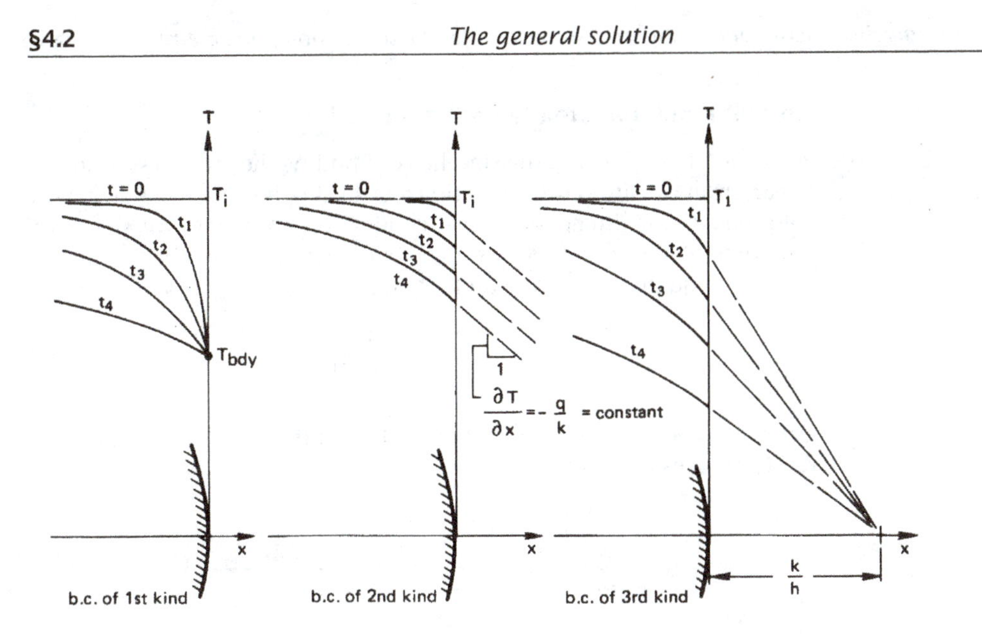

Figure 4.1 The transient cooling of a body as it might occur, subject to boundary conditions of the first, second, and third kinds.

Figure 4.1 shows the transient cooling of body from a constant initial temperature, subject to each of the three b.c.'s described above. Notice that the initial temperature distribution is not subject to the boundary condition, as pointed out previously under 2(a).

The eight-point procedure that was outlined in Section 2.2 for solving the heat diffusion equation assures that a problem will meet the preceding requirements and will be well posed.

4.2 The general solution

Once the heat conduction problem has been posed properly, the first step in solving it is to find the general solution of the heat diffusion equation. We have remarked that this is usually the easiest part of the problem. Let us consider some examples of general solutions.

One-dimensional steady heat conduction

Problem 4.1 emphasizes the simplicity of finding the general solutions of linear ordinary differential equations, by asking for a table of all general solutions of one-dimensional heat conduction problems. We shall work out some of those results to show what is involved. We begin the heat diffusion equation with constant k and $\dot{q}$:

$$\nabla^2 T + \frac{\dot{q}}{k} = \frac{1}{\alpha}\frac{\partial T}{\partial t} \tag{2.11}$$

Cartesian coordinates: Steady conduction in the y-direction. Equation (2.11) reduces as follows:

$$\underbrace{\frac{\partial^2 T}{\partial x^2}}_{=0} + \frac{\partial^2 T}{\partial y^2} + \underbrace{\frac{\partial^2 T}{\partial z^2}}_{=0} + \frac{\dot{q}}{k} = \underbrace{\frac{1}{\alpha}\frac{\partial T}{\partial t}}_{=\,0,\text{ since steady}}$$

Therefore,

$$\frac{d^2 T}{dy^2} = -\frac{\dot{q}}{k}$$

which we integrate twice to get

$$T = -\frac{\dot{q}}{2k}y^2 + C_1 y + C_2$$

or, if $\dot{q} = 0$,

$$T = C_1 y + C_2$$

Cylindrical coordinates with a heat source: Tangential conduction. This time, we look at the heat flow that results in a ring when two points are held at different temperatures. We now express eqn. (2.11) in cylindrical coordinates with the help of eqn. (2.13):

$$\underbrace{\frac{1}{r}\frac{\partial}{\partial r}\left(r\frac{\partial T}{\partial r}\right)}_{=0} + \underbrace{\frac{1}{r^2}\frac{\partial^2 T}{\partial \phi^2}}_{r=\text{constant}} + \underbrace{\frac{\partial^2 T}{\partial z^2}}_{=0} + \frac{\dot{q}}{k} = \underbrace{\frac{1}{\alpha}\frac{\partial T}{\partial t}}_{=\,0,\text{ since steady}}$$

Two integrations give

$$T = -\frac{r^2\dot{q}}{2k}\phi^2 + C_1\phi + C_2 \tag{4.1}$$

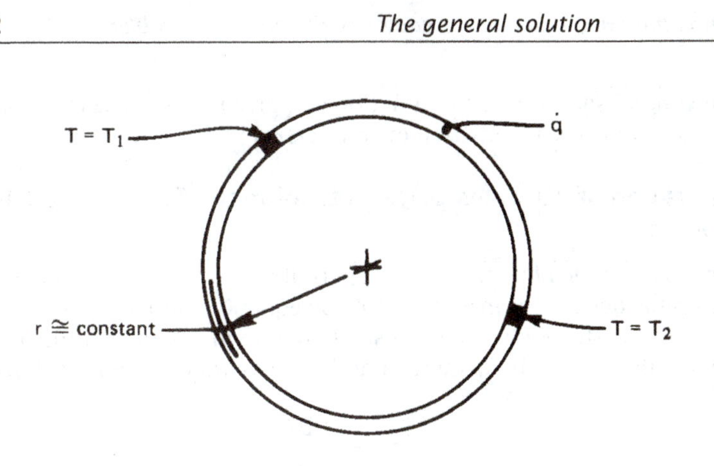

Figure 4.2 One-dimensional heat conduction in a ring.

This would describe, for example, the temperature distribution in the thin ring shown in Fig. 4.2. Here the b.c.'s might consist of temperatures specified at two angular locations, as shown.

T = T(t only)

If T is spatially uniform, it can still vary with time. In such cases

$$\underbrace{\nabla^2 T}_{=0} + \frac{\dot{q}}{k} = \frac{1}{\alpha}\frac{\partial T}{\partial t}$$

and $\partial T/\partial t$ becomes an ordinary derivative. Then, since $\alpha = k/\rho c$,

$$\frac{dT}{dt} = \frac{\dot{q}}{\rho c} \tag{4.2}$$

This result is consistent with the lumped-capacity solution described in Section 1.3. If the Biot number is low and internal resistance is unimportant, the convective removal of heat from the boundary of a body can be *prorated* over the volume of the body and interpreted as

$$\dot{q}_{\text{effective}} = -\frac{\overline{h}(T_{\text{body}} - T_\infty)A}{\text{volume}} \text{ W/m}^3 \tag{4.3}$$

and the heat diffusion equation for this case, eqn. (4.2), becomes

$$\frac{dT}{dt} = -\frac{\overline{h}A}{\rho c V}(T - T_\infty) \tag{4.4}$$

The general solution in this situation was given in eqn. (1.21). [A particular solution was also written in eqn. (1.22).]

Separation of variables: A general solution of multidimensional problems

Suppose that the physical situation permits us to throw out all but one of the spatial derivatives in a heat diffusion equation. Suppose, for example, that we wish to predict the transient cooling in a slab as a function of the location within it. If there is no heat generation, the heat diffusion equation is

$$\frac{\partial^2 T}{\partial x^2} = \frac{1}{\alpha}\frac{\partial T}{\partial t} \tag{4.5}$$

A common trick is to ask: "Can we find a solution in the form of a product of functions of t and x: $T = \mathcal{T}(t) \cdot \mathcal{X}(x)$?" To find the answer, we substitute this in eqn. (4.5) and get

$$\mathcal{X}''\mathcal{T} = \frac{1}{\alpha}\mathcal{T}'\mathcal{X} \tag{4.6}$$

where each prime denotes one differentiation of a function with respect to its argument. Thus $\mathcal{T}' = d\mathcal{T}/dt$ and $\mathcal{X}'' = d^2\mathcal{X}/dx^2$. Rearranging eqn. (4.6), we get

$$\frac{\mathcal{X}''}{\mathcal{X}} = \frac{1}{\alpha}\frac{\mathcal{T}'}{\mathcal{T}} \tag{4.7a}$$

This is an interesting result in that the left-hand side depends only upon x and the right-hand side depends only upon t. Thus, we set *both* sides equal to the same constant, which we call $-\lambda^2$, instead of, say, λ, for reasons that will be clear in a moment:

$$\frac{\mathcal{X}''}{\mathcal{X}} = \frac{1}{\alpha}\frac{\mathcal{T}'}{\mathcal{T}} = -\lambda^2 \quad \text{a constant} \tag{4.7b}$$

It follows that the differential eqn. (4.7a) can be resolved into two ordinary differential equations:

$$\mathcal{X}'' = -\lambda^2\mathcal{X} \quad \text{and} \quad \mathcal{T}' = -\alpha\lambda^2\mathcal{T} \tag{4.8}$$

The general solution of both of these equations are well known and are among the first ones dealt with in any study of differential equations. They are:

$$\begin{aligned} \mathcal{X}(x) &= A\sin\lambda x + B\cos\lambda x \quad &&\text{for} \quad \lambda \neq 0 \\ \mathcal{X}(x) &= Ax + B &&\text{for} \quad \lambda = 0 \end{aligned} \tag{4.9}$$

and

$$\mathcal{T}(t) = Ce^{-\alpha\lambda^2 t} \quad \text{for} \quad \lambda \neq 0$$
$$\mathcal{T}(t) = C \quad \text{for} \quad \lambda = 0 \tag{4.10}$$

where we use capital letters to denote constants of integration. [In either case, these solutions can be verified by substituting them back into eqn. (4.8).] Thus the general solution of eqn. (4.5) can indeed be written in the form of a product, and that product is

$$T = \mathcal{X}\mathcal{T} = e^{-\alpha\lambda^2 t}(D\sin\lambda x + E\cos\lambda x) \quad \text{for} \quad \lambda \neq 0$$
$$T = \mathcal{X}\mathcal{T} = Dx + E \quad \text{for} \quad \lambda = 0 \tag{4.11}$$

The usefulness of this result depends on whether or not it can be fit to the b.c.'s and the i.c. In this case, we made the function $\mathcal{X}(t)$ take the form of sines and cosines (instead of exponential functions) by placing a minus sign in front of λ^2. The sines and cosines make it possible to fit the b.c.'s using Fourier series methods. These general methods are not developed in this book; however, a complete Fourier series solution is presented for one problem in Section 5.3.

The preceding simple methods for obtaining general solutions of linear partial d.e.'s is called the method of *separation of variables.* It can be applied to all kinds of linear d.e.'s. Consider, for example, two-dimensional steady heat conduction without heat sources:

$$\frac{\partial^2 T}{\partial x^2} + \frac{\partial^2 T}{\partial y^2} = 0 \tag{4.12}$$

Set $T = \mathcal{X}\mathcal{Y}$ and get

$$\frac{\mathcal{X}''}{\mathcal{X}} = -\frac{\mathcal{Y}''}{\mathcal{Y}} = -\lambda^2$$

where λ can be an imaginary number. Then

$$\left.\begin{aligned} \mathcal{X} &= A\sin\lambda x + B\cos\lambda x \\ \mathcal{Y} &= Ce^{\lambda y} + De^{-\lambda y} \end{aligned}\right\} \text{for } \lambda \neq 0$$

$$\left.\begin{aligned} \mathcal{X} &= Ax + B \\ \mathcal{Y} &= Cy + D \end{aligned}\right\} \text{for } \lambda = 0$$

The general solution is

$$T = (E\sin\lambda x + F\cos\lambda x)(e^{-\lambda y} + Ge^{\lambda y}) \quad \text{for } \lambda \neq 0$$
$$T = (Ex + F)(y + G) \quad \text{for } \lambda = 0 \tag{4.13}$$

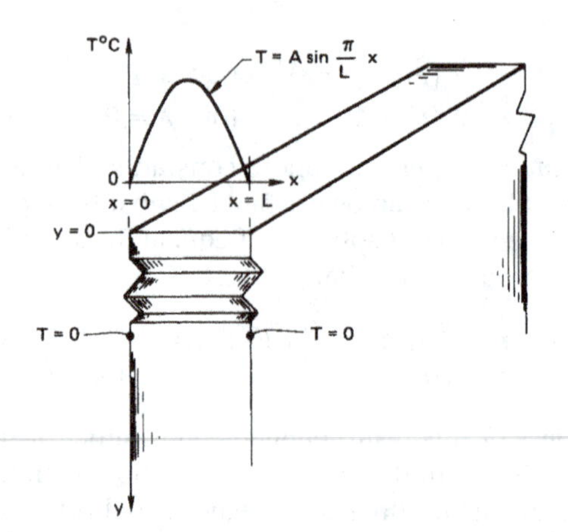

Figure 4.3 A two-dimensional slab maintained at a constant temperature on the sides and subjected to a sinusoidal variation of temperature on one face.

Example 4.1

A long slab is cooled to 0°C on both sides and a blowtorch is turned on the top edge, giving an approximately sinusoidal temperature distribution along the top, as shown in Fig. 4.3. Find the temperature distribution within the slab.

Solution. The general solution is given by eqn. (4.13). We must therefore identify the appropriate b.c.'s and then fit the general solution to it. Those b.c.'s are:

$$\text{on the top surface}: \quad T(x,0) = A\sin\pi\frac{x}{L}$$

$$\text{on the sides}: \quad T(0 \text{ or } L, y) = 0$$

$$\text{as } y \longrightarrow \infty: \quad T(x, y \to \infty) = 0$$

Substitute eqn. (4.13) in the third b.c.:

$$(E\sin\lambda x + F\cos\lambda x)(0 + G\cdot\infty) = 0$$

The only way that this can be true for all x is if $G = 0$. Substitute eqn. (4.13), with $G = 0$, into the second b.c.:

$$(O + F)e^{-\lambda y} = 0$$

so F also equals 0. Substitute eqn. (4.13) with $G = F = 0$, into the first b.c.:

$$E(\sin \lambda x) = A \sin \pi \frac{x}{L}$$

It follows that $A = E$ and $\lambda = \pi/L$. Then eqn. (4.13) becomes the particular solution that satisfies the b.c.'s:

$$T = A\left(\sin \pi \frac{x}{L}\right) e^{-\pi y/L}$$

Thus, the sinusoidal variation of temperature at the top of the slab is attenuated exponentially at lower positions in the slab. At a position of $y = 2L$ below the top, T will be $0.0019 A \sin \pi x/L$. The temperature distribution in the x-direction will still be sinusoidal, but it will have less than $1/500$ of the amplitude at $y = 0$. ■

Consider some important features of this and other solutions:

- The b.c. at $y = 0$ is a special one that works very well with this particular general solution. If we had tried to fit the equation to a general temperature distribution, $T(x, y = 0) = \text{fn}(x)$, it would not have been obvious how to proceed. Actually, this is the kind of problem that Fourier solved with the help of his Fourier series method. We discuss this matter in more detail in Chapter 5.

- Not all forms of general solutions lend themselves to a particular set of boundary and/or initial conditions. In this example, we made the process look simple, but more often than not, *it is in fitting a general solution to a set of boundary conditions that we face difficulties.*

- Normally, on formulating a problem, we must *approximate* real behavior in stating the b.c.'s. It is advisable to consider what kind of assumption will put the b.c.'s in a form compatible with the general solution. The temperature distribution imposed on the slab by the blowtorch in Example 4.1 might just as well have been approximated as a parabola. But as small as the difference between a parabola and a sine function might be, the latter b.c. was far easier to accommodate.

- The twin issues of existence and uniqueness of solutions require a comment here: Mathematicians have established that solutions to all well-posed heat diffusion problems are unique. Furthermore, we know from our experience that if we describe a physical process correctly, a unique outcome exists. Therefore, we are normally safe to ignore these issues in the sort of problems we discuss here.

- Given that a unique solution exists, we accept any solution as correct if we can carve it to fit the boundary conditions. In this sense, the solution of differential equations is often more of an inventive than a formal operation. The person who does it best is often the person who has done it before and so has a large assortment of tricks up his or her sleeve.

4.3 Dimensional analysis

Introduction

Most universities place the first course in heat transfer after an introduction to fluid mechanics: and most fluid mechanics courses include some dimensional analysis. This is normally treated using the familiar *method of indices*, which is seemingly straightforward to teach but is cumbersome and sometimes misleading to use. It is rather well presented in [4.1].

The method we develop here is far simpler to use than the method of indices, and it does much to protect us from the common errors we might fall into. We refer to it as the *method of functional replacement* and strongly recommend its use in place of the method of indices.

The importance of dimensional analysis to heat transfer can be made clearer by recalling Example 2.6, which (like most problems in Part I) involved several variables. Theses variables included the dependent variable of temperature, $(T_\infty - T_i)$;[3] the major independent variable, which was the radius, r; and five system parameters, $r_i, r_o, \overline{h}, k$, and $(T_\infty - T_i)$. By reorganizing the solution into dimensionless groups [eqn. (2.24)], we reduced the total number of variables to only four:

$$\underbrace{\frac{T - T_i}{T_\infty - T_i}}_{\text{dependent variable}} = \text{fn}\Bigg[\underbrace{r/r_i,}_{\text{indep. var.}} \quad \underbrace{r_o/r_i, \qquad \text{Bi}}_{\text{two system parameters}}\Bigg] \tag{2.24a}$$

This solution offered a number of advantages over the dimensional solution. For one thing, it permitted us to plot *all* conceivable solutions

[3]Notice that we do not call T_i a variable. It is simply the reference temperature against which the problem is worked. If it happened to be 0°C, we would not notice its subtraction from the other temperatures.

for a particular shape of cylinder, (r_o/r_i), in a single figure, Fig. 2.13. For another, it allowed us to study the simultaneous roles of $\overline{h}$, k and r_o in defining the character of the solution. By combining them as a Biot number, we were able to say—even before we had solved the problem—whether or not external convection really had to be considered.

The nondimensionalization made it possible for us to consider, simultaneously, the behavior of all *similar* systems of heat conduction through cylinders. Thus a large, highly conducting cylinder might be *similar* in its behavior to a small cylinder with a lower thermal conductivity.

Finally, we shall discover that, by nondimensionalizing a problem *before* we solve it, we can often greatly simplify the process of solving it.

Our next aim is to map out a method for nondimensionalization problems before we have solved then, or, indeed, before we have even written the equations that must be solved. The key to the method is a result called the *Buckingham pi-theorem.*

The Buckingham pi-theorem

The attention of scientific workers was drawn very strongly toward the question of similarity at about the beginning of World War I. Buckingham first organized previous thinking and developed his famous theorem in 1914 in the *Physical Review* [4.2], and he expanded upon the idea in the *Transactions of the ASME* one year later [4.3]. Lord Rayleigh almost simultaneously discussed the problem with great clarity in 1915 [4.4]. To understand Buckingham's theorem, we must first overcome one conceptual hurdle, which, if it is clear to the student, will make everything that follows extremely simple. Let us explain that hurdle first.

Suppose that y depends on r, x, z and so on:

$$y = y(r, x, z, \ldots)$$

We can take any one variable—say, x—and arbitrarily multiply it (or it raised to a power) by any other variables in the equation, without altering the truth of the functional equation. The equation above can thus just as well be written as:

$$\frac{y}{x} = \frac{y}{x}\left(x^2 r, x, xz\right)$$

or an unlimited number of other rearrangements. Many people find such a rearrangement disturbing when they first see it. That is because these are not *algebraic* equations — they are *functional* equations. We have said only that if y depends upon r, x, and z that it will likewise depend upon $x^2 r$, x, and xz. Suppose, for example, that we gave the functional equation the following algebraic form:

$$y = y(r, x, z) = r(\sin x)e^{-z}$$

This need only be rearranged to put it in terms of the desired modified variables and x itself $(y/x, x^2r, x,$ and $xz)$:

$$\frac{y}{x} = \frac{x^2r}{x^3}(\sin x)\exp\left[-\frac{xz}{x}\right]$$

We can do any such multiplying or dividing of powers of any variable we wish without invalidating any functional equation that we choose to write. This simple fact is at the heart of the important example that follows.

Example 4.2

Consider the heat exchanger problem described in Fig. 3.15. The "unknown," or dependent variable, in the problem is either of the exit temperatures. Without any knowledge of heat exchanger analysis, we can write the functional equation on the basis of our physical understanding of the problem:

$$\underbrace{T_{c_{\text{out}}} - T_{c_{\text{in}}}}_{\text{K}} = \text{fn}\left[\underbrace{C_{\text{max}}}_{\text{W/K}}, \underbrace{C_{\text{min}}}_{\text{W/K}}, \underbrace{(T_{h_{\text{in}}} - T_{c_{\text{in}}})}_{\text{K}}, \underbrace{U}_{\text{W/m}^2\text{K}}, \underbrace{A}_{\text{m}^2}\right] \tag{4.14}$$

where the dimensions of each term are noted under the quotation.

We want to know how many dimensionless groups the variables in eqn. (4.14) should reduce to. To determine this number, we use the idea explained above—that is, that we can arbitrarily pick one variable from the equation and divide or multiply it into other variables. Then—one at a time—we select a variable that has one of the dimensions. We divide or multiply it by the other variables in the equation that have that dimension in such a way as to eliminate the dimension from them.

We do this first with the variable $(T_{h_{\text{in}}} - T_{c_{\text{in}}})$, which has the dimension of K.

$$\underbrace{\frac{T_{c_{\text{out}}} - T_{c_{\text{in}}}}{T_{h_{\text{in}}} - T_{c_{\text{in}}}}}_{\text{dimensionless}} = \text{fn}\Bigg[\underbrace{C_{\text{max}}(T_{h_{\text{in}}} - T_{c_{\text{in}}})}_{\text{W}}, \underbrace{C_{\text{min}}(T_{h_{\text{in}}} - T_{c_{\text{in}}})}_{\text{W}}, \underbrace{(T_{h_{\text{in}}} - T_{c_{\text{in}}})}_{\text{K}}, \underbrace{U(T_{h_{\text{in}}} - T_{c_{\text{in}}})}_{\text{W/m}^2}, \underbrace{A}_{\text{m}^2}\Bigg]$$

The interesting thing about the equation in this form is that the only remaining term in it with the units of K is $(T_{h_{\text{in}}} - T_{c_{\text{in}}})$. No such term *can* remain in the equation because it is impossible to achieve dimensional homogeneity without another term in K to balance it. Therefore, we must remove it.

$$\underbrace{\frac{T_{c_{\text{out}}} - T_{c_{\text{in}}}}{T_{h_{\text{in}}} - T_{c_{\text{in}}}}}_{\text{dimensionless}} = \text{fn}\Bigg[\underbrace{C_{\text{max}}(T_{h_{\text{in}}} - T_{c_{\text{in}}})}_{\text{W}}, \underbrace{C_{\text{min}}(T_{h_{\text{in}}} - T_{c_{\text{in}}})}_{\text{W}}, \underbrace{U(T_{h_{\text{in}}} - T_{c_{\text{in}}})}_{\text{W/m}^2}, \underbrace{A}_{\text{m}^2}\Bigg]$$

Now the equation has only two dimensions in it—W and m^2. Next, we multiply $U(T_{h_{\text{in}}} - T_{c_{\text{in}}})$ by A to get rid of m^2 in the second-to-last term. Accordingly, the term A (m^2) can no longer stay in the equation, and we have

$$\underbrace{\frac{T_{c_{\text{out}}} - T_{c_{\text{in}}}}{T_{h_{\text{in}}} - T_{c_{\text{in}}}}}_{\text{dimensionless}} = \text{fn}\Bigg[\underbrace{C_{\text{max}}(T_{h_{\text{in}}} - T_{c_{\text{in}}})}_{\text{W}}, \underbrace{C_{\text{min}}(T_{h_{\text{in}}} - T_{c_{\text{in}}})}_{\text{W}}, \underbrace{UA(T_{h_{\text{in}}} - T_{c_{\text{in}}})}_{\text{W}}\Bigg]$$

Finally, we divide the first and third terms on the right by the second. This leaves only $C_{\text{min}}(T_{h_{\text{in}}} - T_{c_{\text{in}}})$, with the dimensions of W. That term must then be removed, and we are left with the completely dimensionless result:

$$\frac{T_{c_{\text{out}}} - T_{c_{\text{in}}}}{T_{h_{\text{in}}} - T_{c_{\text{in}}}} = \text{fn}\left(\frac{C_{\text{max}}}{C_{\text{min}}}, \frac{UA}{C_{\text{min}}}\right) \tag{4.15}$$

■

Equation (4.15) has exactly the same functional form as eqn. (3.21), which we obtained by direct analysis.

Notice that we removed one variable from eqn. (4.14) for each dimension in which the variables are expressed. If there are n variables—including the dependent variable—expressed in m dimensions, we then expect to be able to express the equation in $(n - m)$ dimensionless groups, or *pi-groups*, as Buckingham called them.

This fact is expressed by the *Buckingham pi-theorem*, which we state formally in the following way:

> A physical relationship among n variables, which can be expressed in *a minimum* of m dimensions, can be rearranged into a relationship among $(n - m)$ *independent* dimensionless groups of the original variables.

Two important qualifications have been italicized. They will be explained in detail in subsequent examples.

Buckingham called the dimensionless groups pi-groups and identified them as $\Pi_1, \Pi_2, ..., \Pi_{n-m}$. Normally we call Π_1 the dependent variable and retain $\Pi_{2\to(n-m)}$ as independent variables. Thus, the dimension*al* functional equation reduces to a dimension*less* functional equation of the form

$$\Pi_1 = \text{fn}\,(\Pi_2, \Pi_3, \ldots, \Pi_{n-m}) \tag{4.16}$$

Applications of the pi-theorem

Example 4.3

Is eqn. (2.24) consistent with the pi-theorem?

Solution. To find out, we first write the dimensional functional equation for Example 2.6:

$$\underbrace{T - T_i}_{\text{K}} = \text{fn}\left[\underbrace{r}_{\text{m}}, \underbrace{r_i}_{\text{m}}, \underbrace{r_o}_{\text{m}}, \underbrace{\overline{h}}_{\text{W/m}^2\text{K}}, \underbrace{k}_{\text{W/m·K}}, \underbrace{(T_\infty - T_i)}_{\text{K}}\right]$$

There are seven variables $(n = 7)$ in three dimensions, K, m, and W $(m = 3)$. Therefore, we look for $7 - 3 = 4$ pi-groups. There *are* four pi-groups in eqn. (2.24):

$$\Pi_1 = \frac{T - T_i}{T_\infty - T_i}, \quad \Pi_2 = \frac{r}{r_i}, \quad \Pi_3 = \frac{r_o}{r_i}, \quad \Pi_4 = \frac{\overline{h} r_o}{k} \equiv \text{Bi}. \qquad \blacksquare$$

Consider two features of this result. First, the minimum number of dimensions was three. If we had written watts as J/s, we would have had four dimensions instead. But Joules never appear in that particular problem independently of seconds. They always appear as a ratio and should not be separated. (If we had worked in English units, this would have seemed more confusing, since there is no name for Btu/sec unless we first convert it to horsepower.) The failure to identify dimensions that are consistently grouped together is one of the major errors that the beginner makes in using the pi-theorem.

The second feature is the *independence* of the groups. This means that we may pick any four dimensionless arrangements of variables, so long as no group or groups can be made into any other group by mathematical manipulation. For example, suppose that someone suggested that there was a fifth pi-group in Example 4.3:

$$\Pi_5 = \sqrt{\frac{\overline{h}r}{k}}$$

It is easy to see that Π_5 can be written as

$$\Pi_5 = \sqrt{\frac{\overline{h}r_o}{k}}\sqrt{\frac{r}{r_i}}\sqrt{\frac{r_i}{r_o}} = \sqrt{\text{Bi}\,\frac{\Pi_2}{\Pi_3}}$$

Therefore Π_5 is not independent of the existing groups, nor will we ever find a fifth grouping that is.

Another matter that is frequently emphasized is that of identifying the pi-groups once the variables are identified for a given problem. (The method of indices [4.1] is a cumbersome arithmetic strategy for doing this but it is perfectly correct.) However, we shall instead find the groups by using either of two much simpler methods:

1. The groups can always be obtained formally by repeating the simple elimination-of-dimensions procedure that was used to derive the pi-theorem in Example 4.2.

2. One may often simply arrange the variables into the required number of independent dimensionless groups by inspection.

In any method, one must make judgments as one combines variables. These decisions can lead to different arrangements of the pi-groups. Therefore, if the problem can be solved by inspection, there is no advantage to be gained by the use of a more formal procedure.

The methods of dimensional analysis can be used to help find the solution of many physical problems. We offer the following example, not entirely with tongue in cheek:

Example 4.4

Einstein might well have noted that the energy equivalent, e, of a rest mass, m_o, depended on the velocity of light, c_o, before he developed the special relativity theory. He would then have had the following dimensional functional equation:

$$\left(e \text{ N·m} \quad \text{or} \quad e \, \frac{\text{kg· m}^2}{\text{s}^2}\right) = \text{fn}\,(c_o \,\text{m/s}, \, m_o \,\text{kg})$$

The minimum number of dimensions is only two: kg and m/s, so we look for $3 - 2 = 1$ pi-group. To find it formally, we eliminated the dimension of mass from e by dividing it by m_o (kg). Thus,

$$\frac{e}{m_o} \frac{\text{m}^2}{\text{s}^2} = \text{fn}\Big[\, c_o \text{ m/s}, \underbrace{\quad m_o \text{ kg} \quad}_{\substack{\text{this must be removed}\\ \text{because it is the only}\\ \text{term with mass in it}}} \Big]$$

Then we eliminate the dimension of velocity (m/s) by dividing e/m_o by c_o^2:

$$\frac{e}{m_o c_o^2} = \text{fn}\,(c_o \text{ m/s})$$

This time c_o must be removed from the function on the right, since it is the only term with the dimensions m/s. This gives the result (which could have been written by inspection once it was known that there could only be one pi-group):

$$\Pi_1 = \frac{e}{m_o c_o^2} = \text{fn}\,(\text{no other groups}) = \text{constant}$$

or

$$e = \text{constant} \cdot \left(m_o c_o^2\right)$$

Of course, it required Einstein's relativity theory to tell us that the constant is one. ■

Example 4.5

What is the velocity of efflux of liquid from the tank shown in Fig. 4.4?

Solution. In this case we can guess that the velocity, V, might depend on gravity, g, and the head H. We might be tempted to include

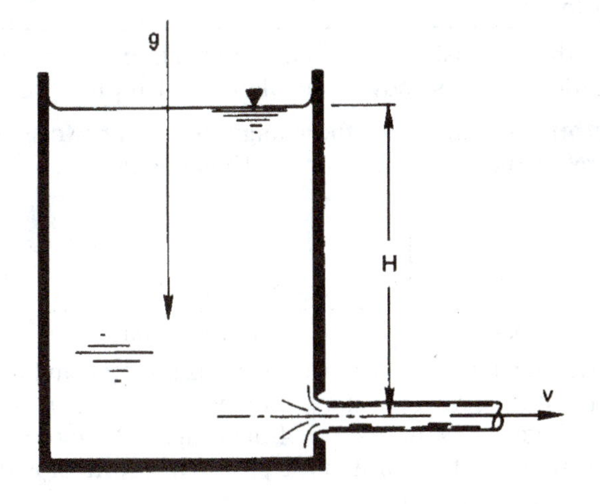

Figure 4.4 Efflux of liquid from a tank.

the density as well until we realize that g is already a *force per unit mass.* To understand this, we can use English units and divide g by the conversion factor,[4] g_c. Thus $(g \text{ ft/s}^2)/(g_c \text{ lb}_\text{m}\cdot\text{ft/lb}_\text{f}\text{ s}^2) = g \text{ lb}_\text{f}/\text{lb}_\text{m}$. Then

$$\underbrace{V}_{\text{m/s}} = \text{fn}\Big[\underbrace{H}_{\text{m}}, \underbrace{g}_{\text{m/s}^2}\Big]$$

so there are three variables in two dimensions, and we look for $3-2=1$ pi-groups. It would have to be

$$\Pi_1 = \frac{V}{\sqrt{gH}} = \text{fn (no other pi-groups)} = \text{constant}$$

or

$$V = \text{constant} \cdot \sqrt{gH}$$

The analytical study of fluid mechanics tells us that this form is correct and that the constant is $\sqrt{2}$. The group V^2/gh, by the way, is called a *Froude number*, Fr (pronounced "Frood"). It compares inertial forces to gravitational forces. Fr is about 1000 for a pitched baseball, and it is between 1 and 10 for the water flowing over the spillway of a dam. ■

[4]One can always divide any variable by a conversion factor without changing it.

Example 4.6

Obtain the dimensionless functional equation for the temperature distribution during steady conduction in a slab with a heat source, $\dot{q}$.

Solution. In such a case, there might be one or two specified temperatures in the problem: T_1 or T_2. Thus the dimensional functional equation is

$$\underbrace{T - T_1}_{\text{K}} = \text{fn}\left[\underbrace{(T_2 - T_1)}_{\text{K}}, \underbrace{x, L}_{\text{m}}, \underbrace{\dot{q}}_{\text{W/m}^3}, \underbrace{k}_{\text{W/m·K}}, \underbrace{\overline{h}}_{\text{W/m}^2\text{K}}\right]$$

where we presume that a convective b.c. is involved and we identify a characteristic length, L, in the x-direction. There are seven variables in three dimensions, or $7 - 3 = 4$ pi-groups. Three of these groups are ones we have dealt with in the past in one form or another:

$$\Pi_1 = \frac{T - T_1}{T_2 - T_1}$$ dimensionless temperature, which we shall give the name Θ

$$\Pi_2 = \frac{x}{L}$$ dimensionless length, which we call ξ

$$\Pi_3 = \frac{\overline{h}L}{k}$$ which we recognize as the Biot number, Bi

The fourth group is new to us:

$$\Pi_4 = \frac{\dot{q}L^2}{k(T_2 - T_1)}$$ which compares the heat generation rate to the rate of heat loss; we call it Γ

Thus, the solution is

$$\Theta = \text{fn}\,(\xi, \text{Bi}, \Gamma) \tag{4.17}$$

■

In Example 2.1, we undertook such a problem, but it differed in two respects. There was no convective boundary condition and hence, no $\overline{h}$, and only one temperature was specified in the problem. In this case, the dimensional functional equation was

$$(T - T_1) = \text{fn}\,(x, L, \dot{q}, k)$$

so there were only five variables in the same three dimensions. The resulting dimensionless functional equation therefore involved only two pi-groups. One was $\xi = x/L$ and the other is a new one equal to Θ/Γ. We call it Φ:

$$\Phi \equiv \frac{T - T_1}{\dot{q}L^2/k} = \text{fn}\left(\frac{x}{L}\right) \tag{4.18}$$

And this is exactly the form of the analytical result, eqn. (2.15).

Finally, we must deal with dimensions that convert into one another. For example, kg and N are defined in terms of one another through Newton's Second Law of Motion. Therefore, they cannot be identified as separate dimensions. The same would appear to be true of J and N·m, since both are dimensions of energy. However, we must discern whether or not a mechanism exists for interchanging them. If mechanical energy remains distinct from thermal energy in a given problem, then J should not be interpreted as N·m.

This issue will prove important when we do the dimensional analysis of several heat transfer problems. See, for example, the analyses of laminar convection problem at the beginning of Section 6.4, of natural convection in Section 8.3, of film condensation in Section 8.5, and of pool boiling burnout in Section 9.3. In all of these cases, heat transfer normally occurs without any conversion of heat to work or work to heat and it would be misleading to break J into N·m.

Additional examples of dimensional analysis appear throughout this book. Dimensional analysis is, indeed, our court of first resort in solving most of the new problems that we undertake.

4.4 An illustration of the use of dimensional analysis in a complex steady conduction problem

Heat conduction problems with convective b.c.s can rapidly grow difficult, even if they start out simple. So we look for ways to avoid making mistakes. For one thing, it is wise to take great care that dimensions are consistent at each stage of the solution. The best way to do this, and to eliminate a great deal of algebra at the same time, is to nondimensionalize the heat conduction equation before we apply the b.c.'s. This nondimensionalization should be consistent with the pi-theorem. We illustrate this idea with an example which, though it is complex, will illustrate several aspects of this idea.

Example 4.7

A slab shown in Fig. 4.5 has different temperatures and different heat transfer coefficients on either side and the heat is generated within it. Calculate the temperature distribution in the slab.

SOLUTION. The differential equation is

$$\frac{d^2T}{dx^2} = -\frac{\dot{q}}{k}$$

and the general solution is

$$T = -\frac{\dot{q}x^2}{2k} + C_1 x + C_2 \tag{4.19}$$

with b.c.'s

$$\overline{h}_1(T_1 - T)_{x=0} = -k\left.\frac{dT}{dx}\right|_{x=0}, \qquad \overline{h}_2(T - T_2)_{x=L} = -k\left.\frac{dT}{dx}\right|_{x=L}. \tag{4.20}$$

There are eight variables involved in the problem: $(T - T_2)$, $(T_1 - T_2)$, x, L, k, $\overline{h}_1$, $\overline{h}_2$, and $\dot{q}$; and there are three dimensions: K, W, and m. This results in 8 – 3 = 5 pi-groups. For these we choose

$$\Pi_1 \equiv \Theta = \frac{T - T_2}{T_1 - T_2}, \qquad \Pi_2 \equiv \xi = \frac{x}{L}, \qquad \Pi_3 \equiv \text{Bi}_1 = \frac{\overline{h}_1 L}{k},$$

$$\Pi_4 \equiv \text{Bi}_2 = \frac{\overline{h}_2 L}{k}, \qquad \text{and} \qquad \Pi_5 \equiv \Gamma = \frac{\dot{q}L^2}{2k(T_1 - T_2)},$$

where Γ can be interpreted as a comparison of the heat generated in the slab to that which could flow through it.

Under this nondimensionalization, eqn. (4.19) becomes[5]

$$\Theta = -\Gamma\,\xi^2 + C_3\xi + C_4 \tag{4.21}$$

and b.c.'s become

$$\text{Bi}_1(1 - \Theta_{\xi=0}) = -\Theta'_{\xi=0}, \qquad \text{Bi}_2\Theta_{\xi=1} = -\Theta'_{\xi=1} \tag{4.22}$$

where the primes denote differentiation with respect to ξ. Substituting eqn. (4.21) in eqn. (4.22), we obtain

$$\text{Bi}_1(1 - C_4) = -C_3, \qquad \text{Bi}_2(-\Gamma + C_3 + C_4) = 2\Gamma - C_3. \tag{4.23}$$

[5]The rearrangement of the dimensional equations into dimensionless form is straightforward algebra. If the results shown here are not immediately obvious, sketch the calculation on a piece of paper.

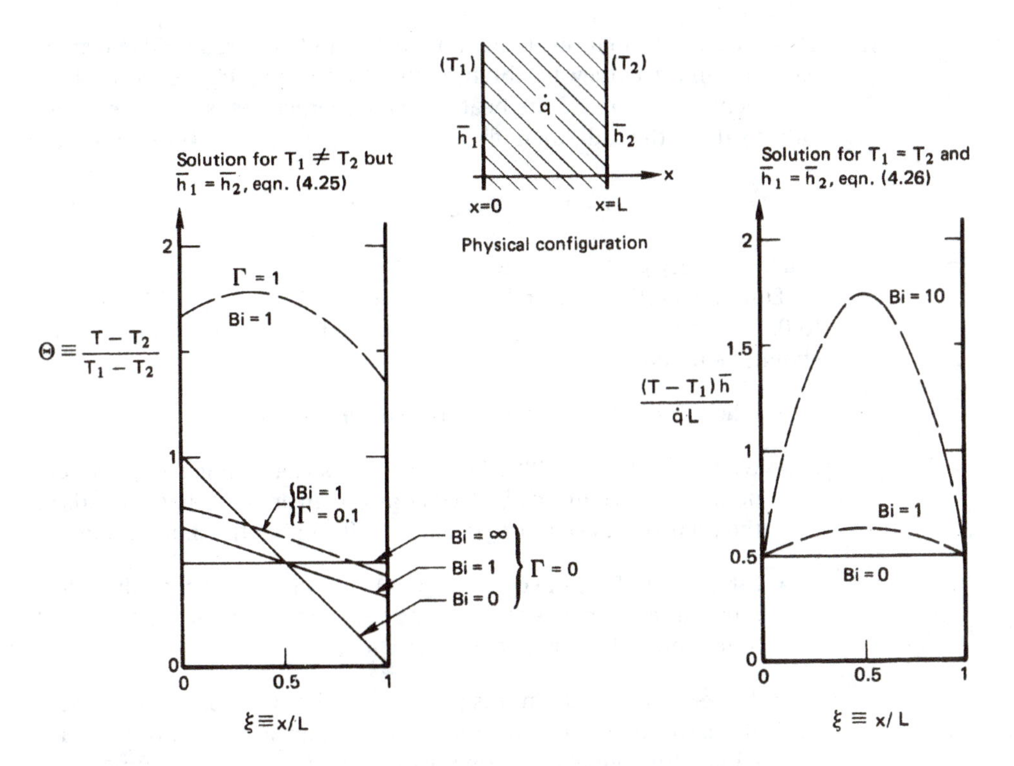

Figure 4.5 Heat conduction through a heat-generating slab with asymmetric boundary conditions.

Substituting the first of eqns. (4.23) in the second we get

$$C_4 = 1 + \frac{-\text{Bi}_1 + 2(\text{Bi}_1/\text{Bi}_2)\Gamma + \text{Bi}_1\Gamma}{\text{Bi}_1 + \text{Bi}_1^2/\text{Bi}_2 + \text{Bi}_1^2}$$

$$C_3 = \text{Bi}_1(C_4 - 1)$$

Thus, eqn. (4.21) becomes

$$\Theta = 1 + \Gamma\left[\frac{2(\text{Bi}_1/\text{Bi}_2) + \text{Bi}_1}{1 + \text{Bi}_1/\text{Bi}_2 + \text{Bi}_1}\xi - \xi^2 + \frac{2(\text{Bi}_1/\text{Bi}_2) + \text{Bi}_1}{\text{Bi}_1 + \text{Bi}_1^2/\text{Bi}_2 + \text{Bi}_1^2}\right] - \frac{\text{Bi}_1}{1 + \text{Bi}_1/\text{Bi}_2 + \text{Bi}_1}\xi - \frac{\text{Bi}_1}{\text{Bi}_1 + \text{Bi}_1^2/\text{Bi}_2 + \text{Bi}_1^2} \tag{4.24}$$

■

This is a complicated result and one that would have required enormous patience and accuracy to obtain without first simplifying the problem statement as we did. If the heat transfer coefficients were the same on either side of the wall, then $\mathrm{Bi}_1 = \mathrm{Bi}_2 \equiv \mathrm{Bi}$, and eqn. (4.24) would reduce to

$$\Theta = 1 + \Gamma\left(\xi - \xi^2 + 1/\mathrm{Bi}\right) - \frac{\xi + 1/\mathrm{Bi}}{1 + 2/\mathrm{Bi}} \tag{4.25}$$

which is a very great simplification.

Equation (4.25) is plotted on the left-hand side of Fig. 4.5 for Bi equal to 0, 1, and ∞ and for Γ equal to 0, 0.1, and 1. The following features should be noted:

- When $\Gamma \ll 0.1$, the heat generation can be ignored.
- When $\Gamma \gg 1, \Theta \to \Gamma/\mathrm{Bi} + \Gamma(\xi - \xi^2)$. This is a simple parabolic temperature distribution displaced upward an amount that depends on the relative external resistance, as reflected in the Biot number.
- If both Γ and $1/\mathrm{Bi}$ become large, $\Theta \to \Gamma/\mathrm{Bi}$. This means that when internal resistance is low and the heat generation is great, the slab temperature is constant and quite high.

If T_2 were equal to T_1 in this problem, Γ would go to infinity. In such a situation, we should redo the dimensional analysis of the problem. The dimensional functional equation now shows $(T - T_1)$ to be a function of x, L, k, $\overline{h}$, and $\dot{q}$. There are six variables in three dimensions, so there are three pi-groups

$$\frac{T - T_1}{\dot{q}L/h} = \mathrm{fn}\,(\xi, \mathrm{Bi})$$

where the dependent variable is like Φ [recall eqn. (4.18)] multiplied by Bi. We can put eqn. (4.25) in this form by multiplying both sides of it by $\overline{h}(T_1 - T_2)/\dot{q}\delta$. The result is

$$\frac{\overline{h}(T - T_1)}{\dot{q}L} = \frac{1}{2}\mathrm{Bi}\left(\xi - \xi^2\right) + \frac{1}{2} \tag{4.26}$$

The result is plotted on the right-hand side of Fig. 4.5. The following features of the graph are of interest:

- Heat generation is the only "force" giving rise to temperature nonuniformity. Since it is symmetric, the graph is also symmetric.

- When $\mathrm{Bi} \ll 1$, the slab temperature approaches a uniform value equal to $T_1 + \dot{q}L/2h$. (In this case, we would have solved the problem with far greater ease by using a simple lumped-capacity heat balance, since it is no longer a heat conduction problem.)

- When $\mathrm{Bi} > 100$, the temperature distribution is a very large parabola with ½ added to it. In this case, the problem could have been solved using boundary conditions of the first kind because the surface temperature stays very close to T_∞ (recall Fig. 1.11).

4.5 Fin design

The purpose of fins

We can substantially improve the convective removal of heat from a surface by putting extensions on that surface to increase its area. These extensions can take many forms. Figure 4.6, for example, shows just some of the ways in which the surface of commercial heat exchanger tubing can be extended with protrusions of a kind we call *fins.*

Figure 4.7 shows another intriguing application of fins in a heat exchanger design. This picture is taken from an issue of *Science* magazine in which Farlow et al. [4.5] present evidence suggesting that the strange rows of fins on the back of the *Stegosaurus* were used to shed excess body heat after strenuous activity.

These examples involve some rather complicated fins. But the analysis of a straight fin protruding from a wall displays the essential features of all fin behavior. This analysis has direct application to a host of problems.

Analysis of a one-dimensional fin

The equations. Figure 4.8 shows a one-dimensional fin protruding from a wall. The wall—and the roots of the fin—are at a temperature T_0, which is either greater or less than the ambient temperature, T_∞. The length of the fin is cooled or heated through a heat transfer coefficient, $\overline{h}$, by the ambient fluid. The heat transfer coefficient will be assumed uniform, although (as we see in Part III) that can introduce serious error in boiling, condensing, or other natural convection situations, and will not be strictly accurate even in forced convection.

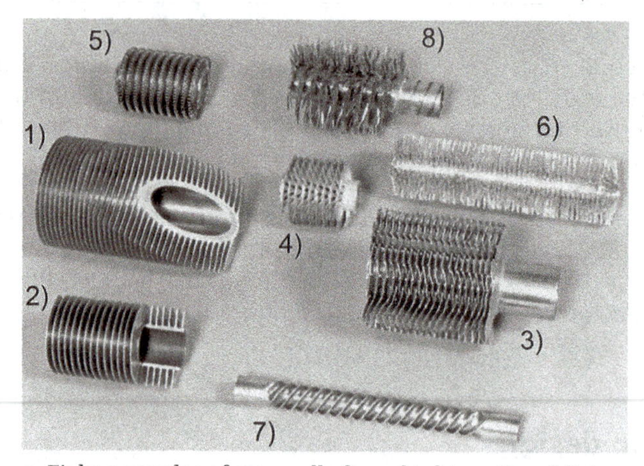

a. Eight examples of externally finned tubing: 1) and 2) typical commercial circular fins of constant thickness; 3) and 4) serrated circular fins and dimpled spirally-wound circular fins, both intended to improve convection; 5) spirally-wound copper coils outside and inside; 6) and 8) bristle fins, spirally wound and machined from base metal; 7) a spirally indented tube to improve convection and increase surface area.

b. An array of commercial internally finned tubing (photo courtesy of Noranda Metal Industries, Inc.)

Figure 4.6 Some of the many varieties of finned tubes.

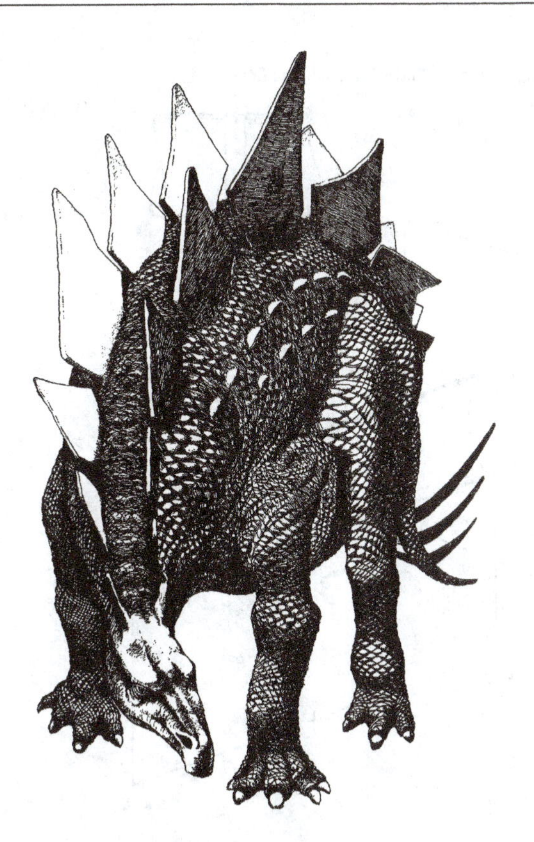

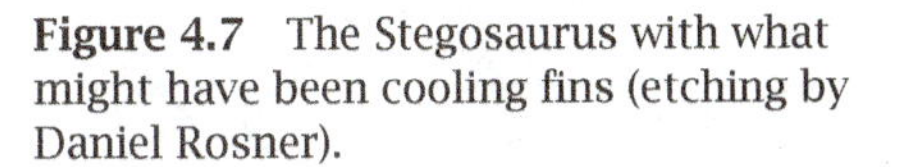

Figure 4.7 The Stegosaurus with what might have been cooling fins (etching by Daniel Rosner).

The tip may or may not exchange heat with the surroundings through a heat transfer coefficient, $\overline{h}_L$, which would generally differ from $\overline{h}$. The length of the fin is L, its uniform cross-sectional area is A, and its circumferential perimeter is P.

The characteristic dimension of the fin in the transverse direction (normal to the x-axis) is taken to be A/P. Thus, for a circular cylindrical fin, $A/P = \pi(\text{radius})^2/(2\pi \text{ radius}) = (\text{radius}/2)$. We define a Biot number for conduction in the transverse direction, based on this dimension, and require that it be small:

$$\text{Bi}_{\text{fin}} = \frac{\overline{h}(A/P)}{k} \ll 1 \tag{4.27}$$

This condition means that the transverse variation of T at any axial position, x, is much less than $(T_{\text{surface}} - T_\infty)$. Thus, $T \simeq T(x \text{ only})$ and the

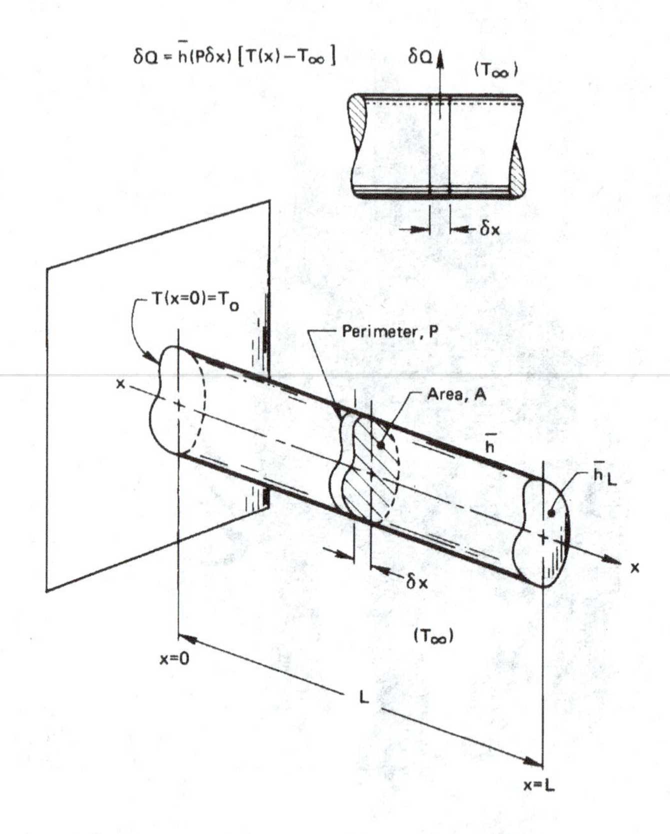

Figure 4.8 The analysis of a one-dimensional fin.

heat flow can be treated as one-dimensional.

An energy balance on the thin slice of the fin shown in Fig. 4.8 gives

$$-kA \left.\frac{dT}{dx}\right|_{x+\delta x} + kA \left.\frac{dT}{dx}\right|_{x} + \overline{h}(P\delta x)(T - T_\infty)_x = 0 \tag{4.28}$$

but

$$\frac{dT/dx|_{x+\delta x} - dT/dx|_x}{\delta x} \longrightarrow \frac{d^2T}{dx^2} = \frac{d^2(T - T_\infty)}{dx^2} \tag{4.29}$$

so

$$\frac{d^2(T - T_\infty)}{dx^2} = \frac{\overline{h}P}{kA}(T - T_\infty) \tag{4.30}$$

The b.c.'s for this equation are

$$(T - T_\infty)_{x=0} = T_0 - T_\infty$$
$$-kA \left.\frac{d(T - T_\infty)}{dx}\right|_{x=L} = \overline{h}_L A(T - T_\infty)_{x=L} \tag{4.31a}$$

Alternatively, if the tip is insulated, or if we can guess that $\overline{h}_L$ is small enough to be unimportant, the b.c.'s are

$$(T - T_\infty)_{x=0} = T_0 - T_\infty \quad \text{and} \quad \left.\frac{d(T - T_\infty)}{dx}\right|_{x=L} = 0 \tag{4.31b}$$

Before we solve this problem, it will pay to do a dimensional analysis of it. The dimensional functional equation is

$$T - T_\infty = \text{fn}\left[(T_0 - T_\infty), x, L, kA, \overline{h}P, \overline{h}_L A\right] \tag{4.32}$$

Notice that we have written kA, $\overline{h}P$, and $\overline{h}_L A$ as single variables. The reason for doing so is subtle but important. Setting $h(A/P)/k \ll 1$, erases any geometric detail of the cross section from the problem. The *only* place where P and A enter the problem is as product of $k, \overline{h}, \text{or} \overline{h}_L$. If they showed up elsewhere, they would have to do so in a physically incorrect way. Thus, we have just seven variables in W, K, and m. This gives four pi-groups if the tip is uninsulated:

$$\frac{T - T_\infty}{T_0 - T_\infty} = \text{fn}\left(\frac{x}{L}, \sqrt{\frac{\overline{h}P}{kA}L^2}, \underbrace{\frac{\overline{h}_L AL}{kA}}_{=\overline{h}_L L/k}\right)$$

or if we rename the groups,

$$\Theta = \text{fn}\left(\xi, mL, \text{Bi}_{\text{axial}}\right) \tag{4.33a}$$

where we call $\sqrt{\overline{h}PL^2/kA} \equiv mL$ because that terminology is common in the literature on fins.

If the tip of the fin is insulated, $\overline{h}_L$ will not appear in eqn. (4.32). There is one less variable but the same number of dimensions; hence, there will be only three pi-groups. The one that is removed is Bi_{axial}, which involves $\overline{h}_L$. Thus, for the insulated fin,

$$\Theta = \text{fn}(\xi, mL) \tag{4.33b}$$

We put eqn. (4.30) in these terms by multiplying it by $L^2/(T_0 - T_\infty)$. The result is

$$\boxed{\frac{d^2\Theta}{d\xi^2} = (mL)^2\Theta} \tag{4.34}$$

This equation is satisfied by $\Theta = Ce^{\pm(mL)\xi}$. The sum of these two solutions forms the general solution of eqn. (4.34):

$$\boxed{\Theta = C_1 e^{mL\xi} + C_2 e^{-mL\xi}} \tag{4.35}$$

Temperature distribution in a one-dimensional fin with the tip insulated The b.c.'s [eqn. (4.31b)] can be written as

$$\Theta_{\xi=0} = 1 \quad \text{and} \quad \left.\frac{d\Theta}{d\xi}\right|_{\xi=1} = 0 \tag{4.36}$$

Substituting eqn. (4.35) into both eqns. (4.36), we get

$$C_1 + C_2 = 1 \qquad \text{and} \qquad C_1 e^{mL} - C_2 e^{-mL} = 0 \tag{4.37}$$

Mathematical Digression 4.1

To put the solution of eqn. (4.37) for C_1 and C_2 in the simplest form, we need to recall a few properties of hyperbolic functions. The four basic functions that we need are defined as

$$\begin{aligned} \sinh x &\equiv \frac{e^x - e^{-x}}{2} \\ \cosh x &\equiv \frac{e^x + e^{-x}}{2} \\ \tanh x &\equiv \frac{\sinh x}{\cosh x} = \frac{e^x - e^{-x}}{e^x + e^{-x}} \\ \coth x &\equiv \frac{e^x + e^{-x}}{e^x - e^{-x}} \end{aligned} \tag{4.38}$$

where x is the independent variable. Additional functions are defined by analogy to the trigonometric counterparts. The differential relations can be written out formally, and they also resemble their trigonometric counterparts.

$$\begin{aligned} \frac{d}{dx}\sinh x &= \frac{1}{2}\left[e^x - (-e^{-x})\right] = \cosh x \\ \frac{d}{dx}\cosh x &= \frac{1}{2}\left[e^x + (-e^{-x})\right] = \sinh x \end{aligned} \tag{4.39}$$

These are analogous to the familiar results, $d\sin x/dx = \cos x$ and $d\cos x/dx = -\sin x$, but without the latter minus sign.

The solution of eqns. (4.37) is then

$$C_1 = \frac{e^{-mL}}{2\cosh mL} \quad \text{and} \quad C_2 = 1 - \frac{e^{-mL}}{2\cosh mL} \tag{4.40}$$

Therefore, eqn. (4.35) becomes

$$\Theta = \frac{e^{-mL(1-\xi)} + (2\cosh mL)e^{-mL\xi} - e^{-mL(1+\xi)}}{2\cosh mL}$$

which simplifies to

$$\boxed{\Theta = \frac{\cosh mL(1-\xi)}{\cosh mL}} \tag{4.41}$$

for a one-dimensional fin with its tip insulated.

One of the most important design variables for a fin is the rate at which it removes (or delivers) heat the wall. To calculate this, we write Fourier's law for the heat flow into the base of the fin:[6]

$$Q = -kA\left.\frac{d(T - T_\infty)}{dx}\right|_{x=0} \tag{4.42}$$

We multiply eqn. (4.42) by $L/kA(T_0 - T_\infty)$ and obtain, after substituting eqn. (4.41) on the right-hand side,

$$\frac{QL}{kA(T_0 - T_\infty)} = mL\frac{\sinh mL}{\cosh mL} = mL\tanh mL \tag{4.43}$$

which can be written

$$\frac{Q}{\sqrt{kA\overline{h}P}(T_0 - T_\infty)} = \tanh mL \tag{4.44}$$

Figure 4.9 includes two graphs showing the behavior of one-dimensional fin with an insulated tip. The top graph shows how the heat removal increases with mL to a virtual maximum at $mL \simeq 3$. This means that no such fin should have a length in excess of $2/m$ or $3/m$ if it is being used to cool (or heat) a wall. Additional length would simply increase the cost without doing any good.

---

[6]We could also integrate $\overline{h}(T - T_\infty)$ over the outside area of the fin to get Q. The answer would be the same, but the calculation would be a little more complicated.

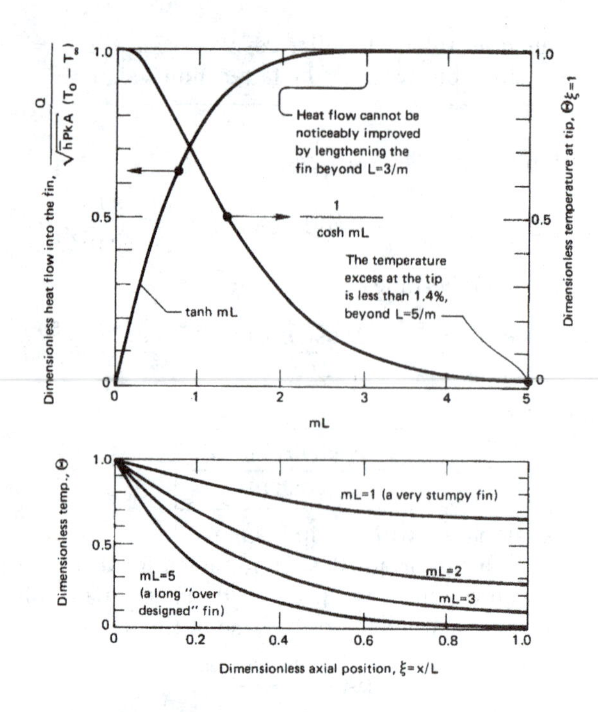

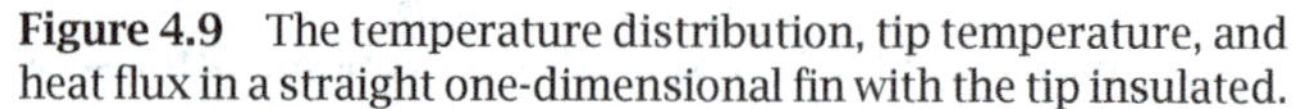

Figure 4.9 The temperature distribution, tip temperature, and heat flux in a straight one-dimensional fin with the tip insulated.

Also shown in the top graph is the temperature of the tip of such a fin. Setting $\xi = 1$ in eqn. (4.41), we discover that

$$\Theta_{\text{tip}} = \frac{1}{\cosh mL} \tag{4.45}$$

This dimensionless temperature drops to about 0.014 at the tip when mL reaches 5. This means that the end is $0.014(T_0 - T_\infty)$ K above T_∞ at the end. Thus, if the fin is actually functioning as a holder for a thermometer or a thermocouple that is intended to read T_∞, the reading will be in error if mL is not significantly greater than five.

The lower graph in Fig. 4.9 shows how the temperature is distributed in insulated-tip fins for various values of mL.

Experiment 4.1

Clamp a 20 cm or so length of copper rod by one end in a horizontal position. Put a candle flame very near the other end and let the arrangement come to a steady state. Run your finger along the rod. How does what you feel correspond to Fig. 4.9? (The diameter for the rod should not exceed about 3 mm. A larger rod of metal with a lower conductivity will also work.)

Exact temperature distribution in a fin with an uninsulated tip. The approximation of an insulated tip may be avoided using the b.c's given in eqn. (4.31a), which take the following dimensionless form:

$$\Theta_{\xi=0} = 1 \quad \text{and} \quad -\left.\frac{d\Theta}{d\xi}\right|_{\xi=1} = \text{Bi}_{\text{ax}}\Theta_{\xi=1} \tag{4.46}$$

Substitution of the general solution, eqn. (4.35), in these b.c.'s yields

$$\begin{aligned} C_1 + C_2 &= 1 \\ -mL(C_1e^{mL} - C_2e^{-mL}) &= \text{Bi}_{\text{ax}}(C_1e^{mL} + C_2e^{-mL}) \end{aligned} \tag{4.47}$$

It requires some manipulation to solve eqn. (4.47) for C_1 and C_2 and to substitute the results in eqn. (4.35). We leave this as an exercise (Problem 4.11). The result is

$$\Theta = \frac{\cosh mL(1-\xi) + (\text{Bi}_{\text{ax}}/mL)\sinh mL(1-\xi)}{\cosh mL + (\text{Bi}_{\text{ax}}/mL)\sinh mL} \tag{4.48}$$

which is the form of eqn. (4.33a), as we anticipated. The corresponding heat flux equation is

$$\frac{Q}{\sqrt{(kA)(\overline{h}P)}\,(T_0 - T_\infty)} = \frac{(\text{Bi}_{\text{ax}}/mL) + \tanh mL}{1 + (\text{Bi}_{\text{ax}}/mL)\tanh mL} \tag{4.49}$$

We have seen that mL is not too much greater than one in a well-designed fin with an insulated tip. Furthermore, when $\overline{h}_L$ is small (as it might be in natural convection), Bi_{ax} is normally much less than one. Therefore, in such cases, we expect to be justified in neglecting terms multiplied by Bi_{ax}. Then eqn. (4.48) reduces to

$$\Theta = \frac{\cosh mL(1-\xi)}{\cosh mL} \tag{4.41}$$

which we obtained by analyzing an insulated fin.

It is worth pointing out that we are in serious difficulty if $\overline{h}_L$ is so large that we cannot assume the tip to be insulated. The reason is that $\overline{h}_L$ is nearly impossible to predict in most practical cases.

Example 4.8

A 2 cm diameter aluminum rod with $k = 205$ W/m·K, 8 cm in length, protrudes from a 150°C wall. Air at 26°C flows by it, and $\overline{h} = 120$ W/m^{2}K. Determine whether or not tip conduction is important in this problem. To do this, make the very crude assumption that $\overline{h} \simeq \overline{h}_L$. Then compare the tip temperatures as calculated with and without considering heat transfer from the tip.

Solution.

$$mL = \sqrt{\frac{\overline{h}PL^2}{kA}} = \sqrt{\frac{120(0.08)^2}{205(0.01/2)}} = 0.8656$$

$$\text{Bi}_{\text{ax}} = \frac{\overline{h}L}{k} = \frac{120(0.08)}{205} = 0.0468$$

Therefore, eqn. (4.48) becomes

$$\Theta(\xi = 1) = \Theta_{\text{tip}} = \frac{\cosh 0 + (0.0468/0.8656)\sinh 0}{\cosh(0.8656) + (0.0468/0.8656)\sinh(0.8656)}$$

$$= \frac{1}{1.3986 + 0.0529} = 0.6886$$

so the exact tip temperature is

$$T_{\text{tip}} = T_\infty + 0.6886(T_0 - T_\infty)$$

$$= 26 + 0.6886(150 - 26) = 111.43°\text{C}$$

Equation (4.41) or Fig. 4.9, on the other hand, gives

$$\Theta_{\text{tip}} = \frac{1}{1.3986} = 0.7150$$

so the approximate tip temperature is

$$T_{\text{tip}} = 26 + 0.715(150 - 26) = 114.66°\text{C}$$

Thus the insulated-tip approximation is adequate for the computation in this case. ■

Very long fin. If a fin is so long that $mL \gg 1$, then eqn. (4.41) becomes

$$\lim_{mL\to\infty} \Theta = \lim_{mL\to\infty} \frac{e^{mL(1-\xi)} + e^{-mL(1-\xi)}}{e^{mL} + e^{-mL}} = \frac{e^{mL(1-\xi)}}{e^{mL}}$$

or

$$\lim_{mL\to\text{large}} \Theta = e^{-mL\xi} \tag{4.50}$$

Substituting this result in eqn. (4.42), we obtain [cf. eqn. (4.44)]

$$Q = \sqrt{(kA\overline{h}P)}\,(T_0 - T_\infty) \tag{4.51}$$

A heating or cooling fin would have to be terribly overdesigned for these results to apply—that is, mL would have been made much larger than necessary. Very long fins are common, however, in a variety of situations related to undesired heat losses. In practice, a fin may be regarded as "infinitely long" in computing its temperature if $mL \gtrsim 5$; in computing Q, $mL \gtrsim 3$ is sufficient for the infinite fin approximation.

Physical significance of mL. The group mL has thus far proved to be extremely useful in the analysis and design of fins. We should therefore say a brief word about its physical significance. Notice that

$$(mL)^2 = \frac{L/kA}{1/\overline{h}(PL)} = \frac{\text{internal resistance in } x\text{-direction}}{\text{gross external resistance}}$$

Thus $(mL)^2$ is a hybrid Biot number. When it is big, $\Theta|_{\xi=1} \to 0$ and we can neglect tip convection. When it is small, the temperature drop along the axis of the fin becomes small (see the lower graph in Fig. 4.9).

The group $(mL)^2$ also has a peculiar similarity to the NTU (Chapter 3) and the dimensionless time, t/T, that appears in the lumped-capacity solution (Chapter 1). Thus,

$$\frac{\overline{h}(PL)}{kA/L} \text{ is like } \frac{UA}{C_{\min}} \text{ is like } \frac{\overline{h}A}{\rho cV/t}$$

In each case a convective heat rate is compared with a heat rate that characterizes the capacity of a system; and in each case the system temperature asymptotically approaches its limit as the numerator becomes large. This was true in eqn. (1.22), eqn. (3.21), eqn. (3.22), and eqn. (4.50).

The problem of specifying the root temperature

Thus far, we have assumed the root temperature of a fin to be given information. There really are many circumstances in which it might be known; however, if a fin protrudes from a wall of the same material, as sketched in Fig. 4.10a, it is clear that for heat to flow, there must be a temperature gradient in the neighborhood of the root.

Consider the situation in which the surface of a wall is kept at a temperature T_s. Then a fin is placed on the wall as shown in the figure. If $T_\infty < T_s$, the wall temperature will be depressed in the neighborhood of the root as heat flows into the fin. The fin's performance should then be predicted using the lowered root temperature, T_{root}.

This heat conduction problem has been analyzed for several fin arrangements by Sparrow and co-workers. Fig. 4.10b is the result of Sparrow and Hennecke's [4.6] analysis for a single circular cylinder. They give

$$1 - \frac{Q_{\text{actual}}}{Q_{\text{no temp. depression}}} = \frac{T_s - T_{\text{root}}}{T_s - T_\infty} = \text{fn}\left[\frac{\overline{h}r}{k}, (mr)\tanh(mL)\right] \tag{4.52}$$

where r is the radius of the fin. From the figure we see that the actual heat flux into the fin, Q_{actual}, and the actual root temperature are both reduced when the Biot number, $\overline{h}r/k$, is large and the fin constant, m, is small.

Example 4.9

Neglect the tip convection from the fin in Example 4.8 and suppose that it is embedded in a wall of the same material. Calculate the error in Q and the actual temperature of the root if the wall is kept at 150°C.

Solution. From Example 4.8 we have $mL = 0.8656$ and $\overline{h}r/k = 120(0.010)/205 = 0.00586$. Then, with $mr = mL(r/L)$, we have $(mr)\tanh(mL) = 0.8656(0.010/0.080)\tanh(0.8656) = 0.0756$. The lower portion of Fig. 4.10b then gives

$$1 - \frac{Q_{\text{actual}}}{Q_{\text{no temp. depression}}} = \frac{T_s - T_{\text{root}}}{T_s - T_\infty} = 0.05$$

so the heat flow is reduced by 5% and the actual root temperature is

$$T_{\text{root}} = 150 - (150 - 26)0.05 = 143.8°\text{C}$$

The correction is modest in this case. ■

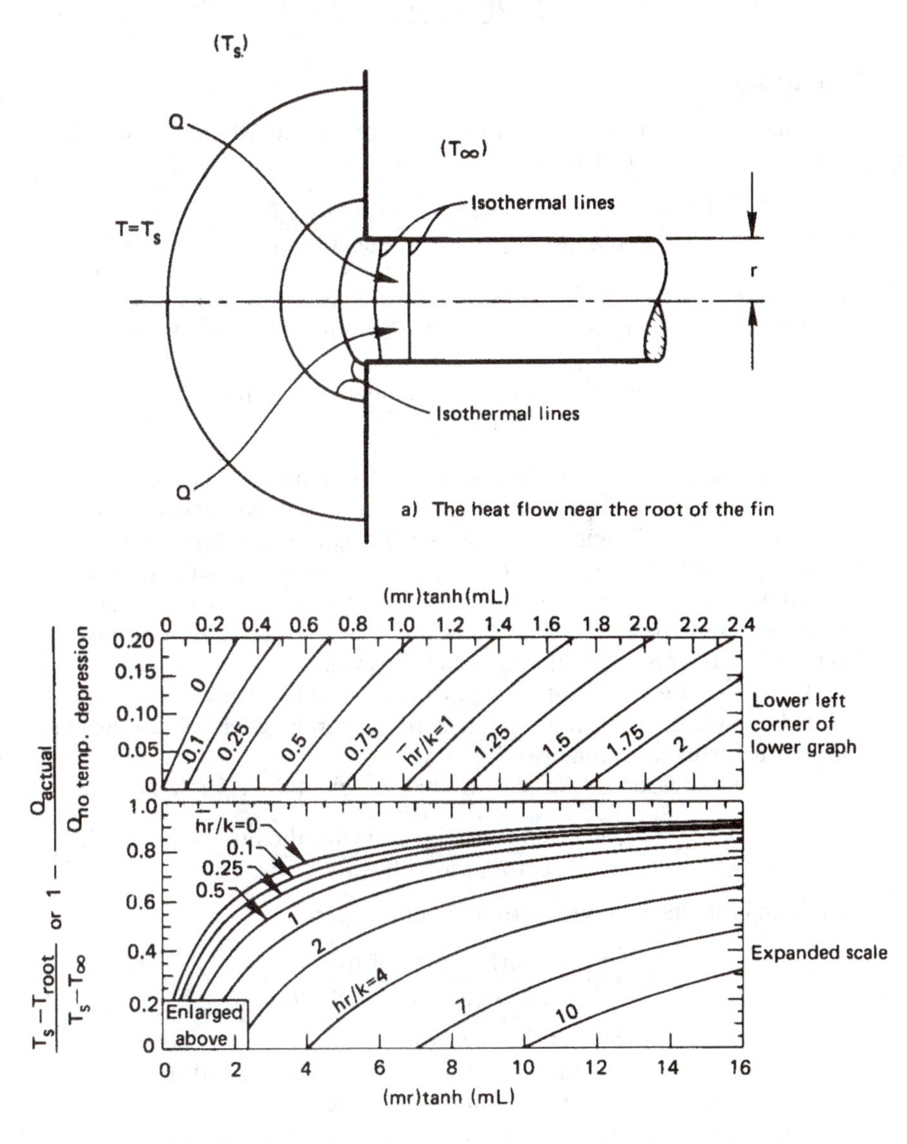

a) The heat flow near the root of the fin

b) Predicted deviations of root temperature and heat flux resulting from local temperature distortion near the root

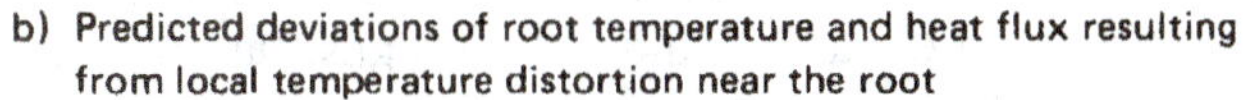

Figure 4.10 The influence of heat flow into the root of circular cylindrical fins [4.6]. This will be useful in fin design when $\overline{h}r/k$ is less than about 1 (or $\overline{h}A/kP$ is less than 1/2). The lines for larger values will not be needed for cooling fins.

Fin design

Two basic measures of fin performance are particularly useful in a fin design. The first is called the *efficiency*, η_f.

$$\eta_f \equiv \frac{\text{actual heat transferred by a fin}}{\text{heat that would be transferred if the entire fin were at } T = T_0} \tag{4.53}$$

where the word "efficiency" used in a rather loose way. To see how this works, we evaluate η_f for a one-dimensional fin with an insulated tip:

$$\eta_f = \frac{\sqrt{(\overline{h}P)(kA)}(T_0 - T_\infty)\tanh mL}{\overline{h}(PL)(T_0 - T_\infty)} = \frac{\tanh mL}{mL} \tag{4.54}$$

This says that, under the definition of efficiency, a very long fin will give $\tanh(mL)/mL \to 1/\text{large number}$, so the fin will be inefficient. On the other hand, the efficiency goes up to 100% as the length is reduced to zero, because $\tanh(mL) \to mL$ as $mL \to 0$. While a fin of zero length would accomplish little, a fin of small m might be designed in order to keep the tip temperature near the root temperature; this, for example, is desirable if the fin is the tip of a soldering iron.

It is therefore clear that, while η_f provides some useful information as to how well a fin is contrived, it is not generally advisable to design toward a particular value of η_f.

A second measure of fin performance is called the *effectiveness*, ε_f:

$$\varepsilon_f \equiv \frac{\text{heat flux from the wall with the fin}}{\text{heat flux from the wall without the fin}} \tag{4.55}$$

This can easily be computed from the efficiency:

$$\varepsilon_f = \eta_f \frac{\text{surface area of the fin}}{\text{cross-sectional area of the fin}} \tag{4.56}$$

Normally, we want the effectiveness to be as high as possible, But this can always be done by extending the length of the fin, and that—as we have seen—rapidly becomes a losing proposition.

The measures η_f and ε_f probably attract the interest of designers not because their absolute values guide the designs, but because they are useful in characterizing fins with more complex shapes. In such cases the solutions are often so complex that η_f and ε_f plots serve as labor-saving graphical solutions. We deal with some of these curves later in this section.

The design of a fin thus becomes an open-ended matter of optimizing, subject to many factors. Some of the factors that have to be considered include:

- The weight of material added by the fin. This might be a cost factor or it might be an important consideration in its own right.
- The possible dependence of $\overline{h}$ on $(T - T_\infty)$, flow velocity past the fin, or other influences.
- The influence of the fin (or fins) on the heat transfer coefficient, $\overline{h}$, as the fluid moves around it (or them).
- The geometric configuration of the channel that the fin lies in.
- The cost and complexity of manufacturing fins.
- The pressure drop introduced by the fins.

Fin thermal resistance

When fins occur in combination with other thermal elements, it can simplify calculations to treat them as a thermal resistance between the root and the surrounding fluid. Specifically, for a straight fin with an insulated tip, we can rearrange eqn. (4.44) as

$$Q = \frac{(T_0 - T_\infty)}{\left(\sqrt{kA\overline{h}P}\,\tanh mL\right)^{-1}} \equiv \frac{(T_0 - T_\infty)}{R_{t_\text{fin}}} \tag{4.57}$$

where

$$R_{t_\text{fin}} = \frac{1}{\sqrt{kA\overline{h}P}\,\tanh mL} \quad \text{for a straight fin} \tag{4.58}$$

In general, for a fin of any shape, fin thermal resistance can be written in terms of fin efficiency and fin effectiveness. From eqns. (4.53) and (4.55), we obtain

$$R_{t_\text{fin}} = \frac{1}{\eta_\text{f} A_\text{surface}\overline{h}} = \frac{1}{\varepsilon_\text{f} A_\text{root}\overline{h}} \tag{4.59}$$

Example 4.10

Consider again the resistor described in Examples 2.8 and 2.9, starting on page 75. Suppose that the two electrical leads are long straight wires 0.62 mm in diameter with $k = 16$ W/m·K and $h_{\text{eff}} = 23$ W/m^{2}K. Recalculate the resistor's temperature taking account of heat conducted into the leads.

Solution. The wires act as very long fins connected to the resistor, so that $\tanh mL \cong 1$ (see Prob. 4.44). Each has a fin resistance of

$$R_{t_{\text{fin}}} = \frac{1}{\sqrt{kA\overline{h}P}} = \frac{1}{\sqrt{(16)(23)(\pi)^2(0.00062)^3/4}} = 2,150 \text{ K/W}$$

These two thermal resistances are in parallel to the thermal resistances for natural convection and thermal radiation from the resistor surface found in Example 2.8. The equivalent thermal resistance is now

$$\begin{aligned} R_{t_{\text{equiv}}} &= \left(\frac{1}{R_{t_{\text{fin}}}} + \frac{1}{R_{t_{\text{fin}}}} + \frac{1}{R_{t_{\text{rad}}}} + \frac{1}{R_{t_{\text{conv}}}} \right)^{-1} \\ &= \left[\frac{2}{2,150} + (1.33 \times 10^{-4})(7.17) + (1.33 \times 10^{-4})(13) \right]^{-1} \\ &= 276.8 \text{ K/W} \end{aligned}$$

The leads reduce the equivalent resistance by about 30% from the value found before. The resistor temperature becomes

$$T_{\text{resistor}} = T_{\text{air}} + Q \cdot R_{t_{\text{equiv}}} = 35 + (0.1)(276.8) = 62.68 \text{ °C}$$

or about 10°C lower than before. ■

Fin Arrays

Fins are often arrayed in banks that are machined, cast, or extruded from single pieces of metal, with a thick base that holds the fin array. The base is fixed to the device to be cooled—a power transistor, a microprocessor, a computer video card—anything that generates a lot of heat that must be removed. Figure 4.11 shows several such typical arrays.

Manufacturers will sometimes simply specify a single thermal resistance for a fin array (or *heat sink*) as a function of the air velocity in the vicinity of the array. Or one might estimate the resistance of the array using the techniques introduced here, taking into account the airflow conditions between the fins and the heat loss from the exposed base between the fins. The detailed treatment of fin arrays becomes highly specialized. We recommend [4.8].

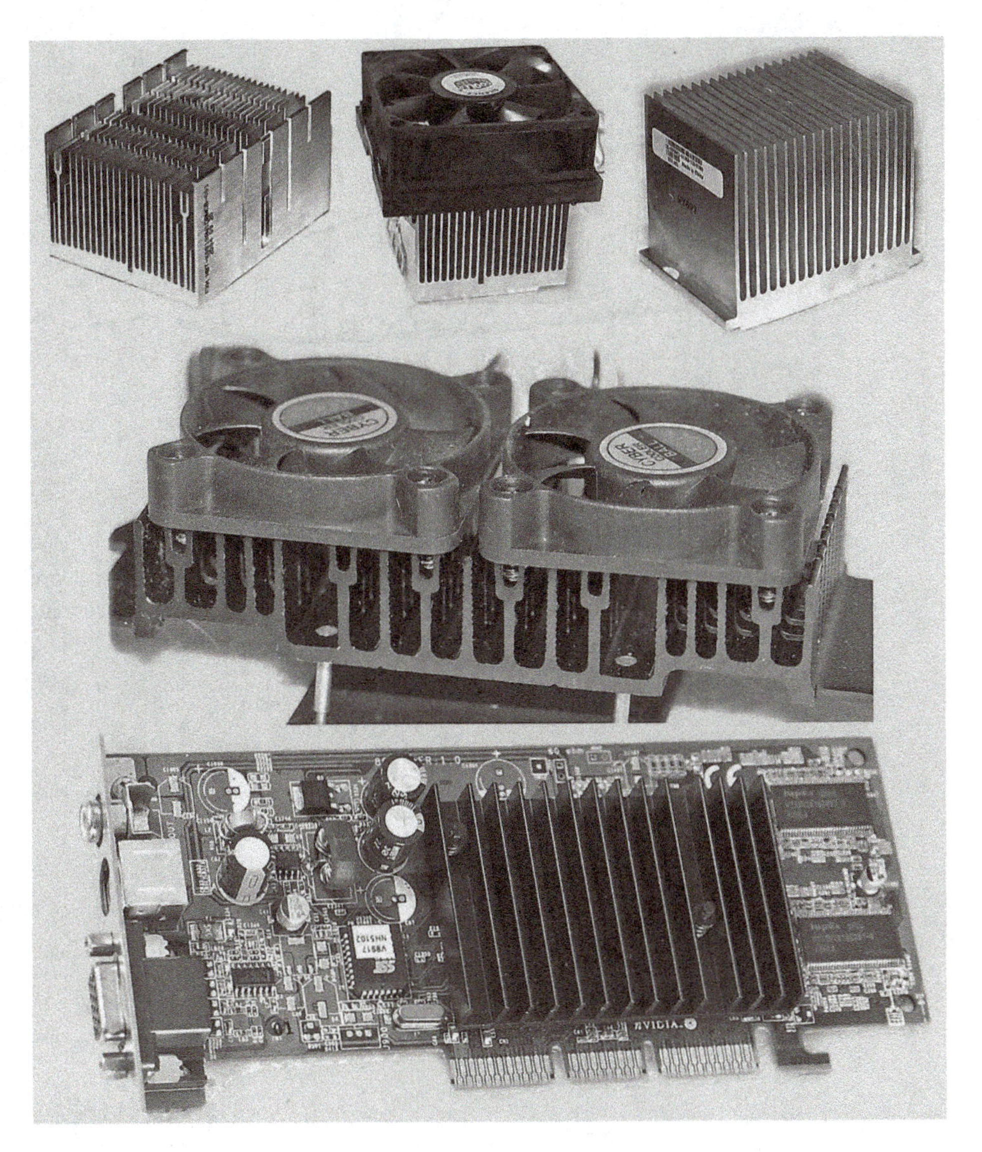

Figure 4.11 Several fin arrays of the kind used to cool computer elements. The top-center and middle arrays are fan-cooled. The other four are cooled by natural convection. Courtesy of Gene Quach, PC&C Computers, Houston, TX.

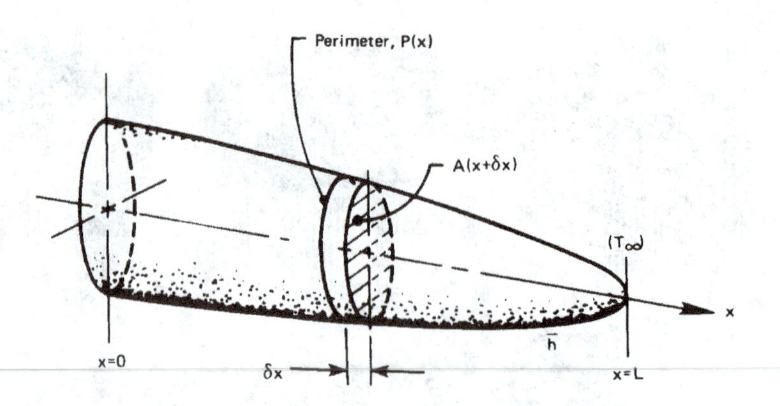

Figure 4.12 A general fin of variable cross section.

Fins of variable cross section

Let us consider what is involved is the design of a fin for which A and P are functions of x. Such a fin is shown in Fig. 4.12. We restrict our attention to fins for which

$$\frac{\overline{h}(A/P)}{k} \ll 1 \quad \text{and} \quad \frac{d(a/P)}{d(x)} \ll 1$$

so the heat flow will be approximately one-dimensional in x.

We begin the analysis, as always, with the First Law statement:

$$Q_{\text{net}} = Q_{\text{cond}} - Q_{\text{conv}} = \frac{dU}{dt}$$

or[7]

$$\underbrace{\left[kA(x+\delta x)\left.\frac{dT}{dx}\right|_{x=\delta x} - kA(x)\left.\frac{dT}{dx}\right|_{x}\right]}_{= \frac{d}{dx}kA(x)\frac{dT}{dx}\,\delta x} - \overline{h}P\,\delta x\,(T - T_\infty)$$

$$= \underbrace{\rho c A(x)\delta x \frac{dT}{dt}}_{=0,\ \text{since steady}}$$

[7]Note that we approximate the external area of the fin as horizontal when we write it as $P\,\delta x$. The actual area is negligibly larger than this in most cases. An exception would be the tip of the fin in Fig. 4.12.

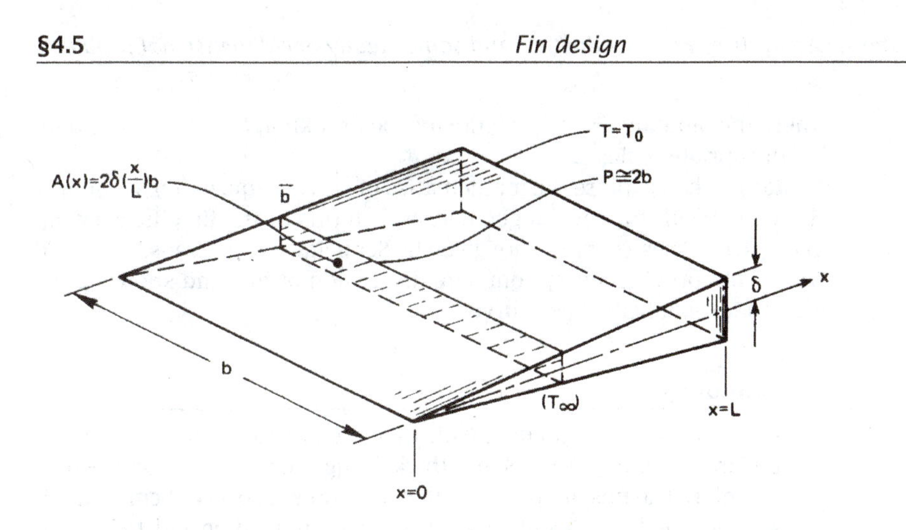

Figure 4.13 A two-dimensional wedge-shaped fin.

Therefore,

$$\boxed{\frac{d}{dx}\left[A(x)\frac{d(T-T_\infty)}{dx}\right]-\frac{\overline{h}P}{k}(T-T_\infty)=0} \tag{4.60}$$

If $A(x) = \text{constant}$, this reduces to $\Theta'' - (mL)^2\Theta = 0$, which is the straight fin equation.

To see how eqn. (4.60) works, consider the triangular fin shown in Fig. 4.13. In this case eqn. (4.60) becomes

$$\frac{d}{dx}\left[2\delta\left(\frac{x}{L}\right)b\,\frac{d(T-T_\infty)}{dx}\right]-\frac{2\overline{h}b}{k}(T-T_\infty)=0$$

or

$$\xi\frac{d^2\Theta}{d\xi^2}+\frac{d\Theta}{d\xi}-\underbrace{\frac{\overline{h}L^2}{k\delta}}_{\substack{\text{a kind}\\ \text{of }(mL)^2}}\Theta=0 \tag{4.61}$$

This second-order linear differential equation is difficult to solve because it has a variable coefficient. Its solution is expressible in Bessel functions:

$$\Theta=\frac{I_o\left(2\sqrt{\overline{h}Lx/k\delta}\right)}{I_o\left(2\sqrt{\overline{h}L^2/k\delta}\right)} \tag{4.62}$$

where the modified Bessel function of the first kind, I_o, can be looked up in appropriate tables.

Rather than explore the mathematics of solving eqn. (4.60), we simply show the result for several geometries in terms of the fin efficiency, η_f, in Fig. 4.14. These curves were given by Schneider [4.7]. Kraus, Aziz, and Welty [4.8] provide a very complete discussion of fins and show a great many additional efficiency curves.

Example 4.11

A thin brass pipe, 3 cm in outside diameter, carries hot water at 85°C. It is proposed to place 0.8 mm thick straight circular fins on the pipe to cool it. The fins are 8 cm in diameter and are spaced 2 cm apart. It is determined that $\overline{h}$ will equal 20 W/m²K on the pipe and 15 W/m²K on the fins, when they have been added. If $T_\infty = 22$°C, compute the heat loss per meter of pipe before and after the fins are added.

Solution. Before the fins are added,

$$Q = \pi(0.03 \text{ m})(20 \text{ W/m}^2\text{K})[(85-22) \text{ K}] = 199 \text{ W/m}$$

where we set $T_{\text{wall}} = T_{\text{water}}$ since the pipe is thin. Notice that, since the wall is constantly heated by the water, we should not have a root-temperature depression problem after the fins are added. Then we can enter Fig. 4.14a with

$$\frac{r_2}{r_1} = 2.67 \quad \text{and} \quad mL\sqrt{\frac{L}{P}} = \sqrt{\frac{\overline{h}L^3}{kA}} = \sqrt{\frac{15(0.04-0.15)^3}{125(0.025)(0.0008)}} = 0.306$$

and we obtain $\eta_f = 89\%$. Thus, the actual heat transfer given by

$$\underbrace{Q_{\text{without fin}}}_{119 \text{ W/m}} \underbrace{\left(\frac{0.02-0.0008}{0.02}\right)}_{\text{fraction of unfinned area}}$$
$$+\,0.89\underbrace{[2\pi(0.04^2-0.015^2)]}_{\text{area per fin (both sides), m}^2}\left(50\frac{\text{fins}}{\text{m}}\right)\left(15\,\frac{\text{W}}{\text{m}^2\text{K}}\right)[(85-22)\text{ K}]$$

so

$$Q_{\text{net}} = 478 \text{ W/m} = 4.02\, Q_{\text{without fins}} \qquad \blacksquare$$

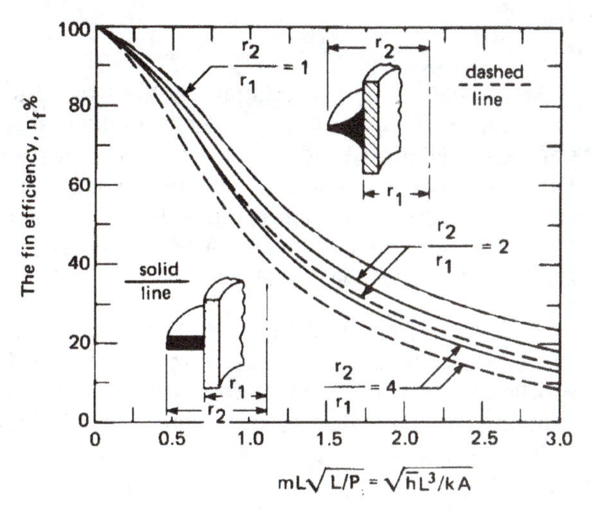

Comparison of a constant thickness circular fin, and a hyperbolic fin with thickness inversely proportional to radius. ($L \equiv r_2 - r_1$ and m is based on the area, A, shown in black.)

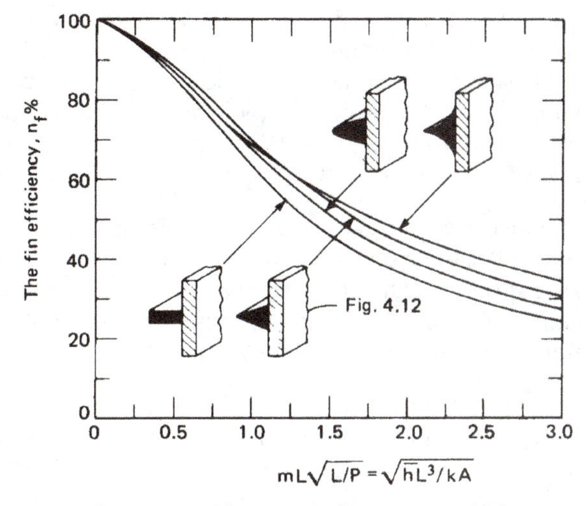

Comparison of four straight fins: constant thickness, triangular, parabolic, and hyperbolic. (m is based on A shown in black.)

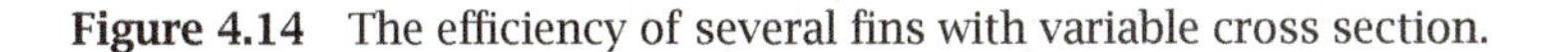

Figure 4.14 The efficiency of several fins with variable cross section.

Problems

4.1 Make a table listing the general solutions of all steady, unidimensional constant-properties heat conduction problems in Cartesian, cylindrical and spherical coordinates, with and without uniform heat generation. This table should prove to be a very useful tool in future problem solving. It should include a total of 18 solutions. State any restrictions on your solutions. Do not include calculations.

4.2 The left side of a slab of thickness L is kept at 0°C. The right side is cooled by air at T_∞°C blowing on it. $\overline{h}_{\text{RHS}}$ is known. An exothermic reaction takes place in the slab such that heat is generated at $A(T - T_\infty)$ W/m^3, where A is a constant. Find a fully dimensionless expression for the temperature distribution in the wall.

4.3 A long, wide plate of known size, material, and thickness L is connected across the terminals of a power supply and serves as a resistance heater. The voltage, current and T_∞ are known. The plate is insulated on the bottom and transfers heat out the top by convection. The temperature, T_{tc}, of the botton is measured with a thermocouple. Obtain expressions for (a) temperature distribution in the plate; (b) $\overline{h}$ at the top; (c) temperature at the top. (Note that your answers must depend on *known information* only.) [$T_{\text{top}} = T_{\text{tc}} - EIL^2/(2k \cdot \text{volume})$]

4.4 The heat transfer coefficient, $\overline{h}$, resulting from a forced flow over a flat plate depends on the fluid velocity, viscosity, density, specific heat, and thermal conductivity, as well as on the length of the plate. Develop the dimensionless functional equation for the heat transfer coefficient (cf. Section 6.5).

4.5 Water vapor condenses on a cold pipe and drips off the bottom in regularly spaced nodes as sketched in Fig. 3.9. The wavelength of these nodes, λ, depends on the liquid-vapor density difference, $\rho_f - \rho_g$, the surface tension, σ, and the gravity, g. Find how λ varies with its dependent variables.

4.6 A thick film flows down a vertical wall. The local film velocity at any distance from the wall depends on that distance, gravity, the liquid kinematic viscosity, and the film thickness. Obtain the dimensionless functional equation for the local velocity (cf. Section 8.5).

4.7 A steam preheater consists of a thick, electrically conducting, cylindrical shell insulated on the outside, with wet stream flowing down the middle. The inside heat transfer coefficient is highly variable, depending on the velocity, quality, and so on, but the flow temperature is constant. Heat is released at $\dot{q}$ J/m^{3}s within the cylinder wall. Evaluate the temperature within the cylinder as a function of position. Plot Θ against ρ, where Θ is an appropriate dimensionless temperature and $\rho = r/r_o$. Use $\rho_i = 2/3$ and note that Bi will be the parameter of a family of solutions. *On the basis of this plot*, recommend criteria (in terms of Bi) for (a) replacing the convective boundary condition on the inside with a constant temperature condition; (b) neglecting temperature variations within the cylinder.

4.8 Steam condenses on the inside of a small pipe, keeping it at a specified temperature, T_i. The pipe is heated by electrical resistance at a rate $\dot{q}$ W/m^3. The outside temperature is T_∞ and there is a natural convection heat transfer coefficient, $\overline{h}$ around the outside. (a) Derive an expression for the dimensionless expression temperature distribution, $\Theta = (T - T_\infty)/(T_i - T_\infty)$, as a function of the radius ratios, $\rho = r/r_o$ and $\rho_i = r_i/r_o$; a heat generation number, $\Gamma = \dot{q}r_o^2/k(T_i - T_\infty)$; and the Biot number. (b) Plot this result for the case $\rho_i = 2/3$, Bi = 1, and for several values of Γ. (c) Discuss any interesting aspects of your result.

4.9 Solve Problem 2.5 if you have not already done so, putting it in dimensionless form before you begin. Then let the Biot numbers approach infinity in the solution. You should get the same solution we got in Example 2.5, using b.c.'s of the first kind. Do you?

4.10 Complete the algebra that is missing between eqns. (4.30) and eqn. (4.31b) and eqn. (4.41).

4.11 Complete the algebra that is missing between eqns. (4.30) and eqn. (4.31a) and eqn. (4.48).

4.12 Obtain eqn. (4.50) from the general solution for a fin [eqn. (4.35)], using the b.c.'s $T(x=0) = T_0$ and $T(x=L) = T_\infty$. Comment on the significance of the computation.

4.13 What is the minimum length, l, of a thermometer well necessary to ensure an error less than 0.5% of the difference between the pipe wall temperature and the temperature of fluid flowing in a pipe? The well consists of a tube reaching into the pipe, with its end closed. It has a 2 cm O.D. and a 1.88 cm I.D. The material is type 304 stainless steel. Assume that the fluid is steam at 260°C and that the heat transfer coefficient between the steam and the tube wall is 300 W/m^{2}K. [3.44 cm.]

4.14 Thin fins with a 0.002 m by 0.02 m rectangular cross section and a thermal conductivity of 50 W/m·K protrude from a wall and have $\overline{h} \simeq 600$ W/m^{2}K and $T_0 = 170$°C. What is the heat flow rate into each fin and what is the effectiveness? $T_\infty =$ 20°C.

4.15 A thin rod is anchored at a wall at $T = T_0$ on one end. It is insulated at the other end. Plot the dimensionless temperature distribution in the rod as a function of dimensionless length: (a) if the rod is exposed to an environment at T_∞ through a heat transfer coefficient; (b) if the rod is insulated but heat is removed from the fin material at the uniform rate $-\dot{q} = \overline{h}P(T_0 - T_\infty)/A$. Comment on the implications of the comparison.

4.16 A tube of outside diameter d_o and inside diameter d_i carries fluid at $T = T_1$ from one wall at temperature T_1 to another wall a distance L away, at T_r. Outside the tube $\overline{h}_o$ is negligible, and inside the tube $\overline{h}_i$ is substantial. Treat the tube as a fin and plot the dimensionless temperature distribution in it as a function of dimensionless length. (*Hint:* The convective heating acts like a heat generation term that varies along the length of the fin.)

4.17 (If you have had some applied mathematics beyond the usual two years of calculus, this problem will not be difficult.) The shape of the fin in Fig. 4.13 is changed so that $A(x) = 2\delta(x/L)^2 b$ instead of $2\delta(x/L)b$. Calculate the temperature distribution and the heat flux at the base. Plot the temperature distribution and fin thickness against x/L. Derive an expression for η_f.

4.18 Work Problem 2.21, if you have not already done so, nondimensionalizing the problem before you attempt to solve it. It should now be much simpler.

4.19 One end of a copper rod 30 cm long is held at 200°C, and the other end is held at 93°C. The heat transfer coefficient in between is 17 W/m^2K (including both convection and radiation). If T_∞ = 38°C and the diameter of the rod is 1.25 cm, what is the net heat removed by the air around the rod? [19.13 W.]

4.20 How much error will the insulated-tip assumption give rise to in the calculation of the heat flow into the fin in Example 4.8?

4.21 A straight cylindrical fin 0.6 cm in diameter and 6 cm long protrudes from a magnesium block at 300°C. Air at 35°C is forced past the fin so that $\overline{h}$ is 130 W/m^2K. Calculate the heat removed by the fin, considering the temperature depression of the root.

4.22 Work Problem 4.19 considering the temperature depression in both roots. To do this, find mL for the two fins with insulated tips that would give the same temperature gradient at each wall. Base the correction on these values of mL.

4.23 A fin of triangular axial section (cf. Fig. 4.13) 0.1 m in length and 0.02 m wide at its base is used to extend the surface area of a 0.5% carbon steel wall. If the wall is at 40°C and heated gas flows past at 200°C ($\overline{h}$ = 230 W/m^2K), compute the heat removed by the fin per meter of breadth, b, of the fin. Neglect temperature distortion at the root.

4.24 Consider the concrete slab in Example 2.1. Suppose that the heat generation were to cease abruptly at time t = 0 and the slab were to start cooling back toward T_w. Predict $T = T_w$ as a function of time, noting that the initial parabolic temperature profile can be nicely approximated as a sine function. (Without the sine approximation, this problem would require the series methods of Chapter 5.)

4.25 Steam condenses in a 2 cm I.D. thin-walled tube of 99% aluminum at 10 atm pressure. There are circular fins of constant thickness, 3.5 cm in diameter, every 0.5 cm on the outside. The fins are 0.8 mm thick and the heat transfer coefficient from them $\overline{h} = 6$ W/m²K (including both convection and radiation). What is the mass rate of condensation if the pipe is 1.5 m in length, the ambient temperature is 18°C, and $\overline{h}$ for condensation is very large? [$\dot{m}_{cond} = 0.802$ kg/hr.]

4.26 How long must a copper fin, 0.4 cm in diameter, be if the temperature of its insulated tip is to exceed the surrounding air temperature by 20% of $(T_0 - T_\infty)$? $T_{air} = 20$°C and $\overline{h} = 28$ W/m²K (including both convection and radiation).

4.27 A 2 cm ice cube sits on a shelf of widely spaced aluminum rods, 3 mm in diameter, in a refrigerator at 10°C. How rapidly, in mm/min, do the rods melt their way through the ice cube if $\overline{h}$ at the surface of the rods is 10 W/m²K (including both convection and radiation). Be sure that you understand the physical mechanism before you make the calculation. Check your result experimentally. $h_{sf} = 333,300$ J/kg.

4.28 The highest heat flux that can be achieved in nucleate boiling (called q_{max}—see the qualitative discussion in Section 9.1) depends upon ρ_g, the saturated vapor density; h_{fg}, the latent heat vaporization; σ, the surface tension; a characteristic length, l; and the gravity force per unit volume, $g(\rho_f - \rho_g)$, where ρ_f is the saturated liquid density. Develop the dimensionless functional equation for q_{max} in terms of dimensionless length.

4.29 You want to rig a handle for a door in the wall of a furnace. The door is at 160°C. You consider bending a 40 cm length of 6.35 mm diam. 0.5% carbon steel rod into a U-shape and welding the ends to the door. Surrounding air at 24°C will cool the handle ($\overline{h} = 12$ W/m²K including both convection and radiation). What is the coolest temperature of the handle? How close to the door can you grasp the handle without getting burned if $T_{burn} = 65$°C? How might you improve the design?

4.30 A 14 cm long by 1 cm square brass rod is supplied with 25 W at its base. The other end is insulated. It is cooled by air at 20°C, with $\overline{h} = 68$ W/m^{2}K. Develop a dimensionless expression for Θ as a function of ε_f and other known information. Calculate the base temperature.

4.31 A cylindrical fin has a constant imposed heat flux of q_1 at one end and q_2 at the other end, and it is cooled convectively along its length. Develop the dimensionless temperature distribution in the fin. Specialize this result for $q_2 = 0$ and $L \to \infty$, and compare it with eqn. (4.50).

4.32 A thin metal cylinder of radius r_o serves as an electrical resistance heater. The temperature along an axial line in one side is kept at T_1. Another line, θ_2 radians away, is kept at T_2. Develop dimensionless expressions for the temperature distributions in the two sections.

4.33 Heat transfer is augmented, in a particular heat exchanger, with a field of 0.007 m diameter fins protruding 0.02 m into a flow. The fins are arranged in a hexagonal array, with a minimum spacing of 1.8 cm. The fins are bronze, and $\overline{h}_f$ around the fins is 168 W/m^{2}K. On the wall itself, $\overline{h}_w$ is only 54 W/m^{2}K. Calculate $\overline{h}_{\text{eff}}$ for the wall with its fins. ($\overline{h}_{\text{eff}} = Q_{\text{wall}}$ divided by A_{wall} and $[T_{\text{wall}} - T_\infty]$.)

4.34 Evaluate $d(\tanh x)/dx$.

4.35 An engineer seeks to study the effect of temperature on the curing of concrete by controlling the temperature of curing in the following way. A sample slab of thickness L is subjected to a heat flux, q_w, on one side, and it is cooled to temperature T_1 on the other. Derive a dimensionless expression for the steady temperature in the slab. Plot the expression and offer a criterion for neglecting the internal heat generation in the slab.

4.36 Develop the dimensionless temperature distribution in a spherical shell with the inside wall kept at one temperature and the outside wall at a second temperature. Reduce your solution to the limiting cases in which $r_{\text{outside}} \gg r_{\text{inside}}$ and in which r_{outside} is very close to r_{inside}. Discuss these limits.

4.37 Does the temperature distribution during steady heat transfer in an object with b.c.'s of only the first kind depend on k? Explain.

4.38 A long, 0.005 m diameter duralumin rod is wrapped with an electrical resistor over 3 cm of its length. The resistor imparts a surface flux of 40 kW/m^2. Evaluate the temperature of the rod in either side of the heated section if $\overline{h}$ = 150 W/m^{2}K around the unheated rod, and T_{ambient} = 27°C.

4.39 The heat transfer coefficient between a cool surface and a saturated vapor, when the vapor condenses in a film on the surface, depends on the liquid density and specific heat, the temperature difference, the buoyant force per unit volume ($g[\rho_f - \rho_g]$), the latent heat, the liquid conductivity and the kinematic viscosity, and the position (x) on the cooler. Develop the dimensionless functional equation for $\overline{h}$.

4.40 A duralumin pipe through a cold room has a 4 cm I.D. and a 5 cm O.D. It carries water that sometimes sits stationary. It is proposed to put electric heating rings around the pipe to protect it against freezing during cold periods of −7°C. The heat transfer coefficient outside the pipe is 9 W/m^{2}K (including both convection and radiation). Neglect the presence of the water in the conduction calculation, and determine how far apart the heaters would have to be if they brought the pipe temperature to 40°C locally. How much heat do they require?

4.41 The specific entropy of an ideal gas depends on its specific heat at constant pressure, its temperature and pressure, the ideal gas constant and reference values of the temperature and pressure. Obtain the dimensionless functional equation for the specific entropy and compare it with the known equation.

4.42 A large freezer's door has a 2.5 cm thick layer of insulation (k_{in} = 0.04 W/m·K) covered on the inside, outside, and edges with a continuous aluminum skin 3.2 mm thick (k_{Al} = 165 W/m·K). The door closes against a nonconducting seal 1 cm wide. Heat gain through the door can result from conduction straight through the insulation and skins (normal to the plane of the door) and from conduction in the aluminum skin only, going from the skin outside, around the edge skin, and to the inside skin. The heat transfer coefficients to the inside, $\overline{h}_i$, and outside, $\overline{h}_o$, are each 12 W/m^{2}K, accounting for both convection and radiation. The temperature outside the freezer is 25°C, and the temperature inside is −15°C.

a. If the door is 1 m wide, estimate the one-dimensional heat gain through the door, neglecting any conduction around the edges of the skin. Your answer will be in watts per meter of door height.

b. Now estimate the heat gain by conduction around the edges of the door, assuming that the insulation is perfectly adiabatic so that all heat flows through the skin. This answer will also be per meter of door height.

4.43 A thermocouple epoxied onto a high conductivity surface is intended to measure the surface temperature. The thermocouple consists of two each bare, 0.51 mm diameter wires. One wire is made of Chromel (Ni-10% Cr with k_{cr} = 17 W/m·K) and the other of constantan (Ni-45% Cu with k_{cn} = 23 W/m·K). The ends of the wires are welded together to create a measuring junction having has dimensions of D_w by $2D_w$. The wires extend perpendicularly away from the surface and do not touch one another. A layer of epoxy (k_{ep} = 0.5 W/m·K separates the thermocouple junction from the surface by 0.2 mm. Air at 20°C surrounds the wires. The heat transfer coefficient between each wire and the surroundings is $\overline{h}$ = 28 W/m^{2}K, including both convection and radiation. If the thermocouple reads T_{tc} = 40°C, estimate the actual temperature T_s of the surface and suggest a better arrangement of the wires.

4.44 The resistor leads in Example 4.10 were assumed to be "infinitely long" fins. What is the minimum length they each must have if they are to be modeled this way? What are the effectiveness, ε_{f}, and efficiency, η_{f}, of the wires?

References

[4.1] V. L. Streeter and E. B. Wylie. *Fluid Mechanics.* McGraw-Hill Book Company, New York, 7th edition, 1979. Chapter 4.

[4.2] E. Buckingham. *Phy. Rev.*, 4:345, 1914.

[4.3] E. Buckingham. Model experiments and the forms of empirical equations. *Trans. ASME*, 37:263–296, 1915.

[4.4] Lord Rayleigh, John Wm. Strutt. The principle of similitude. *Nature*, 95:66–68, 1915.

[4.5] J. O. Farlow, C. V. Thompson, and D. E. Rosner. Plates of the dinosaur stegosaurus: Forced convection heat loss fins? *Science*, 192(4244): 1123–1125 and cover, 1976.

[4.6] D. K. Hennecke and E. M. Sparrow. Local heat sink on a convectively cooled surface—application to temperature measurement error. *Int. J. Heat Mass Transfer*, 13:287–304, 1970.

[4.7] P. J. Schneider. *Conduction Heat Transfer*. Addison-Wesley Publishing Co., Inc., Reading, Mass., 1955.

[4.8] A. D. Kraus, A. Aziz, and J.R. Welty. *Extended Surface Heat Transfer*. John Wiley & Sons, Inc., New York, 2001.

5. Transient and multidimensional heat conduction

When I was a lad, winter was really cold. It would get so cold that if you went outside with a cup of hot coffee it would freeze. I mean it would freeze fast. That cup of hot coffee would freeze so fast that it would still be hot after it froze. Now that's cold! ***Old North-woods tall-tale***

5.1 Introduction

James Watt, of course, did not invent the steam engine. What he did do was to eliminate a destructive transient heating and cooling process that wasted a great amount of energy. By 1763, the great puffing engines of Savery and Newcomen had been used for over half a century to pump the water out of Cornish mines and to do other tasks. What has that to do with our subject? Well, consider what happened that same year, when the young instrument maker, Watt, was called upon to renovate the Newcomen engine model at the University of Glasgow. The Glasgow engine was then being used as a demonstration in the course on natural philosophy. Watt did much more than just renovate the machine—he first recognized, and eventually eliminated, its major shortcoming.

The cylinder of Newcomen's engine was cold when steam entered it and nudged the piston outward. A great deal of steam was wastefully condensed on the cylinder walls until they were warm enough to accommodate it. When the cylinder was filled, the steam valve was closed and jets of water were activated inside the cylinder to cool it again and condense the steam. This created a powerful vacuum, which sucked the piston back in on its working stroke. First, Watt tried to eliminate the wasteful initial condensation of steam by insulating the cylinder. But that simply reduced the vacuum and cut the power of the working stroke. Then he realized that, if he led the steam outside to a *separate condenser*, the cylinder could stay hot while the vacuum was created.

The separate condenser was the main issue in Watt's first patent (1769), and it immediately doubled the thermal efficiency of steam engines from a maximum of 1.1% to 2.2%. By the time Watt died in 1819, his invention had led to efficiencies of 5.7%, and his engine had altered the face of the world by powering the Industrial Revolution. And from 1769 until today, the steam power cycles that engineers study in their thermodynamics courses are accurately represented as steady flow—rather than transient—processes.

The repeated transient heating and cooling that occurred in Newcomen's engine was the kind of process that today's design engineer might still carelessly ignore, but the lesson that we learn from history is that transient heat transfer can be of overwhelming importance. Today, for example, designers of food storage enclosures know that such systems need relatively little energy to keep food cold at steady conditions. The real cost of operating them results from the consumption of energy needed to bring the food down to a low temperature and the losses resulting from people entering and leaving the system with food. The *transient* heat transfer processes are a dominant concern in the design of food storage units.

We therefore turn our attention, first, to the analysis of unsteady heat transfer. And we begin with a more detailed consideration of the lumped-capacity system that we looked at in Section 1.3. And our starting point is the dimensional analysis of such a system.

5.2 Lumped-capacity solutions

Dimensional analysis of transient heat conduction

Consider a fairly representative problem of one-dimensional transient heat conduction:

$$\frac{\partial^2 T}{\partial x^2} = \frac{1}{\alpha}\frac{\partial T}{\partial t} \quad \text{with} \quad \begin{cases} \text{i.c.:} & T(t=0) = T_i \\ \text{b.c.:} & T(t>0, x=0) = T_1 \\ \text{b.c.:} & -k\left.\dfrac{\partial T}{\partial x}\right|_{x=L} = \overline{h}\,(T-T_1)_{x=L} \end{cases}$$

The solution of this problem must take the form of the following dimensional functional equation:

$$T - T_1 = \text{fn}\left[(T_i - T_1), x, L, t, \alpha, \overline{h}, k\right]$$

There are eight variables in four dimensions (K, s, m, W), so we look for $8-4 = 4$ pi-groups. We anticipate, from Section 4.3, that they will include

$$\Theta \equiv \frac{(T - T_1)}{(T_i - T_1)}, \quad \xi \equiv \frac{x}{L}, \quad \text{and Bi} \equiv \frac{\overline{h}L}{k},$$

and we write

$$\Theta = \text{fn}\,(\xi, \text{Bi}, \Pi_4) \tag{5.1}$$

One possible candidate for Π_4, which is independent of the other three, is

$$\Pi_4 \equiv \text{Fo} = \alpha t / L^2 \tag{5.2}$$

where Fo is the *Fourier number.* Another candidate that we use later is

$$\Pi_4 \equiv \zeta = \frac{x}{\sqrt{\alpha t}} \quad \left(\text{this is exactly } \frac{\xi}{\sqrt{\text{Fo}}}\right) \tag{5.3}$$

If the problem involved b.c.'s of only the first kind, the heat transfer coefficient, $\overline{h}$—and hence the Biot number—would go out of the problem. Then the dimensionless function eqn. (5.1) is

$$\Theta = \text{fn}\,(\xi, \text{Fo}) \tag{5.4}$$

By the same token, if the b.c.'s had introduced different values of $\overline{h}$ at $x = 0$ and $x = L$, *two* Biot numbers would appear in the solution as they did in eqn. (4.24).

Dimensional analysis is particularly revealing in the case of the lumped-capacity problem [see eqns. (1.19)–(1.22)]. Neither k nor x enters the problem because we do not retain any features of the internal conduction problem. Therefore, we have ρc rather than α. Furthermore, we do not have to separate ρ and c because they only appear as a product. Finally, we use the volume-to-external-area ratio, V/A, as a characteristic length since no one linear dimension has any significance. Thus, for the transient lumped-capacity problem, the dimensional equation is

$$T - T_\infty = \text{fn}\left[(T_i - T_\infty), \rho c, V/A, \overline{h}, t\right] \tag{5.5}$$

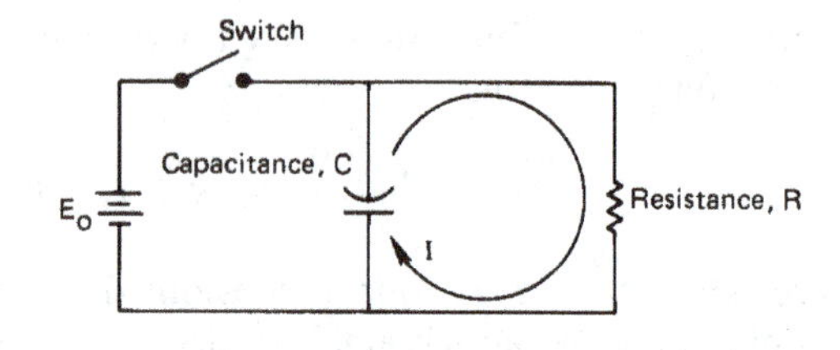

Figure 5.1 A simple resistance-capacitance circuit.

With six variables in the dimensions J, K, m, and s, only two pi-groups will appear in the dimensionless function equation.

$$\Theta = \text{fn}\left(\frac{\overline{h}At}{\rho cV}\right) = \text{fn}\left(\frac{t}{T}\right) \tag{5.6}$$

This is exactly the form of the simple lumped-capacity solution, eqn. (1.22). Notice, too, that the group t/T can be viewed as

$$\frac{t}{T} = \frac{hk(V/A)t}{\rho c(V/A)^2 k} = \frac{\overline{h}(V/A)}{k} \cdot \frac{\alpha t}{(V/A)^2} = \text{Bi Fo} \tag{5.7}$$

Electrical and mechanical analogies to the lumped-thermal-capacity problem

We take the term *capacitance* from electrical circuit theory and can sketch the simple analogous resistance-capacitance circuit in Fig. 5.1. Here, the electrical capacitor is initially charged to a voltage, E_o. When the switch is suddenly opened, the capacitor discharges through the resistor and the voltage drops according to the relation

$$\frac{dE}{dt} + \frac{E}{RC} = 0 \tag{5.8}$$

The solution of eqn. (5.8) with the i.c. $E(t = 0) = E_o$ is

$$E = E_o e^{-t/RC} \tag{5.9}$$

and the current can be computed from Ohm's law, once $E(t)$ is known.

$$I = \frac{E}{R} \tag{5.10}$$

Normally, in a heat conduction problem the *thermal* capacitance, ρcV, is distributed in space. But when the Biot number is small, $T(t)$

is uniform in the body and we can *lump* the capacitance into a single circuit element. The thermal resistance is $1/\overline{h}A$, and the temperature difference $(T - T_\infty)$ is analogous to $E(t)$. Thus, the thermal response, analogous to eqn. (5.9), is [see eqn. (1.22)]

$$T - T_\infty = (T_i - T_\infty)\exp\left(-\frac{\overline{h}At}{\rho cV}\right)$$

Notice that the electrical time constant, analogous to $\rho cV/\overline{h}A$, is RC.

Now consider a slightly more complex system that is also analogous to slightly more complex lumped capacity heat transfer. Figure 5.2 shows a spring-mass-damper system. The well-known response equation (actually, a force balance) for this system is

$$\underbrace{m}\frac{d^2x}{dt^2} + \underbrace{c}\frac{dx}{dt} + \underbrace{k}x = F(t) \tag{5.11}$$

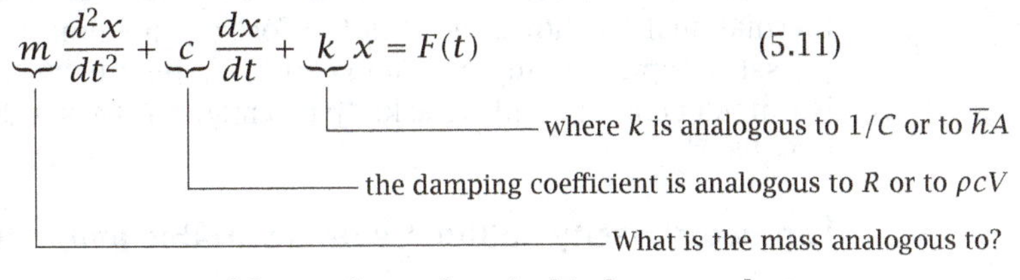

A term analogous to mass would arise from electrical inductance, but we

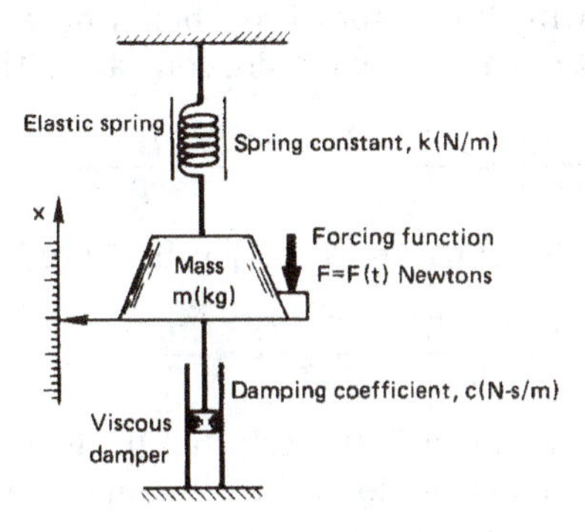

Figure 5.2 A spring-mass-damper system with a forcing function.

did not include it in the electrical circuit. Mass has the effect of carrying the system beyond its final equilibrium point. Thus, in an underdamped mechanical system, we might obtain the sort of response shown in Fig. 5.3 if we specified the velocity at $x = 0$ and provided no forcing function. Electrical inductance provides a similar effect. But the Second Law of Thermodynamics does not permit temperatures to overshoot their equilibrium values spontaneously. *There are no physical elements analogous to mass or inductance in thermal systems.*

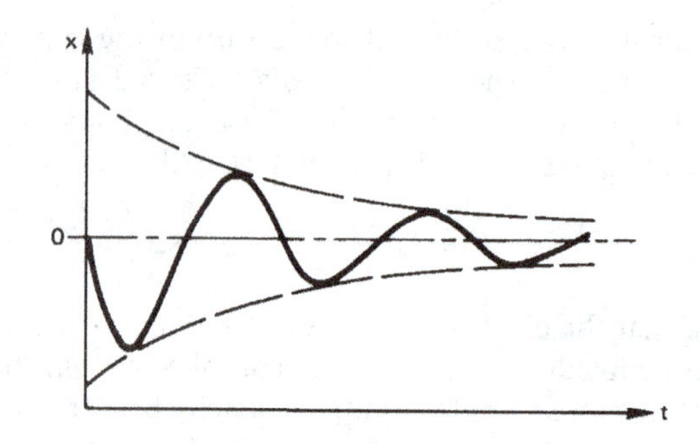

Figure 5.3 Response of an unforced spring-mass-damper system with an initial velocity.

Another mechanical element that we've introduced here does have a thermal analogy, however. It is the forcing function, F. We consider a (massless) spring-damper system with a forcing function F that probably is time-dependent, and we ask: "What might a thermal forcing function look like?"

Lumped-capacity solution with a variable ambient temperature

To answer the preceding question, let us suddenly immerse an object at a temperature $T = T_i$, with $\mathrm{Bi} \ll 1$, into a cool bath whose temperature is rising as $T_\infty(t) = T_i + bt$, where T_i and b are constants. Then eqn. (1.20) becomes

$$\frac{d(T - T_i)}{dt} = -\frac{T - T_\infty}{\mathrm{T}} = -\frac{T - T_i - bt}{\mathrm{T}}$$

where we have arbitrarily subtracted T_i under the differential. Then

$$\frac{d(T - T_i)}{dt} + \frac{T - T_i}{\mathrm{T}} = \frac{bt}{\mathrm{T}} \tag{5.12}$$

To solve eqn. (5.12) we must first recall that the general solution of a linear ordinary differential equation with constant coefficients is equal to the sum of any particular integral of the complete equation and the general solution of the homogeneous equation. We know the latter; it is $T - T_i = (\text{constant}) \exp(-t/\mathrm{T})$. A particular integral of the complete equation can often be formed by guessing solutions and trying them in the complete equation. Here we discover that

$$T - T_i = bt - b\mathrm{T}$$

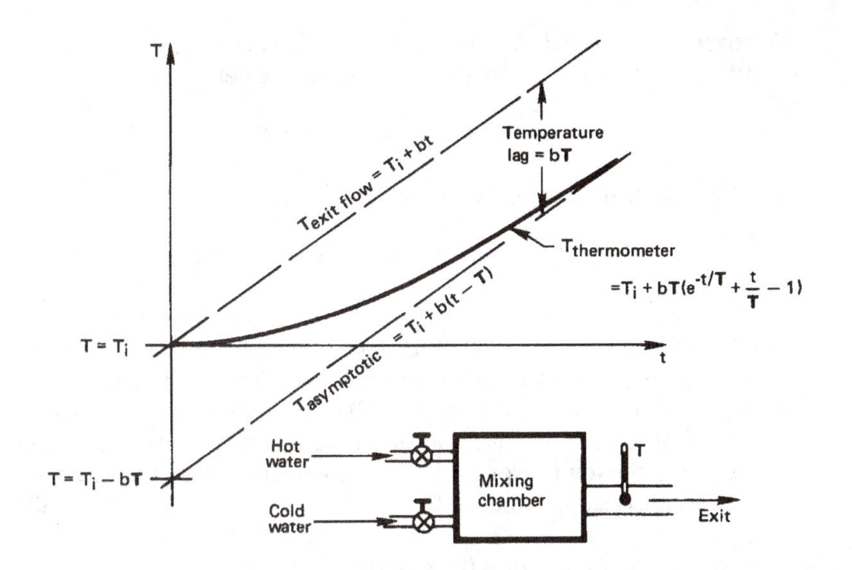

Figure 5.4 Response of a thermometer to a linearly increasing ambient temperature.

satisfies eqn. (5.12). Thus, the general solution of eqn. (5.12) is the sum of these general and particular solutions:

$$T - T_i = C_1 e^{-t/T} + b(t - T) \tag{5.13}$$

The solution for arbitrary variations of $T_\infty(t)$ may be obtained by working Problem 5.52 (see also Problems 5.3, 5.53, and 5.54, as well as the example that follows.).

Example 5.1

The flow rates of hot and cold water are regulated into a mixing chamber. We measure the temperature of the water as it leaves, using a thermometer with a time constant, T. On a particular day, the system started with cold water at $T = T_i$ in the mixing chamber. Then hot water is added in such a way that the outflow temperature rises linearly, as shown in Fig. 5.4, with $T_{\text{exit flow}} = T_i + bt$. How will the thermometer report the temperature variation?

Solution. The initial condition for eqn. (5.13) in this case is $T - T_i = 0$ at $t = 0$. Substituting eqn. (5.13) in the i.c., we get

$$0 = C_1 - b\mathrm{T} \quad \text{so} \quad C_1 = b\mathrm{T}$$

and the response equation is

$$T - (T_i + bt) = b\mathrm{T}\left(e^{-t/\mathrm{T}} - 1\right) \tag{5.14}$$

This result is plotted in Fig. 5.4. Notice that the thermometer reading reflects a transient portion, $b\mathrm{T}e^{-t/\mathrm{T}}$, which decays for a few time constants and then can be neglected, and a steady portion, $T_i + b(t - \mathrm{T})$, which persists thereafter. When the steady response is established, the thermometer follows the bath with a temperature lag of $b\mathrm{T}$. This constant error is reduced when either T or the rate the temperature rises, b, is reduced. ■

Second-order lumped-capacity systems

Now we look at situations in which two lumped-thermal-capacity systems are connected in series. Such an arrangement is shown in Fig. 5.5. Heat is transferred through two slabs with an interfacial resistance, h_c^{-1} between them. We shall require that $h_c L_1/k_1, h_c L_2/k_2$, and $\overline{h}L_2/k_2$ are all much less than one, so we can lump the thermal capacitance of each slab. The differential equations for the temperature response of each slab are then

$$\text{slab 1}: \quad -(\rho c V)_1 \frac{dT_1}{dt} = h_c A(T_1 - T_2) \tag{5.15}$$

$$\text{slab 2}: \quad -(\rho c V)_2 \frac{dT_2}{dt} = \overline{h}A(T_2 - T_\infty) - h_c A(T_1 - T_2) \tag{5.16}$$

and the initial conditions on the temperatures T_1 and T_2 are

$$T_1(t = 0) = T_2(t = 0) = T_i \tag{5.17}$$

We next identify two time constants for this problem:[1]

$$\mathrm{T}_1 \equiv (\rho c V)_1/h_c A \quad \text{and} \quad \mathrm{T}_2 \equiv (\rho c V)_2/\overline{h}A$$

[1]Notice that we could also have used $(\rho c V)_2/h_c A$ for T_2 since both h_c and $\overline{h}$ act on slab 2. The choice is arbitrary.

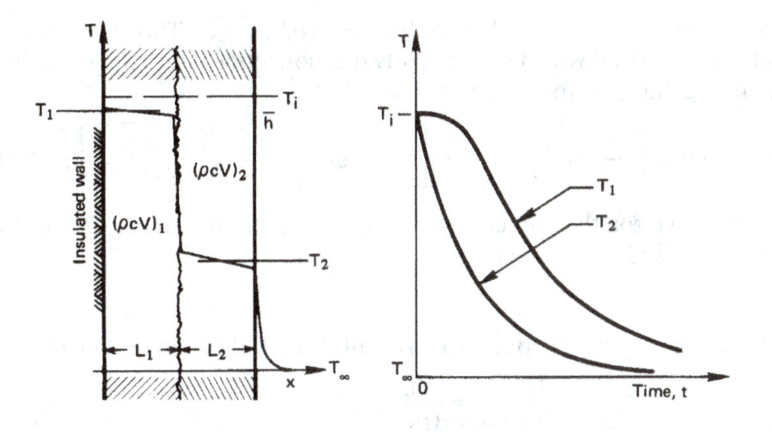

Figure 5.5 Two slabs conducting in series through an interfacial resistance.

Then eqn. (5.15) becomes

$$T_2 = \mathrm{T}_1 \frac{dT_1}{dt} + T_1 \tag{5.18}$$

which we substitute in eqn. (5.16) to get

$$\left(\mathrm{T}_1 \frac{dT_1}{dt} + T_1 - T_\infty\right) + \frac{h_c}{\overline{h}} \mathrm{T}_1 \frac{dT_1}{dt} = \mathrm{T}_1 \mathrm{T}_2 \frac{d^2 T_1}{dt^2} - \mathrm{T}_2 \frac{dT_1}{dt}$$

or

$$\frac{d^2 T_1}{dt^2} + \underbrace{\left[\frac{1}{\mathrm{T}_1} + \frac{1}{\mathrm{T}_2} + \frac{h_c}{\overline{h}\mathrm{T}_2}\right]}_{\equiv b} \frac{dT_1}{dt} + \underbrace{\frac{T_1 - T_\infty}{\mathrm{T}_1 \mathrm{T}_2}}_{c(T_1 - T_\infty)} = 0 \tag{5.19a}$$

if we call $T_1 - T_\infty \equiv \theta$, then eqn. (5.19a) can be written as

$$\frac{d^2\theta}{dt^2} + b\frac{d\theta}{dt} + c\theta = 0 \tag{5.19b}$$

Thus we have reduced the pair of first-order equations, eqn. (5.15) and eqn. (5.16), to a single second-order equation, eqn. (5.19b).

The general solution of eqn. (5.19b) is obtained by guessing a solution of the form $\theta = C_1 e^{Dt}$. Substitution of this guess into eqn. (5.19b) gives

$$D^2 + bD + c = 0 \tag{5.20}$$

from which we find that $D = -(b/2) \pm \sqrt{(b/2)^2 - c}$. This gives us two values of D, from which we can get two exponential solutions. By adding them together, we form a general solution:

$$\theta = C_1 \exp\left[-\frac{b}{2} + \sqrt{\left(\frac{b}{2}\right)^2 - c}\,\right] t + C_2 \exp\left[-\frac{b}{2} - \sqrt{\left(\frac{b}{2}\right)^2 - c}\,\right] t \quad (5.21)$$

To solve for the two constants we first substitute eqn. (5.21) in the first of i.c.'s (5.17) and get

$$T_i - T_\infty = \theta_i = C_1 + C_2 \quad (5.22)$$

The second i.c. can be put into terms of T_1 with the help of eqn. (5.15):

$$-\left.\frac{dT_1}{dt}\right|_{t=0} = \frac{h_c A}{(\rho c V)_1}(T_1 - T_2)_{t=0} = 0$$

We substitute eqn. (5.21) in this and obtain

$$0 = \left[-\frac{b}{2} + \sqrt{\left(\frac{b}{2}\right)^2 - c}\,\right] C_1 + \left[-\frac{b}{2} - \sqrt{\left(\frac{b}{2}\right)^2 - c}\,\right] \underbrace{C_2}_{=\,\theta_i - C_1}$$

so

$$C_1 = -\theta_i \left[\frac{-b/2 - \sqrt{(b/2)^2 - c}}{2\sqrt{(b/2)^2 - c}}\right]$$

and

$$C_2 = \theta_i \left[\frac{-b/2 + \sqrt{(b/2)^2 - c}}{2\sqrt{(b/2)^2 - c}}\right]$$

So we obtain at last:

$$\boxed{\begin{aligned} \frac{T_1 - T_\infty}{T_i - T_\infty} \equiv \frac{\theta}{\theta_i} = \frac{b/2 + \sqrt{(b/2)^2 - c}}{2\sqrt{(b/2)^2 - c}} \exp\left[-\frac{b}{2} + \sqrt{\left(\frac{b}{2}\right)^2 - c}\,\right] t \\ + \frac{-b/2 + \sqrt{(b/2)^2 - c}}{2\sqrt{(b/2)^2 - c}} \exp\left[-\frac{b}{2} - \sqrt{\left(\frac{b}{2}\right)^2 - c}\,\right] t \end{aligned}} \quad (5.23)$$

This is a pretty complicated result—all the more complicated when we remember that b involves three algebraic terms [recall eqn. (5.19a)]. Yet there is nothing very sophisticated about it; it is easy to understand. A system involving three capacitances in series would similarly yield a third-order equation of correspondingly higher complexity, and so forth.

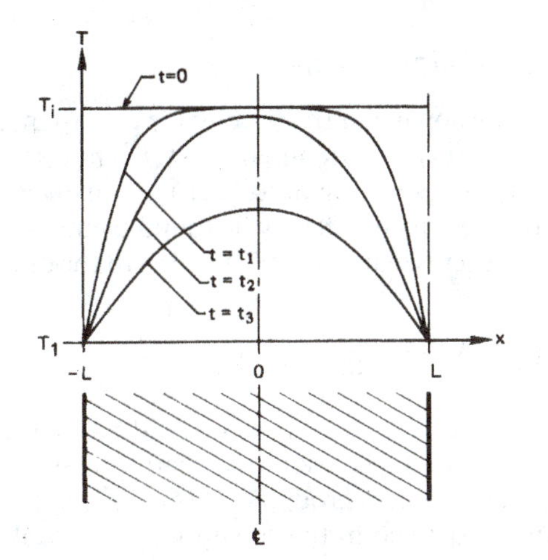

Figure 5.6 The transient cooling of a slab; $\xi = (x/L) + 1$.

5.3 Transient conduction in a one-dimensional slab

We next extend consideration to heat flow in bodies whose internal resistance is significant—to situations in which the lumped capacitance assumption is no longer appropriate. When the temperature within, say, a one-dimensional body varies with position as well as time, we must solve the heat diffusion equation for $T(x, t)$. We shall do this somewhat complicated task for the simplest case and then look at the results of such calculations in other situations.

A simple slab, shown in Fig. 5.6, is initially at a temperature T_i. The temperature of the surface of the slab is suddenly changed to T_1, and we wish to calculate the interior temperature profile as a function of time. The heat conduction equation is

$$\frac{\partial^2 T}{\partial x^2} = \frac{1}{\alpha}\frac{\partial T}{\partial t} \tag{5.24}$$

with the following b.c.'s and i.c.:

$$T(-L, t > 0) = T(L, t > 0) = T_1 \quad \text{and} \quad T(x, t = 0) = T_i \tag{5.25}$$

In fully dimensionless form, eqn. (5.24) and eqn. (5.25) are

$$\frac{\partial^2 \Theta}{\partial \xi^2} = \frac{\partial \Theta}{\partial \text{Fo}} \tag{5.26}$$

and

$$\Theta(0, \mathrm{Fo}) = \Theta(2, \mathrm{Fo}) = 0 \quad \text{and} \quad \Theta(\xi, 0) = 1 \tag{5.27}$$

where we have nondimensionalized the problem in accordance with eqn. (5.4), using $\Theta \equiv (T - T_1)/(T_i - T_1)$ and $\mathrm{Fo} \equiv \alpha t/L^2$; but, to get a nicer looking final result, we have set ξ equal to $(x/L) + 1$ instead of x/L.

The general solution of eqn. (5.26) may be found using the separation of variables technique described in Sect. 4.2, leading to the dimensionless form of eqn. (4.11):

$$\Theta = e^{-\hat{\lambda}^2 \mathrm{Fo}} \left[G \sin(\hat{\lambda}\xi) + E \cos(\hat{\lambda}\xi) \right] \tag{5.28}$$

Direct nondimensionalization of eqn. (4.11) would show that $\hat{\lambda} \equiv \lambda L$, since λ had units of (length)$^{-1}$. The solution therefore appears to have introduced a fourth dimensionless group, $\hat{\lambda}$. This needs explanation. The number λ, which was introduced in the separation-of-variables process, is called an *eigenvalue*.[2] In the present problem, $\hat{\lambda} = \lambda L$ will turn out to be a number—or rather a sequence of numbers—that is independent of system parameters.

Substituting the general solution, eqn. (5.28), in the first b.c. gives

$$0 = e^{-\hat{\lambda}^2 \mathrm{Fo}} \, (0 + E) \quad \text{so} \quad E = 0$$

and substituting it in the second yields

$$0 = e^{-\hat{\lambda}^2 \mathrm{Fo}} [G \sin 2\hat{\lambda}] \quad \text{so either} \quad G = 0$$

or

$$2\hat{\lambda} = 2\hat{\lambda}_n = n\pi, \quad n = 0, 1, 2, \ldots$$

In the second case, we are presented with two choices. The first, $G = 0$, would give $\Theta \equiv 0$ in all situations, so that the initial condition could never be accommodated. (This is what mathematicians call a *trivial* solution.) The second choice, $\hat{\lambda}_n = n\pi/2$, actually yields a string of solutions, each of the form

$$\Theta = G_n \, e^{-n^2\pi^2 \mathrm{Fo}/4} \sin\left(\frac{n\pi}{2} \xi \right) \tag{5.29}$$

where G_n is the constant appropriate to the nth one of these solutions.

[2] The word *eigenvalue* is a curious hybrid of the German term *eigenwert* and its English translation, *characteristic value*.

We still face the problem that none of eqns. (5.29) will fit the initial condition, $\Theta(\xi,0) = 1$. To get around this, we remember that the sum of any number of solutions of a linear differential equation is also a solution. Then we write

$$\Theta = \sum_{n=1}^{\infty} G_n \, e^{-n^2\pi^2 \mathrm{Fo}/4} \sin\left(n\frac{\pi}{2}\xi\right) \tag{5.30}$$

where we drop $n = 0$ since it gives zero contribution to the series. And we arrive, at last, at the problem of choosing the G_n's so that eqn. (5.30) will fit the initial condition.

$$\Theta\,(\xi,0) = \sum_{n=1}^{\infty} G_n \sin\left(n\frac{\pi}{2}\xi\right) = 1 \tag{5.31}$$

The problem of picking the values of G_n that will make this equation true is called "making a Fourier series expansion" of the function $f(\xi) = 1$. We shall not pursue strategies for making Fourier series expansions in any general way. Instead, we merely show how to accomplish the task for the particular problem at hand. We begin with a mathematical trick. We multiply eqn. (5.31) by $\sin(m\pi/2)$, where m may or may not equal n, and we integrate the result between $\xi = 0$ and 2.

$$\int_0^2 \sin\left(\frac{m\pi}{2}\xi\right) d\xi = \sum_{n=1}^{\infty} G_n \int_0^2 \sin\left(\frac{m\pi}{2}\xi\right) \sin\left(\frac{n\pi}{2}\xi\right) d\xi \tag{5.32}$$

(The interchange of summation and integration turns out to be legitimate, although we have not proved, here, that it is.[3]) With the help of a table of integrals, we find that

$$\int_0^2 \sin\left(\frac{m\pi}{2}\xi\right) \sin\left(\frac{n\pi}{2}\xi\right) d\xi = \begin{cases} 0 & \text{for } n \neq m \\ 1 & \text{for } n = m \end{cases}$$

Thus, when we complete the integration of eqn. (5.32), we get

$$-\frac{2}{m\pi} \cos\left(\frac{m\pi}{2}\xi\right)\bigg|_0^2 = \sum_{n=1}^{\infty} G_n \times \begin{cases} 0 & \text{for } n \neq m \\ 1 & \text{for } n = m \end{cases}$$

This reduces to

$$-\frac{2}{m\pi}\left[(-1)^n - 1\right] = G_n$$

[3]What is normally required is that the series in eqn. (5.31) be *uniformly convergent.*

so

$$G_n = \frac{4}{n\pi} \quad \text{where } n \text{ is an odd number}$$

Substituting this result into eqn. (5.30), we finally obtain the solution to the problem:

$$\Theta\,(\xi, \text{Fo}) = \frac{4}{\pi} \sum_{n=\text{odd}}^{\infty} \frac{1}{n}\, e^{-(n\pi/2)^2 \text{Fo}} \sin\left(\frac{n\pi}{2}\xi\right) \tag{5.33}$$

Equation (5.33) admits a very nice simplification for large time (or at large Fo). Suppose that we wish to evaluate Θ at the outer center of the slab—at $x = 0$ or $\xi = 1$. Then

$$\Theta\,(0, \text{Fo}) = \frac{4}{\pi} \times$$

$$\left\{ \underbrace{\exp\left[-\left(\frac{\pi}{2}\right)^2 \text{Fo}\right]}_{\substack{= 0.085 \text{ at Fo} = 1 \\ = 0.781 \text{ at Fo} = 0.1 \\ = 0.976 \text{ at Fo} = 0.01}} - \underbrace{\frac{1}{3}\exp\left[-\left(\frac{3\pi}{2}\right)^2 \text{Fo}\right]}_{\substack{\simeq 10^{-10} \text{ at Fo} = 1 \\ = 0.036 \text{ at Fo} = 0.1 \\ = 0.267 \text{ at Fo} = 0.01}} + \underbrace{\frac{1}{5}\exp\left[-\left(\frac{5\pi}{2}\right)^2 \text{Fo}\right]}_{\substack{\simeq 10^{-27} \text{ at Fo} = 1 \\ = 0.0004 \text{ at Fo} = 0.1 \\ = 0.108 \text{ at Fo} = 0.01}} + \cdots \right\}$$

Thus for values of Fo somewhat greater than 0.1, only the first term in the series need be used in the solution (except at points very close to the boundaries). We discuss these one-term solutions in Sect. 5.5. But first, let us see what happens if the slab had been subjected to b.c.'s of the third kind.

Suppose that the walls of the slab had been cooled by symmetrical convection such that the b.c.'s were

$$\overline{h}(T_\infty - T)_{x=-L} = -k\left.\frac{\partial T}{\partial x}\right|_{x=-L} \quad \text{and} \quad \overline{h}(T - T_\infty)_{x=L} = -k\left.\frac{\partial T}{\partial x}\right|_{x=L}$$

or in dimensionless form, using $\Theta \equiv (T - T_\infty)/(T_i - T_\infty)$ and $\xi = (x/L) + 1$,

$$-\Theta\Big|_{\xi=0} = -\frac{1}{\text{Bi}}\left.\frac{\partial\Theta}{\partial\xi}\right|_{\xi=0} \quad \text{and} \quad \left.\frac{\partial\Theta}{\partial\xi}\right|_{\xi=1} = 0$$

The solution is somewhat harder to find than eqn. (5.33) was, but the result is[4]

$$\Theta = \sum_{n=1}^{\infty} \exp\left(-\hat{\lambda}_n^2 \,\text{Fo}\right)\left(\frac{2\sin\hat{\lambda}_n \cos[\hat{\lambda}_n(\xi - 1)]}{\hat{\lambda}_n + \sin\hat{\lambda}_n \cos\hat{\lambda}_n}\right) \tag{5.34}$$

[4]See, for example, [5.1, §2.3.4] or [5.2, §3.4.3] for details of this calculation.

Table 5.1 Terms of series solutions for slabs, cylinders, and spheres. J_0 and J_1 are Bessel functions of the first kind.

	A_n	f_n	Equation for $\hat{\lambda}_n$
Slab	$\dfrac{2\sin\hat{\lambda}_n}{\hat{\lambda}_n + \sin\hat{\lambda}_n\cos\hat{\lambda}_n}$	$\cos\left(\hat{\lambda}_n \dfrac{x}{L}\right)$	$\cot\hat{\lambda}_n = \dfrac{\hat{\lambda}_n}{\mathrm{Bi}_L}$
Cylinder	$\dfrac{2J_1(\hat{\lambda}_n)}{\hat{\lambda}_n\left[J_0^2(\hat{\lambda}_n) + J_1^2(\hat{\lambda}_n)\right]}$	$J_0\left(\hat{\lambda}_n \dfrac{r}{r_o}\right)$	$\hat{\lambda}_n J_1(\hat{\lambda}_n) = \mathrm{Bi}_{r_o} J_0(\hat{\lambda}_n)$
Sphere	$2\dfrac{\sin\hat{\lambda}_n - \hat{\lambda}_n\cos\hat{\lambda}_n}{\hat{\lambda}_n - \sin\hat{\lambda}_n\cos\hat{\lambda}_n}$	$\left(\dfrac{r_o}{\hat{\lambda}_n r}\right)\sin\left(\dfrac{\hat{\lambda}_n r}{r_o}\right)$	$\hat{\lambda}_n\cot\hat{\lambda}_n = 1 - \mathrm{Bi}_{r_o}$

where the values of $\hat{\lambda}_n$ are given as a function of n and $\mathrm{Bi} = \overline{h}L/k$ by the transcendental equation

$$\cot\hat{\lambda}_n = \frac{\hat{\lambda}_n}{\mathrm{Bi}} \tag{5.35}$$

The successive positive roots of this equation, which are $\hat{\lambda}_n = \hat{\lambda}_1, \hat{\lambda}_2, \hat{\lambda}_3, \ldots$, depend upon Bi. Thus, $\Theta = \mathrm{fn}(\xi, \mathrm{Fo}, \mathrm{Bi})$, as we would expect. This result, although more complicated than the result for b.c.'s of the first kind, still reduces to a single term for $\mathrm{Fo} \gtrsim 0.2$.

Similar series solutions can be constructed for cylinders and spheres that are convectively cooled at their outer surface, $r = r_o$. The solutions for slab, cylinders, and spheres all have the form

$$\boxed{\Theta = \frac{T - T_\infty}{T_i - T_\infty} = \sum_{n=1}^{\infty} A_n \exp\left(-\hat{\lambda}_n^2\,\mathrm{Fo}\right) f_n} \tag{5.36}$$

where the coefficients A_n, the functions f_n, and the equations for the dimensionless eigenvalues $\hat{\lambda}_n$ are given in Table 5.1.

5.4 Temperature-response charts

Methods of solution of the kind described in Section 5.3 were once the norm in solving heat conduction problems. Direct numerical solution of the differential equations have now largely replaced them. Available software will even select the equation and requires only that we provide the boundary and initial conditions, material properties, and numerical parameters. The old series solutions, however, led to the creation of graphical solutions which are often much easier to use than software and which also tell us a great deal about the behavior of heat conduction in simple configurations.

We therefore turn next to the construction and the use of such graphs. Figure 5.7 is a graphical presentation of eqn. (5.34) for $0 \leqslant \text{Fo} \leqslant 1.5$ and for six x-planes in the slab. (Remember that the x-coordinate goes from zero in the center to L on the boundary, while ξ goes from 0 up to 2 in the preceding solution.)

Notice that, with the exception of points for which $1/\text{Bi} < 0.25$ on the outside boundary, the curves are all straight lines when $\text{Fo} \gtrsim 0.2$. Since the coordinates are semilogarithmic, this portion of the graph corresponds to the lead term—the only term that retains any importance—in eqn. (5.34). When we take the logarithm of the one-term version of eqn. (5.34), the result is

$$\ln\Theta \cong \underbrace{\ln\left[\frac{2\sin\hat{\lambda}_1\cos[\hat{\lambda}_1(\xi-1)]}{\hat{\lambda}_1+\sin\hat{\lambda}_1\cos\hat{\lambda}_1}\right]}_{\substack{\Theta\text{-intercept at Fo = 0 of}\\ \text{the straight portion of}\\ \text{the curve}}} - \underbrace{\hat{\lambda}_1^2\,\text{Fo}}_{\substack{\text{slope of the}\\ \text{straight portion}\\ \text{of the curve}}}$$

If Fo is greater than 1.5, the following options are then available to us for solving the problem:

- Extrapolate the given curves using a straightedge.
- Evaluate Θ using the first term of eqn. (5.34), as discussed in Sect. 5.5.
- If Bi is small, use a lumped-capacity result.

Figure 5.8 and Fig. 5.9 are similar graphs for cylinders and spheres. Everything that we have said in general about Fig. 5.7 is also true for these graphs. They were simply calculated from different solutions. The numerical values on them are therefore somewhat different while all their general features remain very similar. These charts are from [5.3, Chap. 5], although such charts are often called Heisler charts, after a collection of related charts subsequently published by Heisler [5.4].

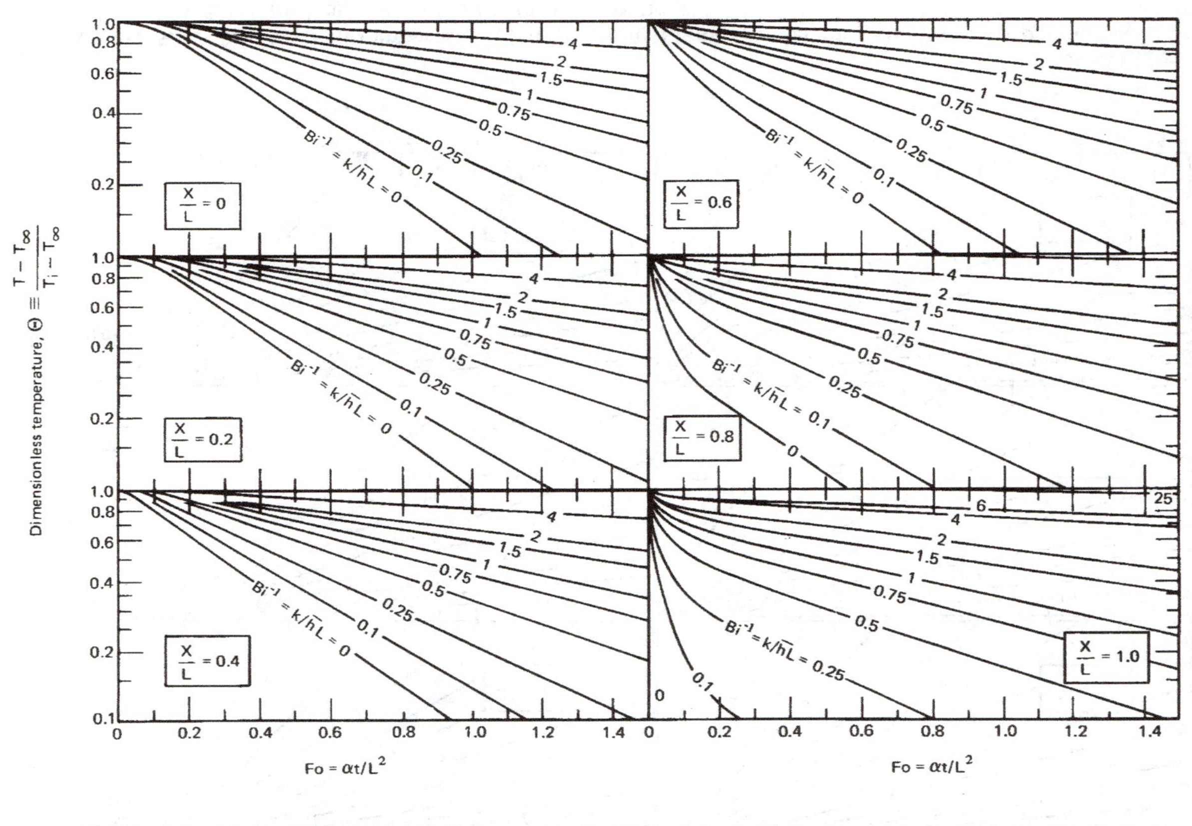

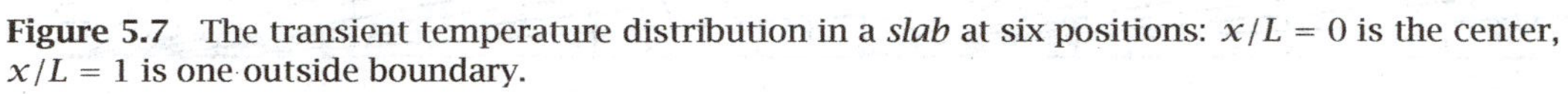

Figure 5.7 The transient temperature distribution in a *slab* at six positions: $x/L = 0$ is the center, $x/L = 1$ is one outside boundary.

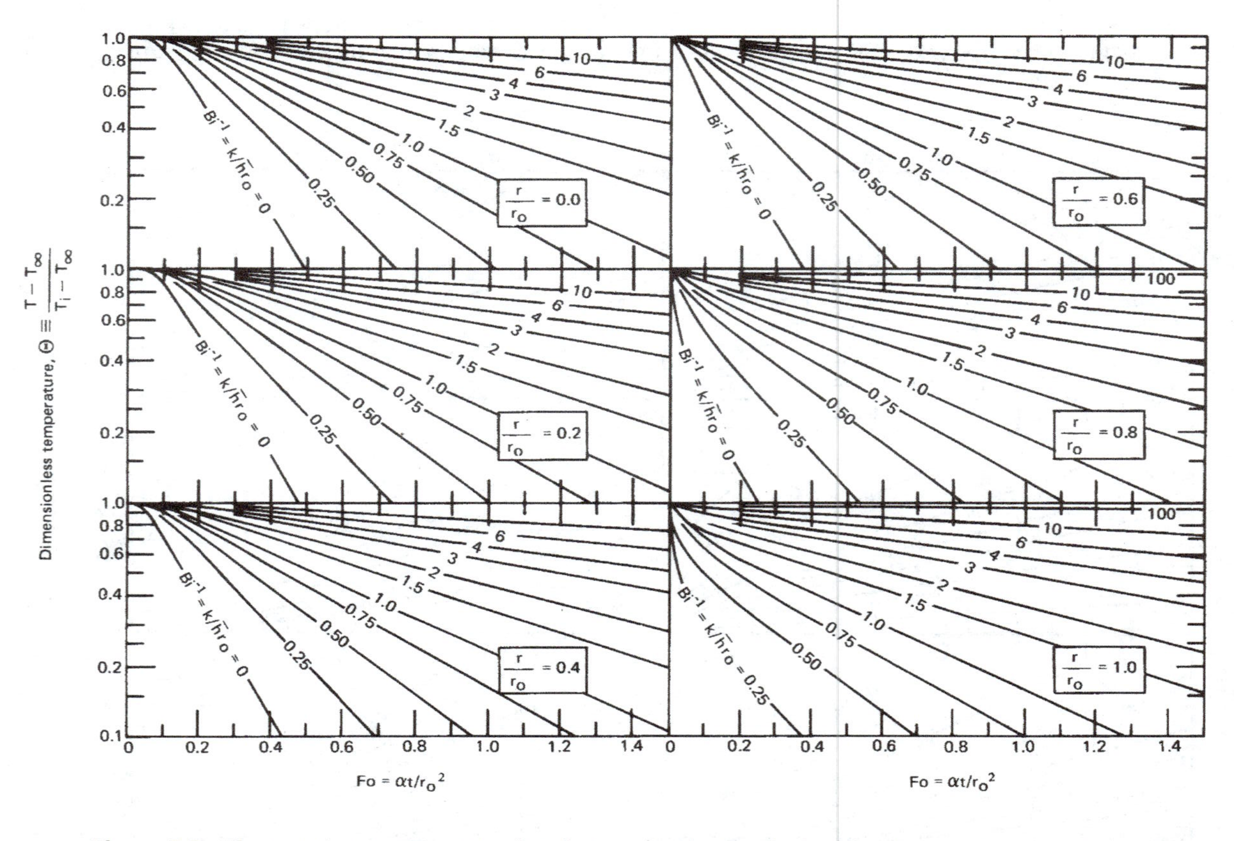

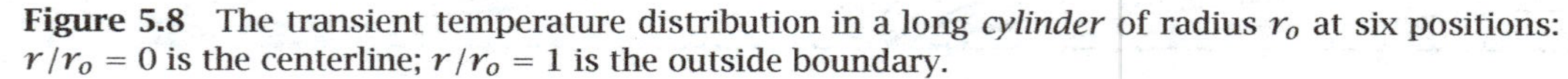
Figure 5.8 The transient temperature distribution in a long *cylinder* of radius r_o at six positions: $r/r_o = 0$ is the centerline; $r/r_o = 1$ is the outside boundary.

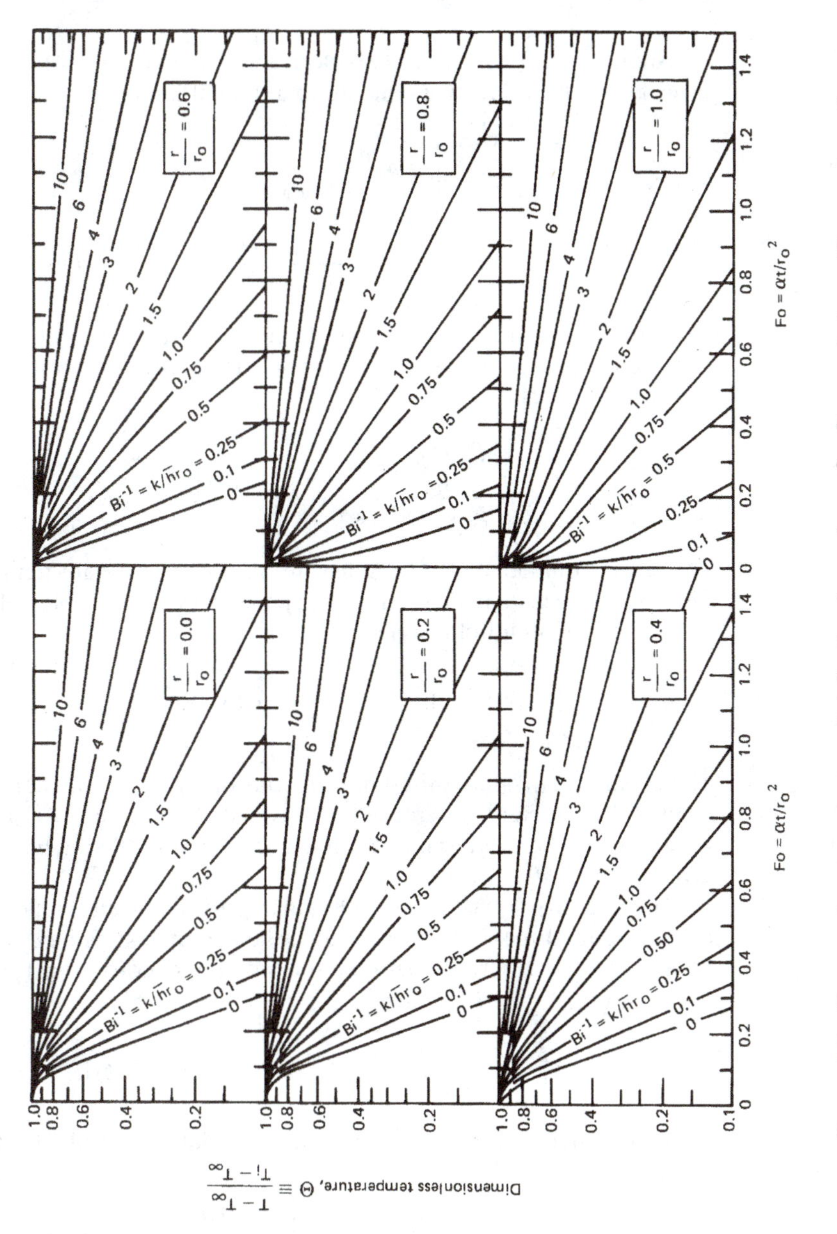

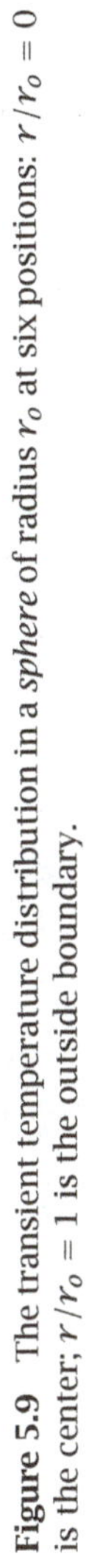
Figure 5.9 The transient temperature distribution in a *sphere* of radius r_o at six positions: $r/r_o = 0$ is the center; $r/r_o = 1$ is the outside boundary.

Another useful kind of chart derivable from eqn. (5.34) is one that gives heat removal from a body up to a time of interest:

$$\int_0^t Q\,dt = -\int_0^t kA\left.\frac{\partial T}{\partial x}\right|_{\text{surface}} dt$$
$$= -\int_0^{\text{Fo}} kA\,\frac{T_i - T_\infty}{L}\left.\frac{\partial \Theta}{\partial \xi}\right|_{\text{surface}}\left(\frac{L^2}{\alpha}\right) d\text{Fo}$$

Dividing this by the total energy of the body above T_∞, we get a quantity, Φ, which approaches one as $t \to \infty$ and the energy is all transferred to the surroundings:

$$\Phi \equiv \frac{\displaystyle\int_0^t Q\,dt}{\rho c V(T_i - T_\infty)} = -\int_0^{\text{Fo}} \left.\frac{\partial \Theta}{\partial \xi}\right|_{\text{surface}} d\text{Fo} \tag{5.37}$$

where the volume, $V = AL$. Substituting the appropriate temperature distribution [e.g., eqn. (5.34) for a slab] in eqn. (5.37), we obtain $\Phi(\text{Fo}, \text{Bi})$ in the form of an infinite series

$$\Phi\,(\text{Fo}, \text{Bi}) = 1 - \sum_{n=1}^{\infty} D_n \exp\left(-\hat{\lambda}_n^2\,\text{Fo}\right) \tag{5.38}$$

The coefficients D_n are different functions of $\hat{\lambda}_n$ — and thus of Bi — for slabs, cylinders, and spheres (e.g., for a slab $D_n = A_n \sin\hat{\lambda}_n/\hat{\lambda}_n$). These functions can be used to plot $\Phi(\text{Fo}, \text{Bi})$ once and for all. Such curves are given in Fig. 5.10.

The quantity Φ has a close relationship to the mean temperature of a body at any time, $\overline{T}(t)$. Specifically, the energy lost as heat by time t determines the difference between the initial temperature and the mean temperature at time t

$$\int_0^t Q\,dt = [U(0) - U(t)] = \rho c V\,[T_i - \overline{T}(t)]. \tag{5.39}$$

Thus, if we define $\overline{\Theta}$ as follows, we find the relationship of $\overline{T}(t)$ to Φ

$$\overline{\Theta} \equiv \frac{\overline{T}(t) - T_\infty}{T_i - T_\infty} = 1 - \frac{\displaystyle\int_0^t Q(t)\,dt}{\rho c V(T_i - T_\infty)} = 1 - \Phi. \tag{5.40}$$

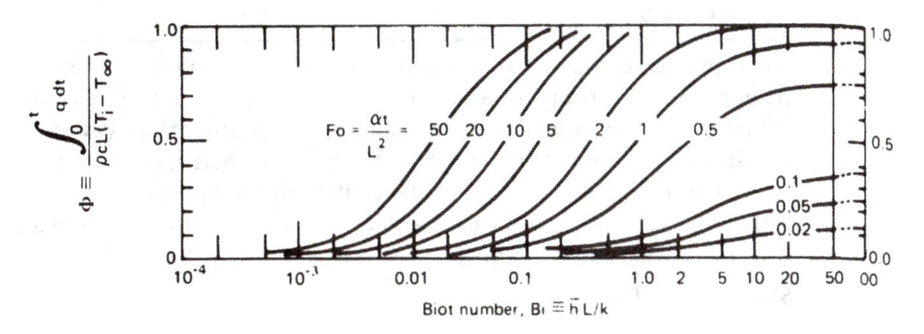

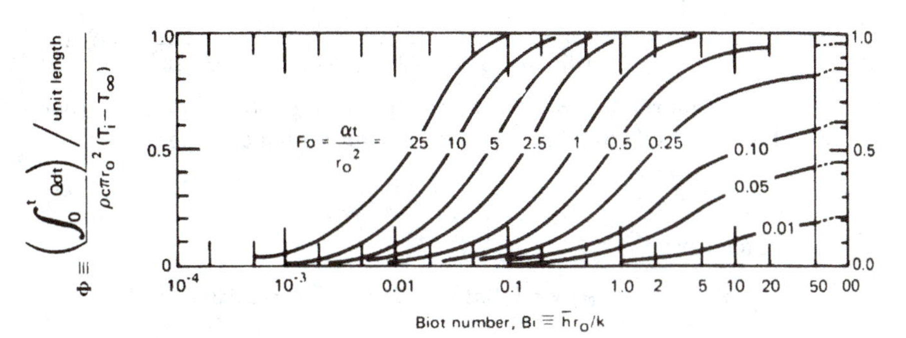

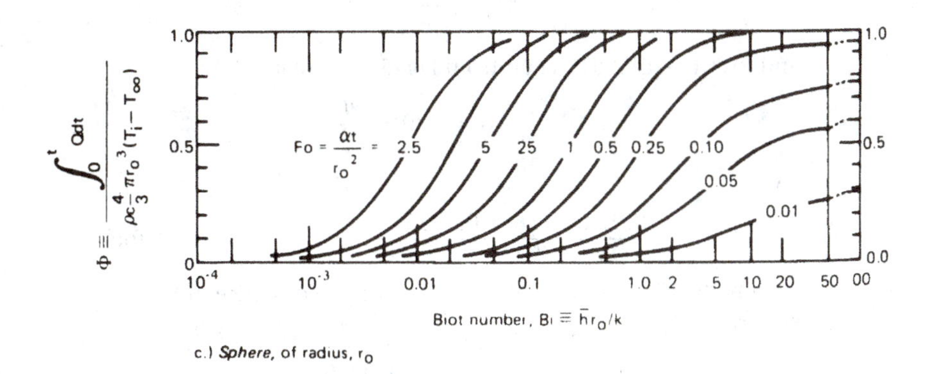

Figure 5.10 The heat removal from suddenly-cooled bodies as a function of $\overline{h}$ and time.

Example 5.2

A dozen approximately spherical apples, 10 cm in diameter are taken from a 30°C environment and laid out on a rack in a refrigerator at 5°C. They have approximately the same physical properties as water, and $\overline{h}$ is approximately 6 W/m²K as the result of natural convection. What will be the temperature of the centers of the apples after 1 hr? How long will it take to bring the centers to 10°C? How much heat will the refrigerator have to carry away to get the centers to 10°C?

Solution. After 1 hr, or 3600 s:

$$\text{Fo} = \frac{\alpha t}{r_o^2} = \left(\frac{k}{\rho c}\right)_{20^\circ\text{C}} \frac{3600 \text{ s}}{(0.05 \text{ m})^2}$$

$$= \frac{(0.603 \text{ J/m·s·K})(3600 \text{ s})}{(997.6 \text{ kg/m}^3)(4180 \text{ J/kg·K})(0.0025 \text{ m}^2)} = 0.208$$

Furthermore, $\text{Bi}^{-1} = (\overline{h}r_o/k)^{-1} = [6(0.05)/0.603]^{-1} = 2.01$. Therefore, we read from Fig. 5.9 in the upper left-hand corner:

$$\Theta = 0.85$$

After 1 hr:

$$T_{\text{center}} = 0.85(30 - 5)^\circ\text{C} + 5^\circ\text{C} = 26.3^\circ\text{C}$$

To find the time to bring the center to 10°C, we first calculate

$$\Theta = \frac{10 - 5}{30 - 5} = 0.2$$

and Bi^{-1} is still 2.01. Then from Fig. 5.9 we read

$$\text{Fo} = 1.29 = \frac{\alpha t}{r_o^2}$$

so

$$t = \frac{1.29(997.6)(4180)(0.0025)}{0.603} = 22,300 \text{ s} = 6 \text{ hr } 12 \text{ min}$$

Finally, we look up Φ at Bi = 1/2.01 and Fo = 1.29 in Fig. 5.10, for spheres:

$$\Phi = 0.80 = \frac{\displaystyle\int_0^t Q\,dt}{\rho c\left(\frac{4}{3}\pi r_0^3\right)(T_i - T_\infty)}$$

so

$$\int_0^t Q\,dt = 997.6(4180)\left(\frac{4}{3}\pi(0.05)^3\right)(25)(0.80) = 43{,}668 \text{ J/apple}$$

Therefore, for the 12 apples,

$$\text{total energy removal} = 12(43.67) = 524 \text{ kJ}$$ ■

The temperature-response charts in Fig. 5.7 through Fig. 5.10 are without doubt among the most useful available since they can be adapted to a host of physical situations. Nevertheless, hundreds of such charts have been formed for other situations, a number of which were cataloged by Schneider [5.5]. Analytical solutions are available for hundreds more problems, and any reader faced with a complex heat conduction calculation would do well consult the literature before trying to solve it. An excellent place to begin is Carslaw and Jaeger's comprehensive treatise on heat conduction [5.6].

Example 5.3

A 1 mm diameter Nichrome (20% Ni, 80% Cr) wire is simultaneously being used as an electric resistance heater and as a resistance thermometer in a liquid flow. The laboratory workers who operate it are attempting to measure the boiling heat transfer coefficient, $\overline{h}$, by supplying an alternating current and measuring the difference between the average temperature of the heater, T_{av}, and the liquid temperature, T_∞. They get $\overline{h} = 30{,}000$ W/m²K at a wire temperature of 100°C and are delighted with such a high value. Then a colleague suggests that $\overline{h}$ is so high because the surface temperature is rapidly oscillating as a result of the alternating current. Is this hypothesis correct?

Solution. Heat is being generated in proportion to the product of voltage and current, or as $\sin^2 \omega t$, where ω is the frequency of the current in rad/s. If the boiling action removes heat rapidly enough in comparison with the heat capacity of the wire, the surface temperature may well vary significantly. This transient conduction problem was first solved by Jeglic in 1962 [5.7]. It was redone in a different form two years later by Switzer and Lienhard (see, e.g. [5.8]), who gave response curves in the form

$$\frac{T_{max} - T_{av}}{T_{av} - T_\infty} = \text{fn}(\text{Bi}, \psi) \tag{5.41}$$

where the left-hand side is the dimensionless range of the temperature oscillation, and $\psi = \omega\delta^2/\alpha$, where δ is a characteristic length [see Problem 5.56]. Because this problem is common and the solution is not widely available, we include the curves for flat plates and cylinders in Fig. 5.11 and Fig. 5.12 respectively.

In the present case:

$$\text{Bi} = \frac{\overline{h}\,\text{radius}}{k} = \frac{30{,}000(0.0005)}{13.8} = 1.09$$

$$\frac{\omega r^2}{\alpha} = \frac{[2\pi(60)](0.0005)^2}{0.00000343} = 27.5$$

and from the chart for cylinders, Fig. 5.12, we find that

$$\frac{T_{\text{max}} - T_{\text{av}}}{T_{\text{av}} - T_\infty} \simeq 0.04$$

A temperature fluctuation of only 4% is probably not serious. It therefore appears that the experiment was valid. ■

5.5 One-term solutions

We have noted previously that when the Fourier number is greater than 0.2 or so, the series solutions from eqn. (5.36) may be approximated using only the first term:

$$\Theta \approx A_1 \cdot f_1 \cdot \exp\left(-\hat{\lambda}_1^2\,\text{Fo}\right). \tag{5.42}$$

Likewise, the fractional heat loss, Φ, or the mean temperature $\overline{\Theta}$ from eqn. (5.40), can be approximated using just the first term of eqn. (5.38):

$$\overline{\Theta} = 1 - \Phi \approx D_1 \exp\left(-\hat{\lambda}_1^2\,\text{Fo}\right). \tag{5.43}$$

Table 5.2 on page 219 lists the values of $\hat{\lambda}_1$, A_1, and D_1 for slabs, cylinders, and spheres as a function of the Biot number. The one-term solution's error in Θ is less than 0.1% for a sphere with Fo $\geq$ 0.28 and for a slab with Fo $\geq$ 0.43. These errors are largest for Biot numbers near one. If high accuracy is not required, these one-term approximations may generally be used whenever Fo $\geq$ 0.2.

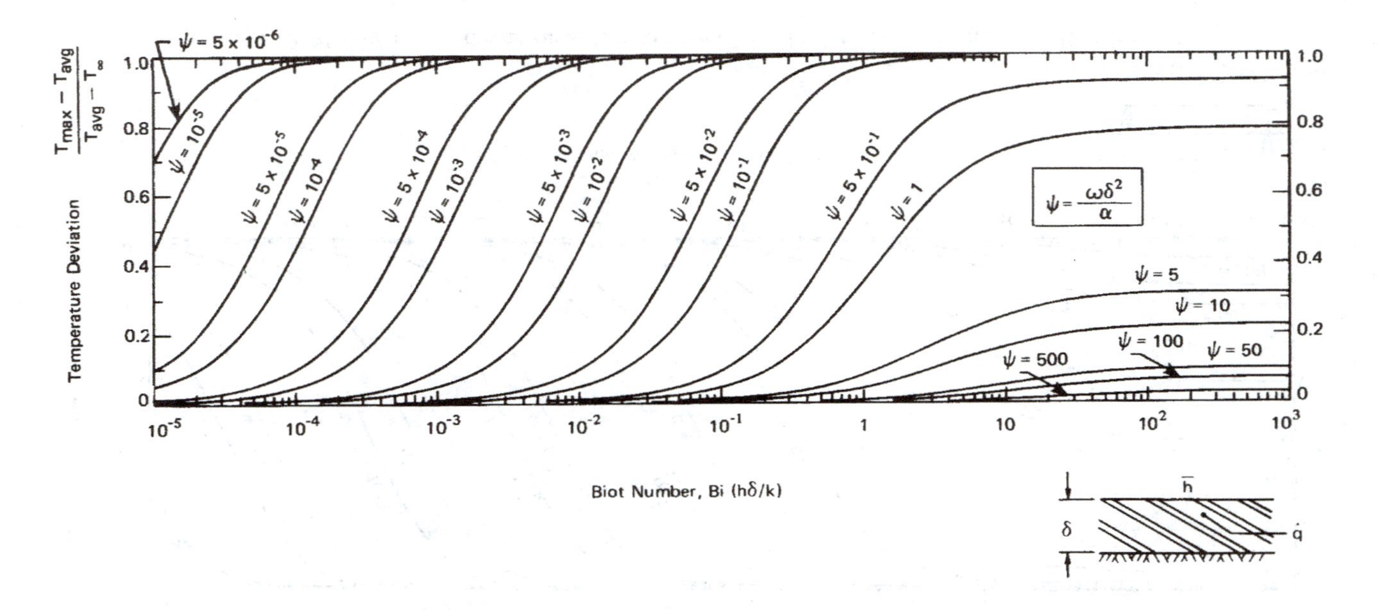

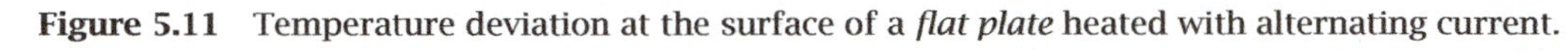

Figure 5.11 Temperature deviation at the surface of a *flat plate* heated with alternating current.

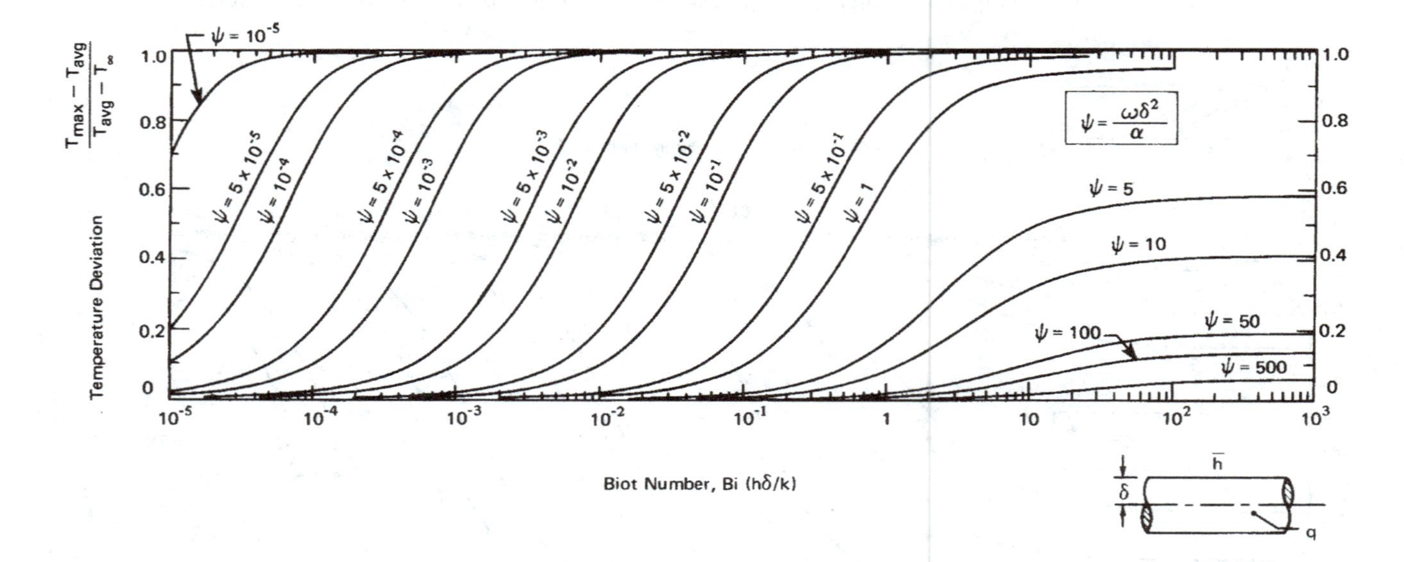

Figure 5.12 Temperature deviation at the surface of a *cylinder* heated with alternating current.

Table 5.2 One-term coefficients for convective cooling [5.1].

Bi	Plate			Cylinder			Sphere		
	$\hat{\lambda}_1$	A_1	D_1	$\hat{\lambda}_1$	A_1	D_1	$\hat{\lambda}_1$	A_1	D_1
0.01	0.09983	1.0017	1.0000	0.14124	1.0025	1.0000	0.17303	1.0030	1.0000
0.02	0.14095	1.0033	1.0000	0.19950	1.0050	1.0000	0.24446	1.0060	1.0000
0.05	0.22176	1.0082	0.9999	0.31426	1.0124	0.9999	0.38537	1.0150	1.0000
0.10	0.31105	1.0161	0.9998	0.44168	1.0246	0.9998	0.54228	1.0298	0.9998
0.15	0.37788	1.0237	0.9995	0.53761	1.0365	0.9995	0.66086	1.0445	0.9996
0.20	0.43284	1.0311	0.9992	0.61697	1.0483	0.9992	0.75931	1.0592	0.9993
0.30	0.52179	1.0450	0.9983	0.74646	1.0712	0.9983	0.92079	1.0880	0.9985
0.40	0.59324	1.0580	0.9971	0.85158	1.0931	0.9970	1.05279	1.1164	0.9974
0.50	0.65327	1.0701	0.9956	0.94077	1.1143	0.9954	1.16556	1.1441	0.9960
0.60	0.70507	1.0814	0.9940	1.01844	1.1345	0.9936	1.26440	1.1713	0.9944
0.70	0.75056	1.0918	0.9922	1.08725	1.1539	0.9916	1.35252	1.1978	0.9925
0.80	0.79103	1.1016	0.9903	1.14897	1.1724	0.9893	1.43203	1.2236	0.9904
0.90	0.82740	1.1107	0.9882	1.20484	1.1902	0.9869	1.50442	1.2488	0.9880
1.00	0.86033	1.1191	0.9861	1.25578	1.2071	0.9843	1.57080	1.2732	0.9855
1.10	0.89035	1.1270	0.9839	1.30251	1.2232	0.9815	1.63199	1.2970	0.9828
1.20	0.91785	1.1344	0.9817	1.34558	1.2387	0.9787	1.68868	1.3201	0.9800
1.30	0.94316	1.1412	0.9794	1.38543	1.2533	0.9757	1.74140	1.3424	0.9770
1.40	0.96655	1.1477	0.9771	1.42246	1.2673	0.9727	1.79058	1.3640	0.9739
1.50	0.98824	1.1537	0.9748	1.45695	1.2807	0.9696	1.83660	1.3850	0.9707
1.60	1.00842	1.1593	0.9726	1.48917	1.2934	0.9665	1.87976	1.4052	0.9674
1.80	1.04486	1.1695	0.9680	1.54769	1.3170	0.9601	1.95857	1.4436	0.9605
2.00	1.07687	1.1785	0.9635	1.59945	1.3384	0.9537	2.02876	1.4793	0.9534
2.20	1.10524	1.1864	0.9592	1.64557	1.3578	0.9472	2.09166	1.5125	0.9462
2.40	1.13056	1.1934	0.9549	1.68691	1.3754	0.9408	2.14834	1.5433	0.9389
3.00	1.19246	1.2102	0.9431	1.78866	1.4191	0.9224	2.28893	1.6227	0.9171
4.00	1.26459	1.2287	0.9264	1.90808	1.4698	0.8950	2.45564	1.7202	0.8830
5.00	1.31384	1.2402	0.9130	1.98981	1.5029	0.8721	2.57043	1.7870	0.8533
6.00	1.34955	1.2479	0.9021	2.04901	1.5253	0.8532	2.65366	1.8338	0.8281
8.00	1.39782	1.2570	0.8858	2.12864	1.5526	0.8244	2.76536	1.8920	0.7889
10.00	1.42887	1.2620	0.8743	2.17950	1.5677	0.8039	2.83630	1.9249	0.7607
20.00	1.49613	1.2699	0.8464	2.28805	1.5919	0.7542	2.98572	1.9781	0.6922
50.00	1.54001	1.2727	0.8260	2.35724	1.6002	0.7183	3.07884	1.9962	0.6434
100.00	1.55525	1.2731	0.8185	2.38090	1.6015	0.7052	3.11019	1.9990	0.6259
∞	1.57080	1.2732	0.8106	2.40483	1.6020	0.6917	3.14159	2.0000	0.6079

5.6 Transient heat conduction to a semi-infinite region

Introduction

Bronowksi's classic television series, *The Ascent of Man* [5.9], included a brilliant reenactment of the ancient ceremonial procedure by which the Japanese forged Samurai swords (see Fig. 5.13). The metal is heated, folded, beaten, and formed, over and over, to create a blade of remarkable toughness and flexibility. When the blade is formed to its final configuration, a tapered sheath of clay is baked on the outside of it, so the cross section is as shown in Fig. 5.13. The red-hot blade with the clay sheath is then subjected to a rapid quenching, which cools the uninsulated cutting edge quickly and the back part of the blade very slowly. The result is a layer of case-hardening that is hardest at the edge and less hard at points farther from the edge.

Figure 5.13 The ceremonial case-hardening of a Samurai sword.

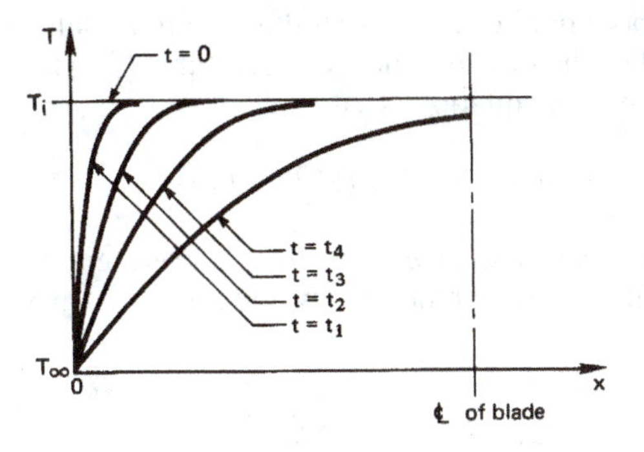

Figure 5.14 The initial cooling of a thin sword blade. Prior to $t = t_4$, the blade might as well be infinitely thick insofar as cooling is concerned.

The blade is then tough and ductile, so it will not break, but has a fine hard outer shell that can be honed to sharpness. We need only look a little way up the side of the clay sheath to find a cross section that was thick enough to prevent the blade from experiencing the sudden effects of the cooling quench. The success of the process actually relies on the *failure* of the cooling to penetrate the clay very deeply in a short time.

Now we ask how we can say whether or not the influence of a heating or cooling process is restricted to the surface of a body. Or if we turn the question around: "Under what conditions can we view the depth of a body as *infinite* with respect to the thickness of the region that has felt the heat transfer process?"

Consider next the cooling process within the blade in the absence of the clay retardant and when $\overline{h}$ is very large. Actually, our considerations will apply initially to any finite body whose boundary suddenly changes temperature. The temperature distribution, in this case, is sketched in Fig. 5.14 for four sequential times. Only the fourth curve—that for which $t = t_4$—is noticeably influenced by the opposite wall. Up to that time, the wall might as well have infinite depth.

Since any body subjected to a sudden change of temperature is infinitely large in comparison with the initial region of temperature change, we must learn how to treat heat transfer in this period.

Solution aided by dimensional analysis

The calculation of the temperature distribution in a semi-infinite region poses a difficulty: we can impose a definite b.c. at only one position—the exposed boundary. We get around that difficulty in a nice way with the help of dimensional analysis.

When the one boundary of a semi-infinite region, initially at $T = T_i$, is suddenly cooled (or heated) to a new temperature, T_∞, as in Fig. 5.14, the dimensional function equation is

$$T - T_\infty = \text{fn}\left[t, x, \alpha, (T_i - T_\infty)\right]$$

where there is *no characteristic length or time.* Since there are five variables in °C, s, and m, we should look for two dimensional groups.

$$\underbrace{\frac{T - T_\infty}{T_i - T_\infty}}_{\Theta} = \text{fn}\bigg(\underbrace{\frac{x}{\sqrt{\alpha t}}}_{\zeta}\bigg) \tag{5.44}$$

The very important thing that we learn from this exercise in dimensional analysis is that position and time collapse into one independent variable. This means that the heat conduction equation and its b.c.s must transform from a partial differential equation into a simpler ordinary differential equation in the single variable, $\zeta = x/\sqrt{\alpha t}$. Thus, we transform each side of

$$\frac{\partial^2 T}{\partial x^2} = \frac{1}{\alpha}\frac{\partial T}{\partial t}$$

as follows, where we call $T_i - T_\infty \equiv \Delta T$:

$$\frac{\partial T}{\partial t} = (T_i - T_\infty)\frac{\partial \Theta}{\partial t} = \Delta T \frac{\partial \Theta}{\partial \zeta}\frac{\partial \zeta}{\partial t} = \Delta T\left(-\frac{x}{2t\sqrt{\alpha t}}\right)\frac{\partial \Theta}{\partial \zeta};$$

$$\frac{\partial T}{\partial x} = \Delta T \frac{\partial \Theta}{\partial \zeta}\frac{\partial \zeta}{\partial x} = \frac{\Delta T}{\sqrt{\alpha t}}\frac{\partial \Theta}{\partial \zeta};$$

$$\text{and} \quad \frac{\partial^2 T}{\partial x^2} = \frac{\Delta T}{\sqrt{\alpha t}}\frac{\partial^2 \Theta}{\partial \zeta^2}\frac{\partial \zeta}{\partial x} = \frac{\Delta T}{\alpha t}\frac{\partial^2 \Theta}{\partial \zeta^2}.$$

Substituting the first and last of these derivatives in the heat conduction equation, we get the ordinary differential equation

$$\frac{d^2\Theta}{d\zeta^2} = -\frac{\zeta}{2}\frac{d\Theta}{d\zeta} \tag{5.45}$$

Notice that we changed from partial to total derivative notation since Θ now depends solely on ζ. The i.c. for eqn. (5.45) is

$$T(t = 0) = T_i \quad \text{or} \quad \Theta(\zeta \to \infty) = 1 \tag{5.46}$$

and the one known b.c. is

$$T(x=0) = T_\infty \quad \text{or} \quad \Theta\,(\zeta = 0) = 0 \tag{5.47}$$

If we call $d\Theta/d\zeta \equiv \chi$, then eqn. (5.45) becomes the first-order equation

$$\frac{d\chi}{d\zeta} = -\frac{\zeta}{2}\chi$$

which can be integrated once to get

$$\chi \equiv \frac{d\Theta}{d\zeta} = C_1\, e^{-\zeta^2/4} \tag{5.48}$$

and we integrate this a second time to get

$$\Theta = C_1 \int_0^\zeta e^{-\zeta^2/4}\, d\zeta + \underbrace{\Theta(0)}_{\substack{=0 \text{ according}\\ \text{to the b.c.}}} \tag{5.49}$$

The b.c. is now satisfied, and we need only substitute eqn. (5.49) in the i.c., eqn. (5.46), to solve for C_1:

$$1 = C_1 \int_0^\infty e^{-\zeta^2/4}\, d\zeta$$

This particular definite integral is given by integral tables as $\sqrt{\pi}$, so

$$C_1 = \frac{1}{\sqrt{\pi}}$$

Thus the solution to the problem of conduction in a semi-infinite region, subject to a b.c. of the first kind is

$$\boxed{\Theta = \frac{1}{\sqrt{\pi}} \int_0^\zeta e^{-\zeta^2/4}\, d\zeta = \frac{2}{\sqrt{\pi}} \int_0^{\zeta/2} e^{-s^2}\, ds \equiv \operatorname{erf}(\zeta/2)} \tag{5.50}$$

The second integral in eqn. (5.50), obtained by a change of variables, is called the *error function* (erf). Its name arises from its relationship to certain statistical problems related to the Gaussian distribution, which describes random errors. In Table 5.3, we list values of the error function and the complementary error function, $\operatorname{erfc}(x) \equiv 1 - \operatorname{erf}(x)$. Equation (5.50) is also plotted in Fig. 5.15.

Table 5.3 Error function and complementary error function.

$\zeta/2$	$\mathrm{erf}(\zeta/2)$	$\mathrm{erfc}(\zeta/2)$	$\zeta/2$	$\mathrm{erf}(\zeta/2)$	$\mathrm{erfc}(\zeta/2)$
0.00	0.00000	1.00000	1.10	0.88021	0.11980
0.05	0.05637	0.94363	1.20	0.91031	0.08969
0.10	0.11246	0.88754	1.30	0.93401	0.06599
0.15	0.16800	0.83200	1.40	0.95229	0.04771
0.20	0.22270	0.77730	1.50	0.96611	0.03389
0.30	0.32863	0.67137	1.60	0.97635	0.02365
0.40	0.42839	0.57161	1.70	0.98379	0.01621
0.50	0.52050	0.47950	1.80	0.98909	0.01091
0.60	0.60386	0.39614	1.8214	*0.99000*	*0.01000*
0.70	0.67780	0.32220	1.90	0.99279	0.00721
0.80	0.74210	0.25790	2.00	0.99532	0.00468
0.90	0.79691	0.20309	2.50	0.99959	0.00041
1.00	0.84270	0.15730	3.00	0.99998	0.00002

In Fig. 5.15 we see that the early-time curves shown in Fig. 5.14 have collapsed into a single curve. This was accomplished by the *similarity transformation*, as we call it[5]: $\zeta/2 = x/2\sqrt{\alpha t}$. From the figure or from Table 5.3, we see that $\Theta \geq 0.99$ when

$$\frac{\zeta}{2} = \frac{x}{2\sqrt{\alpha t}} \geq 1.8214 \quad \text{or} \quad x \geq \delta_{99} \equiv 3.64\sqrt{\alpha t} \tag{5.51}$$

In other words, the local value of $(T - T_\infty)$ is more than 99% of $(T_i - T_\infty)$ for positions in the slab beyond farther from the surface than $\delta_{99} = 3.64\sqrt{\alpha t}$.

Example 5.4

For what maximum time can a samurai sword be analyzed as a semi-infinite region after it is quenched, if it has no clay coating and $\overline{h}_{\text{external}} \cong \infty$?

Solution. First, we must guess the half-thickness of the sword (say, 3 mm) and its material (probably wrought iron with an average α

[5]The transformation is based upon the "similarity" of spatial an temporal changes in this problem.

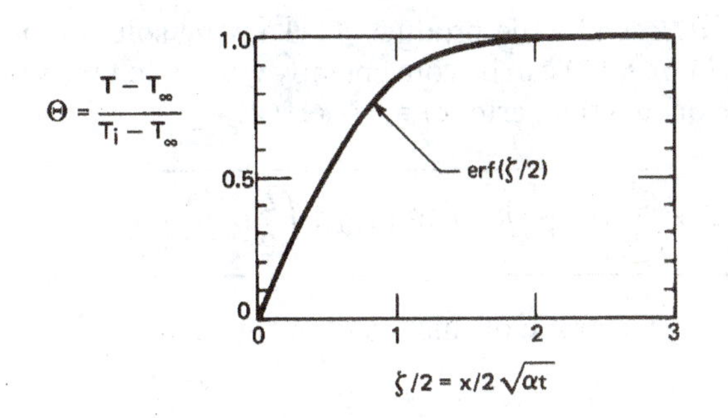

Figure 5.15 Temperature distribution in a semi-infinite region.

around 1.5×10^{-5} m^2/s). The sword will be semi-infinite until δ_{99} equals the half-thickness. Inverting eqn. (5.51), we find

$$t \leqslant \frac{\delta_{99}^2}{3.64^2\alpha} = \frac{(0.003 \text{ m})^2}{13.3(1.5)(10)^{-5} \text{ m}^2/\text{s}} = 0.045 \text{ s}$$

Thus the quench would be felt at the centerline of the sword within only 1/20 s. The thermal diffusivity of clay is smaller than that of steel by a factor of about 30, so the quench time of the coated steel must continue for over 1 s before the temperature of the steel is affected at all, if the clay and the sword thicknesses are comparable. ■

Equation (5.51) provides an interesting foretaste of the notion of a fluid boundary layer. In the context of Fig. 1.9 and Fig. 1.10, we observe that free stream flow around an object is disturbed in a thick layer near the object because the fluid adheres to it. It turns out that the thickness of this boundary layer of altered flow velocity increases in the downstream direction. For flow over a flat plate, this thickness is approximately $4.92\sqrt{\nu t}$, where t is the time required for an element of the stream fluid to move from the leading edge of the plate to a point of interest. This is quite similar to eqn. (5.51), except that the thermal diffusivity, α, has been replaced by its counterpart, the kinematic viscosity, ν, and the constant is a bit larger. The velocity profile will resemble Fig. 5.15.

If we repeated the problem with a boundary condition of the third kind, we would expect to get $\Theta = \Theta(\text{Bi}, \zeta)$, except that there is no length, L, upon which to build a Biot number. Therefore, we must replace L with $\sqrt{\alpha t}$, which has the dimension of length, so

$$\Theta = \Theta\left(\zeta, \frac{\overline{h}\sqrt{\alpha t}}{k}\right) \equiv \Theta(\zeta, \beta) \tag{5.52}$$

The term $\beta \equiv \overline{h}\sqrt{\alpha t}/k$ is like the product: $\text{Bi}\sqrt{\text{Fo}}$. The solution of this problem (see, e.g., [5.6], §2.7) can be conveniently written in terms of the complementary error function, $\text{erfc}(x) \equiv 1 - \text{erf}(x)$:

$$\boxed{\Theta = \text{erf}\,\frac{\zeta}{2} + \exp\left(\beta\zeta + \beta^2\right)\left[\text{erfc}\left(\frac{\zeta}{2} + \beta\right)\right]} \tag{5.53}$$

We offer our own original graph of this result in Fig. 5.16.

Example 5.5

Most of us have passed our finger through an 800°C candle flame and know that if we limit exposure to about 1/4 s we will not be burned. Why not?

Solution. The short exposure to the flame causes only a *very* superficial heating, so we consider the finger to be a semi-infinite region and go to eqn. (5.53) to calculate $(T_{\text{burn}} - T_{\text{flame}})/(T_i - T_{\text{flame}})$. It turns out that the burn threshold of human skin, T_{burn}, is about 65°C. (That is why 140°F or 60°C tap water is considered to be "scalding.") Therefore, we shall calculate how long it will take for the surface temperature of the finger to rise from body temperature (37°C) to 65°C, when it is protected by an assumed $\overline{h} \cong 100$ W/m²K. We shall assume that the thermal conductivity of human flesh equals that of its major component—water—and that the thermal diffusivity is equal to the known value for beef. Then

$$\Theta = \frac{65 - 800}{37 - 800} = 0.963$$

$$\beta\zeta = \frac{\overline{h}x}{k} = 0 \quad \text{since } x = 0 \text{ at the surface}$$

$$\beta^2 = \frac{\overline{h}^2 \alpha t}{k^2} = \frac{100^2(0.135 \times 10^{-6})t}{0.63^2} = 0.0034(t \text{ s})$$

The situation is quite far into the corner of Fig. 5.16. We read $\beta^2 \cong 0.001$, which corresponds with $t \cong 0.3$ s. For greater accuracy, we must go to eqn. (5.53):

$$0.963 = \underbrace{\text{erf}\,0}_{=0} + e^{0.0034t}\left[\text{erfc}\left(0 + \sqrt{0.0034\,t}\right)\right]$$

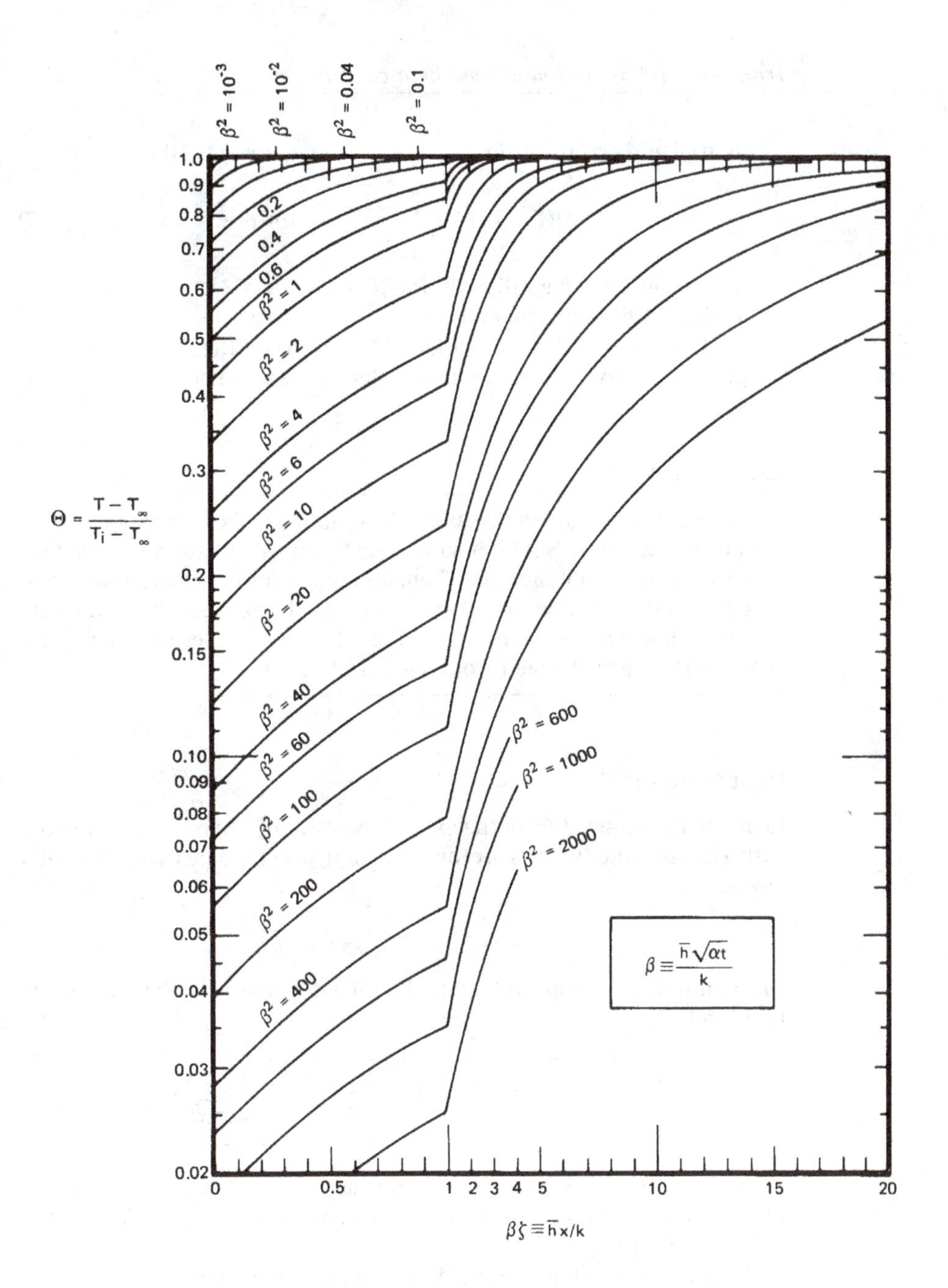

Figure 5.16 The cooling of a semi-infinite region by an environment at T_∞, through a heat transfer coefficient, $\overline{h}$.

By trial and error, we get $t \cong 0.33$ s. In fact, it can be shown that

$$\Theta(\zeta = 0, \beta) \cong 1 - \frac{2\beta}{\sqrt{\pi}} \qquad \text{for } \beta \ll 1 \qquad \blacksquare$$

which can be solved directly for $\beta = (1 - 0.963)\sqrt{\pi}/2 = 0.03279$, leading to the same answer.

Thus, it would require about $1/3$ s to bring the skin to the burn point if we have chosen a correct value of the heat transfer coefficient.

Experiment 5.1

Immerse your hand in the subfreezing air in the freezer compartment of your refrigerator. Next immerse your finger in a mixture of ice cubes and water, but do not move it. Then, immerse your finger in a mixture of ice cubes and water , swirling it around as you do so. Describe your initial sensation in each case, and explain the differences in terms of Fig. 5.16. What variable has changed from one case to another?

Heat transfer

Heat will be removed from the exposed surface of a semi-infinite region, with a b.c. of either the first or the third kind, in accordance with Fourier's law:

$$q = -k \left.\frac{\partial T}{\partial x}\right|_{x=0} = \frac{k(T_\infty - T_i)}{\sqrt{\alpha t}} \left.\frac{d\Theta}{d\zeta}\right|_{\zeta=0}$$

Differentiating Θ as given by eqn. (5.50), we obtain, for the b.c. of the first kind,

$$\boxed{q = \frac{k(T_\infty - T_i)}{\sqrt{\alpha t}} \left(\frac{1}{\sqrt{\pi}} e^{-\zeta^2/4}\right)_{\zeta=0} = \frac{k(T_\infty - T_i)}{\sqrt{\pi \alpha t}}} \tag{5.54}$$

Thus, q decreases with increasing time, as $t^{-1/2}$. When the temperature of the surface is first changed, the heat removal rate is enormous. Then it drops off rapidly.

It often occurs that we suddenly apply a specified input heat flux, q_w, at the boundary of a semi-infinite region. In such a case, we can differentiate the heat diffusion equation with respect to x, so

$$\alpha \frac{\partial^3 T}{\partial x^3} = \frac{\partial^2 T}{\partial t \partial x}$$

When we substitute $q = -k\,\partial T/\partial x$ in this, we obtain

$$\alpha \frac{\partial^2 q}{\partial x^2} = \frac{\partial q}{\partial t}$$

with the b.c.'s:

$$q(x = 0, t > 0) = q_w \qquad \text{or} \qquad \left.\frac{q_w - q}{q_w}\right|_{x=0} = 0$$

$$q(x \geqslant 0, t = 0) = 0 \qquad \text{or} \qquad \left.\frac{q_w - q}{q_w}\right|_{t=0} = 1$$

What we have done here is quite elegant. We have made the problem of predicting the local heat flux q into exactly the same form as that of predicting the local temperature in a semi-infinite region subjected to a step change of wall temperature. Therefore, the solution must be the same:

$$\frac{q_w - q}{q_w} = \text{erf}\left(\frac{x}{2\sqrt{\alpha t}}\right). \tag{5.55}$$

The temperature distribution is obtained by integrating Fourier's law. At the wall, for example:

$$\int_{T_i}^{T_w} dT = -\int_{\infty}^{0} \frac{q}{k}\, dx$$

where $T_i = T(x \to \infty)$ and $T_w = T(x = 0)$. Then

$$T_w = T_i + \frac{q_w}{k} \int_0^\infty \text{erfc}(x/2\sqrt{\alpha t})\, dx$$

This becomes

$$T_w = T_i + \frac{q_w}{k} \sqrt{\alpha t} \underbrace{\int_0^\infty \text{erfc}(\zeta/2)\, d\zeta}_{=2/\sqrt{\pi}}$$

so

$$T_w(t) = T_i + 2\frac{q_w}{k}\sqrt{\frac{\alpha t}{\pi}} \tag{5.56}$$

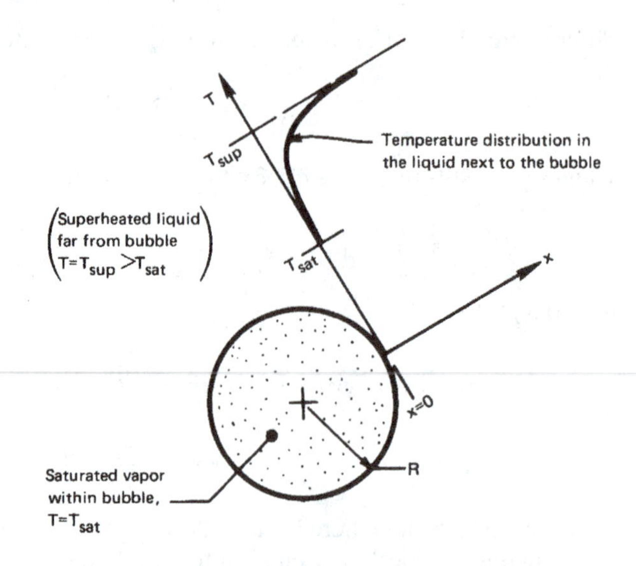

Figure 5.17 A bubble growing in a superheated liquid.

Example 5.6 Predicting the Growth Rate of a Vapor Bubble in an Infinite Superheated Liquid

This prediction is relevant to a large variety of processes, ranging from nuclear thermohydraulics to direct-contact heat exchange. It was originally presented by Max Jakob and others in the early 1930s (see, e.g., [5.10, Chap. I]). Jakob (pronounced Yah′-kob) was an important figure in heat transfer during the 1920s and 1930s. He left Nazi Germany in 1936 to come to the United States. We encounter his name again later.

Figure 5.17 shows how growth occurs. When a liquid is superheated to a temperature somewhat above its boiling point, a small gas or vapor cavity in that liquid will grow. (That is what happens in the superheated water at the bottom of a teakettle.)

This bubble grows into the surrounding liquid because its boundary is kept at the saturation temperature, T_{sat}, by the near-equilibrium coexistence of liquid and vapor. Therefore, heat must flow from the superheated surroundings to the interface, where evaporation occurs. So long as the layer of cooled liquid is thin, we should not suffer too much error by using the one-dimensional semi-infinite region solution to predict the heat flow.

Thus, we can write the energy balance at the bubble interface:

$$\underbrace{\left(-q\,\frac{\text{W}}{\text{m}^2}\right)\left(4\pi R^2\ \text{m}^2\right)}_{Q\text{ into bubble}} = \underbrace{\left(\rho_g h_{fg}\,\frac{\text{J}}{\text{m}^3}\right)\left(\frac{dV}{dt}\,\frac{\text{m}^3}{\text{s}}\right)}_{\substack{\text{rate of energy increase}\\ \text{of the bubble}}}$$

and then substitute eqn. (5.54) for q and $4\pi R^3/3$ for the volume, V. This gives

$$\frac{k(T_{\text{sup}} - T_{\text{sat}})}{\sqrt{\alpha\pi t}} = \rho_g h_{fg}\,\frac{dR}{dt} \tag{5.57}$$

Integrating eqn. (5.57) from $R = 0$ at $t = 0$ up to R at t, we obtain Jakob's prediction:

$$R = \frac{2}{\sqrt{\pi}}\,\frac{k\Delta T}{\rho_g h_{fg}\sqrt{\alpha}}\,\sqrt{t} \tag{5.58}$$

■

This analysis was done without assuming the curved bubble interface to be plane, 24 years after Jakob's work, by Plesset and Zwick [5.11]. It was verified in a more exact way after another 5 years by Scriven [5.12]. These calculations are more complicated, but they lead to a very similar result:

$$R = \frac{2\sqrt{3}}{\sqrt{\pi}}\,\frac{k\Delta T}{\rho_g h_{fg}\sqrt{\alpha}}\,\sqrt{t} = \sqrt{3}\,R_{\text{Jakob}}. \tag{5.59}$$

Both predictions are compared with some of the data of Dergarabedian [5.13] in Fig. 5.18. The data and the exact theory match almost perfectly. The simple theory of Jakob et al. shows the correct dependence on R on all its variables, but it shows growth rates that are low by a factor of $\sqrt{3}$. This is because the expansion of the spherical bubble causes a relative motion of liquid toward the bubble surface, which helps to thin the region of thermal influence in the radial direction. Consequently, the temperature gradient and heat transfer rate are higher than in Jakob's model, which neglected the liquid motion. Therefore, the temperature profile flattens out more slowly than Jakob predicts, and the bubble grows more rapidly.

Experiment 5.2

Touch various objects in the room around you: glass, wood, corkboard, paper, steel, and gold or diamond, if available. Rank them in order of which feels coldest at the first instant of contact (see Problem 5.29).

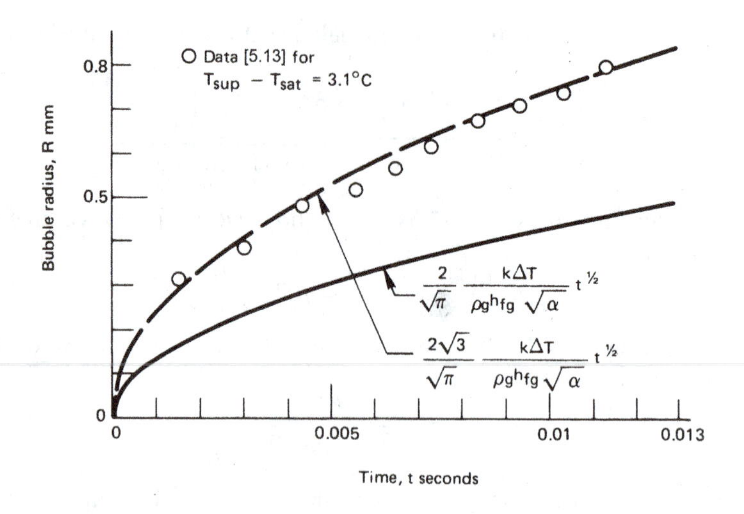

Figure 5.18 The growth of a vapor bubble—predictions and measurements.

The more advanced theory of heat conduction (see, e.g., [5.6]) shows that if two semi-infinite regions at uniform temperatures T_1 and T_2 are placed together suddenly, their interface temperature, T_s, is given by[6]

$$\frac{T_s - T_2}{T_1 - T_2} = \frac{\sqrt{(k\rho c_p)_1}}{\sqrt{(k\rho c_p)_1} + \sqrt{(k\rho c_p)_2}}$$

If we identify one region with your body ($T_1 \simeq 37°C$) and the other with the object being touched ($T_2 \simeq 20°C$), we can determine the temperature, T_s, that the surface of your finger will reach upon contact. Compare the ranking you obtain experimentally with the ranking given by this equation.

Notice that your bloodstream and capillary system provide a heat source in your finger, so the equation is valid only for a moment. Then you start replacing heat lost to the objects. If you included a diamond

[6]For semi-infinite regions, initially at uniform temperatures, T_s does not vary with time. For finite bodies, T_s will eventually change. A constant value of T_s means that each of the two bodies independently behaves as a semi-infinite body whose surface temperature has been changed to T_s at time zero. Consequently, our previous results—eqns. (5.50), (5.51), and (5.54)—apply to each of these bodies while they may be treated as semi-infinite. We need only replace T_∞ by T_s in those equations.

among the objects that you touched, you will notice that it warmed up almost instantly. Most diamonds are quite small but are possessed of the highest known value of α. Therefore, they can behave as a semi-infinite region only for an instant. Immediately after, they feel warm to the touch.

Conduction to a semi-infinite region with a harmonically oscillating temperature at the boundary

Suppose that we approximate the annual variation of the ambient temperature as sinusoidal; then we may ask what the influence of this variation will be beneath the ground. We want to calculate $T - \overline{T}$ (where $\overline{T}$ is the time-average surface temperature) as a function of: depth, x; thermal diffusivity, α; frequency of oscillation, ω; amplitude of oscillation, ΔT; and time, t. There are six variables in K, m, and s, so the problem can be represented in three dimensionless variables:

$$\Theta \equiv \frac{T - \overline{T}}{\Delta T}; \qquad \Omega \equiv \omega t; \qquad \xi \equiv x\sqrt{\frac{\omega}{2\alpha}}.$$

We pose the problem as follows in these variables. The heat conduction equation is

$$\frac{1}{2}\frac{\partial^2 \Theta}{\partial \xi^2} = \frac{\partial \Theta}{\partial \Omega} \tag{5.60}$$

and the b.c.'s are

$$\Theta\bigg|_{\xi=0} = \cos \omega t \qquad \text{and} \qquad \Theta\bigg|_{\xi>0} = \text{finite} \tag{5.61}$$

No i.c. is needed because, after the initial transient decays, the remaining steady oscillation must be periodic.

The solution is given by Carslaw and Jaeger (see [5.6, §2.6] or work Problem 5.16). It is

$$\Theta\left(\xi, \Omega\right) = e^{-\xi} \cos\left(\Omega - \xi\right) \tag{5.62}$$

This result is plotted in Fig. 5.19. It shows that the surface temperature variation decays exponentially into the region and suffers a phase shift as it does so.

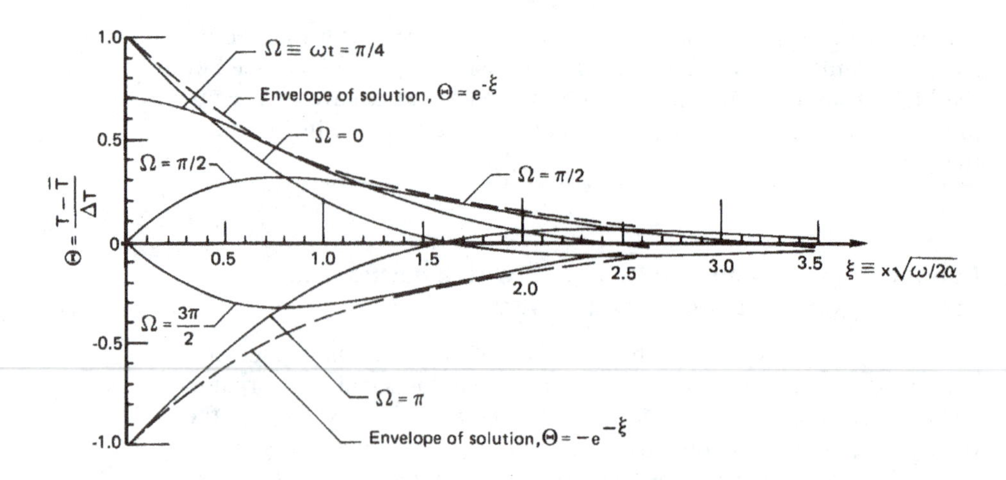

Figure 5.19 The temperature variation within a semi-infinite region whose temperature varies harmonically at the boundary.

Example 5.7

How deep in the earth must we dig to find the temperature wave that was launched by the coldest part of the last winter if it is now high summer?

Solution. $\omega = 2\pi$ rad/yr, and $\Omega = \omega t = 0$ at the present. First, we must find the depths at which the $\Omega = 0$ curve reaches its local extrema. (We pick the $\Omega = 0$ curve because it gives the highest temperature at $t = 0$.)

$$\left.\frac{d\Theta}{d\xi}\right|_{\Omega=0} = -e^{-\xi}\cos(0-\xi) + e^{-\xi}\sin(0-\xi) = 0$$

This gives

$$\tan(0-\xi) = 1 \quad \text{so} \quad \xi = \frac{3\pi}{4}, \frac{7\pi}{4}, \ldots$$

and the first minimum occurs where $\xi = 3\pi/4 = 2.356$, as we can see in Fig. 5.19. Thus,

$$\xi = x\sqrt{\omega/2\alpha} = 2.356$$

or, if we take $\alpha = 0.139 \times 10^{-6}$ m^2/s (given in [5.14] for coarse, gravelly earth),

$$x = 2.356 \bigg/ \sqrt{\frac{2\pi}{2\,(0.139 \times 10^{-6})} \frac{1}{365(24)(3600)}} = 2.783 \text{ m}$$

If we dug in the earth, we would find it growing colder and colder until it reached a maximum coldness at a depth of about 2.8 m. Farther down, it would begin to warm up again, but not much. In midwinter ($\Omega = \pi$), the reverse would be true. ■

5.7 Steady multidimensional heat conduction

Introduction

The general equation for $T(\vec{r})$ during steady conduction in a region of constant thermal conductivity, without heat sources, is called Laplace's equation:

$$\nabla^2 T = 0 \tag{5.63}$$

It looks easier to solve than it is, since [recall eqn. (2.12) and eqn. (2.14)] the Laplacian, $\nabla^2 T$, is a sum of several second partial derivatives. We solved one two-dimensional heat conduction problem in Example 4.1, but this was not difficult because the boundary conditions were made to order. Depending upon one's mathematical background and the specific problem, the analytical solution of multidimensional problems can be anything from straightforward calculation to a considerable challenge. The reader who wishes to study such analyses in depth should refer to [5.6] or [5.15], where such calculations are discussed in detail.

Faced with a steady multidimensional problem, three routes are open to us:

- Find out whether or not the analytical solution is already available in a heat conduction text or in other published literature.

- Solve the problem.

  (a) Analytically.

  (b) Numerically.

- Obtain the solution graphically if the problem is two-dimensional.

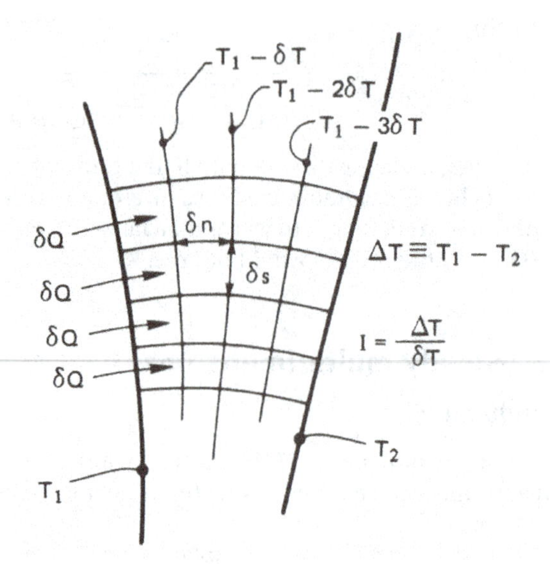

Figure 5.20 The two-dimensional flow of heat between two isothermal walls.

The last of these options is out of style as a solution method, yet it is remarkably simple and effective. We turn to it next since anyone who takes the trouble to master it will develop an uncommonly strong intuitive understanding of multidimensional heat transfer along the way.

The flux plot

The method of *flux plotting* will solve all steady planar problems in which all boundaries are held at either of two temperatures or are insulated. With a little skill, it provides accuracies of a few percent—almost always greater than the accuracy with which the b.c.'s and k can be specified. And it reveals the physics of the problem very clearly.

Figure 5.20 shows heat flowing from one isothermal wall to another in a regime that does not conform to any convenient coordinate scheme. We identify a series of channels, each which carries the same heat flow, δQ W/m. We also include a set of equally spaced isotherms, δT apart, between the walls. Since the heat fluxes in all channels are the same,

$$\left|\delta Q\right| = k\frac{\delta T}{\delta n}\,\delta s \tag{5.64}$$

Notice that if we arrange things so that δQ, δT, and k are the same for flow through each rectangle in the flow field, then $\delta s/\delta n$ must be the same for each rectangle. We therefore arbitrarily set the ratio equal to one, so all the elements appear as distorted *squares.*

The objective then is to sketch the isothermal lines and the adiabatic,[7] or heat flow, lines which run perpendicular to them. This sketch is to be done subject to two constraints

- Isothermal and adiabatic lines must intersect at right angles.
- They must subdivide the flow field into elements that are nearly square—"nearly" because they have slightly curved sides.

Once the grid has been sketched, the temperature anywhere in the field can be read directly from the sketch. And the heat flow per unit depth into the paper is

$$\boxed{Q \text{ W/m} = Nk\,\delta T\,\frac{\delta s}{\delta n} = \frac{N}{I}\,k\Delta T} \tag{5.65}$$

where N is the number of heat flow channels and I is the number of temperature increments, $\Delta T/\delta T$.

The first step in constructing a flux plot is to draw the boundaries of the region accurately *in ink*, using drafting software or drafting instruments. The next is to obtain a soft pencil (such as a no. 2 grade) and a soft eraser. We begin with an example that was executed nicely in the influential *Heat Transfer Notes* [5.3] of the mid-twentieth century. This example is shown in Fig. 5.21.

The particular example happens to have an axis of symmetry in it. We immediately interpret this as an adiabatic boundary because heat cannot cross it. The problem therefore reduces to the simpler one of sketching lines in only one half of the area. We illustrate this process in four steps. Notice the following steps and features in this plot:

- Begin by dividing the region, by sketching in either a single isothermal or adiabatic line.
- Fill in the lines perpendicular to the original line so as to make squares. Allow the original line to move in such a way as to accommodate squares. This will *always* require some erasing. Therefore:
- *Never* make the original lines dark and firm.

[7] Adiabatic lines are ones *in the direction of heat flow*: Since by definition there can be no component of heat flow normal to them, they *must* be adiabatic.

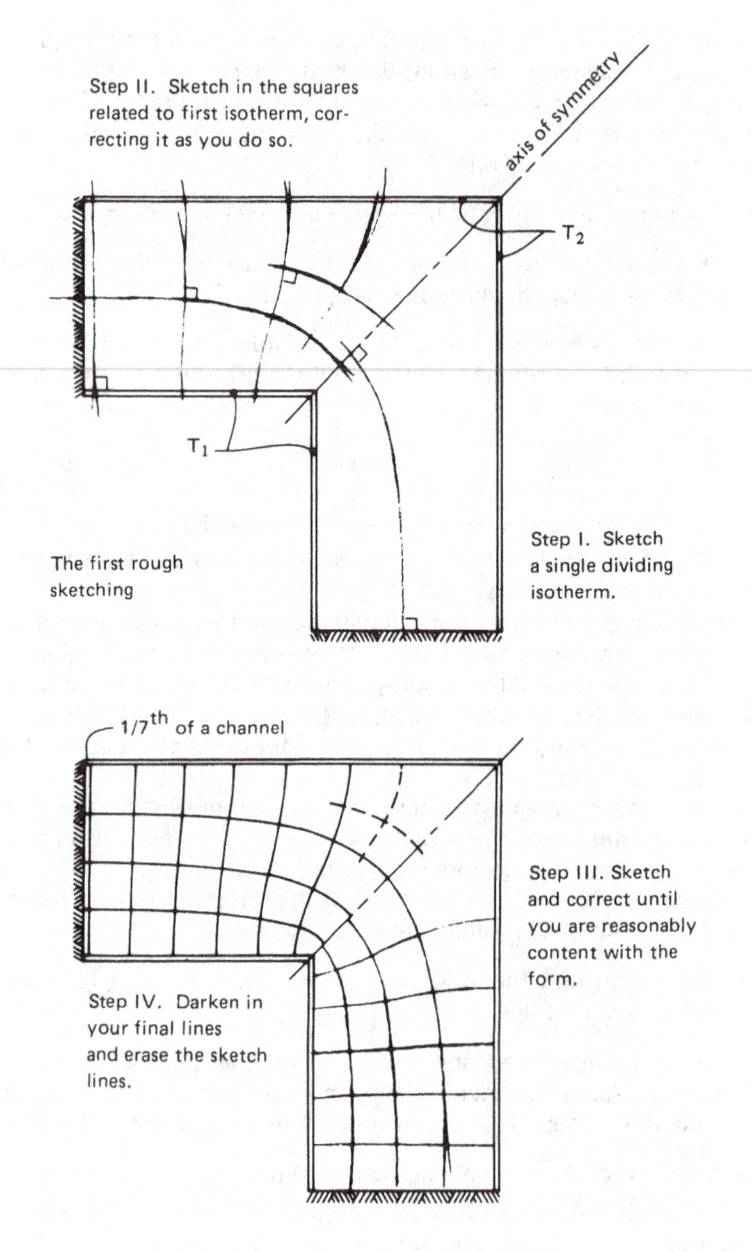

Figure 5.21 The evolution of a flux plot.

- By successive subdividing of the squares, make the final grid. *Do not make the grid very fine.* If you do, you will lose accuracy because the lack of perpendicularity and squareness will be less evident to the eye. Step IV in Fig. 5.21 is as fine a grid as should ever be made.

- If you have doubts about whether any large, ill-shaped regions are correct, fill them in with an extra isotherm and adiabatic line to be sure that they resolve into appropriate squares (see the dashed lines in Fig. 5.21).

- Fill in the final grid, when you are sure of it, either in hard pencil or pen, and erase any lingering background sketch lines.

- Your flow channels need not come out even. Notice that there is an extra 1/7 of a channel in Fig. 5.21. This is simply counted as 1/7 of a square in eqn. (5.65).

- Never allow isotherms or adiabatic lines to intersect themselves.

When the sketch is complete, we return to eqn. (5.65) to compute the heat flux. In this case

$$Q = \frac{N}{I}\,k\Delta T = \frac{2(6.14)}{4}\,k\Delta T = 3.07\,k\Delta T$$

When the authors of [5.3] did this problem, they obtained $N/I = 3.00$—a value only 2% below ours. This kind of agreement is typical when flux plotting is done with care.

One must be careful not to grasp at a false axis of symmetry. Figure 5.22 shows a shape similar to the one that we just treated, but with unequal legs. In this case, no lines must enter (or leave) the corners A and B. The reason is that since there *is* no symmetry, we have no guidance as to the direction of the lines at these corners. In particular, we know that a line leaving A will no longer arrive at B.

Example 5.8

A structure consists of metal walls, 8 cm apart, with insulating material ($k = 0.12$ W/m·K) between. Ribs 4 cm long protrude from one wall every 14 cm. They can be assumed to stay at the temperature of that wall. Find the heat flux through the wall if the first wall is at 40°C and the one with ribs is at 0°C. Find the temperature in the middle of the wall, 2 cm from a rib, as well.

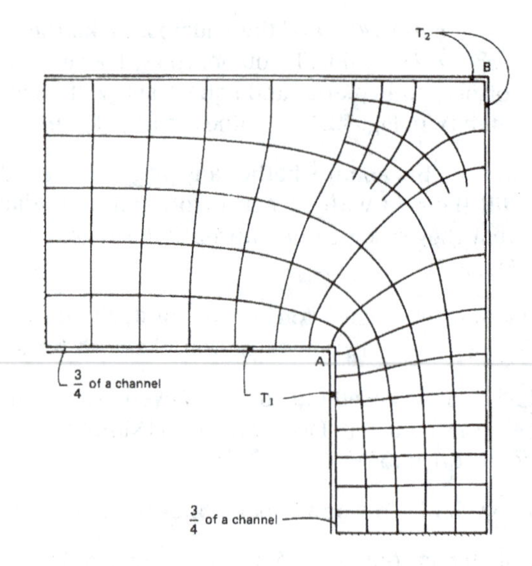

Figure 5.22 A flux plot with no axis of symmetry to guide construction.

SOLUTION. The flux plot for this configuration is shown in Fig. 5.23. For a typical section, there are approximately 5.6 isothermal increments and 6.15 heat flow channels, so

$$Q = \frac{N}{I} k\Delta T = \frac{2(6.15)}{5.6}(0.12)(40-0) = 10.54 \text{ W/m}$$

where the factor of 2 accounts for the fact that there are two halves in the section. We deduce the temperature for the point of interest, A, by a simple proportionality:

$$T_{\text{point } A} = \frac{2.1}{5.6}(40-0) = 15°\text{C}$$

■

The shape factor

A heat conduction *shape factor* S may be defined for steady problems involving two isothermal surfaces as follows:

$$Q \equiv S\, k\Delta T. \tag{5.66}$$

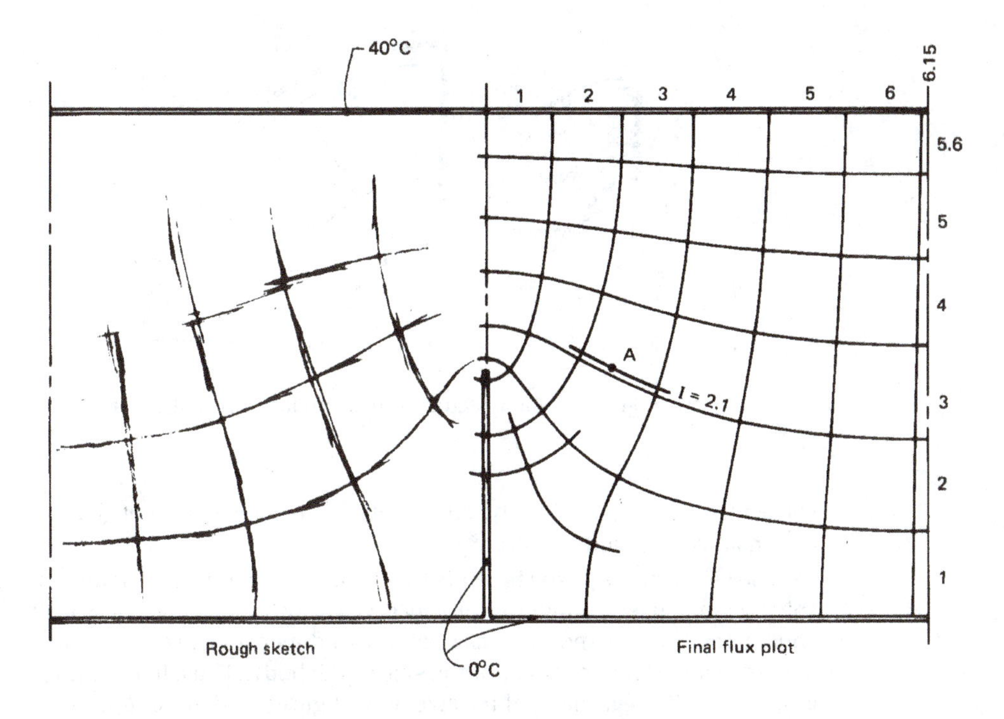

Figure 5.23 Heat transfer through a wall with isothermal ribs.

Thus far, every steady heat conduction problem we have done has taken this form. For these situations, the heat flow always equals a function of the geometric shape of the body multiplied by $k\Delta T$.

The shape factor can be obtained analytically, numerically, or through flux plotting. For example, let us compare eqn. (5.65) and eqn. (5.66):

$$Q\,\frac{\mathrm{W}}{\mathrm{m}} = (S \text{ dimensionless})\left(k\Delta T\,\frac{\mathrm{W}}{\mathrm{m}}\right) = \frac{N}{I}\,k\Delta T \tag{5.67}$$

This shows S to be dimensionless in a two-dimensional problem, but in three dimensions S has units of meters:

$$Q\;\mathrm{W} = (S\;\mathrm{m})\left(k\Delta T\,\frac{\mathrm{W}}{\mathrm{m}}\right). \tag{5.68}$$

It also follows that the thermal resistance of a two-dimensional body is

$$R_t = \frac{1}{kS} \qquad \text{where} \qquad Q = \frac{\Delta T}{R_t} \tag{5.69}$$

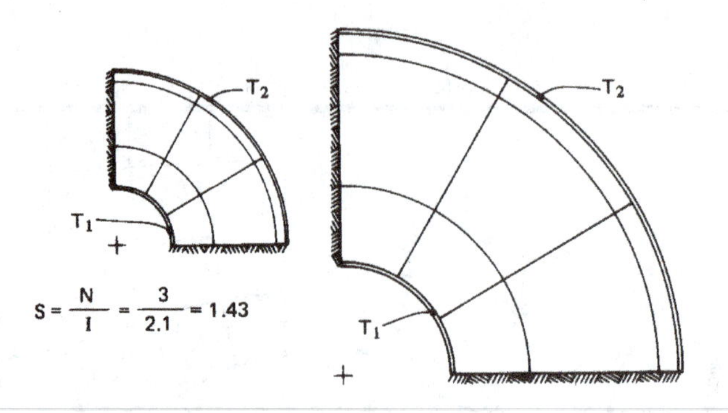

Figure 5.24 The shape factor for two similar bodies of different size.

For a three-dimensional body, eqn. (5.69) is unchanged except that the dimensions of Q and R_t differ.[8]

The virtue of the shape factor is that it summarizes a heat conduction solution in a given configuration. Once S is known, it can be used again and again. That S is nondimensional in two-dimensional configurations means that Q is independent of the size of the body. Thus, in Fig. 5.21, S is always 3.07—regardless of the size of the figure—and in Example 5.8, S is $2(6.15)/5.6 = 2.196$, whether or not the wall is made larger or smaller. When a body's breadth is increased so as to increase Q, its thickness in the direction of heat flow is also increased so as to decrease Q by the same factor.

Example 5.9

Calculate the shape factor for a one-quarter section of a thick cylinder.

Solution. We already know R_t for a thick cylinder. It is given by eqn. (2.22). From it we compute

$$S_{\text{cyl}} = \frac{1}{kR_t} = \frac{2\pi}{\ln(r_o/r_i)}$$

[8]Recall that we noted after eqn. (2.22) that the dimensions of R_t changed, depending on whether or not Q was expressed in a unit-length basis.

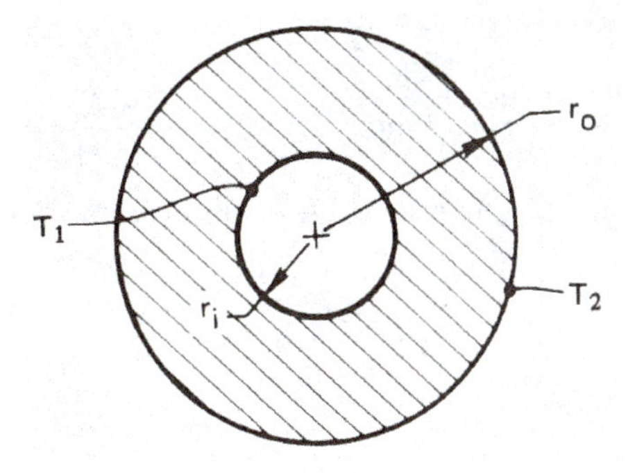

Figure 5.25 Heat transfer through a thick, hollow sphere.

so on the case of a quarter-cylinder,

$$S = \frac{\pi}{2\ln(r_o/r_i)}$$

The quarter-cylinder is pictured in Fig. 5.24 for a radius ratio, $r_o/r_i = 3$, but for two different sizes. In both cases $S = 1.43$. (Note that the same S is also given by the flux plot shown.) ■

Example 5.10

Calculate S for a three-dimensional object—a thick hollow sphere, as shown in Fig. 5.25. Notice that, in this case, since we expect the shape factor to have the dimension of length, it should increase linearly with the size of the sphere.

Solution. The general solution of the heat diffusion equation in spherical coordinates for purely radial heat flow (see Problem 4.1) is:

$$T = \frac{C_1}{r} + C_2$$

when $T = \text{fn}(r \text{ only})$. The b.c.'s are

$$T(r = r_i) = T_i \quad \text{and} \quad T(r = r_o) = T_o$$

substituting the general solution in the b.c.'s we get

$$\frac{C_1}{r_i} + C_2 = T_i \quad \text{and} \quad \frac{C_1}{r_o} + C_1 = T_o$$

Therefore,

$$C_1 = \frac{T_i - T_o}{r_o - r_i} r_i r_o \quad \text{and} \quad C_2 = T_i - \frac{T_i - T_o}{r_o - r_i} r_o$$

Putting C_1 and C_2 in the general solution, and calling $T_i - T_o \equiv \Delta T$, we get

$$T = T_i + \Delta T \left[\frac{r_i r_o}{r(r_o - r_i)} - \frac{r_o}{r_o - r_i} \right]$$

Then

$$Q = -kA \frac{dT}{dr} = \frac{4\pi(r_i r_o)}{r_o - r_i} k\Delta T$$

$$S = \frac{4\pi(r_i r_o)}{r_o - r_i} \text{ m}$$

where S does indeed have the dimension of m and is hence size dependent. ■

Table 5.4 includes a number of analytically derived shape factors for use in calculating the heat flux in different configurations. Notice that these results will not give local temperatures. To obtain that information, one must solve the Laplace equation, $\nabla^2 T = 0$, by one of the methods listed at the beginning of this section. Notice, too, that this table is restricted to bodies with isothermal and insulated boundaries.

In the two-dimensional cases, both a hot and a cold surface must be present in order to have a steady-state solution; if only a single hot (or cold) body is present, steady state is never reached. For example, a hot isothermal cylinder in a cooler, infinite medium never reaches steady state with that medium. Likewise, in situations 5, 6, and 7 in the table, the medium far from the isothermal plane must also be at temperature T_2 in order for steady state to occur; otherwise the isothermal plane and the medium below it would behave as an unsteady, semi-infinite body. Of course, since no real medium is truly infinite, what this means in practice is that steady state only occurs after the medium "at infinity" comes to a temperature T_2. Conversely, in three-dimensional situations (such as 4, 8, 12, and 13), a body *can* come to steady state with a surrounding infinite or semi-infinite medium at a different temperature.

Example 5.11

A spherical heat source of 6 cm in diameter is buried 30 cm below the surface of a very large box of soil and kept at 35°C. The surface of

Table 5.4 Conduction shape factors: $Q = S\,k\Delta T$, $R_t = 1/(kS)$.

Situation	*Shape factor, S*	*Dimensions*	*Source*
1. Conduction through a slab	A/L	meter	Example 2.2
2. Conduction through wall of a long thick cylinder	$\dfrac{2\pi}{\ln(r_o/r_i)}$	none	Example 5.9
3. Conduction through a thick-walled hollow sphere	$\dfrac{4\pi\,(r_o r_i)}{r_o - r_i}$	meter	Example 5.10
4. The boundary of a spherical hole of radius R conducting into an infinite medium 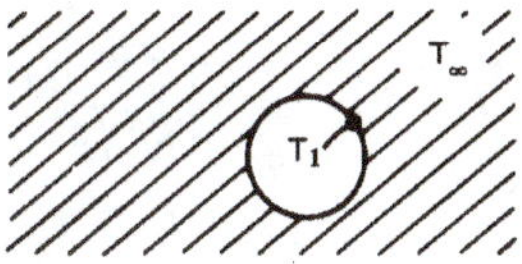	$4\pi R$	meter	Problems 5.19 and 2.15
5. Cylinder of radius R and length L, transferring heat to a parallel isothermal plane; $h \ll L$ 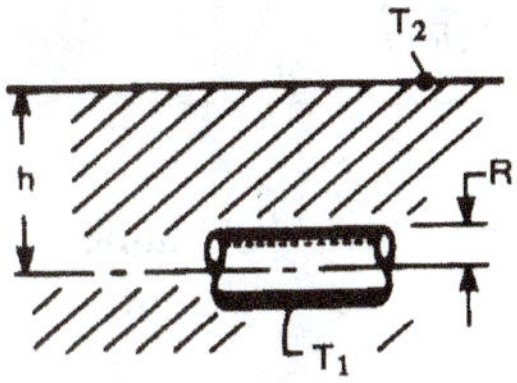	$\dfrac{2\pi L}{\cosh^{-1}(h/R)}$	meter	[5.16]
6. Same as item 5, but with $L \longrightarrow \infty$ (two-dimensional conduction)	$\dfrac{2\pi}{\cosh^{-1}(h/R)}$	none	[5.16]
7. An isothermal sphere of radius R transfers heat to an isothermal plane; $R/h < 0.8$ (see item 4) 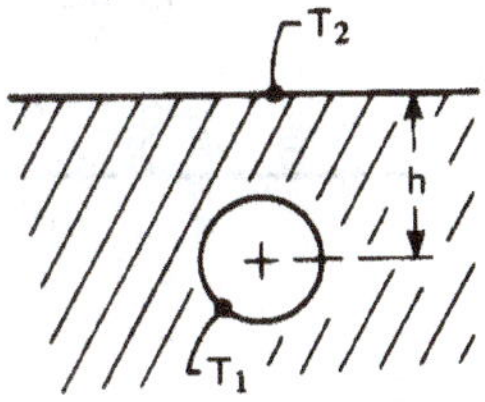	$\dfrac{4\pi R}{1 - R/2h}$	meter	[5.16, 5.17]

Table 5.4 Conduction shape factors: $Q = S\,k\Delta T$, $R_t = 1/(kS)$ *(con't).*

Situation	*Shape factor, S*	*Dimensions*	*Source*
8. An isothermal sphere of radius R, near an insulated plane, transfers heat to a semi-infinite medium at T_∞ (see items 4 and 7) 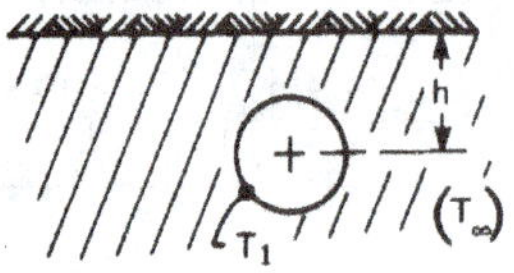	$\dfrac{4\pi R}{1 + R/2h}$	meter	[5.18]
9. Parallel cylinders exchange heat in an infinite conducting medium 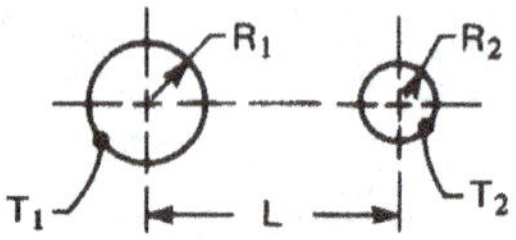	$\dfrac{2\pi}{\cosh^{-1}\left(\dfrac{L^2 - R_1^2 - R_2^2}{2R_1R_2}\right)}$	none	[5.6]
10. Same as 9, but with cylinders widely spaced; $L \gg R_1$ and R_2	$\dfrac{2\pi}{\cosh^{-1}\left(\dfrac{L}{2R_1}\right) + \cosh^{-1}\left(\dfrac{L}{2R_2}\right)}$	none	[5.16]
11. Cylinder of radius R_i surrounded by eccentric cylinder of radius $R_o > R_i$; centerlines a distance L apart (see item 2)	$\dfrac{2\pi}{\cosh^{-1}\left(\dfrac{R_o^2 + R_i^2 - L^2}{2R_oR_i}\right)}$	none	[5.6]
12. Isothermal disc of radius R on an otherwise insulated plane conducts heat into a semi-infinite medium at T_∞ below it	$4R$	meter	[5.6]
13. Isothermal ellipsoid of semimajor axis b and semiminor axes a conducts heat into an infinite medium at T_∞; $b > a$ (see 4)	$\dfrac{4\pi b\sqrt{1 - a^2/b^2}}{\tanh^{-1}\left(\sqrt{1 - a^2/b^2}\right)}$	meter	[5.16]

the soil is kept at 21°C. If the steady heat transfer rate is 14 W, what is the thermal conductivity of this sample of soil?

SOLUTION.

$$Q = S\,k\Delta T = \left(\frac{4\pi R}{1 - R/2h}\right) k\Delta T$$

where S is that for situation 7 in Table 5.4. Then

$$k = \frac{14\ \text{W}}{(35-21)\text{K}}\ \frac{1-(0.06/2)/2(0.3)}{4\pi(0.06/2)\ \text{m}} = 2.545\ \text{W/m·K} \qquad \blacksquare$$

Readers who desire a broader catalogue of shape factors should refer to [5.16], [5.18], or [5.19].

The problem of locally vanishing resistance

Suppose that two different temperatures are specified on adjacent sides of a square, as shown in Fig. 5.26. The shape factor in this case is

$$S = \frac{N}{I} = \frac{\infty}{4} = \infty$$

(It is futile to try and count channels beyond $N \simeq 10$, but it is clear that they multiply without limit in the lower left corner.) The problem is that we have violated our rule that isotherms cannot intersect and have created a $1/r$ singularity. If we actually tried to sustain such a situation,

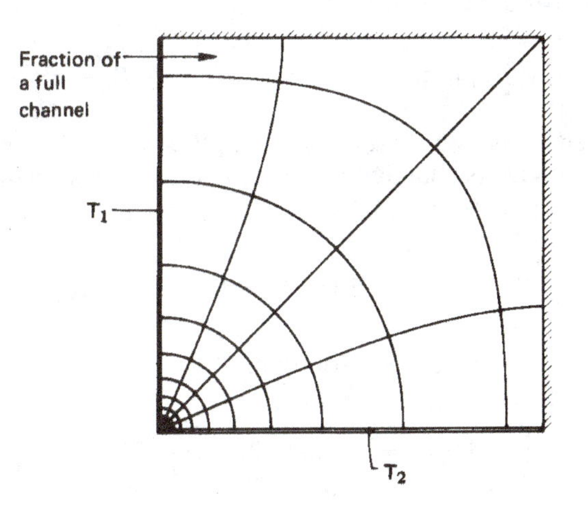

Figure 5.26 Resistance vanishes where two isothermal boundaries intersect.

the figure would be correct at some distance from the corner. However, where the isotherms are close to one another, they will necessarily influence and distort one another in such a way as to avoid intersecting. And S would never really be infinite, as it appears to be in the figure.

5.8 Transient multidimensional heat conduction—The tactic of superposition

Consider the cooling of a stubby cylinder, such as the one shown in Fig. 5.27a. The cylinder is initially at $T = T_i$, and it is suddenly subjected to a common b.c. on all sides. It has a length $2L$ and a radius r_o. Finding the temperature field in this situation is inherently complicated. It requires solving the heat conduction equation for $T = \text{fn}(r, z, t)$ with b.c.'s of the first, second, or third kind.

However, Fig. 5.27a suggests that this can somehow be viewed as a combination of an infinite cylinder and an infinite slab. It turns out that the problem *can be analyzed* from that point of view.

If the body is subject to uniform b.c.'s of the first, second, or third kind, and if it has a uniform initial temperature, then its temperature response is simply the product of an infinite slab solution and an infinite cylinder solution each having the same boundary and initial conditions. For the case shown in Fig. 5.27a, if the cylinder begins convective cooling into a medium at temperature T_∞ at time $t = 0$, the dimensional temperature response is

$$T(r, z, t) - T_\infty = \left[T_{\text{slab}}(z, t) - T_\infty\right] \times \left[T_{\text{cyl}}(r, t) - T_\infty\right] \tag{5.70a}$$

Observe that the slab has as a characteristic length L, its half thickness, while the cylinder has as its characteristic length R, its radius. In dimensionless form, we may write eqn. (5.70a) as

$$\Theta \equiv \frac{T(r, z, t) - T_\infty}{T_i - T_\infty} = \left[\Theta_{\text{inf slab}}(\xi, \text{Fo}_s, \text{Bi}_s)\right]\left[\Theta_{\text{inf cyl}}(\rho, \text{Fo}_c, \text{Bi}_c)\right] \tag{5.70b}$$

For the cylindrical component of the solution,

$$\rho = \frac{r}{r_o}, \quad \text{Fo}_c = \frac{\alpha t}{r_o^2}, \quad \text{and} \quad \text{Bi}_c = \frac{\overline{h} r_o}{k},$$

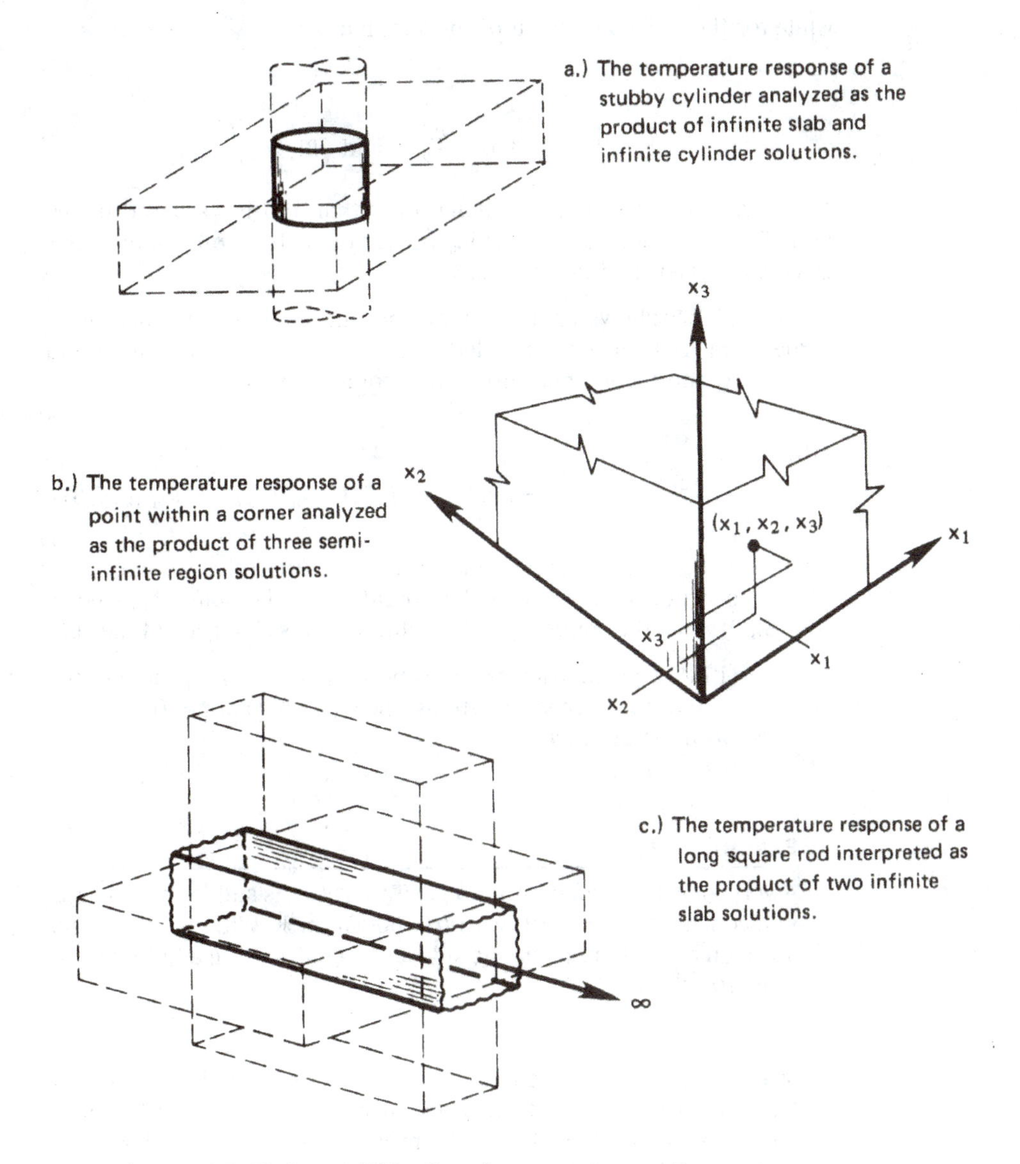

Figure 5.27 Various solid bodies whose transient cooling can be treated as the product of one-dimensional solutions.

while for the slab component of the solution

$$\xi = \frac{z}{L} + 1, \quad \text{Fo}_s = \frac{\alpha t}{L^2}, \quad \text{and} \quad \text{Bi}_s = \frac{\overline{h}L}{k}.$$

The component solutions are none other than those discussed in Sections 5.3–5.5. The proof of the legitimacy of such product solutions is given by Carlsaw and Jaeger [5.6, §1.15].

Figure 5.27b shows a point inside a one-eighth-infinite region, near the corner. This case may be regarded as the product of three semi-infinite bodies. To find the temperature at this point we write

$$\Theta \equiv \frac{T(x_1, x_2, x_3, t) - T_\infty}{T_i - T_\infty} = [\Theta_{\text{semi}}(\zeta_1, \beta)]\,[\Theta_{\text{semi}}(\zeta_2, \beta)]\,[\Theta_{\text{semi}}(\zeta_3, \beta)] \tag{5.71}$$

in which Θ_{semi} is either the semi-infinite body solution given by eqn. (5.53) when convection is present at the boundary or the solution given by eqn. (5.50) when the boundary temperature itself is changed at time zero.

Several other geometries can also be represented by product solutions. Note that for of these solutions, the value of Θ at $t = 0$ is one for each factor in the product.

Example 5.12

A very long 4 cm square iron rod at $T_i = 100$°C is suddenly immersed in a coolant at $T_\infty = 20$°C with $\overline{h} = 800$ W/m²K. What is the temperature on a line 1 cm from one side and 2 cm from the adjoining side, after 10 s?

Solution. With reference to Fig. 5.27c, see that the bar may be treated as the product of two slabs, each 4 cm thick. We first evaluate $\text{Fo}_1 = \text{Fo}_2 = \alpha t/L^2 = (0.0000226 \text{ m}^2\text{/s})(10 \text{ s})/(0.04 \text{ m}/2)^2 = 0.565$, and $\text{Bi}_1 = \text{Bi}_2 = \overline{h}L/k = 800(0.04/2)/76 = 0.2105$, and we then write

$$\Theta\left[\left(\frac{x}{L}\right)_1 = 0,\ \left(\frac{x}{L}\right)_2 = \frac{1}{2},\ \text{Fo}_1, \text{Fo}_2, \text{Bi}_1^{-1}, \text{Bi}_2^{-1}\right]$$

$$= \underbrace{\Theta_1\left[\left(\frac{x}{L}\right)_1 = 0,\ \text{Fo}_1 = 0.565,\ \text{Bi}_1^{-1} = 4.75\right]}_{\substack{= 0.93 \text{ from upper left-hand} \\ \text{side of Fig. 5.7}}}$$

$$\times\, \underbrace{\Theta_2\left[\left(\frac{x}{L}\right)_2 = \frac{1}{2},\ \text{Fo}_2 = 0.565,\ \text{Bi}_2^{-1} = 4.75\right]}_{\substack{= 0.91 \text{ from interpolation} \\ \text{between lower lefthand side and} \\ \text{upper righthand side of Fig. 5.7}}}$$

Thus, at the axial line of interest,

$$\Theta = (0.93)(0.91) = 0.846$$

so

$$\frac{T - 20}{100 - 20} = 0.846 \quad \text{or} \quad T = 87.7°\text{C}$$ ■

Product solutions can also be used to determine the mean temperature, $\overline{\Theta}$, and the total heat removal, Φ, from a multidimensional object. For example, when two or three solutions (Θ_1, Θ_2, and perhaps Θ_3) are multiplied to obtain Θ, the corresponding mean temperature of the multidimensional object is simply the product of the one-dimensional mean temperatures from eqn. (5.40)

$$\overline{\Theta} = \overline{\Theta}_1(\text{Fo}_1, \text{Bi}_1) \times \overline{\Theta}_2(\text{Fo}_2, \text{Bi}_2) \quad \text{for two factors} \tag{5.72a}$$

$$\overline{\Theta} = \overline{\Theta}_1(\text{Fo}_1, \text{Bi}_1) \times \overline{\Theta}_2(\text{Fo}_2, \text{Bi}_2) \times \overline{\Theta}_3(\text{Fo}_3, \text{Bi}_3) \quad \text{for three factors.} \tag{5.72b}$$

Since $\Phi = 1 - \overline{\Theta}$, a simple calculation shows that Φ can found from Φ_1, Φ_2, and Φ_3 as follows:

$$\Phi = \Phi_1 + \Phi_2(1 - \Phi_1) \quad \text{for two factors} \tag{5.73a}$$

$$\Phi = \Phi_1 + \Phi_2(1 - \Phi_1) + \Phi_3(1 - \Phi_2)(1 - \Phi_1) \quad \text{for three factors.} \tag{5.73b}$$

Example 5.13

For the bar described in Example 5.12, what is the mean temperature after 10 s and how much heat has been lost at that time?

Solution. For the Biot and Fourier numbers given in Example 5.12, we find from Fig. 5.10a

$$\Phi_1\,(\mathrm{Fo}_1 = 0.565, \mathrm{Bi}_1 = 0.2105) = 0.10$$
$$\Phi_2\,(\mathrm{Fo}_2 = 0.565, \mathrm{Bi}_2 = 0.2105) = 0.10$$

and, with eqn. (5.73a),

$$\Phi = \Phi_1 + \Phi_2\,(1 - \Phi_1) = 0.19$$

The mean temperature is

$$\overline{\Theta} = \frac{\overline{T} - 20}{100 - 20} = 1 - \Phi = 0.81$$

so

$$\overline{T} = 20 + 80(0.81) = 84.8^\circ\mathrm{C}$$

■

Problems

5.1 Rework Example 5.1, and replot the solution, with one change. This time, insert the thermometer at zero time, at an initial temperature $< (T_i - bT)$.

5.2 A body of known volume and surface area and temperature T_i is suddenly immersed in a bath whose temperature is rising as $T_{\mathrm{bath}} = T_i + (T_0 - T_i)e^{t/\tau}$. If $\overline{h}$ is known, $\tau = 10\rho cV/\overline{h}A$, t is measured from the time of immersion, and the Biot number of the body is small, find the temperature response of the body. Plot it and the bath temperature against time up to $t = 2\tau$.

5.3 A body of known volume and surface area is immersed in a bath whose temperature is varying sinusoidally with a frequency ω about an average value. The heat transfer coefficient is known and the Biot number is small. Find the temperature variation of the body after a long time has passed, and plot it along with the bath temperature. Comment on any interesting aspects of the solution.

A suggested program for solving this problem:

- Write the differential equation of response.
- To get the particular integral of the complete equation, guess that $T - T_{\text{mean}} = C_1 \cos \omega t + C_2 \sin \omega t$. Substitute this in the differential equation and find C_1 and C_2 values that will make the resulting equation valid.
- Write the general solution of the complete equation. It will have one unknown constant in it.
- Write any initial condition you wish—the simplest one you can think of—and use it to get rid of the constant.
- Let the time be large and note which terms vanish from the solution. Throw them away.
- Combine two trigonometric terms in the solution into a term involving $\sin(\omega t - \beta)$, where $\beta = \text{fn}(\omega T)$ is the phase lag of the body temperature.

5.4 A block of copper floats within a large region of well-stirred mercury. The system is initially at a uniform temperature, T_i. There is a heat transfer coefficient, $\overline{h}_m$, on the inside of the thin metal container of the mercury and another one, $\overline{h}_c$, between the copper block and the mercury. The container is then suddenly subjected to a change in ambient temperature from T_i to $T_s < T_i$. Predict the temperature response of the copper block, neglecting the internal resistance of both the copper and the mercury. Check your result by seeing that it fits both initial conditions and that it gives the expected behavior at $t \to \infty$.

5.5 Sketch the electrical circuit that is analogous to the second-order lumped capacity system treated in the context of Fig. 5.5 and explain it fully.

5.6 A one-inch diameter copper sphere with a thermocouple in its center is mounted as shown in Fig. 5.28 and immersed in water that is saturated at 211°F. The figure shows an actual thermocouple reading as a function of time during the quenching process. If the Biot number is small, the center temperature can be

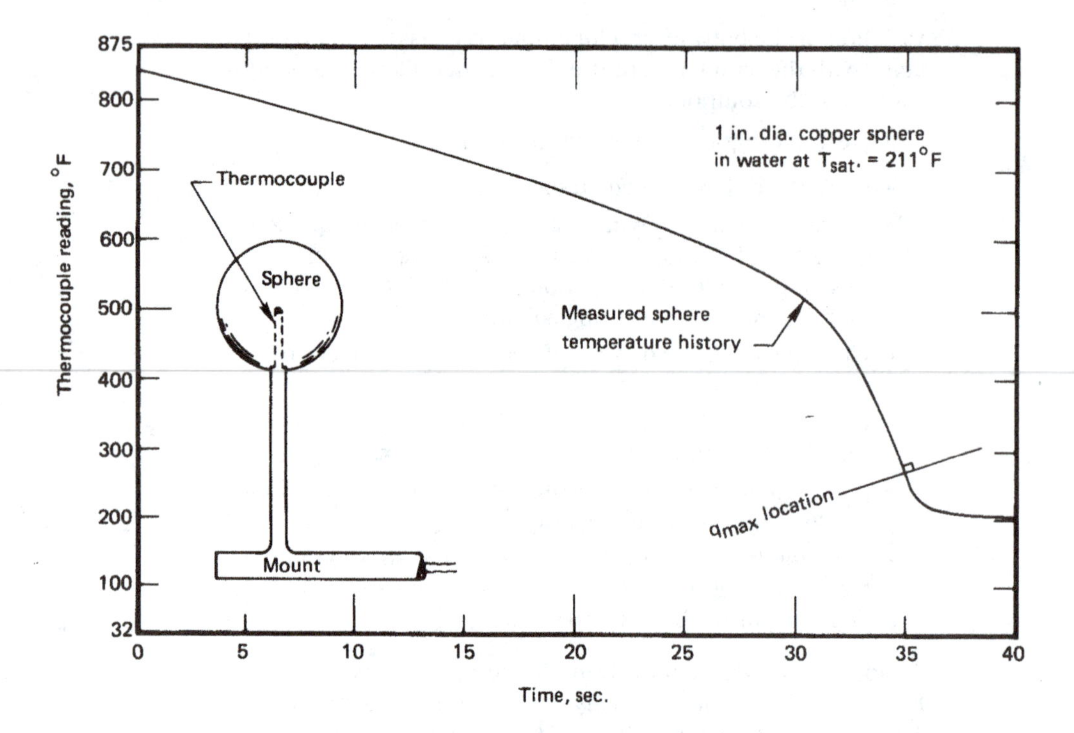

Figure 5.28 Configuration and temperature response for Problem 5.6

interpreted as the uniform temperature of the sphere during the quench. First draw tangents to the curve, and graphically differentiate it. Then use the resulting values of dT/dt to construct a graph of the heat transfer coefficient as a function of $(T_{\text{sphere}} - T_{\text{sat}})$. Check to see whether or not the largest value of the Biot number is too great to permit the use of lumped-capacity methods.

5.7 A butt-welded 36-gage thermocouple is placed in a gas flow whose temperature rises at the rate 20°C/s. The thermocouple steadily records a temperature 2.4°C below the known gas flow temperature. If ρc is 3800 kJ/m^{3}K for the thermocouple material, what is $\overline{h}$ on the thermocouple? [$\overline{h}$ = 1006 W/m^{2}K.]

5.8 Check the point on Fig. 5.7 at Fo = 0.2, Bi = 10, and $x/L = 0$ analytically.

5.9 Prove that when Bi is large, eqn. (5.34) reduces to eqn. (5.33).

5.10 Check the point at Bi = 0.1 and Fo = 2.5 on the slab curve in Fig. 5.10 analytically.

5.11 Sketch one of the curves in Fig. 5.7, 5.8, or 5.9 and identify:

- The region in which b.c.'s of the third kind can be replaced with b.c.'s of the first kind.
- The region in which a lumped-capacity response can be assumed.
- The region in which the solid can be viewed as a semi-infinite region.

5.12 Water flows over a flat slab of Nichrome, 0.05 mm thick, which serves as a resistance heater using AC power. The apparent value of $\overline{h}$ is 2000 W/m^{2}K. How much surface temperature fluctuation will there be?

5.13 Put Jakob's bubble growth formula in dimensionless form, identifying a "Jakob number", Ja $\equiv c_p(T_{\text{sup}} - T_{\text{sat}})/h_{fg}$ as one of the groups. (Ja is the ratio of sensible heat to latent heat.) Be certain that your nondimensionalization is consistent with the Buckingham pi-theorem.

5.14 A 7 cm long vertical glass tube is filled with water that is uniformly at a temperature of $T = 102$°C. The top is suddenly opened to the air at 1 atm pressure. Plot the decrease of the height of water in the tube by evaporation as a function of time until the bottom of the tube has cooled by 0.05°C.

5.15 A slab is cooled convectively on both sides from a known initial temperature. Compare the variation of surface temperature with time as given in Fig. 5.7 with that given by eqn. (5.53) if Bi = 2. Discuss the meaning of your comparisons.

5.16 To obtain eqn. (5.62), assume a complex solution of the type $\Theta = \text{fn}(\xi)\exp(i\Omega)$, where $i \equiv \sqrt{-1}$. This will assure that the real part of your solution has the required periodicity and, when you substitute it in eqn. (5.60), you will get an easy-to-solve ordinary d.e. in $\text{fn}(\xi)$.

5.17 A certain steel cylinder wall is subjected to a temperature oscillation that we approximate at $T = 650°\text{C} + (300°\text{C}) \cos \omega t$, where the piston fires eight times per second. For stress design purposes, plot the amplitude of the temperature variation in the steel as a function of depth. If the cylinder is 1 cm thick, can we view it as having infinite depth?

5.18 A 40 cm diameter pipe at 75°C is buried in a large block of Portland cement. It runs parallel with a 15°C isothermal surface at a depth of 1 m. Plot the temperature distribution along the line normal to the 15°C surface that passes through the center of the pipe. Compute the heat loss from the pipe both graphically and analytically.

5.19 Derive shape factor 4 in Table 5.4.

5.20 Verify shape factor 9 in Table 5.4 with a flux plot. Use $R_1/R_2 = 2$ and $R_1/L = ½$. (Be sure to start out with enough blank paper surrounding the cylinders.)

5.21 A copper block 1 in. thick and 3 in. square is held at 100°F on one 1 in. by 3 in. surface. The opposing 1 in. by 3 in. surface is adiabatic for 2 in. and 90°F for 1 inch. The remaining surfaces are adiabatic. Find the rate of heat transfer. [$Q = 36.8$ W.]

5.22 Obtain the shape factor for any or all of the situations pictured in Fig. 5.29a through j on pages 257–258. In each case, present a well-drawn flux plot. You may optionally check these results using numerical simulation software. [$S_b \simeq 1.03$, $S_c \gg S_d$, $S_g = 1$.]

5.23 Two copper slabs, 3 cm thick and insulated on the outside, are suddenly slapped tightly together. The one on the left side is initially at 100°C and the one on the right side at 0°C. Determine the left-hand adiabatic boundary's temperature after 2.3 s have elapsed. [$T_{\text{wall}} \simeq 80.5°\text{C}$]

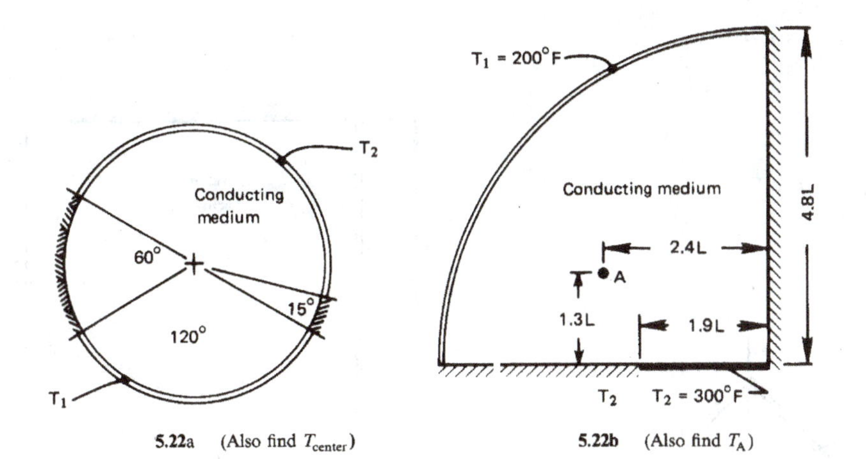

5.22a (Also find T_{center}) **5.22b** (Also find T_A)

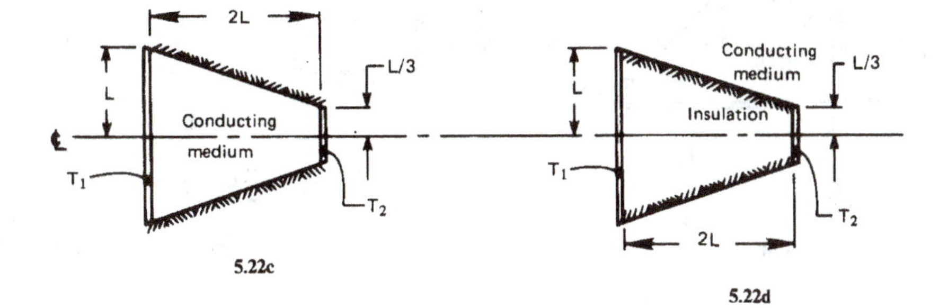

5.22c **5.22d**

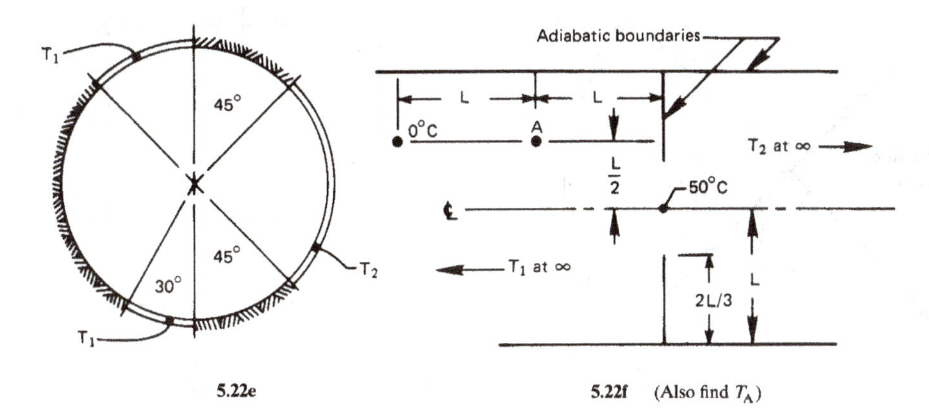

5.22e **5.22f** (Also find T_A)

Figure 5.29 Configurations for Problem 5.22

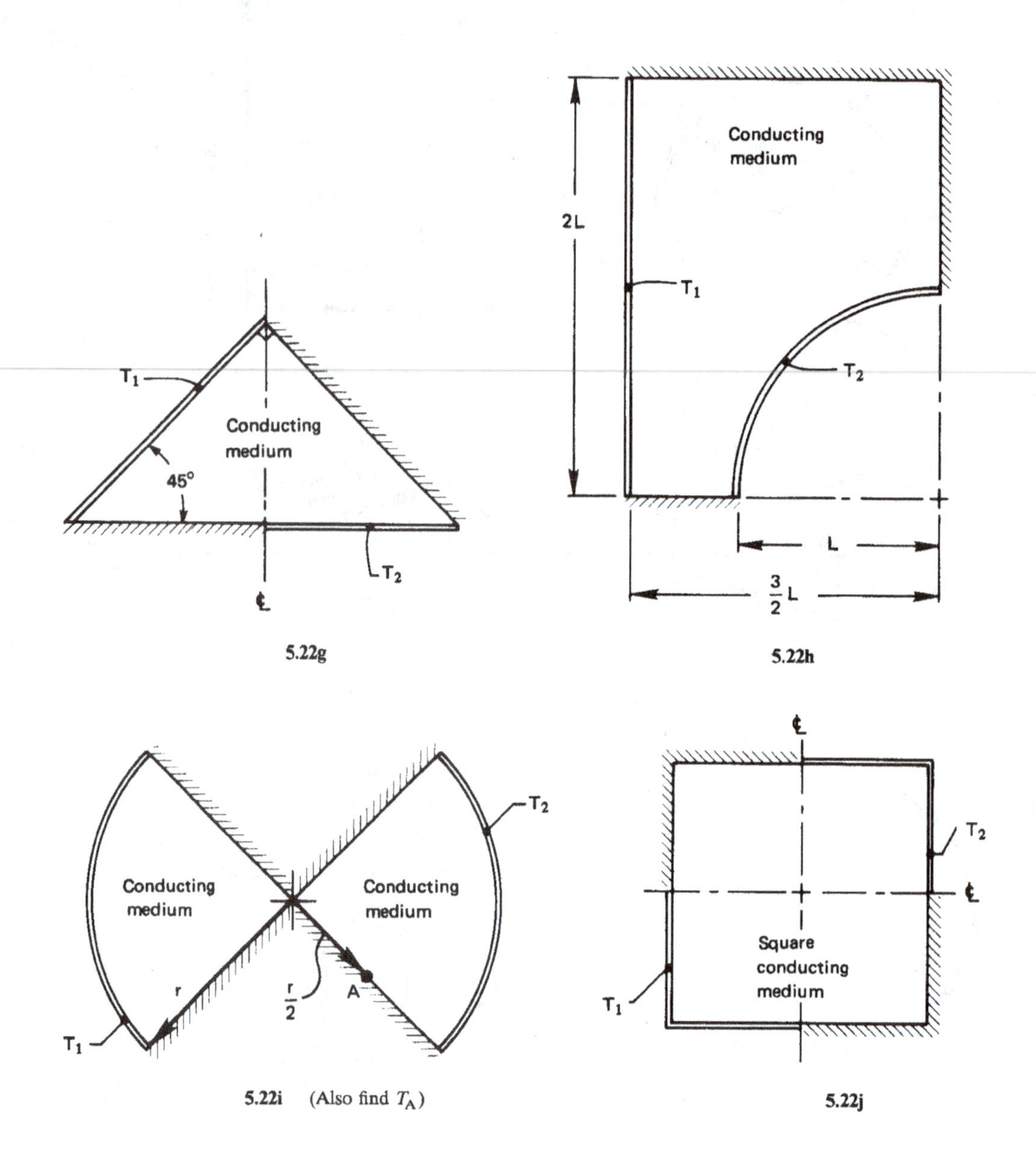

Figure 5.29 Configurations for Problem 5.22 *(con't)*

5.24 Estimate the time required to hard-cook an egg if:

- The minor diameter is 45 mm.
- k for the entire egg is about the same as for egg white. No significant heat release or change of properties occurs during cooking.
- $\overline{h}$ between the egg and the water is 1000 W/m^{2}K.
- The egg has a uniform temperature of 20°C when it is put into simmering water at 85°C.
- The egg is done when the center reaches 75°C.

Eggs cook as their proteins denature and coagulate. The time to cook depends on whether a soft or hard cooked egg desired. Eggs may be cooked by placing them (cold or warm) into cold water before heating starts or by placing warm eggs directly into simmering water [5.20].

5.25 Prove that T_1 in Fig. 5.5 cannot oscillate.

5.26 Show that when isothermal and adiabatic lines are interchanged in a two-dimenisonal body, the new shape factor is the inverse of the original one.

5.27 A 0.5 cm diameter cylinder at 300°C is suddenly immersed in saturated water at 1 atm. If $\overline{h}$ = 10,000 W/m^{2}K, find the centerline and surface temperatures after 0.2 s:

a. If the cylinder is copper.

b. If the cylinder is Nichrome V. [$T_{\text{sfc}} \simeq 200$°C.]

c. If the cylinder is Nichrome V, obtain the most accurate value of the temperatures after 0.04 s that you can.

5.28 A large, flat electrical resistance strip heater is fastened to a firebrick wall, unformly at 15°C. When it is suddenly turned on, it releases heat at the uniform rate of 4000 W/m^2. Plot the temperature of the brick immediately under the heater as a function of time if the other side of the heater is insulated. What is the heat flux at a depth of 1 cm when the surface reaches 200°C.

5.29 Do Experiment 5.2 and submit a report on the results.

5.30 An approximately spherical container, 2 cm in diameter, containing electronic equipment is placed in wet mineral soil with its center 2 m below the surface. The soil surface is kept at 0°C. What is the maximum rate at which energy can be released by the equipment if the surface of the sphere is not to exceed 30°C?

5.31 A semi-infinite slab of ice at $-10°$C is exposed to air at 15°C through a heat transfer coefficient of 10 W/m^{2}K. What is the initial rate of melting of ice in kg/m^{2}s? What is the asymptotic rate? Describe the melting process in physical terms. (The latent heat of fusion of ice, $h_{sf} = 333,300$ J/kg.)

5.32 One side of an insulating firebrick wall, 10 cm thick, initially at 20°C is exposed to 1000°C flame through a heat transfer coefficient of 230 W/m^{2}K. How long will it be before the other side is too hot to touch, say at 65°C? (Estimate properties at 500°C, and assume that $\overline{h}$ is quite low on the cool side.)

5.33 A particular lead bullet travels for 0.5 sec within a shock wave that heats the air near the bullet to 300°C. Approximate the bullet as a cylinder 0.8 cm in diameter. What is its surface temperature at impact if $\overline{h} = 600$ W/m^{2}K and if the bullet was initially at 20°C? What is its center temperature?

5.34 A loaf of bread is removed from an oven at 125°C and set on the (insulating) counter to cool in a kitchen at 25°C. The loaf is 30 cm long, 15 cm high, and 12 cm wide. If $k = 0.05$ W/m·K and $\alpha = 5 \times 10^{-7}$ m^2/s for bread, and $\overline{h} = 10$ W/m^{2}K, when will the hottest part of the loaf have cooled to 60°C? [About 1 h 5 min.]

5.35 A lead cube, 50 cm on each side, is initially at 20°C. The surroundings are suddenly raised to 200°C and $\overline{h}$ around the cube is 272 W/m^{2}K. Plot the cube temperature along a line from the center to the middle of one face after 20 minutes have elapsed.

5.36 A jet of clean water superheated to 150°C issues from a 1/16 inch diameter sharp-edged orifice into air at 1 atm, moving at 27 m/s. The coefficient of contraction of the jet is 0.611. Evaporation at $T = T_{\text{sat}}$ begins immediately on the outside of the jet. Plot the centerline temperature of the jet and $T(r/r_o = 0.6)$ as functions of distance from the orifice up to about 5 m. Neglect any axial conduction and any dynamic interactions between the jet and the air.

5.37 A 3 cm thick slab of aluminum (initially at 50°C) is slapped tightly against a 5 cm slab of copper (initially at 20°C). The outsides are both insulated and the contact resistance is neglible. What is the initial interfacial temperature? Estimate how long the interface will keep its initial temperature.

5.38 A cylindrical underground gasoline tank, 2 m in diameter and 4 m long, is embedded in 10°C soil with $k = 0.8$ W/m^{2}K and $\alpha = 1.3 \times 10^{-6}$ m^2/s. water at 27°C is injected into the tank to test it for leaks. It is well-stirred with a submerged ½ kW pump. We observe the water level in a 10 cm I.D. transparent standpipe and measure its rate of rise and fall. What rate of change of height will occur after one hour if there is no leakage? Will the level rise or fall? Neglect thermal expansion and deformation of the tank, which should be complete by the time the tank is filled.

5.39 A 47°C copper cylinder, 3 cm in diameter, is suddenly immersed horizontally in water at 27°C in a reduced gravity environment. Plot T_{cyl} as a function of time if $g = 0.76$ m/s^2 and if $\overline{h} = [2.733 + 10.448(\Delta T °\text{C})^{1/6}]^2$ W/m^{2}K. (Do it numerically if you cannot integrate the resulting equation analytically.)

5.40 The mechanical engineers at the University of Utah end spring semester by roasting a pig and having a picnic. The pig is roughly cylindrical and about 26 cm in diameter. It is roasted over a propane flame, whose products have properties similar to those of air, at 280°C. The hot gas flows across the pig at about 2 m/s. If the meat is cooked when it reaches 95°C, and if it is to be served at 2:00 pm, what time should cooking commence? Assume Bi to be large, but note Problem 7.40. The pig is initially at 25°C.

5.41 People from cold northern climates know not to grasp metal with their bare hands in subzero weather. A very slightly frosted peice of, say, cast iron will stick to your hand like glue in, say, −20°C weather and might tear off patches of skin. Explain this *quantitatively.*

5.42 A 4 cm diameter rod of type 304 stainless steel has a very small hole down its center. The hole is clogged with wax that has a melting point of 60°C. The rod is at 20°C. In an attempt to free the hole, a workman swirls the end of the rod—and about a meter of its length—in a tank of water at 80°C. If $\overline{h}$ is 688 W/m^{2}K on both the end and the sides of the rod, plot the depth of the melt front as a function of time up to say, 4 cm.

5.43 A cylindrical insulator contains a single, very thin electrical resistor wire that runs along a line halfway between the center and the outside. The wire liberates 480 W/m. The thermal conductivity of the insulation is 3 W/m^{2}K, and the outside perimeter is held at 20°C. Develop a flux plot for the cross section, considering carefully how the field should look in the neighborhood of the point through which the wire passes. Evaluate the temperature at the center of the insulation.

5.44 A long, 10 cm square copper bar is bounded by 260°C gas flows on two opposing sides. These flows impose heat transfer coefficients of 46 W/m^{2}K. The two intervening sides are cooled by natural convection to water at 15°C, with a heat transfer coefficient of 30 W/m^{2}K. What is the heat flow through the block and the temperature at the center of the block? (This could be a pretty complicated problem, but take the trouble to think about Biot numbers before you begin.)

5.45 Lord Kelvin made an interesting estimate of the age of the earth in 1864. He assumed that the earth originated as a mass of molten rock at 4144 K (7000°F) and that it had been cooled by outer space at 0 K ever since. To do this, he assumed that Bi for the earth is very large and that cooling had thus far penetrated through only a relatively thin (one-dimensional) layer. Using $\alpha_{\text{rock}} = 1.18 \times 10^{-6}$ m^2/s and the measured surface temperature gradient of the earth, $\frac{1}{27}$°C/m, Find Kelvin's value of Earth's age. (Kelvin's result turns out to be much less than the accepted value of 4 billion years. His calculation fails because Earth is not solid. Rather, the molten core is convectively stirred below the solid lithosphere. Consequently, the surface gradient has little to do with its age.)

5.46 A pure aluminum cylinder, 4 cm diam. by 8 cm long, is initially at 300°C. It is plunged into a liquid bath at 40°C with $\overline{h}$ = 500 W/m^{2}K. Calculate the hottest and coldest temperatures in the cylinder after one minute. Compare these results with the lumped capacity calculation, and discuss the comparison.

5.47 When Ivan cleaned his freezer, he accidentally put a large can of frozen juice into the refrigerator. The juice can is 17.8 cm tall and has an 8.9 cm I.D. The can was at -15°C in the freezer, but the refrigerator is at 4°C. The can now lies on a shelf of widely-spaced plastic rods, and air circulates freely over it. Thermal interactions with the rods can be ignored. The effective heat transfer coefficient to the can (for simultaneous convection and thermal radiation) is 8 W/m^{2}K. The can has a 1.0 mm thick cardboard skin with $k = 0.2$ W/m·K. The frozen juice has approximately the same physical properties as ice.

a. How important is the cardboard skin to the thermal response of the juice? Justify your answer quantitatively.

b. If Ivan finds the can in the refrigerator 30 minutes after putting it in, will the juice have begun to melt?

5.48 A cleaning crew accidentally switches off the heating system in a warehouse one Friday night during the winter, just ahead of the holidays. When the staff return two weeks later, the warehouse is quite cold. In some sections, moisture that condensed has formed a layer of ice 1 to 2 mm thick on the concrete floor. The concrete floor is 25 cm thick and sits on compacted earth. Both the slab and the ground below it are now at 20°F. The building operator turns on the heating system, quickly warming the air to 60°F. If the heat transfer coefficient between the air and the floor is 15 W/m^{2}K, how long will it take for the ice to start melting? Take $\alpha_{\text{concr}} = 7.0 \times 10^{-7}$ m^2/s and $k_{\text{concr}} = 1.4$ W/m·K, and make justifiable approximations as appropriate.

5.49 A thick wooden wall, initially at 25°C, is made of fir. It is suddenly exposed to flames at 800°C. If the effective heat transfer coefficient for convection and radiation between the wall and the flames is 80 W/m^{2}K, how long will it take the wooden wall to reach its ignition temperature of 430°C?

5.50 Cold butter does not spread as well as warm butter. A small tub of whipped butter bears a label suggesting that, before use, it be allowed to warm up in room air for 30 minutes after being removed from the refrigerator. The tub has a diameter of 9.1 cm with a height of 5.6 cm, and the properties of whipped butter are: $k = 0.125$ W/m·K, $c_p = 2520$ J/kg·K, and $\rho = 620$ kg/m^3. Assume that the tub's cardboard walls offer negligible thermal resistance, that $\overline{h} = 10$ W/m^{2}K outside the tub. Negligible heat is gained through the low conductivity lip around the bottom of the tub. If the refrigerator temperature was 5°C and the tub has warmed for 30 minutes in a room at 20°C, find: the temperature in the center of the butter tub, the temperature around the edge of the top surface of the butter, and the total energy (in J) absorbed by the butter tub.

5.51 A two-dimensional, 90° annular sector has an adiabatic inner arc, $r = r_i$, and an adiabatic outer arc, $r = r_o$. The flat surface along $\theta = 0$ is isothermal at T_1, and the flat surface along $\theta = \pi/2$ is isothermal at T_2. Show that the shape factor is $S = (2/\pi)\ln(r_o/r_i)$.

5.52 Suppose that $T_\infty(t)$ is the time-dependent environmental temperature surrounding a convectively-cooled, lumped object.

a. Show that eqn. (1.20) leads to

$$\frac{d}{dt}(T - T_\infty) + \frac{(T - T_\infty)}{\boldsymbol{T}} = -\frac{dT_\infty}{dt}$$

where the time constant $\boldsymbol{T}$ is defined as usual.

b. If the initial temperature of the object is T_i, use either an integrating factor or a Laplace transform to show that $T(t)$ is

$$T(t) = T_\infty(t) + [T_i - T_\infty(0)]\, e^{-t/\boldsymbol{T}} - e^{-t/\boldsymbol{T}} \int_0^t e^{s/\boldsymbol{T}} \frac{d}{ds} T_\infty(s)\, ds.$$

5.53 Use the result of Problem 5.52 to verify eqn. (5.13).

5.54 Suppose that a thermocouple with an initial temperature T_i is placed into an airflow for which its Bi $\ll 1$ and its time constant is $\boldsymbol{T}$. Suppose also that the temperature of the airflow varies harmonically as $T_\infty(t) = T_i + \Delta T \cos(\omega t)$.

a. Use the result of Problem 5.52 to find the temperature of the thermocouple, $T_{tc}(t)$, for $t > 0$. (If you wish, note that the real part of $e^{i\omega t}$ is $\text{Re}\left\{e^{i\omega t}\right\} = \cos\omega t$ and use complex variables to do the integration.)

b. Approximate your result for $t \gg T$. Then determine the value of $T_{tc}(t)$ for $\omega T \ll 1$ and for $\omega T \gg 1$. Explain in physical terms the relevance of these limits to the frequency response of the thermocouple.

c. If the thermocouple has a time constant of $T = 0.1$ sec, estimate the highest frequency temperature variation that it will measure accurately.

5.55 A particular tungsten lamp filament has a diameter of 100 μm and sits inside a glass bulb filled with inert gas. The effective heat transfer coefficient for conduction and radiation is 750 W/m·K and the electrical current is at 60 Hz. How much does the filament's surface temperature fluctuate if the gas temperature is 200°C and the average wire temperature is 2900°C?

5.56 The consider the parameter ψ in eqn. (5.41).

a. If the timescale for heat to diffuse a distance δ is δ^2/α, explain the physical significance of ψ and the consequence of large or small values of ψ.

b. Show that the timescale for the thermal response of a wire with $\text{Bi} \ll 1$ is $\rho c_p \delta/(2\overline{h})$. Then explain the meaning of the new parameter $\phi = \rho c_p \omega\delta/(4\pi\overline{h})$.

c. When $\text{Bi} \ll 1$, is ϕ or ψ a more relevant parameter?

References

[5.1] H. D. Baehr and K. Stephan. *Heat and Mass Transfer.* Springer-Verlag, Berlin, 1998.

[5.2] A. F. Mills. *Basic Heat and Mass Transfer.* Prentice-Hall, Inc., Upper Saddle River, NJ, 2nd edition, 1999.

[5.3] L. M. K. Boelter, V. H. Cherry, H. A. Johnson, and R. C. Martinelli. *Heat Transfer Notes.* McGraw-Hill Book Company, New York, 1965.

[5.4] M. P. Heisler. Temperature charts for induction and constant temperature heating. *Trans. ASME*, 69:227–236, 1947.

[5.5] P. J. Schneider. *Temperature Response Charts.* John Wiley & Sons, Inc., New York, 1963.

[5.6] H. S. Carslaw and J. C. Jaeger. *Conduction of Heat in Solids.* Oxford University Press, New York, 2nd edition, 1959.

[5.7] F. A. Jeglic. An analytical determination of temperature oscillations in wall heated by alternating current. NASA TN D-1286, July 1962.

[5.8] F. A. Jeglic, K. A. Switzer, and J. H. Lienhard. Surface temperature oscillations of electric resistance heaters supplied with alternating current. *J. Heat Transfer*, 102(2):392–393, 1980.

[5.9] J. Bronowski. *The Ascent of Man.* Chapter 4. Little, Brown and Company, Boston, 1973.

[5.10] N. Zuber. Hydrodynamic aspects of boiling heat transfer. AEC Report AECU-4439, Physics and Mathematics, June 1959.

[5.11] M. S. Plesset and S. A. Zwick. The growth of vapor bubbles in superheated liquids. *J. Appl. Phys.*, 25:493–500, 1954.

[5.12] L. E. Scriven. On the dynamics of phase growth. *Chem. Eng. Sci.*, 10:1–13, 1959.

[5.13] P. Dergarabedian. The rate of growth of bubbles in superheated water. *J. Appl. Mech., Trans. ASME*, 75:537, 1953.

[5.14] E. R. G. Eckert and R. M. Drake, Jr. *Analysis of Heat and Mass Transfer.* Hemisphere Publishing Corp., Washington, D.C., 1987.

[5.15] V. S. Arpaci. *Conduction Heat Transfer.* Ginn Press/Pearson Custom Publishing, Needham Heights, Mass., 1991.

[5.16] E. Hahne and U. Grigull. Formfactor and formwiderstand der stationären mehrdimensionalen wärmeleitung. *Int. J. Heat Mass Transfer*, 18:751–767, 1975.

[5.17] P. M. Morse and H. Feshbach. *Methods of Theoretical Physics.* McGraw-Hill Book Company, New York, 1953.

[5.18] R. Rüdenberg. Die ausbreitung der luft—und erdfelder um hochspannungsleitungen besonders bei erd—und kurzschlüssen. *Electrotech. Z.*, 36:1342-1346, 1925.

[5.19] M. M. Yovanovich. Conduction and thermal contact resistances (conductances). In W. M. Rohsenow, J. P. Hartnett, and Y. I. Cho, editors, *Handbook of Heat Transfer*, chapter 3. McGraw-Hill, New York, 3rd edition, 1998.

[5.20] S. H. Corriher. *Cookwise: the hows and whys of successful cooking.* Wm. Morrow and Company, New York, 1997. Includes excellent desciptions of the physical and chemical processes of cooking. *The* cookbook for those who enjoyed freshman chemistry.

PART III

CONVECTIVE HEAT TRANSFER

6. Laminar and turbulent boundary layers

In cold weather, if the air is calm, we are not so much chilled as when there is wind along with the cold; for in calm weather, our clothes and the air entangled in them receive heat from our bodies; this heat...brings them nearer than the surrounding air to the temperature of our skin. But in windy weather, this heat is prevented...from accumulating; the cold air, by its impulse...both cools our clothes faster and carries away the warm air that was entangled in them.

notes on "The General Effects of Heat", **Joseph Black, c. 1790s**

6.1 Some introductory ideas

Joseph Black's perception about forced convection (above) represents a very correct understanding of the way forced convective cooling works. When cold air moves past a warm body, it constantly sweeps away warm air that has become, as Black put it, "entangled" with the body and replaces it with cold air. In this chapter we learn to form analytical descriptions of these convective heating (or cooling) processes.

Our aim is to predict h and $\overline{h}$, and it is clear that such predictions must begin in the motion of fluid around the bodies that they heat or cool. Once we understand these fluid motions, we can begin the process of predicting how much heat they add or remove.

Flow boundary layer

Fluids flowing past solid bodies adhere to them, so a region of variable velocity must be built up between the body and the free fluid stream, as indicated in Fig. 6.1. This region is called a *boundary layer*, which

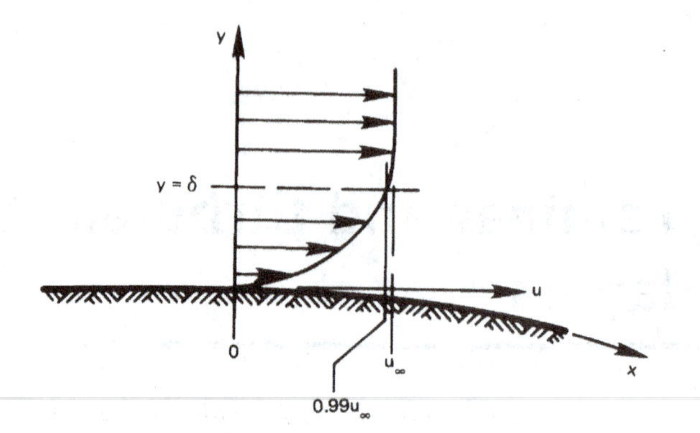

Figure 6.1 A boundary layer of thickness δ.

we abbreviate as b.l. The b.l. has a thickness, δ. The boundary layer thickness is arbitrarily defined as the distance from the wall at which the flow velocity approaches to within 1% of u_∞. The boundary layer is normally very thin in comparison with the dimensions of the body immersed in the flow.[1]

The first step we must take before we can predict h is the mathematical description of the boundary layer. This was first done by Prandtl[2] (see Fig. 6.2) and his students, starting in 1904, and it depended upon simplifications he could make after he recognized how thin the layer must be.

The dimensional functional equation for the boundary layer thickness on a flat surface is

$$\delta = \text{fn}(u_\infty, \rho, \mu, x)$$

where x is the length along the surface and ρ and μ are the fluid density in kg/m^3 and the dynamic viscosity in kg/m·s. We have five variables in

[1] We qualify this remark when we treat the b.l. quantitatively.

[2] Prandtl was educated at the Technical University in Munich and finished his doctorate there in 1900. He was given a chair in a new fluid mechanics institute at Göttingen University in 1904—the same year that he presented his historic paper explaining the boundary layer. His work at Göttingen during the early 20th century set the course of modern fluid mechanics and aerodynamics and laid the foundations for the analysis of heat convection.

Figure 6.2 Ludwig Prandtl (1875–1953). (Courtesy of *Appl. Mech. Rev.* [6.1])

kg, m, and s, so we anticipate two pi-groups:

$$\frac{\delta}{x} = \text{fn}(\text{Re}_x) \qquad \text{Re}_x \equiv \frac{\rho u_\infty x}{\mu} = \frac{u_\infty x}{\nu} \tag{6.1}$$

where ν is the kinematic viscosity μ/ρ and Re_x is called the *Reynolds number.* It characterizes the relative influences of inertial and viscous forces in a fluid problem. The subscript on Re—x in this case—tells what length it is based upon.

We discover shortly that the actual form of eqn. (6.1) for a flat surface, where u_∞ remains constant, is

$$\boxed{\frac{\delta}{x} = \frac{4.92}{\sqrt{\text{Re}_x}}} \tag{6.2}$$

which means that if the velocity is great or the viscosity is low, δ/x will be relatively small. Heat transfer will be relatively high in such cases. If the velocity is low, the b.l. will be relatively thick. A good deal of nearly

Osborne Reynolds (1842 to 1912)
Reynolds was born in Ireland but he taught at the University of Manchester. He was a significant contributor to the subject of fluid mechanics in the late 19th C. His original laminar-to-turbulent flow transition experiment, pictured below, was still being used as a student experiment at the University of Manchester in the 1970s.

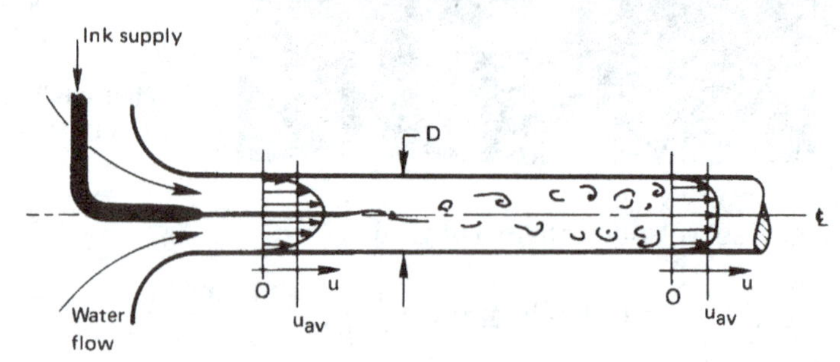

Figure 6.3 Osborne Reynolds and his laminar-turbulent flow transition experiment. (Detail from a portrait at the University of Manchester.)

stagnant fluid will accumulate near the surface and be "entangled" with the body, although in a different way than Black envisioned it to be.

The Reynolds number is named after Osborne Reynolds (see Fig. 6.3), who discovered the laminar-turbulent transition during fluid flow in a tube. He injected ink into a steady and undisturbed flow of water and found that, beyond a certain average velocity, u_{av}, the liquid streamline marked with ink would become wobbly and then break up into increasingly disorderly eddies, and it would finally be completely mixed into the

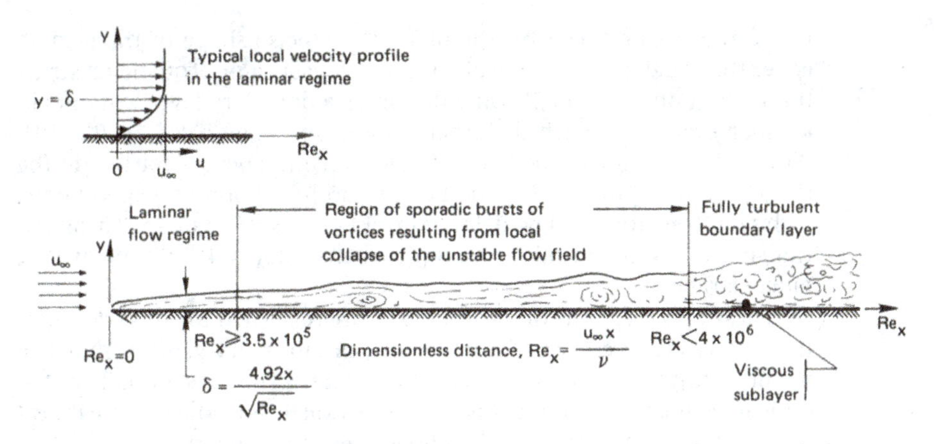

Figure 6.4 Boundary layer on a long, flat surface with a sharp leading edge.

water, as is suggested in the sketch.

To define the transition, we first note that $(u_{\text{av}})_{\text{crit}}$, the transitional value of the average velocity, must depend on the pipe diameter, D, on μ, and on ρ—four variables in kg, m, and s. There is therefore only one pi-group:

$$\text{Re}_{\text{critical}} \equiv \frac{\rho D (u_{\text{av}})_{\text{crit}}}{\mu} \tag{6.3}$$

The maximum Reynolds number for which fully developed laminar flow in a pipe will always be stable, regardless of the level of background noise, is 2100. In a reasonably careful experiment, laminar flow can be made to persist up to Re = 10,000. With enormous care it can be increased still another order of magnitude. But the value below which the flow will *always* be laminar—the critical value of Re—is 2100.

Much the same sort of thing happens in a boundary layer. Figure 6.4 shows fluid flowing over a plate with a sharp leading edge. The flow is laminar up to a transitional Reynolds number based on x:

$$\text{Re}_{x_{\text{critical}}} = \frac{u_\infty x_{\text{crit}}}{\nu} \tag{6.4}$$

At larger values of x the b.l. exhibits sporadic vortexlike instabilities over a fairly long range, and it finally settles into a fully turbulent b.l.

For the boundary layer shown, $\text{Re}_{x_{\text{critical}}} = 3.5 \times 10^5$, but in general the critical Reynolds number depends strongly on the amount of turbulence in the freestream flow over the plate, the precise shape of the leading edge, the roughness of the wall, and the presence of acoustic or structural vibrations [6.2, §5.5]. On a flat plate, a boundary layer will remain laminar even when such disturbances are very large if $\text{Re}_x \le 6 \times 10^4$. With relatively undisturbed conditions, transition occurs for Re_x in the range of 3×10^5 to 5×10^5, and in very careful laboratory experiments, turbulent transition can be delayed until $\text{Re}_x \approx 3 \times 10^6$ or so. Turbulent transition is essentially always complete before $\text{Re}_x = 4 \times 10^6$ and usually much earlier.

These specifications of the critical Re are restricted to flat surfaces. If the surface is curved away from the flow, as shown in Fig. 6.1, turbulence might be triggered at much lower values of Re_x. The susceptibility of the critical Reynolds to minor disturbances obviously introduces a great deal of troublesome uncertainty when flow rates are in this range.

Thermal boundary layer

When a wall is at a temperature T_w, different from that of the free stream, T_∞, a *thermal boundary layer* is present, and it has a thickness, δ_t, different from the *flow* b.l. thickness, δ. A thermal b.l. is pictured in Fig. 6.5. Now, with reference to this picture, we equate the heat conducted away from the wall by the fluid to the same heat transfer expressed in terms of a convective heat transfer coefficient:

$$\underbrace{-k_f \left.\frac{\partial T}{\partial y}\right|_{y=0}}_{\substack{\text{conduction} \\ \text{into the fluid}}} = h(T_w - T_\infty) \tag{6.5}$$

where k_f is the conductivity of the fluid. Notice two things about this result. In the first place, it is correct to express heat removal *at the wall* using Fourier's law of conduction, because no fluid moves in the direction of q. Also, while eqn. (6.5) looks like a b.c. of the third kind, it is not. This condition *defines h within the fluid* instead of specifying it as known information on the boundary. Equation (6.5) can be arranged in the form

$$\left.\frac{\partial\left(\dfrac{T_w - T}{T_w - T_\infty}\right)}{\partial(y/L)}\right|_{y/L=0} = \frac{hL}{k_f} = \text{Nu}_L, \text{ the Nusselt number} \tag{6.5a}$$

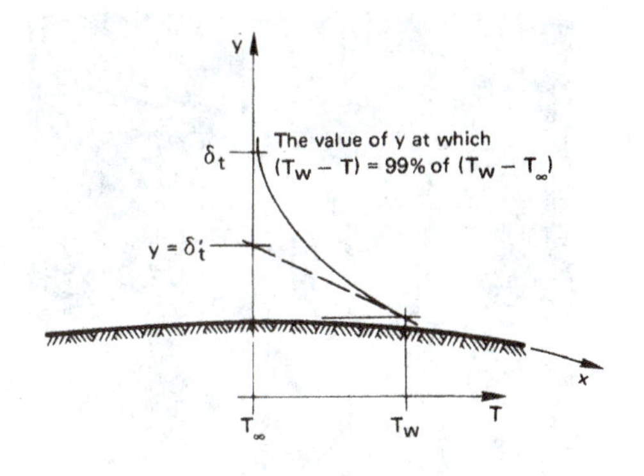

Figure 6.5 The thermal boundary layer during the flow of cool fluid over a warm plate.

where L is a characteristic dimension of the body under consideration—the length of a plate, the diameter of a cylinder, or [if we write eqn. (6.5) at a point, x, of interest along a flat surface] $\mathrm{Nu}_x \equiv hx/k_f$. From Fig. 6.5 we see immediately that the physical significance of Nu is given by

$$\mathrm{Nu}_L = \frac{L}{\delta_t'} \tag{6.6}$$

In other words, the Nusselt number is inversely proportional to the thickness of the thermal b.l.

The Nusselt number is named after Wilhelm Nusselt,[3] whose work on convective heat transfer was as fundamental as Prandtl's was in analyzing the related fluid dynamics (see Fig. 6.6).

We now turn to the detailed evaluation of h. And, as the preceding remarks make very clear, this evaluation will have to start with a development of the flow field in the boundary layer.

[3]Nusselt finished his doctorate in mechanical engineering at the Technical University in Munich in 1907. During an indefinite teaching appointment at Dresden (1913 to 1917) he made two of his most important contributions: He did the dimensional analysis of heat convection before he had access to Buckingham and Rayleigh's work. In so doing, he showed how to generalize limited data, and he set the pattern of subsequent analysis. He also showed how to predict convective heat transfer during film condensation. After moving about Germany and Switzerland from 1907 until 1925, he was named to the important Chair of Theoretical Mechanics at Munich. During his early years in this post, he made seminal contributions to heat exchanger design methodology. He held this position until 1952, during which time his, and Germany's, great influence in heat transfer and fluid mechanics waned. He was succeeded in the chair by another of Germany's heat transfer luminaries, Ernst Schmidt.

Figure 6.6 Ernst Kraft Wilhelm Nusselt (1882–1957). This photograph, provided by his student, G. Lück, shows Nusselt at the Kesselberg waterfall in 1912. He was an avid mountain climber.

6.2 Laminar incompressible boundary layer on a flat surface

We predict the boundary layer flow field by solving the equations that express conservation of mass and momentum in the b.l. Thus, the first order of business is to develop these equations.

Conservation of mass—The continuity equation

A two- or three-dimensional velocity field can be expressed in vectorial form:

$$\vec{u} = \vec{i}u + \vec{j}v + \vec{k}w$$

where u, v, and w are the x, y, and z components of velocity. Figure 6.7 shows a two-dimensional velocity flow field. If the flow is steady, the paths of individual particles appear as steady *streamlines.* The streamlines can be expressed in terms of a *stream function,* $\psi(x,y)$ = constant, where each value of the constant identifies a separate streamline, as shown in the figure.

The velocity, $\vec{u}$, is directed along the streamlines. Since no flow can cross a streamline, any pair of adjacent streamlines resembles a heat

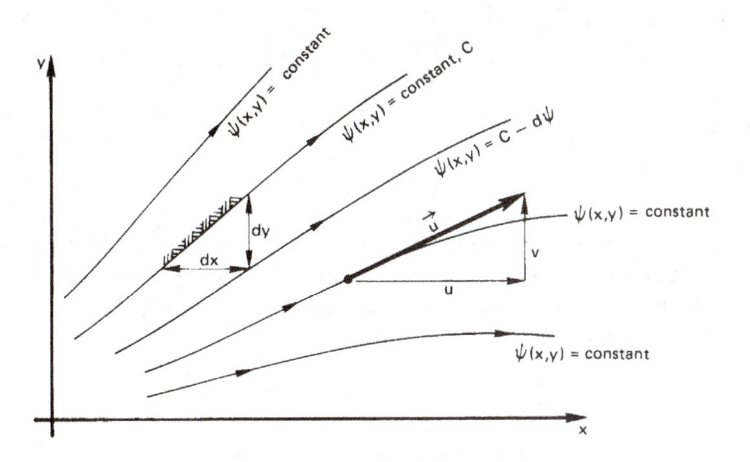

Figure 6.7 A steady, incompressible, two-dimensional flow field represented by streamlines, or lines of constant ψ.

flow channel in a flux plot (Section 5.7); such channels are adiabatic—no heat flow can cross them. Therefore, we write the equation for the conservation of mass by summing the inflow and outflow of mass on two faces of a triangular element of unit depth, as shown in Fig. 6.7:

$$\rho v\, dx - \rho u\, dy = 0 \tag{6.7}$$

If the fluid is incompressible, so that ρ = constant along each streamline, then

$$-v\, dx + u\, dy = 0 \tag{6.8}$$

But we can also differentiate the stream function along any streamline, $\psi(x, y)$ = constant, in Fig. 6.7:

$$d\psi = \left.\frac{\partial \psi}{\partial x}\right|_y dx + \left.\frac{\partial \psi}{\partial y}\right|_x dy = 0 \tag{6.9}$$

Comparison of eqns. (6.8) and (6.9) suggests a definition of the stream function in terms of velocities—a definition from which we get a more useful mass conservation equation than eqn. (6.8). By requiring that eqns. (6.8) and (6.9) are equivalent, we get the following pair of equations which define psi:

$$v \equiv -\left.\frac{\partial \psi}{\partial x}\right|_y \quad \text{and} \quad u \equiv \left.\frac{\partial \psi}{\partial y}\right|_x \tag{6.10}$$

Furthermore,

$$\frac{\partial^2 \psi}{\partial y \partial x} = \frac{\partial^2 \psi}{\partial x \partial y}$$

so that

$$\boxed{\frac{\partial u}{\partial x} + \frac{\partial v}{\partial y} = 0} \tag{6.11}$$

This is called the two-dimensional *continuity equation* for incompressible flow, because it expresses mathematically the fact that the flow is *continuous*; it has no breaks in it. In three dimensions, the continuity equation for an incompressible fluid is

$$\boxed{\nabla \cdot \vec{u} = \frac{\partial u}{\partial x} + \frac{\partial v}{\partial y} + \frac{\partial w}{\partial z} = 0}$$

Example 6.1

Fluid moves with a uniform velocity, u_∞, in the x-direction. Find the stream function and see if it gives plausible behavior (see Fig. 6.8).

Solution. $u = u_\infty$ and $v = 0$. Therefore, from eqns. (6.10)

$$u_\infty = \left.\frac{\partial \psi}{\partial y}\right|_x \quad \text{and} \quad 0 = \left.\frac{\partial \psi}{\partial x}\right|_y$$

Integrating these equations, we get

$$\psi = u_\infty y + \text{fn}(x) \quad \text{and} \quad \psi = 0 + \text{fn}(y)$$

Comparing these equations, we get $\text{fn}(x)$ = constant and $\text{fn}(y) = u_\infty y+$ constant, so

$$\psi = u_\infty y + \text{constant}$$ ■

This gives a series of equally spaced, horizontal streamlines, as we would expect (see Fig. 6.8). We set the arbitrary constant equal to zero in the figure so that the stream function is zero at the wall.

Conservation of momentum

The momentum equation in a viscous flow is a complicated vectorial expression called the Navier-Stokes equation. Its derivation is carried out

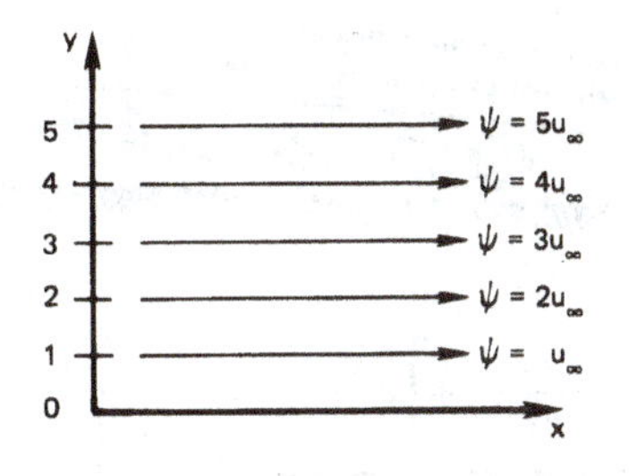

Figure 6.8 Streamlines in a uniform horizontal flow field, $\psi = u_\infty y$.

in any advanced fluid mechanics text (see, e.g., [6.3, Chap. III]). We shall offer a very restrictive derivation of the equation—one that applies only to a two-dimensional incompressible b.l. flow, as shown in Fig. 6.9.

Here we see that shear stresses act upon any element such as to continuously distort and rotate it. In the lower part of the figure, one such element is enlarged, so we can see the horizontal shear stresses[4] and the pressure forces that act upon it. They are shown as heavy arrows. We also display, as lighter arrows, the momentum fluxes entering and leaving the element.

Notice that both x- and y-directed momentum enters and leaves the element. To understand this, one can envision a boxcar moving down the railroad track with a man standing, facing its open door. A child standing at a crossing throws him a baseball as the car passes. When he catches the ball, its momentum will push him back, but a component of momentum will also jar him toward the rear of the train, because of the relative motion. Particles of fluid entering element A will likewise influence its motion, with their x components of momentum carried into the element by both components of flow.

The velocities must adjust themselves to satisfy the principle of conservation of linear momentum. To evaluate this, we consider a stationary control volume surrounding the particle A at one instant in time, say the instant when it is rectangular as shown in the figure. We require that the sum of the external forces in the x-direction, which act on the control volume, must be balanced by the rate at which the control volume forces

[4]The stress, τ, is often given two subscripts. The first one identifies the direction normal to the plane on which it acts, and the second one identifies the line along which it acts. Thus, if both subscripts are the same, the stress must act normal to a surface—it must be a pressure or tension instead of a shear stress.

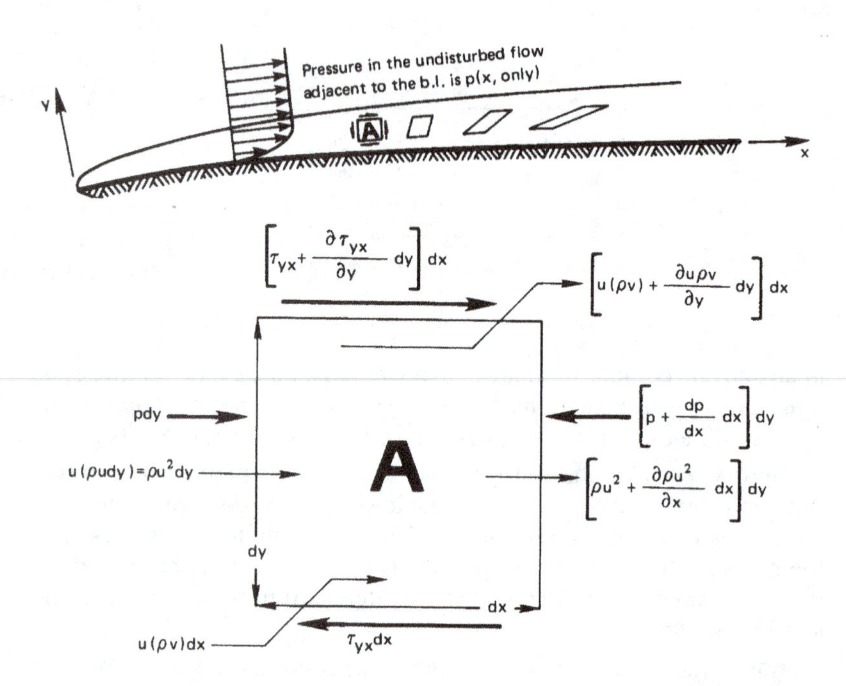

Figure 6.9 Forces acting in a two-dimensional incompressible boundary layer.

x-directed momentum out. The external forces, shown in Fig. 6.9, are

$$\left(\tau_{yx} + \frac{\partial \tau_{yx}}{\partial y} dy\right) dx - \tau_{yx}\, dx + p\, dy - \left(p + \frac{dp}{dx} dx\right) dy$$
$$= \left(\frac{\partial \tau_{yx}}{\partial y} - \frac{dp}{dx}\right) dx\, dy$$

The rate at which A loses x-directed momentum to its surroundings is

$$\left(\rho u^2 + \frac{\partial \rho u^2}{\partial x} dx\right) dy - \rho u^2\, dy + \left[u(\rho v) + \frac{\partial \rho u v}{\partial y} dy\right] dx$$
$$- \rho u v\, dx = \left(\frac{\partial \rho u^2}{\partial x} + \frac{\partial \rho u v}{\partial y}\right) dx\, dy$$

We equate these results and obtain the basic statement of conservation of x-directed momentum for the b.l.:

$$\frac{\partial \tau_{yx}}{\partial y}\,dy\,dx - \frac{dp}{dx}\,dx\,dy = \left(\frac{\partial \rho u^2}{\partial x} + \frac{\partial \rho u v}{\partial y}\right) dx\,dy$$

The shear stress in this result can be eliminated with the help of Newton's law of viscous shear:

$$\tau_{yx} = \mu \frac{\partial u}{\partial y}$$

so the momentum equation becomes

$$\frac{\partial}{\partial y}\left(\mu \frac{\partial u}{\partial y}\right) - \frac{dp}{dx} = \left(\frac{\partial \rho u^2}{\partial x} + \frac{\partial \rho u v}{\partial y}\right)$$

Finally, we remember that the analysis is limited to $\rho \simeq$ constant, and we limit use of the equation to temperature ranges in which $\mu \simeq$ constant. Then

$$\boxed{\frac{\partial u^2}{\partial x} + \frac{\partial u v}{\partial y} = -\frac{1}{\rho}\frac{dp}{dx} + \nu \frac{\partial^2 u}{\partial y^2}} \tag{6.12}$$

This is one form of the steady, two-dimensional, incompressible boundary layer momentum equation. Although we have taken $\rho \simeq$ constant, a more complete derivation reveals that the result is valid for compressible flow as well. If we multiply eqn. (6.11) by u and subtract the result from the left-hand side of eqn. (6.12), we obtain a second form of the momentum equation:

$$\boxed{u\frac{\partial u}{\partial x} + v\frac{\partial u}{\partial y} = -\frac{1}{\rho}\frac{dp}{dx} + \nu \frac{\partial^2 u}{\partial y^2}} \tag{6.13}$$

Equation (6.13) has a number of so-called *boundary layer approximations* built into it:

- $|\partial u/\partial x|$ is generally $\ll |\partial u/\partial y|$.
- v is generally $\ll u$.
- $p \neq \text{fn}(y)$

We obtain the pressure gradient by writing the Bernoulli equation for the free stream flow just above the boundary layer where there is no viscous shear. Thus,

$$\frac{p}{\rho} + \frac{u_\infty^2}{2} = \text{constant}$$

Differentiate this and use it to eliminate the pressure gradient,

$$\frac{1}{\rho}\frac{dp}{dx} = -u_\infty \frac{du_\infty}{dx}$$

so from eqn. (6.12):

$$\frac{\partial u^2}{\partial x} + \frac{\partial (uv)}{\partial y} = u_\infty \frac{du_\infty}{dx} + \nu \frac{\partial^2 u}{\partial y^2} \tag{6.14}$$

And if there is no pressure gradient in the flow—if p and u_∞ are constant as they would be for flow past a flat plate—then eqns. (6.12), (6.13), and (6.14) become

$$\boxed{\frac{\partial u^2}{\partial x} + \frac{\partial (uv)}{\partial y} = u\frac{\partial u}{\partial x} + v\frac{\partial u}{\partial y} = \nu \frac{\partial^2 u}{\partial y^2}} \tag{6.15}$$

Predicting the velocity profile in the laminar boundary layer without a pressure gradient

Exact solution. Two strategies for solving eqn. (6.15) for the velocity profile have long been widely used. The first was developed by Prandtl's student, H. Blasius,[5] before World War I. It is exact, and we shall sketch it only briefly. First we introduce the stream function, ψ, into eqn. (6.15). This reduces the number of dependent variables from two (u and v) to just one—namely, ψ. We do this by substituting eqns. (6.10) in eqn. (6.15):

$$\frac{\partial \psi}{\partial y}\frac{\partial^2 \psi}{\partial y \partial x} - \frac{\partial \psi}{\partial x}\frac{\partial^2 \psi}{\partial y^2} = \nu \frac{\partial^3 \psi}{\partial y^3} \tag{6.16}$$

It turns out that eqn. (6.16) can be converted into an ordinary d.e. with the following change of variables:

$$\psi(x, y) \equiv \sqrt{u_\infty \nu x}\, f(\eta) \quad \text{where} \quad \eta \equiv \sqrt{\frac{u_\infty}{\nu x}}\, y \tag{6.17}$$

where $f(\eta)$ is an as-yet-undertermined function. [This transformation is rather similar to the one that we used to make an ordinary d.e. of the

[5]Blasius achieved great fame for many accomplishments in fluid mechanics and then gave it up. Despite how much he achieved, he was eventually quoted as saying: "I decided that I had no gift for it; all of my ideas came from Prandtl."

Table 6.1 Exact velocity profile in the boundary layer on a flat surface with no pressure gradient

$y\sqrt{u_\infty/\nu x}$		u/u_∞	$v\sqrt{x/\nu u_\infty}$	
η	$f(\eta)$	$f'(\eta)$	$(\eta f' - f)/2$	$f''(\eta)$
0.00	0.00000	0.00000	0.00000	0.33206
0.20	0.00664	0.06641	0.00332	0.33199
0.40	0.02656	0.13277	0.01322	0.33147
0.60	0.05974	0.19894	0.02981	0.33008
0.80	0.10611	0.26471	0.05283	0.32739
1.00	0.16557	0.32979	0.08211	0.32301
2.00	0.65003	0.62977	0.30476	0.26675
3.00	1.39682	0.84605	0.57067	0.16136
4.00	2.30576	0.95552	0.75816	0.06424
4.918	3.20169	*0.99000*	0.83344	0.01837
6.00	4.27964	0.99898	0.85712	0.00240
8.00	6.27923	1.00000^-	0.86039	0.00001

heat conduction equation, between eqns. (5.44) and (5.45).] After some manipulation of partial derivatives, this substitution gives (Problem 6.2)

$$f\frac{d^2f}{d\eta^2} + 2\frac{d^3f}{d\eta^3} = 0 \tag{6.18}$$

and

$$\frac{u}{u_\infty} = \frac{df}{d\eta} \qquad \frac{v}{\sqrt{u_\infty \nu/x}} = \frac{1}{2}\left(\eta\frac{df}{d\eta} - f\right) \tag{6.19}$$

The boundary conditions for this flow are

$$\left.\begin{aligned} u(y=0) &= 0 \quad &\text{or}\quad \left.\frac{df}{d\eta}\right|_{\eta=0} &= 0 \\ u(y=\infty) &= u_\infty \quad &\text{or}\quad \left.\frac{df}{d\eta}\right|_{\eta=\infty} &= 1 \\ v(y=0) &= 0 \quad &\text{or}\quad f(\eta=0) &= 0 \end{aligned}\right\} \tag{6.20}$$

The solution of eqn. (6.18) subject to these b.c.'s must be done numerically. (See Problem 6.3.)

The solution of the Blasius problem is listed in Table 6.1, and the dimensionless velocity components are plotted in Fig. 6.10. The u component increases from zero at the wall ($\eta = 0$) to 99% of u_∞ at $\eta = 4.92$.

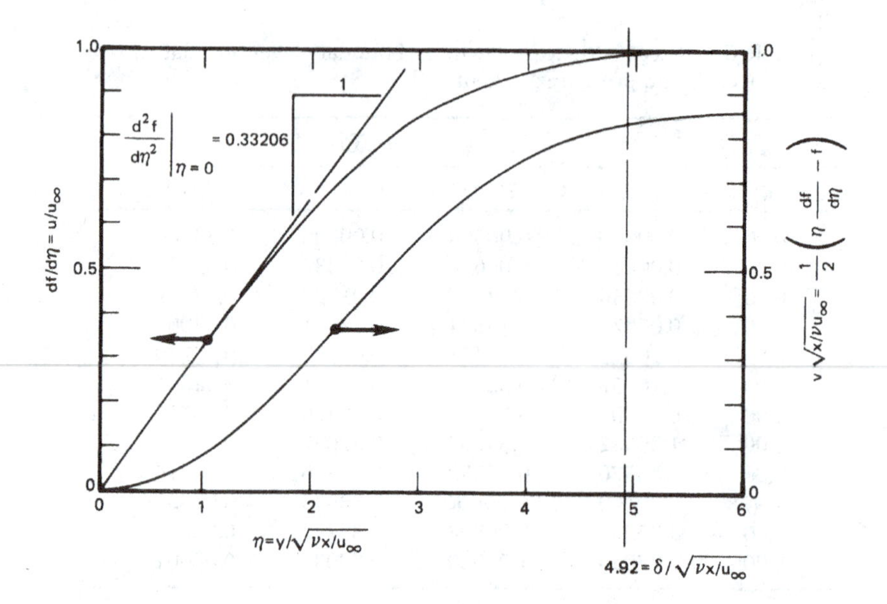

Figure 6.10 The dimensionless velocity components in a laminar boundary layer.

Thus, the b.l. thickness is given by

$$4.92 = \frac{\delta}{\sqrt{\nu x/u_\infty}}$$

or, as we anticipated earlier [eqn. (6.2)],

$$\frac{\delta}{x} = \frac{4.92}{\sqrt{u_\infty x/\nu}} = \frac{4.92}{\sqrt{\mathrm{Re}_x}}$$

Concept of similarity. The exact solution for $u(x, y)$ reveals a most useful fact—namely, that u can be expressed as a function of a single variable, η:

$$\frac{u}{u_\infty} = f'(\eta) = f'\left(y\sqrt{\frac{u_\infty}{\nu x}}\right)$$

This is called a *similarity solution.* To see why, we solve eqn. (6.2) for

$$\sqrt{\frac{u_\infty}{\nu x}} = \frac{4.92}{\delta(x)}$$

and substitute this in $f'(y\sqrt{u_\infty/\nu x})$. The result is

$$f' = \frac{u}{u_\infty} = \text{fn}\left[\frac{y}{\delta(x)}\right] \tag{6.21}$$

The velocity profile thus has the same shape with respect to the b.l. thickness at each x-station. We say, in other words, that the profile is *similar* at each station. This is what we found to be true for conduction into a semi-infinite region. In that case [recall eqn. (5.51)], $x/\sqrt{t}$ always had the same value at the outer limit of the thermally disturbed region.

Boundary layer similarity makes it especially easy to use a simple approximate method for solving other b.l. problems. This method, called the *momentum integral method*, is the subject of the next subsection.

Example 6.2

Air at 27°C blows over a flat surface with a sharp leading edge at 1.5 m/s. Find the b.l. thickness $\frac{1}{2}$ m from the leading edge. Check the b.l. assumption that $u \gg v$ at the trailing edge.

Solution. The dynamic and kinematic viscosities are $\mu = 1.853 \times 10^{-5}$ kg/m·s and $\nu = 1.566 \times 10^{-5}$ m²/s. Then

$$\text{Re}_x = \frac{u_\infty x}{\nu} = \frac{1.5(0.5)}{1.566 \times 10^{-5}} = 47{,}893$$

The Reynolds number is low enough to permit the use of a laminar flow analysis. Then

$$\delta = \frac{4.92x}{\sqrt{\text{Re}_x}} = \frac{4.92(0.5)}{\sqrt{47{,}893}} = 0.01124 = 1.124 \text{ cm}$$

(Remember that the b.l. analysis is only valid if $\delta/x \ll 1$. In this case, $\delta/x = 1.124/50 = 0.0225$.) From Fig. 6.10 or Table 6.1, we observe that v/u is greatest beyond the outside edge of the b.l, at large η. Using data from Table 6.1 at $\eta = 8$, v at $x = 0.5$ m is

$$\begin{aligned} v = \frac{0.8604}{\sqrt{x/\nu u_\infty}} &= 0.8604\sqrt{\frac{(1.566)(10^{-5})(1.5)}{(0.5)}} \\ &= 0.00590 \text{ m/s} \end{aligned}$$

or, since $u/u_\infty \to 1$ at large η

$$\frac{v}{u} = \frac{v}{u_\infty} = \frac{0.00590}{1.5} = 0.00393$$

■

Since v grows larger as x grows smaller, the condition $v \ll u$ is not satisfied very near the leading edge. There, the b.l. approximations themselves break down. We say more about this breakdown after eqn. (6.34).

Momentum integral method.[6] A second method for solving the b.l. momentum equation is approximate and much easier to apply to a wide range of problems than is any exact method of solution. The idea is this: We are not really interested in the details of the velocity or temperature profiles in the b.l., beyond learning their slopes at the wall. [These slopes give us the shear stress at the wall, $\tau_w = \mu(\partial u/\partial y)_{y=0}$, and the heat flux at the wall, $q_w = -k(\partial T/\partial y)_{y=0}$.] Therefore, we integrate the b.l. equations from the wall, $y = 0$, to the b.l. thickness, $y = \delta$, to make ordinary d.e.'s of them. It turns out that while these much simpler equations do not reveal anything new about the temperature and velocity profiles, they do give quite accurate explicit equations for τ_w and q_w.

Let us see how this procedure works with the b.l. momentum equation. We integrate eqn. (6.15), as follows, for the case in which there is no pressure gradient ($dp/dx = 0$):

$$\int_0^\delta \frac{\partial u^2}{\partial x}\,dy + \int_0^\delta \frac{\partial (uv)}{\partial y}\,dy = \nu \int_0^\delta \frac{\partial^2 u}{\partial y^2}\,dy$$

At $y = \delta$, u can be approximated as the free stream value, u_∞ and other quantities can be evaluated just as though y were infinite:

$$\int_0^\delta \frac{\partial u^2}{\partial x}\,dy + \Big[\underbrace{(uv)_{y=\delta}}_{=u_\infty v_\infty} - \underbrace{(uv)_{y=0}}_{=0}\Big] = \nu\Bigg[\underbrace{\left(\frac{\partial u}{\partial y}\right)_{y=\delta}}_{\simeq 0} - \left(\frac{\partial u}{\partial y}\right)_{y=0}\Bigg] \tag{6.22}$$

The continuity equation (6.11) can be integrated thus:

$$v_\infty - \underbrace{v_{y=0}}_{=0} = -\int_0^\delta \frac{\partial u}{\partial x}\,dy \tag{6.23}$$

Multiplying this by u_∞ gives

$$u_\infty v_\infty = -\int_0^\delta \frac{\partial u u_\infty}{\partial x}\,dy$$

[6] This method was developed by Pohlhausen, von Kármán, and others. See the discussion in [6.3, Chap. XII].

Using this result in eqn. (6.22), we obtain

$$\int_0^\delta \frac{\partial}{\partial x}[u(u-u_\infty)]\,dy = -\nu \left.\frac{\partial u}{\partial y}\right|_{y=0}$$

Finally, since $\mu(\partial u/\partial y)_{y=0}$ is the shear stress on the wall, $\tau_w = \tau_w$ (x only), this becomes[7]

$$\boxed{\frac{d}{dx}\int_0^{\delta(x)} u(u-u_\infty)\,dy = -\frac{\tau_w}{\rho}} \tag{6.24}$$

Equation (6.24) expresses the conservation of linear momentum in integrated form. It shows that the rate of momentum loss caused by the b.l. is balanced by the shear force on the wall. When we use it in place of eqn. (6.15), we are said to be *using an integral method.* To make use of eqn. (6.24), we first nondimensionalize it as follows:

$$\begin{aligned}\frac{d}{dx}\left[\delta\int_0^1 \frac{u}{u_\infty}\left(\frac{u}{u_\infty}-1\right)d\left(\frac{y}{\delta}\right)\right] &= -\frac{\nu}{u_\infty\delta}\left.\frac{\partial(u/u_\infty)}{\partial(y/\delta)}\right|_{y=0}\\ &= -\frac{\tau_w(x)}{\rho u_\infty^2} \equiv -\frac{1}{2}C_f(x)\end{aligned} \tag{6.25}$$

where $\tau_w/(\rho u_\infty^2/2)$ is the *skin friction coefficient,* C_f.

Equation (6.25) will be satisfied precisely by the exact solution (Problem 6.4) for u/u_∞. However, the point is to use eqn. (6.25) to determine u/u_∞ when we do not already have an exact solution. To do this, we recall that the exact solution exhibits *similarity.* First, we guess the solution in the form of eqn. (6.21): $u/u_\infty = \text{fn}(y/\delta)$. This guess is made in such a way that it will fit the following four things that are true of the velocity profile:

- $u/u_\infty = 0$ at $y/\delta = 0$
- $u/u_\infty \cong 1$ at $y/\delta = 1$
- $d\left(\dfrac{u}{u_\infty}\right)\Big/d\left(\dfrac{y}{\delta}\right) \cong 0$ at $y/\delta = 1$

(6.26) — applies to the three conditions above

- and from eqn. (6.15), we know that at $y/\delta = 0$:

$$\underbrace{u\frac{\partial u}{\partial x}}_{=0} + \underbrace{v\frac{\partial u}{\partial y}}_{=0} = \nu\left.\frac{\partial^2 u}{\partial y^2}\right|_{y=0}.$$

[7]The interchange of integration and differentiation is consistent with Leibnitz's rule for differentiation of an integral (Problem 6.14).

so

$$\left.\frac{\partial^2(u/u_\infty)}{\partial(y/\delta)^2}\right|_{y/\delta=0} = 0 \tag{6.27}$$

If $\text{fn}(y/\delta)$ is written as a polynomial with four constants—a, b, c, and d—in it,

$$\frac{u}{u_\infty} = a + b\frac{y}{\delta} + c\left(\frac{y}{\delta}\right)^2 + d\left(\frac{y}{\delta}\right)^3 \tag{6.28}$$

the four things that are known about the profile give

- $0 = a$, which eliminates a immediately
- $1 = 0 + b + c + d$
- $0 = b + 2c + 3d$
- $0 = 2c$, which eliminates c as well

Solving the middle two equations (above) for b and d, we obtain $d = -\frac{1}{2}$ and $b = +\frac{3}{2}$, so

$$\frac{u}{u_\infty} = \frac{3}{2}\frac{y}{\delta} - \frac{1}{2}\left(\frac{y}{\delta}\right)^3 \tag{6.29}$$

We compare this approximate velocity profile to the exact Blasius profile in Fig. 6.11, and they prove to be equal within a maximum error of 8%. The only remaining problem is calculating $\delta(x)$. To do this, we substitute eqn. (6.29) in eqn. (6.25) and get, after integration (see Problem 6.5):

$$-\frac{d}{dx}\left[\delta\left(\frac{39}{280}\right)\right] = -\frac{\nu}{u_\infty \delta}\left(\frac{3}{2}\right) \tag{6.30}$$

or

$$-\frac{39}{280}\left(\frac{2}{3}\right)\left(\frac{1}{2}\right)\frac{d\delta^2}{dx} = -\frac{\nu}{u_\infty}$$

We integrate this using the b.c. $\delta^2 = 0$ at $x = 0$:

$$\delta^2 = \frac{280}{13}\frac{\nu x}{u_\infty} \tag{6.31a}$$

or

$$\frac{\delta}{x} = \frac{4.64}{\sqrt{\text{Re}_x}} \tag{6.31b}$$

This b.l. thickness is of the correct functional form, and the constant is low by only 5.6%.

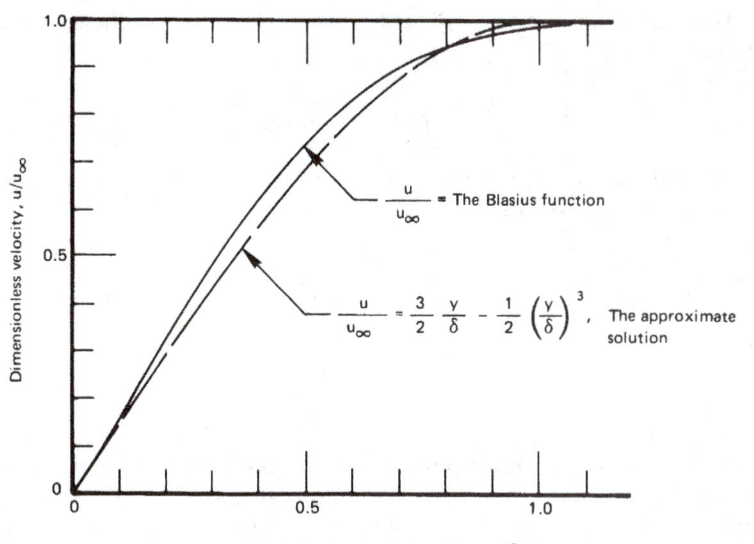

Figure 6.11 Comparison of the third-degree polynomial fit with the exact b.l. velocity profile. (Notice that the approximate result has been forced to $u/u_\infty = 1$ instead of 0.99 at $y = \delta$.)

The skin friction coefficient

Since the function $f(\eta)$ gives all information about flow in the b.l., the shear stress at the wall can be obtained from it by using Newton's law of viscous shear:

$$\begin{aligned}\tau_w &= \mu \left.\frac{\partial u}{\partial y}\right|_{y=0} = \mu \left.\frac{\partial}{\partial y}(u_\infty f')\right|_{y=0} = \mu u_\infty \left(\frac{df'}{d\eta}\frac{\partial \eta}{\partial y}\right)_{y=0} \\ &= \mu u_\infty \frac{\sqrt{u_\infty}}{\sqrt{\nu x}} \left.\frac{d^2 f}{d\eta^2}\right|_{\eta=0}\end{aligned}$$

But from Fig. 6.10 and Table 6.1, we see that $(d^2 f/d\eta^2)_{\eta=0} = 0.33206$, so

$$\tau_w = 0.332 \frac{\mu u_\infty}{x} \sqrt{\mathrm{Re}_x} \tag{6.32}$$

The integral method that we just outlined would have given 0.323 for the constant in eqn. (6.32) instead of 0.332 (Problem 6.6).

The *local skin friction coefficient*, or local skin drag coefficient, is defined as

$$C_f \equiv \frac{\tau_w}{\rho u_\infty^2/2} = \frac{0.664}{\sqrt{\mathrm{Re}_x}} \tag{6.33}$$

The *overall skin friction coefficient*, $\overline{C}_f$, is based on the average of the shear stress, τ_w, over the length, L, of the plate

$$\overline{\tau}_w = \frac{1}{L}\int_0^L \tau_w\,dx = \frac{\rho u_\infty^2}{2L}\int_0^L \frac{0.664}{\sqrt{u_\infty x/\nu}}\,dx = 1.328\,\frac{\rho u_\infty^2}{2}\sqrt{\frac{\nu}{u_\infty L}}$$

so

$$\overline{C}_f = \frac{1.328}{\sqrt{\mathrm{Re}_L}} \tag{6.34}$$

Notice that $C_f(x)$ approaches infinity at the leading edge of the flat surface. This means that to stop the fluid that first touches the front of the plate—dead in its tracks—would require infinite shear stress right at that point. Nature, of course, will not allow such a thing to happen; and it turns out that the boundary layer analysis is not really valid right at the leading edge.

In fact, the range $x \lesssim 5\delta$ is too close to the edge to use this analysis with accuracy because the b.l. is relatively thick and v is no longer $\ll u$. With eqn. (6.2), this converts to

$$x > 600\,\nu/u_\infty \quad \text{for a boundary layer to exist}$$

or $\mathrm{Re}_x \gtrsim 600$. In Example 6.2, this condition is satisfied for all x's greater than about 6 mm. This region is usually very small.

Example 6.3

Calculate the average shear stress and the overall friction coefficient for the surface in Example 6.2 if its total length is $L = 0.5$ m. Compare $\overline{\tau}_w$ with τ_w at the trailing edge. At what point on the surface does $\tau_w = \overline{\tau}_w$? Finally, estimate what fraction of the surface can legitimately be analyzed using boundary layer theory.

Solution.

$$\overline{C}_f = \frac{1.328}{\sqrt{\mathrm{Re}_{0.5}}} = \frac{1.328}{\sqrt{47{,}893}} = 0.00607$$

and

$$\overline{\tau}_w = \frac{\rho u_\infty^2}{2}\overline{C}_f = \frac{1.183(1.5)^2}{2}\,0.00607 = 0.00808\ \underbrace{\text{kg/m}\cdot\text{s}^2}_{\text{N/m}^2}$$

(This is very little drag. It amounts only to about 1/50 ounce/m^2.)

At $x = L$,

$$\left.\frac{\tau_w(x)}{\overline{\tau}_w}\right|_{x=L} = \frac{\rho u_\infty^2/2}{\rho u_\infty^2/2}\left[\frac{0.664/\sqrt{\text{Re}_L}}{1.328/\sqrt{\text{Re}_L}}\right] = \frac{1}{2}$$

and

$$\tau_w(x) = \overline{\tau}_w \quad \text{where} \quad \frac{0.664}{\sqrt{x}} = \frac{1.328}{\sqrt{0.5}}$$

so the local shear stress equals the average value, where

$$x = \tfrac{1}{8}\text{ m} \qquad \text{or} \qquad \frac{x}{L} = \frac{1}{4}$$

Thus, the shear stress, which is initially infinite, plummets to $\overline{\tau}_w$ one-fourth of the way from the leading edge and drops only to one-half of $\overline{\tau}_w$ in the remaining 75% of the plate.

The boundary layer assumptions fail when

$$x < 600\,\frac{\nu}{u_\infty} = 600\,\frac{1.566\times 10^{-5}}{1.5} = 0.0063\text{ m}$$

Thus, the preceding analysis should be good over almost 99% of the 0.5 m length of the surface. ■

6.3 The energy equation

Derivation

We now know how fluid moves in the b.l. Next, we must extend the heat conduction equation to allow for the motion of the fluid. It can be solved for the temperature field in the b.l. and its solution can be used to calculate h, using Fourier's law:

$$\boxed{h = \frac{q}{T_w - T_\infty} = -\frac{k}{T_w - T_\infty}\left.\frac{\partial T}{\partial y}\right|_{y=0}} \tag{6.35}$$

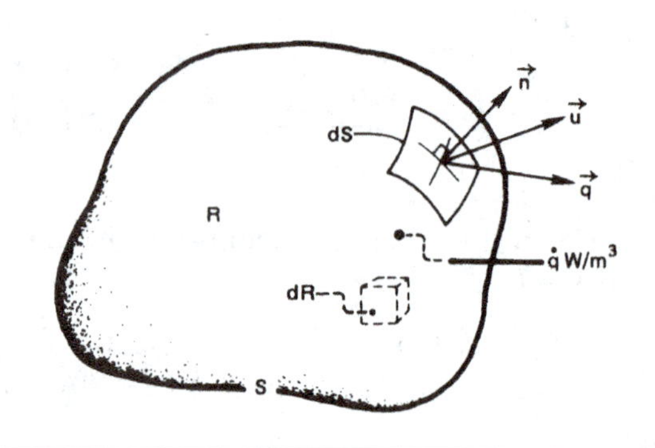

Figure 6.12 Control volume in a heat-flow and fluid-flow field.

To predict T, we extend the analysis of Section 2.1. Figure 2.4 shows a volume containing a solid subjected to a temperature field. We now allow this volume to contain a moving fluid with a velocity field $\vec{u}(x, y, z)$, as shown in Fig. 6.12. We make the following restrictive approximations:

- Pressure variations in the flow are not large enough to affect thermodynamic properties. From thermodynamics, we know that the specific internal energy, $\hat{u}$, is related to the specific enthalpy as $\hat{h} = \hat{u} + p/\rho$, and that $d\hat{h} = c_p\, dT + (\partial\hat{h}/\partial p)_T\, dp$. We shall neglect the effect of dp on enthalpy, internal energy, and density. This approximation is reasonable for most liquid flows and for gas flows moving at speeds less than about 1/3 the speed of sound.

- Under these conditions, density changes result only from temperature changes and will also be small; and the flow will behave as if incompressible. For such flows, $\nabla \cdot \vec{u} = 0$ (Sect. 6.2).

- Temperature variations in the flow are not large enough to change k significantly. When we consider the flow field, we will also presume μ to be unaffected by temperature change.

- Potential and kinetic energy changes are negligible in comparison to thermal energy changes. Since the kinetic energy of a fluid can change owing to pressure gradients, this again means that pressure variations may not be too large.

- The viscous stresses do not dissipate enough energy to warm the fluid significantly.

Just as we wrote eqn. (2.7) in Section 2.1, we now write conservation of energy in the form

$$\underbrace{\frac{d}{dt}\int_R \rho\hat{u}\,dR}_{\substack{\text{rate of internal}\\\text{energy increase}\\\text{in } R}} = -\underbrace{\int_S (\rho\hat{h})\,\vec{u}\cdot\vec{n}\,dS}_{\substack{\text{rate of internal energy and}\\\text{flow work out of } R}}$$

$$-\underbrace{\int_S (-k\nabla T)\cdot\vec{n}\,dS}_{\substack{\text{net heat conduction}\\\text{rate out of } R}} + \underbrace{\int_R \dot{q}\,dR}_{\substack{\text{rate of heat}\\\text{generation in } R}} \tag{6.36}$$

In the second integral, $\vec{u}\cdot\vec{n}\,dS$ represents the volume flow rate through an element dS of the control surface. The position of R is not changing in time, so we can bring the time derivative inside the first integral. If we then we call in Gauss's theorem [eqn. (2.8)] to make volume integrals of the surface integrals, eqn. (6.36) becomes

$$\int_R \left(\frac{\partial(\rho\hat{u})}{\partial t} + \nabla\cdot(\rho\vec{u}\hat{h}) - \nabla\cdot k\nabla T - \dot{q}\right) dR = 0$$

Because the integrand must vanish identically (recall the footnote on pg. 55 in Chap. 2) and because k depends weakly on T,

$$\frac{\partial(\rho\hat{u})}{\partial t} + \underbrace{\nabla\cdot(\rho\vec{u}\hat{h})}_{=\rho\vec{u}\cdot\nabla\hat{h}+\hat{h}\nabla\cdot(\rho\vec{u})} - k\nabla^2 T - \dot{q} = 0$$

Since we are neglecting pressure effects, we may introduce the following approximation:

$$d(\rho\hat{u}) = d(\rho\hat{h}) - dp \approx d(\rho\hat{h}) = \rho d\hat{h} + \hat{h}\,d\rho$$

Thus, collecting and rearranging terms

$$\rho\left(\frac{\partial\hat{h}}{\partial t} + \vec{u}\cdot\nabla\hat{h}\right) + \hat{h}\underbrace{\left(\frac{\partial\rho}{\partial t} + \nabla\cdot(\rho\vec{u})\right)}_{\text{neglect}} = k\nabla^2 T + \dot{q}$$

where we neglect the term involving density derivatives on the basis that density changes are small and the flow is nearly incompressible (but see Problem 6.36 for a more general result).

Upon substituting $d\hat{h} \approx c_p\, dT$, we obtain our final result:

$$\rho c_p \Bigg(\underbrace{\frac{\partial T}{\partial t}}_{\substack{\text{energy}\\ \text{storage}}} + \underbrace{\vec{u} \cdot \nabla T}_{\substack{\text{enthalpy}\\ \text{convection}}} \Bigg) = \underbrace{k \nabla^2 T}_{\substack{\text{heat}\\ \text{conduction}}} + \underbrace{\dot{q}}_{\substack{\text{heat}\\ \text{generation}}} \tag{6.37}$$

This is the energy equation for a constant pressure flow field. It is the same as the corresponding equation (2.11) for a solid body, except for the enthalpy transport, or convection, term, $\rho c_p \vec{u} \cdot \nabla T$.

Consider the term in parentheses in eqn. (6.37):

$$\frac{\partial T}{\partial t} + \vec{u} \cdot \nabla T = \frac{\partial T}{\partial t} + u\frac{\partial T}{\partial x} + v\frac{\partial T}{\partial y} + w\frac{\partial T}{\partial z} \equiv \frac{DT}{Dt} \tag{6.38}$$

DT/Dt is exactly the so-called *material derivative*, which is treated in some detail in every fluid mechanics course. DT/Dt is the rate of change of the temperature of a fluid particle as it moves in a flow field.

In a steady two-dimensional flow field without heat sources, eqn. (6.37) takes the form

$$u\frac{\partial T}{\partial x} + v\frac{\partial T}{\partial y} = \alpha \left(\frac{\partial^2 T}{\partial x^2} + \frac{\partial^2 T}{\partial y^2} \right) \tag{6.39}$$

Furthermore, in a b.l., $\partial^2 T/\partial x^2 \ll \partial^2 T/\partial y^2$, so

$$\boxed{u\frac{\partial T}{\partial x} + v\frac{\partial T}{\partial y} = \alpha\frac{\partial^2 T}{\partial y^2}} \tag{6.40}$$

Heat and momentum transfer analogy

Consider a b.l. in a fluid of bulk temperature T_∞, flowing over a flat surface at temperature T_w. The momentum equation and its b.c.'s can be written as

$$u\frac{\partial}{\partial x}\left(\frac{u}{u_\infty}\right) + v\frac{\partial}{\partial y}\left(\frac{u}{u_\infty}\right) = \nu\frac{\partial^2}{\partial y^2}\left(\frac{u}{u_\infty}\right) \quad \begin{cases} \left.\dfrac{u}{u_\infty}\right|_{y=0} = 0 \\ \left.\dfrac{u}{u_\infty}\right|_{y=\infty} = 1 \\ \left.\dfrac{\partial}{\partial y}\left(\dfrac{u}{u_\infty}\right)\right|_{y=\infty} = 0 \end{cases} \tag{6.41}$$

And the energy equation (6.40) can be expressed in terms of a dimensionless temperature, $\Theta = (T - T_w)/(T_\infty - T_w)$, as

$$u\frac{\partial \Theta}{\partial x} + v\frac{\partial \Theta}{\partial y} = \alpha\frac{\partial^2 \Theta}{\partial y^2} \qquad \begin{cases} \Theta(y=0) = 0 \\ \Theta(y=\infty) = 1 \\ \left.\dfrac{\partial \Theta}{\partial y}\right|_{y=\infty} = 0 \end{cases} \tag{6.42}$$

Notice that the problems of predicting u/u_∞ and Θ are *identical*, with one exception: eqn. (6.41) has ν in it whereas eqn. (6.42) has α. If ν and α should happen to be equal, the temperature distribution in the b.l. is

$$\text{for } \nu = \alpha: \frac{T - T_w}{T_\infty - T_w} = f'(\eta) \quad \text{derivative of the Blasius function}$$

since the two problems must have the same solution.

In this case, we can immediately calculate the heat transfer coefficient using eqn. (6.5):

$$h = \frac{k}{T_\infty - T_w} \left.\frac{\partial (T - T_w)}{\partial y}\right|_{y=0} = k\left(\frac{\partial f'}{\partial \eta}\frac{\partial \eta}{\partial y}\right)_{\eta=0}$$

but $(\partial^2 f/\partial \eta^2)_{\eta=0} = 0.33206$ (see Fig. 6.10) and $\partial\eta/\partial y = \sqrt{u_\infty/\nu x}$, so

$$\frac{hx}{k} = \text{Nu}_x = 0.33206\sqrt{\text{Re}_x} \quad \text{for } \nu = \alpha \tag{6.43}$$

Normally, in using eqn. (6.43) or any other forced convection equation, properties should be evaluated at the *film temperature*, $T_f = (T_w + T_\infty)/2$.

Example 6.4

Water flows over a flat heater, 0.06 m in length, at 15 atm pressure and 440 K. The free stream velocity is 2 m/s and the heater is held at 460 K. What is the average heat flux?

Solution. At $T_f = (460 + 440)/2 = 450$ K:

$$\nu = 1.725 \times 10^{-7} \text{ m}^2/\text{s}$$
$$\alpha = 1.724 \times 10^{-7} \text{ m}^2/\text{s}$$

Therefore, $\nu \simeq \alpha$, and we can use eqn. (6.43). First, we must calculate the average heat flux, $\overline{q}$. To do this, we set $\Delta T \equiv T_w - T_\infty$ and write

$$\overline{q} = \frac{1}{L}\int_0^L (h\Delta T)\,dx = \frac{\Delta T}{L}\int_0^L \frac{k}{x}\text{Nu}_x\,dx = 0.332\frac{k\Delta T}{L}\underbrace{\int_0^L \sqrt{\frac{u_\infty}{\nu x}}\,dx}_{=2\sqrt{u_\infty L/\nu}}$$

so

$$\overline{q} = 2\left(0.332\frac{k}{L}\sqrt{\text{Re}_L}\right)\Delta T = 2q_{x=L}$$

Note that the average heat flux is twice that at the trailing edge, $x = L$. Using $k = 0.674$ W/m·K for water at the film temperature,

$$\begin{aligned}\overline{q} &= 2(0.332)\frac{0.674}{0.06}\sqrt{\frac{2(0.06)}{1.72\times10^{-7}}}(460-440)\\ &= 124{,}604\ \text{W/m}^2 = 125\ \text{kW/m}^2\end{aligned}$$ ■

Equation (6.43) is clearly a very restrictive heat transfer solution. We now want to find how to evaluate q when ν does not equal α.

6.4 The Prandtl number and the boundary layer thicknesses

Dimensional analysis

Let us look more closely at implications of the similarity between the velocity and thermal boundary layers. We first ask what dimensional analysis reveals about heat transfer in the laminar b.l. We know that the dimensional functional equation for the heat transfer coefficient, h, should be

$$h = \text{fn}(k, x, \rho, c_p, \mu, u_\infty)$$

We have excluded $T_w - T_\infty$ on the basis of Newton's original hypothesis, borne out in eqn. (6.43)—that $h \neq \text{fn}(\Delta T)$ during forced convection. This gives seven variables in J/K, m, kg, and s, or $7 - 4 = 3$ pi-groups. Note that, as we indicated at the end of Section 4.3, this is a situation in which heat and work do not convert into one another. That means we should not regard J as N·m, but rather as a separate unit. The dimensionless groups are then:

$$\Pi_1 = \frac{hx}{k} \equiv \text{Nu}_x \qquad \Pi_2 = \frac{\rho u_\infty x}{\mu} \equiv \text{Re}_x$$

and a new group:

$$\Pi_3 = \frac{\mu c_p}{k} \equiv \frac{\nu}{\alpha} \equiv \text{Pr, Prandtl number}$$

Thus,

$$\text{Nu}_x = \text{fn}(\text{Re}_x, \text{Pr}) \tag{6.44}$$

in forced convection flow situations. Equation (6.43) was developed for the case in which $\nu = \alpha$ or $\text{Pr} = 1$; therefore, it is of the same form as eqn. (6.44), although it does not display the Pr dependence of Nu_x.

We can better understand the physical meaning of the Prandtl number if we briefly consider how to predict its value in a gas.

Kinetic theory of μ and k

Figure 6.13 shows a small neighborhood of a point in a gas where there exists a velocity or temperature gradient. We identify the *mean free path* of molecules between collisions as ℓ and indicate planes at $y \pm \ell/2$ which bracket the average travel of those molecules found at plane y. (Actually, these planes should be located closer to $y \pm \ell$ for a variety of subtle reasons. This and other fine points of these arguments

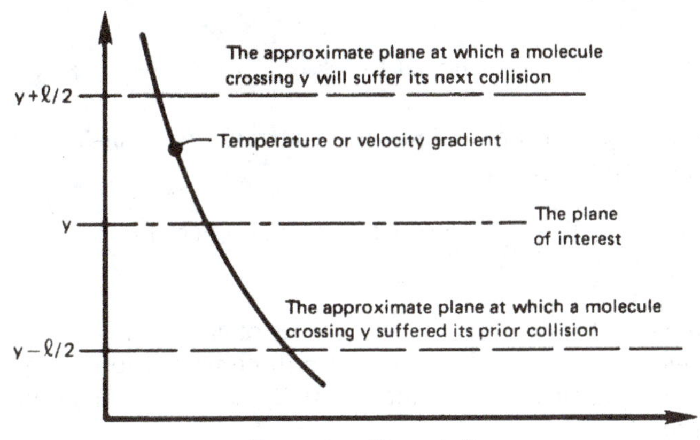

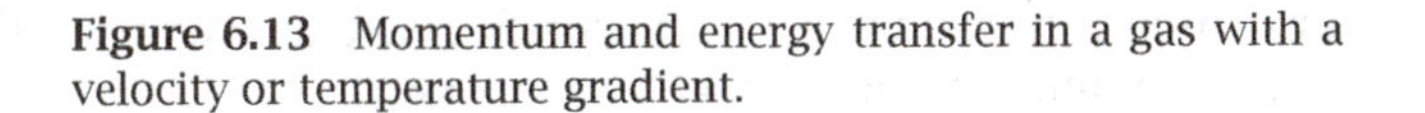

Figure 6.13 Momentum and energy transfer in a gas with a velocity or temperature gradient.

are explained in detail in [6.4].)

The shear stress, τ_{yx}, can be expressed as the change of momentum of all molecules that pass through the y-plane of interest, per unit area:

$$\tau_{yx} = \begin{pmatrix} \text{mass flux of molecules} \\ \text{from } y - \ell/2 \text{ to } y + \ell/2 \end{pmatrix} \cdot \begin{pmatrix} \text{change in fluid} \\ \text{velocity} \end{pmatrix}$$

The mass flux from top to bottom is proportional to $\rho\overline{C}$, where $\overline{C}$, the mean molecular speed of the stationary fluid, is $\gg u$ or v in incompressible flow. Thus,

$$\tau_{yx} = C_1 \left(\rho\overline{C}\right) \left(\ell \frac{du}{dy}\right) \frac{\text{N}}{\text{m}^2} \text{ and this also equals } \mu \frac{du}{dy} \tag{6.45}$$

By the same token,

$$q_y = C_2 \left(\rho c_v \overline{C}\right) \left(\ell \frac{dT}{dy}\right) \text{ and this also equals } -k \frac{dT}{dy}$$

where c_v is the specific heat at constant volume. The constants, C_1 and C_2, are on the order of one. It follows immediately that

$$\mu = C_1 \left(\rho\overline{C}\ell\right) \qquad \text{so} \qquad \nu = C_1 \left(\overline{C}\ell\right)$$

and

$$k = C_2 \left(\rho c_v \overline{C}\ell\right) \qquad \text{so} \qquad \alpha = C_2 \frac{\overline{C}\ell}{\gamma}$$

where $\gamma \equiv c_p/c_v$ is approximately a constant on the order of one for a given gas. Thus, for a gas,

$$\Pr \equiv \frac{\nu}{\alpha} = \text{ a constant on the order of one}$$

More detailed use of the kinetic theory of gases reveals more specific information as to the value of the Prandtl number, and these points are borne out reasonably well experimentally, as you can determine from Appendix A:

- For simple monatomic gases, $\Pr = \frac{2}{3}$.
- For diatomic gases in which vibration is unexcited (such as N_2 and O_2 at room temperature), $\Pr = \frac{5}{7}$.

- As the complexity of gas molecules increases, Pr approaches an upper value of one.
- Pr is least sensitive to temperature in gases made up of the simplest molecules because their structure is least responsive to temperature changes.

In a liquid, the physical mechanisms of molecular momentum and energy transport are much more complicated and Pr can be far from one. For example (cf. Table A.3):

- For liquids composed of fairly simple molecules, excluding metals, Pr is of the order of magnitude of 1 to 10.
- For liquid metals, Pr is of the order of magnitude of 10^{-2} or less.
- If the molecular structure of a liquid is very complex, Pr might reach values on the order of 10^5. This is true of oils made of long-chain hydrocarbons, for example.

Thus, while Pr can vary over almost eight orders of magnitude in common fluids, it is still the result of analogous mechanisms of heat and momentum transfer. The numerical values of Pr, as well as the analogy itself, have their origins in similar molecular transport processes.

Boundary layer thicknesses, δ and δ_t, and the Prandtl number

We have seen that the exact solution of the b.l. equations gives $\delta = \delta_t$ for Pr = 1, and it gives dimensionless velocity and temperature profiles that are identical on a flat surface. Also:

- When Pr > 1, $\delta > \delta_t$, and when Pr < 1, $\delta < \delta_t$. This is true because high viscosity leads to a thick velocity b.l., and a high thermal diffusivity should give a thick thermal b.l.
- Since the exact governing equations (6.41) and (6.42) are identical for either b.l., except for the appearance of α in one and ν in the other, we expect that

$$\frac{\delta_t}{\delta} = \text{fn}\left(\frac{\nu}{\alpha} \text{ only}\right)$$

Therefore, we can combine these two observations, defining $\delta_t/\delta \equiv \phi$, and get

$$\phi = \text{monotonically decreasing function of Pr only} \tag{6.46}$$

The exact solution of the thermal b.l. equations proves this to be precisely true.

The fact that ϕ is independent of x will greatly simplify the use of the integral method. We shall establish the correct form of eqn. (6.46) in the following section.

6.5 Heat transfer coefficient for laminar, incompressible flow over a flat surface

The integral method for solving the energy equation

Integrating the b.l. energy equation in the same way as the momentum equation gives

$$\int_0^{\delta_t} u\frac{\partial T}{\partial x}\,dy + \int_0^{\delta_t} v\frac{\partial T}{\partial y}\,dy = \alpha\int_0^{\delta_t}\frac{\partial^2 T}{\partial y^2}\,dy$$

And the chain rule of differentiation in the form $x dy \equiv dxy - y dx$, reduces this to

$$\int_0^{\delta_t}\frac{\partial uT}{\partial x}\,dy - \int_0^{\delta_t} T\frac{\partial u}{\partial x}\,dy + \int_0^{\delta_t}\frac{\partial vT}{\partial y}\,dy - \int_0^{\delta_t} T\frac{\partial v}{\partial y}\,dy = \alpha\left.\frac{\partial T}{\partial y}\right|_0^{\delta_t}$$

or

$$\int_0^{\delta_t}\frac{\partial uT}{\partial x}\,dy + \underbrace{\left. vT\right|_0^{\delta_t}}_{=T_\infty v|_{y=\delta_t}-0} - \int_0^{\delta_t} T\underbrace{\left(\frac{\partial u}{\partial x}+\frac{\partial v}{\partial y}\right)}_{=\,0,\text{ eqn. (6.11)}}dy$$

$$= \alpha\left[\underbrace{\left.\frac{\partial T}{\partial y}\right|_{\delta_t}}_{=0} - \left.\frac{\partial T}{\partial y}\right|_0\right]$$

We evaluate v at $y = \delta_t$, using the continuity equation in the form of eqn. (6.23), in the preceeding expression:

$$\int_0^{\delta_t}\frac{\partial}{\partial x}u(T-T_\infty)\,dy = \frac{1}{\rho c_p}\left(-k\left.\frac{\partial T}{\partial y}\right|_0\right) = \text{fn}(x\text{ only})$$

or

$$\boxed{\frac{d}{dx}\int_0^{\delta_t} u(T-T_\infty)\,dy = \frac{q_w}{\rho c_p}} \tag{6.47}$$

Equation (6.47) expresses the conservation of thermal energy in integrated form. It shows the rate thermal energy is carried away by the b.l. flow being matched by the rate heat is transferred in at the wall.

Predicting the temperature distribution in the laminar thermal boundary layer

We can continue to paraphrase the development of the velocity profile in the laminar b.l., from the preceding section. We previously guessed the velocity profile in such a way as to make it match what we know to be true. We also know certain things to be true of the temperature profile. The temperatures at the wall and at the outer edge of the b.l. are known. Furthermore, the temperature distribution should be smooth as it blends into T_∞ for $y > \delta_t$. This condition is imposed by setting $\partial T/\partial y$ equal to zero at $y = \delta_t$. A fourth condition is obtained by writing eqn. (6.40) at the wall, where $u = v = 0$. This gives $(\partial^2 T/\partial y^2)_{y=0} = 0$. These four conditions take the following dimensionless form:

$$\left.\begin{aligned}
\frac{T - T_\infty}{T_w - T_\infty} &= 1 \quad \text{at } y/\delta_t = 0 \\
\frac{T - T_\infty}{T_w - T_\infty} &= 0 \quad \text{at } y/\delta_t = 1 \\
\frac{\partial[(T - T_\infty)/(T_w - T_\infty)]}{\partial(y/\delta_t)} &= 0 \quad \text{at } y/\delta_t = 1 \\
\frac{\partial^2[(T - T_\infty)/(T_w - T_\infty)]}{\partial(y/\delta_t)^2} &= 0 \quad \text{at } y/\delta_t = 0
\end{aligned}\right\} \tag{6.48}$$

Equations (6.48) provide enough information to approximate the temperature profile with a cubic function.

$$\frac{T - T_\infty}{T_w - T_\infty} = a + b\frac{y}{\delta_t} + c\left(\frac{y}{\delta_t}\right)^2 + d\left(\frac{y}{\delta_t}\right)^3 \tag{6.49}$$

Substituting eqn. (6.49) into eqns. (6.48), we get

$$a = 1 \qquad -1 = b + c + d \qquad 0 = b + 2c + 3d \qquad 0 = 2c$$

which gives

$$a = 1 \qquad b = -\tfrac{3}{2} \qquad c = 0 \qquad d = \tfrac{1}{2}$$

so the temperature profile is

$$\boxed{\frac{T - T_\infty}{T_w - T_\infty} = 1 - \frac{3}{2}\frac{y}{\delta_t} + \frac{1}{2}\left(\frac{y}{\delta_t}\right)^3} \tag{6.50}$$

Predicting the heat flux in the laminar boundary layer

Equation (6.47) contains an as-yet-unknown quantity—the thermal b.l. thickness, δ_t. To calculate δ_t, we substitute the temperature profile, eqn. (6.50), and the velocity profile, eqn. (6.29), in the integral form of the energy equation, (6.47), which we first express as

$$u_\infty(T_w - T_\infty)\frac{d}{dx}\left[\delta_t \int_0^1 \frac{u}{u_\infty}\left(\frac{T - T_\infty}{T_w - T_\infty}\right) d\left(\frac{y}{\delta_t}\right)\right]$$

$$= -\frac{\alpha(T_w - T_\infty)}{\delta_t} \left.\frac{d\left(\dfrac{T - T_\infty}{T_w - T_\infty}\right)}{d(y/\delta_t)}\right|_{y/\delta_t=0} \tag{6.51}$$

This will work fine as long as $\delta_t < \delta$. But, if $\delta_t > \delta$, the velocity will be given by $u/u_\infty = 1$, instead of eqn. (6.29), beyond $y = \delta$. Let us proceed for the moment in the hope that the requirement that $\delta_t \leqslant \delta$ will be satisfied. Introducing $\phi \equiv \delta_t/\delta$ in eqn. (6.51) and calling $y/\delta_t \equiv \eta$, we get

$$\delta_t \frac{d}{dx}\left[\delta_t \underbrace{\int_0^1 \left(\frac{3}{2}\eta\phi - \frac{1}{2}\eta^3\phi^3\right)\left(1 - \frac{3}{2}\eta + \frac{1}{2}\eta^3\right) d\eta}_{=\frac{3}{20}\phi - \frac{3}{280}\phi^3}\right] = \frac{3\alpha}{2u_\infty} \tag{6.52}$$

Since ϕ is a constant for any Pr [recall eqn. (6.46)], we separate variables:

$$2\delta_t \frac{d\delta_t}{dx} = \frac{d\delta_t^2}{dx} = \frac{3\alpha/u_\infty}{\left(\dfrac{3}{20}\phi - \dfrac{3}{280}\phi^3\right)}$$

Integrating this result with respect to x and taking $\delta_t = 0$ at $x = 0$, we get

$$\delta_t = \sqrt{\frac{3\alpha x}{u_\infty}} \bigg/ \sqrt{\frac{3}{20}\phi - \frac{3}{280}\phi^3} \tag{6.53}$$

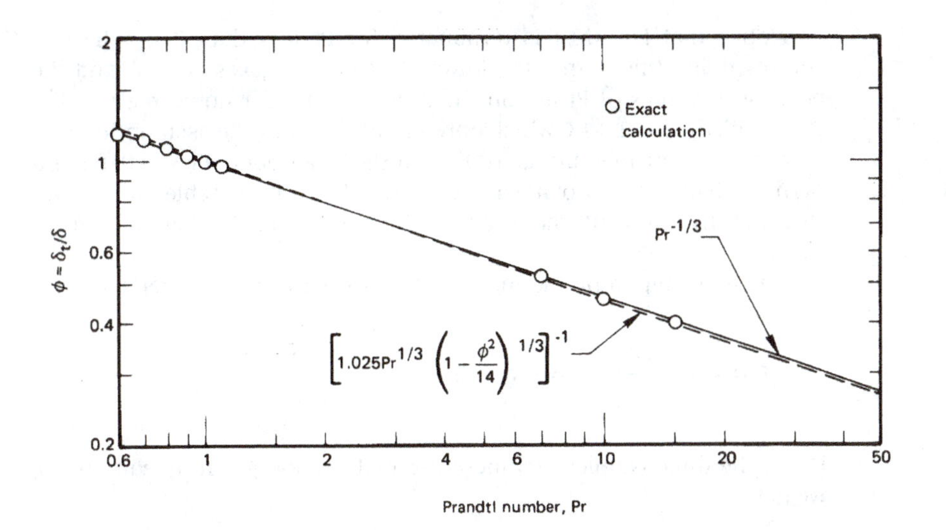

Figure 6.14 The exact and approximate Prandtl number influence on the ratio of b.l. thicknesses.

But $\delta = 4.64x/\sqrt{\mathrm{Re}_x}$ in the integral formulation [eqn. (6.31b)]. We divide by this value of δ to be consistent and obtain

$$\frac{\delta_t}{\delta} \equiv \phi = 0.9638 \Big/ \sqrt{\mathrm{Pr}\,\phi\,(1-\phi^2/14)}$$

Rearranging this gives

$$\frac{\delta_t}{\delta} = \frac{1}{1.025\,\mathrm{Pr}^{1/3}\left[1-(\delta_t^2/14\delta^2)\right]^{1/3}} \simeq \frac{1}{1.025\,\mathrm{Pr}^{1/3}} \tag{6.54}$$

The unapproximated result above is shown in Fig. 6.14, along with the results of Pohlhausen's precise calculation (see Schlichting [6.3, Chap. 14]). It turns out that the exact ratio, δ/δ_t, is represented with great accuracy by

$$\boxed{\frac{\delta_t}{\delta} = \mathrm{Pr}^{-1/3}} \qquad 0.6 \leqslant \mathrm{Pr} \leqslant 50 \tag{6.55}$$

So the integral method is accurate within 2.5% in the Prandtl number range indicated.

Notice that Fig. 6.14 is terminated for Pr less than 0.6. The reason for doing this is that the lowest Pr for pure gases is 0.67, and the next lower values of Pr are on the order of 10^{-2} for liquid metals. For Pr = 0.67, $\delta_t/\delta = 1.143$, which only slightly violates the assumption that $\delta_t \leqslant \delta$. For, say, mercury at 100°C, on the other hand, Pr = 0.0162 and $\delta_t/\delta = 3.952$, which violates the condition by an intolerable margin. We therefore have a result that is acceptable for gases and all liquids except the metallic ones.

The final step in predicting the heat flux is to write Fourier's law:

$$q = -k\left.\frac{\partial T}{\partial y}\right|_{y=0} = -k\frac{T_w - T_\infty}{\delta_t}\left.\frac{\partial\left(\dfrac{T - T_\infty}{T_w - T_\infty}\right)}{\partial(y/\delta_t)}\right|_{y/\delta_t=0} \tag{6.56}$$

Using the dimensionless temperature distribution given by eqn. (6.50), we get

$$q = +k\frac{T_w - T_\infty}{\delta_t}\frac{3}{2}$$

or

$$h \equiv \frac{q}{\Delta T} = \frac{3k}{2\delta_t} = \frac{3}{2}\frac{k}{\delta}\frac{\delta}{\delta_t} \tag{6.57}$$

and substituting eqns. (6.54) and (6.31b) for δ/δ_t and δ, we obtain

$$\mathrm{Nu}_x \equiv \frac{hx}{k} = \frac{3}{2}\frac{\sqrt{\mathrm{Re}_x}}{4.64}\,1.025\,\mathrm{Pr}^{1/3} = 0.3314\,\mathrm{Re}_x^{1/2}\,\mathrm{Pr}^{1/3}$$

Considering the various approximations, this is very close to the result of the exact calculation, which turns out to be

$$\boxed{\mathrm{Nu}_x = 0.332\,\mathrm{Re}_x^{1/2}\,\mathrm{Pr}^{1/3}} \qquad 0.6 \leqslant \mathrm{Pr} \leqslant 50 \tag{6.58}$$

This expression gives very accurate results under the assumptions on which it is based, namely a laminar two-dimensional b.l. on a flat surface, with T_w = constant and $0.6 \leqslant \mathrm{Pr} \leqslant 50$.

Some other laminar boundary layer heat transfer equations

High Pr. At high Pr, eqn. (6.58) is still close to correct. The exact solution is

$$\mathrm{Nu}_x \longrightarrow 0.339\,\mathrm{Re}_x^{1/2}\,\mathrm{Pr}^{1/3}, \quad \mathrm{Pr} \longrightarrow \infty \tag{6.59}$$

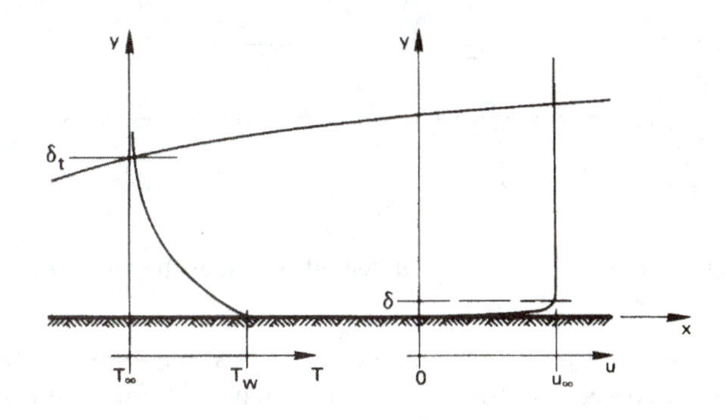

Figure 6.15 A laminar b.l. in a low-Pr liquid. The velocity b.l. is so thin that $u \simeq u_\infty$ in the thermal b.l.

Low Pr. Figure 6.15 shows a low-Pr liquid flowing over a flat plate. In this case $\delta_t \gg \delta$, and for all practical purposes $u = u_\infty$ everywhere within the thermal b.l. It is as though the no-slip condition $[u(y=0)=0]$ and the influence of viscosity were removed from the problem. Thus, the dimensional functional equation for h becomes

$$h = \text{fn}\left(x, k, \rho c_p, u_\infty\right) \tag{6.60}$$

There are five variables in J/K, m, and s, so there are only two pi-groups. They are

$$\text{Nu}_x = \frac{hx}{k} \quad \text{and} \quad \Pi_2 \equiv \text{Re}_x\text{Pr} = \frac{u_\infty x}{\alpha}$$

The new group, Π_2, is called a *Péclét number*, Pe_x, where the subscript identifies the length upon which it is based. It can be interpreted as follows:

$$\text{Pe}_x \equiv \frac{u_\infty x}{\alpha} = \frac{\rho c_p u_\infty \Delta T}{k\Delta T} = \frac{\text{heat capacity rate of fluid in the b.l.}}{\text{axial heat conductance of the b.l.}} \tag{6.61}$$

So long as Pe_x is large, the b.l. assumption that $\partial^2 T/\partial x^2 \ll \partial^2 T/\partial y^2$ will be valid; but for small Pe_x (i.e., $\text{Pe}_x \ll 100$), it will be violated and a boundary layer solution cannot be used.

The exact solution of the b.l. equations gives, in this case:

$$\text{Nu}_x = 0.565\,\text{Pe}_x^{1/2} \qquad \begin{cases} \text{Pe}_x \geq 100 & \text{and} \\ \text{Pr} \lesssim \frac{1}{100} & \text{or} \\ \text{Re}_x \geq 10^4 \end{cases} \tag{6.62}$$

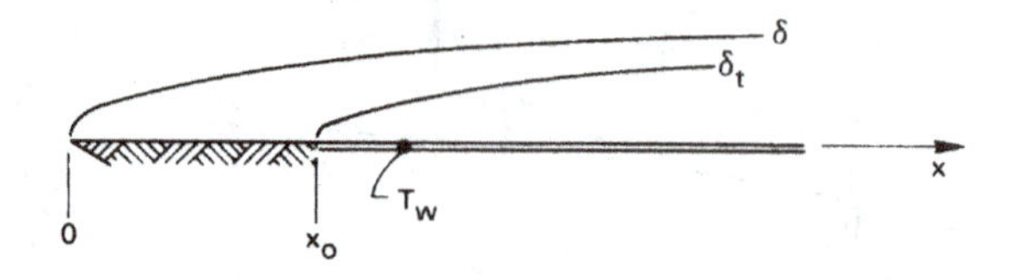

Figure 6.16 A b.l. with an unheated region at the leading edge.

General relationship. Churchill and Ozoe [6.5] recommend the following empirical correlation for laminar flow on a constant-temperature flat surface for the entire range of Pr:

$$\boxed{\mathrm{Nu}_x = \frac{0.3387\,\mathrm{Re}_x^{1/2}\,\mathrm{Pr}^{1/3}}{\left[1 + (0.0468/\mathrm{Pr})^{2/3}\right]^{1/4}}} \qquad \mathrm{Pe}_x > 100 \tag{6.63}$$

This relationship proves to be quite accurate, and it approximates eqns. (6.59) and (6.62), respectively, in the high- and low-Pr limits. The calculations of an average Nusselt number for the general case is left as an exercise (Problem 6.10).

Boundary layer with an unheated starting length Figure 6.16 shows a b.l. with a heated region that starts at a distance x_0 from the leading edge. The heat transfer in this instance is easily obtained using integral methods (see Prob. 6.41).

$$\mathrm{Nu}_x = \frac{0.332\,\mathrm{Re}_x^{1/2}\,\mathrm{Pr}^{1/3}}{\left[1 - (x_0/x)^{3/4}\right]^{1/3}}, \qquad x > x_0 \tag{6.64}$$

Average heat transfer coefficient, $\overline{h}$. The heat transfer coefficient h, is the ratio of two quantities, q and ΔT, either of which might vary with x. So far, we have only dealt with the *uniform wall temperature problem.* Equations (6.58), (6.59), (6.62), and (6.63), for example, can all be used to calculate $q(x)$ when $(T_w - T_\infty) \equiv \Delta T$ is a specified constant. In the next subsection, we discuss the problem of predicting $[T(x) - T_\infty]$ when q is a specified constant. That is called the *uniform wall heat flux problem.*

The term $\overline{h}$ is used to designate either $\overline{q}/\Delta T$ in the uniform wall temperature problem or $q/\overline{\Delta T}$ in the uniform wall heat flux problem. Thus,

$$\text{uniform wall temp.:}\quad \overline{h} \equiv \frac{\overline{q}}{\Delta T} = \frac{1}{\Delta T}\left[\frac{1}{L}\int_0^L q\,dx\right] = \frac{1}{L}\int_0^L h(x)\,dx \tag{6.65}$$

$$\text{uniform heat flux:}\quad \overline{h} \equiv \frac{q}{\overline{\Delta T}} = \frac{q}{\dfrac{1}{L}\displaystyle\int_0^L \Delta T(x)\,dx} \tag{6.66}$$

The Nusselt number based on $\overline{h}$ and a characteristic length, L, is designated $\overline{\mathrm{Nu}}_L$. This is not to be construed as an average of Nu_x, which would be meaningless in either of these cases.

Thus, for a flat surface (with $x_0 = 0$), we use eqn. (6.58) in eqn. (6.65) to get

$$\overline{h} = \frac{1}{L}\int_0^L \underbrace{h(x)\,dx}_{\frac{k}{x}\mathrm{Nu}_x} = \frac{0.332\,k\,\mathrm{Pr}^{1/3}}{L}\sqrt{\frac{u_\infty}{\nu}}\int_0^L \frac{\sqrt{x}\,dx}{x}$$

$$= 0.664\,\mathrm{Re}_L^{1/2}\,\mathrm{Pr}^{1/3}\left(\frac{k}{L}\right) \tag{6.67}$$

Thus, $\overline{h} = 2h(x = L)$ in a laminar flow, and

$$\boxed{\overline{\mathrm{Nu}}_L = \frac{\overline{h}L}{k} = 0.664\,\mathrm{Re}_L^{1/2}\,\mathrm{Pr}^{1/3}} \tag{6.68}$$

Likewise for liquid metal flows:

$$\overline{\mathrm{Nu}}_L = 1.13\,\mathrm{Pe}_L^{1/2} \tag{6.69}$$

Some final observations. The preceding results are restricted to the two-dimensional, incompressible, laminar b.l. on a flat isothermal wall at velocities that are not too high. These conditions are usually met if:

- Re_x or Re_L is not above the turbulent transition value, which is typically a few hundred thousand.
- The Mach number of the flow, $\mathrm{Ma} \equiv u_\infty/(\text{sound speed})$, is less than about 0.3. (Even gaseous flows behave incompressibly at velocities well below sonic.) A related condition is:

- The *Eckert number*, $\text{Ec} \equiv u_\infty^2/c_p(T_w - T_\infty)$, is substantially less than one. (This means that heating by viscous dissipation—which we have neglected—does not play any role in the problem. We made this assumption implicitly when we treated J as an independent unit in the dimensional analysis of this problem.)

And we note how h and Nu depend on their independent variables:

$$\begin{aligned} h \text{ or } \overline{h} \propto \frac{1}{\sqrt{x}} \text{ or } \frac{1}{\sqrt{L}}, &\quad \sqrt{u_\infty},\ \nu^{-1/6},\ (\rho c_p)^{1/3},\ k^{2/3} \\ \text{Nu}_x \text{ or } \overline{\text{Nu}}_L \propto \sqrt{x} \text{ or } L, &\quad \sqrt{u_\infty},\ \nu^{-1/6},\ (\rho c_p)^{1/3},\ k^{-1/3} \end{aligned} \tag{6.70}$$

Thus, $h \longrightarrow \infty$ and Nu_x vanishes at the leading edge, $x = 0$. Of course, an infinite value of h, like infinite shear stress, will not really occur at the leading edge because the b.l. description will actually break down in a small neighborhood of $x = 0$.

In all of the preceding considerations, we have assumed the fluid properties to be constant. Actually, k, ρc_p, and especially μ might all vary noticeably with T within the b.l. It turns out that if properties are all evaluated at the average temperature of the b.l. or film temperature $T_f = (T_w + T_\infty)/2$, the results will normally be quite accurate. It is also worth noting that, although properties are given only at one pressure in Appendix A, μ, k, and c_p change very little with pressure, especially in liquids.

Example 6.5

Air at 20°C and moving at 15 m/s is warmed by an isothermal steam-heated plate at 110°C, ½ m in length and ½ m in width. Find the average heat transfer coefficient and the total heat transferred. What are h, δ_t, and δ at the trailing edge?

SOLUTION. We evaluate properties at $T_f = (110+20)/2 = 65$°C. Then

$$\text{Pr} = 0.707 \qquad \text{and} \qquad \text{Re}_L = \frac{u_\infty L}{\nu} = \frac{15(0.5)}{0.0000194} = 386{,}600$$

so the flow ought to be laminar up to the trailing edge. The Nusselt number is then

$$\overline{\text{Nu}}_L = 0.664\ \text{Re}_L^{1/2}\ \text{Pr}^{1/3} = 367.8$$

and

$$\overline{h} = 367.8\frac{k}{L} = \frac{367.8(0.02885)}{0.5} = 21.2 \text{ W/m}^2\text{K}$$

The value is quite low because of the low conductivity of air. The total heat flux is then

$$Q = \overline{h}A\,\Delta T = 21.2(0.5)^2(110-20) = 477 \text{ W}$$

By comparing eqns. (6.58) and (6.68), we see that $h(x = L) = \frac{1}{2}\overline{h}$, so

$$h(\text{trailing edge}) = \tfrac{1}{2}(21.2) = 10.6 \text{ W/m}^2\text{K}$$

And finally,

$$\delta(x = L) = 4.92L/\sqrt{\text{Re}_L} = \frac{4.92(0.5)}{\sqrt{386,600}} = 0.00396 \text{ m}$$
$$= 3.96 \text{ mm}$$

and

$$\delta_t = \frac{\delta}{\sqrt[3]{\text{Pr}}} = \frac{3.96}{\sqrt[3]{0.707}} = 4.44 \text{ mm}$$ ■

The problem of uniform wall heat flux

When the heat flux at the heater wall, q_w, is specified instead of the temperature, it is T_w that we need to know. We leave the problem of finding Nu_x for q_w = constant as an exercise (Problem 6.11). The exact result is

$$\boxed{\text{Nu}_x = 0.453\,\text{Re}_x^{1/2}\,\text{Pr}^{1/3}} \qquad \text{for} \quad \text{Pr} \geqslant 0.6 \tag{6.71}$$

where $\text{Nu}_x = hx/k = q_w x/k(T_w - T_\infty)$. The integral method gives the same result with a slightly lower constant (0.417).

We must be very careful in discussing *average* results in the constant heat flux case. The problem now might be that of finding an average temperature difference [cf. eqn. (6.66)]:

$$\overline{T_w - T_\infty} = \frac{1}{L}\int_0^L (T_w - T_\infty)\,dx = \frac{1}{L}\int_0^L \frac{q_w x}{k(0.453\sqrt{u_\infty/\nu}\,\text{Pr}^{1/3})}\frac{dx}{\sqrt{x}}$$

or

$$\overline{T_w - T_\infty} = \frac{q_w L/k}{0.6795\,\text{Re}_L^{1/2}\,\text{Pr}^{1/3}} \tag{6.72}$$

which can be put into the form $\overline{\mathrm{Nu}}_L = \overline{h}L/k = 0.6795\ \mathrm{Re}_L^{1/2}\mathrm{Pr}^{1/3}$ for $\overline{h} = q_w/(\overline{T_w - T_\infty})$.

Churchill and Ozoe [6.5] have pointed out that their eqn. (6.63) will describe $(T_w - T_\infty)$ with high accuracy over the full range of Pr if the constants are changed as follows:

$$\mathrm{Nu}_x = \frac{0.4637\ \mathrm{Re}_x^{1/2}\ \mathrm{Pr}^{1/3}}{\left[1 + (0.02052/\mathrm{Pr})^{2/3}\right]^{1/4}} \qquad \mathrm{Pe}_x > 100 \qquad (6.73)$$

Example 6.6

Air at 15°C flows at 1.8 m/s over a 0.6 m-long heating panel. The panel is intended to supply 420 W/m^2 to the air, but the surface can sustain only about 105°C without being damaged. Is it safe? What is the *average* temperature of the plate?

Solution. In accordance with eqn. (6.71),

$$\Delta T_{\mathrm{max}} = \Delta T_{x=L} = \frac{qL}{k\,\mathrm{Nu}_{x=L}} = \frac{qL/k}{0.453\,\mathrm{Re}_x^{1/2}\ \mathrm{Pr}^{1/3}}$$

or if we evaluate properties at $(85 + 15)/2 = 50$°C, for the moment,

$$\Delta T_{\mathrm{max}} = \frac{420(0.6)/0.0278}{0.453\left[0.6(1.8)/1.794\times10^{-5}\right]^{1/2}(0.709)^{1/3}} = 91.5°\mathrm{C}$$

This will give $T_{w_{\mathrm{max}}} = 15 + 91.5 = 106.5$°C. This is very close to 105°C. If 105°C is at all conservative, $q = 420$ W/m^2 should be safe—particularly since it only occurs over a very small distance at the end of the plate.

From eqn. (6.72) we find that

$$\overline{\Delta T} = \frac{0.453}{0.6795}\Delta T_{\mathrm{max}} = 61.0°\mathrm{C}$$

so

$$\overline{T_w} = 15 + 61.0 = 76.0°\mathrm{C} \qquad \blacksquare$$

6.6 The Reynolds analogy

The analogy between heat and momentum transfer can now be generalized to provide a very useful result. We begin with eqn. (6.25), which is restricted to a flat surface with no pressure gradient:

$$\frac{d}{dx}\left[\delta\int_0^1 \frac{u}{u_\infty}\left(\frac{u}{u_\infty}-1\right)d\left(\frac{y}{\delta}\right)\right] = -\frac{C_f}{2} \tag{6.25}$$

and we rewrite eqns. (6.47) and (6.51) to obtain for the constant wall temperature case:

$$\frac{d}{dx}\left[\phi\,\delta\int_0^1 \frac{u}{u_\infty}\left(\frac{T-T_\infty}{T_w-T_\infty}\right)d\left(\frac{y}{\delta_t}\right)\right] = \frac{q_w}{\rho c_p u_\infty (T_w-T_\infty)} \tag{6.74}$$

But the similarity of temperature and flow boundary layers to one another [see, e.g., eqns. (6.29) and (6.50)], suggests the following approximation, which becomes exact only when Pr = 1:

$$\frac{T-T_\infty}{T_w-T_\infty}\,\delta = \left(1-\frac{u}{u_\infty}\right)\delta_t$$

Substituting this result in eqn. (6.74) and comparing it to eqn. (6.25), we get

$$-\frac{d}{dx}\left[\delta\int_0^1 \frac{u}{u_\infty}\left(\frac{u}{u_\infty}-1\right)d\left(\frac{y}{\delta}\right)\right] = -\frac{C_f}{2} = -\frac{q_w}{\rho c_p u_\infty (T_w-T_\infty)\phi^2} \tag{6.75}$$

Finally, we substitute eqn. (6.55) to eliminate ϕ from eqn. (6.75). The result is one instance of the *Reynolds-Colburn analogy*:[8]

$$\frac{h}{\rho c_p u_\infty}\,\mathrm{Pr}^{2/3} = \frac{C_f}{2} \tag{6.76}$$

For use in Reynolds' analogy, C_f must be a pure *skin* friction coefficient. The profile drag that results from the variation of pressure around the body is unrelated to heat transfer. The analogy does not apply when profile drag is included in C_f.

The dimensionless group $h/\rho c_p u_\infty$ is called the *Stanton number.*, defined as follows:

$$\text{St, Stanton number} \equiv \frac{h}{\rho c_p u_\infty} = \frac{\mathrm{Nu}_x}{\mathrm{Re}_x\mathrm{Pr}}$$

The physical significance of the Stanton number is

$$\mathrm{St} = \frac{h\Delta T}{\rho c_p u_\infty \Delta T} = \frac{\text{actual heat flux to the fluid}}{\text{heat flux capacity of the fluid flow}} \tag{6.77}$$

[8]Reynolds [6.6] developed the analogy in 1874. Colburn made important use of it in this century. The form given is for flat plates with $0.6 \le \mathrm{Pr} \le 50$. The Prandtl number factor is usually a little different for other flows or other ranges of Pr.

The group St $\mathrm{Pr}^{2/3}$ was dealt with by the chemical engineer Colburn, who gave it a special symbol:

$$j \equiv \text{Colburn } j\text{-factor} = \mathrm{St}\,\mathrm{Pr}^{2/3} = \frac{\mathrm{Nu}_x}{\mathrm{Re}_x \mathrm{Pr}^{1/3}} \tag{6.78}$$

Example 6.7

Does the equation for the Nusselt number on an isothermal flat surface in laminar flow satisfy the Reynolds analogy?

Solution. If we rewrite eqn. (6.58), we obtain

$$\frac{\mathrm{Nu}_x}{\mathrm{Re}_x \mathrm{Pr}^{1/3}} = \mathrm{St}\,\mathrm{Pr}^{2/3} = \frac{0.332}{\sqrt{\mathrm{Re}_x}} \tag{6.79}$$

But comparison with eqn. (6.33) reveals that the left-hand side of eqn. (6.79) is precisely $C_f/2$, so the analogy is satisfied perfectly. Likewise, from eqns. (6.68) and (6.34), we get

$$\frac{\overline{\mathrm{Nu}}_L}{\mathrm{Re}_L \mathrm{Pr}^{1/3}} \equiv \overline{\mathrm{St}}\,\mathrm{Pr}^{2/3} = \frac{0.664}{\sqrt{\mathrm{Re}_L}} = \frac{\overline{C}_f}{2} \tag{6.80}$$

■

The Reynolds-Colburn analogy can be used directly to infer heat transfer data from measurements of the shear stress, or vice versa. It can also be extended to turbulent flow, which is much harder to predict analytically. We undertake that problem in Sect. 6.8.

Example 6.8

How much drag force does the air flow in Example 6.5 exert on the heat transfer surface?

Solution. From eqn. (6.80) in Example 6.7, we obtain

$$\overline{C}_f = \frac{2\,\overline{\mathrm{Nu}}_L}{\mathrm{Re}_L\,\mathrm{Pr}^{1/3}}$$

From Example 6.5 we obtain $\overline{\mathrm{Nu}}_L$, Re_L, and $\mathrm{Pr}^{1/3}$:

$$\overline{C}_f = \frac{2(367.8)}{(386{,}600)(0.707)^{1/3}} = 0.002135$$

so

$$\overline{\tau_{yx}} = (0.002135)\,\frac{1}{2}\rho u_\infty^2 = \frac{(0.002135)(1.05)(15)^2}{2}$$
$$= 0.2522 \text{ kg/m·s}^2$$

and the force is

$$\overline{\tau_{yx}}A = 0.2522(0.5)^2 = 0.06305 \text{ kg·m/s}^2 = 0.06305 \text{ N}$$
$$= 0.23 \text{ oz}$$ ■

6.7 Turbulent boundary layers

Turbulence

> Big whirls have little whirls,
> That feed on their velocity.
> Little whirls have littler whirls,
> And so on, to viscosity.

This bit of doggerel by the British fluid mechanic, L. F. Richardson, tells us a great deal about the nature of turbulence. Turbulence in a fluid can be viewed as a spectrum of coexisting vortices in which kinetic energy from the larger ones is dissipated to successively smaller ones until the very smallest of these vortices (or "whirls") are damped out by viscous shear stresses.

Notice the cloud patterns when the weatherman shows a satellite photograph of North America on the evening news. One or two enormous vortices have continental proportions. They in turn feed smaller "weather-making" vortices hundreds of kilometers in diameter. These further dissipate into vortices of cyclone and tornado proportions—sometimes with that level of violence but more often not. These dissipate into still smaller whirls as they interact with the ground and its various protrusions. The next time the wind blows, stand behind any tree and *feel* the vortices. In the great plains, where there are few vortex generators (such as trees), one sees small cyclonic eddies called "dust devils." The process continues right on down to millimeter or even micrometer scales. There, momentum exchange is no longer identifiable as turbulence but appears simply as viscous stretching of the fluid.

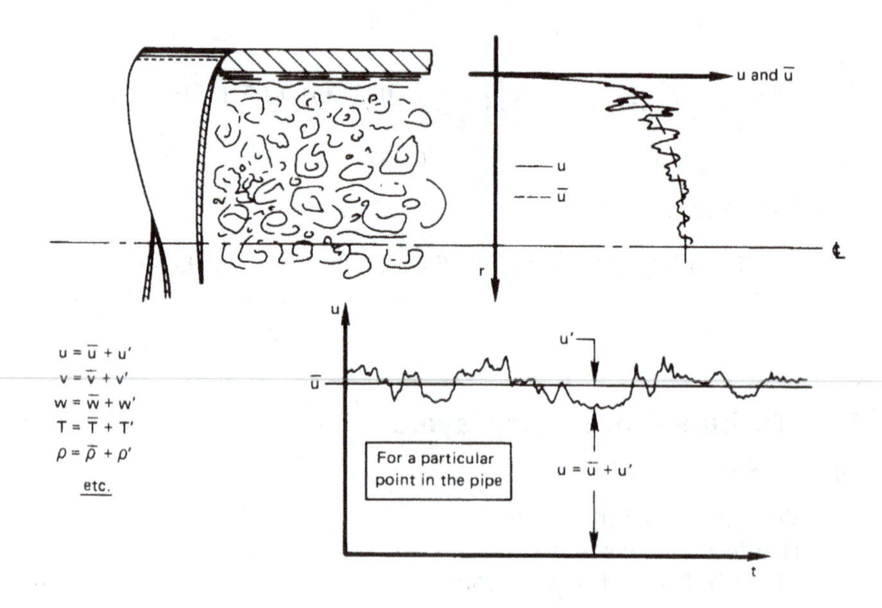

Figure 6.17 Fluctuation of u and other quantities in a turbulent pipe flow.

The same kind of process exists within, say, a turbulent pipe flow at high Reynolds number. Such a flow is shown in Fig. 6.17. Turbulence in such a case consists of coexisting vortices which vary in size from a substantial fraction of the pipe radius down to micrometer dimensions. The spectrum of sizes varies with location in the pipe. The size and intensity of vortices at the wall must clearly approach zero, since the fluid velocity goes to zero at the wall.

Figure 6.17 shows the fluctuation of a typical flow variable—namely, velocity—both with location in the pipe and with time. This fluctuation arises because of the turbulent motions that are superposed on the average local flow. Other flow variables, such as T or ρ, can vary in the same manner. For any variable we can write a local time-average value as

$$\overline{u} \equiv \frac{1}{\mathrm{T}} \int_0^{\mathrm{T}} u \, dt \tag{6.81}$$

where T is a time that is much longer than the period of typical fluctuations.[9] Equation (6.81) is most useful for so-called *stationary processes*—ones for which $\overline{u}$ is nearly time-independent.

If we substitute $u = \overline{u} + u'$ in eqn. (6.81), where u is the actual local velocity and u' is the instantaneous magnitude of the fluctuation, we obtain

$$\overline{u} = \underbrace{\frac{1}{\mathbf{T}}\int_0^{\mathbf{T}} \overline{u}\,dt}_{=\overline{u}} + \underbrace{\frac{1}{\mathbf{T}}\int_0^{\mathbf{T}} u'\,dt}_{=\overline{u'}} \tag{6.82}$$

This is consistent with the fact that

$$\overline{u'} \text{ or any other average fluctuation} = 0 \tag{6.83}$$

since the fluctuations are defined as deviations from the average.

We now want to create a measure of the size, or *lengthscale*, of turbulent vortices. This might be done experimentally by placing two velocity-measuring devices very close to one another in a turbulent flow field. When the probes are close, their measurements will be very highly correlated with one one another. Then, suppose that the two velocity probes are moved apart until the measurements first become unrelated to one another. That spacing gives an indication of the average size of the turbulent motions.

Prandtl invented a slightly different (although related) measure of the lengthscale of turbulence, called the *mixing length*, ℓ. He saw ℓ as an average distance that a parcel of fluid moves between interactions. It has a physical significance similar to that of the molecular mean free path. It is harder to devise a clean experimental measure of ℓ than of the correlation lengthscale of turbulence. But we can still use the concept of ℓ to examine the notion of a turbulent shear stress.

The shear stresses of turbulence arise from the same kind of momentum exchange process that gives rise to the molecular viscosity. Recall that, in the latter case, a kinetic calculation gave eqn. (6.45) for the laminar shear stress

$$\tau_{yx} = (\text{constant})\left(\rho\overline{C}\right)\underbrace{\left(\ell\frac{\partial u}{\partial y}\right)}_{=u'} \tag{6.45}$$

where ℓ was the molecular mean free path and u' was the velocity difference for a molecule that had travelled a distance ℓ in the mean velocity

[9]Take care not to interpret this **T** as the thermal time constant that we introduced in Chapter 1; we denote time constants with italics as T.

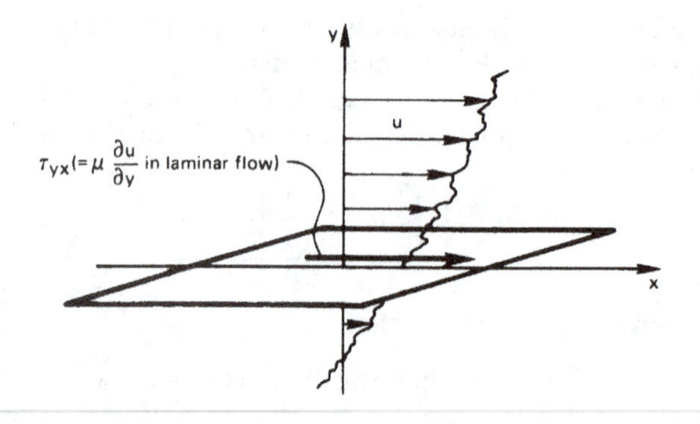

Figure 6.18 The shear stress, τ_{yx}, in a laminar or turbulent flow.

gradient. In the turbulent flow case, pictured in Fig. 6.18, we can think of Prandtl's parcels of fluid (rather than individual molecules) as carrying the x-momentum. Let us rewrite eqn. (6.45) in the following way:

- The shear stress τ_{yx} becomes a fluctuation in shear stress, τ'_{yx}, resulting from the turbulent movement of a parcel of fluid
- ℓ changes from the mean free path to the mixing length
- $\overline{C}$ is replaced by $v = \overline{v} + v'$, the instantaneous vertical speed of the fluid parcel
- The velocity fluctuation, u', is for a fluid parcel that moves a distance ℓ through the mean velocity gradient, $\partial\overline{u}/\partial y$. It is given by $\ell(\partial\overline{u}/\partial y)$.

Then

$$\tau'_{yx} = (\text{constant})\left[\rho\,(\overline{v} + v')\right] u' \tag{6.84}$$

Equation (6.84) can also be derived formally and precisely with the help of the Navier-Stokes equation. When this is done, the constant comes out equal to -1. The average of the fluctuating shear stress is

$$\overline{\tau'_{yx}} = -\frac{\rho}{\mathrm{T}}\int_0^{\mathrm{T}} (\overline{v}u' + v'u')\,dt = -\rho\overline{v}\,\underbrace{\overline{u'}}_{=0} - \rho\overline{v'u'} \tag{6.85}$$

Notice that, while $\overline{u'} = \overline{v'} = 0$, averages of cross products of fluctuations (such as $\overline{u'v'}$ or $\overline{u'^2}$) do not generally vanish. Thus, the time average of the fluctuating component of shear stress is

$$\overline{\tau'_{yx}} = -\rho\overline{v'u'} \tag{6.86}$$

In addition to the fluctuating shear stress, the flow will have a mean shear stress associated with the mean velocity gradient, $\partial\overline{u}/\partial y$. That stress is $\mu(\partial\overline{u}/\partial y)$, just as in Newton's law of viscous shear.

It is not obvious how to calculate $\overline{v'u'}$ (although it can be measured), so we shall not make direct use of eqn. (6.86). Instead, we can try to model $\overline{v'u'}$. From the preceding discussion, we see that $\overline{v'u'}$ should go to zero when the velocity gradient, $(\partial\overline{u}/\partial y)$, is zero, and that it should increase when the velocity gradient increases. We might therefore assume it to be proportional to $(\partial\overline{u}/\partial y)$. Then the total time-average shear stress, τ_{yx}, can be expressed as a sum of the mean flow and turbulent contributions that are each proportional to the mean velocity gradient. Specifically,

$$\tau_{yx} = \mu\frac{\partial\overline{u}}{\partial y} - \rho\overline{v'u'} \tag{6.87a}$$

$$\tau_{yx} = \mu\frac{\partial\overline{u}}{\partial y} + \underbrace{\begin{pmatrix}\text{some other factor, which}\\ \text{reflects turbulent mixing}\end{pmatrix}}_{\equiv\,\rho\,\cdot\,\varepsilon_m}\frac{\partial\overline{u}}{\partial y} \tag{6.87b}$$

or

$$\tau_{yx} = \rho\,(\nu + \varepsilon_m)\,\frac{\partial\overline{u}}{\partial y} \tag{6.87c}$$

where ε_m is called the *eddy diffusivity for momentum.* We shall use this characterization in examining the flow field and the heat transfer.

The eddy diffusivity itself may be expressed in terms of the mixing length. Suppose that $\overline{u}$ increases in the y-direction (i.e., $\partial\overline{u}/\partial y > 0$). Then, when a fluid parcel moves downward into slower moving fluid, it has $u' \cong \ell(\partial\overline{u}/\partial y)$. If that parcel moves upward into faster fluid, the sign changes. The vertical velocity fluctation, v', is positive for an upward moving parcel and negative for a downward motion. On average, u' and v' for the eddies should be about the same size. Hence, we expect that

$$\rho\varepsilon_m\frac{\partial\overline{u}}{\partial y} = -\rho\overline{v'u'} = -\rho(\text{constant})\left(\pm\ell\left|\frac{\partial\overline{u}}{\partial y}\right|\right)\left(\mp\ell\frac{\partial\overline{u}}{\partial y}\right) \tag{6.88a}$$

$$= \rho(\text{constant})\,\ell^2\left|\frac{\partial\overline{u}}{\partial y}\right|\frac{\partial\overline{u}}{\partial y} \tag{6.88b}$$

where the absolute value is needed to get the right sign when $\partial\overline{u}/\partial y < 0$. Both $\partial\overline{u}/\partial y$ and $\overline{v'u'}$ can be measured, so we may arbitrarily set the constant in eqn. (6.88) to one to obtain a *measurable* definition of the mixing length. We also obtain an expression for the eddy diffusivity:

$$\varepsilon_m = \ell^2 \left| \frac{\partial \overline{u}}{\partial y} \right| . \tag{6.89}$$

Turbulence near walls

The most important convective heat transfer issue is how flowing fluids cool solid surfaces. Thus, we are principally interested in turbulence near walls. In a turbulent boundary layer, the gradients are very steep near the wall and weaker farther from the wall where the eddies are larger and turbulent mixing is more efficient. This is in contrast to the gradual variation of velocity and temperature in a laminar boundary layer, where heat and momentum are transferred by molecular diffusion rather than the vertical motion of vortices. In fact,the most important processes in turbulent convection occur very close to walls, perhaps within only a fraction of a millimeter. The outer part of the b.l. is less significant.

Let us consider the turbulent flow close to a wall. When the boundary layer momentum equation is time-averaged for turbulent flow, the result is

$$\underbrace{\rho\left(\overline{u}\frac{\partial \overline{u}}{\partial x} + \overline{v}\frac{\partial \overline{u}}{\partial y}\right)}_{\text{neglect very near wall}} = \frac{\partial}{\partial y}\left(\mu\frac{\partial \overline{u}}{\partial y} - \rho\overline{v'u'}\right) \tag{6.90a}$$

$$= \frac{\partial}{\partial y}\tau_{yx} \tag{6.90b}$$

$$= \frac{\partial}{\partial y}\left[\rho\left(\nu + \varepsilon_m\right)\frac{\partial \overline{u}}{\partial y}\right] \tag{6.90c}$$

In the innermost region of a turbulent boundary layer — $y/\delta \lesssim 0.2$, where δ is the b.l. thickness — the mean velocities are small enough that the convective terms in eqn. (6.90a) can be neglected. As a result, $\partial\tau_{yx}/\partial y \cong 0$. The total shear stress is thus essentially constant in y and must equal the wall shear stress:

$$\tau_w \cong \tau_{yx} = \rho\left(\nu + \varepsilon_m\right)\frac{\partial \overline{u}}{\partial y} \tag{6.91}$$

Equation (6.91) shows that the near-wall velocity profile does not depend directly upon x. In functional form

$$\overline{u} = \text{fn}\left(\tau_w, \rho, \nu, y\right) \tag{6.92}$$

(Note that ε_m does not appear as an explicit variable since it is defined by the velocity field.) The effect of the streamwise position is likewise carried in τ_w, which varies slowly with x. As a result, the flow field near the wall is not very sensitive to upstream conditions, except through their effect on τ_w. When the velocity profile is scaled in terms of the local value τ_w, essentially the same velocity profile is obtained in *every* turbulent boundary layer.

Equation (6.92) involves five variables in three dimensions (kg, m, s), so just two dimensionless groups are needed to describe the velocity profile:

$$\frac{\overline{u}}{u^*} = \text{fn}\left(\frac{u^* y}{\nu}\right) \tag{6.93}$$

where the velocity scale $u^* \equiv \sqrt{\tau_w/\rho}$ is called the *friction velocity*. The friction velocity is a speed that is characteristic of the turbulent fluctuations in the boundary layer.

Equation (6.91) can be integrated to find the near wall velocity profile:

$$\underbrace{\int_0^{\overline{u}} d\overline{u}}_{=\overline{u}(y)} = \frac{\tau_w}{\rho}\int_0^y \frac{dy}{\nu + \varepsilon_m} \tag{6.94}$$

To complete the integration, we need an equation for $\varepsilon_m(y)$. Measurements show that the mixing length varies linearly with the distance from the wall for small y

$$\ell = \kappa y \quad \text{for} \quad y/\delta \lesssim 0.2 \tag{6.95}$$

where $\kappa = 0.41$ is called the *von Kármán constant*. Physically, this says that the turbulent eddies at a location y must be no bigger that the distance to wall. That makes sense, since eddies cannot cross into the wall.

The viscous sublayer. Very near the wall, the eddies must become tiny; ℓ and thus ε_m will tend to zero, so that $\nu \gg \varepsilon_m$. In other words, in this region turbulent shear stress is negligible compared to viscous shear stress. If we integrate eqn. (6.94) in that range, we find

$$\begin{aligned} \overline{u}(y) &= \frac{\tau_w}{\rho} \int_0^y \frac{dy}{\nu} = \frac{\tau_w}{\rho} \frac{y}{\nu} \\ &= \frac{(u^*)^2 y}{\nu} \end{aligned} \tag{6.96}$$

Experimentally, eqn. (6.96) is found to apply for $(u^* y/\nu) \lesssim 7$. We call this thin region the *viscous sublayer.* The sublayer is on the order of tens to hundreds of micrometers thick, depending upon the fluid and the shear stress. Because turbulent mixing is ineffective in the sublayer, the sublayer is responsible for a major fraction of the thermal resistance of a turbulent boundary layer. Even a small wall roughness can disrupt this thin sublayer, causing a large decrease in the thermal resistance (but also a large increase in the wall shear stress).

The log layer. Farther away from the wall, ℓ is larger and turbulent shear stress is dominant: $\varepsilon_m \gg \nu$. Then, from eqns. (6.91) and (6.89)

$$\tau_w \cong \rho \varepsilon_m \frac{\partial \overline{u}}{\partial y} = \rho \ell^2 \left| \frac{\partial \overline{u}}{\partial y} \right| \frac{\partial \overline{u}}{\partial y} \tag{6.97}$$

Assuming the velocity gradient to be positive, we may take the square root of eqn. (6.97), rearrange, and integrate it:

$$\int d\overline{u} = \sqrt{\frac{\tau_w}{\rho}} \int \frac{dy}{\ell} \tag{6.98a}$$

$$\overline{u}(y) = u^* \int \frac{dy}{\kappa y} + \text{constant} \tag{6.98b}$$

$$= \frac{u^*}{\kappa} \ln y + \text{constant} \tag{6.98c}$$

Experimental data may be used to fix the constant, with the result that

$$\frac{\overline{u}(y)}{u^*} = \frac{1}{\kappa} \ln \left(\frac{u^* y}{\nu} \right) + B \tag{6.99}$$

for $B \cong 5.5$. Equation (6.99) is sometimes called the *log law.* Experiments show it to apply for $(u^* y/\nu) \gtrsim 30$ and $y/\delta \lesssim 0.2$.

Other regions of the turbulent b.l. For the range $7 < (u^* y/\nu) < 30$, the so-called *buffer layer*, more complicated equations for ℓ, ε_m, or $\overline{u}$ are used to connect the viscous sublayer to the log layer [6.7, 6.8]. Here, ℓ actually decreases a little faster than shown by eqn. (6.95), as $y^{3/2}$ [6.9].

In contrast, for the outer part of the turbulent boundary layer ($y/\delta \gtrsim 0.2$), the mixing length is approximately constant: $\ell \cong 0.09\delta$. Gradients in this part of the boundary layer are weak and do not directly affect transport at the wall. This part of the b.l. is nevertheless essential to the streamwise momentum balance that determines how τ_w and δ vary along the wall. Analysis of that momentum balance [6.2] leads to the following expressions for the boundary thickness and the skin friction coefficient as a function of x:

$$\frac{\delta(x)}{x} = \frac{0.16}{\mathrm{Re}_x^{1/7}} \tag{6.100}$$

$$C_f(x) = \frac{0.027}{\mathrm{Re}_x^{1/7}} \tag{6.101}$$

These expressions are based on an assumption that the turbulent b.l. begins at $x = 0$, neglecting the initial laminar region. They are reasonably accurate for Reynolds numbers ranging from about 10^6 to 10^9. A more accurate formula for C_f, valid for all turbulent Re_x, was given by White [6.10]:

$$\boxed{C_f(x) = \frac{0.455}{\left[\ln(0.06\,\mathrm{Re}_x)\right]^2}} \tag{6.102}$$

6.8 Heat transfer in turbulent boundary layers

The turbulent thermal boundary layer, like the turbulent momentum boundary layer, is characterized by inner and outer regions. In the inner part of the thermal boundary layer, turbulent mixing is increasingly weak; there, heat transport is controlled by heat conduction in the sublayer. Farther from the wall, the temperature profile is logarithmic, and turbulent mixing is the dominant mode of transport in the outermost parts of the boundary layer.

The boundary layer ends where turbulence dies out and uniform freestream conditions prevail, with the result that the thermal and momentum boundary layer thicknesses are the same. At first, this might seem to suggest that Prandtl number does not affect turbulent heat transfer, but actually it does. Its effect is now found in the sublayers near the wall, where molecular viscosity and thermal conductivity still control the transport of heat and momentum.

The Reynolds-Colburn analogy for turbulent flow

The eddy diffusivity for momentum was introduced by Boussinesq [6.11] in 1877. It was subsequently proposed that Fourier's law might likewise be modified for turbulent flow as follows:

$$q = -k\frac{\partial \overline{T}}{\partial y} - \underbrace{\begin{pmatrix}\text{another constant, which}\\ \text{reflects turbulent mixing}\end{pmatrix}}_{\equiv \rho c_p \cdot \varepsilon_h}\frac{\partial \overline{T}}{\partial y}$$

where $\overline{T}$ is the local time-average value of the temperature. Therefore,

$$q = -\rho c_p\,(\alpha + \varepsilon_h)\,\frac{\partial \overline{T}}{\partial y} \tag{6.103}$$

where ε_h is called the *eddy diffusivity of heat.* This immediately suggests yet another definition:

$$\text{turbulent Prandtl number, } \Pr_t \equiv \frac{\varepsilon_m}{\varepsilon_h} \tag{6.104}$$

Equation (6.103) can be written in terms of ν and ε_m by introducing Pr and $\Pr_t$ into it. Thus,

$$q = -\rho c_p\left(\frac{\nu}{\Pr} + \frac{\varepsilon_m}{\Pr_t}\right)\frac{\partial \overline{T}}{\partial y} \tag{6.105}$$

Before we try to build a form of the Reynolds analogy for turbulent flow, we must note the behavior of Pr and $\Pr_t$:

- Pr is a physical property of the fluid. It is both theoretically and actually near one for ideal gases, while for liquids it may differ from one by orders of magnitude.

- $\Pr_t$ is a property of the flow field more than of the fluid. The numerical value of $\Pr_t$ is normally well within a factor of 2 of one. It varies with location in the b.l., but, for nonmetallic fluids, it is often near 0.85.

The time-average boundary-layer energy equation is similar to the time-average momentum equation [eqn. (6.90a)]

$$\underbrace{\rho c_p \left(\overline{u} \frac{\partial \overline{T}}{\partial x} + \overline{v} \frac{\partial \overline{T}}{\partial y} \right)}_{\text{neglect very near wall}} = -\frac{\partial}{\partial y} q = \frac{\partial}{\partial y} \left[\rho c_p \left(\frac{\nu}{\mathrm{Pr}} + \frac{\varepsilon_m}{\mathrm{Pr}_t} \right) \frac{\partial \overline{T}}{\partial y} \right] \tag{6.106}$$

and in the near wall region the convective terms are again negligible. This means that $\partial q / \partial y \cong 0$ near the wall, so that the heat flux is constant in y and equal to the wall heat flux:

$$q = q_w = -\rho c_p \left(\frac{\nu}{\mathrm{Pr}} + \frac{\varepsilon_m}{\mathrm{Pr}_t} \right) \frac{\partial \overline{T}}{\partial y} \tag{6.107}$$

We may integrate this equation as we did eqn. (6.91), with the result that

$$\frac{T_w - \overline{T}(y)}{q_w / (\rho c_p u^*)} = \begin{cases} \mathrm{Pr} \left(\dfrac{u^* y}{\nu} \right) & \text{thermal sublayer} \\[2ex] \dfrac{1}{\kappa} \ln \left(\dfrac{u^* y}{\nu} \right) + A(\mathrm{Pr}) & \text{thermal log layer} \end{cases} \tag{6.108}$$

where the thermal sublayer extends to $(u^* y/\nu) \lesssim 7$ and the thermal log layer applies for $(u^* y/\nu) \gtrsim 30$ and $y/\delta \lesssim 0.2$. The constant A depends upon the Prandtl number. It reflects the thermal resistance of the sublayer near the wall. As was done for the constant B in the velocity profile, experimental data or numerical simulation may be used to determine $A(\mathrm{Pr})$ [6.12, 6.13]. For $\mathrm{Pr} \geq 0.5$,

$$A(\mathrm{Pr}) = 12.8\,\mathrm{Pr}^{0.68} - 7.3 \tag{6.109}$$

To obtain the Reynolds analogy, we subtract the dimensionless log-law, eqn. (6.99), from its thermal counterpart, eqn. (6.108):

$$\frac{T_w - \overline{T}(y)}{q_w / (\rho c_p u^*)} - \frac{\overline{u}(y)}{u^*} = A(\mathrm{Pr}) - B \tag{6.110a}$$

In the outer part of the boundary layer, $\overline{T}(y) \cong T_\infty$ and $\overline{u}(y) \cong u_\infty$, so

$$\frac{T_w - T_\infty}{q_w / (\rho c_p u^*)} - \frac{u_\infty}{u^*} = A(\mathrm{Pr}) - B \tag{6.110b}$$

We eliminate the friction velocity in favor of the skin friction coefficient by using the definitions of each:

$$\frac{u^*}{u_\infty} = \sqrt{\frac{\tau_w}{\rho u_\infty^2}} = \sqrt{\frac{C_f}{2}} \tag{6.110c}$$

Hence,

$$\frac{T_w - T_\infty}{q_w/(\rho c_p u_\infty)}\sqrt{\frac{C_f}{2}} - \sqrt{\frac{2}{C_f}} = A(\text{Pr}) - B \tag{6.110d}$$

Rearrangment of the last equation gives

$$\frac{q_w}{(\rho c_p u_\infty)(T_w - T_\infty)} = \frac{C_f/2}{1 + [A(\text{Pr}) - B]\sqrt{C_f/2}} \tag{6.110e}$$

The lefthand side is simply the Stanton number, $\text{St} = h/(\rho c_p u_\infty)$. Upon substituting $B = 5.5$ and eqn. (6.109) for $A(\text{Pr})$, we obtain the Reynolds-Colburn analogy for turbulent flow:

$$\boxed{\text{St}_x = \frac{C_f/2}{1 + 12.8\left(\text{Pr}^{0.68} - 1\right)\sqrt{C_f/2}}} \qquad \text{Pr} \geq 0.5 \tag{6.111}$$

This result can be used with eqn. (6.102) for C_f, or with data for C_f, to calculate the local heat transfer coefficient in a turbulent boundary layer. It works for either uniform T_w or uniform q_w. This is because the thin, near-wall part of the boundary layer controls most of the thermal resistance and that thin layer is not strongly dependent on upstream history of the flow.

Equation (6.111) is valid for smooth walls with a mild or a zero pressure gradient. The factor $12.8\,(\text{Pr}^{0.68} - 1)$ in the denominator accounts for the thermal resistance of the sublayer. If the walls are rough, the sublayer will be disrupted and that term must be replaced by one that takes account of the roughness (see Sect. 7.3).

Other equations for heat transfer in the turbulent b.l.

Although eqn. (6.111) gives an excellent prediction of the local value of h in a turbulent boundary layer, a number of simplified approximations to it have been suggested in the literature. For example, for Prandtl numbers not too far from one and Reynolds numbers not too far above transition, the laminar flow Reynolds-Colburn analogy can be used

$$\mathrm{St}_x = \left(\frac{C_f}{2}\right) \mathrm{Pr}^{-2/3} \qquad \text{for Pr near 1} \tag{6.76}$$

The best exponent for the Prandtl number in such an equation actually depends upon the Reynolds and Prandtl numbers. For gases, an exponent of −0.4 gives somewhat better results.

A more wide-ranging approximation can be obtained after introducing a simplifed expression for C_f. For example, Schlichting [6.3, Chap. XXI] shows that, for turbulent flow over a smooth flat plate in the low-Re range,

$$C_f \cong \frac{0.0592}{\mathrm{Re}_x^{1/5}}, \qquad 5 \times 10^5 \leqslant \mathrm{Re}_x \leqslant 10^7 \tag{6.112}$$

With this Reynolds number dependence, Žukauskas and coworkers [6.14, 6.15] found that

$$\mathrm{St}_x = \left(\frac{C_f}{2}\right) \mathrm{Pr}^{-0.57}, \qquad 0.7 \leq \mathrm{Pr} \leq 380 \tag{6.113}$$

so that when eqn. (6.112) is used to eliminate C_f

$$\mathrm{Nu}_x = 0.0296\, \mathrm{Re}_x^{0.8}\, \mathrm{Pr}^{0.43} \tag{6.114}$$

Somewhat better agreement with data, for $2 \times 10^5 \leqslant \mathrm{Re}_x \leqslant 5 \times 10^6$, is obtained by adjusting the constant [6.15]:

$$\boxed{\mathrm{Nu}_x = 0.032\, \mathrm{Re}_x^{0.8}\, \mathrm{Pr}^{0.43}} \tag{6.115}$$

The average Nusselt number for uniform T_w is obtained from eqn. (6.114) as follows:

$$\overline{\mathrm{Nu}}_L = \frac{L}{k}\overline{h} = \frac{0.0296\, \mathrm{Pr}^{0.43} L}{k} \left[\frac{k}{L}\int_0^L \left(\frac{1}{x}\mathrm{Re}_x^{0.8}\right) dx\right]$$

where we ignore the fact that there is a laminar region at the front of the plate. Thus,

$$\overline{\mathrm{Nu}}_L = 0.0370\, \mathrm{Re}_L^{0.8}\, \mathrm{Pr}^{0.43} \tag{6.116}$$

This equation may be used for either uniform T_w or uniform q_w, and for Re_L up to about 3×10^7 [6.14, 6.15].

A flat heater with a turbulent b.l. on it actually has a laminar b.l. between $x = 0$ and $x = x_{\text{trans}}$, as is indicated in Fig. 6.4. The obvious way to calculate $\overline{h}$ in this case is to write

$$\begin{aligned} \overline{h} &= \frac{1}{L\Delta T}\int_0^L q\,dx \\ &= \frac{1}{L}\left[\int_0^{x_{\text{trans}}} h_{\text{laminar}}\,dx + \int_{x_{\text{trans}}}^L h_{\text{turbulent}}\,dx\right] \end{aligned} \tag{6.117}$$

where $x_{\text{trans}} = (\nu/u_\infty)\text{Re}_{\text{trans}}$. Thus, we substitute eqns. (6.58) and (6.114) in eqn. (6.117) and obtain, for $0.6 \leqslant \Pr \leqslant 50$,

$$\boxed{\overline{\text{Nu}}_L = 0.037\,\Pr^{0.43}\left\{\text{Re}_L^{0.8} - \left[\text{Re}_{\text{trans}}^{0.8} - 17.95\,\Pr^{-0.097}\left(\text{Re}_{\text{trans}}\right)^{1/2}\right]\right\}} \tag{6.118}$$

If $\text{Re}_L \gg \text{Re}_{\text{trans}}$, this result reduces to eqn. (6.116).

Whitaker [6.16] suggested setting $\Pr^{-0.097} \approx 1$ and $\text{Re}_{\text{trans}} \approx 200,000$ in eqn. (6.118):

$$\boxed{\overline{\text{Nu}}_L = 0.037\,\Pr^{0.43}\left(\text{Re}_L^{0.8} - 9200\right)\left(\frac{\mu_\infty}{\mu_w}\right)^{1/4}} \qquad 0.6 \le \Pr \le 380 \tag{6.119}$$

This expression has been corrected to account for the variability of liquid viscosity with the factor $(\mu_\infty/\mu_w)^{1/4}$, where μ_∞ is the viscosity at the free-stream temperature, T_∞, and μ_w is that at the wall temperature, T_w; other physical properties should be evaluated at T_∞. If eqn. (6.119) is used to predict heat transfer to a gaseous flow, the viscosity-ratio correction term should not be used and properties should be evaluated at the film temperature. This is because the viscosity of a gas rises with temperature instead of dropping, and the correction will be incorrect.

Finally, it is important to remember that eqns. (6.118) and (6.119) should be used only when Re_L is substantially above the transitional value.

A correlation for laminar, transitional, and turbulent flow

A problem with the two preceding relations is that they do not really deal with the question of heat transfer in the rather lengthy transition region. Both eqns. (6.118) and (6.119) are based on the assumption that flow abruptly passes from laminar to turbulent at a critical value of x, and we have noted in the context of Fig. 6.4 that this is *not* what occurs. The location of the transition depends upon such variables as surface roughness and the turbulence, or lack of it, in the stream approaching the heater.

Churchill [6.17] suggests correlating any *particular set* of data with

$$\text{Nu}_x = 0.45 + \left(0.3387\,\phi^{1/2}\right)\left\{1 + \frac{(\phi/2{,}600)^{3/5}}{\left[1 + (\phi_u/\phi)^{7/2}\right]^{2/5}}\right\}^{1/2} \qquad (6.120\text{a})$$

where

$$\phi \equiv \text{Re}_x \text{Pr}^{2/3}\left[1 + \left(\frac{0.0468}{\text{Pr}}\right)^{2/3}\right]^{-1/2} \qquad (6.120\text{b})$$

and ϕ_u is a number between about 10^5 and 10^7. The actual value of ϕ_u must be fit to the particular set of data. In a very "clean" system, ϕ_u will be larger; in a very "noisy" one, it will be smaller. If the Reynolds number at the end of the turbulent transition region is Re_u, an estimate is $\phi_u \approx \phi(\text{Re}_x = \text{Re}_u)$.

The equation is for uniform T_w, but it may be used for uniform q_w if the constants 0.3387 and 0.0468 are replaced by 0.4637 and 0.02052, respectively.

Churchill also gave an expression for the average Nusselt number:

$$\overline{\text{Nu}}_L = 0.45 + \left(0.6774\,\phi^{1/2}\right)\left\{1 + \frac{(\phi/12{,}500)^{3/5}}{\left[1 + (\phi_{\text{um}}/\phi)^{7/2}\right]^{2/5}}\right\}^{1/2} \qquad (6.120\text{c})$$

where ϕ is defined as in eqn. (6.120b), using Re_L in place of Re_x, and $\phi_{\text{um}} \approx 1.875\,\phi(\text{Re}_L = \text{Re}_u)$. This equation may be used for either uniform T_w or uniform q_w.

The advantage of eqns. (6.120a) or (6.120c) is that, once ϕ_u or ϕ_{um} is known, they will predict heat transfer from the laminar region, *through* the transition regime, and into the turbulent regime.

Example 6.9

After loading its passengers, a ship sails out of the mouth of a river, where the water temperature is 24°C, into 10°C ocean water. The forward end of the ship's hull is sharp and relatively flat. If the ship travels at 5 knots, find C_f and h at a distance of 1 m from the forward edge of the hull.

Solution. If we assume that the hull's heat capacity holds it at the river temperature for a time, we can take the properties of water at $T_f = (10+24)/2 = 17°\text{C}$: $\nu = 1.085\times10^{-6}$ m^2/s, $k = 0.5927$ W/m·K, $\rho = 998.8$ kg/m^3, $c_p = 4187$ J/kg·K, and Pr $= 7.66$.

One knot equals 0.5144 m/s, so $u_\infty = 5(0.5144) = 2.572$ m/s. Then, $\text{Re}_x = (2.572)(1)/(1.085\times10^{-6}) = 2.371\times10^6$, indicating that the flow is turbulent at this location.

We have given several different equations for C_f in a turbulent boundary layer, but the most accurate of these is eqn. (6.102):

$$C_f(x) = \frac{0.455}{[\ln(0.06\,\text{Re}_x)]^2} = \frac{0.455}{\{\ln[0.06(2.371\times10^6)]\}^2} = 0.003232$$

For the heat transfer coefficient, we can use either eqn. (6.115)

$$\begin{aligned} h(x) &= \frac{k}{x}\cdot 0.032\,\text{Re}_x^{0.8}\,\text{Pr}^{0.43} \\ &= \frac{(0.5927)(0.032)(2.371\times10^6)^{0.8}(7.66)^{0.43}}{(1.0)} \\ &= 5{,}729\ \text{W/m}^2\text{K} \end{aligned}$$

or its more complex counterpart, eqn. (6.111):

$$\begin{aligned} h(x) &= \rho c_p u_\infty \cdot \frac{C_f/2}{1+12.8\left(\text{Pr}^{0.68}-1\right)\sqrt{C_f/2}} \\ &= \frac{998.8(4187)(2.572)(0.003232/2)}{1+12.8\left[(7.66)^{0.68}-1\right]\sqrt{0.003232/2}} \\ &= 6{,}843\ \text{W/m}^2\text{K} \end{aligned}$$

The two values of h differ by about 18%, which is within the uncertainty of eqn. (6.115). ■

Example 6.10

In a wind tunnel experiment, an aluminum plate 2.0 m in length is electrically heated at a power density of 1 kW/m^2 and is cooled on one surface by air flowing at 10 m/s. The air in the wind tunnel has a temperature of 290 K and is at 1 atm pressure, and the Reynolds number at the end of turbulent transition regime is observed to be 400,000. Estimate the average temperature of the plate.

Solution. For this low heat flux, we expect the plate temperature to be near the air temperature, so we evaluate properties at 300 K: $\nu = 1.578 \times 10^{-5}$ m^2/s, $k = 0.02623$ W/m·K, and Pr = 0.713. At 10 m/s, the plate Reynolds number is $\mathrm{Re}_L = (10)(2)/(1.578 \times 10^{-5}) = 1.267 \times 10^6$. From eqn. (6.118), we get

$$\overline{\mathrm{Nu}}_L = 0.037(0.713)^{0.43}\Big\{(1.267\times10^6)^{0.8} - \Big[(400{,}000)^{0.8} - 17.95(0.713)^{-0.097}(400{,}000)^{1/2}\Big]\Big\} = 1{,}845$$

so

$$\overline{h} = \frac{1845\,k}{L} = \frac{1845(0.02623)}{2.0} = 24.19\ \mathrm{W/m^2K}$$

It follows that the average plate temperature is

$$\overline{T}_w = 290\ K + \frac{10^3\ \mathrm{W/m^2}}{24.19\ \mathrm{W/m^2K}} = 331\ \mathrm{K}.$$

The film temperature is $(331+290)/2 = 311$ K; if we recalculate using properties at 311 K, the $\overline{h}$ changes by less than 4%, and $\overline{T}_w$ by 1.3 K.

To take better account of the transition regime, we can use Churchill's equation, (6.120c). First, we evaluate ϕ:

$$\phi = \frac{(1.267\times10^6)(0.713)^{2/3}}{\left[1 + (0.0468/0.713)^{2/3}\right]^{1/2}} = 9.38\times10^5$$

We then estimate

$$\begin{aligned}\phi_{\mathrm{um}} &= 1.875\cdot\phi(\mathrm{Re}_L = 400{,}000)\\ &= \frac{(1.875)(400{,}000)(0.713)^{2/3}}{\left[1 + (0.0468/0.713)^{2/3}\right]^{1/2}} = 5.55\times10^5\end{aligned}$$

Finally,

$$\overline{\text{Nu}}_L = 0.45 + (0.6774)\left(9.38 \times 10^5\right)^{1/2} \times \left\{1 + \frac{\left(9.38 \times 10^5/12,500\right)^{3/5}}{\left[1 + (5.55 \times 10^5/9.38 \times 10^5)^{7/2}\right]^{2/5}}\right\}^{1/2}$$

$$= 2,418$$

which leads to

$$\overline{h} = \frac{2418\,k}{L} = \frac{2418(0.02623)}{2.0} = 31.71 \text{ W/m}^2\text{K}$$

and

$$\overline{T}_w = 290\ K + \frac{10^3 \text{ W/m}^2}{31.71 \text{ W/m}^2\text{K}} = 322 \text{ K.}$$

Thus, in this case, the average heat transfer coefficient is 33% higher when the transition regime is included. ■

A word about the analysis of turbulent boundary layers

The preceding discussion has circumvented serious *analysis* of heat transfer in turbulent boundary layers. In the past, boundary layer heat transfer has been analyzed in many flows (with and without pressure gradients, dp/dx) using sophisticated integral methods. In recent decades, however, computational techniques have largely replaced integral analyses. Various computational schemes, particularly those based on turbulent kinetic energy and viscous dissipation (so-called k-ε methods), are widely-used and have been implemented in a variety of commercial fluid-dynamics codes. These methods are described in the technical literature and in monographs on turbulence [6.18, 6.19].

We have found our way around analysis by presenting some correlations for the simple plane surface. In the next chapter, we deal with more complicated configurations. A few of these configurations will be amenable to elementary analyses, but for others we shall only be able to present the best data correlations available.

Problems

6.1 Verify that eqn. (6.13) follows from eqns. (6.11) and (6.12).

6.2 The student with some analytical ability (or some assistance from the instructor) should complete the algebra between eqns. (6.16) and (6.20).

6.3 Use a computer to solve eqn. (6.18) subject to b.c.'s (6.20). To do this you need all three b.c.'s at $\eta = 0$, but one is presently at $\eta = \infty$. There are three ways to get around this:

- Start out by guessing a value of $\partial f'/\partial\eta$ at $\eta = 0$—say, $\partial f'/\partial\eta = 1$. When η is large—say, 6 or 10—$\partial f'/\partial\eta$ will asymptotically approach a constant. If the constant > 1, go back and guess a lower value of $\partial f'/\partial\eta$, or vice versa, until the constant converges on one. (One might invent a number of means to automate the successive guesses.)
- The correct value of $df'/d\eta$ is approximately 0.33206 at $\eta = 0$. You might cheat and begin with it, but where is the fun in that?
- There exists a clever way to map $df/d\eta = 1$ at $\eta = \infty$ back into the origin. (Consult your instructor.)

6.4 Verify that the Blasius solution (Table 6.1) satisfies eqn. (6.25). To do this, carry out the required numerical and/or graphical integration.

6.5 Verify eqn. (6.30).

6.6 Obtain the counterpart of eqn. (6.32) based on the velocity profile given by the integral method.

6.7 Assume a laminar b.l. velocity profile of the simple form $u/u_\infty = y/\delta$ and calculate δ and C_f on the basis of this very rough estimate, using the momentum integral method. How accurate is each? [C_f is about 13% low.]

6.8 In a certain flow of water at 40°C over a flat plate $\delta = 0.005\sqrt{x}$, for δ and x measured in meters. Plot *to scale* on a common graph (with an appropriately expanded y-scale):

- δ and δ_t for the water.
- δ and δ_t for air at the same temperature and velocity.

6.9 A thin film of liquid with a constant thickness, δ_0, falls down a vertical plate. It has reached its terminal velocity so that viscous shear and weight are in balance and the flow is steady. The b.l. equation for such a flow is the same as eqn. (6.13), except that it has a gravity force in it. Thus,

$$u\frac{\partial u}{\partial x} + v\frac{\partial u}{\partial y} = -\frac{1}{\rho}\frac{dp}{dx} + g + \nu\frac{\partial^2 u}{\partial y^2}$$

where x increases in the downward direction and y is normal to the wall. Assume that the surrounding air density $\simeq 0$, so there is no hydrostatic pressure gradient in the surrounding air. Then:

- Simplify the equation to describe this situation.
- Write the b.c.'s for the equation, neglecting any air drag on the film.
- Solve for the velocity distribution in the film, assuming that you know δ_0 (cf. Chap. 8).

(This solution is the starting point in the study of many heat and mass transfer proceses.)

6.10 Develop an equation for $\overline{\text{Nu}}_L$ that is valid over the entire range of Pr for a laminar b.l. over a flat, isothermal surface.

6.11 Use an integral method to develop a prediction of Nu_x for a laminar b.l. over a uniform heat flux surface. Compare your result with eqn. (6.71). What temperature difference does this give at the leading edge of the surface?

6.12 Verify eqn. (6.118).

6.13 It is known from flow measurements that the transition to turbulence occurs when the Reynolds number based on mean velocity and diameter exceeds 4000 in a certain pipe. Use the fact that the laminar boundary layer on a flat plate grows according to the relation

$$\frac{\delta}{x} = 4.92\sqrt{\frac{\nu}{u_{\text{max}}x}}$$

to find an equivalent value for the Reynolds number of transition based on distance from the leading edge of the plate and u_{max}. (Note that $u_{\text{max}} = 2\overline{u}_{\text{av}}$ during laminar flow in a pipe.)

6.14 Execute the differentiation in eqn. (6.24) with the help of Leibnitz's rule for the differentiation of an integral and show that the equation preceding it results.

6.15 Liquid at 23°C flows at 2 m/s over a smooth, sharp-edged, flat surface 12 cm in length which is kept at 57°C. Calculate h at the trailing edge (a) if the fluid is water; (b) if the fluid is glycerin ($h = 346$ W/m^{2}K). (c) Compare the drag forces in the two cases. [There is 23.4 times as much drag in the glycerin.]

6.16 Air at −10°C flows over a smooth, sharp-edged, almost-flat, aerodynamic surface at 240 km/hr. The surface is at 10°C. Find (a) the approximate location of the laminar turbulent transition; (b) the overall $\overline{h}$ for a 2 m chord; (c) h at the trailing edge for a 2 m chord; (d) δ and h at the beginning of the transition region. [δ_{x_t} should be around half a millimeter.]

6.17 Find $\overline{h}$ in Example 6.10 using eqn. (6.120c) with $\mathrm{Re}_u = 10^5$ and 2×10^5. Discuss the results.

6.18 For system described in Example 6.10, plot the local value of h over the whole length of the plate using eqn. (6.120c). On the same graph, plot h from eqn. (6.71) for $\mathrm{Re}_x < 400{,}000$ and from eqn. (6.115) for $\mathrm{Re}_x > 200{,}000$. Discuss the results.

6.19 Mercury at 25°C flows at 0.7 m/s over a 4 cm-long flat heater at 60°C. Find $\overline{h}$, $\overline{\tau}_w$, $h(x = 0.04\text{ m})$, and $\delta(x = 0.04\text{ m})$.

6.20 A large plate is at rest in water at 15°C. The plate is suddenly translated parallel to itself, at 1.5 m/s. The resulting fluid movement is not exactly like that in a b.l. because the velocity profile builds up uniformly, all over, instead of from an edge. The governing transient momentum equation, $Du/Dt = \nu(\partial^2 u/\partial y^2)$, takes the form

$$\frac{1}{\nu}\frac{\partial u}{\partial t} = \frac{\partial^2 u}{\partial y^2}$$

Determine u at 0.015 m from the plate for $t = 1$, 10, and 1000 s. Do this by first posing the problem fully and then comparing it with the solution in Section 5.6. [$u \simeq 0.003$ m/s after 10 s.]

6.21 Notice that, when Pr is large, the velocity b.l. on an isothermal, flat heater is much larger than δ_t. The small part of the velocity b.l. inside the thermal b.l. is approximately $u/u_\infty = \frac{3}{2}y/\delta = \frac{3}{2}\phi(y/\delta_t)$. Derive Nu_x for this case based on this velocity profile.

6.22 Plot the ratio of $h(x)_{\text{laminar}}$ to $h(x)_{\text{turbulent}}$ against Re_x in the range of Re_x that might be either laminar or turbulent. What does the plot suggest about heat transfer design?

6.23 Water at 7°C flows at 0.38 m/s across the top of a 0.207 m-long, thin copper plate. Methanol at 87°C flows across the bottom of the same plate, at the same speed but in the opposite direction. Make the obvious first guess as to the temperature at which to evaluate physical properties. Then plot the plate temperature as a function of position. (Do not bother to correct the physical properties in this problem, but note Problem 6.24.) With everything that varies along the plate, determine where the local heat flux would be least.

6.24 Work Problem 6.23 taking full account of property variations.

6.25 If the wall temperature in Example 6.6 (with a uniform q_w = 420 W/m^2) were instead fixed at its average value of 76°C, what would the *average* wall heat flux be?

6.26 A cold, 20 mph westerly wind at 20°F cools a rectangular building, 35 ft by 35 ft by 22 ft high, with a flat roof. The outer walls are at 27°F. Find the heat loss, conservatively assuming that the east and west faces have the same $\overline{h}$ as the north, south, and top faces. Estimate U for the walls.

6.27 A 2 ft-square slab of mild steel leaves a forging operation 0.25 in. thick at 1000°C. It is laid flat on an insulating bed and 27°C air is blown over it at 30 m/s. How long will it take to cool to 200°C. (State your assumptions about property evaluation.)

6.28 Do Problem 6.27 numerically, recalculating properties at successive points. If you did Problem 6.27, compare results.

6.29 Plot T_w against x for the situation described in Example 6.10.

6.30 Consider the plate in Example 6.10. Suppose that instead of specifying $q_w = 1000$ W/m^2, we specified $T_w = 200$°C. Plot q_w against x for this case.

6.31 A thin metal sheet separates air at 44°C, flowing at 48 m/s, from water at 4°C, flowing at 0.2 m/s. Both fluids start at a leading edge and move in the same direction. Plot T_{plate} and q as a function of x up to $x = 0.1$ m.

6.32 A mixture of 60% glycerin and 40% water flows over a 1-m-long flat plate. The glycerin is at 20°C and the plate is at 40°. A thermocouple 1 mm above the trailing edge records 35°C. What is u_∞, and what is u at the thermocouple?

6.33 What is the maximum $\overline{h}$ that can be achieved in laminar flow over a 5 m plate, based on data from Table A.3? What physical circumstances give this result?

6.34 A 17°C sheet of water, Δ_1 m thick and moving at a constant speed u_∞ m/s, impacts a horizontal plate at 45°, turns, and flows along it. Develop a dimensionless equation for the thickness Δ_2 at a distance L from the point of impact. Assume that $\delta \ll \Delta_2$. Evaluate the result for $u_\infty = 1$ m/s, $\Delta_1 = 0.01$ m, and $L = 0.1$ m, in water at 27°C.

6.35 A good approximation to the temperature dependence of μ in gases is given by the Sutherland formula:

$$\frac{\mu}{\mu_{\text{ref}}} = \left(\frac{T}{T_{\text{ref}}}\right)^{1.5} \frac{T_{\text{ref}} + S}{T + S},$$

where the reference state can be chosen anywhere. Use data for air at two points to evaluate S for air. Use this value to predict a third point. (T and T_{ref} are expressed in kelvin.)

6.36 We have derived a steady-state continuity equation in Section 6.3. Now derive the time-dependent, compressible, three-dimensional version of the equation:

$$\frac{\partial \rho}{\partial t} + \nabla \cdot (\rho \vec{u}) = 0$$

To do this, paraphrase the development of equation (2.10), requiring that mass be conserved instead of energy.

6.37 Various considerations show that the smallest-scale motions in a turbulent flow have no preferred spatial orientation at large enough values of Re. Moreover, these small eddies are responsible for most of the viscous dissipation of kinetic energy. The dissipation rate, ε (W/kg), may be regarded as given information about the small-scale motion, since it is set by the larger-scale motion. Both ε and ν are governing parameters of the small-scale motion.

a. Find the characteristic length and velocity scales of the small-scale motion. These are called the *Kolmogorov scales* of the flow.

b. Compute Re for the small-scale motion and interpret the result.

c. The Kolmogorov length scale characterizes the smallest motions found in a turbulent flow. If ε is 10 W/kg and the mean free path is 7×10^{-8} m, show that turbulent motion is a continuum phenomenon and thus is properly governed by the equations of this chapter.

6.38 The temperature outside is 35°F, but with the wind chill it's −15°F. And you forgot your hat. If you go outdoors for long, are you in danger of freezing your ears?

6.39 To heat the airflow in a wind tunnel, an experimenter uses an array of electrically heated, horizontal Nichrome V strips. The strips are perpendicular to the flow. They are 20 cm long, very thin, 2.54 cm wide (in the flow direction), with the flat sides parallel to the flow. They are spaced vertically, each 1 cm above the next. Air at 1 atm and 20°C passes over them at 10 m/s.

a. How much power must each strip deliver to raise the mean temperature of the airstream to 30°C?

b. What is the heat flux if the electrical heating in the strips is uniformly distributed?

c. What are the average and maximum temperatures of the strips?

6.40 An airflow sensor consists of a 5 cm long, heated copper slug that is smoothly embedded 10 cm from the leading edge of a flat plate. The overall length of the plate is 15 cm, and the width of the plate and the slug are both 10 cm. The slug is electrically heated by an internal heating element, but, owing to its high thermal conductivity, the slug has an essentially uniform temperature along its airside surface. The heater's controller adjusts its power to keep the slug surface at a fixed temperature. The air velocity is found from measurements of the slug temperature, the air temperature, and the heating power needed to hold the slug at the set temperature.

a. If the air is at 280 K, the slug is at 300 K, and the heater power is 5.0 W, find the airspeed assuming the flow is laminar. *Hint:* For $x_1/x_0 = 1.5$

$$\int_{x_0}^{x_1} x^{-1/2}\left[1-(x_0/x)^{3/4}\right]^{-1/3} dx = 1.0035\sqrt{x_0}$$

b. Suppose that a disturbance trips the boundary layer near the leading edge, causing it to become turbulent over the whole plate. The air speed, air temperature, and the slug's set-point temperature remain the same. Make a very rough estimate of the heater power that the controller now delivers, without doing a lot of analysis.

6.41 Equation (6.64) gives Nu_x for a flat plate with an unheated starting length. This equation may be derived using the integral energy equation [eqn. (6.47)], modelling the velocity and temperature profiles with eqns. (6.29) and (6.50), respectively, and taking $\delta(x)$ from eqn. (6.31a). Equation (6.52) is again obtained; however, in this case, $\phi = \delta_t/\delta$ is a function of x for $x > x_0$. Derive eqn. (6.64) by starting with eqn. (6.52), neglecting the term $3\phi^3/280$, and replacing δ_t by $\phi\delta$. After some manipulation, you will obtain

$$x\frac{4}{3}\frac{d}{dx}\phi^3 + \phi^3 = \frac{13}{14\,\mathrm{Pr}}$$

Show that its solution is

$$\phi^3 = Cx^{-3/4} + \frac{13}{14\,\mathrm{Pr}}$$

for an unknown constant C. Then apply an appropriate initial condition and the definition of q_w and Nu_x to obtain eqn. (6.64).

References

[6.1] S. Juhasz. Notes on Applied Mechanics Reviews - Referativnyi Zhurnal Mekhanika exhibit at XIII IUTAM, Moscow 1972. *Appl. Mech. Rev.*, 26(2):145–160, 1973.

[6.2] F.M. White. *Viscous Fluid Flow.* McGraw-Hill, Inc., New York, 2nd edition, 1991.

[6.3] H. Schlichting. *Boundary-Layer Theory.* (trans. J. Kestin). McGraw-Hill Book Company, New York, 6th edition, 1968.

[6.4] C. L. Tien and J. H. Lienhard. *Statistical Thermodynamics.* Hemisphere Publishing Corp., Washington, D.C., rev. edition, 1978.

[6.5] S. W. Churchill and H. Ozoe. Correlations for laminar forced convection in flow over an isothermal flat plate and in developing and fully developed flow in an isothermal tube. *J. Heat Trans., Trans. ASME, Ser. C*, 95:78, 1973.

[6.6] O. Reynolds. On the extent and action of the heating surface for steam boilers. *Proc. Manchester Lit. Phil. Soc.*, 14:7–12, 1874.

[6.7] J.A. Schetz. *Foundations of Boundary Layer Theory for Momentum, Heat, and Mass Transfer.* Prentice-Hall, Inc., Englewood Cliffs, NJ, 1984.

[6.8] P. S. Granville. A modified Van Driest formula for the mixing length of turbulent boundary layers in pressure gradients. *J. Fluids Engr.*, 111(1):94–97, 1989.

[6.9] P. S. Granville. A near-wall eddy viscosity formula for turbulent boundary layers in pressure gradients suitable for momentum, heat, or mass transfer. *J. Fluids Engr.*, 112(2):240–243, 1990.

[6.10] F. M. White. A new integral method for analyzing the turbulent boundary layer with arbitrary pressure gradient. *J. Basic Engr.*, 91: 371–378, 1969.

[6.11] J. Boussinesq. Théorie de l'écoulement tourbillant. *Mem. Pres. Acad. Sci.*, (Paris), 23:46, 1877.

[6.12] F. M. White. *Viscous Fluid Flow.* McGraw-Hill Book Company, New York, 1974.

[6.13] B. S. Petukhov. Heat transfer and friction in turbulent pipe flow with variable physical properties. In T.F. Irvine, Jr. and J. P. Hartnett, editors, *Advances in Heat Transfer*, volume 6, pages 504–564. Academic Press, Inc., New York, 1970.

[6.14] A. A. Žukauskas and A. B. Ambrazyavichyus. Heat transfer from a plate in a liquid flow. *Int. J. Heat Mass Transfer*, 3(4):305–309, 1961.

[6.15] A. Žukauskas and A. Šlanciauskas. *Heat Transfer in Turbulent Fluid Flows.* Hemisphere Publishing Corp., Washington, 1987.

[6.16] S. Whitaker. Forced convection heat transfer correlation for flow in pipes past flat plates, single cylinders, single spheres, and for flow in packed beds and tube bundles. *AIChE J.*, 18:361, 1972.

[6.17] S. W. Churchill. A comprehensive correlating equation for forced convection from flat plates. *AIChE J.*, 22:264–268, 1976.

[6.18] S. B. Pope. *Turbulent Flows.* Cambridge University Press, Cambridge, 2000.

[6.19] P. A. Libby. *Introduction to Turbulence.* Taylor & Francis, Washington D.C., 1996.

7. Forced convection in a variety of configurations

The bed was soft enough to suit me...But I soon found that there came such a draught of cold air over me from the sill of the window that this plan would never do at all, especially as another current from the rickety door met the one from the window and both together formed a series of small whirlwinds in the immediate vicinity of the spot where I had thought to spend the night. ***Moby Dick,*** **H. Melville, 1851**

7.1 Introduction

Consider for a moment the fluid flow pattern within a shell-and-tube heat exchanger, such as shown in Fig. 3.5. The shell-pass flow moves up and down across the tube bundle from one baffle to the next. The flow around each pipe is determined by the complexities of the one before it, and the direction of the mean flow relative to each pipe can vary. Yet the problem of determining the heat transfer in this situation, however difficult it appears to be, is a task that must be undertaken.

The flow within the tubes of the exchanger is somewhat more tractable, but it, too, brings with it several problems that do not arise in the flow of fluids over a flat surface. Heat exchangers thus present a kind of microcosm of internal and external forced convection problems. Other such problems arise everywhere that energy is delivered, controlled, utilized, or produced. They arise in the complex flow of water through nuclear heating elements or in the liquid heating tubes of a solar collector—in the flow of a cryogenic liquid coolant in certain digital computers or in the circulation of refrigerant in the spacesuit of a lunar astronaut.

We dealt with the simple configuration of flow over a flat surface in

Chapter 6. This situation has considerable importance in its own right, and it also reveals a number of analytical methods that apply to other configurations. Now we wish to undertake a sequence of progressively harder problems of forced convection heat transfer in more complicated flow configurations.

Incompressible forced convection heat transfer problems normally admit an extremely important simplification: the fluid flow problem can be solved without reference to the temperature distribution in the fluid. Thus, we can first find the velocity distribution and then put it in the energy equation as known information and solve for the temperature distribution. Two things can impede this procedure, however:

- If the fluid properties (especially μ and ρ) vary significantly with temperature, we cannot predict the velocity without knowing the temperature, and vice versa. The problems of predicting velocity and temperature become intertwined and harder to solve. We encounter such a situation later in the study of natural convection, where the fluid is driven by thermally induced density changes.

- Either the fluid flow solution or the temperature solution can, itself, become prohibitively hard to find. When that happens, we resort to the correlation of experimental data with the help of dimensional analysis.

Our aim in this chapter is to present the analysis of a few simple problems and to show the progression toward increasingly empirical solutions as the problems become progressively more unwieldy. We begin this undertaking with one of the simplest problems: that of predicting laminar convection in a pipe.

7.2 Heat transfer to and from laminar flows in pipes

Not many industrial pipe flows are laminar, but laminar heating and cooling does occur in an increasing variety of modern instruments and equipment: micro-electro-mechanical systems (MEMS), laser coolant lines, and many compact heat exchangers, for example. As in any forced convection problem, we first describe the flow field. This description will include a number of ideas that apply to turbulent as well as laminar flow.

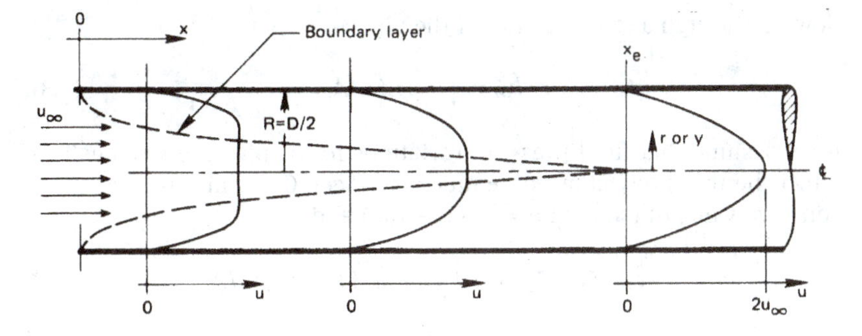

Figure 7.1 The development of a laminar velocity profile in a pipe.

Development of a laminar flow

Figure 7.1 shows the evolution of a laminar velocity profile from the entrance of a pipe. Throughout the length of the pipe, the mass flow rate, $\dot{m}$ (kg/s), is constant, of course, and the *average*, or *bulk*, velocity u_{av} is also constant:

$$\dot{m} = \int_{A_c} \rho u \, dA_c = \rho u_{\text{av}} A_c \tag{7.1}$$

where A_c is the cross-sectional area of the pipe. The velocity profile, on the other hand, changes greatly near the inlet to the pipe. A b.l. builds up from the front, generally accelerating the otherwise undisturbed core. The b.l. eventually occupies the entire flow area and defines a velocity profile that changes very little thereafter. We call such a flow *fully developed*. A flow is fully developed from the hydrodynamic standpoint when

$$\frac{\partial u}{\partial x} = 0 \quad \text{or} \quad v = 0 \tag{7.2}$$

at each radial location in the cross section. An attribute of a dynamically fully developed flow is that the streamlines are all parallel to one another.

The concept of a fully developed flow, from the thermal standpoint, is a little more complicated. We must first understand the notion of the *mixing-cup*, or *bulk*, enthalpy and temperature, $\hat{h}_b$ and T_b. The enthalpy is of interest because we use it in writing the First Law of Thermodynamics when calculating the inflow of thermal energy and flow work to open control volumes. The bulk enthalpy is an average enthalpy for the fluid flowing through a cross section of the pipe:

$$\dot{m}\,\hat{h}_b \equiv \int_{A_c} \rho u \hat{h}\, dA_c \tag{7.3}$$

If we assume that fluid pressure variations in the pipe are too small to affect the thermodynamic state much (see Sect. 6.3) and if we assume a constant value of c_p, then $\hat{h} = c_p(T - T_{\text{ref}})$ and

$$\dot{m}\, c_p\,(T_b - T_{\text{ref}}) = \int_{A_c} \rho c_p u\,(T - T_{\text{ref}})\; dA_c \tag{7.4}$$

or simply

$$T_b = \frac{\displaystyle\int_{A_c} \rho c_p u T\, dA_c}{\dot{m} c_p} \tag{7.5}$$

In words, then,

$$T_b \equiv \frac{\text{rate of flow of enthalpy through a cross section}}{\text{rate of flow of heat capacity through a cross section}}$$

Thus, if the pipe were broken at any x-station and allowed to discharge into a mixing cup, the enthalpy of the mixed fluid in the cup would equal the average enthalpy of the fluid flowing through the cross section, and the temperature of the fluid in the cup would be T_b. This definition of T_b is perfectly general and applies to either laminar or turbulent flow. For a circular pipe, with $dA_c = 2\pi r\, dr$, eqn. (7.5) becomes

$$T_b = \frac{\displaystyle\int_0^R \rho c_p u T\, 2\pi r\, dr}{\displaystyle\int_0^R \rho c_p u\, 2\pi r\, dr} \tag{7.6}$$

A fully developed flow, from the thermal standpoint, is one for which the relative shape of the temperature profile does not change with x. We state this mathematically as

$$\frac{\partial}{\partial x}\left(\frac{T_w - T}{T_w - T_b}\right) = 0 \tag{7.7}$$

where T generally depends on x and r. This means that the profile can be scaled up or down with $T_w - T_b$. Of course, a flow must be hydrodynamically developed if it is to be thermally developed.

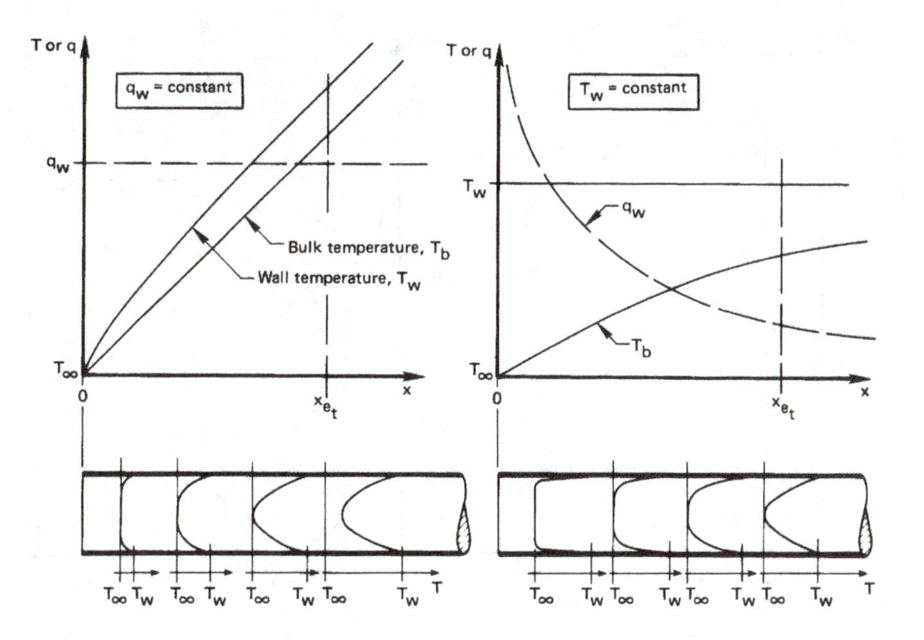

Figure 7.2 The thermal development of flows in tubes with a uniform wall heat flux and with a uniform wall temperature (the *entrance region*).

Figures 7.2 and 7.3 show the development of two flows and their subsequent behavior. The two flows are subjected to either a uniform wall heat flux or a uniform wall temperature. In Fig. 7.2 we see each flow develop until its temperature profile achieves a shape which, except for a linear stretching, it will retain thereafter. If we consider a small length of pipe, dx long with perimeter P, then its surface area is $P\,dx$ (e.g., $2\pi R\,dx$ for a circular pipe) and an energy balance on it is[1]

$$dQ = q_w\,P dx = \dot{m} d\hat{h}_b \tag{7.8}$$

$$= \dot{m} c_p\,dT_b \tag{7.9}$$

so that

$$\frac{dT_b}{dx} = \frac{q_w P}{\dot{m} c_p} \tag{7.10}$$

[1]Here we make the same approximations as were made in deriving the energy equation in Sect. 6.3.

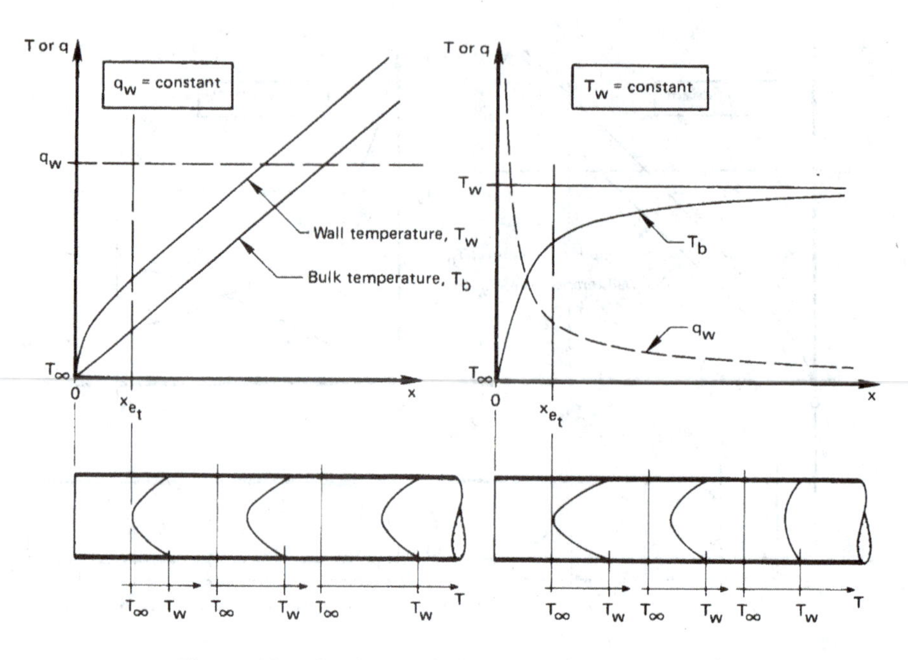

Figure 7.3 The thermal behavior of flows in tubes with a uniform wall heat flux and with a uniform temperature (the *thermally developed region*).

This result is also valid for the bulk temperature in a turbulent flow.

In Fig. 7.3 we see the fully developed variation of the temperature profile. If the flow is fully developed, the boundary layers are no longer growing thicker, and we expect that h will become constant. When q_w is constant, then $T_w - T_b$ will be constant in fully developed flow, so that the temperature profile will retain the same shape while the temperature rises at a constant rate at all values of r. Thus, at any radial position,

$$\frac{\partial T}{\partial x} = \frac{dT_b}{dx} = \frac{q_w P}{\dot{m} c_p} = \text{constant} \tag{7.11}$$

In the uniform wall temperature case, the temperature profile keeps the same shape, but its amplitude decreases with x, as does q_w. The lower right-hand corner of Fig. 7.3 has been drawn to conform with this requirement, as expressed in eqn. (7.7).

The velocity profile in laminar tube flows

The Buckingham pi-theorem tells us that if the hydrodynamic *entry length*, x_e, required to establish a fully developed velocity profile depends on u_{av}, μ, ρ, and D in three dimensions (kg, m, and s), then we expect to find two pi-groups:

$$\frac{x_e}{D} = \text{fn}\,(\text{Re}_D)$$

where $\text{Re}_D \equiv u_{av}D/\nu$. The matter of entry length is discussed by White [7.1, Chap. 4], who quotes

$$\frac{x_e}{D} \simeq 0.03\,\text{Re}_D \tag{7.12}$$

The constant, 0.03, guarantees that the laminar shear stress on the pipe wall will be within 5% of the value for fully developed flow when $x > x_e$. The number 0.05 can be used, instead, if a deviation of just 1.4% is desired. The thermal entry length, x_{e_t}, turns out to be different from x_e. We deal with it shortly.

The hydrodynamic entry length for a pipe carrying fluid at speeds near the transitional Reynolds number (2100) will extend beyond 100 diameters. Since heat transfer in pipes shorter than this is very often important, we will eventually have to deal with the entry region.

The velocity profile for a fully developed laminar incompressible pipe flow can be derived from the momentum equation for an axisymmetric flow. It turns out that the b.l. assumptions all happen to be valid for a fully developed pipe flow:

- The pressure is constant across any section.
- $\partial^2 u/\partial x^2$ is exactly zero.
- The radial velocity is not just small, but it is zero.
- The term $\partial u/\partial x$ is not just small, but it is zero.

The boundary layer equation for cylindrically symmetrical flows is quite similar to that for a flat surface, eqn. (6.13):

$$u\frac{\partial u}{\partial x} + v\frac{\partial u}{\partial r} = -\frac{1}{\rho}\frac{dp}{dx} + \frac{\nu}{r}\frac{\partial}{\partial r}\left(r\frac{\partial u}{\partial r}\right) \tag{7.13}$$

For fully developed flows, we go beyond the b.l. assumptions and set v and $\partial u/\partial x$ equal to zero as well, so eqn. (7.13) becomes

$$\frac{1}{r}\frac{d}{dr}\left(r\frac{du}{dr}\right) = \frac{1}{\mu}\frac{dp}{dx}$$

We integrate this twice and get

$$u = \left(\frac{1}{4\mu}\frac{dp}{dx}\right) r^2 + C_1 \ln r + C_2$$

The two b.c.'s on u express the no-slip (or zero-velocity) condition at the wall and the fact that u must be symmetrical in r:

$$u(r = R) = 0 \quad \text{and} \quad \left.\frac{du}{dr}\right|_{r=0} = 0$$

They give $C_1 = 0$ and $C_2 = (-dp/dx)R^2/4\mu$, so

$$u = \frac{R^2}{4\mu}\left(-\frac{dp}{dx}\right)\left[1 - \left(\frac{r}{R}\right)^2\right] \tag{7.14}$$

This is the familiar Hagen-Poiseuille[2] parabolic velocity profile. We can identify the lead constant $(-dp/dx)R^2/4\mu$ as the maximum centerline velocity, u_{max}. In accordance with the conservation of mass (see Problem 7.1), $2u_{\text{av}} = u_{\text{max}}$, so

$$\frac{u}{u_{\text{av}}} = 2\left[1 - \left(\frac{r}{R}\right)^2\right] \tag{7.15}$$

Thermal behavior of a flow with a uniform heat flux at the wall

The b.l. energy equation for a fully developed laminar incompressible flow, eqn. (6.40), takes the following simple form in a pipe flow where the radial velocity is equal to zero:

$$u\frac{\partial T}{\partial x} = \alpha\frac{1}{r}\frac{\partial}{\partial r}\left(r\frac{\partial T}{\partial r}\right) \tag{7.16}$$

[2]The German scientist G. Hagen showed experimentally how u varied with r, dp/dx, μ, and R, in 1839. J. Poiseuille (pronounced Pwa-zói or, more precisely, Pwä-zəē) did the same thing, almost simultaneously (1840), in France. Poiseuille was a physician interested in blood flow, and we find today that if medical students know nothing else about fluid flow, they know "Poiseuille's law."

For a fully developed flow with q_w = constant, T_w and T_b increase linearly with x. In particular, by integrating eqn. (7.10), we find

$$T_b(x) - T_{b_{\text{in}}} = \int_0^x \frac{q_w P}{\dot{m} c_p}\, dx = \frac{q_w P x}{\dot{m} c_p} \tag{7.17}$$

Then, from eqns. (7.11) and (7.1), we get

$$\frac{\partial T}{\partial x} = \frac{dT_b}{dx} = \frac{q_w P}{\dot{m} c_p} = \frac{q_w (2\pi R)}{\rho c_p u_{\text{av}} (\pi R^2)} = \frac{2 q_w \alpha}{u_{\text{av}} R k}$$

Using this result and eqn. (7.15) in eqn. (7.16), we obtain

$$4\left[1 - \left(\frac{r}{R}\right)^2\right] \frac{q_w}{Rk} = \frac{1}{r}\frac{d}{dr}\left(r \frac{dT}{dr}\right) \tag{7.18}$$

This ordinary d.e. in r can be integrated twice to obtain

$$T = \frac{4 q_w}{Rk}\left(\frac{r^2}{4} - \frac{r^4}{16R^2}\right) + C_1 \ln r + C_2 \tag{7.19}$$

The first b.c. on this equation is the symmetry condition, $\partial T/\partial r = 0$ at $r = 0$, and it gives $C_1 = 0$. The second b.c. is the definition of the mixing-cup temperature, eqn. (7.6). Substituting eqn. (7.19) with $C_1 = 0$ into eqn. (7.6) and carrying out the indicated integrations, we get

$$C_2 = T_b - \frac{7}{24}\frac{q_w R}{k}$$

so

$$T - T_b = \frac{q_w R}{k}\left[\left(\frac{r}{R}\right)^2 - \frac{1}{4}\left(\frac{r}{R}\right)^4 - \frac{7}{24}\right] \tag{7.20}$$

and at $r = R$, eqn. (7.20) gives

$$T_w - T_b = \frac{11}{24}\frac{q_w R}{k} = \frac{11}{48}\frac{q_w D}{k} \tag{7.21}$$

so the local Nu_D for fully developed flow, based on $h(x) = q_w/[T_w(x) - T_b(x)]$, is

$$\text{Nu}_D \equiv \frac{q_w D}{(T_w - T_b)k} = \frac{48}{11} = 4.364 \tag{7.22}$$

Equation (7.22) is surprisingly simple. Indeed, the fact that there is only one dimensionless group in it is predictable by dimensional analysis. In this case the dimensional functional equation is merely

$$h = \text{fn}(D, k)$$

We exclude ΔT, because h should be independent of ΔT in forced convection; μ, because the flow is parallel regardless of the viscosity; and ρu_{av}^2, because there is no influence of momentum in a laminar incompressible flow that never changes direction. This gives three variables, effectively in only two dimensions, W/K and m, resulting in just one dimensionless group, Nu_D, which must therefore be a constant.

Example 7.1

Water at 20°C flows through a small-bore tube 1 mm in diameter at a uniform speed of 0.2 m/s. The flow is fully developed at a point beyond which a constant heat flux of 6000 W/m^2 is imposed. How much farther down the tube will the water reach 74°C at its hottest point?

Solution. As a fairly rough approximation, we evaluate properties at $(74 + 20)/2 = 47$°C: $k = 0.6367$ W/m·K, $\alpha = 1.541 \times 10^{-7}$, and $\nu = 0.556\times10^{-6}$ m^2/s. Therefore, $\text{Re}_D = (0.001 \text{ m})(0.2 \text{ m/s})/0.556\times 10^{-6} \text{ m}^2/\text{s} = 360$, and the flow is laminar. Then, noting that T is greatest at the wall and setting $x = L$ at the point where $T_{\text{wall}} = 74$°C, eqn. (7.17) gives:

$$T_b(x = L) = 20 + \frac{q_w P}{\dot{m} c_p} L = 20 + \frac{4 q_w \alpha}{u_{\text{av}} D k} L$$

And eqn. (7.21) gives

$$74 = T_b(x = L) + \frac{11}{48}\frac{q_w D}{k} = 20 + \frac{4 q_w \alpha}{u_{\text{av}} D k} L + \frac{11}{48}\frac{q_w D}{k}$$

so

$$\frac{L}{D} = \left(54 - \frac{11}{48}\frac{q_w D}{k}\right)\frac{u_{\text{av}} k}{4 q_w \alpha}$$

or

$$\frac{L}{D} = \left[54 - \frac{11}{48}\frac{6000(0.001)}{0.6367}\right]\frac{0.2(0.6367)}{4(6000)1.541(10)^{-7}} = 1785$$

so the wall temperature reaches the limiting temperature of 74°C at

$$L = 1785(0.001 \text{ m}) = 1.785 \text{ m}$$

While we did not evaluate the thermal entry length here, it may be shown to be much, much less than 1785 diameters. ■

In the preceding example, the heat transfer coefficient is actually rather large

$$h = \mathrm{Nu}_D \frac{k}{D} = 4.364 \frac{0.6367}{0.001} = 2,778 \text{ W/m}^2\text{K}$$

The high h is a direct result of the small tube diameter, which keeps the thermal boundary layer thin and the thermal resistance low. This trend leads directly to the notion of a *microchannel heat exchanger.* Small scale fabrication technologies, such as have been developed in the semiconductor industry, allow us to create channels whose characteristic diameter is in the range of 100 μm. These yield heat transfer coefficients in the range of 10^4 W/m^{2}K for water [7.2]. If, instead, we use liquid sodium ($k \approx 80$ W/m·K) as the working fluid, the laminar flow heat transfer coefficient is on the order of 10^6 W/m^{2}K — a range usually associated with boiling processes!

Thermal behavior of the flow in an isothermal pipe

The dimensional analysis that showed Nu_D = constant for flow with a uniform heat flux at the wall is unchanged when the pipe wall is isothermal. Thus, Nu_D should still be constant. But this time (see, e.g., [7.3, Chap. 8]) the constant changes to

$$\mathrm{Nu}_D = 3.657, \qquad T_w = \text{constant} \tag{7.23}$$

for fully developed flow. The behavior of the bulk temperature is discussed in Sect. 7.4.

The thermal entrance region

The thermal entrance region is of great importance in laminar flow because the thermally undeveloped region becomes extremely long for higher-Pr fluids. The entry-length equation (7.12) takes the following form for the thermal entry region, where the velocity profile is assumed to be fully developed before heat transfer starts at $x = 0$[3]:

$$\frac{x_{e_t}}{D} \simeq \begin{cases} 0.034 \,\mathrm{Re}_D\mathrm{Pr} & \text{for } T_w = \text{constant} \\ 0.043 \,\mathrm{Re}_D\mathrm{Pr} & \text{for } q_w = \text{constant} \end{cases} \tag{7.24}$$

[3]The Nusselt number will be within 5% of the fully developed value beyond x_{e_t}. When the velocity and temperature profiles develop simultaneously, the coefficient next to $\mathrm{Re}_D\mathrm{Pr}$ ranges between about 0.028 and 0.053 depending upon the Prandtl number and the wall boundary condition [7.4, 7.5].

Entry lengths can become very long in certain cases. For the flow of cold water ($\Pr \simeq 10$), the entry length can reach more than 600 diameters as we approach the transitional Reynolds number. For Pr on the order of 10^4 (oil flows, for example), a fully developed profile is virtually unobtainable.

A complete analysis of the heat transfer rate in the thermal entry region becomes quite complicated. The reader interested in details should look at [7.3, Chap. 8]. Dimensional analysis of the entry problem shows that the local value of h depends on u_{av}, μ, ρ, D, c_p, k, and x—eight variables in m, s, kg, and J/K. This means that we should anticipate four pi-groups:

$$\text{Nu}_D = \text{fn}\left(\text{Re}_D, \Pr, x/D\right) \tag{7.25}$$

In other words, to the already familiar Nu_D, Re_D, and Pr, we add a new length parameter, x/D. The solution of the constant wall temperature problem, originally formulated by Graetz in 1885 [7.6] and solved in convenient form by Sellars, Tribus, and Klein in 1956 [7.7], includes an arrangement of these dimensionless groups, called the Graetz number:

$$\text{Graetz number, Gz} \equiv \frac{\text{Re}_D \Pr D}{x} \tag{7.26}$$

Figure 7.4 shows values of $\text{Nu}_D \equiv hD/k$ for both the uniform wall temperature and uniform wall heat flux cases. The independent variable in the figure is a dimensionless length equal to $2/\text{Gz}$. The figure also presents an average Nusselt number, $\overline{\text{Nu}}_D$ for the isothermal wall case:

$$\overline{\text{Nu}}_D \equiv \frac{\overline{h}D}{k} = \frac{D}{k}\left(\frac{1}{L}\int_0^L h\,dx\right) = \frac{1}{L}\int_0^L \text{Nu}_D\,dx \tag{7.27}$$

where, since $h = q(x)/[T_w - T_b(x)]$, we cannot simply average just q or ΔT. We show how to find the change in T_b using $\overline{h}$ for an isothermal wall in Sect. 7.4. For a fixed heat flux, the change in T_b is given by eqn. (7.17), and a value of $\overline{h}$ is not needed.

For an isothermal wall, the following curve fits are available for the Nusselt number in thermally developing flow [7.4]:

$$\text{Nu}_D = 3.657 + \frac{0.0018\ \text{Gz}^{1/3}}{\left(0.04 + \text{Gz}^{-2/3}\right)^2} \tag{7.28}$$

$$\overline{\text{Nu}}_D = 3.657 + \frac{0.0668\ \text{Gz}^{1/3}}{0.04 + \text{Gz}^{-2/3}} \tag{7.29}$$

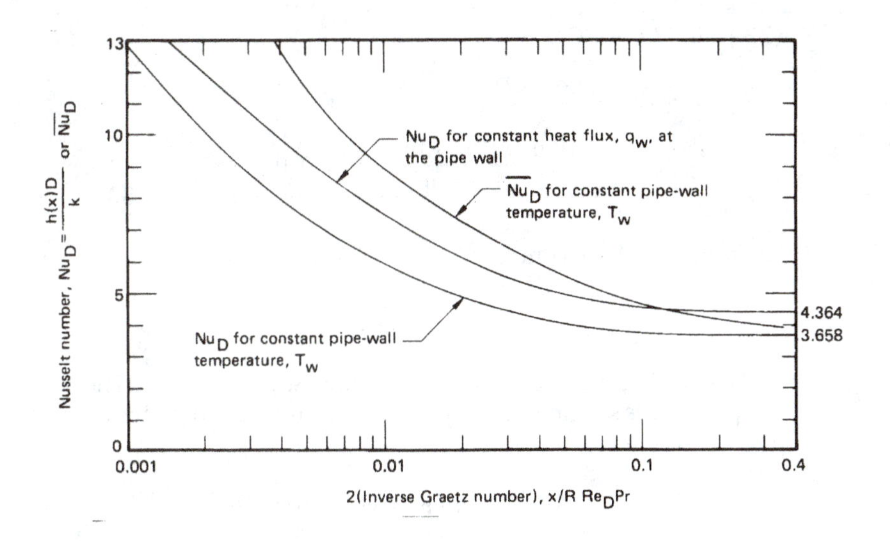

Figure 7.4 Local and average Nusselt numbers for the thermal entry region in a hydrodynamically developed laminar pipe flow.

The error is less than 14% for Gz > 1000 and less than 7% for Gz < 1000. For fixed q_w, a more complicated formula reproduces the exact result for local Nusselt number to within 1%:

$$\mathrm{Nu}_D = \begin{cases} 1.302\,\mathrm{Gz}^{1/3} - 1 & \text{for } 2 \times 10^4 \le \mathrm{Gz} \\ 1.302\,\mathrm{Gz}^{1/3} - 0.5 & \text{for } 667 \le \mathrm{Gz} \le 2 \times 10^4 \\ 4.364 + 0.263\,\mathrm{Gz}^{0.506}\, e^{-41/\mathrm{Gz}} & \text{for } 0 \le \mathrm{Gz} \le 667 \end{cases} \tag{7.30}$$

Example 7.2

A fully developed flow of air at 27°C moves at 2 m/s in a 1 cm I.D. pipe. An electric resistance heater surrounds the last 20 cm of the pipe and supplies a constant heat flux to bring the air out at $T_b = 40$°C. What power input is needed to do this? What will be the wall temperature at the exit?

Solution. This is a case in which the wall heat flux is uniform along the pipe. We first must compute $\mathrm{Gz}_{20\ \mathrm{cm}}$, evaluating properties at

$(27 + 40)/2 \simeq 34°\text{C}$.

$$\text{Gz}_{20\text{ cm}} = \frac{\text{Re}_D \text{Pr}\, D}{x} = \frac{\dfrac{(2\text{ m/s})(0.01\text{ m})}{16.4 \times 10^{-6}\text{ m}^2/\text{s}}(0.711)(0.01\text{ m})}{0.2\text{ m}} = 43.38$$

From eqn. 7.30, we compute $\text{Nu}_D = 5.05$, so

$$T_{w_{\text{exit}}} - T_b = \frac{q_w D}{5.05\, k}$$

Notice that we still have two unknowns, q_w and T_w. The bulk temperature is specified as 40°C, and q_w is obtained from this number by a simple energy balance:

$$q_w(2\pi R x) = \rho c_p u_{\text{av}}(T_b - T_{\text{entry}})\pi R^2$$

so

$$q_w = 1.159\frac{\text{kg}}{\text{m}^3} \cdot 1004\frac{\text{J}}{\text{kg·K}} \cdot 2\frac{\text{m}}{\text{s}} \cdot (40 - 27)°\text{C} \cdot \underbrace{\frac{R}{2x}}_{1/80} = 378\text{ W/m}^2$$

Then

$$T_{w_{\text{exit}}} = 40°\text{C} + \frac{(378\text{ W/m}^2)(0.01\text{ m})}{5.05(0.0266\text{ W/m·K})} = 68.1°\text{C}$$ ■

7.3 Turbulent pipe flow

Turbulent entry length

The entry lengths x_e and x_{e_t} are generally shorter in turbulent flow than in laminar flow. Table 7.1 gives the thermal entry length for various values of Pr and Re_D. These will give Nu_D values within 5% of the fully developed values. These results are for a uniform wall heat flux imposed on a hydrodynamically fully developed flow. Similar results have been obtained for a uniform wall temperature.

For Prandtl numbers typical of gases and nonmetallic liquids, the entry length is not strongly sensitive to the Reynolds number. For $\text{Pr} > 1$ in particular, the entry length is just a few diameters. This is because the

Table 7.1 Thermal entry lengths, x_{et}/D, for which Nu_D will be no more than 5% above its fully developed value in turbulent flow

Pr	Re_D		
	20,000	100,000	500,000
0.01	7	22	32
0.7	10	12	14
3.0	4	3	3

heat transfer rate is controlled by the thin thermal sublayer on the wall, and it develops very quickly.

Only liquid metals give fairly long thermal entrance lengths, and, for these fluids, x_{e_t} depends on both Re and Pr in a complicated way. Since liquid metals have very high thermal conductivities, the heat transfer rate is also more strongly affected by the temperature distribution in the center of the pipe. We discuss liquid metals in more detail at the end of this section.

When heat transfer begins at the pipe inlet, the velocity and temperature profiles develop simultaneously. The entry length is then very strongly affected by the shape of the inlet. For example, an inlet that induces vortices in the pipe—a sharp bend or contraction—can create a much longer entry length than occurs for a thermally developing flow. These vortices may require 20 to 40 diameters to die out. For various types of inlets, Bhatti and Shah [7.8] provide the following correlation for $\overline{\mathrm{Nu}}_D$ with $L/D > 3$ for air (or other fluids with $\mathrm{Pr} \approx 0.7$)

$$\frac{\overline{\mathrm{Nu}}_D}{\mathrm{Nu}_\infty} = 1 + \frac{C}{(L/D)^n} \qquad \text{for Pr} = 0.7 \tag{7.31}$$

where Nu_∞ is the fully developed value of the Nusselt number, and C and n depend on the inlet configuration as shown in Table 7.2.

Whereas the entry effect on the local Nusselt number is confined to a few ten's of diameters, the effect on the average Nusselt number may persist for a hundred diameters. This is because much additional length is needed to average out the higher heat transfer rates near the entry.

Table 7.2 Constants for the gas-flow simultaneous entry length correlation, eqn. (7.31), for various inlet configurations

Inlet configuration	C	n
Long, straight pipe	0.9756	0.760
Square-edged inlet	2.4254	0.676
180° circular bend	0.9759	0.700
90° circular bend	1.0517	0.629
90° sharp elbow	2.0152	0.614

Illustrative experiment

Figure 7.5 shows average heat transfer data given by Kreith [7.9, Chap. 8] for air flowing in a 1 in. I.D. isothermal pipe 60 in. in length. Let us see how these data compare with what we know about pipe flows thus far.

The data are plotted for a single Prandtl number on $\overline{\text{Nu}}_D$ vs. Re_D coordinates. This format is consistent with eqn. (7.25) in the fully developed range, but the actual pipe incorporates a significant entry region. Therefore, the data will reflect entry behavior.

For laminar flow, $\overline{\text{Nu}}_D \simeq 3.66$ at $\text{Re}_D = 750$. This is the correct value for an isothermal pipe. However, the pipe is too short for flow to be fully developed over much, if any, of its length. Therefore $\overline{\text{Nu}}_D$ is not constant in the laminar range. The rate of rise of $\overline{\text{Nu}}_D$ with Re_D becomes very great in the transitional range, which lies between $\text{Re}_D = 2100$ and about 5000 in this case. Above $\text{Re}_D \simeq 5000$, the flow is turbulent and it turns out that $\overline{\text{Nu}}_D \simeq \text{Re}_D^{0.8}$.

The Reynolds analogy and heat transfer

A form of the Reynolds analogy appropriate to fully developed turbulent pipe flow can be obtained from eqn. (6.111)

$$\text{St}_x = \frac{h}{\rho c_p u_\infty} = \frac{C_f(x)/2}{1 + 12.8\left(\text{Pr}^{0.68} - 1\right)\sqrt{C_f(x)/2}} \tag{6.111}$$

where h, in a pipe flow, is defined as $q_w/(T_w - T_b)$. We merely replace u_∞ with u_{av} and $C_f(x)$ with the friction coefficient for fully developed

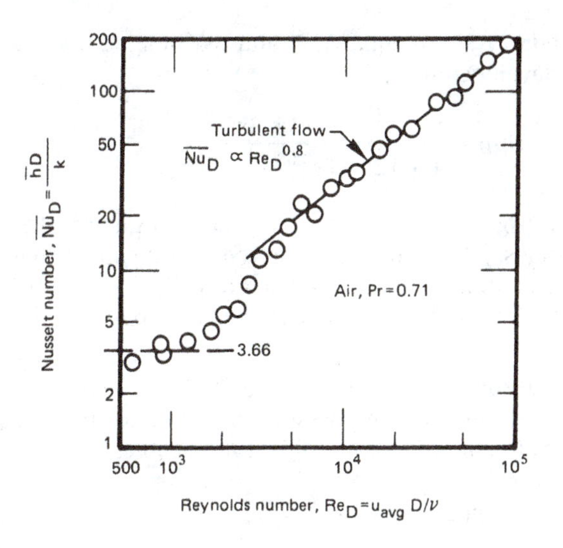

Figure 7.5 Heat transfer to air flowing in a 1 in. I.D., 60 in. long pipe (after Kreith [7.9]).

pipe flow, C_f (which is constant), to get

$$\text{St} = \frac{h}{\rho c_p u_{\text{av}}} = \frac{C_f/2}{1 + 12.8\left(\text{Pr}^{0.68} - 1\right)\sqrt{C_f/2}} \tag{7.32}$$

This equation is not accurate at very low Pr's, but it has the advantage of applying to either uniform q_w or uniform T_w situations. It is limited to smooth walls.

The frictional resistance to flow in a pipe is normally expressed in terms of the Darcy-Weisbach friction factor, f [recall eqn. (3.25)]:

$$f \equiv \frac{\text{head loss}}{\left(\dfrac{\text{pipe length}}{D}\,\dfrac{u_{\text{av}}^2}{2}\right)} = \frac{\Delta p}{\left(\dfrac{L}{D}\,\dfrac{\rho u_{\text{av}}^2}{2}\right)} \tag{7.33}$$

where Δp is the pressure drop in a pipe of length L. However,

$$\tau_w = \frac{\text{frictional force on liquid}}{\text{surface area of pipe}} = \frac{\Delta p\left[(\pi/4)D^2\right]}{\pi D L} = \frac{\Delta p D}{4L}$$

so

$$f = \frac{\tau_w}{\rho u_{\text{av}}^2/8} = 4C_f \tag{7.34}$$

Substituting eqn. (7.34) in eqn. (7.32) and rearranging the result, we obtain, for fully developed flow,

$$\mathrm{Nu}_D = \frac{(f/8)\mathrm{Re}_D\,\mathrm{Pr}}{1 + 12.8\left(\mathrm{Pr}^{0.68} - 1\right)\sqrt{f/8}} \tag{7.35}$$

The friction factor is given graphically in Fig. 7.6 as a function of Re_D and the relative roughness, ε/D, where ε is the root-mean-square roughness of the pipe wall. Equation (7.35) can be used directly along with Fig. 7.6 to calculate the Nusselt number for smooth-walled pipes ($\varepsilon/D = 0$).

Historical formulations. A number of the earliest equations for the Nusselt number in turbulent pipe flow were based on Reynolds analogy in the form of eqn. (6.76), which for a pipe flow becomes

$$\mathrm{St} = \frac{C_f}{2}\,\mathrm{Pr}^{-2/3} = \frac{f}{8}\,\mathrm{Pr}^{-2/3} \tag{7.36}$$

or

$$\mathrm{Nu}_D = \mathrm{Re}_D\,\mathrm{Pr}^{1/3}(f/8) \tag{7.37}$$

For smooth pipes, the curve $\varepsilon/D = 0$ in Fig. 7.6 is approximately given by this equation:

$$\frac{f}{4} = C_f = \frac{0.046}{\mathrm{Re}_D^{0.2}} \tag{7.38}$$

in the range $20{,}000 < \mathrm{Re}_D < 300{,}000$, so eqn. (7.37) becomes

$$\mathrm{Nu}_D = 0.023\,\mathrm{Pr}^{1/3}\,\mathrm{Re}_D^{0.8}$$

for smooth pipes. This result was given by Colburn [7.10] in 1933. Actually, it is quite similar to an earlier result developed by Dittus and Boelter in 1930 (see [7.11, pg. 552]) for smooth pipes:

$$\mathrm{Nu}_D = 0.0243\,\mathrm{Pr}^{0.4}\,\mathrm{Re}_D^{0.8} \tag{7.39}$$

These equations are intended for reasonably low temperature differences under which properties can be evaluated at a mean temperature $(T_b + T_w)/2$. In 1936, a study by Sieder and Tate [7.12] showed that when $|T_w - T_b|$ is large enough to cause significant changes of μ, the Colburn

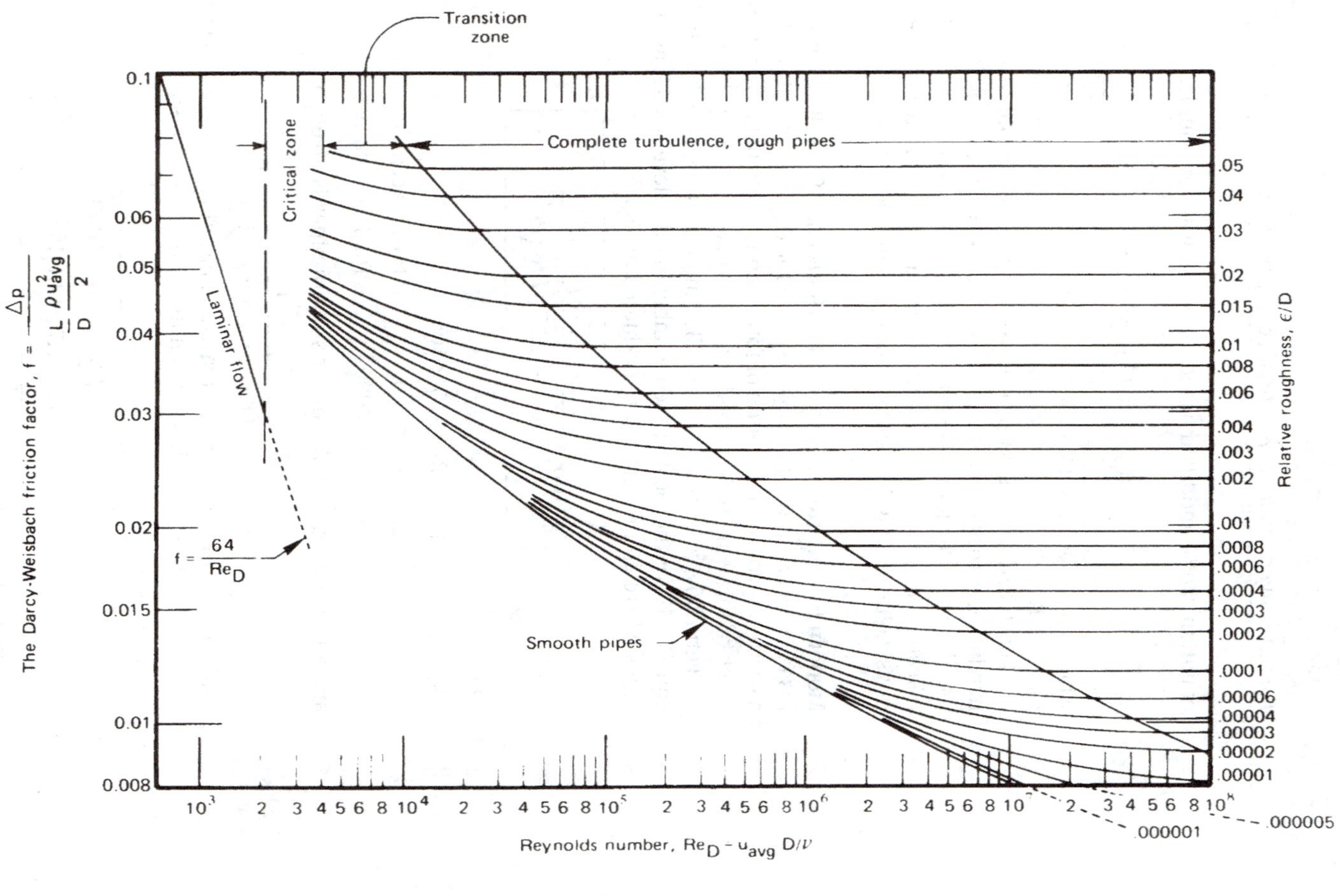

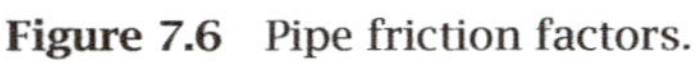

Figure 7.6 Pipe friction factors.

equation can be modified in the following way for liquids:

$$\mathrm{Nu}_D = 0.023\,\mathrm{Re}_D^{0.8}\,\mathrm{Pr}^{1/3}\left(\frac{\mu_b}{\mu_w}\right)^{0.14} \tag{7.40}$$

where all properties are evaluated at the local bulk temperature except μ_w, which is the viscosity evaluated at the wall temperature.

These early relations proved to be fair approximations. They gave maximum errors of +25% and −40% in the range $0.67 \leqslant \mathrm{Pr} < 100$ and usually were considerably more accurate than this. However, subsequent research has provided far more data, and a much better theoretical and physical understanding of how to represent them accurately.

Modern formulations. During the 1950s and 1960s, B. S. Petukhov and his co-workers at the Moscow Institute for High Temperature developed a vastly improved description of forced convection heat transfer in pipes. Much of this work is described in a 1970 survey article by Petukhov [7.13].

Petukhov recommends the following equation, which is built from eqn. (7.35), for the local Nusselt number in fully developed flow in smooth pipes where all properties are evaluated at T_b.

$$\boxed{\mathrm{Nu}_D = \frac{(f/8)\,\mathrm{Re}_D\,\mathrm{Pr}}{1.07 + 12.7\sqrt{f/8}\left(\mathrm{Pr}^{2/3} - 1\right)}} \tag{7.41}$$

where

$$10^4 < \mathrm{Re}_D < 5 \times 10^6$$

$$0.5 < \mathrm{Pr} < 200 \qquad \text{for 6\% accuracy}$$

$$200 \leqslant \mathrm{Pr} < 2000 \qquad \text{for 10\% accuracy}$$

and where the friction factor for smooth pipes is given by

$$f = \frac{1}{\left(1.82 \log_{10} \mathrm{Re}_D - 1.64\right)^2} \tag{7.42}$$

Gnielinski [7.14] later showed that the range of validity could be extended down to the transition Reynolds number by making a small adjustment to eqn. (7.41):

$$\boxed{\mathrm{Nu}_D = \frac{(f/8)\,(\mathrm{Re}_D - 1000)\,\mathrm{Pr}}{1 + 12.7\sqrt{f/8}\left(\mathrm{Pr}^{2/3} - 1\right)}} \tag{7.43}$$

for $2300 \le \mathrm{Re}_D \le 5 \times 10^6$.

Variations in physical properties. Sieder and Tate's work on property variations was also refined in later years [7.13]. The effect of variable physical properties is dealt with differently for liquids and gases. In both cases, the Nusselt number is first calculated with all properties evaluated at T_b using eqn. (7.41) or (7.43). For liquids, one then corrects by multiplying with a viscosity ratio. Over the interval $0.025 \leq (\mu_b/\mu_w) \leq 12.5$,

$$\mathrm{Nu}_D = \mathrm{Nu}_D\Big|_{T_b}\left(\frac{\mu_b}{\mu_w}\right)^n \qquad \text{where } n = \begin{cases} 0.11 & \text{for } T_w > T_b \\ 0.25 & \text{for } T_w < T_b \end{cases} \tag{7.44}$$

For gases and a temperatures ratio *in kelvins* within $0.27 \leq (T_b/T_w) \leq 2.7$,

$$\mathrm{Nu}_D = \mathrm{Nu}_D\Big|_{T_b}\left(\frac{T_b}{T_w}\right)^n \qquad \text{where } n = \begin{cases} 0.47 & \text{for } T_w > T_b \\ 0 & \text{for } T_w < T_b \end{cases} \tag{7.45}$$

After eqn. (7.43) is used to calculate Nu_D, it should also be corrected for the effect of variable viscosity. For liquids, with $0.5 \leq (\mu_b/\mu_w) \leq 3$

$$f = f\Big|_{T_b} \times K \qquad \text{where } K = \begin{cases} (7 - \mu_b/\mu_w)/6 & \text{for } T_w > T_b \\ (\mu_b/\mu_w)^{-0.24} & \text{for } T_w < T_b \end{cases} \tag{7.46}$$

For gases, the data are much weaker [7.15, 7.16]. For $0.14 \leq (T_b/T_w) \leq 3.3$

$$f = f\Big|_{T_b}\left(\frac{T_b}{T_w}\right)^m \qquad \text{where } m \approx \begin{cases} 0.23 & \text{for } T_w > T_b \\ 0.23 & \text{for } T_w < T_b \end{cases} \tag{7.47}$$

Example 7.3

A 21.5 kg/s flow of water is dynamically and thermally developed in a 12 cm I.D. pipe. The pipe is held at 90°C and $\varepsilon/D = 0$. Find h and f where the bulk temperature of the fluid has reached 50°C.

SOLUTION.

$$u_{\mathrm{av}} = \frac{\dot{m}}{\rho A_c} = \frac{21.5}{977\pi(0.06)^2} = 1.946 \text{ m/s}$$

so

$$\mathrm{Re}_D = \frac{u_{\mathrm{av}} D}{\nu} = \frac{1.946(0.12)}{4.07 \times 10^{-7}} = 573{,}700$$

and

$$\text{Pr} = 2.47, \qquad \frac{\mu_b}{\mu_w} = \frac{5.38 \times 10^{-4}}{3.10 \times 10^{-4}} = 1.74$$

From eqn. (7.42), $f = 0.0128$ at T_b, and since $T_w > T_b$, $n = 0.11$ in eqn. (7.44). Thus, with eqn. (7.41) we have

$$\text{Nu}_D = \frac{(0.0128/8)(5.74 \times 10^5)(2.47)}{1.07 + 12.7\sqrt{0.0128/8}\,(2.47^{2/3} - 1)}(1.74)^{0.11} = 1617$$

or

$$h = \text{Nu}_D \frac{k}{D} = 1617 \frac{0.661}{0.12} = 8{,}907 \text{ W/m}^2\text{K}$$

The corrected friction factor, with eqn. (7.46), is

$$f = (0.0128)\,(7 - 1.74)/6 = 0.0122$$ ■

Rough-walled pipes. Roughness on a pipe wall can disrupt the viscous and thermal sublayers if it is sufficiently large. Figure 7.6 shows the effect of increasing root-mean-square roughness height ε on the friction factor, f. As the Reynolds number increases, the viscous sublayer becomes thinner and smaller levels of roughness influence f. Some typical pipe roughnesses are given in Table 7.3.

The importance of a given level of roughness on friction and heat transfer can determined by comparing ε to the sublayer thickness. We saw in Sect. 6.7 that the thickness of the sublayer is around 30 times ν/u^*, where $u^* = \sqrt{\tau_w/\rho}$ was the friction velocity. We can define the ratio of ε and ν/u^* as the *roughness Reynolds number*, Re_ε

$$\text{Re}_\varepsilon \equiv \frac{u^*\varepsilon}{\nu} = \text{Re}_D \frac{\varepsilon}{D}\sqrt{\frac{f}{8}} \tag{7.48}$$

where the second equality follows from the definitions of u^* and f (and a little algebra). Experimental data then show that the smooth, transitional, and fully rough regions seen in Fig. 7.6 correspond to the following ranges of Re_ε:

$\text{Re}_\varepsilon < 5$	hydraulically smooth
$5 \le \text{Re}_\varepsilon \le 70$	transitionally rough
$70 < \text{Re}_\varepsilon$	fully rough

Table 7.3 Typical wall roughness of commercially available pipes when new.

Pipe	*ε (μm)*	*Pipe*	*ε (μm)*
Glass	0.31	Asphalted cast iron	120.
Drawn tubing	1.5	Galvanized iron	150.
Steel or wrought iron	46.	Cast iron	260.

In the fully rough regime, Bhatti and Shah [7.8] provide the following correlation for the local Nusselt number

$$\mathrm{Nu}_D = \frac{(f/8)\,\mathrm{Re}_D\,\mathrm{Pr}}{1+\sqrt{f/8}\left(4.5\,\mathrm{Re}_\varepsilon^{0.2}\mathrm{Pr}^{0.5}-8.48\right)} \tag{7.49}$$

which applies for the ranges

$$10^4 \leqslant \mathrm{Re}_D, \quad 0.5 \leqslant \mathrm{Pr} \leqslant 10, \quad \text{and } 0.002 \leqslant \frac{\varepsilon}{D} \leqslant 0.05$$

The corresponding friction factor may be computed from Haaland's equation [7.17]:

$$f = \frac{1}{\left\{1.8\log_{10}\left[\dfrac{6.9}{\mathrm{Re}_D}+\left(\dfrac{\varepsilon/D}{3.7}\right)^{1.11}\right]\right\}^2} \tag{7.50}$$

The heat transfer coefficient on a rough wall can be several times that for a smooth wall at the same Reynolds number. The friction factor, and thus the pressure drop and pumping power, will also be higher. Nevertheless, designers sometimes deliberately roughen tube walls so as to raise h and reduce the surface area needed for heat transfer. Several manufacturers offer tubing that has had some pattern of roughness impressed upon its interior surface. Periodic ribs are one common configuration. Specialized correlations have been developed for a number of such configurations [7.18, 7.19].

Example 7.4

Repeat Example 7.3, now assuming the pipe to be cast iron with a wall roughness of $\varepsilon = 260$ μm.

Solution. The Reynolds number and physical properties are unchanged. From eqn. (7.50)

$$f = \left\{1.8 \log_{10}\left[\frac{6.9}{573,700} + \left(\frac{260 \times 10^{-6}/0.12}{3.7}\right)^{1.11}\right]\right\}^{-2}$$

$$= 0.02424$$

The roughness Reynolds number is then

$$\mathrm{Re}_\varepsilon = (573,700)\frac{260 \times 10^{-6}}{0.12}\sqrt{\frac{0.02424}{8}} = 68.4$$

This corresponds to fully rough flow. With eqn. (7.49) we have

$$\mathrm{Nu}_D = \frac{(0.02424/8)(5.74 \times 10^5)(2.47)}{1 + \sqrt{0.02424/8}\,[4.5(68.4)^{0.2}(2.47)^{0.5} - 8.48]}$$

$$= 2,985$$

so

$$h = 2985\,\frac{0.661}{0.12} = 16.4 \text{ kW/m}^2\text{K}$$

In this case, wall roughness causes a factor of 1.8 increase in h and a factor of 2.0 increase in f and the pumping power. We have omitted the variable properties corrections used in Example 7.3 because they apply only to smooth-walled pipes. ■

Heat transfer to fully developed liquid-metal flows in tubes

A dimensional analysis of the forced convection flow of a liquid metal over a flat surface [recall eqn. (6.60) et seq.] showed that

$$\mathrm{Nu} = \mathrm{fn(Pe)} \tag{7.51}$$

because viscous influences were confined to a region very close to the wall. Thus, the thermal b.l., which extends far beyond δ, is hardly influenced by the dynamic b.l. or by viscosity. During heat transfer to liquid metals in pipes, the same thing occurs as is illustrated in Fig. 7.7. The region of thermal influence extends far beyond the laminar sublayer, when $\mathrm{Pr} \ll 1$, and the temperature profile is not influenced by the sublayer.

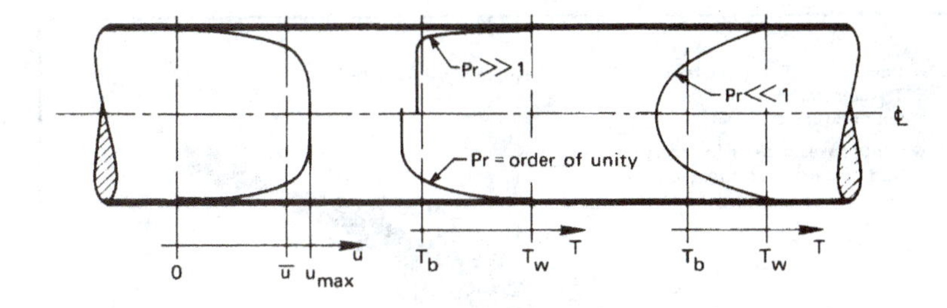

Figure 7.7 Velocity and temperature profiles during fully developed turbulent flow in a pipe.

Conversely, if Pr $\gg$ 1, the temperature profile is largely shaped within the laminar sublayer. At high or even moderate Pr's, ν is therefore very important, but at low Pr's it vanishes from the functional equation. Equation (7.51) thus applies to pipe flows as well as to flow over a flat surface.

Lubarsky and Kaufman [7.20] collected measured values of Nu_D for liquid metals flowing in pipes with a constant wall heat flux, q_w (see Fig. 7.8). It is clear that while most of the data correlate fairly well on Nu_D vs. Pe coordinates, certain sets of data are badly scattered. This occurs in part because liquid metal experiments are hard to carry out. Temperature differences are small and must often be measured at high temperatures. Some of the very low data might possibly result from a failure of the metals to wet the inner surface of the pipe.

Another problem that besets liquid metal heat transfer measurements is the very great difficulty involved in keeping such liquids pure. Most impurities tend to result in lower values of h. Thus, most of the Nusselt numbers in Fig. 7.8 have probably been lowered by impurities in the liquids; the few high values are probably the more correct ones for pure liquids.

A body of theory for turbulent liquid metal heat transfer yields a prediction of the form

$$\text{Nu}_D = C_1 + C_2 \,\text{Pe}_D^{0.8} \tag{7.52}$$

where this Péclét number is defined as $\text{Pe}_D = u_{\text{av}}D/\alpha$. The constants, however, are a problem. They usually lie in the ranges $2 \leqslant C_1 \leqslant 7$ and $0.0185 \leqslant C_2 \leqslant 0.386$ according to the test circumstances. Using the few reliable data sets available for uniform wall temperature conditions,

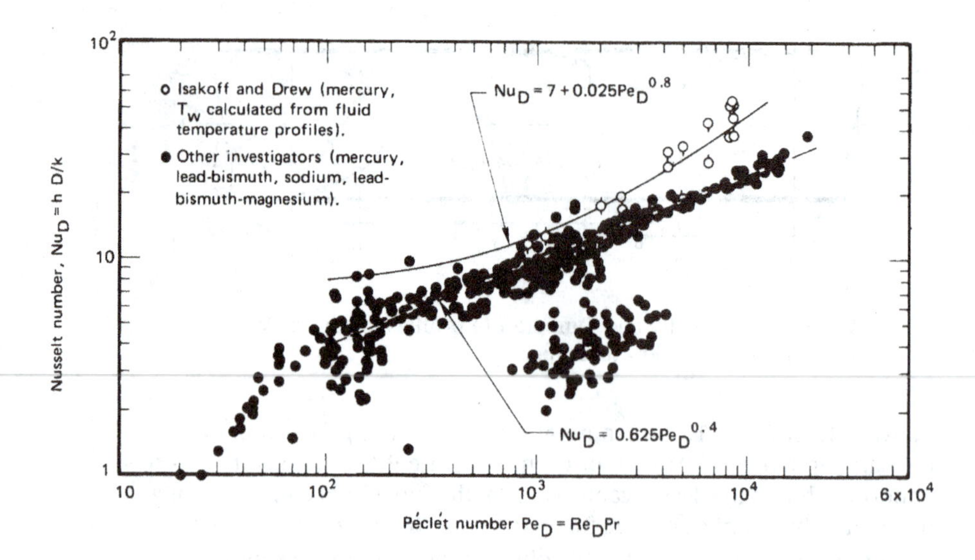

Figure 7.8 Comparison of measured and predicted Nusselt numbers for liquid metals heated in long tubes with uniform wall heat flux, q_w. (See NACA TN 336, 1955, for details and data source references.)

Reed [7.21] recommends

$$\mathrm{Nu}_D = 3.3 + 0.02\,\mathrm{Pe}_D^{0.8} \tag{7.53}$$

(Earlier work by Seban and Shimazaki [7.22] had suggested $C_1 = 4.8$ and $C_2 = 0.025$.) For uniform wall heat flux, many more data are available, and Lyon [7.23] recommends the following equation, shown in Fig. 7.8:

$$\mathrm{Nu}_D = 7 + 0.025\,\mathrm{Pe}_D^{0.8} \tag{7.54}$$

In both these equations, properties should be evaluated at the average of the inlet and outlet bulk temperatures and the pipe flow should have $L/D > 60$ and $\mathrm{Pe}_D > 100$. For lower Pe_D, axial heat conduction in the liquid metal may become significant.

Although eqns. (7.53) and (7.54) are probably correct for pure liquids, we cannot overlook the fact that the liquid metals in actual use are seldom pure. Lubarsky and Kaufman [7.20] put the following line through the bulk of the data in Fig. 7.8:

$$\mathrm{Nu}_D = 0.625\,\mathrm{Pe}_D^{0.4} \tag{7.55}$$

So what is a design engineer to do? The use of eqn. (7.55) for q_w = constant is far less optimistic than the use of eqn. (7.54). We should probably use it if it is safer to err on the low side.

7.4 Heat transfer surface viewed as a heat exchanger

Let us reconsider the problem of a fluid flowing through a pipe with a uniform wall temperature. By now we can predict $\overline{h}$ for a pretty wide range of conditions. Suppose that we need to know the net heat transfer to a pipe of known length once $\overline{h}$ is known. This problem is complicated by the fact that the bulk temperature, T_b, is varying along its length.

However, we need only recognize that such a section of pipe is a heat exchanger whose overall heat transfer coefficient, U (between the wall and the bulk), is just $\overline{h}$. Thus, if we wish to know how much pipe surface area is needed to raise the bulk temperature from T_{b_in} to T_{b_out}, we can calculate it as follows:

$$Q = (\dot{m}c_p)_b\left(T_{b_\text{out}} - T_{b_\text{in}}\right) = \overline{h}A(\text{LMTD})$$

or

$$A = \frac{(\dot{m}c_p)_b\left(T_{b_\text{out}} - T_{b_\text{in}}\right)}{\overline{h}} \frac{\ln\left(\dfrac{T_{b_\text{out}} - T_w}{T_{b_\text{in}} - T_w}\right)}{\left(T_{b_\text{out}} - T_w\right) - \left(T_{b_\text{in}} - T_w\right)} \tag{7.56}$$

By the same token, heat transfer in a duct can be analyzed with the effectiveness method (Sect. 3.3) if the exiting fluid temperature is unknown. Suppose that we do not know T_{b_out} in the example above. Then we can write an energy balance at any cross section, as we did in eqn. (7.8):

$$dQ = q_w P\,dx = hP\left(T_w - T_b\right)dx = \dot{m}c_p\,dT_b$$

Integration can be done from $T_b(x = 0) = T_{b_\text{in}}$ to $T_b(x = L) = T_{b_\text{out}}$

$$\int_0^L \frac{hP}{\dot{m}c_p}\,dx = -\int_{T_{b_\text{in}}}^{T_{b_\text{out}}} \frac{d(T_w - T_b)}{(T_w - T_b)}$$

$$\frac{P}{\dot{m}c_p}\int_0^L h\,dx = -\ln\left(\frac{T_w - T_{b_\text{out}}}{T_w - T_{b_\text{in}}}\right)$$

We recognize in this the definition of $\overline{h}$ from eqn. (7.27). Hence,

$$\frac{\overline{h}PL}{\dot{m}c_p} = -\ln\left(\frac{T_w - T_{b_\text{out}}}{T_w - T_{b_\text{in}}}\right)$$

which can be rearranged as

$$\frac{T_{b_\text{out}} - T_{b_\text{in}}}{T_w - T_{b_\text{in}}} = 1 - \exp\left(-\frac{\overline{h}PL}{\dot{m}c_p}\right) \tag{7.57}$$

This equation applies to either laminar or turbulent flow. It will give the variation of bulk temperature if T_{b_out} is replaced by $T_b(x)$, L is replaced by x, and $\overline{h}$ is adjusted accordingly.

The left-hand side of eqn. (7.57) is the heat exchanger effectiveness. On the right-hand side we replace U with $\overline{h}$; we note that $PL = A$, the exchanger surface area; and we write $C_\text{min} = \dot{m}c_p$. Since T_w is uniform, the stream that it represents must have a very large capacity rate, so that $C_\text{min}/C_\text{max} = 0$. Under these substitutions, we identify the argument of the exponential as $\text{NTU} = UA/C_\text{min}$, and eqn. (7.57) becomes

$$\varepsilon = 1 - \exp(-\text{NTU}) \tag{7.58}$$

which we could have obtained directly, from either eqn. (3.20) or (3.21), by setting $C_\text{min}/C_\text{max} = 0$. A heat exchanger for which one stream is isothermal, so that $C_\text{min}/C_\text{max} = 0$, is sometimes called a *single-stream* heat exchanger.

Equation (7.57) applies to ducts of any cross-sectional shape. We can cast it in terms of the *hydraulic diameter*, $D_h = 4A_c/P$, by substituting $\dot{m} = \rho u_\text{av} A_c$:

$$\frac{T_{b_\text{out}} - T_{b_\text{in}}}{T_w - T_{b_\text{in}}} = 1 - \exp\left(-\frac{\overline{h}PL}{\rho u_\text{av} c_p A_c}\right) \tag{7.59a}$$

$$= 1 - \exp\left(-\frac{\overline{h}}{\rho u_\text{av} c_p}\frac{4L}{D_h}\right) \tag{7.59b}$$

For a circular tube, with $A_c = \pi D^2/4$ and $P = \pi D$, $D_h = 4(\pi D^2/4)/(\pi D) = D$. To use eqn. (7.59b) for a noncircular duct, of course, we will need the value of $\overline{h}$ for its more complex geometry. We consider this issue in the next section.

Example 7.5

Air at 20°C is hydrodynamically fully developed as it flows in a 1 cm I.D. pipe. The average velocity is 0.7 m/s. If it enters a section where the pipe wall is at 60°C, what is the temperature 0.25 m farther downstream?

Solution.

$$\mathrm{Re}_D = \frac{u_{av}D}{\nu} = \frac{(0.7)(0.01)}{1.66\times10^{-5}} = 422$$

The flow is therefore laminar. To account for the thermal entry region, we compute the Graetz number from eqn. (7.26)

$$\mathrm{Gz} = \frac{\mathrm{Re}_D \mathrm{Pr}\, D}{x} = \frac{(422)(0.709)(0.01)}{0.25} = 12.0$$

Substituting this value into eqn. (7.29), we find $\overline{\mathrm{Nu}}_D = 4.32$. Thus,

$$\overline{h} = \frac{3.657(0.0268)}{0.01} = 11.6\ \mathrm{W/m^2K}$$

Then, using eqn. (7.59b),

$$\frac{T_{b_{out}} - T_{b_{in}}}{T_w - T_{b_{in}}} = 1 - \exp\left[-\frac{11.6}{1.14(1007)(0.7)}\,\frac{4(0.25)}{0.01}\right]$$

so that

$$\frac{T_b - 20}{60 - 20} = 0.764 \qquad \text{or} \qquad T_b = 50.6°\mathrm{C}$$ ■

7.5 Heat transfer coefficients for noncircular ducts

So far, we have focused on flows within circular tubes, which are by far the most common configuration. Nevertheless, other cross-sectional shapes often occur. For example, the fins of a heat exchanger may form a rectangular passage through which air flows. Sometimes, the passage cross-section is very irregular, as might happen when fluid passes through a clearance between other objects. In situations like these, all the qualitative ideas that we developed in Sections 7.1–7.3 still apply, but the Nusselt numbers for circular tubes cannot be used in calculating heat transfer rates.

The hydraulic diameter, which was introduced in connection with eqn. (7.59b), provides a basis for approximating heat transfer coefficients in noncircular ducts. Recall that the hydraulic diameter is defined as

$$D_h \equiv \frac{4A_c}{P} \tag{7.60}$$

where A_c is the cross-sectional area and P is the passage's wetted perimeter (Fig. 7.9). The hydraulic diameter measures the fluid area per unit

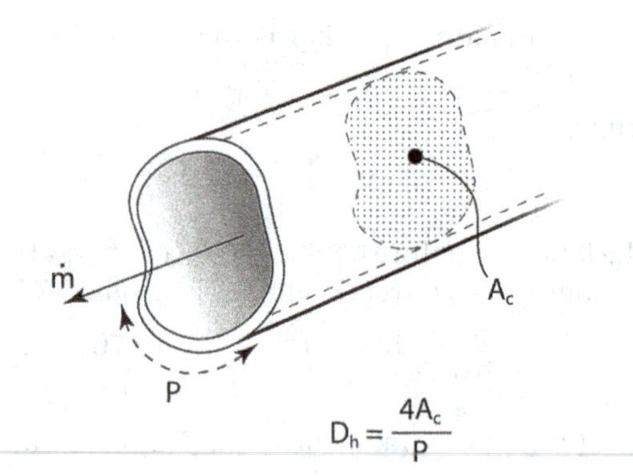

Figure 7.9 Flow in a noncircular duct.

length of wall. In turbulent flow, where most of the convection resistance is in the sublayer on the wall, this ratio determines the heat transfer coefficient to within about ±20% across a broad range of duct shapes. In fully-developed laminar flow, where the thermal resistance extends into the core of the duct, the heat transfer coefficient depends on the details of the duct shape, and D_h alone cannot define the heat transfer coefficient. Nevertheless, the hydraulic diameter provides an appropriate characteristic length for cataloging laminar Nusselt numbers.

The factor of four in the definition of D_h ensures that it gives the actual diameter of a circular tube. We noted in the preceding section that, for a circular tube of diameter D, $D_h = D$. Some other important cases include:

a rectangular duct of width a and height b

$$D_h = \frac{4\,ab}{2a + 2b} = \frac{2ab}{a + b} \tag{7.61a}$$

an annular duct of inner diameter D_i and outer diameter D_o

$$\begin{aligned} D_h &= \frac{4\,(\pi D_o^2/4 - \pi D_i^2/4)}{\pi\,(D_o + D_i)} \\ &= (D_o - D_i) \end{aligned} \tag{7.61b}$$

and, for very wide parallel plates, eqn. (7.61a) with $a \gg b$ gives

two parallel plates a distance b apart

$$D_h = 2b \tag{7.61c}$$

Turbulent flow in noncircular ducts

With some caution, we may use D_h directly in place of the circular tube diameter when calculating turbulent heat transfer coefficients and bulk temperature changes. Specifically, D_h replaces D in the Reynolds number, which is then used to calculate f and Nu_{D_h} from the circular tube formulas. The mass flow rate and the bulk velocity must be based on the true cross-sectional area, which does *not* usually equal $\pi D_h^2/4$ (see Problem 7.46). The following example illustrates the procedure.

Example 7.6

An air duct carries chilled air at an inlet bulk temperature of $T_{b_{\text{in}}} = 17°\text{C}$ and a speed of 1 m/s. The duct is made of thin galvanized steel, has a square cross-section of 0.3 m by 0.3 m, and is not insulated. A length of the duct 15 m long runs outdoors through warm air at $T_\infty = 37°\text{C}$. The heat transfer coefficient on the outside surface, due to natural convection and thermal radiation, is 5 $\text{W/m}^2\text{K}$. Find the bulk temperature change of the air over this length.

Solution. The hydraulic diameter, from eqn. (7.61a) with $a = b$, is simply

$$D_h = a = 0.3 \text{ m}$$

Using properties of air at the inlet temperature (290 K), the Reynolds number is

$$\mathrm{Re}_{D_h} = \frac{u_{\text{av}} D_h}{\nu} = \frac{(1)(0.3)}{(1.578 \times 10^{-5})} = 19{,}011$$

The Reynolds number for turbulent transition in a noncircular duct is typically approximated by the circular tube value of about 2300, so this flow is turbulent. The friction factor is obtained from eqn. (7.42)

$$f = \left[1.82 \log_{10}(19{,}011) - 1.64\right]^{-2} = 0.02646$$

and the Nusselt number is found with Gnielinski's equation, (7.43)

$$\mathrm{Nu}_{D_h} = \frac{(0.02646/8)(19{,}011 - 1{,}000)(0.713)}{1 + 12.7\sqrt{0.02646/8}\left[(0.713)^{2/3} - 1\right]} = 49.82$$

The heat transfer coefficient is

$$h = \mathrm{Nu}_{D_h} \frac{k}{D_h} = \frac{(49.82)(0.02623)}{0.3} = 4.371 \text{ W/m}^2\text{K}$$

The remaining problem is to find the bulk temperature change. The thin metal duct wall offers little thermal resistance, but convection resistance outside the duct must be considered. Heat travels first from the air at T_∞ through the outside heat transfer coefficient to the duct wall, through the duct wall, and then through the inside heat transfer coefficient to the flowing air — effectively through three resistances in series from the fixed temperature T_∞ to the rising temperature T_b. We have seen in Section 2.4 that an overall heat transfer coefficient may be used to describe such series resistances. Here, with $A_{\text{inside}} \simeq A_{\text{outside}}$, we find U based on inside area to be

$$U = \frac{1}{A_{\text{inside}}}\left[\frac{1}{(hA)_{\text{inside}}} + \underbrace{R_{t\,\text{wall}}}_{\text{neglect}} + \frac{1}{(hA)_{\text{outside}}}\right]^{-1}$$

$$= \left(\frac{1}{4.371} + \frac{1}{5}\right)^{-1} = 2.332\ \text{W/m}^2\text{K}$$

We then adapt eqn. (7.59b) by replacing $\overline{h}$ by U and T_w by T_∞:

$$\frac{T_{b_{\text{out}}} - T_{b_{\text{in}}}}{T_\infty - T_{b_{\text{in}}}} = 1 - \exp\left(-\frac{U}{\rho u_{\text{av}} c_p}\frac{4L}{D_h}\right)$$

$$= 1 - \exp\left[-\frac{2.332}{(1.217)(1)(1007)}\frac{4(15)}{0.3}\right] = 0.3165$$

The outlet bulk temperature is therefore

$$T_{b_{\text{out}}} = [17 + (37 - 17)(0.3165)]\ ^\circ\text{C} = 23.3\ ^\circ\text{C}$$ ■

The results obtained by substituting D_h for D in turbulent circular tube formulæ are generally accurate to within ±20% and are often within ±10%. Worse results are obtained for duct cross-sections having sharp corners—say an acute triangle. Specialized equations for "effective" hydraulic diameters have been developed for specific geometries and can improve the accuracy to 5 or 10% [7.8].

When only a portion of the duct cross-section is heated — one wall of a rectangle, for example — the procedure for finding h is the same. The hydraulic diameter is based upon the entire *wetted* perimeter, not simply the heated part. However, in eqn. (7.59a) P is the *heated* perimeter: eqn. (7.59b) does *not* apply for nonuniform heating.

One situation in which one-sided or unequal heating often occurs is an annular duct, with the inner tube serving as a heating element. The hydraulic diameter procedure will typically predict the heat transfer coefficient on the outer tube to within ±10%, irrespective of the heating configuration. The heat transfer coefficient on the inner surface, however, is sensitive to both the diameter ratio and the heating configuration. For that surface, the hydraulic diameter approach is not very accurate, especially if $D_i \ll D_o$; other methods have been developed to accurately predict heat transfer in annular ducts (see [7.3] or [7.8]).

Laminar flow in noncircular ducts

Laminar velocity profiles in noncircular ducts develop in essentially the same way as for circular tubes, and the fully developed velocity profiles are generally paraboloidal in shape. For example, for fully developed flow between parallel plates located at $y = b/2$ and $y = -b/2$,

$$\frac{u}{u_{\text{av}}} = \frac{3}{2}\left[1 - 4\left(\frac{y}{b}\right)^2\right] \tag{7.62}$$

for u_{av} the bulk velocity. This should be compared to eqn. (7.15) for a circular tube. The constants and coordinates differ, but the equations are otherwise identical. Likewise, an analysis of the temperature profiles between parallel plates leads to constant Nusselt numbers, which may be expressed in terms of the hydraulic diameter for various boundary conditions:

$$\text{Nu}_{D_h} = \frac{hD_h}{k} = \begin{cases} 7.541 & \text{for fixed plate temperatures} \\ 8.235 & \text{for fixed flux at both plates} \\ 5.385 & \text{one plate fixed flux, one adiabatic} \end{cases} \tag{7.63}$$

Some other cases are summarized in Table 7.4. Many more have been considered in the literature (see, especially, [7.5]). The latter include different wall boundary conditions and a wide variety cross-sectional shapes, both practical and ridiculous: triangles, circular sectors, trapezoids, rhomboids, hexagons, limaçons, and even crescent moons! The boundary conditions, in particular, should be considered when the duct is small (so that h will be large): if the conduction resistance of the tube wall is comparable to the convective resistance within the duct, then temperature or flux variations around the tube perimeter must be expected. This will significantly affect the laminar Nusselt number. The rectangular duct values in Table 7.4 for fixed wall flux, for example, assume a uniform temperature around the perimeter of the tube, as if the wall has

Table 7.4 Laminar, fully developed Nusselt numbers based on hydraulic diameters given in eqn. (7.61)

Cross-section	T_w *fixed*	q_w *fixed*
Circular	3.657	4.364
Square	2.976	3.608
Rectangular		
$a = 2b$	3.391	4.123
$a = 4b$	4.439	5.331
$a = 8b$	5.597	6.490
Parallel plates	7.541	8.235

no conduction resistance around its perimeter. This might be true for a copper duct heated at a fixed rate in watts per meter of duct length.

Laminar entry length formulæ for noncircular ducts are also given by Shah and London [7.5].

7.6 Heat transfer during cross flow over cylinders

Fluid flow pattern

It will help us to understand the complexity of heat transfer from bodies in a cross flow if we first look in detail at the fluid flow patterns that occur in one cross-flow configuration—a cylinder with fluid flowing normal to it. Figure 7.10 shows how the flow develops as $\text{Re} \equiv u_\infty D/\nu$ is increased from below 5 to near 10^7. An interesting feature of this evolving flow pattern is the fairly continuous way in which one flow transition follows another. The flow field degenerates to greater and greater degrees of disorder with each successive transition until, rather strangely, it regains order at the highest values of Re_D.

An important reflection of the complexity of the flow field is the vortex-shedding frequency, f_v. Dimensional analysis shows that a dimensionless frequency called the Strouhal number, Str, depends on the Reynolds number of the flow:

$$\text{Str} \equiv \frac{f_v D}{u_\infty} = \text{fn}\,(\text{Re}_D) \tag{7.64}$$

Figure 7.11 defines this relationship experimentally on the basis of about

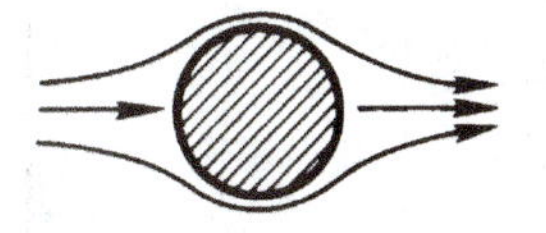

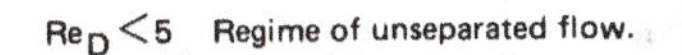

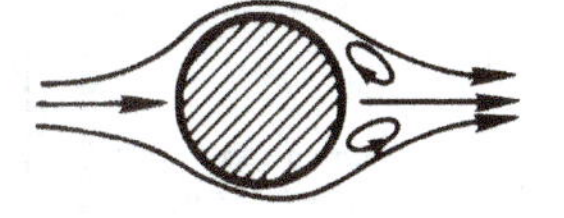

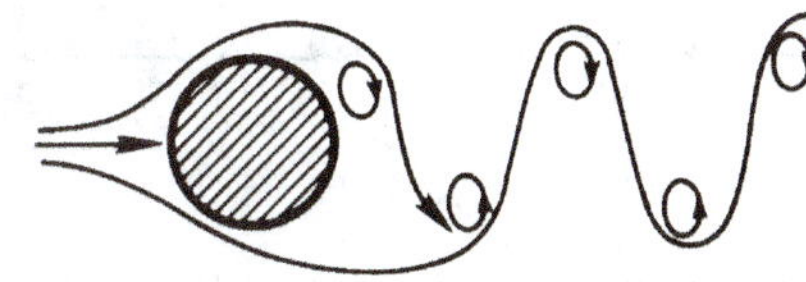

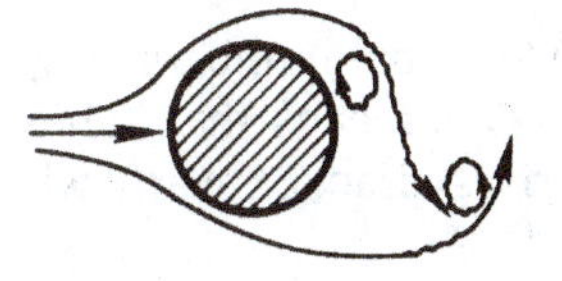

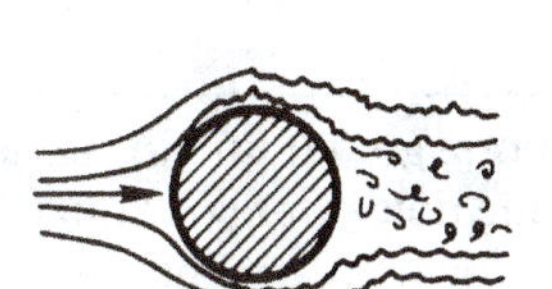

Figure 7.10 Regimes of fluid flow across circular cylinders [7.24].

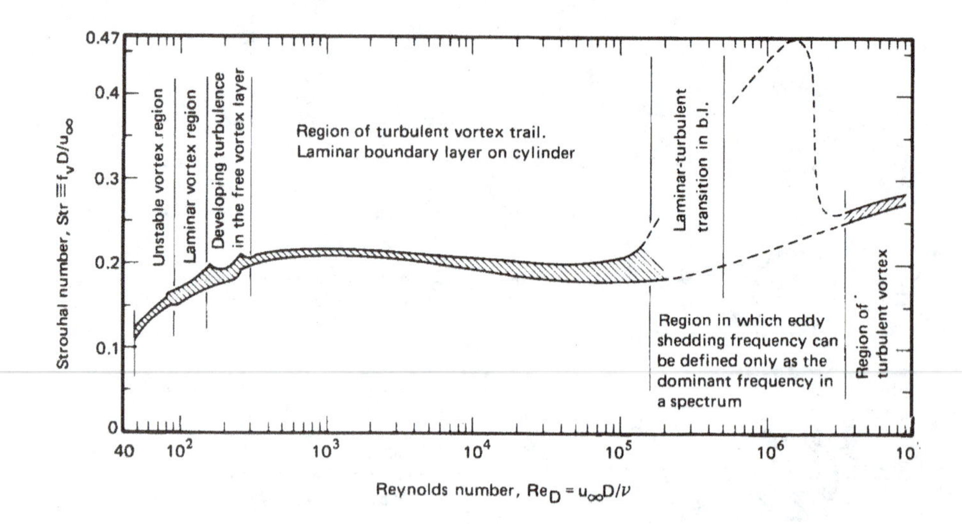

Figure 7.11 The Strouhal-Reynolds number relationship for circular cylinders, as defined by existing data [7.24].

550 of the best data available (see [7.24]). The Strouhal numbers stay a little over 0.2 over most of the range of Re_D. This means that behind a given object, the vortex-shedding frequency rises almost linearly with velocity.

Experiment 7.1

When there is a gentle breeze blowing outdoors, go out and locate a large tree with a straight trunk or, say, a telephone pole. Wet your finger and place it in the wake a couple of diameters downstream and about one radius off center. Estimate the vortex-shedding frequency and use $\mathrm{Str} \simeq 0.21$ to estimate u_∞. Is your value of u_∞ reasonable?

Heat transfer

The action of vortex shedding greatly complicates the heat removal process. Giedt's data [7.25] in Fig. 7.12 show how the heat removal changes as the constantly fluctuating motion of the fluid to the rear of the cylinder changes with Re_D. Notice, for example, that Nu_D is near its minimum

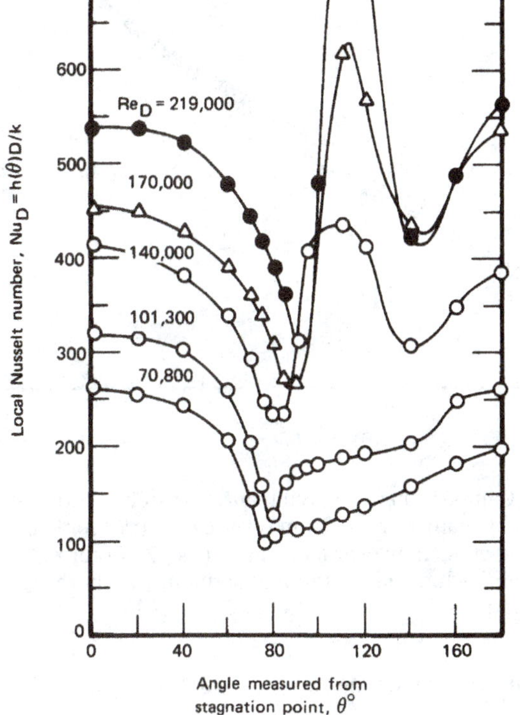
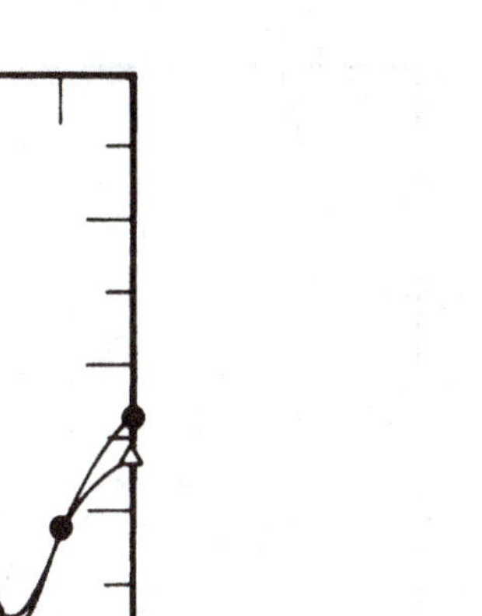

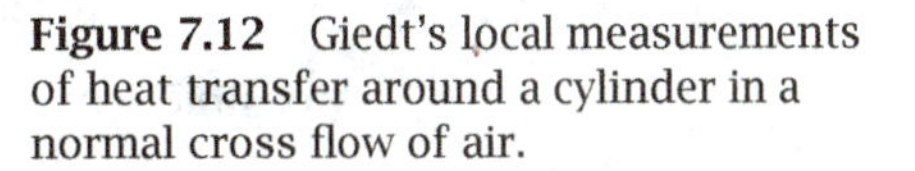

Figure 7.12 Giedt's local measurements of heat transfer around a cylinder in a normal cross flow of air.

at 110° when $Re_D = 71,000$, but it maximizes at the same place when $Re_D = 140,000$. Direct prediction by the sort of b.l. methods that we discussed in Chapter 6 is out of the question. However, a great deal can be done with the data using relations of the form

$$\overline{Nu}_D = \text{fn}\,(Re_D, Pr)$$

The broad study of Churchill and Bernstein [7.26] probably brings the correlation of heat transfer data from cylinders about as far as it is

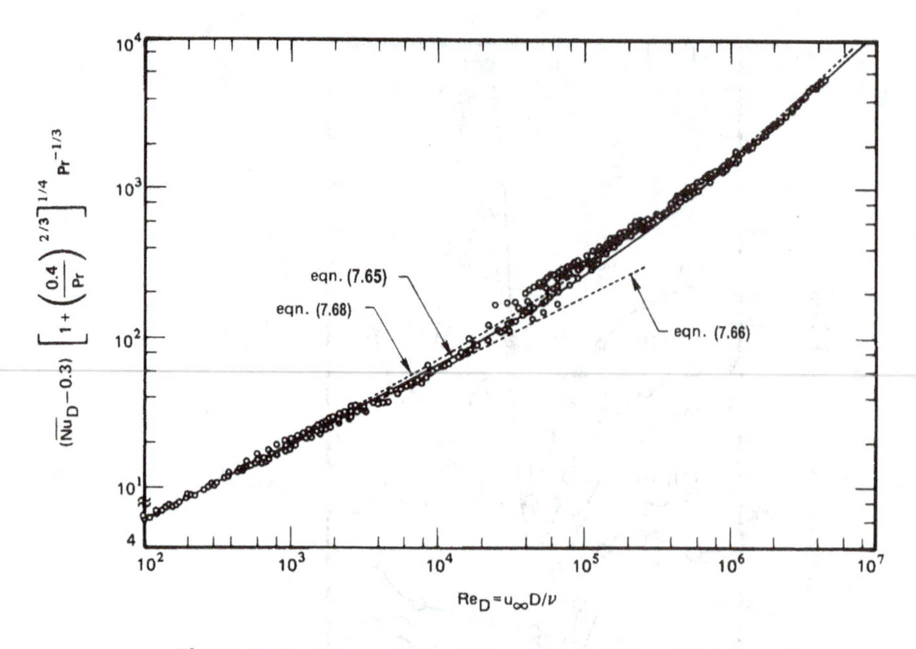

Figure 7.13 Comparison of Churchill and Bernstein's correlation with data by many workers from several countries for heat transfer during cross flow over a cylinder. (See [7.26] for data sources.) Fluids include air, water, and sodium, with both q_w and T_w constant.

possible. For the entire range of the available data, they offer

$$\boxed{\overline{Nu}_D = 0.3 + \frac{0.62\,Re_D^{1/2}\,Pr^{1/3}}{\left[1+(0.4/Pr)^{2/3}\right]^{1/4}}\left[1+\left(\frac{Re_D}{282{,}000}\right)^{5/8}\right]^{4/5}} \tag{7.65}$$

This expression underpredicts most of the data by about 20% in the range $20{,}000 < Re_D < 400{,}000$ but is quite good at other Reynolds numbers above $Pe_D \equiv Re_D Pr = 0.2$. This is evident in Fig. 7.13, where eqn. (7.65) is compared with data.

Greater accuracy and, in most cases, greater convenience results from breaking the correlation into component equations:

- Below $Re_D = 4000$, the bracketed term $[1 + (Re_D/282{,}000)^{5/8}]^{4/5}$

is $\simeq 1$, so

$$\overline{\text{Nu}}_D = 0.3 + \frac{0.62\,\text{Re}_D^{1/2}\text{Pr}^{1/3}}{\left[1 + (0.4/\text{Pr})^{2/3}\right]^{1/4}} \tag{7.66}$$

- Below Pe = 0.2, the Nakai-Okazaki [7.27] relation

$$\overline{\text{Nu}}_D = \frac{1}{0.8237 - \ln(\text{Pe}^{1/2})} \tag{7.67}$$

  should be used.

- In the range $20{,}000 < \text{Re}_D < 400{,}000$, somewhat better results are given by

$$\overline{\text{Nu}}_D = 0.3 + \frac{0.62\,\text{Re}_D^{1/2}\,\text{Pr}^{1/3}}{\left[1 + (0.4/\text{Pr})^{2/3}\right]^{1/4}}\left[1 + \left(\frac{\text{Re}_D}{282{,}000}\right)^{1/2}\right] \tag{7.68}$$

  than by eqn. (7.65).

All properties in eqns. (7.65) to (7.68) are to be evaluated at a film temperature $T_f = (T_w + T_\infty)/2$.

Example 7.7

An electric resistance wire heater 0.0001 m in diameter is placed perpendicular to an air flow. It holds a temperature of 40°C in a 20°C air flow while it dissipates 17.8 W/m of heat to the flow. How fast is the air flowing?

Solution. $\overline{h} = (17.8 \text{ W/m})/[\pi(0.0001 \text{ m})(40 - 20)\ K] = 2833$ W/m²K. Therefore, $\overline{\text{Nu}}_D = 2833(0.0001)/0.0264 = 10.75$, where we have evaluated $k = 0.0264$ at $T = 30$°C. We now want to find the Re_D for which $\overline{\text{Nu}}_D$ is 10.75. From Fig. 7.13 we see that Re_D is around 300 when the ordinate is on the order of 10. This means that we can solve eqn. (7.66) to get an accurate value of Re_D:

$$\text{Re}_D = \left\{(\overline{\text{Nu}}_D - 0.3)\left[1 + \left(\frac{0.4}{\text{Pr}}\right)^{2/3}\right]^{1/4} \Big/ 0.62\,\text{Pr}^{1/3}\right\}^2$$

but Pr = 0.71, so

$$\text{Re}_D = \left\{(10.75 - 0.3)\left[1 + \left(\frac{0.40}{0.71}\right)^{2/3}\right]^{1/4} \Big/ 0.62(0.71)^{1/3}\right\}^2 = 463$$

Then

$$u_\infty = \frac{\nu}{D}\,\text{Re}_D = \left(\frac{1.596 \times 10^{-5}}{10^{-4}}\right) 463 = 73.9 \text{ m/s}$$

The data scatter in Re_D is quite small—less than 10%, it would appear—in Fig. 7.13. Therefore, this method can be used to measure local velocities with good accuracy. If the device is calibrated, its accuracy is improved further. Such an air speed indicator is called a *hot-wire anemometer*, as discussed further in Problem 7.45. ■

Heat transfer during flow across tube bundles

A rod or tube bundle is an arrangement of parallel cylinders that heat, or are being heated by, a fluid that might flow normal to them, parallel with them, or at some angle in between. The flow of coolant through the fuel elements of all nuclear reactors being used in this country is parallel to the heating rods. The flow on the shell side of most shell-and-tube heat exchangers is generally normal to the tube bundles.

Figure 7.14 shows the two basic configurations of a tube bundle in a cross flow. In one, the tubes are in a line with the flow; in the other, the tubes are staggered in alternating rows. For either of these configurations, heat transfer data can be correlated reasonably well with power-law relations of the form

$$\overline{\text{Nu}}_D = C\,\text{Re}_D^n\,\text{Pr}^{1/3} \tag{7.69}$$

but in which the Reynolds number is based on the maximum velocity,

$$u_{\text{max}} = \overline{u}_{\text{av}} \text{ in the narrowest transverse area of the passage}$$

Thus, the Nusselt number based on the average heat transfer coefficient over any particular isothermal tube is

$$\overline{\text{Nu}}_D = \frac{\overline{h}D}{k} \quad \text{and} \quad \text{Re}_D = \frac{u_{\text{max}}D}{\nu}$$

Žukauskas at the Lithuanian Academy of Sciences Institute in Vilnius has written two comprehensive review articles on tube-bundle heat transfer [7.28, 7.29]. In these he summarizes his work and that of other Soviet workers, together with earlier work from the West. He was able to correlate data over very large ranges of Pr, Re_D, S_T/D, and S_L/D (see Fig. 7.14)

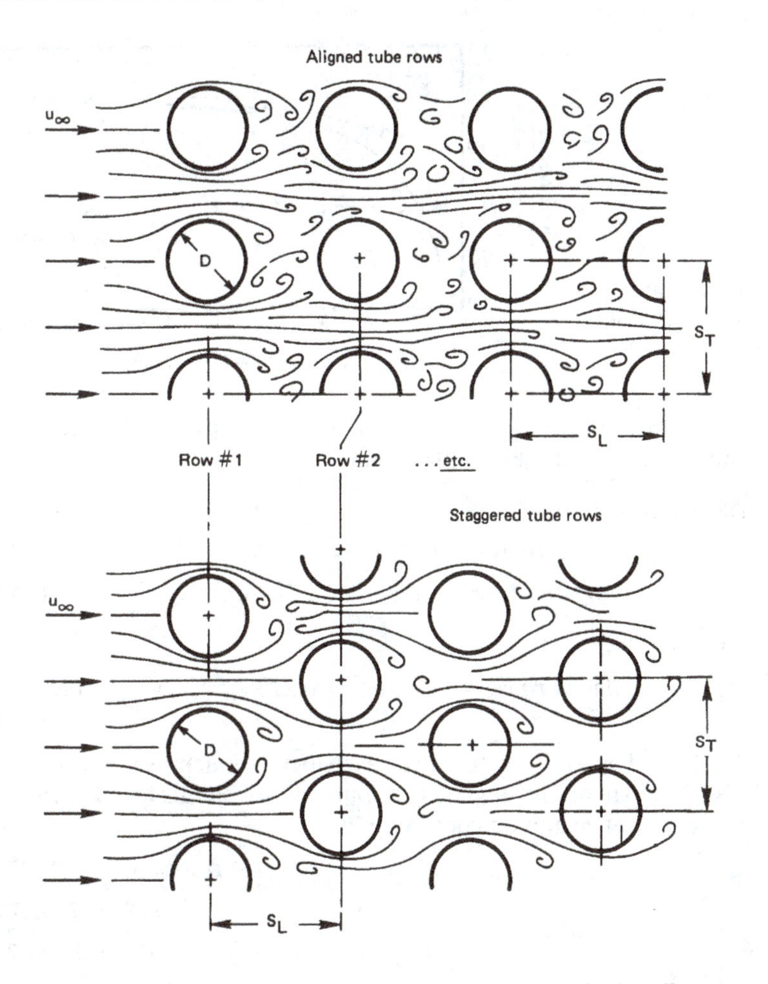

Figure 7.14 Aligned and staggered tube rows in tube bundles.

with an expression of the form

$$\overline{\text{Nu}}_D = \text{Pr}^{0.36}\,(\text{Pr}/\text{Pr}_w)^n\ \text{fn}\,(\text{Re}_D) \quad \text{with } n = \begin{cases} 0 & \text{for gases} \\ \frac{1}{4} & \text{for liquids} \end{cases} \tag{7.70}$$

where properties are to be evaluated at the local fluid bulk temperature, except for Pr_w, which is evaluated at the uniform tube wall temperature, T_w.

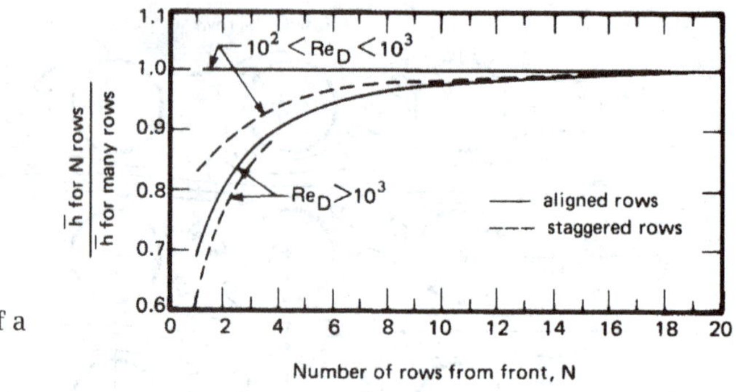

Figure 7.15 Correction for the heat transfer coefficients in the front rows of a tube bundle [7.28].

The function $\text{fn}(\text{Re}_D)$ takes the following form for the various circumstances of flow and tube configuration:

$100 \leqslant \text{Re}_D \leqslant 10^3$:

aligned rows: $\text{fn}(\text{Re}_D) = 0.52\,\text{Re}_D^{0.5}$ (7.71a)

staggered rows: $\text{fn}(\text{Re}_D) = 0.71\,\text{Re}_D^{0.5}$ (7.71b)

$10^3 \leqslant \text{Re}_D \leqslant 2 \times 10^5$:

aligned rows: $\text{fn}(\text{Re}_D) = 0.27\,\text{Re}_D^{0.63}, \quad S_T/S_L \geqslant 0.7$ (7.71c)

For $S_T/S_L < 0.7$, heat exchange is much less effective. Therefore, aligned tube bundles are not designed in this range and no correlation is given.

staggered rows: $\text{fn}(\text{Re}_D) = 0.35\,(S_T/S_L)^{0.2}\,\text{Re}_D^{0.6}, \quad S_T/S_L \leqslant 2$ (7.71d)

$\text{fn}(\text{Re}_D) = 0.40\,\text{Re}_D^{0.6}, \quad S_T/S_L > 2$ (7.71e)

$\text{Re}_D > 2 \times 10^5$:

aligned rows: $\text{fn}(\text{Re}_D) = 0.033\,\text{Re}_D^{0.8}$ (7.71f)

staggered rows: $\text{fn}(\text{Re}_D) = 0.031\,(S_T/S_L)^{0.2}\,\text{Re}_D^{0.8}, \quad \text{Pr} > 1$ (7.71g)

$\overline{\text{Nu}}_D = 0.027\,(S_T/S_L)^{0.2}\,\text{Re}_D^{0.8}, \quad \text{Pr} = 0.7$ (7.71h)

All of the preceding relations apply to the inner rows of tube bundles.

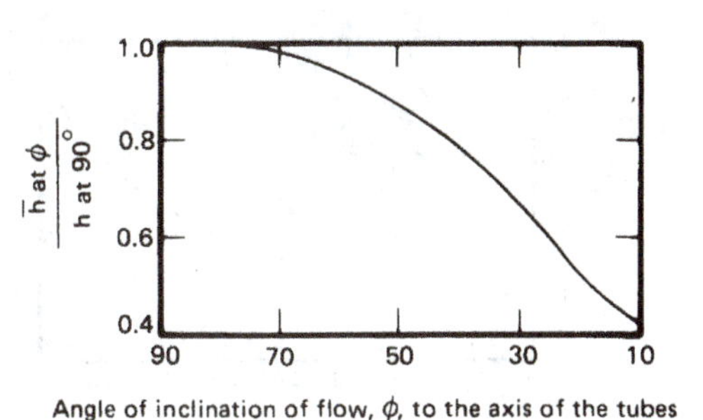

Figure 7.16 Correction for the heat transfer coefficient in flows that are not perfectly perpendicular to heat exchanger tubes [7.28].

The heat transfer coefficient is smaller in the rows at the front of a bundle, facing the oncoming flow. The heat transfer coefficient can be corrected so that it will apply to any of the front rows using Fig. 7.15.

Early in this chapter we alluded to the problem of predicting the heat transfer coefficient during the flow of a fluid at an angle other than 90° to the axes of the tubes in a bundle. Žukauskas provides the empirical corrections in Fig. 7.16 to account for this problem.

The work of Žukauskas does not extend to liquid metals. However, Kalish and Dwyer [7.30] present the results of an experimental study of heat transfer to the liquid eutectic mixture of 77.2% potassium and 22.8% sodium (called NaK). NaK is a fairly popular low-melting-point metallic coolant which has received a good deal of attention for its potential use in certain kinds of nuclear reactors. For isothermal tubes in an equilateral triangular array, as shown in Fig. 7.17, Kalish and Dwyer give

$$\mathrm{Nu}_D = \left(5.44 + 0.228\,\mathrm{Pe}^{0.614}\right)\sqrt{C\,\frac{P-D}{P}\left(\frac{\sin\phi + \sin^2\phi}{1+\sin^2\phi}\right)} \tag{7.72}$$

where

- ϕ is the angle between the flow direction and the rod axis.
- P is the "pitch" of the tube array, as shown in Fig. 7.17, and D is the tube diameter.
- C is the constant given in Fig. 7.17.
- Pe_D is the Péclét number based on the mean flow velocity through the narrowest opening between the tubes.

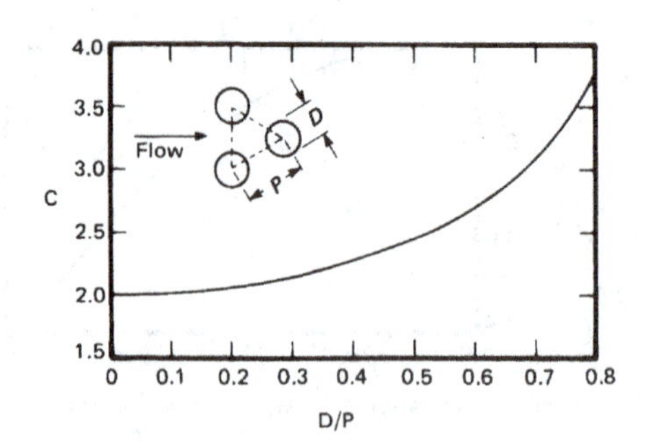

Figure 7.17 Geometric correction for the Kalish-Dwyer equation (7.72).

- For the same uniform heat flux around each tube, the constants in eqn. (7.72) change as follows: 5.44 becomes 4.60; 0.228 becomes 0.193.

7.7 Other configurations

At the outset, we noted that this chapter would move further and further beyond the reach of analysis in the heat convection problems that it dealt with. However, we must not forget that even the most completely empirical relations in Section 7.6 were devised by people who were keenly aware of the theoretical framework into which these relations had to fit. Notice, for example, that eqn. (7.66) reduces to $\mathrm{Nu}_D \propto \sqrt{\mathrm{Pe}_D}$ as Pr becomes small. That sort of theoretical requirement did not just pop out of a data plot. Instead, it was a consideration that led the authors to select an empirical equation that agreed with theory at low Pr.

Thus, the theoretical considerations in Chapter 6 guide us in correlating limited data in situations that cannot be analyzed. Such correlations can be found for all kinds of situations, but all must be viewed critically. Many are based on limited data, and many incorporate systematic errors of one kind or another.

In the face of a heat transfer situation that has to be predicted, one can often find a correlation of data from similar systems. This might involve flow in or across noncircular ducts; axial flow through tube or rod bundles; flow over such bluff bodies as spheres, cubes, or cones; or flow in circular and noncircular annuli. The *Handbook of Heat Transfer* [7.31], the shelf of heat transfer texts in your library, or the journals referred to by the *Engineering Index* are among the first places to look for a correlation curve or equation. When you find a correlation, there are many questions that you should ask yourself:

- Is my case included within the range of dimensionless parameters upon which the correlation is based, or must I extrapolate to reach my case?
- What geometric differences exist between the situation represented in the correlation and the one I am dealing with? (Such elements as these might differ:
  (a) inlet flow conditions;
  (b) small but important differences in hardware, mounting brackets, and so on;
  (c) minor aspect ratio or other geometric nonsimilarities
- Does the form of the correlating equation that represents the data, if there is one, have any basis in theory? (If it is only a curve fit to the existing data, one might be unjustified in using it for more than interpolation of those data.)
- What nuisance variables might make our systems different? For example:
  (a) surface roughness;
  (b) fluid purity;
  (c) problems of surface wetting
- To what extend do the data scatter around the correlation line? Are error limits reported? Can I actually see the data points? (In this regard, you must notice whether you are looking at a correlation on linear or logarithmic coordinates. Errors usually appear smaller than they really are on logarithmic coordinates. Compare, for example, the data of Figs. 8.3 and 8.10.)
- Are the ranges of physical variables large enough to guarantee that I can rely on the correlation for the full range of dimensionless groups that it purports to embrace?
- Am I looking at a primary or secondary source (i.e., is this the author's original presentation or someone's report of the original)? If it is a secondary source, have I been given enough information to question it?

- Has the correlation been signed by the persons who formulated it? (If not, why *haven't* the authors taken responsibility for the work?) Has it been subjected to critical review by independent experts in the field? This is a matter that we must be particularly aware of when we go to the Internet for information. The wiki concept is that open review and editing of information will make such unsigned material effectively self-correcting. But even then the material must be meticulously backed up with references for which individuals take responsibility and which include the necessary supporting data.

Problems

7.1 Prove that in fully developed laminar pipe flow, $(-dp/dx)R^2/4\mu$ is twice the average velocity in the pipe. To do this, set the mass flow rate through the pipe equal to (ρu_{av})(area).

7.2 A flow of air at 27°C and 1 atm is hydrodynamically fully developed in a 1 cm I.D. pipe with $u_{\text{av}} = 2$ m/s. Plot (to scale) T_w, q_w, and T_b as a function of the distance x after T_w is changed or q_w is imposed:

   a. In the case for which $T_w = 68.4$°C = constant.

   b. In the case for which $q_w = 378$ W/m^2 = constant.

Indicate x_{e_t} on your graphs.

7.3 Prove that C_f is $16/\text{Re}_D$ in fully developed laminar pipe flow.

7.4 Air at 200°C flows at 4 m/s over a 3 cm O.D. pipe that is kept at 240°C. (a) Find $\overline{h}$. (b) If the flow were pressurized water at 200°C, what velocities would give the same $\overline{h}$, the same $\overline{\text{Nu}}_D$, and the same Re_D? (c) If someone asked if you could model the water flow with an air experiment, how would you answer? [$u_\infty = 0.0156$ m/s for same $\overline{\text{Nu}}_D$.]

7.5 Compare the h value calculated in Example 7.3 with those calculated from the Dittus-Boelter, Colburn, and Sieder-Tate equations. Comment on the comparison.

7.6 Water at $T_{b_{\text{local}}} = 10$°C flows in a 3 cm I.D. pipe at 1 m/s. The pipe walls are kept at 70°C and the flow is fully developed.

Evaluate h and the local value of dT_b/dx at the point of interest. The relative roughness is 0.001.

7.7 Water at 10°C flows over a 3 cm O.D. cylinder at 70°C. The velocity is 1 m/s. Evaluate $\overline{h}$.

7.8 Consider the hot wire anemometer in Example 7.7. Suppose that 17.8 W/m is the constant heat input, and plot u_∞ vs. T_{wire} over a reasonable range of variables. Must you deal with any changes in the flow regime over the range of interest?

7.9 Water at 20°C flows at 2 m/s over a 2 m length of pipe, 10 cm in diameter, at 60°C. Compare $\overline{h}$ for flow normal to the pipe with that for flow parallel to the pipe. What does the comparison suggest about baffling in a heat exchanger?

7.10 A thermally fully developed flow of NaK in a 5 cm I.D. pipe moves at $u_{\text{av}} = 8$ m/s. If $T_b = 395$°C and T_w is constant at 403°C, what is the local heat transfer coefficient? Is the flow laminar or turbulent?

7.11 Water enters a 7 cm I.D. pipe at 5°C and moves through it at an average speed of 0.86 m/s. The pipe wall is kept at 73°C. Plot T_b against the position in the pipe until $(T_w - T_b)/68 = 0.01$. Neglect the entry problem and consider property variations.

7.12 Air at 20°C flows over a very large bank of 2 cm O.D. tubes that are kept at 100°C. The air approaches at an angle 15° off normal to the tubes. The tube array is staggered, with $S_L =$ 3.5 cm and $S_T = 2.8$ cm. Find $\overline{h}$ on the first tubes and on the tubes deep in the array if the air velocity is 4.3 m/s before it enters the array. [$\overline{h}_{\text{deep}} = 118$ W/m^{2}K.]

7.13 Rework Problem 7.11 using a single value of $\overline{h}$ evaluated at $3(73 - 5)/4 = 51$°C and treating the pipe as a heat exchanger. At what length would you judge that the pipe is no longer efficient as an exchanger? Explain.

7.14 Go to the periodical engineering literature in your library. Find a correlation of heat transfer data. Evaluate the applicability of the correlation according to the criteria outlined in Section 7.7.

7.15 Water at 24°C flows at 0.8 m/s in a smooth, 1.5 cm I.D. tube that is kept at 27°C. The system is extremely clean and quiet, and the flow stays laminar until a noisy air compressor is turned on in the laboratory. Then it suddenly goes turbulent. Calculate the ratio of the turbulent h to the laminar h. [h_{turb} = 4429 W/m^{2}K.]

7.16 Laboratory observations of heat transfer during the forced flow of air at 27°C over a bluff body, 12 cm wide, kept at 77°C yield q = 646 W/m^2 when the air moves 2 m/s and q = 3590 W/m^2 when it moves 18 m/s. In another test, everything else is the same, but now 17°C water flowing 0.4 m/s yields 131,000 W/m^2. The correlations in Chapter 7 suggest that, with such limited data, we can probably create a fairly good correlation in the form: $\overline{\text{Nu}}_L = C\text{Re}^a\text{Pr}^b$. Estimate the constants C, a, and b. One easy way to do this is by by cross-plotting the data on log-log paper.

7.17 Air at 200 psia flows at 12 m/s in an 11 cm I.D. duct. Its bulk temperature is 40°C and the pipe wall is at 268°C. Evaluate h if ε/D = 0.00006.

7.18 How does $\overline{h}$ during cross flow over a cylindrical heater vary with the diameter when Re_D is very large?

7.19 Air enters a 0.8 cm I.D. tube at 20°C with an average velocity of 0.8 m/s. The tube wall is kept at 40°C. Plot $T_b(x)$ until it reaches 39°C. Use properties evaluated at [(20 + 40)/2]°C for the whole problem, but report the local error in h at the end to get a sense of the error incurred by the simplification.

7.20 Write Re_D in terms of $\dot{m}$ in pipe flow and explain why this representation could be particularly useful in dealing with compressible pipe flows.

7.21 NaK at 394°C flows at 0.57 m/s across a 1.82 m length of 0.036 m O.D. tube. The tube is kept at 404°C. Find $\overline{h}$ and the heat removal rate from the tube.

7.22 Verify the value of h specified in Problem 3.22.

7.23 Check the value of h given in Example 7.3 by using Reynolds's analogy *directly* to calculate it. Which h do you deem to be in error, and by what percent?

7.24 A homemade heat exchanger consists of a copper plate, 0.5 m square, with twenty 1.5 cm I.D. copper tubes soldered to it. The ten tubes on top are evenly spaced across the top and parallel with two sides. The ten on the bottom are also evenly spaced, but they run at 90° to the top tubes. The exchanger is used to cool methanol flowing at 0.48 m/s in the tubes from an initial temperature of 73°C, using water flowing at 0.91 m/s and entering at 7°C. What is the temperature of the methanol when it is mixed in a header on the outlet side? Make a judgement of the heat exchanger.

7.25 Given that $\overline{\mathrm{Nu}}_D = 12.7$ at $(2/\mathrm{Gz}) = 0.004$, evaluate $\overline{\mathrm{Nu}}_D$ at $(2/\mathrm{Gz}) = 0.02$ numerically, using Fig. 7.4. Compare the result with the value you read from the figure.

7.26 Report the maximum percent scatter of data in Fig. 7.13. What is happening in the fluid flow when the scatter is worst?

7.27 Water at 27°C flows at 2.2 m/s in a 0.04 m I.D. thin-walled pipe. Air at 227°C flows across it at 7.6 m/s. Find the pipe wall temperature.

7.28 Freshly painted aluminum rods, 0.02 m in diameter, are withdrawn from a drying oven at 150°C and cooled in a 3 m/s cross flow of air at 23°C. How long will it take to cool them to 50°C so they can be handled?

7.29 At what speed, u_∞, must 20°C air flow across an insulated tube before the insulation on it will do any good? The tube is at 60°C and is 6 mm in diameter. The insulation is 12 mm in diameter, with $k = 0.08$ W/m·K. (Notice that we do *not* ask for the u_∞ for which the insulation will do the most harm.)

7.30 Water at 37°C flows at 3 m/s across at 6 cm O.D. tube that is held at 97°C. In a second configuration, 37°C water flows at an average velocity of 3 m/s through a bundle of 6 cm O.D. tubes that are held at 97°C. The bundle is staggered, with $S_T/S_L = 2$. Compare the heat transfer coefficients for the two situations.

7.31 It is proposed to cool 64°C air as it flows, fully developed, in a 1 m length of 8 cm I.D. smooth, thin-walled tubing. The coolant is Freon 12 flowing, fully developed, in the opposite direction, in eight smooth 1 cm I.D. tubes equally spaced around the periphery of the large tube. The Freon enters at −15°C and is fully developed over almost the entire length. The average speeds are 30 m/s for the air and 0.5 m/s for the Freon. Determine the exiting air temperature, assuming that soldering provides perfect thermal contact between the entire surface of the small tubes and the surface of the large tube. Criticize the heat exchanger design and propose some design improvement.

7.32 Evaluate $\overline{\text{Nu}}_D$ using Giedt's data for air flowing over a cylinder at $\text{Re}_D = 140,000$. Compare your result with the appropriate correlation and with Fig. 7.13.

7.33 A 25 mph wind blows across a 0.25 in. telephone line. What is the musical note for the pitch of the hum that it emits?

7.34 A large Nichrome V slab, 0.2 m thick, has two parallel 1 cm I.D. holes drilled through it. Their centers are 8 cm apart. One carries liquid CO_2 at 1.2 m/s from a −13°C reservoir below. The other carries methanol at 1.9 m/s from a 47°C reservoir above. Take account of the intervening Nichrome and compute the heat transfer. Will the CO_2 be significantly warmed by the methanol?

7.35 Consider the situation described in Problem 4.38 but suppose that we do not know $\overline{h}$. Suppose, instead, that we know there is a 10 m/s cross flow of 27°C air over the rod. Rework the problem under these conditions.

7.36 A liquid whose properties are not known flows across a 40 cm O.D. tube at 20 m/s. The measured heat transfer coefficient is 8000 W/m²K. We can be fairly confident that Re_D is very large indeed. What would $\overline{h}$ be if D were 53 cm? What would $\overline{h}$ be if u_∞ were 28 m/s?

7.37 Water flows at 4 m/s, at a temperature of 100°C, in a 6 cm I.D. thin-walled tube with a 2 cm layer of 85% magnesia insulation on it. The outside heat transfer coefficient is 6 W/m²K, and the outside temperature is 20°C. Find: (a) U based on the inside area, (b) Q W/m, and (c) the temperature on either side of the insulation.

7.38 Glycerin is added to water in a mixing tank at 20°C. The mixture discharges through a 4 m length of 0.04 m I.D. tubing under a constant 3 m head. Plot the discharge rate in m^3/hr as a function of composition.

7.39 Plot $\overline{h}$ as a function of composition for the discharge pipe in Problem 7.38. Assume a small temperature difference.

7.40 Rework Problem 5.40 without assuming the Bi number to be very large.

7.41 Water enters a 0.5 cm I.D. pipe at 24°C. The pipe walls are held at 30°C. Plot T_b against distance from entry if u_{av} is 0.27 m/s, neglecting entry behavior in your calculation. (Indicate the entry region on your graph, however.)

7.42 Devise a numerical method to find the velocity distribution and friction factor for laminar flow in a square duct of side length a. Set up a square grid of size N by N and solve the difference equations by hand for $N = 2$, 3, and 4. *Hint*: First show that the velocity distribution is given by the solution to the equation

$$\frac{\partial^2 \overline{u}}{\partial \overline{x}^2} + \frac{\partial^2 \overline{u}}{\partial \overline{y}^2} = 1$$

where $u = 0$ on the sides of the square and we define $\overline{u} = u/[(a^2/\mu)(dp/dz)]$, $\overline{x} = (x/a)$, and $\overline{y} = (y/a)$. Then show that the friction factor, f [eqn. (7.34)], is given by

$$f = \frac{-2}{\dfrac{\rho u_{av} a}{\mu} \displaystyle\iint \overline{u}\, d\overline{x}\, d\overline{y}}$$

Note that the area integral can be evaluated as $\sum \overline{u}/N^2$.

7.43 Chilled air at 15°C enters a horizontal duct at a speed of 1 m/s. The duct is made of thin galvanized steel and is not insulated. A 30 m section of the duct runs outdoors through humid air at 30°C. Condensation of moisture on the outside of the duct is undesirable, but it will occur if the duct wall is at or below the dew point temperature of 20°C. For this problem, assume that condensation rates are so low that their thermal effects can be ignored.

a. Suppose that the duct's square cross-section is 0.3 m by 0.3 m and the effective outside heat transfer coefficient is 5 W/m^{2}K in still air. Determine whether condensation occurs.

b. The single duct is replaced by four circular horizontal ducts, each 0.17 m in diameter. The ducts are parallel to one another in a vertical plane with a center-to-center separation of 0.5 m. Each duct is wrapped with a layer of fiberglass insulation 6 cm thick (k_i = 0.04 W/m·K) and carries air at the same inlet temperature and speed as before. If a 15 m/s wind blows perpendicular to the plane of the circular ducts, find the bulk temperature of the air exiting the ducts.

7.44 An x-ray "monochrometer" is a mirror that reflects only a single wavelength from a broadband beam of x-rays. Over 99% of the beam's energy arrives on other wavelengths and is absorbed creating a high heat flux on part of the surface of the monochrometer. Consider a monochrometer made from a silicon block 10 mm long and 3 mm by 3 mm in cross-section which absorbs a flux of 12.5 W/mm^2 over an area of 6 mm^2 on one face (a heat load of 75 W). To control the temperature, it is proposed to pump liquid nitrogen through a circular channel bored down the center of the silicon block. The channel is 10 mm long and 1 mm in diameter. LN_2 enters the channel at 80 K and a pressure of 1.6 MPa (T_{sat} = 111.5 K). The entry to this channel is a long, straight, unheated passage of the same diameter.

a. For what range of mass flow rates will the LN_2 have a bulk temperature rise of less than 1.5 K over the length of the channel?

b. At your minimum flow rate, estimate the maximum wall temperature in the channel. As a first approximation, assume that the silicon conducts heat well enough to distribute the 75 W heat load uniformly over the channel surface. Could boiling occur in the channel? Discuss the influence of entry length and variable property effects.

7.45 Turbulent fluid velocities are sometimes measured with a *constant temperature hot-wire anemometer*, which consists of a long, fine wire (typically platinum, 4 μm in diameter and 1.25 mm long) supported between two much larger needles. The needles are connected to an electronic bridge circuit which electrically heats the wire while adjusting the heating voltage, V_w, so that the wire's temperature — and thus its resistance, R_w — stays constant. The electrical power dissipated in the wire, V_w^2/R_w, is convected away at the surface of the wire. Analyze the heat loss from the wire to show

$$V_w^2 = (T_{wire} - T_{flow})\left(A + Bu^{1/2}\right)$$

where u is the instantaneous flow speed perpendicular to the wire. Assume that u is between 2 and 100 m/s and that the fluid is an isothermal gas. The constants A and B depend on properties, dimensions, and resistance; they are usually found by calibration of the anemometer. This result is called *King's law*.

7.46 (a) Show that the Reynolds number for a circular tube may be written in terms of the mass flow rate as $\text{Re}_D = 4\dot{m}/\pi\mu D$. (b) Show that this result does not apply to a noncircular tube, specifically $\text{Re}_{D_h} \neq 4\dot{m}/\pi\mu D_h$.

7.47 Olive oil is introduced into a helical coil at 20°C. The coil is made of 1 cm I.D. copper tubing with a 1 mm wall thickness, and consists of ten turns with a diameter of 50 cm. The total length of tubing is 17 m. It is surrounded by a bath of water at 50°C, stirred in such a way as to give an external heat transfer coefficient of 3000 W/m²K. The through-flow of oil is 0.3 kg/s. (a) Determine the bulk temperature of the oil leaving the coil. (b) An engineer suggests raising the outlet temperature by adding fins to the coil. Is his suggestion a good one? Explain.

References

[7.1] F. M. White. *Viscous Fluid Flow.* McGraw-Hill Book Company, New York, 1974.

[7.2] S. S. Mehendale, A. M. Jacobi, and R. K. Shah. Fluid flow and heat transfer at micro- and meso-scales with application to heat exchanger design. *Appl. Mech. Revs.*, 53(7):175–193, 2000.

[7.3] W. M. Kays and M. E. Crawford. *Convective Heat and Mass Transfer.* McGraw-Hill Book Company, New York, 3rd edition, 1993.

[7.4] R. K. Shah and M. S. Bhatti. Laminar convective heat transfer in ducts. In S. Kakaç, R. K. Shah, and W. Aung, editors, *Handbook of Single-Phase Convective Heat Transfer*, chapter 3. Wiley-Interscience, New York, 1987.

[7.5] R. K. Shah and A. L. London. *Laminar Flow Forced Convection in Ducts.* Academic Press, Inc., New York, 1978. Supplement 1 to the series *Advances in Heat Transfer.*

[7.6] L. Graetz. Über die wärmeleitfähigkeit von flüssigkeiten. *Ann. Phys.*, 25:337, 1885.

[7.7] S. R. Sellars, M. Tribus, and J. S. Klein. Heat transfer to laminar flow in a round tube or a flat plate—the Graetz problem extended. *Trans. ASME*, 78:441–448, 1956.

[7.8] M. S. Bhatti and R. K. Shah. Turbulent and transition flow convective heat transfer in ducts. In S. Kakaç, R. K. Shah, and W. Aung, editors, *Handbook of Single-Phase Convective Heat Transfer*, chapter 4. Wiley-Interscience, New York, 1987.

[7.9] F. Kreith. *Principles of Heat Transfer.* Intext Press, Inc., New York, 3rd edition, 1973.

[7.10] A. P. Colburn. A method of correlating forced convection heat transfer data and a comparison with fluid friction. *Trans. AIChE*, 29:174, 1933.

[7.11] L. M. K. Boelter, V. H. Cherry, H. A. Johnson, and R. C. Martinelli. *Heat Transfer Notes.* McGraw-Hill Book Company, New York, 1965.

[7.12] E. N. Sieder and G. E. Tate. Heat transfer and pressure drop of liquids in tubes. *Ind. Eng. Chem.*, 28:1429, 1936.

[7.13] B. S. Petukhov. Heat transfer and friction in turbulent pipe flow with variable physical properties. In T.F. Irvine, Jr. and J. P. Hartnett, editors, *Advances in Heat Transfer*, volume 6, pages 504–564. Academic Press, Inc., New York, 1970.

[7.14] V. Gnielinski. New equations for heat and mass transfer in turbulent pipe and channel flow. *Int. Chemical Engineering*, 16:359-368, 1976.

[7.15] D. M. McEligot. Convective heat transfer in internal gas flows with temperature-dependent properties. In A. S. Majumdar and R. A. Mashelkar, editors, *Advances in Transport Processes*, volume IV, pages 113–200. Wiley, New York, 1986.

[7.16] M. F. Taylor. Prediction of friction and heat-transfer coefficients with large variations in fluid properties. NASA TM X-2145, December 1970.

[7.17] S. E. Haaland. Simple and explicit formulas for the friction factor in turbulent pipe flow. *J. Fluids Engr.*, 105:89–90, 1983.

[7.18] T. S. Ravigururajan and A. E. Bergles. Development and verification of general correlations for pressure drop and heat transfer in single-phase turbulent flow in enhanced tubes. *Exptl. Thermal Fluid Sci.*, 13:55–70, 1996.

[7.19] R. L. Webb. Enhancement of single-phase heat transfer. In S. Kakaç, R. K. Shah, and W. Aung, editors, *Handbook of Single-Phase Convective Heat Transfer*, chapter 17. Wiley-Interscience, New York, 1987.

[7.20] B. Lubarsky and S. J. Kaufman. Review of experimental investigations of liquid-metal heat transfer. NACA Tech. Note 3336, 1955.

[7.21] C. B. Reed. Convective heat transfer in liquid metals. In S. Kakaç, R. K. Shah, and W. Aung, editors, *Handbook of Single-Phase Convective Heat Transfer*, chapter 8. Wiley-Interscience, New York, 1987.

[7.22] R. A. Seban and T. T. Shimazaki. Heat transfer to a fluid flowing turbulently in a smooth pipe with walls at a constant temperature. *Trans. ASME*, 73:803, 1951.

[7.23] R. N. Lyon, editor. *Liquid Metals Handbook*. A.E.C. and Dept. of the Navy, Washington, D.C., 3rd edition, 1952.

[7.24] J. H. Lienhard. Synopsis of lift, drag, and vortex frequency data for rigid circular cylinders. Bull. 300. Wash. State Univ., Pullman, 1966. May be downloaded as a 2.3 MB pdf file from http://www.uh.edu/engines/vortexcylinders.pdf.

[7.25] W. H. Giedt. Investigation of variation of point unit-heat-transfer coefficient around a cylinder normal to an air stream. *Trans. ASME*, 71:375–381, 1949.

[7.26] S. W. Churchill and M. Bernstein. A correlating equation for forced convection from gases and liquids to a circular cylinder in crossflow. *J. Heat Transfer, Trans. ASME, Ser. C*, 99:300–306, 1977.

[7.27] S. Nakai and T. Okazaki. Heat transfer from a horizontal circular wire at small Reynolds and Grashof numbers—1 pure convection. *Int. J. Heat Mass Transfer*, 18:387–396, 1975.

[7.28] A. Žukauskas. Heat transfer from tubes in crossflow. In T.F. Irvine, Jr. and J. P. Hartnett, editors, *Advances in Heat Transfer*, volume 8, pages 93–160. Academic Press, Inc., New York, 1972.

[7.29] A. Žukauskas. Heat transfer from tubes in crossflow. In T. F. Irvine, Jr. and J. P. Hartnett, editors, *Advances in Heat Transfer*, volume 18, pages 87–159. Academic Press, Inc., New York, 1987.

[7.30] S. Kalish and O. E. Dwyer. Heat transfer to NaK flowing through unbaffled rod bundles. *Int. J. Heat Mass Transfer*, 10:1533–1558, 1967.

[7.31] W. M. Rohsenow, J. P. Hartnett, and Y. I. Cho, editors. *Handbook of Heat Transfer*. McGraw-Hill, New York, 3rd edition, 1998.

8. Natural convection in single-phase fluids and during film condensation

There is a natural place for everything to seek, as:
Heavy things go downward, fire upward, and rivers to the sea.
The Anatomy of Melancholy, **R. Burton, 1621**

8.1 Scope

The remaining convection processes that we deal with are largely gravity-driven. Unlike forced convection, in which the driving force is external to the fluid, these so-called natural convection processes are driven by body forces exerted directly within the fluid as the result of heating or cooling. Two such mechanisms are remarkably similar. They are single-phase *natural convection* and *film condensation*. Because these processes have so much in common, we deal with both mechanisms in this chapter. We develop the governing equations side by side in two brief opening sections. Then we treat each mechanism independently in Sections 8.3 and 8.4 and in Section 8.5, respectively.

Chapter 9 deals with other natural convection heat transfer processes that involve phase change—for example:

- *Nucleate boiling.* This heat transfer process is highly disordered as opposed to the processes described in Chapter 8.
- *Film boiling.* This is so similar to film condensation that we can often just modify film condensation predictions.
- *Dropwise condensation.* This bears some similarity to nucleate boiling.

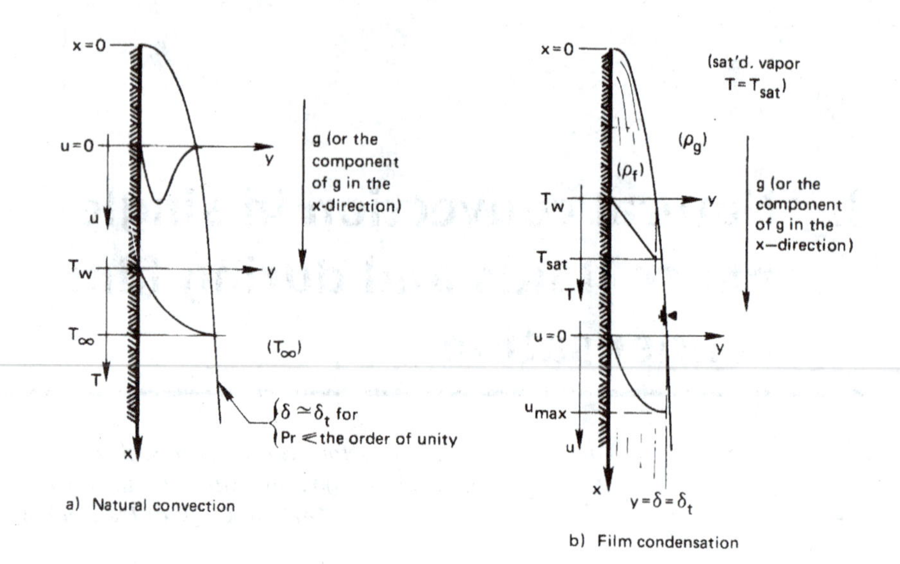

Figure 8.1 The convective boundary layers for natural convection and film condensation. In both sketches, but particularly in that for film condensation, the y-coordinate has been stretched making each layer look thicker than it really is.

8.2 The nature of the problems of film condensation and of natural convection

Description

The natural convection problem is sketched in its simplest form on the left-hand side of Fig. 8.1. Here a vertical isothermal plate cools the fluid adjacent to it. The cooled fluid sinks downward to form a b.l. The figure would be inverted if the plate were warmer than the fluid next to it. Then the fluid would buoy upward.

The corresponding film condensation problem is shown in its simplest form in Fig. 8.1. An isothermal vertical plate cools an adjacent vapor, which condenses and forms a liquid film on the wall.[1] The film is normally very thin and flows off, rather like a b.l., as the figure suggests.

[1] It might instead condense into individual droplets, which roll of without forming into a film. This process is *dropwise condensation.* See Section 9.9.

While natural convection can carry fluid either upward or downward, a condensate film can only move downward. The temperature in the condensing film rises from T_w at the cool wall to T_{sat} at the outer edge of the film.

In both problems, but particularly in film condensation, the b.l. and the film are normally thin enough to accommodate the b.l. assumptions [recall the discussion following eqn. (6.13)]. A second feature of both problems is that δ and δ_t are closely related. In the condensing film they are equal, since the edge of the condensate film forms the edge of both b.l.'s. In natural convection, δ and δ_t are approximately equal when Pr is on the order of one or less, because all cooled (or heated) fluid must buoy downward (or upward). When Pr is large, not just the cooled (or heated) fluid will fall (or rise). Owing to its high viscosity, it will also drag unheated liquid with it. In this case, δ can exceed δ_t. The analysis that follows below is for cases for which $\delta \cong \delta_t$.

Governing equations

To describe laminar film condensation and laminar natural convection, we must add a gravity term to the momentum equation. First we examine the dimensions of the terms in the momentum equation (6.13):

$$\left(u\frac{\partial u}{\partial x} + v\frac{\partial u}{\partial y}\right)\underbrace{\frac{\text{m}}{\text{s}^2}}_{=\frac{\text{kg}\cdot\text{m}}{\text{kg}\cdot\text{s}^2}=\frac{\text{N}}{\text{kg}}} = -\frac{1}{\rho}\frac{dp}{dx}\underbrace{\frac{\text{m}^3}{\text{kg}}\frac{\text{N}}{\text{m}^2\cdot\text{m}}}_{=\frac{\text{N}}{\text{kg}}} + \nu\frac{\partial^2 u}{\partial y^2}\underbrace{\frac{\text{m}^2}{\text{s}}\frac{\text{m}}{\text{s}\cdot\text{m}^2}}_{=\frac{\text{m}}{\text{s}^2}=\frac{\text{N}}{\text{kg}}}$$

where $\partial p/\partial x \simeq dp/dx$ in the b.l. since the pressure does not vary with y, and where $\mu \simeq$ constant. Thus, every term in the equation has units of acceleration or (equivalently) force per unit mass. The component of gravity in the x-direction therefore enters the momentum balance as $(+g)$. This is because x and g point in the same direction. Gravity would enter as $-g$ if it acted opposite the x-direction.

$$u\frac{\partial u}{\partial x} + v\frac{\partial u}{\partial y} = -\frac{1}{\rho}\frac{dp}{dx} + g + \nu\frac{\partial^2 u}{\partial y^2} \tag{8.1}$$

The pressure gradient for both problems is the hydrostatic gradient outside the b.l. Thus,

$$\underbrace{\frac{dp}{dx} = \rho_\infty g}_{\substack{\text{natural}\\\text{convection}}} \qquad \underbrace{\frac{dp}{dx} = \rho_g g}_{\substack{\text{film}\\\text{condensation}}} \tag{8.2}$$

where ρ_∞ is the density of the undisturbed fluid and ρ_g (and ρ_f below) are the saturated vapor and liquid densities. Equation (8.1) then becomes

$$u\frac{\partial u}{\partial x} + v\frac{\partial u}{\partial y} = \left(1 - \frac{\rho_\infty}{\rho}\right) g + \nu\frac{\partial^2 u}{\partial y^2} \qquad \text{for natural convection} \qquad (8.3)$$

$$u\frac{\partial u}{\partial x} + v\frac{\partial u}{\partial y} = \left(1 - \frac{\rho_g}{\rho_f}\right) g + \nu\frac{\partial^2 u}{\partial y^2} \qquad \text{for film condensation} \qquad (8.4)$$

Two boundary conditions, which apply to *both* problems, are

$$\left.\begin{aligned} u\,(y=0) &= 0 \qquad \text{the no-slip condition} \\ v\,(y=0) &= 0 \qquad \text{no flow into the wall} \end{aligned}\right\} \qquad (8.5a)$$

The third b.c. is different for the film condensation and natural convection problems:

$$\left.\begin{aligned} \left.\frac{\partial u}{\partial y}\right|_{y=\delta} &= 0 \qquad \text{condensation: no shear at the edge of the film} \\ u\,(y=\delta) &= 0 \qquad \text{natural convection: undisturbed fluid outside the b.l.} \end{aligned}\right\} \qquad (8.5b)$$

The energy equation for either of the two cases is eqn. (6.40):

$$u\frac{\partial T}{\partial x} + v\frac{\partial T}{\partial y} = \alpha\frac{\partial^2 T}{\partial y^2}$$

We leave the identification of the b.c.'s for temperature until later.

The crucial thing we must recognize about the momentum equation at the moment is that it is coupled to the energy equation. Let us consider how that occurs:

In natural convection: The velocity, u, is driven by buoyancy, which is reflected in the term $(1 - \rho_\infty/\rho)g$ in the momentum equation. The density, $\rho = \rho(T)$, varies with T, so it is impossible to solve the momentum and energy equations independently of one another.

In film condensation: The third boundary condition (8.5b) for the momentum equation involves the film thickness, δ. But to calculate δ we must make an energy balance on the film to find out how much latent heat—and thus how much condensate—it has absorbed. This will bring $(T_{\text{sat}} - T_w)$ into the solution of the momentum equation.

The boundary layer on a flat surface, during forced convection, was easy to analyze because the momentum equation could be solved completely before any consideration of the energy equation was attempted. Now we no longer have that advantage in predicting either natural convection or film condensation.

8.3 Laminar natural convection on a vertical isothermal surface

Dimensional analysis and experimental data

Before we attempt a dimensional analysis of the natural convection problem, let us simplify the buoyancy term, $(\rho - \rho_\infty)g/\rho$, in eqn. (8.3). We derived it for incompressible flow, but we modified it by admitting a small variation of density with temperature in this term only. Now we wish to eliminate $(\rho - \rho_\infty)$ in favor of $(T - T_\infty)$ with the help of the coefficient of thermal expansion, β:

$$\beta \equiv \frac{1}{v}\frac{\partial v}{\partial T}\bigg|_p = -\frac{1}{\rho}\frac{\partial \rho}{\partial T}\bigg|_p \simeq -\frac{1}{\rho}\frac{\rho - \rho_\infty}{T - T_\infty} = -\frac{(1 - \rho_\infty/\rho)}{T - T_\infty} \tag{8.6}$$

where v designates the specific volume here—not a velocity component.

Figure 8.2 shows natural convection from a vertical surface that is hotter than its surroundings. In either this case or the cold plate shown in Fig. 8.1, we replace $(1 - \rho_\infty/\rho)g$ with $-g\beta(T - T_\infty)$. The sign of this term (see Fig. 8.2) is the same in either case. The direction in which it acts will depend upon whether T is greater or less than T_∞. Then

$$u\frac{\partial u}{\partial x} + v\frac{\partial u}{\partial y} = -g\beta(T - T_\infty) + \nu\frac{\partial^2 u}{\partial y^2} \tag{8.7}$$

where the minus sign corresponds to plate orientation in Fig. 8.1a. This conveniently removes ρ from the equation and makes the coupling of the momentum and energy equations very clear.

The functional equation for the heat transfer coefficient, h, in natural convection is now (cf. Section 6.4)

$$h \text{ or } \overline{h} = \text{fn}\left(k, |T_w - T_\infty|, x \text{ or } L, \nu, \alpha, g, \beta\right)$$

where L is a length that must be specified for a given problem. Notice that while we could take h to be independent of ΔT in the forced convection

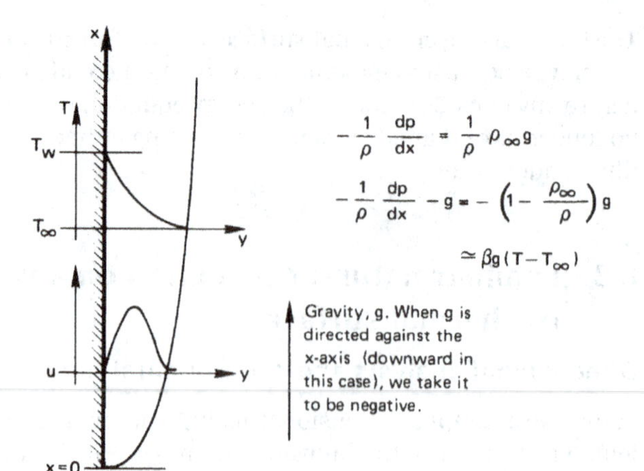

Figure 8.2 Natural convection from a vertical heated plate.

problem (Section 6.4), the explicit appearance of $(T - T_\infty)$ in eqn. (8.7) suggests that we cannot do so here. There are thus eight variables in W, m, s, and °C (where we again regard J as a unit independent of N and m); so we look for 8 − 4 = 4 pi-groups. For $\overline{h}$ and a characteristic length, L, the groups may be chosen as

$$\overline{\text{Nu}}_L \equiv \frac{\overline{h}L}{k}, \qquad \text{Pr} \equiv \frac{\nu}{\alpha}, \qquad \Pi_3 \equiv \frac{L^3}{\nu^2}\,|g|, \qquad \Pi_4 \equiv \beta\,|T_w - T_\infty| = \beta\,\Delta T$$

where we set $\Delta T \equiv |T_w - T_\infty|$. Two of these groups are new to us:

- $\Pi_3 \equiv gL^3/\nu^2$: This characterizes the importance of buoyant forces relative to viscous forces.[2]

- $\Pi_4 \equiv \beta\Delta T$ characterizes the thermal expansion of the fluid. For an ideal gas,

$$\beta = \frac{1}{v}\frac{\partial}{\partial T}\left(\frac{RT}{p}\right)_p = \frac{1}{T_\infty}$$

where R is the gas constant. Therefore, for ideal gases

$$\beta\,\Delta T = \frac{\Delta T}{T_\infty} \tag{8.8}$$

[2]Note that gL is dimensionally the same as a velocity squared—say, u^2. Then $\sqrt{\Pi_3}$ can be interpreted as a Reynolds number: uL/ν. In a laminar b.l. we recall that $\text{Nu} \propto \text{Re}^{1/2}$; so here we might anticipate that $\text{Nu} \propto \Pi_3^{1/4}$.

It turns out that Π_3 and Π_4 (which do not bear the names of famous people) usually appear as a product. This product is called the Grashof (pronounced Gráhs-hoff) number[3], Gr_L, where the subscript designates the length on which it is based:

$$\Pi_3\Pi_4 \equiv \mathrm{Gr}_L = \frac{g\beta\Delta T L^3}{\nu^2} \tag{8.9}$$

Two exceptions in which Π_3 and Π_4 appear independently are rotating systems (where Coriolis forces are part of the body force) and situations in which $\beta\Delta T$ is no longer $\ll 1$ but instead approaches one. We therefore expect to correlate data in most other situations with functional equations of the form

$$\mathrm{Nu} = \mathrm{fn}(\mathrm{Gr}, \mathrm{Pr}) \tag{8.10}$$

Another attribute of the dimensionless functional equation is that the most influential independent variable is usually the product of Gr and Pr. This is called the Rayleigh number, Ra_L, where the subscript designates the length on which it is based:

$$\mathrm{Ra}_L \equiv \mathrm{Gr}_L \mathrm{Pr} = \frac{g\beta\Delta T L^3}{\alpha\nu} \tag{8.11}$$

Thus, most (but not all) analyses and correlations of natural convection take the form:

$$\mathrm{Nu} = \mathrm{fn}(\underbrace{\mathrm{Ra}}_{\text{most important independent variable}}, \underbrace{\mathrm{Pr}}_{\text{secondary parameter}}) \tag{8.12}$$

Figure 8.3 is a careful selection of the best data available for natural convection from vertical isothermal surfaces. These data were organized by Churchill and Chu [8.3] and they span 13 orders of magnitude of the Rayleigh number. The correlation of these data in the coordinates of Fig. 8.2 is exactly in the form of eqn. (8.12), and it shows the dominant influence of Ra_L, while any influence of Pr is small.

---

[3] Nu, Pr, Π_3, Π_4, and Gr were all suggested by Nusselt in his pioneering paper on convective heat transfer [8.1]. Grashof was a notable nineteenth-century mechanical engineering professor who was simply given the honor of having a dimensionless group named after him posthumously (see, e.g., [8.2]). He did not work with natural convection.

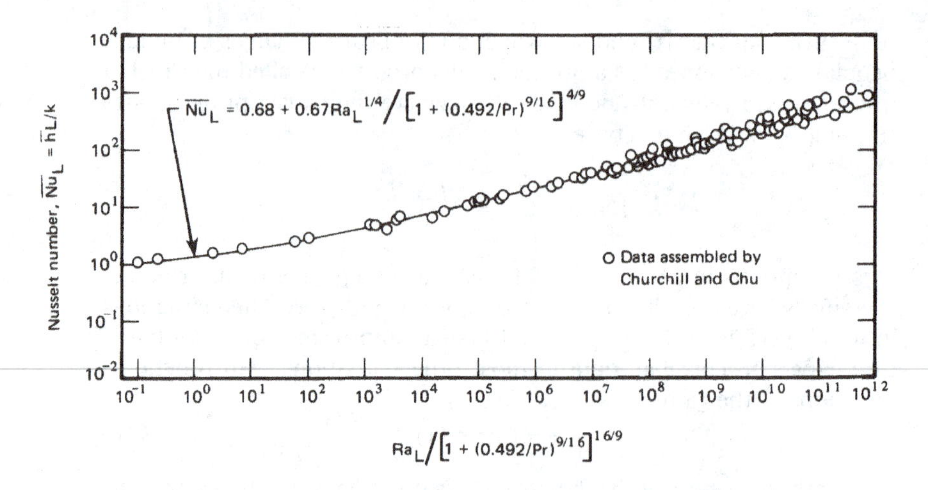

Figure 8.3 The correlation of $\overline{h}$ data for vertical isothermal surfaces by Churchill and Chu [8.3], using $\text{Nu}_L = \text{fn}(\text{Ra}_L, \text{Pr})$. (Applies to full range of Pr.)

The data correlate on these coordinates within a few percent up to $\text{Ra}_L/[1+(0.492/\text{Pr}^{9/16})]^{16/9} \simeq 10^8$. That is about where the b.l. starts exhibiting turbulent behavior. Beyond that point, the overall Nusselt number, Nu_L, rises more sharply, and the data scatter increases as the heat transfer mechanisms become much more complex.

Prediction of h in natural convection on a vertical surface

The analysis of natural convection using an integral method was done independently by Squire (see [8.4]) and by Eckert [8.5] in the 1930s. We refer to this important analysis as the Squire-Eckert formulation.

It begins with the integrated momentum and energy equations. We assume $\delta = \delta_t$ and integrate both equations to the same value of δ:

$$\frac{d}{dx}\int_0^\delta \Big(u^2 - \underbrace{uu_\infty}_{\substack{=0,\text{ since}\\ u_\infty = 0}}\Big)dy = -\nu\left.\frac{\partial u}{\partial y}\right|_{y=0} + g\beta\int_0^\delta (T - T_\infty)\,dy \qquad (8.13)$$

and [eqn. (6.47)]

$$\frac{d}{dx}\int_0^\delta u\,(T - T_\infty)\,dy = \frac{q_w}{\rho c_p} = -\alpha\left.\frac{\partial T}{\partial y}\right|_{y=0}$$

The integrated momentum equation is the same as eqn. (6.24) except that it includes the buoyancy term, which was added to the differential momentum equation in eqn. (8.7).

We now must estimate the temperature and velocity profiles for use in eqns. (8.13) and (6.47). We do this here in much the same way as we did in Sections 6.2 and 6.3 for forced convection. We write a set of known facts about the profiles, then use these facts to evaluate the constants in power-series expressions for u and T.

Since the temperature profile has a fairly simple shape, we choose a simple quadratic expression:

$$\frac{T - T_\infty}{T_w - T_\infty} = a + b\left(\frac{y}{\delta}\right) + c\left(\frac{y}{\delta}\right)^2 \tag{8.14}$$

Notice that the thermal boundary layer thickness, δ_t, should be roughly equal to δ in eqn. (8.14). This would seemingly limit the results to Prandtl numbers not too much larger than one. Actually, the analysis will also prove useful for large Pr's because the velocity profile exerts diminishing influence on the temperature profile as Pr increases. We require the following things to be true of this profile:

- $T(y = 0) = T_w$ or $\left.\dfrac{T - T_\infty}{T_w - T_\infty}\right|_{y/\delta=0} = 1 = a$
- $T(y = \delta) = T_\infty$ or $\left.\dfrac{T - T_\infty}{T_w - T_\infty}\right|_{y/\delta=1} = 0 = 1 + b + c$
- $\left.\dfrac{\partial T}{\partial y}\right|_{y=\delta} = 0$ or $\dfrac{d}{d(y/\delta)}\left(\dfrac{T - T_\infty}{T_w - T_\infty}\right)_{y/\delta=1} = 0 = b + 2c$

so $a = 1$, $b = -2$, and $c = 1$. This gives the following dimensionless temperature profile:

$$\frac{T - T_\infty}{T_w - T_\infty} = 1 - 2\left(\frac{y}{\delta}\right) + \left(\frac{y}{\delta}\right)^2 = \left(1 - \frac{y}{\delta}\right)^2 \tag{8.15}$$

We anticipate a somewhat complicated velocity profile (recall Fig. 8.1) and seek to represent it with a cubic function:

$$u = u_c(x)\left[\left(\frac{y}{\delta}\right) + c\left(\frac{y}{\delta}\right)^2 + d\left(\frac{y}{\delta}\right)^3\right] \tag{8.16}$$

where, since there is no obvious characteristic velocity in the problem, we write u_c as an as-yet-unknown function. (u_c will have to increase with

x, since u must increase with x.) We know three things about u:

- $u(y=0)=0$ $\quad \left\{\begin{array}{l}\text{we have already satisfied this condition by} \\ \text{writing eqn. (8.16) with no lead constant}\end{array}\right.$
- $u(y=\delta)=0 \quad$ or $\quad \dfrac{u}{u_c}=0=(1+c+d)$
- $\left.\dfrac{\partial u}{\partial y}\right|_{y=\delta}=0 \quad$ or $\quad \left.\dfrac{\partial u}{\partial(y/\delta)}\right|_{y/\delta=1}=0=(1+2c+3d)\,u_c$

These give $c=-2$ and $d=1$, so

$$\frac{u}{u_c(x)}=\frac{y}{\delta}\left(1-\frac{y}{\delta}\right)^2 \tag{8.17}$$

We could also have written the momentum equation (8.7) at the wall, where $u=v=0$, and created a fourth condition:

$$\left.\frac{\partial^2 u}{\partial y^2}\right|_{y=0}=-\frac{g\beta\,(T_w-T_\infty)}{\nu}$$

Then we could have evaluated $u_c(x)$ as $\beta g|T_w-T_\infty|\delta^2/4\nu$. A correct expression for u_c will eventually depend upon these variables, but we will not attempt to make u_c fit this particular condition. Doing so would yield two equations, (8.13) and (6.47), in a single unknown, $\delta(x)$. It would be impossible to satisfy both of them. Instead, we allow the velocity profile to violate this condition slightly and write

$$u_c(x)=C_1\frac{\beta g\,|T_w-T_\infty|}{\nu}\,\delta^2(x) \tag{8.18}$$

Then we solve the *two* integrated conservation equations for the two unknowns, C_1 (which should $\simeq$¼) and $\delta(x)$.

The dimensionless temperature and velocity profiles are plotted in Fig. 8.4. With them are included Schmidt and Beckmann's exact calculation for air (Pr = 0.7), as presented in [8.4]. Notice that the integral approximation to the temperature profile is better than the approximation to the velocity profile. That is fortunate, since the temperature profile ultimately determines the heat transfer solution.

When we substitute eqns. (8.15) and (8.17) in the momentum equa-

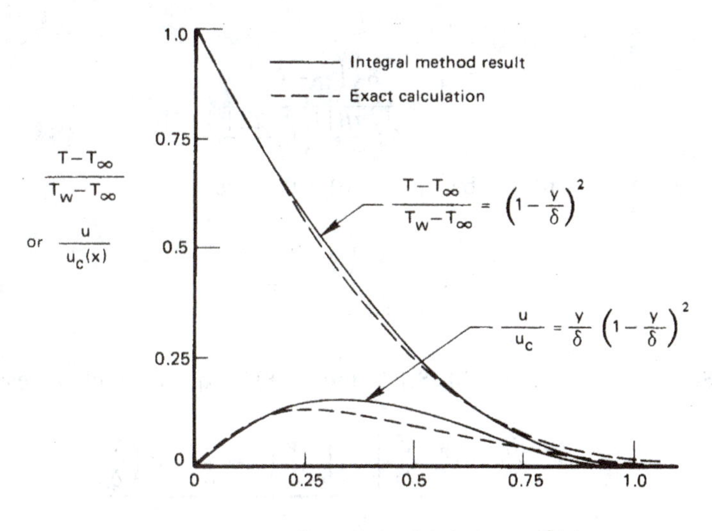

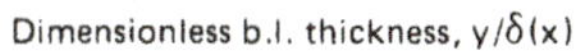

Figure 8.4 The temperature and velocity profiles for air (Pr = 0.7) in a laminar convection b.l.

tion (8.13), using eqn. (8.18) for $u_c(x)$, we get

$$C_1^2\left(\frac{g\beta\,|T_w - T_\infty|}{\nu}\right)^2 \frac{d}{dx}\left[\delta^5 \underbrace{\int_0^1 \left(\frac{y}{\delta}\right)^2\left(1-\frac{y}{\delta}\right)^4 d\left(\frac{y}{\delta}\right)}_{=\frac{1}{105}}\right]$$

$$= g\beta\,|T_w - T_\infty|\,\delta \underbrace{\int_0^1 \left(1-\frac{y}{\delta}\right)^2 d\left(\frac{y}{\delta}\right)}_{=\frac{1}{3}}$$

$$- C_1 g\beta\,|T_w - T_\infty|\,\delta(x)\underbrace{\frac{\partial}{\partial\,(y/\delta)}\left[\frac{y}{\delta}\left(1-\frac{y}{\delta}\right)^2\right]_{\frac{y}{\delta}=0}}_{=1} \tag{8.19}$$

where we change the sign of the terms on the left by replacing $(T_w - T_\infty)$ with its absolute value. Equation (8.19) then becomes

$$\left(\frac{1}{21}C_1^2\,\frac{g\beta\,|T_w - T_\infty|}{\nu^2}\right)\delta^3\frac{d\delta}{dx} = \frac{1}{3} - C_1$$

or

$$\frac{d\delta^4}{dx} = \frac{84\left(\dfrac{1}{3} - C_1\right)}{C_1^2 \dfrac{g\beta\,|T_w - T_\infty|}{\nu^2}}$$

Integrating this with the b.c., $\delta(x = 0) = 0$, gives

$$\delta^4 = \frac{84\left(\dfrac{1}{3} - C_1\right)}{C_1^2 \dfrac{g\beta\,|T_w - T_\infty|}{\nu^2}}\, x \tag{8.20}$$

Substituting eqns. (8.15), (8.17), and (8.18) in eqn. (6.47) likewise gives

$$(T_w - T_\infty)\, C_1 \frac{g\beta\,|T_w - T_\infty|}{\nu} \frac{d}{dx}\left[\delta^3 \underbrace{\int_0^1 \frac{y}{\delta}\left(1 - \frac{y}{\delta}\right)^4 d\left(\frac{y}{\delta}\right)}_{=\frac{1}{30}}\right]$$

$$= -\alpha \frac{T_w - T_\infty}{\delta} \underbrace{\frac{d}{d(y/\delta)}\left[\left(1 - \frac{y}{\delta}\right)^2\right]_{y/\delta=0}}_{=-2}$$

or

$$3\frac{C_1}{30}\delta^3 \frac{d\delta}{dx} = \frac{C_1}{40}\frac{d\delta^4}{dx} = \frac{2}{\mathrm{Pr}\,\dfrac{g\beta\,|T_w - T_\infty|}{\nu^2}}$$

Integrating this with the b.c., $\delta(x = 0) = 0$, we get

$$\delta^4 = \frac{80}{C_1 \mathrm{Pr}\,\dfrac{g\beta|T_w - T_\infty|}{\nu^2}}\, x \tag{8.21}$$

Equating eqns. (8.20) and (8.21) for δ^4, we then obtain

$$\frac{21}{20}\,\frac{\dfrac{1}{3} - C_1}{C_1 \dfrac{g\beta\,|T_w - T_\infty|}{\nu^2}}\, x = \frac{1}{\mathrm{Pr}\,\dfrac{g\beta\,|T_w - T_\infty|}{\nu^2}}\, x$$

or

$$C_1 = \frac{\mathrm{Pr}}{3\left(\dfrac{20}{21} + \mathrm{Pr}\right)} \tag{8.22}$$

Then, from eqn. (8.21):

$$\delta^4 = \frac{240\left(\dfrac{20}{21} + \Pr\right)}{\Pr^2 \dfrac{g\beta\,|T_w - T_\infty|}{\nu^2}}\, x$$

or

$$\frac{\delta}{x} = 3.936\left(\frac{0.952 + \Pr}{\Pr^2}\right)^{1/4} \frac{1}{\mathrm{Gr}_x^{1/4}} \tag{8.23}$$

Equation (8.23) can be combined with the known temperature profile, eqn. (8.15), and substituted in Fourier's law to find q:

$$q = -k \left.\frac{\partial T}{\partial y}\right|_{y=0} = -\frac{k(T_w - T_\infty)}{\delta} \underbrace{\left.\frac{d\left(\dfrac{T - T_\infty}{T_w - T_\infty}\right)}{d\left(\dfrac{y}{\delta}\right)}\right|_{y/\delta=0}}_{=-2} = 2\,\frac{k\Delta T}{\delta} \tag{8.24}$$

so, writing $h = q/|T_w - T_\infty| \equiv q/\Delta T$, we obtain[4]

$$\mathrm{Nu}_x \equiv \frac{qx}{\Delta T k} = 2\,\frac{x}{\delta} = \frac{2}{3.936}\,(\Pr\mathrm{Gr}_x)^{1/4}\left(\frac{\Pr}{0.952 + \Pr}\right)^{1/4}$$

or

$$\mathrm{Nu}_x = 0.508\,\mathrm{Ra}_x^{1/4}\left(\frac{\Pr}{0.952 + \Pr}\right)^{1/4} \tag{8.25}$$

This is the Squire-Eckert result for the local heat transfer from a vertical isothermal wall during laminar natural convection. It applies for either $T_w > T_\infty$ or $T_w < T_\infty$.

The average heat transfer coefficient can be obtained from

$$\overline{h} = \frac{\displaystyle\int_0^L q(x)\,dx}{L\Delta T} = \frac{\displaystyle\int_0^L h(x)\,dx}{L}$$

Thus,

$$\overline{\mathrm{Nu}}_L = \frac{\overline{h}L}{k} = \frac{1}{k}\int_0^L \frac{k}{x}\,\mathrm{Nu}_x\,dx = \frac{4}{3}\,\mathrm{Nu}_x\Big|_{x=L}$$

[4]Recall that, in footnote 2, we wondered if Nu would vary as $\mathrm{Gr}^{1/4}$. We now see that this is the case.

or

$$\boxed{\overline{\text{Nu}}_L = 0.678\ \text{Ra}_L^{1/4}\left(\frac{\text{Pr}}{0.952+\text{Pr}}\right)^{1/4}} \tag{8.26}$$

All properties in eqn. (8.26) and the preceding equations should be evaluated at $T = (T_w + T_\infty)/2$ except in gases, where β should be evaluated at T_∞.

Example 8.1

A thin-walled metal tank containing fluid at 40°C cools in air at 14°C; $\overline{h}$ is very large inside the tank. If the sides are 0.4 m high, compute $\overline{h}$, $\overline{q}$, and δ at the top. Are the b.l. assumptions reasonable?

Solution.

$$\beta_{\text{air}} = 1/T_\infty = 1/(273+14) = 0.00348\ \text{K}^{-1}. \quad \text{Then}$$

$$\text{Ra}_L = \frac{g\beta\Delta T L^3}{\nu\alpha} = \frac{9.8(0.00348)(40-14)(0.4)^3}{(1.566\times10^{-5})\,(2.203\times10^{-5})} = 1.645\times10^8$$

and Pr = 0.711, where the properties are evaluated at 300 K = 27°C. Then, from eqn. (8.26),

$$\overline{\text{Nu}}_L = 0.678\left(1.645\times10^8\right)^{1/4}\left(\frac{0.711}{0.952+0.711}\right)^{1/4} = 62.1$$

so

$$\overline{h} = \frac{62.1k}{L} = \frac{62.1(0.02614)}{0.4} = 4.06\ \text{W/m}^2\text{K}$$

and

$$\overline{q} = \overline{h}\,\Delta T = 4.06(40-14) = 105.5\ \text{W/m}^2$$

The b.l. thickness at the top of the tank is given by eqn. (8.23) at $x = L$:

$$\frac{\delta}{L} = 3.936\left(\frac{0.952+0.711}{0.711^2}\right)^{1/4}\frac{1}{(\text{Ra}_L/\text{Pr})^{1/4}} = 0.0430$$

Thus, the b.l. thickness at the end of the plate is only 4% of the height, or 1.72 cm thick. This is thicker than typical forced convection b.l.'s, but it is still reasonably thin. ■

Example 8.2

Large thin metal sheets of length L are dipped in an electroplating bath in the vertical position. Their average temperature is initially cooler than the liquid in the bath. How rapidly will they come up to bath temperature, T_b?

Solution. We can probably take Bi $\ll 1$ and use the lumped-capacity response equation (1.20). We obtain $\overline{h}$ for use in eqn. (1.20) from eqn. (8.26):

$$\overline{h} = \underbrace{0.678\,\frac{k}{L}\left(\frac{\text{Pr}}{0.952+\text{Pr}}\right)^{1/4}\left(\frac{g\beta L^3}{\alpha\nu}\right)^{1/4}}_{\text{call this } B}\Delta T^{1/4}$$

Since $\overline{h} \propto \Delta T^{1/4}$, with $\Delta T = T_b - T$, eqn. (1.20) becomes

$$\frac{d(T_b - T)}{dt} = -\frac{BA}{\rho c V}(T_b - T)^{5/4}$$

where V/A = the half-thickness of the plate, w. Integrating this between the initial temperature of the plate, T_i, and the temperature at time t, we get

$$\int_{T_i}^{T} \frac{d(T_b - T)}{(T_b - T)^{5/4}} = -\int_0^t \frac{B}{\rho c w}\,dt$$

so

$$T_b - T = \left[\frac{1}{(T_b - T_i)^{1/4}} + \frac{B}{4\rho c w}\,t\right]^{-4}$$

(Before we use this result, we should check Bi $= Bw\Delta T^{1/4}/k$ to be certain that it is, in fact, less than one.) The temperature can be put in dimensionless form as

$$\frac{T_b - T}{T_b - T_i} = \left[1 + \frac{B\,(T_b - T_i)^{1/4}}{4\rho c w}\,t\right]^{-4}$$

where the coefficient of t is a kind of inverse time constant of the response. Thus, the temperature dependence of $\overline{h}$ in natural convection leads to a solution quite different from the exponential response that resulted from a constant $\overline{h}$ [eqn. (1.22)]. ■

Comparison of analysis and correlations with experimental data

Churchill and Chu [8.3] have proposed two equations for the data correlated in Fig. 8.3. And here we see the way in which analysis and correlation work hand in hand. Churchill and Chu have begun with the general form of the Squire-Eckert formulation, then modified it to fit a vast collection of experimental data. And the fact that they have a rational nondimensionalization of the problem allows them to treat a vast range of situations in a single graph.

In any case, the simpler of the two Churchill-Chu formulations is shown in the figure. It is

$$\overline{\mathrm{Nu}}_L = 0.68 + 0.67\,\mathrm{Ra}_L^{1/4}\left[1 + \left(\frac{0.492}{\mathrm{Pr}}\right)^{9/16}\right]^{-4/9} \tag{8.27}$$

which applies for all Pr and for the range of Ra shown in the figure. The Squire–Eckert prediction is within 1.2% of this correlation for high Pr and high Ra_L, and it differs by only 5.5% if the fluid is a gas and $\mathrm{Ra}_L > 10^5$. Typical Rayleigh numbers usually exceed 10^5, so the Squire-Eckert prediction is remarkably accurate in the range of practical interest, despite the approximations upon which it is built. The additive constant of 0.68 in eqn. (8.27) is a correction for low Ra_L, where the b.l. assumptions are inaccurate and $\overline{\mathrm{Nu}}_L$ is no longer proportional to $\mathrm{Ra}_L^{1/4}$.

The Squire-Eckert prediction fails also fails at low Prandtl numbers, and eqn. (8.27) has to be used. In the turbulent regime, $\mathrm{Gr} \gtrsim 10^9$ [8.6], eqn. (8.27) predicts a lower bound on the data (see Fig. 8.3), although it is really intended only for laminar boundary layers. In this correlation, as in eqn. (8.26), the thermal properties should all be evaluated at a film temperature, $T_f = (T_\infty + T_w)/2$, except for β, which is to be evaluated at T_∞ if the fluid is a gas.

Example 8.3

Verify the first heat transfer coefficient in Table 1.1. It is for air at 20°C next to a 0.3 m high wall at 50°C.

Solution. At $T = 35$°C $= 308$ K, we find $\mathrm{Pr} = 0.71$, $\nu = 16.45 \times 10^{-6}\,\mathrm{m^2/s}$, $\alpha = 2.318\times10^{-5}\,\mathrm{m^2/s}$, and $\beta = 1/(273{+}20) = 0.00341\,\mathrm{K^{-1}}$.

Then

$$\mathrm{Ra}_L = \frac{g\beta\Delta T L^3}{\alpha\nu} = \frac{9.8(0.00341)(30)(0.3)^3}{(16.45)(0.2318)10^{-10}} = 7.10 \times 10^7$$

The Squire-Eckert prediction gives

$$\overline{\mathrm{Nu}_L} = 0.678\left(7.10 \times 10^7\right)^{1/4}\left(\frac{0.71}{0.952 + 0.71}\right)^{1/4} = 50.3$$

so

$$\overline{h} = 50.3\,\frac{k}{L} = 50.3\left(\frac{0.0267}{0.3}\right) = 4.48 \text{ W/m}^2\text{K}.$$

And the Churchill-Chu correlation gives

$$\overline{\mathrm{Nu}_L} = 0.68 + 0.67\,\frac{(7.10 \times 10^7)^{1/4}}{\left[1 + (0.492/0.71)^{9/16}\right]^{4/9}} = 47.88$$

so

$$\overline{h} = 47.88\left(\frac{0.0267}{0.3}\right) = 4.26 \text{ W/m}^2\text{K}$$

The prediction is therefore within 5% of the correlation. We should use the latter result in preference to the theoretical one, although the difference is slight. ■

Variable-properties problem

Sparrow and Gregg [8.7] provide an extended discussion of the influence of physical property variations on predicted values of Nu. They found that while β for gases should be evaluated at T_∞, all other properties should be evaluated at a temperature T_r, where

$$T_r = T_w - C\,(T_w - T_\infty) \tag{8.28}$$

and where $C = 0.38$ for gases. Most books recommend that a simple mean between T_w and T_∞ (or $C = 0.50$) be used. A simple mean seldom differs much from the more precise result above, of course.

It has also been shown by Barrow and Sitharamarao [8.8] that when $\beta\Delta T$ is no longer $\ll 1$, the Squire-Eckert formula should be corrected as follows:

$$\mathrm{Nu} = \mathrm{Nu}_{\text{sq-Ek}}\left[1 + \tfrac{3}{5}\beta\Delta T + \mathcal{O}(\beta\Delta T)^2\right]^{1/4} \tag{8.29}$$

This same correction can be applied to the Churchill-Chu correlation or to other expressions for Nu. Since $\beta = 1/T_\infty$ for an ideal gas, eqn. (8.29) gives only about a 1.5% correction for a 330 K plate heating 300 K air.

Note on the validity of the boundary layer approximations

The boundary layer approximations are sometimes put to a rather severe test in natural convection problems. Thermal b.l. thicknesses are often fairly large, and the usual analyses that take the b.l. to be thin can be significantly in error. This is particularly true as Gr becomes small. Figure 8.5 includes three pictures that illustrate this. These pictures are interferograms (or in the case of Fig. 8.5c, data deduced from interferograms). An interferogram is a photograph made in a kind of lighting that causes regions of uniform density to appear as alternating light and dark bands.

Figure 8.5a was made at the University of Kentucky by G.S. Wang and R. Eichhorn. The Grashof number based on the radius of the leading edge is 2250 in this case. This is low enough to result in a b.l. that is larger than the radius near the leading edge. Figure 8.5b and c are from Kraus's classic study of natural convection visualization methods [8.9]. Figure 8.5c shows that, at Gr = 585, the b.l. assumptions are quite unreasonable since the cylinder is small in comparison with the large region of thermal disturbance.

The analysis of free convection becomes a far more complicated problem at low Gr's, since the b.l. equations can no longer be used. We shall not discuss any of the numerical solutions of the full Navier-Stokes equations that have been carried out in this regime. We shall instead note that correlations of data using functional equations of the form

$$\mathrm{Nu} = \mathrm{fn}(\mathrm{Ra}, \mathrm{Pr})$$

will be the first thing that we resort to in such cases. Indeed, Fig. 8.3 reveals that Churchill and Chu's equation (8.27) already serves this purpose in the case of the vertical isothermal plate, at low values of $\mathrm{Ra} \equiv \mathrm{Gr}\,\mathrm{Pr}$.

8.4 Natural convection in other situations

Natural convection from horizontal isothermal cylinders

Churchill and Chu [8.10] provide yet another comprehensive correlation of existing data. For natural convection from horizontal isothermal cylinders, they find that an equation with the same form as eqn. (8.27) correlates the data for horizontal cylinders as well. Data from a variety of sources, over about 24 orders of magnitude of the Rayleigh number based

a. A 1.34 cm wide flat plate with a rounded leading edge in air. $T_w = 46.5°\text{C}$, $\Delta T = 17.0°\text{C}$, $\text{Gr}_{\text{radius}} \simeq 2250$

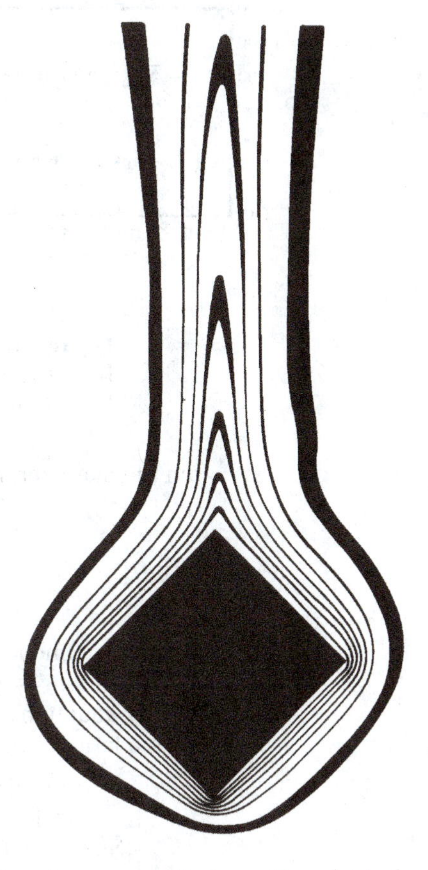

b. A square cylinder with a fairly low value of Gr. (Rendering of an interferogram shown in [8.9].)

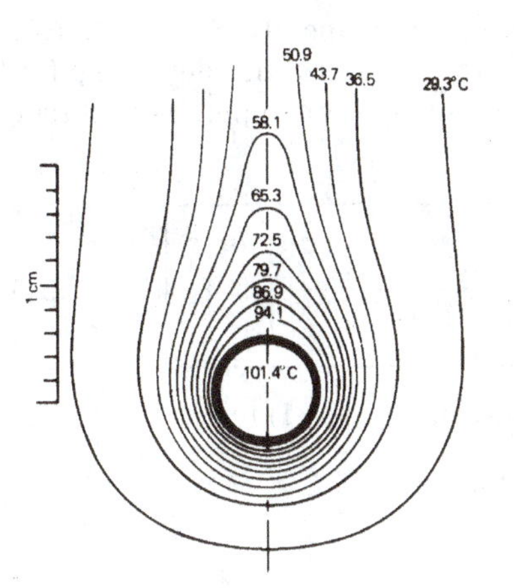

c. Measured isotherms around a cylinder in airwhen $\text{Gr}_D \approx 585$ (from [8.9]).

Figure 8.5 The thickening of the b.l. during natural convection at low Gr, as illustrated by interferograms made on two-dimensional bodies. (The dark lines in the pictures are isotherms.)

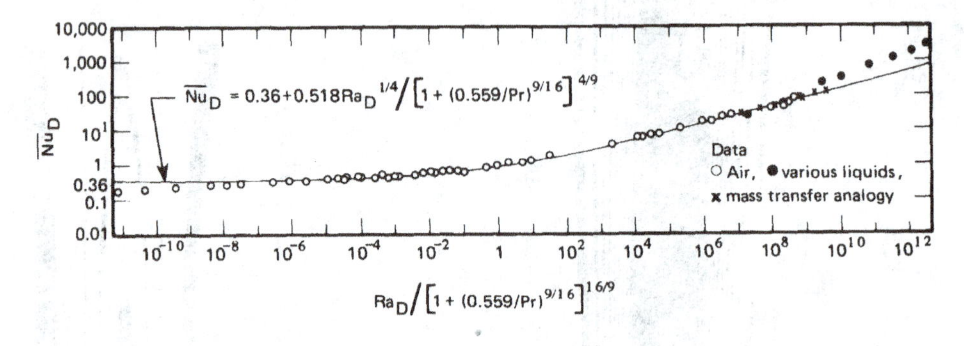

Figure 8.6 The data of many investigators for heat transfer from isothermal horizontal cylinders during natural convection, as correlated by Churchill and Chu [8.10].

on the diameter, Ra_D, are shown in Fig. 8.6. The correlating equation is

$$\overline{Nu}_D = 0.36 + \frac{0.518\, Ra_D^{1/4}}{\left[1 + (0.559/Pr)^{9/16}\right]^{4/9}} \tag{8.30}$$

They recommend that eqn. (8.30) be used in the range $10^{-6} \leqslant Ra_D \leqslant 10^9$.

When Ra_D is greater than 10^9, the flow becomes turbulent. The following equation is a little more complex, but it gives comparable accuracy over a larger range:

$$\overline{Nu}_D = \left\{0.60 + 0.387\left[\frac{Ra_D}{\left[1 + (0.559/Pr)^{9/16}\right]^{16/9}}\right]^{1/6}\right\}^2 \tag{8.31}$$

The recommended range of applicability of eqn. (8.31) is

$$10^{-6} \leqslant Ra_D$$

Example 8.4

Space vehicles are subject to a "g-jitter," or background variation of acceleration, on the order of 10^{-6} or 10^{-5} earth gravities. Brief periods of gravity up to 10^{-4} or 10^{-2} earth gravities can be exerted by accelerating the whole vehicle. A certain line carrying hot oil is ½ cm in diameter and it is at 127°C. How does Q vary with g-level if T_∞ = 27°C in the air around the tube?

Solution. The average b.l. temperature is 350 K. We evaluate properties at this temperature and write g as $g_e \times$ (g-level), where g_e is g at the earth's surface and the g-level is the fraction of g_e in the space vehicle. With $\beta = 1/T_\infty$ for an ideal gas

$$\mathrm{Ra}_D = \frac{g\beta\Delta T\, D^3}{\nu\alpha} = \frac{9.8\left(\dfrac{400-300}{300}\right)(0.005)^3}{2.062(10)^{-5}2.92(10)^{-5}}(g\text{-level})$$

$$= (678.2)\,(g\text{-level})$$

From eqn. (8.31), with Pr = 0.706, we compute

$$\overline{\mathrm{Nu}}_D = \left\{0.6 + \underbrace{0.387\left[\frac{678.2}{\left[1+(0.559/0.706)^{9/16}\right]^{16/9}}\right]^{1/6}}_{=0.952}(g\text{-level})^{1/6}\right\}^2$$

so

g-level	$\overline{\mathrm{Nu}}_D$	$\overline{h} = \overline{\mathrm{Nu}}_D\left(\dfrac{0.0297}{0.005}\right)$	$Q = \pi D\overline{h}\Delta T$
10^{-6}	0.483	2.87 W/m²K	4.51 W/m of tube
10^{-5}	0.547	3.25 W/m²K	5.10 W/m of tube
10^{-4}	0.648	3.85 W/m²K	6.05 W/m of tube
10^{-2}	1.086	6.45 W/m²K	10.1 W/m of tube

The numbers in the rightmost column are quite low. Cooling is extremely inefficient at these low gravities. ■

Natural convection from vertical cylinders

The heat transfer from the wall of a cylinder with its axis running vertically is the same as that from a vertical plate, as long as the thermal b.l. is thin. However, if the b.l. is thick, as is indicated in Fig. 8.7, heat transfer will be enhanced by the curvature of the thermal b.l. This correction was first considered some years ago by Sparrow and Gregg, and the analysis was subsequently extended with the help of more powerful numerical methods by Cebeci [8.11].

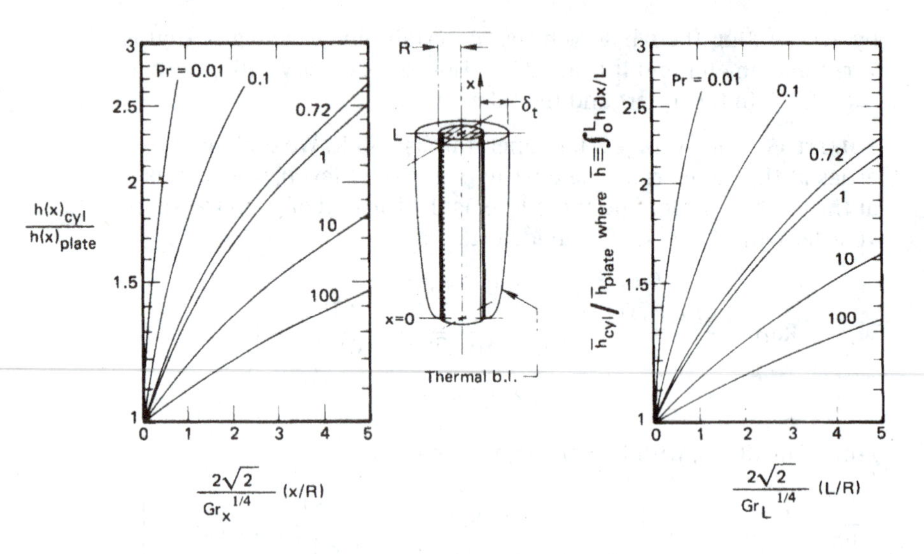

Figure 8.7 Corrections for h and $\overline{h}$ on vertical isothermal plates to adapt them to vertical isothermal cylinders [8.11].

Figure 8.7 includes the corrections to the vertical plate results that were calculated for many Pr's by Cebeci. The left-hand graph gives a correction that must be multiplied by the local flat-plate Nusselt number to get the vertical cylinder result. Notice that the correction increases when the Grashof number decreases. The right-hand curve gives a similar correction for the overall Nusselt number on a cylinder of height L. Notice that in either situation, the correction for all but liquid metals is less than 10% if $(x \text{ or } L)/R < 0.08\,\mathrm{Gr}_{x \text{ or } L}^{1/4}$.

Heat transfer from general submerged bodies

Spheres. The sphere is interesting because it has a clearly specifiable value of Nu_D as $\mathrm{Ra}_D \to 0$. We look first at this limit. When the buoyancy forces approach zero by virtue of:

- low gravity,
- small diameter,
- very high viscosity,
- a very small value of β,

then heated fluid will no longer be buoyed away convectively. In that case, only conduction will serve to remove heat. Using shape factor number 4 in Table 5.4, we compute in this case

$$\lim_{\mathrm{Ra}_D \to 0} \mathrm{Nu}_D = \frac{Q}{A\Delta T}\frac{D}{k} = \frac{k\Delta T(S)D}{4\pi(D/2)^2\Delta T k} = \frac{4\pi(D/2)}{4\pi(D/4)} = 2 \tag{8.32}$$

Every proper correlation of data for heat transfer from spheres therefore has the lead constant, 2, in it.[5] A typical example is that of Yuge [8.12] for spheres immersed in gases:

$$\overline{\mathrm{Nu}}_D = 2 + 0.43\,\mathrm{Ra}_D^{1/4}, \quad \mathrm{Ra}_D < 10^5 \tag{8.33}$$

A more complex expression [8.13] encompasses other Prandtl numbers:

$$\boxed{\overline{\mathrm{Nu}}_D = 2 + \frac{0.589\,\mathrm{Ra}_D^{1/4}}{\left[1 + (0.492/\mathrm{Pr})^{9/16}\right]^{4/9}}} \quad \mathrm{Ra}_D < 10^{12} \tag{8.34}$$

This result has an estimated uncertainty of 5% in air and an rms error of about 10% at higher Prandtl numbers.

Rough estimate of Nu for other bodies. In 1973 Lienhard [8.14] noted that, for laminar convection in which the b.l. does not separate, the expression

$$\boxed{\overline{\mathrm{Nu}}_\tau \simeq 0.52\,\mathrm{Ra}_\tau^{1/4}} \tag{8.35}$$

would predict heat transfer from any submerged body within about 10% if Pr is not $\ll 1$. The characteristic dimension in eqn. (8.35) is the length of travel, τ, of fluid in the unseparated b.l.

In the case of spheres without separation, for example, $\tau = \pi D/2$, the distance from the bottom to the top around the circumference. Thus, for spheres, eqn. (8.35) becomes

$$\frac{\overline{h}\pi D}{2k} = 0.52\left[\frac{g\beta\Delta T(\pi D/2)^3}{\nu\alpha}\right]^{1/4}$$

or

$$\frac{\overline{h}D}{k} = 0.52\left(\frac{2}{\pi}\right)\left(\frac{\pi}{2}\right)^{3/4}\left[\frac{g\beta\Delta T D^3}{\nu\alpha}\right]^{1/4}$$

[5] It is important to note that while Nu_D for spheres approaches a limiting value at small Ra_D, no such limit exists for cylinders or vertical surfaces. The constants in eqns. (8.27) and (8.30) are not valid at extremely low values of Ra_D.

or

$$\overline{\text{Nu}}_D = 0.465\ \text{Ra}_D^{1/4}$$

This is within 8% of Yuge's correlation if Ra_D remains fairly large.

Laminar heat transfer from inclined and horizontal plates

In 1953, Rich [8.15] showed that heat transfer from inclined plates could be predicted by vertical plate formulas if the component of the gravity vector along the surface of the plate was used in the calculation of the Grashof number. Thus, g is replaced by $g\cos\theta$, where θ is the angle of inclination measured from the vertical, as shown in Fig. 8.8. The heat transfer rate decreases as $(\cos\theta)^{1/4}$.

Subsequent studies have shown that Rich's result is substantially correct for the lower surface of a heated plate or the upper surface of a cooled plate. For the upper surface of a heated plate or the lower surface of a cooled plate, the boundary layer becomes unstable and separates at a relatively low value of Gr. Experimental observations of such instability have been reported by Fujii and Imura [8.16], Vliet [8.17], Pera and Gebhart [8.18], and Al-Arabi and El-Riedy [8.19], among others.

In the limit $\theta = 90°$ — a horizontal plate — the fluid flow above a hot plate or below a cold plate must form one or more plumes, as shown in Fig. 8.8c and d. In such cases, the b.l. is unstable for all but small Rayleigh numbers, and even then a plume must leave the center of the plate. The unstable cases can only be represented with empirical correlations.

Theoretical considerations, and experiments, show that the Nusselt number for laminar b.l.s on horizontal and slightly inclined plates varies as $\text{Ra}^{1/5}$ [8.20, 8.21]. For the unstable cases, when the Rayleigh number exceeds 10^4 or so, the experimental variation is as $\text{Ra}^{1/4}$, and once the flow is fully turbulent, for Rayleigh numbers above about 10^7, experiments show a $\text{Ra}^{1/3}$ variation of the Nusselt number [8.22, 8.23]. In the latter case, both Nu_L and $\text{Ra}_L^{1/3}$ are proportional to L, so the heat transfer coefficient is *independent* of L. Moreover, the flow field in these situations is driven mainly by the component of gravity normal to the plate.

Unstable Cases: For the lower side of cold plates and the upper side of hot plates, the boundary layer becomes increasingly unstable as Ra is increased.

- For inclinations $\theta \lesssim 45°$ and $10^5 \leqslant \text{Ra}_L \leqslant 10^9$, replace g with $g\cos\theta$ in eqn. (8.27).

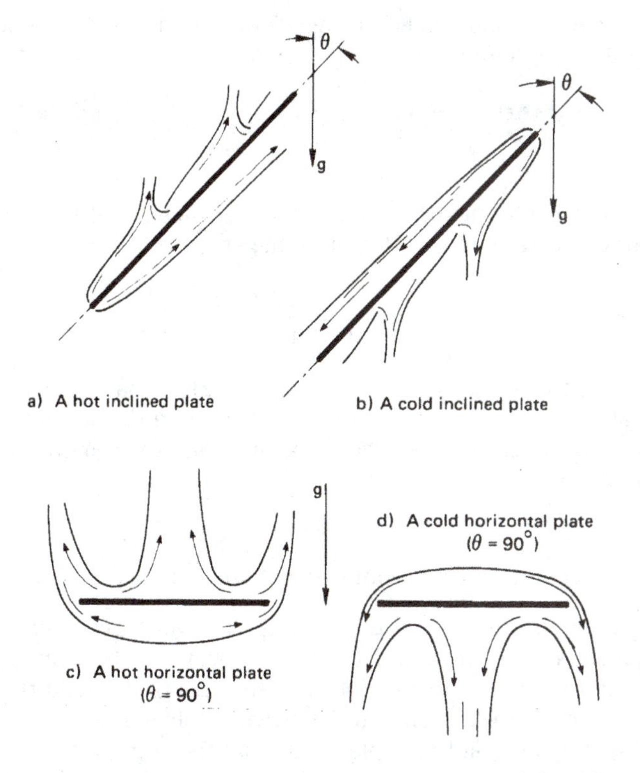

Figure 8.8 Natural convection b.l.'s on some inclined and horizontal surfaces. The b.l. separation, shown here for the unstable cases in (a) and (b), occurs only at sufficiently large values of Gr.

- For horizontal plates with Rayleigh numbers above 10^7, nearly identical results have been obtained by many investigators. From these results, Raithby and Hollands propose [8.13]:

$$\overline{\text{Nu}}_L = 0.14\ \text{Ra}_L^{1/3}\left(\frac{1+0.0107\ \text{Pr}}{1+0.01\ \text{Pr}}\right),\quad 0.024 \leqslant \text{Pr} \leqslant 2000 \tag{8.36}$$

This formula is consistent with available data up to $\text{Ra}_L = 2 \times 10^{11}$, and probably goes higher. As noted before, the choice of lengthscale L is immaterial. Fujii and Imura's results support using the above for $60° \leqslant \theta \leqslant 90°$ with g in the Rayleigh number.

For high Ra in gases, temperature differences and variable properties effects can be large. From experiments on upward facing plates,

Clausing and Berton [8.23] suggest evaluating all gas properties at a reference temperature, in kelvin, of

$$T_{\text{ref}} = T_w - 0.83\,(T_w - T_\infty) \quad \text{for} \quad 1 \leqslant T_w/T_\infty \leqslant 3.$$

- For horizontal plates of area A and perimeter P at lower Rayleigh numbers, Raithby and Hollands suggest [8.13]

$$\overline{\text{Nu}}_{L*} = \frac{0.560\,\text{Ra}_{L*}^{1/4}}{\left[1 + (0.492/\text{Pr})^{9/16}\right]^{4/9}} \tag{8.37a}$$

where, following Lloyd and Moran [8.22], a characteristic length-scale $L* = A/P$, is used in the Rayleigh and Nusselt numbers. If $\overline{\text{Nu}}_{L*} \lesssim 10$, the b.l.s will be thick, and they suggest correcting the result to

$$\overline{\text{Nu}}_{\text{corrected}} = \frac{1.4}{\ln\left(1 + 1.4/\overline{\text{Nu}}_{L*}\right)} \tag{8.37b}$$

These equations are recommended[6] for $1 < \text{Ra}_{L*} < 10^7$.

- In general, for inclined plates in the unstable cases, Raithby and Hollands [8.13] recommend that the heat flow be computed first using the formula for a vertical plate with $g\cos\theta$ and then using the formula for a horizontal plate with $g\sin\theta$ (i.e., the component of gravity normal to the plate) and that the larger value of the heat flow be taken.

Stable Cases: For the upper side of cold plates and the lower side of hot plates, the flow is generally stable. The following results assume that the flow is not obstructed at the edges of the plate; a surrounding adiabatic surface, for example, will lower $\overline{h}$ [8.24, 8.25].

- For $\theta < 88°$ and $10^5 \leqslant \text{Ra}_L \leqslant 10^{11}$, eqn. (8.27) is still valid for the upper side of cold plates and the lower side of hot plates when g is replaced with $g\cos\theta$ in the Rayleigh number [8.16].

---

[6] Raithby and Hollands also suggest using a blending formula for $1 < \text{Ra}_{L*} < 10^{10}$

$$\overline{\text{Nu}}_{\text{blended},L*} = \left[\left(\overline{\text{Nu}}_{\text{corrected}}\right)^{10} + \left(\overline{\text{Nu}}_{\text{turb}}\right)^{10}\right]^{1/10} \tag{8.37c}$$

in which $\overline{\text{Nu}}_{\text{turb}}$ is calculated from eqn. (8.36) using $L*$. The formula is useful for numerical progamming, but its effect on $\overline{h}$ is usually small.

- For downward-facing hot plates and upward-facing cold plates of width L with very slight inclinations, Fujii and Imura give:

$$\overline{\text{Nu}}_L = 0.58\ \text{Ra}_L^{1/5} \tag{8.38}$$

This is valid for $10^6 < \text{Ra}_L < 10^9$ if $87° \leqslant \theta \leqslant 90°$ and for $10^9 \leqslant \text{Ra}_L < 10^{11}$ if $89° \leqslant \theta \leqslant 90°$. Ra_L is based on g (*not* $g\cos\theta$). Fujii and Imura's results are for two-dimensional plates—ones in which infinite breadth has been approximated by suppression of end effects.

For circular plates of diameter D in the stable horizontal configurations, the data of Kadambi and Drake [8.26] suggest that

$$\overline{\text{Nu}}_D = 0.82\ \text{Ra}_D^{1/5}\ \text{Pr}^{0.034} \tag{8.39}$$

Natural convection with uniform heat flux

When q_w is specified instead of $\Delta T \equiv (T_w - T_\infty)$, ΔT becomes the unknown dependent variable. Because $h \equiv q_w/\Delta T$, the dependent variable appears in the Nusselt number; however, for natural convection, it also appears in the Rayleigh number. Thus, the situation is more complicated than in forced convection.

Since Nu often varies as $\text{Ra}^{1/4}$, we may write

$$\text{Nu}_x = \frac{q_w}{\Delta T}\frac{x}{k} \propto \text{Ra}_x^{1/4} \propto \Delta T^{1/4} x^{3/4}$$

The relationship between x and ΔT is then

$$\Delta T = C x^{1/5} \tag{8.40}$$

where the constant of proportionality C involves q_w and the relevant physical properties. The average of ΔT over a heater of length L is

$$\overline{\Delta T} = \frac{1}{L}\int_0^L C x^{1/5}\, dx = \frac{5}{6} C \tag{8.41}$$

We plot $\Delta T/C$ against x/L in Fig. 8.9. Here, $\overline{\Delta T}$ and $\Delta T(x/L = ½)$ are within 4% of each other. This suggests that, if we are interested in *average* values of ΔT, we can use ΔT evaluated at the midpoint of the plate in both the Rayleigh number, Ra_L, and the average Nusselt number, $\overline{\text{Nu}}_L = q_w L/k\overline{\Delta T}$. Churchill and Chu, for example, show that their vertical plate correlation, eqn. (8.27), represents q_w = constant data exceptionally well

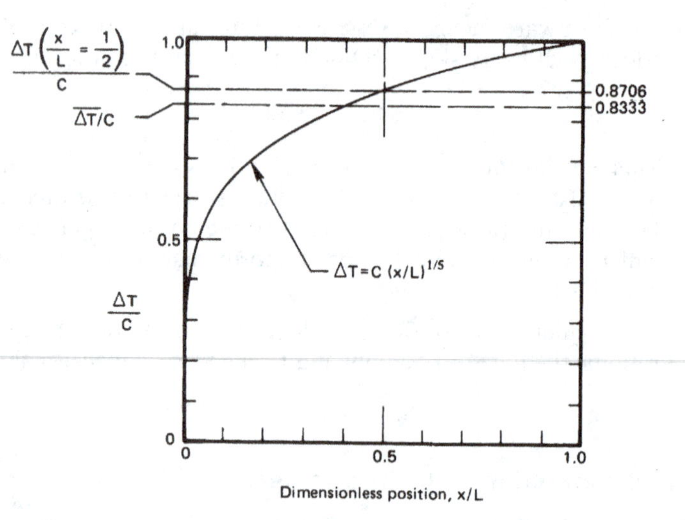

Figure 8.9 The mean value of $\Delta T \equiv T_w - T_\infty$ during natural convection.

in the range $\text{Ra}_L > 1$ when Ra_L is based on ΔT at the middle of the plate. This approach eliminates the variation of ΔT with x from the calculation, but the temperature difference at the middle of the plate must still be found by iteration.

To avoid iterating, we need to eliminate ΔT from the Rayleigh number. We can do this by introducing a modified Rayleigh number, Ra_x^*, defined as

$$\text{Ra}_x^* \equiv \text{Ra}_x \text{Nu}_x \equiv \frac{g\beta\Delta T x^3}{\nu\alpha}\,\frac{q_w x}{\Delta T k} = \frac{g\beta q_w x^4}{k\nu\alpha} \tag{8.42}$$

For example, in eqn. (8.27), we replace Ra_L with $\text{Ra}_L^*/\overline{\text{Nu}}_L$. The result is

$$\overline{\text{Nu}}_L = 0.68 + 0.67\left(\text{Ra}_L^*\right)^{1/4} \Big/ \overline{\text{Nu}}_L^{1/4}\left[1 + \left(\frac{0.492}{\text{Pr}}\right)^{9/16}\right]^{4/9}$$

which may be rearranged as

$$\overline{\text{Nu}}_L^{1/4}\left(\overline{\text{Nu}}_L - 0.68\right) = \frac{0.67\left(\text{Ra}_L^*\right)^{1/4}}{\left[1 + (0.492/\text{Pr})^{9/16}\right]^{4/9}} \tag{8.43a}$$

When $\overline{\text{Nu}}_L \gtrsim 5$, the term 0.68 may be neglected, with the result

$$\overline{\text{Nu}}_L = \frac{0.73\left(\text{Ra}_L^*\right)^{1/5}}{\left[1 + (0.492/\text{Pr})^{9/16}\right]^{16/45}} \tag{8.43b}$$

Raithby and Hollands [8.13] give the following, somewhat simpler correlations for laminar natural convection from vertical plates with a uniform wall heat flux:

$$\text{Nu}_x = 0.630\left(\frac{\text{Ra}_x^*\,\text{Pr}}{4 + 9\sqrt{\text{Pr}} + 10\,\text{Pr}}\right)^{1/5} \tag{8.44a}$$

$$\overline{\text{Nu}}_L = \frac{6}{5}\left(\frac{\text{Ra}_L^*\,\text{Pr}}{4 + 9\sqrt{\text{Pr}} + 10\,\text{Pr}}\right)^{1/5} \tag{8.44b}$$

These equations apply for all Pr and for Nu $\gtrsim 5$ (equations for lower Nu or Ra* are given in [8.13]).

Example 8.5

A horizontal circular disk heater of diameter 0.17 m faces downward in air at 27°C. If it delivers 15 W, estimate its average surface temperature.

Solution. We have no formula for this situation, so the problem calls for some judicious guesswork. Following the lead of Churchill and Chu, we replace Ra_D with $\text{Ra}_D^*/\overline{\text{Nu}}_D$ in eqn. (8.39):

$$\left(\overline{\text{Nu}}_D\right)^{6/5} = \left(\frac{q_w D}{\overline{\Delta T} k}\right)^{6/5} = 0.82\left(\text{Ra}_D^*\right)^{1/5}\text{Pr}^{0.034}$$

so

$$\begin{aligned}
\overline{\Delta T} &= 1.18\,\frac{q_w D/k}{\left(\dfrac{g\beta q_w D^4}{k\nu\alpha}\right)^{1/6}\text{Pr}^{0.028}} \\
&= 1.18\,\frac{\left(\dfrac{15}{\pi(0.085)^2}\right)\dfrac{0.17}{0.02614}}{\left[\dfrac{9.8[15/\pi(0.085)^2]0.17^4}{300(0.02164)(1.566)(2.203)10^{-10}}\right]^{1/6}(0.711)^{0.028}} \\
&= 140\text{ K}
\end{aligned}$$

In the preceding computation, all properties were evaluated at T_∞. Now we must return the calculation, reevaluating all properties except β at 27 + (140/2) = 97°C:

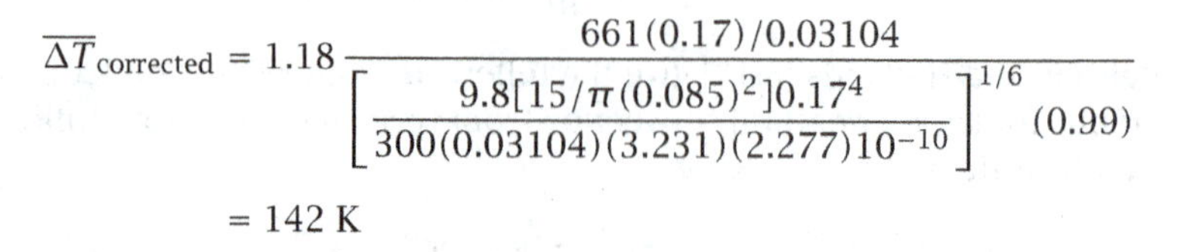

$$\overline{\Delta T}_{\text{corrected}} = 1.18 \frac{661(0.17)/0.03104}{\left[\dfrac{9.8[15/\pi(0.085)^2]0.17^4}{300(0.03104)(3.231)(2.277)10^{-10}}\right]^{1/6}(0.99)}$$

$$= 142 \text{ K}$$

so the surface temperature is 27 + 142 = 169°C.

That is rather hot. Obviously, the cooling process is quite ineffective in this case. ■

Some other natural convection problems

We have clearly moved into the realm of handbook information at this point. And it would be beyond the scope of this book to go much further. Still, two matters deserve at least qualitative mention. They are:

Natural convection in enclosures. When a natural convection process occurs within a confined space, the heated fluid buoys up and then follows the contours of the container, releasing heat and in some way returning to the heater. This recirculation process normally enhances heat transfer beyond that which would occur by conduction through the stationary fluid. These processes are of importance to energy conservation processes in buildings (as in multiply glazed windows, uninsulated walls, and attics), to crystal growth and solidification processes, to hot or cold liquid storage systems, and to countless other configurations. Survey articles on natural convection in enclosures have been written by Yang [8.27], Raithby and Hollands [8.13], and Catton [8.28].

Combined natural and forced convection. When forced convection along, say, a vertical wall occurs at a relatively low velocity but at a relatively high heating rate, the resulting density changes can give rise to a superimposed natural convection process. We saw in footnote 2 on page 404 that $\text{Gr}_L^{1/2}$ plays the role of of a natural convection Reynolds number, it follows that we can estimate of the relative importance of natural and forced convection can be gained by considering the ratio

$$\frac{\mathrm{Gr}_L}{\mathrm{Re}_L^2} = \frac{\text{strength of natural convection flow}}{\text{strength of forced convection flow}} \tag{8.45}$$

where Re_L is for the forced convection along the wall. If this ratio is small compared to one, the flow is essentially driven by forced convection, whereas if it is much larger than one, we have natural convection. When $\mathrm{Gr}_L/\mathrm{Re}_L^2$ is on the order of one, we have a *mixed convection* process.

Of course, the relative orientation of the forced flow and the natural convection flow matters. For example, compare cool air flowing downward past a hot wall to cool air flowing upward along a hot wall. The former situation is called *opposing flow* and the latter is called *assisting flow.* Opposing flow may lead to boundary layer separation and degraded heat transfer.

Churchill [8.29] has provided an extensive discussion of both the conditions that give rise to mixed convection and the prediction of heat transfer for it. Review articles on the subject have been written by Chen and Armaly [8.30] and by Aung [8.31].

8.5 Film condensation

Dimensional analysis and experimental data

The dimensional functional equation for h (or $\overline{h}$) during film condensation is[7]

$$\overline{h} \text{ or } h = \text{fn}\left[c_p, \rho_f, h_{fg}, g(\rho_f - \rho_g), k, \mu, (T_{\text{sat}} - T_w), L \text{ or } x\right]$$

where h_{fg} is the latent heat of vaporization. It does not appear in the differential equations (8.4) and (6.40); however, it is used in the calculation of δ [which enters in the b.c.'s (8.5)]. The film thickness, δ, depends heavily on the latent heat and slightly on the sensible heat, $c_p\Delta T$, which the film must absorb to condense. Notice, too, that $g(\rho_f - \rho_g)$ is included as a product, because gravity only enters the problem as it acts upon the density difference [cf. eqn. (8.4)].

The problem is therefore expressed nine variables in J, kg, m, s, and °C (where we once more avoid resolving J into N · m since heat is not being

[7]Note that, throughout this section, k, μ, c_p, and Pr refer to properties of the liquid, rather than the vapor.

converted into work in this situation). It follows that we look for 9 − 5 = 4 pi-groups. The ones we choose are

$$\Pi_1 = \overline{\mathrm{Nu}}_L \equiv \frac{\overline{h}L}{k} \qquad \Pi_2 = \mathrm{Pr} \equiv \frac{\nu}{\alpha} \tag{8.46}$$

$$\Pi_3 = \mathrm{Ja} \equiv \frac{c_p(T_{\mathrm{sat}} - T_w)}{h_{fg}} \qquad \Pi_4 \equiv \frac{\rho_f(\rho_f - \rho_g)gh_{fg}L^3}{\mu k(T_{\mathrm{sat}} - T_w)} \tag{8.47}$$

Two of these groups are new to us. The group Π_3 is called the *Jakob number*, Ja, to honor Max Jakob's important pioneering work during the 1930s on problems of phase change. It compares the maximum sensible heat absorbed by the liquid to the latent heat absorbed. The group Π_4 does not normally bear anyone's name, but, if it is multiplied by Ja, it may be regarded as a Rayleigh number for the condensate film.

Notice that if we condensed water at 1 atm on a wall 10°C below T_{sat}, then Ja would equal $4.174(10/2257) = 0.0185$. Although 10°C is a fairly large temperature difference in a condensation process, it gives a maximum sensible heat that is less than 2% of the latent heat. The Jakob number is accordingly small in most cases of practical interest, and it turns out that sensible heat can often be neglected. (There are important exceptions to this.) The same is true of the role of the Prandtl number. Therefore, during film condensation

$$\overline{\mathrm{Nu}}_L = \mathrm{fn}\left(\underbrace{\frac{\rho_f(\rho_f - \rho_g)gh_{fg}L^3}{\mu k(T_{\mathrm{sat}} - T_w)}}_{\text{primary independent variable, } \Pi_4}, \underbrace{\mathrm{Pr}, \mathrm{Ja}}_{\text{secondary independent variables}}\right) \tag{8.48}$$

Equation (8.48) is not restricted to any geometrical configuration, since the same variables govern h during film condensation on any body. Figure 8.10, for example, shows laminar film condensation data given for spheres by Dhir[8] [8.32]. They have been correlated according to eqn. (8.12). The data are for only one value of Pr but for a range of Π_4 and Ja. They generally correlate well within ±10%, despite a broad variation of the not-very-influential variable, Ja. A predictive curve [8.32] is included in Fig. 8.10 for future reference.

[8]Vijay K. Dhir very kindly recalculated his data into the form shown in Fig. 8.10 for use here.

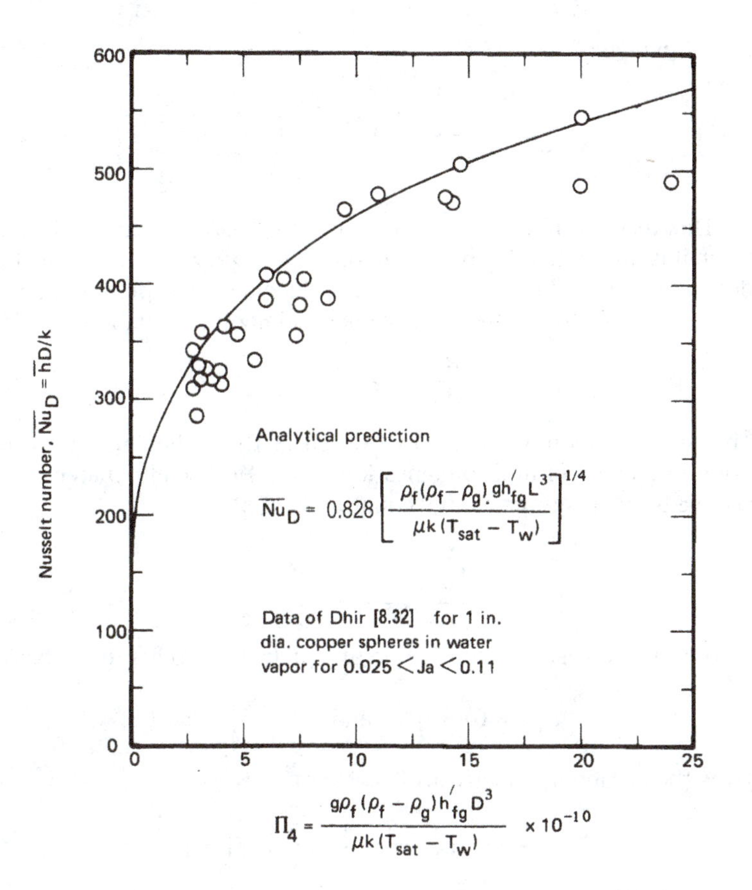

Figure 8.10 Correlation of the data of Dhir [8.32] for laminar film condensation on spheres at one value of Pr and for a range of Π_4 and Ja, with properties evaluated at $(T_{sat} + T_w)/2$. Analytical prediction from [8.33].

Laminar film condensation on a vertical plate

Consider the following feature of film condensation. The latent heat of a liquid is normally a very large number. Therefore, even a high rate of heat transfer will typically result in only very thin films. These films move relatively slowly, so it is safe to ignore the inertia terms in the momentum equation (8.4):

$$\underbrace{u\frac{\partial u}{\partial x}+v\frac{\partial u}{\partial y}}_{\simeq 0}=\left(1-\frac{\rho_g}{\rho_f}\right)g+\nu\underbrace{\frac{\partial^2 u}{\partial y^2}}_{\simeq \frac{d^2u}{dy^2}}$$

This result will give $u = u(y,\delta)$ (where δ is the local b.l. thickness) when it is integrated. We recognize that $\delta = \delta(x)$, so that u is not strictly dependent on y alone. However, the y-dependence is predominant, and it is reasonable to use the approximate momentum equation

$$\frac{d^2u}{dy^2} = -\frac{\rho_f-\rho_g}{\rho_f}\frac{g}{\nu} \tag{8.49}$$

This simplification was made by Nusselt in 1916 when he set down the original analysis of film condensation [8.34]. He also eliminated the convective terms from the energy equation (6.40):

$$\underbrace{u\frac{\partial T}{\partial x}+v\frac{\partial T}{\partial y}}_{\simeq 0}=\alpha\frac{\partial^2 T}{\partial y^2}$$

on the same basis. The integration of eqn. (8.49) subject to the b.c.'s

$$u\,(y=0)=0 \qquad \text{and} \qquad \left.\frac{\partial u}{\partial y}\right|_{y=\delta}=0$$

gives the parabolic velocity profile:

$$u=\frac{(\rho_f-\rho_g)g\delta^2}{2\mu}\left[2\left(\frac{y}{\delta}\right)-\left(\frac{y}{\delta}\right)^2\right] \tag{8.50}$$

And integration of the energy equation subject to the b.c.'s

$$T\,(y=0)=T_w \qquad \text{and} \qquad T\,(y=\delta)=T_{\text{sat}}$$

gives the linear temperature profile:

$$T=T_w+(T_{\text{sat}}-T_w)\,\frac{y}{\delta} \tag{8.51}$$

To complete the analysis, we must calculate δ. We can do this in two steps. First, we express the mass flow rate per unit width of film, $\dot{m}$, in terms of δ, with the help of eqn. (8.50):

$$\dot{m}=\int_0^\delta \rho_f u\,dy=\frac{\rho_f(\rho_f-\rho_g)}{3\mu}g\delta^3 \tag{8.52}$$

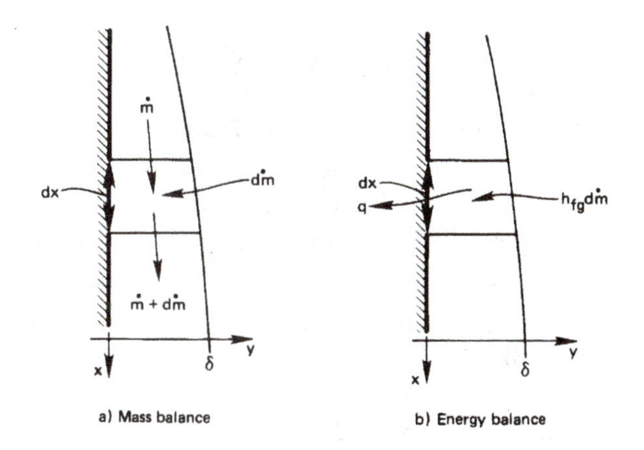

Figure 8.11 Heat and mass flow in an element of a condensing film.

Second, we neglect the sensible heat absorbed by that part of the film cooled below T_{sat} and express the local heat flux in terms of the rate of change of $\dot{m}$ (see Fig. 8.11):

$$|q| = k \left.\frac{\partial T}{\partial y}\right|_{y=0} = k\,\frac{T_{\text{sat}} - T_w}{\delta} = h_{fg}\,\frac{d\dot{m}}{dx} \tag{8.53}$$

Substituting eqn. (8.52) in eqn. (8.53), we obtain a first-order differential equation for δ:

$$k\,\frac{T_{\text{sat}} - T_w}{\delta} = \frac{h_{fg}\rho_f(\rho_f - \rho_g)}{\mu}\,g\delta^2\,\frac{d\delta}{dx} \tag{8.54}$$

This can be integrated directly, subject to the b.c., $\delta(x = 0) = 0$. The result is

$$\delta = \left[\frac{4k(T_{\text{sat}} - T_w)\mu x}{\rho_f(\rho_f - \rho_g)gh_{fg}}\right]^{1/4} \tag{8.55}$$

Both Nusselt and, subsequently, Rohsenow [8.35] suggested replacing h_{fg} with a corrected value, h'_{fg}, which would depend on Ja. We will give an expression for this correction below. For the moment we simply write this yet-to-be-provided h'_{fg} in place of h_{fg} in the equations that follow.

Finally, we calculate the heat transfer coefficient

$$h \equiv \frac{q}{T_{\text{sat}} - T_w} = \frac{1}{T_{\text{sat}} - T_w}\left[\frac{k(T_{\text{sat}} - T_w)}{\delta}\right] = \frac{k}{\delta} \tag{8.56}$$

so

$$\mathrm{Nu}_x = \frac{hx}{k} = \frac{x}{\delta} \tag{8.57}$$

Thus, we substitute eqn. (8.55) in eqn. (8.57) and get

$$\boxed{\mathrm{Nu}_x = 0.707\left[\frac{\rho_f(\rho_f-\rho_g)gh'_{fg}x^3}{\mu k(T_{\text{sat}}-T_w)}\right]^{1/4}} \tag{8.58}$$

This equation carries out the functional dependence that we anticipated in eqn. (8.48):

$$\mathrm{Nu}_x = \mathrm{fn}\left(\underbrace{\Pi_4}, \underbrace{\mathrm{Ja}}, \underbrace{\mathrm{Pr}}\right)$$

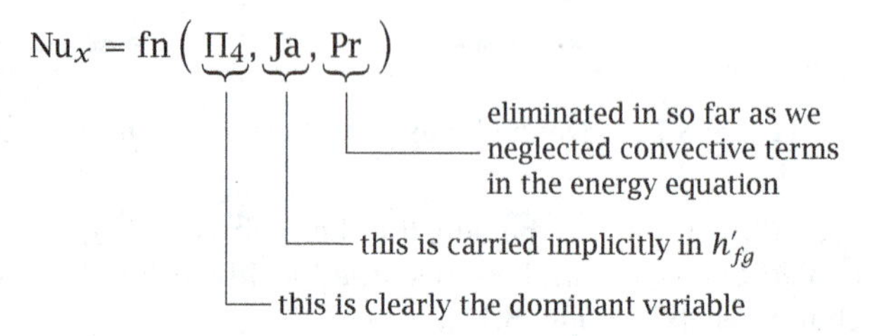

The liquid properties in Π_4, Ja, and Pr (with the exception of h_{fg}) are to be evaluated at the mean film temperature. However, if $T_{\text{sat}} - T_w$ is small—and it often is—one might approximate them at T_{sat}.

At this point we should ask just how great the missing influence of Pr is and what degree of approximation is involved in representing the influence of Ja with the use of h'_{fg}. Sparrow and Gregg [8.36] answered these questions with a complete b.l. analysis of film condensation. They did not introduce Ja in a corrected latent heat but instead showed its influence directly.

Figure 8.12 displays two figures from the Sparrow and Gregg paper. The first shows heat transfer results plotted in the form

$$\frac{\mathrm{Nu}_x}{\sqrt[4]{\Pi_4}} = \mathrm{fn}\,(\mathrm{Ja}, \mathrm{Pr}) \longrightarrow \text{constant as Ja} \longrightarrow 0 \tag{8.59}$$

Notice that the calculation approaches Nusselt's simple result for all Pr as Ja → 0. It also approaches Nusselt's result, even for fairly large values of Ja, if Pr is not small. The second figure shows how the temperature deviates from the linear profile that we assumed to exist in the film in developing eqn. (8.51). If we remember that a Jakob number of 0.02 is about as large as we normally find in laminar condensation, it is

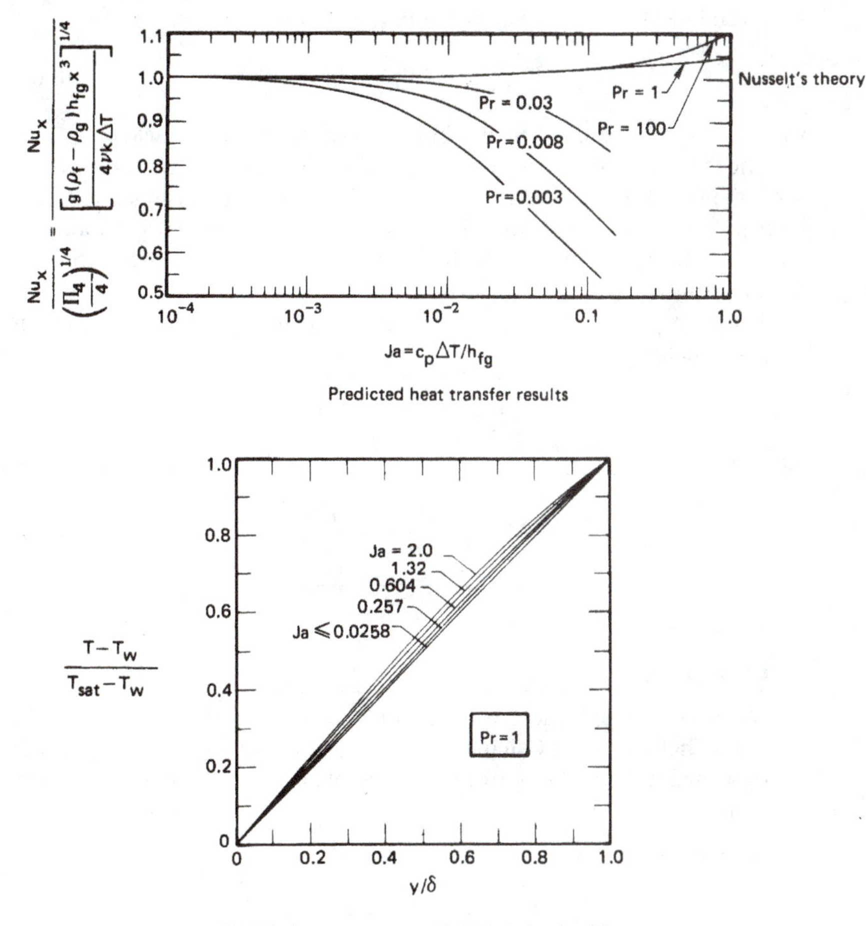

Figure 8.12 Results of the exact b.l. analysis of laminar film condensation on a vertical plate [8.36].

clear that the linear temperature profile is a very sound assumption for nonmetallic liquids.

Sadasivan and Lienhard [8.37] have shown that the Sparrow-Gregg formulation can be expressed with high accuracy, for $\Pr \geqslant 0.6$, by including

Pr in the latent heat correction. Thus they wrote

$$h'_{fg} = h_{fg}\left[1 + (0.683 - 0.228/\text{Pr})\,\text{Ja}\right] \qquad (8.60)$$

which we use in eqn. (8.58) and in the relevant equations below.

The Sparrow and Gregg analysis proves that Nusselt's analysis is quite accurate for all Prandtl numbers above the liquid-metal range. The very high Ja flows, for which Nusselt's theory requires some correction, usually result in thicker films, which become turbulent so the exact analysis no longer applies.

The average heat transfer coefficient is calculated in the usual way for T_{wall} = constant:

$$\overline{h} = \frac{1}{L}\int_0^L h(x)\,dx = \tfrac{4}{3}\,h(L)$$

so

$$\boxed{\overline{\text{Nu}_L} = 0.9428\left[\frac{\rho_f(\rho_f - \rho_g)gh'_{fg}L^3}{\mu k(T_{\text{sat}} - T_w)}\right]^{1/4}} \qquad (8.61)$$

Example 8.6

Water at atmospheric pressure condenses on a strip 30 cm in height that is held at 90°C. Calculate the overall heat transfer per meter, the film thickness at the bottom, and the mass rate of condensation per meter.

Solution.

$$\delta = \left[\frac{4k(T_{\text{sat}} - T_w)\mu x}{\rho_f(\rho_f - \rho_g)gh'_{fg}}\right]^{1/4}$$

where we have replaced h_{fg} with h'_{fg}:

$$h'_{fg} = 2257\left[1 + \left(0.683 - \frac{0.228}{1.86}\right)\frac{4.211(10)}{2257}\right] = 2281 \text{ kJ/kg}$$

so

$$\delta = \left[\frac{4(0.677)(10)(2.99\times10^{-4})\,x}{961.9(961.9 - 0.6)(9.806)(2281\times10^3)}\right]^{1/4} = 0.000141\,x^{1/4}$$

Then

$$\delta(L) = 0.000104 \text{ m} = 0.104 \text{ mm}$$

Notice how thin the film is. Finally, we use eqns. (8.57) and (8.60) to compute

$$\overline{\text{Nu}}_L = \frac{4}{3}\frac{L}{\delta} = \frac{4(0.3)}{3(0.000104)} = 3846$$

so

$$q = \frac{\overline{\text{Nu}}_L\, k\Delta T}{L} = \frac{3846(0.677)(10)}{0.3} = 8.68 \times 10^4 \text{ W/m}^2$$

(This would correspond to a heat flow of 86.8 kW on an area about half the size of a desk top. That is very high for such a small temperature difference.) Then

$$Q = (8.68 \times 10^4)(0.3) = 26{,}040 \text{ W/m} = 26.0 \text{ kW/m}$$

The rate of condensate flow, $\dot{m}$ is

$$\dot{m} = \frac{Q}{h'_{fg}} = \frac{26.0}{2281} = 0.0114 \text{ kg/m·s}$$ ■

Condensation on other bodies

Nusselt himself extended his prediction to certain other bodies but was restricted by the lack of a digital computer from evaluating as many cases as he might have. In 1971 Dhir and Lienhard [8.33] showed how Nusselt's method could be readily extended to a large class of problems. They showed that one need only to replace the gravity, g, with an effective gravity, g_{eff}:

$$g_{\text{eff}} \equiv \frac{x\,(gR)^{4/3}}{\displaystyle\int_0^x g^{1/3}R^{4/3}\,dx} \tag{8.62}$$

in eqns. (8.55) and (8.58), to predict δ and Nu_x for a variety of bodies. The terms in eqn. (8.62) are (see Fig. 8.13):

- x is the distance along the liquid film measured from the upper stagnation point.
- $g = g(x)$, the component of gravity (or other body force) along x; g can vary from point to point as it does in Fig. 8.13b and c.

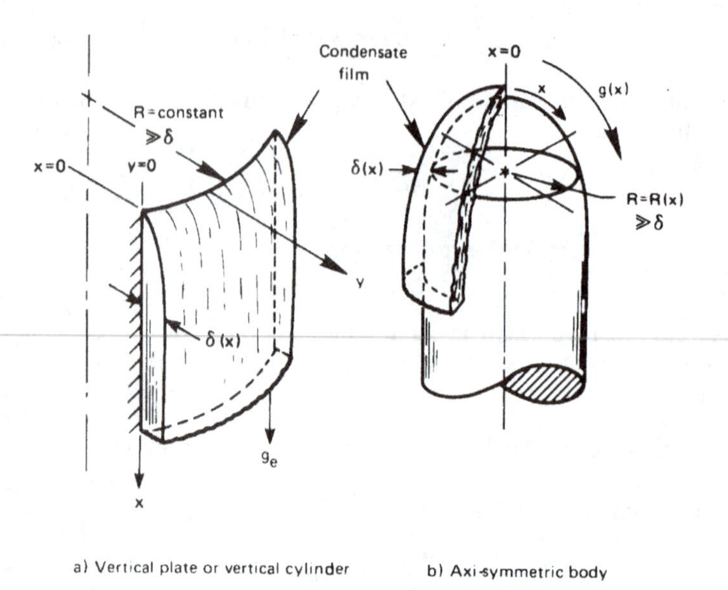

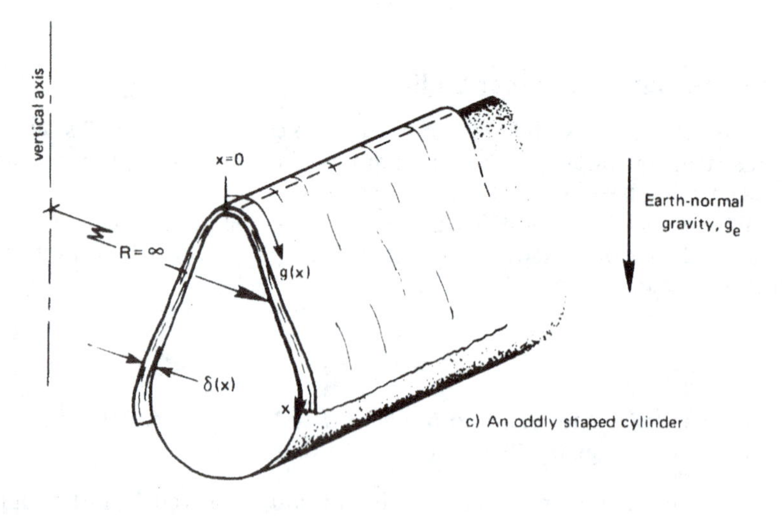

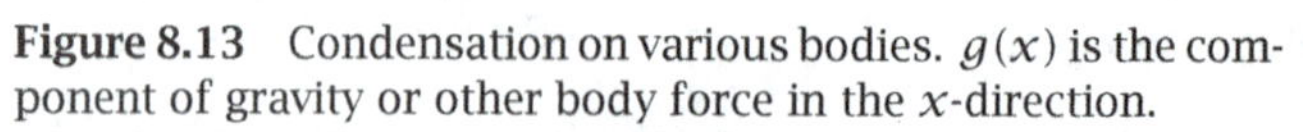

Figure 8.13 Condensation on various bodies. $g(x)$ is the component of gravity or other body force in the x-direction.

- $R(x)$ is a radius of curvature about the vertical axis. In Fig. 8.13a, it is a constant that factors out of eqn. (8.62). In Fig. 8.13c, R is infinite. Since it appears to the same power in both the numerator and the denominator, it again can be factored out of eqn. (8.62). Only in axisymmetric bodies, where R varies with x, need it be included. When it can be factored out,

$$g_{\text{eff}} \text{ reduces to } \frac{x g^{4/3}}{\displaystyle\int_0^x g^{1/3}\, dx} \tag{8.63}$$

- g_e is earth-normal gravity. We introduce g_e at this point to distinguish it from $g(x)$.

Example 8.7

Find Nu_x for laminar film condensation on the top of a flat surface sloping at $\theta°$ from the vertical plane.

Solution. In this case $g = g_e \cos\theta$ and $R = \infty$. Therefore, eqn. (8.62) or (8.63) reduces to

$$g_{\text{eff}} = \frac{x g_e^{4/3} (\cos\theta)^{4/3}}{g_e^{1/3} (\cos\theta)^{1/3} \displaystyle\int_0^x dx} = g_e \cos\theta \tag{8.64}$$

as we might expect. Then, for a slanting plate,

$$\mathrm{Nu}_x = 0.707 \left[\frac{\rho_f(\rho_f - \rho_g)(g_e \cos\theta) h'_{fg} x^3}{\mu k (T_{\text{sat}} - T_w)} \right]^{1/4} \tag{8.65}$$

■

Example 8.8

Find the overall Nusselt number for a horizontal cylinder.

Solution. There is an important conceptual hurdle here. The radius $R(x)$ is infinity, as shown in Fig. 8.13c—it is not the radius of the cylinder. It is also very easy to show that $g(x)$ is equal to $g_e \sin(2x/D)$, where D is the diameter of the cylinder. Then

$$g_{\text{eff}} = \frac{x g_e^{4/3} (\sin 2x/D)^{4/3}}{g_e^{1/3} \displaystyle\int_0^x (\sin 2x/D)^{1/3}\, dx}$$

and, with $h(x)$ from eqn. (8.58),

$$\overline{h} = \frac{2}{\pi D}\int_0^{\pi D/2} \frac{1}{\sqrt{2}}\frac{k}{x}\left[\frac{\rho_f(\rho_f-\rho_g)h'_{fg}x^3}{\mu k\,(T_{\text{sat}}-T_w)}\,\frac{x g_e(\sin 2x/D)^{4/3}}{\int_0^x (\sin 2x/D)^{1/3}\,dx}\right]^{1/4} dx$$

This integral can be evaulated in terms of gamma functions. The result, when it is put back in the form of a Nusselt number, is

$$\overline{\text{Nu}}_D = 0.728\left[\frac{\rho_f(\rho_f-\rho_g)g_e h'_{fg}D^3}{\mu k\,(T_{\text{sat}}-T_w)}\right]^{1/4} \tag{8.66}$$

for a horizontal cylinder. (Nusselt got 0.725 for the lead constant, but he had to approximate the integral with a hand calculation.) ■

Some other results of this calculation include the following cases.

Sphere of diameter D:

$$\overline{\text{Nu}}_D = 0.828\left[\frac{\rho_f(\rho_f-\rho_g)g_e h'_{fg}D^3}{\mu k\,(T_{\text{sat}}-T_w)}\right]^{1/4} \tag{8.67}$$

This result[9] has already been compared with the experimental data in Fig. 8.10.

Vertical cone with the apex on top, the bottom insulated, and a cone angle of $\alpha°$:

$$\text{Nu}_x = 0.874\,[\cos(\alpha/2)]^{1/4}\left[\frac{\rho_f(\rho_f-\rho_g)g_e h'_{fg}x^3}{\mu k\,(T_{\text{sat}}-T_w)}\right]^{1/4} \tag{8.68}$$

Rotating horizontal disk[10]: In this case, $g = \omega^2 x$, where x is the distance from the center and ω is the speed of rotation. The Nusselt number, based on $L = (\mu/\rho_f\omega)^{1/2}$, is

$$\overline{\text{Nu}} = 0.9034\left[\frac{\mu(\rho_f-\rho_g)h'_{fg}}{\rho_f k\,(T_{\text{sat}}-T_w)}\right]^{1/4} = \text{constant} \tag{8.69}$$

[9] There is an error in [8.33]: the constant given there is 0.785. The value of 0.828 given here is correct.

[10] This problem was originally solved by Sparrow and Gregg [8.38].

This result might seem strange at first glance. It says that $\text{Nu} \neq \text{fn}(x \text{ or } \omega)$. The reason is that δ just happens to be independent of x in this configuration.

The Nusselt solution can thus be bent to fit many complicated geometric figures. One of the most complicated ones that have been dealt with is the reflux condenser shown in Fig. 8.14. In such a configuration, cooling water flows through a helically wound tube and vapor condenses on the outside, running downward along the tube. As the condensate flows, centripetal forces sling the liquid outward at a downward angle. This complicated flow was analyzed by Karimi [8.39], who found that

$$\overline{\text{Nu}} \equiv \frac{\overline{h} d \cos\alpha}{k} = \left[\frac{(\rho_f - \rho_g)\rho_f h'_{fg} g (d\cos\alpha)^3}{\mu k \Delta T}\right]^{1/4} \text{fn}\left(\frac{d}{D}, B\right) \tag{8.70}$$

where B is a centripetal parameter:

$$B \equiv \frac{\rho_f - \rho_g}{\rho_f} \frac{c_p \Delta T}{h'_{fg}} \frac{\tan^2\alpha}{\text{Pr}}$$

and α is the helix angle (see Fig. 8.14). The function on the righthand side of eqn. (8.70) was a complicated one that must be evaluated numerically. Karimi's result is plotted in Fig. 8.14.

Laminar–turbulent transition

The mass flow rate of condensate per unit width of film, $\dot{m}$, is more commonly designated as Γ_c (kg/m · s). Its calculation in eqn. (8.52) involved substituting eqn. (8.50) in

$$\dot{m} \text{ or } \Gamma_c = \rho_f \int_0^\delta u\, dy$$

Equation (8.50) gives $u(y)$ independently of any geometric features. [The geometry is characterized by $\delta(x)$.] Thus, the resulting equation for the mass flow rate is still

$$\Gamma_c = \frac{\rho_f(\rho_f - \rho_g) g \delta^3}{3\mu} \tag{8.52a}$$

This expression is valid for any location along any film, regardless of the geometry of the body. The configuration will lead to variations of $g(x)$ and $\delta(x)$, but eqn. (8.52a) still applies.

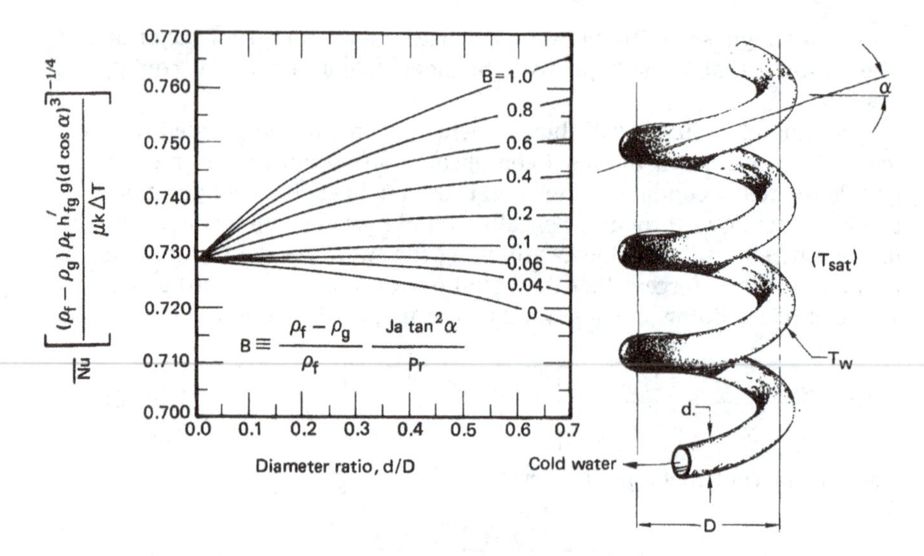

Figure 8.14 Fully developed film condensation heat transfer on a helical reflux condenser [8.39].

It is useful to define a Reynolds number in terms of Γ_c. This is easy to do, because Γ_c is equal to $\rho u_{\text{av}}\delta$.

$$\text{Re}_c = \frac{\Gamma_c}{\mu} = \frac{\rho_f(\rho_f - \rho_g)g\delta^3}{3\mu^2} \tag{8.71}$$

It turns out that the Reynolds number dictates the onset of film instability, just as it dictates the instability of a b.l. or of a pipe flow.[11] When $\text{Re}_c \cong 7$, scallop-shaped ripples become visible on the condensate film. When Re_c reaches about 400, a full-scale laminar-to-turbulent transition occurs.

Gregorig, Kern, and Turek [8.40] reviewed many data for the film condensation of water and added their own measurements. Figure 8.15 shows these data in comparison with Nusselt's theory, eqn. (8.61). The comparison is almost perfect up to $\text{Re}_c \cong 7$. Then the data start yielding somewhat higher heat transfer rates than the prediction. This is because

[11]Two Reynolds numbers are defined for film condensation: Γ_c/μ and $4\Gamma_c/\mu$. The latter one, which is simply four times as large as the one we use, is more common in the American literature.

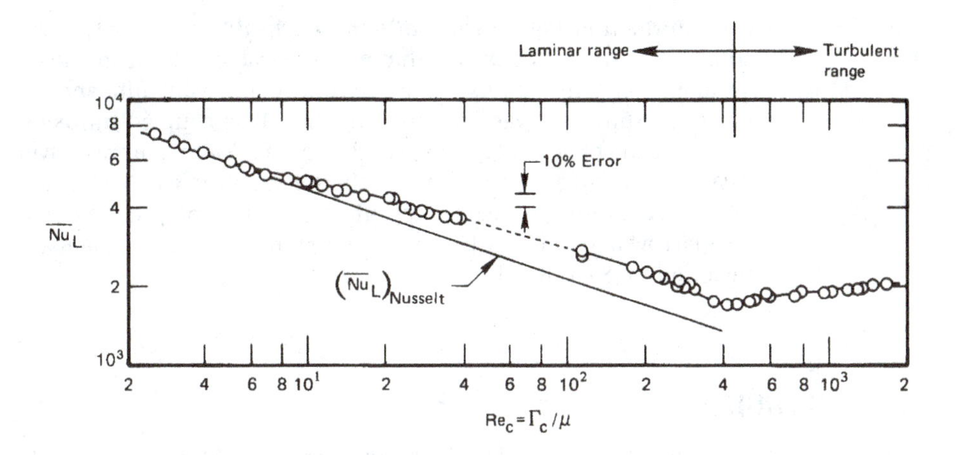

Figure 8.15 Film condensation on vertical plates. Data are for water [8.40].

the ripples improve heat transfer—just a little at first and by about 20% when the full laminar-to-turbulent transition occurs at $\mathrm{Re}_c = 400$.

Above $\mathrm{Re}_c = 400$, $\overline{\mathrm{Nu}}_L$ begins to rise with Re_c. The Nusselt number begins to exhibit an increasingly strong dependence on the Prandtl number in this turbulent regime. Therefore, one can use Fig. 8.15, directly as a data correlation, to predict the heat transfer coefficient for steam condensating at 1 atm. But for other fluids with different Prandtl numbers, one should consult [8.41] or [8.42].

Two final issues in natural convection film condensation

- *Condensation in tube bundles.* Nusselt showed that if n horizontal tubes are arrayed over one another, and if the condensate leaves each one and flows directly onto the one below it without splashing, then

$$\mathrm{Nu}_{D_{\text{for } n \text{ tubes}}} = \frac{\mathrm{Nu}_{D_{1\text{ tube}}}}{n^{1/4}} \tag{8.72}$$

  This is a fairly optimistic extension of the theory, of course. In addition, the effects of vapor shear stress on the condensate and of pressure losses on the saturation temperature are often important in tube bundles. These effects are discussed by Rose et al. [8.42] and Marto [8.41].

- *Condensation in the presence of noncondensable gases.* When the condensing vapor is mixed with noncondensable air, uncondensed air must constantly diffuse away from the condensing film and vapor must diffuse inward toward the film. This coupled diffusion process can considerably slow condensation. The resulting h can easily be cut by a factor of five if there is as little as 5% by mass of air mixed into the steam. This effect was first analyzed in detail by Sparrow and Lin [8.43]. More recent studies of this problem are reviewed in [8.41, 8.42].

Problems

8.1 Show that Π_4 in the film condensation problem can properly be interpreted as $\Pr \mathrm{Re}^2/\mathrm{Ja}$.

8.2 A 20 cm high vertical plate is kept at 34°C in a 20°C room. Plot (*to scale*) δ and h vs. height and the actual temperature and velocity vs. y at the top.

8.3 Redo the Squire-Eckert analysis, neglecting inertia, to get a high-Pr approximation to Nu_x. Compare your result to the Squire-Eckert formula. Explain the difference.

8.4 Assume a linear temperature profile and a simple triangular velocity profile, as shown in Fig. 8.16, for natural convection on a vertical isothermal plate. Derive $\mathrm{Nu}_x = \mathrm{fn}(\Pr, \mathrm{Gr}_x)$, compare your result with the Squire-Eckert result, and discuss the comparison.

8.5 A horizontal cylindrical duct of diamond-shaped cross section (Fig. 8.17) carries air at 35°C. Since almost all thermal resistance is in the natural convection b.l. on the outside, take T_w to be approximately 35°C. T_∞ = 25°C. Estimate the heat loss per meter of duct if the duct is uninsulated. [Q = 24.0 W/m.]

8.6 The heat flux from a 3 m high electrically heated panel in a wall is 75 W/m^2 in an 18°C room. What is the average temperature of the panel? What is the temperature at the top? at the bottom?

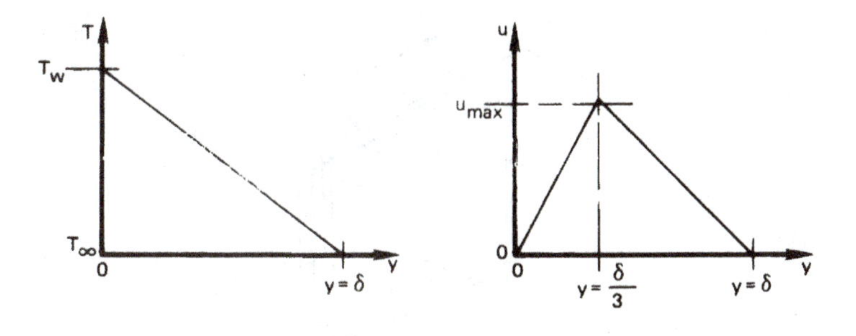

Figure 8.16 Configuration for Problem 8.4.

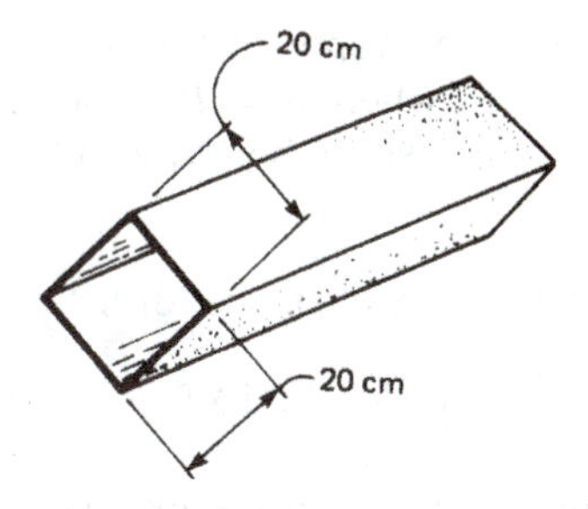

Figure 8.17 Configuration for Problem 8.5.

8.7 Find pipe diameters and wall temperatures for which the film condensation heat transfer coefficients given in Table 1.1 are valid.

8.8 Consider Example 8.6. What value of wall temperature (if any), or what height of the plate, would result in a laminar-to-turbulent transition at the bottom in this example?

8.9 A plate spins, as shown in Fig. 8.18, in a vapor that rotates synchronously with it. Neglect earth-normal gravity and calculate Nu_L as a result of film condensation.

8.10 A laminar liquid film of temperature T_{sat} flows down a vertical wall that is also at T_{sat}. Flow is fully developed and the film thickness is δ_o. Along a particular horizontal line, the wall temperature has a lower value, T_w, and it is kept at that temperature everywhere below that position. Call the line where the wall temperature changes $x = 0$. If the whole system is

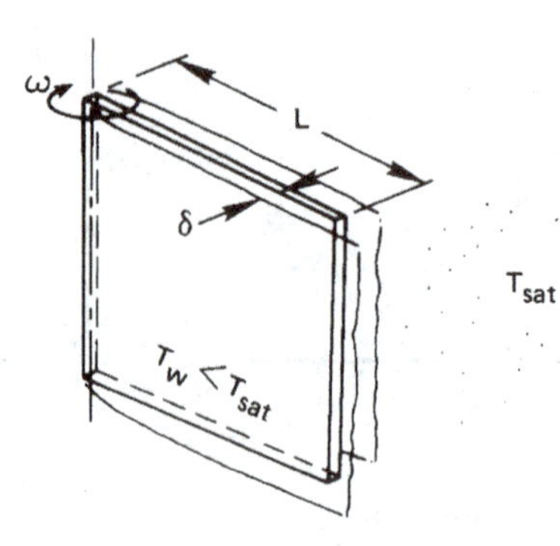

Figure 8.18 Configuration for Problem 8.9.

immersed in saturated vapor of the flowing liquid, calculate $\delta(x)$, Nu_x, and Nu_L, where $x = L$ is the bottom edge of the wall. (Neglect any transition behavior in the neighborhood of $x = 0$.)

8.11 Prepare a table of formulas of the form

$$\overline{h}\,(\mathrm{W/m^2K}) = C\,[\Delta T\,^\circ\mathrm{C}/L\ \mathrm{m}]^{1/4}$$

for natural convection at normal gravity in air and in water at $T_\infty = 27°$C. Assume that T_w is close to 27°C. Your table should include results for vertical plates, horizontal cylinders, spheres, and possibly additional geometries. Do not include your calculations.

8.12 For what value of Pr is the condition

$$\left.\frac{\partial^2 u}{\partial y^2}\right|_{y=0} = \frac{g\beta(T_w - T_\infty)}{\nu}$$

satisfied exactly in the Squire-Eckert b.l. solution? [Pr = 2.86.]

8.13 The overall heat transfer coefficient on the side of a particular house 10 m in height is 2.5 W/m^{2}K, excluding exterior convection. It is a cold, still winter night with $T_{\text{outside}} = -30°$C and $T_{\text{inside air}} = 25°$C. What is $\overline{h}$ on the outside of the house? Is external convection laminar or turbulent?

8.14 Consider Example 8.2. The sheets are mild steel, 2 m long and 6 mm thick. The bath is basically water at 60°C, and the sheets are put in it at 18°C. (a) Plot the sheet temperature as a function of time. (b) Approximate $\overline{h}$ at $\Delta T = [(60 + 18)/2 - 18]$°C and plot the conventional exponential response on the same graph.

8.15 A vertical heater 0.15 m in height is immersed in water at 7°C. Plot $\overline{h}$ against $(T_w - T_\infty)^{1/4}$, where T_w is the heater temperature, in the range $0 < (T_w - T_\infty) < 100$°C. Comment on the result. should the line be straight?

8.16 A 77°C vertical wall heats 27°C air. Evaluate $\delta_{\text{top}}/L, \text{Ra}_L$, and L where the line in Fig. 8.3 ceases to be straight. Comment on the implications of your results. [$\delta_{\text{top}}/L \simeq 0.6$.]

8.17 A horizontal 8 cm O.D. pipe carries steam at 150°C through a room at 17°C. The pipe has a 1.5 cm layer of 85% magnesia insulation on it. Evaluate the heat loss per meter of pipe. [$Q = 97.3$ W/m.]

8.18 What heat rate (in W/m) must be supplied to a 0.01 mm horizontal wire to keep it 30°C above the 10°C water around it?

8.19 A vertical run of copper tubing, 5 mm in diameter and 20 cm long, carries condensation vapor at 60°C through 27°C air. What is the total heat loss?

8.20 A body consists of two cones joined at their bases. The diameter is 10 cm and the overall length of the joined cones is 25 cm. The axis of the body is vertical, and the body is kept at 27°C in 7°C air. What is the rate of heat removal from the body? [$Q = 3.38$ W.]

8.21 Consider the plate dealt with in Example 8.3. Plot $\overline{h}$ as a function of the angle of inclination of the plate as the hot side is tilted both upward and downward. Note that you must make do with discontinuous formulas in different ranges of θ.

8.22 You have been asked to design a vertical wall panel heater, 1.5 m high, for a dwelling. What should the heat flux be if no part of the wall should exceed 33°C? How much heat will be added to the room if the panel is 7 m in width?

8.23 A 14 cm high vertical surface is heated by condensing steam at 1 atm. If the wall is kept at 30°C, how would the average heat transfer coefficient change if ammonia, R22, methanol, or acetone were used instead of steam to heat it? How would the heat flux change? (Data for methanol and acetone must be obtained from sources outside this book.)

8.24 A 1 cm diameter tube extends 27 cm horizontally through a region of saturated steam at 1 atm. The outside of the tube can be maintained at any temperature between 50°C and 150°C. Plot the total heat transfer as a function of tube temperature.

8.25 A 2 m high vertical plate condenses steam at 1 atm. Below what temperature will Nusselt's prediction of $\overline{h}$ be in error? Below what temperature will the condensing film be turbulent?

8.26 A reflux condenser is made of copper tubing 0.8 cm in diameter with a wall temperature of 30°C. It condenses steam at 1 atm. Find $\overline{h}$ if $\alpha = 18°$ and the coil diameter is 7 cm.

8.27 The coil diameter of a helical condenser is 5 cm and the tube diameter is 5 mm. The condenser carries water at 15°C and is in a bath of saturated steam at 1 atm. Specify the number of coils and a reasonable helix angle if 6 kg/hr of steam is to be condensed. $h_{\text{inside}} = 600$ W/m^{2}K.

8.28 A schedule 40 type 304 stainless steam pipe with a 4 in. nominal diameter carries saturated steam at 150 psia in a processing plant. Calculate the heat loss per unit length of pipe if it is bare and the surrounding air is still at 68°F. How much would this heat loss be reduced if the pipe were insulated with a 1 in. layer of 85% magnesia insulation? [$Q_{\text{saved}} \simeq 127$ W/m.]

8.29 What is the maximum speed of air in the natural convection b.l. in Example 8.1?

8.30 All of the uniform-T_w, natural convection formulas for $\overline{\text{Nu}}$ take the same form, within a constant, at high Pr and Ra. What is that form? (Exclude any equation that includes turbulence.)

8.31 A large industrial process requires that water be heated by a large horizontal cylinder using natural convection. The water is at 27°C. The diameter of the cylinder is 5 m, and it is kept at 67°C. First, find $\overline{h}$. Then suppose that D is increased to 10 m.

What is the new $\overline{h}$? Explain the similarity of these answers in the turbulent natural convection regime.

8.32 A vertical jet of liquid of diameter d and moving at velocity u_∞ impinges on a horizontal disk rotating ω rad/s. There is no heat transfer in the system. Develop an expression for $\delta(r)$, where r is the radial coordinate on the disk. Contrast the r dependence of δ with that of a condensing film on a rotating disk and explain the difference qualitatively.

8.33 We have seen that if properties are constant, $h \propto \Delta T^{1/4}$ in natural convection. If we consider the variation of properties as T_w is increased over T_∞, will h depend more or less strongly on ΔT in air? in water?

8.34 A film of liquid falls along a vertical plate. It is initially saturated and it is surrounded by saturated vapor. The film thickness is δ_o. If the wall temperature below a certain point on the wall (call it $x = 0$) is raised to a value of T_w, slightly above T_{sat}, derive expressions for $\delta(x)$, Nu_x, and x_f—the distance at which the plate becomes dry. Calculate x_f if the fluid is water at 1 atm, if $T_w = 105°\text{C}$ and $\delta_o = 0.1$ mm.

8.35 In a particular solar collector, dyed water runs down a vertical plate in a laminar film with thickness δ_o at the top. The sun's rays pass through parallel glass plates (see Section 10.6) and deposit q_s W/m^2 in the film. Assume the water to be saturated at the inlet and the plate behind it to be insulated. Develop an expression for $\delta(x)$ as the water evaporates. Develop an expression for the maximum length of wetted plate, and provide a criterion for the laminar solution to be valid.

8.36 What heat removal flux can be achieved at the surface of a horizontal 0.01 mm diameter electrical resistance wire in still 27°C air if its melting point is 927°C? Neglect radiation.

8.37 A 0.03 m O.D. vertical pipe, 3 m in length, carries refrigerant through a 24°C room. How much heat does it absorb from the room if the pipe wall is at 10°C?

8.38 A 1 cm O.D. tube at 50°C runs horizontally in 20°C air. What is the critical radius of 85% magnesium insulation on the tube?

8.39 A 1 in. cube of ice is suspended in 20°C air. Estimate the drip rate in gm/min. (Neglect ΔT through the departing water film. $h_{sf} = 333,300$ J/kg.)

8.40 A horizontal electrical resistance heater, 1 mm in diameter, releases 100 W/m in water at 17°C. What is the wire temperature?

8.41 Solve Problem 5.39 using the correct formula for the heat transfer coefficient.

8.42 A red-hot vertical rod, 0.02 m in length and 0.005 m in diameter, is used to shunt an electrical current in air at room temperature. How much power can it dissipate if it melts at 1200°C? Note all assumptions and corrections. Include radiation using $\mathcal{F}_{\text{rod-room}} = 0.064$.

8.43 A 0.25 mm diameter platinum wire, 0.2 m long, is to be held horizontally at 1035°C. It is black. How much electric power is needed? Is it legitimate to treat it as a constant-wall-temperature heater in calculating the convective part of the heat transfer? The surroundings are at 20°C and the surrounding room is virtually black.

8.44 A vertical plate, 11.6 m long, condenses saturated steam at 1 atm. We want to be sure that the film stays laminar. What is the lowest allowable plate temperature, and what is $\overline{q}$ at this temperature?

8.45 A straight horizontal fin exchanges heat by laminar natural convection with the surrounding air.

a. Show that

$$\frac{d^2\theta}{d\xi^2} = m^2L^2\theta^{5/4}$$

where m is based on $\overline{h}_o \equiv \overline{h}(T = T_o)$.

b. Develop an iterative numerical method to solve this equation for $T(x = 0) = T_o$ and an insulated tip. (*Hint*: linearize the right side by writing it as $(m^2L^2\theta^{1/4})\theta$, and evaluate the term in parenthesis at the previous iteration step.)

c. Solve the resulting difference equations for m^2L^2 values ranging from 10^{-3} to 10^3. Use Gauss elimination or the tridiagonal algorithm. Express the results as η/η_o where η is the fin efficiency and η_o is the efficiency that would result if $\overline{h}_o$ were the uniform heat transfer coefficient over the entire fin.

8.46 A 2.5 cm black sphere ($\mathcal{F} = 1$) is in radiation-convection equilibrium with air at 20°C. The surroundings are at 1000 K. What is the temperature of the sphere?

8.47 Develop expressions for $\overline{h}(D)$ and $\overline{\text{Nu}}_D$ during condensation on a vertical circular plate.

8.48 A cold copper plate is surrounded by a 5 mm high ridge which forms a shallow container. It is surrounded by saturated water vapor at 100°C. Estimate the steady heat flux and the rate of condensation.

a. When the plate is perfectly horizontal and filled to overflowing with condensate.

b. When the plate is in the vertical position.

c. Did you have to make any idealizations? Would they result in under- or over-estimation of the condensation?

8.49 A proposed design for a nuclear power plant uses molten lead to remove heat from the reactor core. The heated lead is then used to boil water that drives a steam turbine. Water at 5 atm pressure ($T_{\text{sat}} = 152$°C) enters a heated section of a pipe at 60°C with a mass flow rate of $\dot{m} = 2$ kg/s. The pipe is stainless steel ($k_s = 15$ W/m·K) with a wall thickness of 12 mm and an outside diameter of 6.2 cm. The outside surface of the pipe is surrounded by an almost-stationary pool of molten lead at 477°C.

a. At point where the liquid water has a bulk temperature of $T_b = 80$°C, estimate the inside and outside wall temperatures of the pipe, T_{w_i} and T_{w_o}, to within about 5°C. Neglect entry length and variable properties effects and take $\beta \approx 0.000118$ K^{-1} for lead. (*Hint:* Guess an outside wall temperature above 370°C when computing $\overline{h}$ for the lead.)

b. At what distance from the inlet will the inside wall of the pipe reach T_{sat}? What redesign may be needed?

8.50 A flat plate 10 cm long and 40 cm wide is inclined at 30° from the vertical. It is held at a uniform temperature of 250 K. Saturated HCFC-22 vapor at 260 K condenses onto the plate. Determine the following:

a. The ratio h'_{fg}/h_{fg}.

b. The average heat transfer coefficient, $\overline{h}$, and the rate at which the plate must be cooled, Q (watts).

c. The film thickness, δ (μm), at the bottom of the plate, and the plate's rate of condensation in g/s.

8.51 One component in a particular automotive air-conditioning system is a "receiver", a small vertical cylindrical tank that contains a pool of liquid refrigerant, HFC-134a, with vapor above it. The receiver stores extra refrigerant for the system and helps to regulate the pressure. The receiver is at equilibrium with surroundings at 330 K. A 5 mm diameter, spherical thermistor inside the receiver monitors the liquid level. The thermistor is a temperature-sensing resistor driven by a small electric current; it dissipates a power of 0.1 W. When the system is fully charged with refrigerant, the thermistor sits below the liquid surface. When refrigerant leaks from the system, the liquid level drops and the thermistor eventually sits in vapor. The thermistor is small compared to the receiver, and its power is too low to affect the bulk temperature in the receiver.

a. If the system is fully charged, determine the temperature of the thermistor.

b. If enough refrigerant has leaked that the thermistor sits in vapor, find the thermistor's temperature. Neglect thermal radiation.

8.52 Ammonia vapor at 300 K and 1.062 MPa pressure condenses onto the outside of a horizontal tube. The tube has an O.D. of 1.91 cm.

a. Suppose that the outside of the tube has a uniform temperature of 290 K. Determine the average condensation heat transfer cofficient of the tube.

b. The tube is cooled by cold water flowing through it and the thin wall of the copper tube offers negligible thermal resistance. If the bulk temperature of the water is 275 K at a location where the outside surface of the tube is at 290 K, what is the heat transfer coefficient inside the tube?

c. Using the heat transfer coefficients you just found, estimate the largest wall thickness for which the thermal resistance of the tube could be neglected. Discuss the variation the tube wall temperature around the circumference and along the length of the tube.

8.53 An inclined plate in a piece of process equipment is tilted 30° above the horizontal and is 20 cm long and 25 cm wide (in the horizontal direction). The plate is held at 280 K by a stream of liquid flowing past its bottom side; the liquid in turn is cooled by a refrigeration system capable of removing 12 watts from it. If the heat transfer from the plate to the stream exceeds 12 watts, the temperature of both the liquid and the plate will begin to rise. The upper surface of the plate is in contact with gaseous ammonia vapor at 300 K and a varying pressure. An engineer suggests that any rise in the bulk temperature of the liquid will signal that the pressure has exceeded a level of about $p_{\text{crit}} = 551$ kPa.

a. Explain why the gas's pressure will affect the heat transfer to the coolant.

b. Suppose that the pressure is 255.3 kPa. What is the heat transfer (in watts) from gas to the plate, if the plate temperature is $T_w = 280$ K? Will the coolant temperature rise? Data for ammonia are given in App. A.

c. Suppose that the pressure rises to 1062 kPa. What is the heat transfer to the plate if the plate is still at $T_w = 280$ K? Will the coolant temperature rise?

References

[8.1] W. Nusselt. Das grundgesetz des wärmeüberganges. *Gesund. Ing.*, 38:872, 1915.

[8.2] C. J. Sanders and J. P. Holman. Franz Grashof and the Grashof Number. *Int. J. Heat Mass Transfer*, 15:562–563, 1972.

[8.3] S. W. Churchill and H. H. S. Chu. Correlating equations for laminar and turbulent free convection from a vertical plate. *Int. J. Heat Mass Transfer*, 18:1323–1329, 1975.

[8.4] S. Goldstein, editor. *Modern Developments in Fluid Mechanics*, volume 2, chapter 14. Oxford University Press, New York, 1938.

[8.5] E. R. G. Eckert and R. M. Drake, Jr. *Analysis of Heat and Mass Transfer*. Hemisphere Publishing Corp., Washington, D.C., 1987.

[8.6] A. Bejan and J. L. Lage. The Prandtl number effect on the transition in natural convection along a vertical surface. *J. Heat Transfer, Trans. ASME*, 112:787–790, 1990.

[8.7] E. M. Sparrow and J. L. Gregg. The variable fluid-property problem in free convection. In J. P. Hartnett, editor, *Recent Advances in Heat and Mass Transfer*, pages 353–371. McGraw-Hill Book Company, New York, 1961.

[8.8] H. Barrow and T. L. Sitharamarao. The effect of variable β on free convection. *Brit. Chem. Eng.*, 16(8):704, 1971.

[8.9] W. Kraus. *Messungen des Temperatur- und Geschwindigskeitsfeldes bei freier Konvection*. Verlag G. Braun, Karlsruhe, 1955. Chapter F.

[8.10] S. W. Churchill and H. H. S. Chu. Correlating equations for laminar and turbulent free convection from a horizontal cylinder. *Int. J. Heat Mass Transfer*, 18:1049–1053, 1975.

[8.11] T. Cebeci. Laminar-free-convective-heat transfer from the outer surface of a vertical slender circular cylinder. In *Proc. Fifth Intl. Heat Transfer Conf.*, volume 3, pages 15–19. Tokyo, September 3–7 1974.

[8.12] T. Yuge. Experiments on heat transfer from spheres including combined forced and natural convection. *J. Heat Transfer, Trans. ASME, Ser. C*, 82(1):214, 1960.

[8.13] G. D. Raithby and K. G. T. Hollands. Natural convection. In W. M. Rohsenow, J. P. Hartnett, and Y. I. Cho, editors, *Handbook of Heat Transfer*, chapter 4. McGraw-Hill, New York, 3rd edition, 1998.

[8.14] J. H. Lienhard. On the commonality of equations for natural convection from immersed bodies. *Int. J. Heat Mass Transfer*, 16:2121, 1973.

[8.15] B. R. Rich. An investigation of heat transfer from an inclined flat plate in free convection. *Trans. ASME*, 75:489–499, 1953.

[8.16] T. Fujii and H. Imura. Natural convection heat transfer from a plate with arbitrary inclination. *Int. J. Heat Mass Transfer*, 15(4): 755–767, 1972.

[8.17] G. C. Vliet. Natural convection local heat transfer on constant heat transfer inclined surface. *J. Heat Transfer, Trans. ASME, Ser. C*, 91:511–516, 1969.

[8.18] L. Pera and B. Gebhart. On the stability of natural convection boundary layer flow over horizontal and slightly inclined surfaces. *Int. J. Heat Mass Transfer*, 16(6):1147–1163, 1973.

[8.19] M. Al-Arabi and M. K. El-Riedy. Natural convection heat transfer from isothermal horizontal plates of different shapes. *Int. J. Heat Mass Transfer*, 19:1399–1404, 1976.

[8.20] L. Pera and B. Gebhart. Natural convection boundary layer flow over horizontal and slightly inclined surfaces. *Int. J. Heat Mass Transfer*, 16(6):1131–1147, 1973.

[8.21] B. Gebhart, Y. Jaluria, R. L. Mahajan, and B. Sammakia. *Buoyancy-Induced Flows and Transport*. Hemisphere Publishing Corp., Washington, 1988.

[8.22] J. R. Lloyd and W. R. Moran. Natural convection adjacent to horizontal surface of various planforms. *J. Heat Transfer, Trans. ASME, Ser. C*, 96(4):443–447, 1974.

[8.23] A. M. Clausing and J. J. Berton. An experimental investigation of natural convection from an isothermal horizontal plate. *J. Heat Transfer, Trans. ASME*, 111(4):904–908, 1989.

[8.24] F. Restrepo and L. R. Glicksman. The effect of edge conditions on natural convection heat transfer from a horizontal plates. *Int. J. Heat Mass Transfer*, 17(1):135–142, 1974.

[8.25] D. W. Hatfield and D. K. Edwards. Edge and aspect ratio effects on natural convection from the horizontal heated plate facing downwards. *Int. J. Heat Mass Transfer*, 24(6):1019–1024, 1981.

[8.26] V. Kadambi and R. M. Drake, Jr. Free convection heat transfer from horizontal surfaces for prescribed variations in surface temperature and mass flow through the surface. Tech. Rept. Mech. Eng. HT-1, Princeton Univ., June 30 1959.

[8.27] K. T. Yang. Natural convection in enclosures. In S. Kakaç, R. K. Shah, and W. Aung, editors, *Handbook of Single-Phase Convective Heat Transfer*, chapter 13. Wiley-Interscience, New York, 1987.

[8.28] I. Catton. Natural convection in enclosures. In *Proc. Sixth Intl. Heat Transfer Conf.*, volume 6, pages 13–31. Toronto, Aug. 7–11 1978.

[8.29] S. W. Churchill. A comprehensive correlating equation for laminar, assisting, forced and free convection. *AIChE J.*, 23(1):10–16, 1977.

[8.30] T. S. Chen and B. F. Armaly. Mixed convection in external flow. In S. Kakaç, R. K. Shah, and W. Aung, editors, *Handbook of Single-Phase Convective Heat Transfer*, chapter 14. Wiley-Interscience, New York, 1987.

[8.31] W. Aung. Mixed convection in internal flow. In S. Kakaç, R. K. Shah, and W. Aung, editors, *Handbook of Single-Phase Convective Heat Transfer*, chapter 15. Wiley-Interscience, New York, 1987.

[8.32] V. K. Dhir. Quasi-steady laminar film condensation of steam on copper spheres. *J. Heat Transfer, Trans. ASME, Ser. C*, 97(3):347–351, 1975.

[8.33] V. K. Dhir and J. H. Lienhard. Laminar film condensation on plane and axi-symmetric bodies in non-uniform gravity. *J. Heat Transfer, Trans. ASME, Ser. C*, 93(1):97–100, 1971.

[8.34] W. Nusselt. Die oberflächenkondensation des wasserdampfes. *Z. Ver. Dtsch. Ing.*, 60:541 and 569, 1916.

[8.35] W. M. Rohsenow. Heat transfer and temperature distribution in laminar-film condensation. *Trans. ASME*, 78:1645–1648, 1956.

[8.36] E. M. Sparrow and J. L. Gregg. A boundary-layer treatment of laminar-film condensation. *J. Heat Transfer, Trans. ASME, Ser. C*, 81:13–18, 1959.

[8.37] P. Sadasivan and J. H. Lienhard. Sensible heat correction in laminar film boiling and condensation. *J. Heat Transfer, Trans. ASME*, 109: 545–547, 1987.

[8.38] E. M. Sparrow and J. L. Gregg. A theory of rotating condensation. *J. Heat Transfer, Trans. ASME, Ser. C*, 81:113–120, 1959.

[8.39] A. Karimi. Laminar film condensation on helical reflux condensers and related configurations. *Int. J. Heat Mass Transfer*, 20:1137–1144, 1977.

[8.40] R. Gregorig, J. Kern, and K. Turek. Improved correlation of film condensation data based on a more rigorous application of similarity parameters. *Wärme- und Stoffübertragung*, 7:1–13, 1974.

[8.41] P. J. Marto. Condensation. In W. M. Rohsenow, J. P. Hartnett, and Y. I. Cho, editors, *Handbook of Heat Transfer*, chapter 14. McGraw-Hill, New York, 3rd edition, 1998.

[8.42] J. Rose, H. Uehara, S. Koyama, and T. Fujii. Film condensation. In S. G. Kandlikar, M. Shoji, and V. K. Dhir, editors, *Handbook of Phase Change: Boiling and Condensation*, chapter 19. Taylor & Francis, Philadelphia, 1999.

[8.43] E. M. Sparrow and S. H. Lin. Condensation in the presence of a non-condensible gas. *J. Heat Transfer, Trans. ASME, Ser. C*, 86: 430, 1963.

9. Heat transfer in boiling and other phase-change configurations

For a charm of powerful trouble,
like a Hell-broth boil and bubble....
...Cool it with a baboon's blood,
then the charm is firm and good.
Macbeth, **Wm. Shakespeare**

"A watched pot never boils"—the water in a teakettle takes a long time to get hot enough to boil because natural convection initially warms it rather slowly. Once boiling begins, the water is heated the rest of the way to the saturation point very quickly. Boiling is of interest to us because it is remarkably effective in carrying heat from a heater into a liquid. The heater in question might be a red-hot horseshoe quenched in a bucket or the core of a nuclear reactor with coolant flowing through it. Our aim is to learn enough about the boiling process to design systems that use boiling for cooling. We begin by considering pool boiling—the boiling that occurs when a stationary heater transfers heat to an otherwise stationary liquid.

9.1 Nukiyama's experiment and the pool boiling curve

Hysteresis in the q vs. ΔT relation for pool boiling

In 1934, Nukiyama [9.1] did the experiment pictured in Fig. 9.1. He boiled saturated water on a horizontal wire that functioned both as an electric resistance heater and as a resistance thermometer. By calibrating the resistance of a Nichrome wire as a function of temperature before the

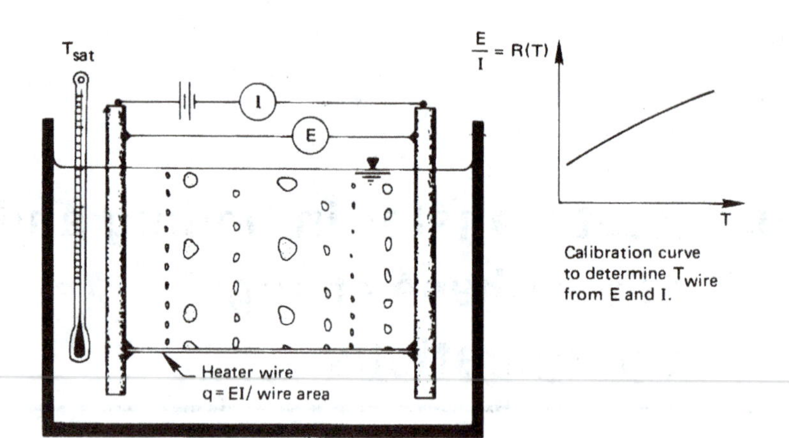

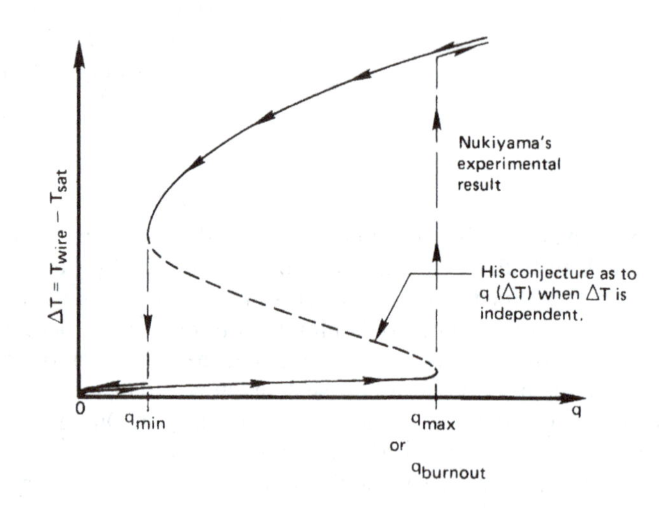

Figure 9.1 Nukiyama's boiling hysteresis loop.

experiment, he was able to obtain both the heat flux and the temperature from the observed current and voltage. He found that, as he increased the power input to the wire, the heat flux rose sharply but the temperature of the wire increased relatively little. Suddenly, at a particular high value of the heat flux, the wire abruptly melted. Nukiyama then obtained a platinum wire and tried again. This time the wire reached the same limiting heat flux, but then turned almost white-hot without melting.

As he reduced the power input to the white-hot wire, the temperature dropped in a continuous way, as shown in Fig. 9.1, until the heat flux was far below the value where the first temperature jump occurred. Then the temperature dropped abruptly to the original q vs. $\Delta T = (T_{\text{wire}} - T_{\text{sat}})$ curve, as shown. Nukiyama suspected that the hysteresis would not occur if ΔT could be specified as the independent controlled variable. He conjectured that such an experiment would result in the connecting line shown between the points where the temperatures jumped.

In 1937, Drew and Mueller [9.2] succeeded in making ΔT the independent variable by boiling organic liquids outside a tube. They allowed steam to condense inside the tube at an elevated pressure. The steam's saturation temperature—and hence the tube-wall temperature—was varied by controlling the steam's pressure. This permitted them to obtain a few scattered data that seemed to bear out Nukiyama's conjecture. Measurements of this kind are inherently hard to make accurately. For the next forty years, the relatively few nucleate boiling data that people obtained were usually—and sometimes imaginatively—interpreted as verifying Nukiyama's suggestion that this part of the boiling curve is continuous.

Figure 9.2 is a completed boiling curve for saturated water at atmospheric pressure on a particular flat horizontal heater. It displays the behavior shown in Fig. 9.1, but it has been rotated to place the independent variable, ΔT, on the abscissa. (We represent Nukiyama's connecting region as two unconnected extensions of the neighboring regions for reasons that we explain subsequently.)

Modes of pool boiling

The boiling curve in Fig. 9.2 is divided into five regimes of behavior. We consider these regimes, and the transitions that divide them, next.

Natural convection. Water that is not in contact with its own vapor does not boil at the so-called normal boiling point,[1] T_{sat}. Instead, it continues to rise in temperature until bubbles finally begin to form. On conventional machined metal surfaces, this occurs when the surface is a few degrees above T_{sat}. Below the bubble inception point, heat is removed by natural convection, and it can be predicted by the methods laid out in Chapter 8.

[1]This notion might be new to some readers. It is explained in Section 9.2.

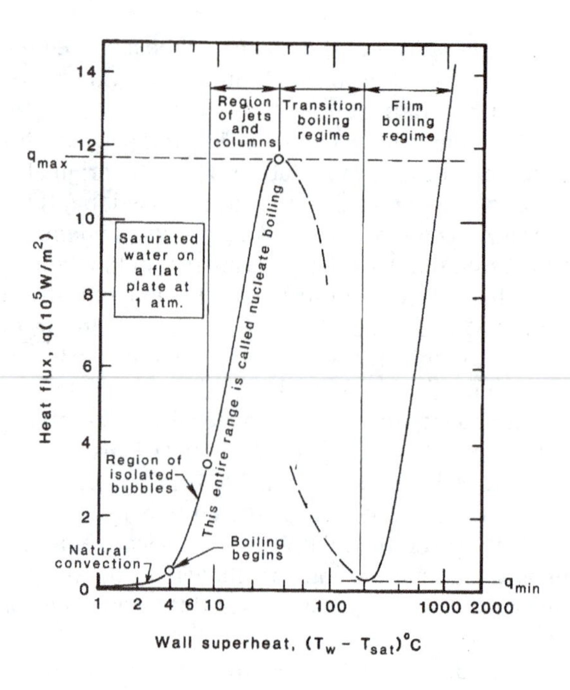

Figure 9.2 Typical boiling curve and regimes of boiling for an unspecified heater surface.

Nucleate boiling. The nucleate boiling regime embraces the two distinct regimes that lie between bubble inception and Nukiyama's first transition point:

1. *The region of isolated bubbles.* In this range, bubbles rise from isolated nucleation sites, more or less as they are sketched in Fig. 9.1. As q and ΔT increase, more and more sites are activated. Figure 9.3a is a photograph of this regime as it appears on a horizontal plate.

2. *The region of slugs and columns.* When the active sites become very numerous, the bubbles start to merge into one another, and an entirely different kind of vapor escape path comes into play. Vapor formed at the surface merges immediately into jets that feed into large overhead bubbles or "slugs" of vapor. This process is shown as it occurs on a horizontal cylinder in Fig. 9.3b.

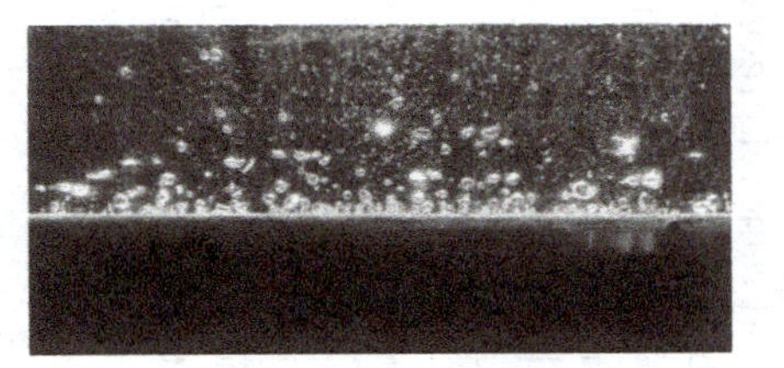

a. Isolated bubble regime—water.

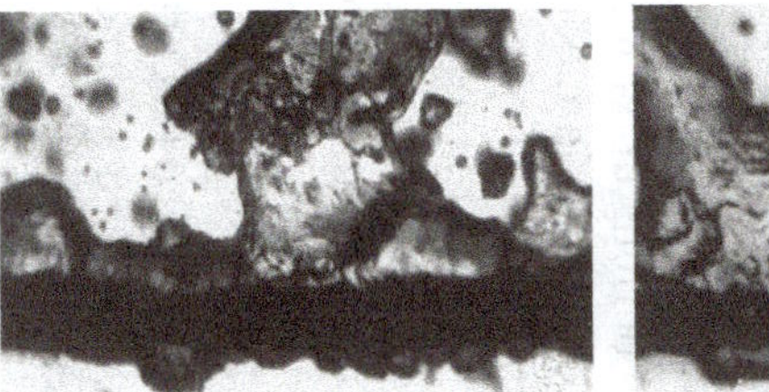

b. Two views of transitional boiling in acetone on a 0.32 cm diam. tube.

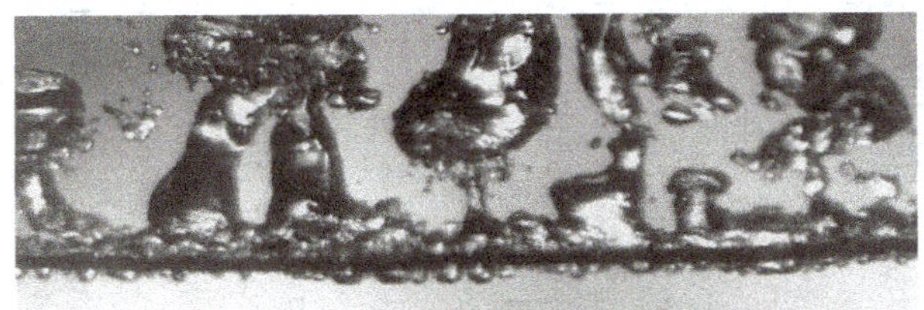

3.45 cm length of 0.0322 cm diam. wire in methanol at 10 earth-normal gravities. q=1.04×10^6 W/m^2

3.75 cm length of 0.164 cm diam. wire in benzene at earth-normal gravity. q=0.35×10^6 W/m^2

c. Two views of the regime of slugs and columns.

d. Film boiling of acetone on a 22 gage wire at earth-normal gravity. The true width of this image is 3.48 cm.

Figure 9.3 Typical photographs of boiling in the four regimes identified in Fig. 9.2.

Peak heat flux. We would clearly like operate heat exchange equipment at the upper end of the region of slugs and columns. Here the temperature difference is low while the heat flux is very high. Heat transfer coefficients in this range are enormous. However, it is very dangerous to run equipment near q_{max} in systems for which q is the independent variable (as in nuclear reactors). If q is raised beyond the upper limit of the nucleate boiling regime, such a system will suffer a sudden and damaging increase of temperature. This transition[2] is known by a variety of names: the *burnout* point (although a complete burning up or melting away does not always accompany it); the *peak heat flux* (a modest descriptive term); the *boiling crisis* (a Russian term); the *DNB*, or *departure from nucleate boiling*, and the *CHF*, or *critical heat flux* (terms more often used in flow boiling); and the *first boiling transition* (which term ignores previous transitions). We shall refer to it as the peak heat flux and designate it q_{max}.

Transitional boiling regime. It might seem odd that the heat flux actually diminishes with ΔT after q_{max} is reached. However, the effectiveness of the vapor escape process in this regime becomes worse and worse. As ΔT is further increased, the hot surface becomes completely blanketed in vapor and q reaches a minimum heat flux which we call q_{min}. Figure 9.3c shows two typical instances of transitional boiling just beyond the peak heat flux.

Film boiling. Once a stable vapor blanket is established, q again increases with increasing ΔT. The mechanics of the heat removal process during film boiling, and the regular removal of bubbles, has a great deal in common with film condensation, but the heat transfer coefficients are much lower because heat must be conducted through a vapor film instead of through a liquid film. We see an instance of film boiling in Fig. 9.3d.

Experiment 9.1

Set an open pan of cold tap water on your stove to boil. Observe the following stages as you watch:

- At first nothing appears to happen; then you notice that numerous small, stationary bubbles have formed over the bottom of the pan.

[2]We defer a proper physical explanation of the transition to Section 9.3.

These bubbles have nothing to do with boiling—they contain air being driven out of solution as the temperature rises.

- Suddenly the pan will begin to "sing" with a somewhat high-pitched buzzing-humming sound as the first vapor bubbles are triggered. They grow at the heated surface and condense very suddenly when their tops encounter the still-cold water above them. This so-called *cavitation* collapse is accompanied by a small "ping" or "click," over and over, as the process is repeated at a fairly high frequency.

- As the temperature of the liquid bulk rises, the singing is increasingly muted. You may then look in the pan and see a number of points on the bottom where a feathery blur appears to be affixed. These blurred images are bubble columns emanating scores of bubbles per second. The bubbles in these columns condense completely at some distance above the heater surface. Notice that the air bubbles are all gradually being swept away.

- The "singing" finally gives way to a full rolling boil, accompanied by a gentle burbling sound. Bubbles no longer condense but now reach the liquid surface, where they spill their vapor into the air.

- A full rolling-boil process, in which the liquid bulk is saturated, is a kind of isolated-bubble process, as plotted in Fig. 9.2. No kitchen stove supplies energy fast enough to boil water in the slugs-and-columns regime. In fact, that gives us some sense of the relative intensity of the slugs-and-columns process.

Experiment 9.2

Repeat Experiment 9.1 with a glass beaker instead of a kitchen pan. Place a strobe light, blinking about 6 to 10 times per second, behind the beaker with a piece of frosted glass or tissue paper between it and the beaker. You can now see the evolution of bubble columns from the first singing mode up to the rolling boil. You will also be able to see natural convection in the refraction of the light before boiling begins.

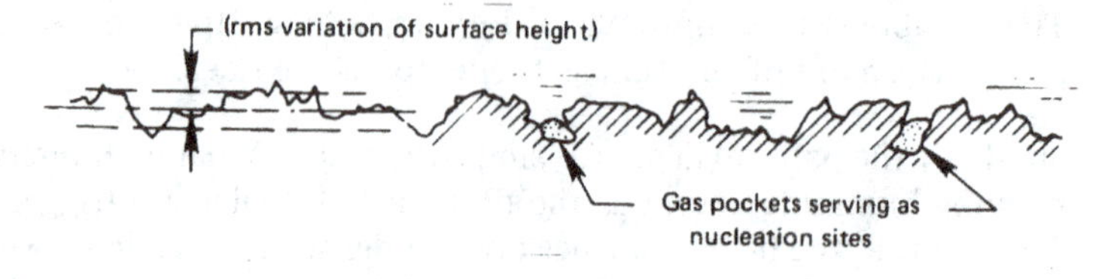

Figure 9.4 Enlarged sketch of a typical metal surface.

9.2 Nucleate boiling

Inception of boiling

Figure 9.4 shows a highly enlarged sketch of a heater surface. Most metal-finishing operations score tiny grooves on the surface, but they also typically involve some *chattering* or bouncing action, which hammers small holes into the surface. When a surface is wetted, liquid is prevented by surface tension from entering these holes, so small gas or vapor pockets are formed. These little pockets are the sites at which bubble nucleation occurs.

To see why vapor pockets serve as nucleation sites, consider Fig. 9.5. Here we see the problem in highly idealized form. Suppose that a spherical bubble of pure saturated steam is at equilibrium with an infinite superheated liquid. To determine the size of such a bubble, we impose the conditions of mechanical and thermal equilibrium.

The bubble will be in *mechanical* equilibrium when the pressure difference between the inside and the outside of the bubble is balanced by the forces of surface tension, σ, as indicated in the cutaway sketch in Fig. 9.5. Since *thermal* equilibrium requires that the temperature must be the same inside and outside the bubble, and since the vapor inside must be saturated at T_{sup} because it is in contact with its liquid, the force balance takes the form

$$R_b = \frac{2\sigma}{(p_{\text{sat}} \text{ at } T_{\text{sup}}) - p_{\text{ambient}}} \tag{9.1}$$

The p–v diagram in Fig. 9.5 shows the state points of the internal vapor and external liquid for a bubble at equilibrium. Notice that the external liquid is superheated to $(T_{\text{sup}} - T_{\text{sat}})$ K above its boiling point at the ambient pressure; but the vapor inside, being held at just the right elevated pressure by surface tension, is just saturated.

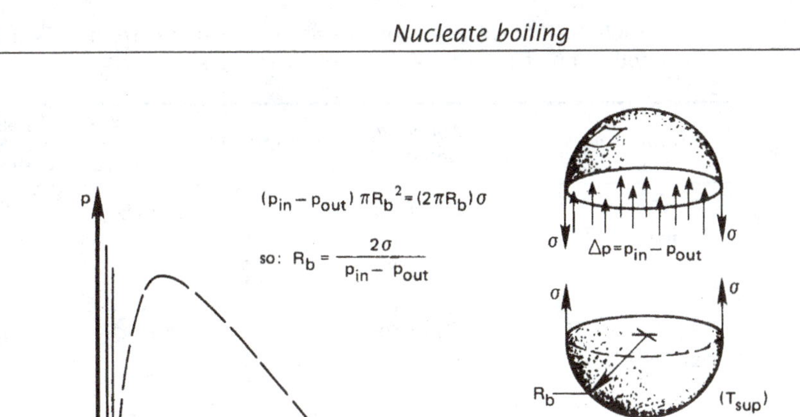

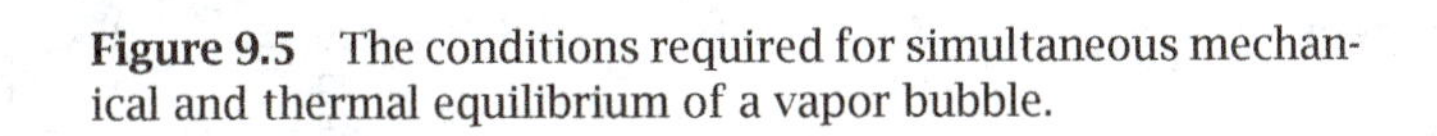

Figure 9.5 The conditions required for simultaneous mechanical and thermal equilibrium of a vapor bubble.

Physical Digression 9.1

The surface tension of water in contact with its vapor is given with great accuracy by [9.3]:

$$\sigma_{\text{water}} = 235.8\left(1 - \frac{T_{\text{sat}}}{T_c}\right)^{1.256}\left[1 - 0.625\left(1 - \frac{T_{\text{sat}}}{T_c}\right)\right]\frac{\text{mN}}{\text{m}} \tag{9.2a}$$

where both T_{sat} and the thermodynamical critical temperature, T_c = 647.096 K, are expressed in K. The units of σ are millinewtons (mN) per meter. Table 9.1 gives additional values of σ for several substances.

Most of the expressions in Table 9.1 are linear curve fits that apply to small ranges of surface tension. However, eqn. (9.2a) is a specialized

Table 9.1 Surface tension of various substances from the collection of Jasper [9.4][a] and other sources.

Substance	*Temperature Range* (°C)	σ (mN/m)	$\sigma = a - bT$ (°C)	
			a (mN/m)	b (mN/m·°C)
Acetone	25 to 50		26.26	0.112
Ammonia	−70	42.39		
	−60	40.25		
	−50	37.91		
	−40	35.38		
Aniline	15 to 90		44.83	0.1085
Benzene	10	30.21		
	30	27.56		
	50	24.96		
	70	22.40		
Butyl alcohol	10 to 100		27.18	0.08983
Carbon tetrachloride	15 to 105		29.49	0.1224
Cyclohexanol	20 to 100		35.33	0.0966
Ethyl alcohol	10 to 100		24.05	0.0832
Ethylene glycol	20 to 140		50.21	0.089
Hydrogen	−258	2.80		
	−255	2.29		
	−253	1.95		
Isopropyl alcohol	10 to 100		22.90	0.0789
Mercury	5 to 200		490.6	0.2049
Methane	90	18.877		
	100	16.328		
	115	12.371		
Methyl alcohol	10 to 60		24.00	0.0773
Naphthalene	100 to 200		42.84	0.1107
Nicotine	−40 to 90		41.07	0.1112
Nitrogen	−195 to −183		26.42	0.2265
Octane	10 to 120		23.52	0.09509
Oxygen	−202 to −184		−33.72	−0.2561
Pentane	10 to 30		18.25	0.11021
Toluene	10 to 100		30.90	0.1189
Water	10 to 100		75.83	0.1477

Substance	*Temperature Range* (°C)	$\sigma = \sigma_o\,[1 - T\,(\mathrm{K})/T_c]^n$		
		σ_o (mN/m)	T_c (K)	n
Carbon dioxide	−56 to 31	75.00	304.26	1.25
CFC-12 (R12) [9.5]	−148 to 112	56.52	385.01	1.27
HCFC-22 (R22) [9.5]	−158 to 96	61.23	369.32	1.23
HFC-134a (R134a) [9.6]	−30 to 101	59.69	374.18	1.266
Propane [9.7]	−173 to 96	53.13	369.85	1.242

[a] The function $\sigma = \sigma(T)$ is not really linear, but Jasper was able to linearize it over modest ranges of temperature [e.g., compare the water equation above with eqn. (9.2a)].

refinement of the following simple, but quite accurate and widely-used, semi-empirical equation for correlating surface tension:

$$\sigma = \sigma_o \left(1 - T_{sat}/T_c\right)^{11/9} \tag{9.2b}$$

We include correlating equations of this form for CO_2, propane, and some refrigerants at the bottom of Table 9.1. Equations of this general form are discussed in Reference [9.8].

It is easy to see that the equilibrium bubble, whose radius is described by eqn. (9.1), is unstable. If its radius is less than this value, surface tension will overbalance $[p_{sat}(T_{sup}) - p_{ambient}]$. When that happens, vapor inside will condense at this higher pressure and the bubble will collapse. If the bubble radius is slightly larger than the equation specifies, liquid at the interface will evaporate and the bubble will begin to grow.

Thus, as the heater surface temperature is increased, higher and higher values of $[p_{sat}(T_{sup}) - p_{ambient}]$ will result and the equilibrium radius, R_b, will decrease in accordance with eqn. (9.1). It follows that smaller and smaller vapor pockets will be triggered into active bubble growth as the temperature is increased. As an approximation, we can use eqn. (9.1) to specify the radius of those vapor pockets that become active nucleation sites. More accurate estimates can be made using Hsu's [9.9] bubble inception theory, the subsequent work by Rohsenow and others (see, e.g., [9.10]), or the still more recent technical literature.

Example 9.1

Estimate the approximate size of active nucleation sites in water at 1 atm on a wall superheated by 8 K and by 16 K. This is roughly in the regime of isolated bubbles indicated in Fig. 9.2.

Solution. $p_{sat} = 1.203 \times 10^5$ N/m^2 at 108°C and 1.769×10^5 N/m^2 at 116°C, and σ is given as 57.36 mN/m at T_{sat} = 108°C and as 55.78 mN/m at T_{sat} = 116°C by eqn. (9.2a). Then, at 108°C, R_b from eqn. (9.1) is

$$R_b = \frac{2(57.36 \times 10^{-3})\ \text{N/m}}{(1.203 \times 10^5 - 1.013 \times 10^5)\ \text{N/m}^2}$$

and similarly for 116°C, so the radius of active nucleation sites is on the order of

$$R_b = 0.0060\ \text{mm at}\ T = 108°\text{C} \quad \text{or} \quad 0.0015\ \text{mm at}\ 116°\text{C}$$

This means that active nucleation sites would be holes with diameters very roughly on the order of magnitude of 0.005 mm or 5 μm—at least on the heater represented by Fig. 9.2. That is within the range of roughness of commercially finished surfaces. ■

Region of isolated bubbles

The mechanism of heat transfer enhancement in the isolated bubble regime was hotly argued in the years following World War II. A few conclusions have emerged from that debate, and we shall attempt to identify them. There is little doubt that bubbles act in some way as small pumps that keep replacing liquid heated at the wall with cool liquid. The question is that of specifying the correct mechanism. Figure 9.6 shows the way bubbles probably act to remove hot liquid from the wall and introduce cold liquid to be heated.

It is apparent that the number of active nucleation sites generating bubbles will strongly influence q. On the basis of his experiments, Yamagata showed in 1955 (see, e.g., [9.11]) that

$$q \propto \Delta T^a n^b \tag{9.3}$$

where $\Delta T \equiv T_w - T_{sat}$ and n is the site density or number of active sites per square meter. A great deal of subsequent work has been done to fix the constant of proportionality and the constant exponents, a and b. The exponents turn out to be approximately $a = 1.2$ and $b = \frac{1}{3}$.

The problem with eqn. (9.3) is that it introduces what engineers call a *nuisance variable*. A nuisance variable is one that varies from system to system and cannot easily be evaluated—the site density, n, in this case. Normally, n increases with ΔT in some way, but how? If all sites were identical in size, all sites would be activated simultaneously, and q would be a discontinuous function of ΔT. When the sites have a typical distribution of sizes, n (and hence q) can increase very strongly with ΔT.

It is a lucky fact that for a large class of factory-finished materials, n varies approximately as $\Delta T^{5 \text{ or } 6}$, so q varies roughly as ΔT^3. This has made it possible for various authors to correlate q approximately for a large variety of materials. One of the first correlations for nucleate

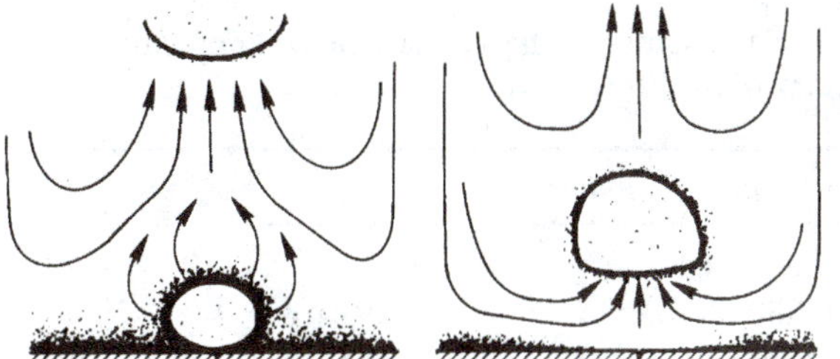

A bubble growing and departing in saturated liquid. The bubble grows, absorbing heat from the superheated liquid on its periphery. As it leaves, it entrains cold liquid onto the plate which then warms up until nucleation occurs and the cycle repeats.

A bubble growing in subcooled liquid. When the bubble protrudes into cold liquid, steam can condense on the top while evaporation continues on the bottom. This provides a short-circuit for cooling the wall. Then, when the bubble caves in, cold liquid is brought to the wall.

Figure 9.6 Heat removal by bubble action during boiling. Dark regions denote locally superheated liquid.

boiling was that of Rohsenow [9.12] in 1952. It is

$$\frac{c_p\,(T_w - T_{\mathrm{sat}})}{h_{fg}\,\mathrm{Pr}^s} = C_{\mathrm{sf}}\left[\frac{q}{\mu h_{fg}}\sqrt{\frac{\sigma}{g(\rho_f - \rho_g)}}\right]^{0.33} \tag{9.4}$$

where all properties, unless otherwise noted, are for liquid at T_{sat}. The constant C_{sf} is an empirical correction for typical surface conditions. Table 9.2 includes a set of values of C_{sf} for common surfaces (taken from [9.12]) as well as the Prandtl number exponent, s. A more extensive compilation of these constants was published by Pioro in 1999 [9.13].

We noted, initially, that there are two nucleate boiling regimes, and the Yamagata equation (9.3) applies only to the first of them. Rohsenow's equation is frankly empirical and does not depend on the rational analysis of either nucleate boiling process. It turns out that it represents $q(\Delta T)$ in both regimes, but it is not terribly accurate in either one. Figure 9.7 shows Rohsenow's original comparison of eqn. (9.4) with data for water over a large range of conditions. It shows typical errors in heat flux of 100% and typical errors in ΔT of about 25%.

Thus, our ability to predict the nucleate pool boiling heat flux is poor. Our ability to predict ΔT is better because, with $q \propto \Delta T^3$, a large error in

Table 9.2 Selected values of the surface correction factor for use with eqn. (9.4) [9.12]

Surface-Fluid Combination	C_{sf}	s
Water-nickel	0.006	1.0
Water-platinum	0.013	1.0
Water-copper	0.013	1.0
Water-brass	0.006	1.0
CCl_4-copper	0.013	1.7
Benzene-chromium	0.010	1.7
n-Pentane-chromium	0.015	1.7
Ethyl alcohol-chromium	0.0027	1.7
Isopropyl alcohol-copper	0.0025	1.7
35% K_2CO_3-copper	0.0054	1.7
50% K_2CO_3-copper	0.0027	1.7
n-Butyl alcohol-copper	0.0030	1.7

q gives a much smaller error in ΔT. It appears that any substantial improvement in this situation will have to wait until someone has managed to deal realistically with the nuisance variable, n.

We are in luck, however, because we do not often have to calculate q, given ΔT, in the nucleate boiling regime. More often, the major problem is to avoid exceeding $q_{\max}$. We turn our attention in the next section to predicting this limit.

Example 9.2

What is C_{sf} for the heater surface in Fig. 9.2?

Solution. From eqn. (9.4) we obtain

$$\frac{q}{\Delta T^3} C_{sf}^3 = \frac{\mu c_p^3}{h_{fg}^2 \mathrm{Pr}^3} \sqrt{\frac{g(\rho_f - \rho_g)}{\sigma}}$$

where, since the liquid is water, we take s to be 1.0. Then, for water at $T_{sat} = 100°\mathrm{C}$: $c_p = 4.22$ kJ/kg·K, Pr = 1.75, $(\rho_f - \rho_g) = 958$ kg/m^3,

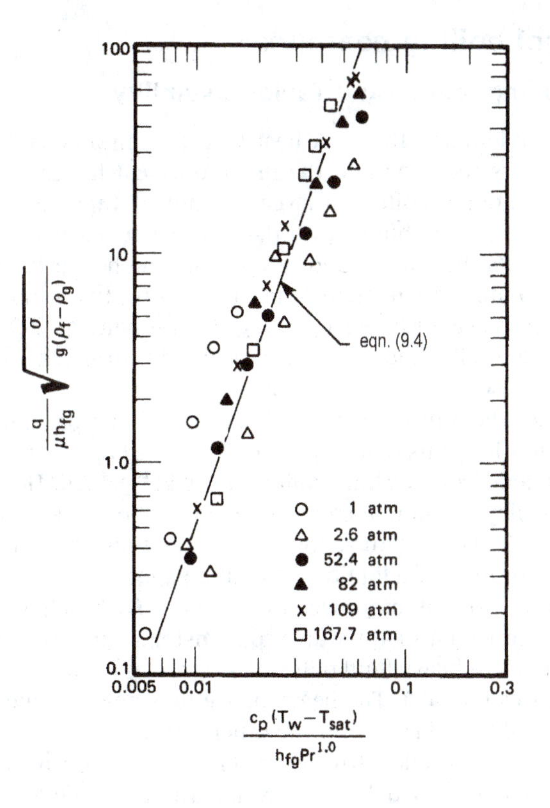

Figure 9.7 Illustration of Rohsenow's [9.12] correlation applied to data for water boiling on 0.61 mm diameter platinum wire.

$\sigma = 0.0589$ N/m or kg/s^2, $h_{fg} = 2257$ kJ/kg, $\mu = 0.000282$ kg/m·s. Thus,

$$\frac{q}{\Delta T^3} C_{\text{sf}}^3 = 3.10 \times 10^{-7} \frac{\text{kW}}{\text{m}^2\text{K}^3}$$

At $q = 800$ kW/m^2, we read $\Delta T = 22$ K from Fig. 9.2. This gives

$$C_{\text{sf}} = \left[\frac{3.10 \times 10^{-7}(22)^3}{800}\right]^{1/3} = 0.016$$

This value compares favorably with C_{sf} for a platinum or copper surface under water. ■

9.3 Peak pool boiling heat flux

Transitional boiling regime and Taylor instability

It will help us to understand the peak heat flux if we first consider the process that connects the peak and the minimum heat fluxes. During high heat flux transitional boiling, a large amount of vapor is glutted about the heater. It wants to buoy upward, but it has no clearly defined escape route. The jets that carry vapor away from the heater in the region of slugs and columns are unstable and cannot serve that function in this regime. Therefore, vapor buoys up in big slugs—then liquid falls in, touches the surface briefly, and a new slug begins to form. Figure 9.3c shows part of this process.

The high and low heat flux transitional boiling regimes are different in character. The low heat flux region does not look like Fig. 9.2c but is almost indistinguishable from the film boiling shown in Fig. 9.2d. However, both processes display a common conceptual key: In both, the heater is almost completely blanketed with vapor. In both, we must contend with the unstable configuration of a liquid on top of a vapor.

Figure 9.8 shows two commonplace examples of such behavior. In either an inverted honey jar or the water condensing from a cold water pipe, we have seen how a heavy fluid falls into a light one (water or honey, in this case, collapses into air). The heavy phase falls down at one node of a wave and the light fluid rises into the other node.

The collapse process is called *Taylor instability* after G. I. Taylor, who first predicted it. The so-called Taylor wavelength, λ_d, is the length of the wave that grows fastest and therefore predominates during the collapse of an infinite plane horizontal interface. It can be predicted using dimensional analysis. The dimensional functional equation for λ_d is

$$\lambda_d = \text{fn}\left[\sigma, g(\rho_f - \rho_g)\right] \tag{9.5}$$

since the wave is formed as a result of the balancing forces of surface tension against inertia and gravity. There are three variables involving m and kg/s^2, so we look for just one dimensionless group:

$$\lambda_d \sqrt{\frac{g(\rho_f - \rho_g)}{\sigma}} = \text{constant}$$

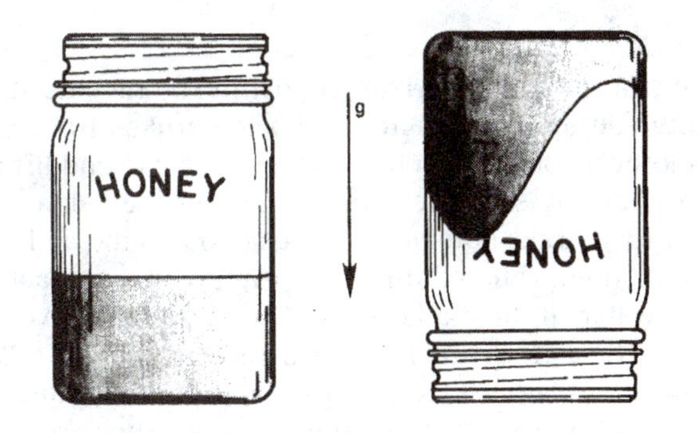

a. Taylor instability in the surface of the honey in an inverted honey jar

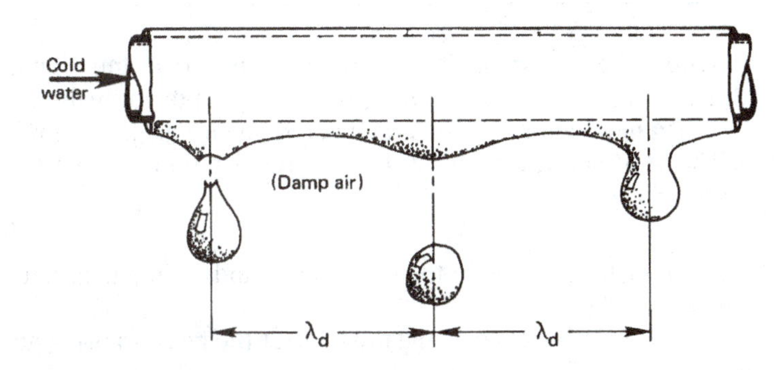

b. Taylor instability in the interface of the water condensing on the underside of a small cold water pipe.

Figure 9.8 Two examples of Taylor instabilities that one might commonly experience.

This relationship was derived analytically by Bellman and Pennington [9.14] for one-dimensional waves and by Sernas [9.15] for the two-dimensional waves that actually occur in a plane horizontal interface. The results were

$$\lambda_d \sqrt{\frac{g(\rho_f - \rho_g)}{\sigma}} = \begin{cases} 2\pi\sqrt{3} & \text{for one-dimensional waves} \\ 2\pi\sqrt{6} & \text{for two-dimensional waves} \end{cases} \tag{9.6}$$

Experiment 9.3

Hang a metal rod in the horizontal position by threads at both ends. The rod should be about 30 cm in length and perhaps 1 to 2 cm in diameter. Pour motor oil or glycerin in a narrow cake pan and lift the pan up under the rod until it is submerged. Then lower the pan and watch the liquid drain into it. Take note of the wave action on the underside of the rod. The same thing can be done in an even more satisfactory way by running cold water through a horizontal copper tube above a beaker of boiling water. The condensing liquid will also come off in a Taylor wave such as is shown in Fig. 9.8. In either case, the waves will approximate λ_{d_1} (the length of a one-dimensional wave, since they are arrayed on a line); however, the wavelength will be influenced by the curvature of the rod.

Throughout the transitional boiling regime, vapor rises into liquid on the nodes of Taylor waves, and at $q_{\max}$ this rising vapor forms into jets. These jets arrange themselves on a staggered square grid, as shown in Fig. 9.9. The basic spacing of the grid is λ_{d_2} (the two-dimensional Taylor wavelength). Since

$$\lambda_{d_2} = \sqrt{2}\,\lambda_{d_1} \tag{9.7}$$

[recall eqn. (9.6)], the spacing of the most basic module of jets is actually λ_{d_1}, as shown in Fig. 9.9.

Next we see how the jets become unstable at the peak, to bring about burnout.

Helmholtz instability of vapor jets

Figure 9.10 shows a commonplace example of what is called *Helmholtz instability.* This is the phenomenon that causes the vapor jets to cave in when the vapor velocity in them reaches a critical value. Any flag in a breeze will constantly be in a state of collapse as the result of relatively high pressures where the velocity is low and relatively low pressures where the velocity is high, as is indicated in the top view.

This same instability is shown as it occurs in a vapor jet wall in Fig. 9.11. This situation differs from the flag in one important particular. Surface tension in the jet walls tends to balance the flow-induced pressure forces that bring about collapse. Thus, while the flag is unstable in *any* breeze, the vapor velocity in the jet must reach a limiting value, u_g, before the jet becomes unstable.

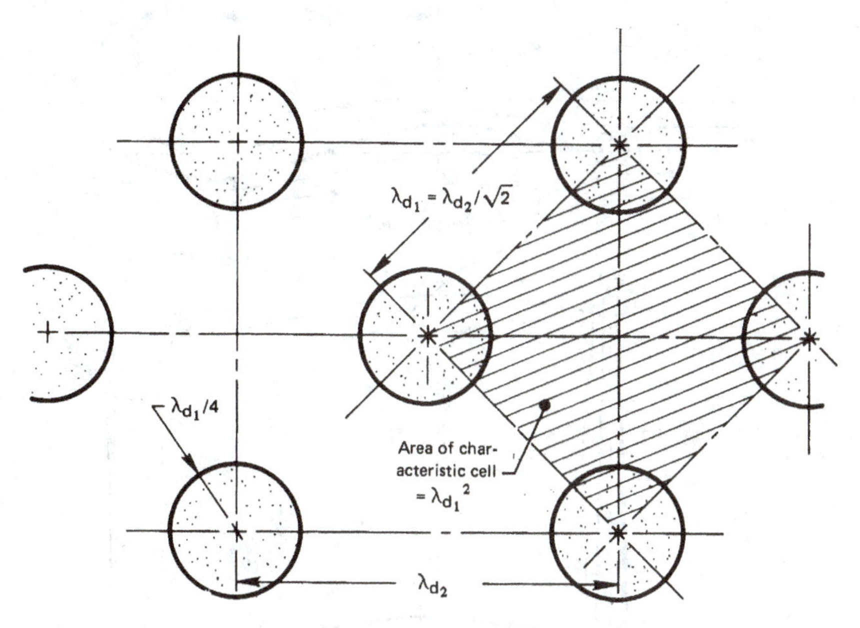

a. Plan view of bubbles rising from surface

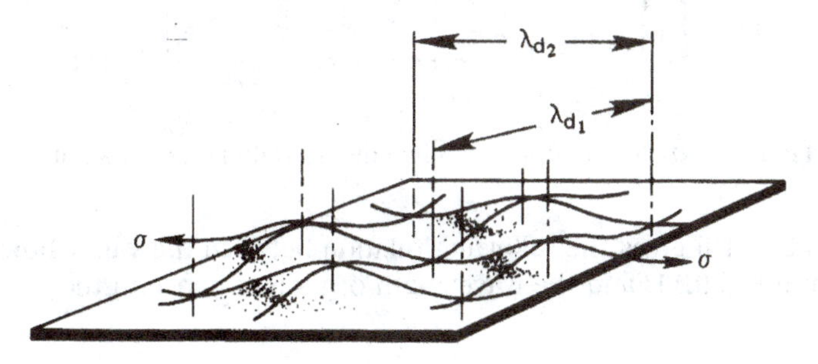

b. Waveform underneath the bubbles shown in **a.**

Figure 9.9 The array of vapor jets as seen on an infinite horizontal heater surface.

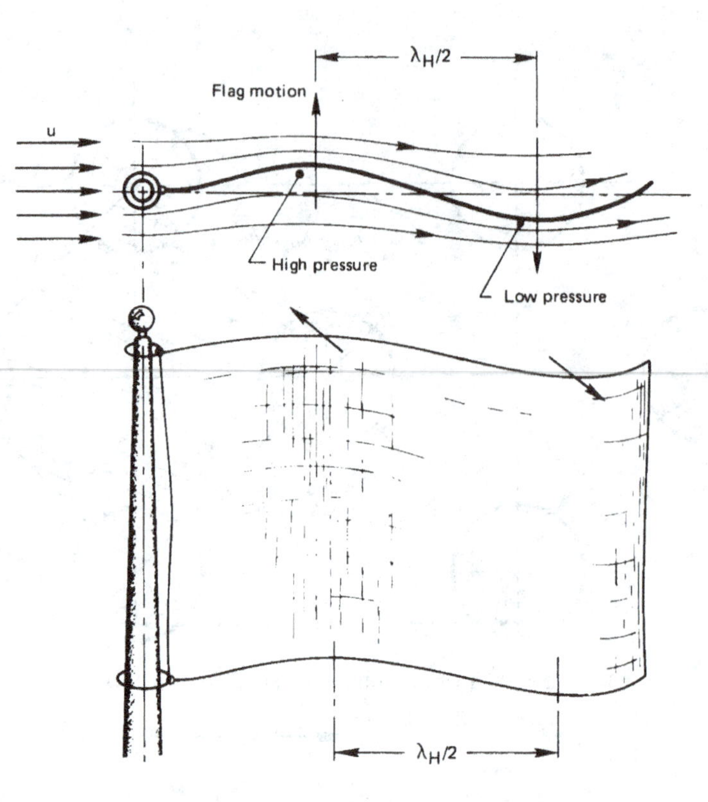

Figure 9.10 The flapping of a flag due to Helmholtz instability.

Lamb [9.16] gives the following relation between the vapor flow u_g, shown in Fig. 9.11, and the wavelength of a disturbance in the jet wall, λ_H:

$$u_g = \sqrt{\frac{2\pi\sigma}{\rho_g \lambda_H}} \tag{9.8}$$

[This result, like eqn. (9.6), can be predicted within a constant using dimensional analysis. See Problem 9.19.] A real liquid–vapor interface will usually be irregular, and therefore it can be viewed as containing all possible sinusoidal wavelengths superposed on one another. One problem we face is that of guessing whether or not one of those wavelengths will be better developed than the others and therefore more liable to

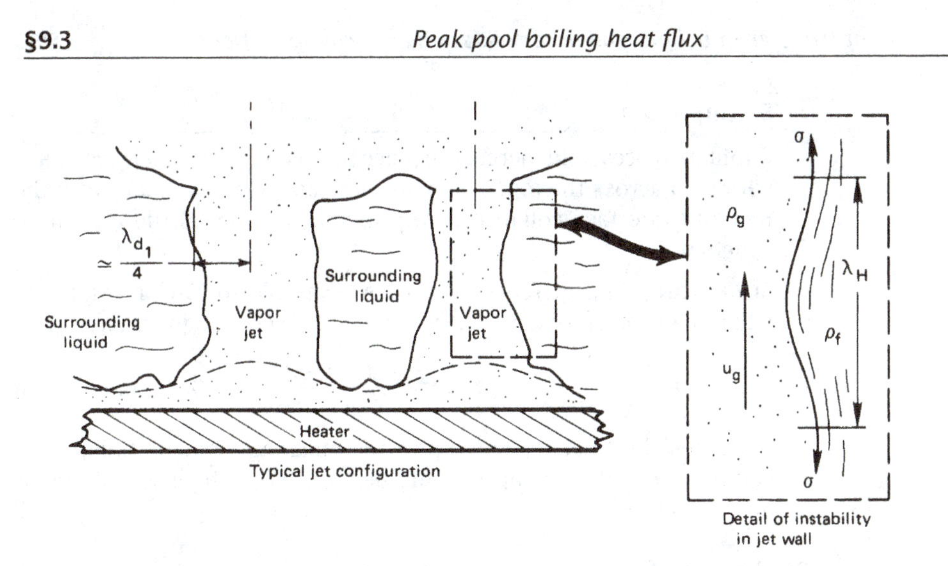

Figure 9.11 Helmholtz instability of vapor jets.

collapse during the brief life of the jet.

Example 9.3

Saturated water at 1 atm flows slowly down the inside wall of a 10 cm I.D. vertical tube. Steam flows rapidly upward in the center. The wall of the pipe has circumferential corrugations in it, with a 4 cm wavelength in the axial direction. Neglect problems raised by radial curvature and the finite thickness of the liquid, and estimate the steam velocity required to destabilize the liquid flow.

Solution. If we can neglect the liquid velocity, the flow will be Helmholtz-stable until the steam velocity reaches the value given by eqn. (9.8):

$$u_g = \sqrt{\frac{2\pi(0.0589)}{0.577(0.04 \text{ m})}}$$

Thus, the maximum stable steam velocity would be $u_g = 4$ m/s. Beyond that, the liquid will form whitecaps and be blown back upward. ■

Example 9.4

Capillary forces hold mercury in place between two parallel steel plates with a lid across the top. The plates are slowly pulled apart until the mercury interface collapses. Approximately what is the maximum spacing?

Solution. The mercury is most susceptible to Taylor instability when the spacing reaches the wavelength given by eqn. (9.6):

$$\lambda_{d_1} = 2\pi\sqrt{3}\sqrt{\frac{\sigma}{g(\rho_f - \rho_g)}} = 2\pi\sqrt{3}\sqrt{\frac{0.487}{9.8(13600)}} = 0.021 \text{ m} = 2.1 \text{ cm}$$

(Actually, this spacing would give the maximum *rate* of collapse. It can be shown that collapse would begin at $1/\sqrt{3}$ times this value, or at 1.2 cm.) ■

Prediction of q_{max}

General expression for q_{max} The heat flux must be balanced by the latent heat carried away in the jets when the liquid is saturated. Thus, we can write immediately

$$q_{max} = \rho_g h_{fg} u_g \left(\frac{A_j}{A_h}\right) \tag{9.9}$$

where A_j is the cross-sectional area of a jet and A_h is the heater area that supplies each jet.

For any heater configuration, two things must be determined. One is the length of the particular disturbance in the jet wall, λ_H, which will trigger Helmholtz instability and fix u_g in eqn. (9.8) for use in eqn. (9.9). The other is the ratio A_j/A_h. The prediction of q_{max} in any pool boiling configuration always comes down to these two problems.

q_{max} on an infinite horizontal plate. The original analysis of this type was done by Zuber in his doctoral dissertation at UCLA in 1958 (see [9.17]). He first guessed that the jet radius was $\lambda_{d_1}/4$. This guess has received corroboration by subsequent investigators, and (with reference to Fig. 9.9) it gives

$$\frac{A_j}{A_h} = \frac{\text{cross-sectional area of circular jet}}{\text{area of the square portion of the heater that feeds the jet}} = \frac{\pi(\lambda_{d_1}/4)^2}{(\lambda_{d_1})^2} = \frac{\pi}{16} \tag{9.10}$$

Lienhard and Dhir ([9.18, 9.19, 9.20]) guessed that the Helmholtz-unstable wavelength might be equal to λ_{d_1}, so eqn. (9.9) became

$$q_{\text{max}} = \rho_g h_{fg} \sqrt{\frac{2\pi\sigma}{\rho_g} \frac{1}{2\pi\sqrt{3}} \sqrt{\frac{g(\rho_f - \rho_g)}{\sigma}}} \times \frac{\pi}{16}$$

or[3]

$$\boxed{q_{\text{max}} = 0.149\, \rho_g^{1/2} h_{fg} \sqrt[4]{g(\rho_f - \rho_g)\sigma}} \tag{9.11}$$

Equation (9.11) is compared with available data for large flat heaters, with vertical sidewalls to prevent any liquid sideflow, in Fig. 9.12. As long as the diameter or width of the heater is more than about $3\lambda_{d_1}$, the prediction is quite accurate. When the width or diameter is less than this, there is a small integral number of jets on a plate which may be larger or smaller in area than $16/\pi$ per jet. When this is the case, the actual q_{max} may be larger or smaller than that predicted by eqn. (9.11) (see Problem 9.13).

The form of the preceding prediction is usually credited to Kutateladze [9.21] and Zuber [9.17]. Kutateladze (then working in Leningrad and later director of the Heat Transfer Laboratory near Novosibirsk, Siberia) recognized that burnout resembled the flooding of a distillation column. At any level in a distillation column, alcohol-rich vapor (for example) rises while water-rich liquid flows downward in counterflow. If the process is driven too far, the flows become Helmholtz-unstable and the process collapses. The liquid then cannot move downward and the column is said to "flood."

Kutateladze did the dimensional analysis of q_{max} based on the flooding mechanism and obtained the following relationship, which, lacking a characteristic length and being of the same form as eqn. (9.11), is really valid only for an infinite horizontal plate:

$$q_{\text{max}} = C\, \rho_g^{1/2} h_{fg} \sqrt[4]{g(\rho_f - \rho_g)\sigma}$$

He then suggested that C was equal to 0.131 on the basis of data from configurations other than infinite flat plates (horizontal cylinders, for example). Zuber's analysis yielded $C = \pi/24 = 0.1309$, which was quite close to Kutateladze's value but lower by 14% than eqn. (9.11). We therefore designate the Zuber-Kutateladze prediction as q_{max_z}. However, we

[3]Readers are reminded that $\sqrt[n]{x} \equiv x^{1/n}$.

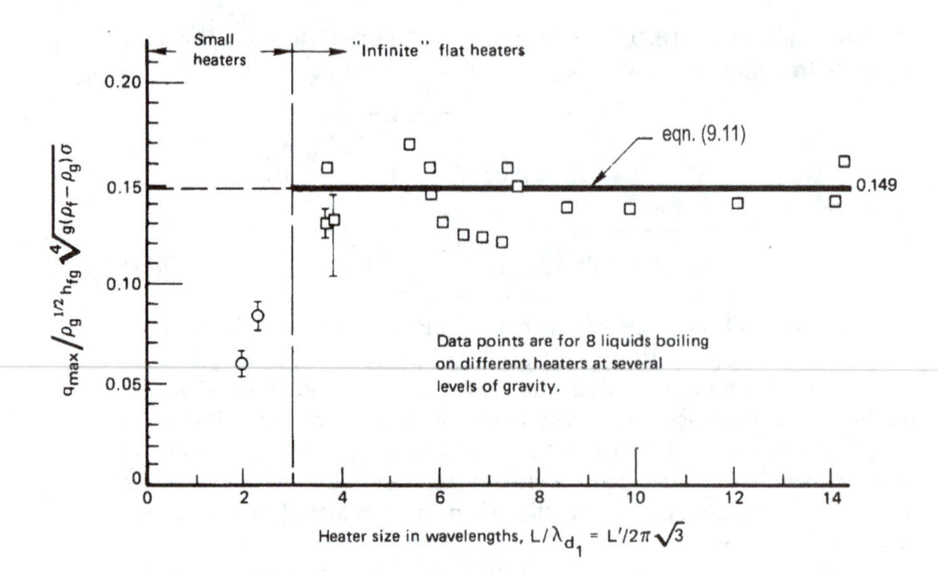

Figure 9.12 Comparison of the q_{max} prediction for infinite horizontal heaters with data reported in [9.18].

shall not use it directly, since it does not predict any actual physical configuration.

$$q_{max_z} \equiv 0.131\, \rho_g^{1/2} h_{fg} \sqrt[4]{g(\rho_f - \rho_g)\sigma} \tag{9.12}$$

It is very interesting that C. F. Bonilla, whose q_{max} experiments in the early 1940s are included in Fig. 9.12, also suggested that q_{max} should be compared with the column-flooding mechanism. He presented these ideas in a paper, but A. P. Colburn wrote to him: "A correlation [of the flooding velocity plots with] boiling data would not serve any great purpose and would perhaps be very misleading." And T. H. Chilton—another eminent chemical engineer of that period—wrote to him: "I venture to suggest that you delete from the manuscript...the relationship between boiling rates and loading velocities in packed towers." Thus, the technical conservativism of the period prevented the idea from gaining acceptance for another decade.

Example 9.5

Predict the peak heat flux for Fig. 9.2.

Solution. We use eqn. (9.11) to evaluate q_{max} for water at 100°C on an infinite flat plate:

$$\begin{aligned} q_{max} &= 0.149\, \rho_g^{1/2} h_{fg} \sqrt[4]{g(\rho_f - \rho_g)\sigma} \\ &= 0.149(0.597)^{1/2}(2,257,000)\sqrt[4]{9.8(958.2-0.6)(0.0589)} \\ &= 1.260 \times 10^6 \text{ W/m}^2 \\ &= 1.260 \text{ MW/m}^2 \end{aligned}$$

Figure 9.2 shows $q_{max} \simeq 1.160$ MW/m^2, which is less by only about 8%. ■

Example 9.6

What is q_{max} in mercury on a large flat plate at 1 atm?

Solution. The normal boiling point of mercury is 355°C. At this temperature, $h_{fg} = 292,500$ J/kg, $\rho_f = 13,400$ kg/m^3, $\rho_g = 4.0$ kg/m^3, and $\sigma \simeq 0.418$ kg/s^2, so

$$\begin{aligned} q_{max} &= 0.149(4.0)^{1/2}(292,500)\sqrt[4]{9.8(13,400-4)(0.418)} \\ &= 1.334 \text{ MW/m}^2 \end{aligned}$$

The result is very close to that for water. The increases in density and surface tension have been compensated by a much lower latent heat. ■

Peak heat flux in other pool boiling configurations

The prediction of q_{max} in configurations other than an infinite flat heater will involve a characteristic length, L. Thus, the dimensional functional equation for q_{max} becomes

$$q_{max} = \text{fn}\left[\rho_g, h_{fg}, \sigma, g(\rho_f - \rho_g), L\right].$$

This involves six variables and four dimensions: J, m, s, and kg, where, once more in accordance with Section 4.3, we note that no significant conversion from work to heat is occurring and J must be retained as a separate unit. There are thus two pi-groups. The first group can arbitrarily be multiplied by $24/\pi$ to give

$$\Pi_1 = \frac{q_{max}}{(\pi/24)\,\rho_g^{1/2} h_{fg} \sqrt[4]{\sigma g(\rho_f - \rho_g)}} = \frac{q_{max}}{q_{max_z}} \tag{9.13}$$

Notice that the factor of $24/\pi$ has served to make the denominator equal to $q_{\max_z}$ (Zuber's expression for $q_{\max}$). Thus, for $q_{\max}$ on a flat plate, Π_1 equals 0.149/0.131, or 1.14. The second pi-group is

$$\Pi_2 = \frac{L}{\sqrt{\sigma/g(\rho_f - \rho_g)}} = 2\pi\sqrt{3}\frac{L}{\lambda_{d_1}} \equiv L' \tag{9.14}$$

The latter group, Π_2, is the square root of the *Bond number*, Bo—a group that has often been used to compare buoyant force with capillary forces.

Predictions and correlations of $q_{\max}$ have been made for several finite geometries in the form

$$\frac{q_{\max}}{q_{\max_z}} = \text{fn}\,(L') \tag{9.15}$$

The dimensionless characteristic length in eqn. (9.15) might be a dimensionless radius (R'), a dimensionless diameter (D'), or a dimensionless height (H'). The graphs in Fig. 9.13 are comparisons of several of the existing predictions and correlations with experimental data. These predictions and others are listed in Table 9.3. Notice that the last three items in Table 9.3 (10, 11, and 12) are general expressions from which several of the preceding expressions in the table can be obtained.

The equations in Table 9.3 are all valid within $\pm 15\%$ or 20%, which is very close to the inherent scatter of $q_{\max}$ data. However, they are subject to the following conditions:

- The bulk liquid is saturated.
- There are no pathological surface imperfections.
- There is no forced convection.

Another limitation on all the equations in Table 9.3 is that neither the size of the heater nor the relative force of gravity can be too small. When $L' < 0.15$ in most configurations, the Bond number is

$$\text{Bo} \equiv L'^2 = \frac{g(\rho_f - \rho_g)L^3}{\sigma L} = \frac{\text{buoyant force}}{\text{capillary force}} < \frac{1}{44}$$

In such cases, the process becomes completely dominated by surface tension and the Taylor-Helmholtz wave mechanisms no longer operate. As L' is reduced, the peak and minimum heat fluxes cease to occur and

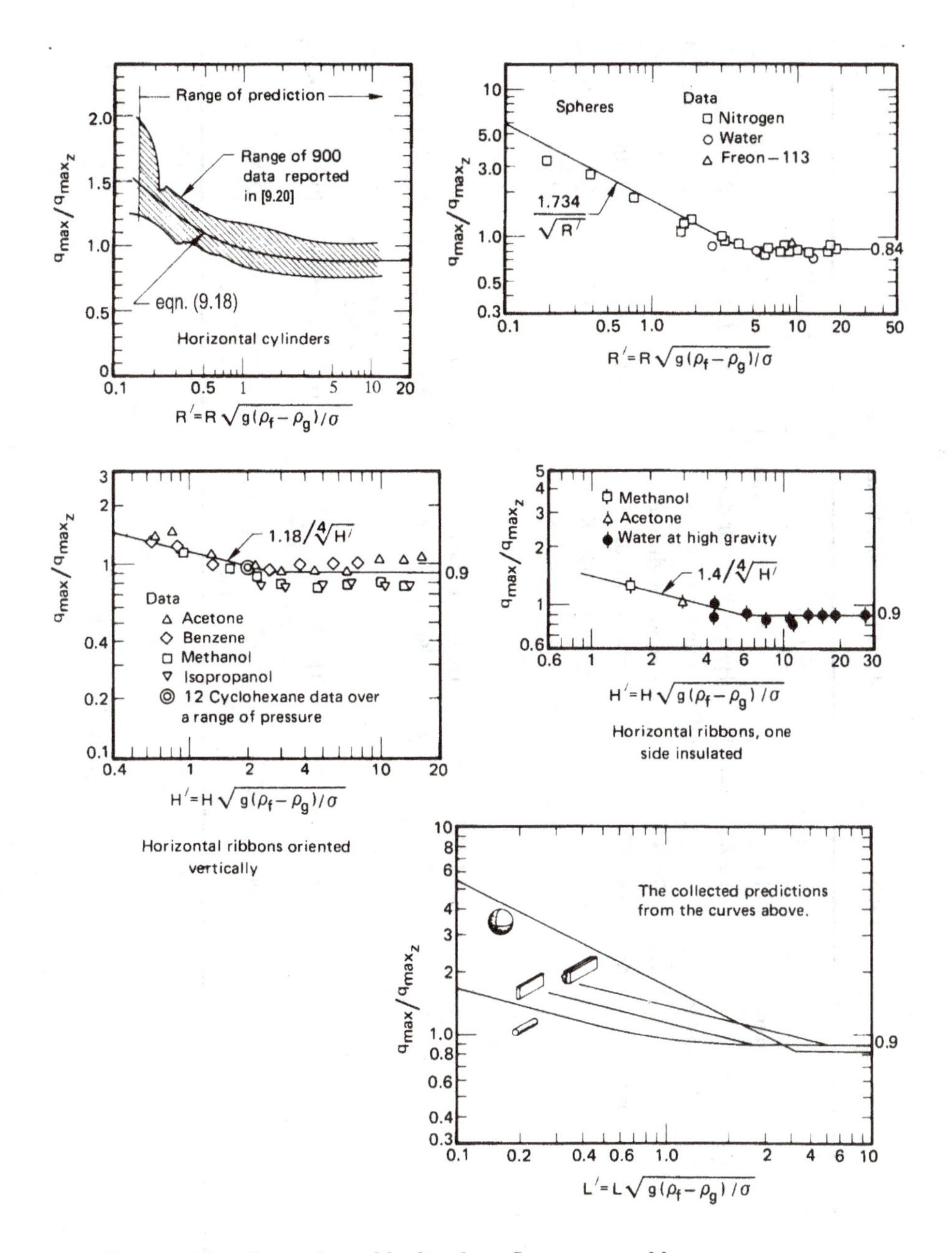

Figure 9.13 The peak pool boiling heat flux on several heaters.

Table 9.3 Predictions of the peak pool boiling heat flux

Situation	q_{max}/q_{max_z}	*Basis for L′*	*Range of L′*	*Source*	*Eqn. No.*
1. Infinite flat heater	1.14	Heater width or diameter	$L' \geq 27$	[9.19]	(9.16)
2. Small flat heater	$1.14(\lambda_{d_1}/A_{heater})$	Heater width or diameter	$9 < L' < 20$	[9.19]	(9.17)
3. Horizontal cylinder	$0.89 + 2.27e^{-3.44\sqrt{R'}}$	Cylinder radius, R	$R' \geq 0.15$	[9.22]	(9.18)
4. Large horizontal cylinder	0.90	Cylinder radius, R	$R' \geq 1.2$	[9.20]	(9.19)
5. Small horizontal cylinder	$0.94/(R')^{1/4}$	Cylinder radius, R	$0.15 \leq R' \leq 1.2$	[9.20]	(9.20)
6. Large sphere	0.84	Sphere radius, R	$R' \geq 4.26$	[9.23]	(9.21)
7. Small sphere	$1.734/(R')^{1/2}$	Sphere radius, R	$0.15 \leq R' \leq 4.26$	[9.23]	(9.22)
Small horizontal ribbon oriented vertically					
8. plain	$1.18/(H')^{1/4}$	Height of side, H	$0.15 \leq H' \leq 2.96$	[9.20]	(9.23)
9. 1 side insulated	$1.4/(H')^{1/4}$	Height of side, H	$0.15 \leq H' \leq 5.86$	[9.20]	(9.24)
10. Any large finite body	~ 0.90	Characteristic length, L	cannot specify generally; $L' \gtrsim 4$	[9.20]	(9.25)
11. Small slender cylinder of any cross section	$1.4/(P')^{1/4}$	Transverse perimeter, P	$0.15 \leq P' \leq 5.86$	[9.20]	(9.26)
12. Small bluff body	$\text{Constant}/(L')^{1/2}$	Characteristic length, L	cannot specify generally; $L' \lesssim 4$	[9.20]	(9.27)

the boiling curve becomes monotonic. When nucleation occurs on a very small wire, for example, the wire is immediately enveloped in vapor and the mechanism of heat removal passes directly from natural convection to film boiling.

Example 9.7

A metal body, only roughly spherical in shape, has a surface area of 400 cm^2 and a volume of 600 cm^3. It is quenched in saturated water at 1 atm. What is the most rapid rate of heat removal during the quench?

Solution. As the cooling process progresses, it goes through the boiling curve from film boiling, through q_{min}, up the transitional boiling regime, through q_{max}, and down the nucleate boiling curve. Cooling is finally completed by natural convection. If you have ever seen a red-hot horseshoe quenched, you might recall recall the great gush of bubbling that occurs as q_{max} is reached. We therefore calculate the required heat flow as $Q = q_{max} A_{spheroid}$, where q_{max} is given by eqn. (9.25) for large bodies in Table 9.3:

$$q_{max} = 0.9\, q_{max_z} = 0.9(0.131)\rho_g^{1/2} h_{fg} \sqrt[4]{g\sigma(\rho_f - \rho_g)}$$

so

$$Q = \left[0.9(0.131)(0.597)^{1/2}(2,257,000)\sqrt[4]{9.8(0.0589)(958)}\ \text{W/m}^2\right] \times \left(400 \times 10^{-4}\ \text{m}^2\right)$$

or

$$Q = 39,900\ \text{W} \quad \text{or} \quad 39.9\ \text{kW}$$

This is a startlingly large rate of energy removal for such a small object.

To complete the calculation, it is necessary to check whether or not R' is large enough to justify the use of eqn. (9.25):

$$R' = \frac{V/A}{\sqrt{\sigma/g(\rho_f - \rho_g)}} = \frac{0.0006}{0.04}\sqrt{\frac{9.8(958)}{0.0589}} = 6.0$$

This is larger than the value of about 4 for which the body must be considered "large." ■

9.4 Film boiling

Film boiling bears an uncanny similarity to film condensation. The similarity is so great that in 1950, Bromley [9.24] was able to use eqn. (8.66) for condensation on cylinders—almost directly—to predict film boiling from cylinders. He observed that the boundary condition $(\partial u/\partial y)_{y=\delta} = 0$ at the liquid-vapor interface in film condensation would have to change to something in between $(\partial u/\partial y)_{y=\delta} = 0$ and $u(y = \delta) = 0$ during film boiling. The reason is that the external liquid is not so easily set into motion. He then redid the film condensation analysis, merely changing k and ν from liquid to vapor properties. The change of boundary conditions gave eqn. (8.66) with the constant changed from 0.729 to 0.512 and with k and ν changed to vapor values. By comparing the equation with experimental data, he fixed the constant at the intermediate value of 0.62. Thus, $\overline{\text{Nu}}_D$ based on k_g became

$$\overline{\text{Nu}}_D = 0.62 \left[\frac{(\rho_f - \rho_g) g h'_{fg} D^3}{\nu_g k_g (T_w - T_{\text{sat}})} \right]^{1/4} \tag{9.28}$$

where vapor and liquid properties should be evaluated at $T_{\text{sat}} + \Delta T/2$ and at T_{sat}, respectively. The latent heat correction in this case is similar in form to that for film condensation, but with different constants in it. Sadasivan and Lienhard [9.25] have shown it to be

$$h'_{fg} = h_{fg} \left[1 + \left(0.968 - 0.163/\text{Pr}_g \right) \text{Ja}_g \right] \tag{9.29}$$

for $\text{Pr}_g \geq 0.6$, where $\text{Ja}_g = c_{p_g}(T_w - T_{\text{sat}})/h_{fg}$.

Dhir and Lienhard [9.26] did the same thing for *spheres*, as Bromley did for *cylinders*, 20 years later. Their result [cf. eqn. (8.67)] was

$$\overline{\text{Nu}}_D = 0.67 \left[\frac{(\rho_f - \rho_g) g h'_{fg} D^3}{\nu_g k_g (T_w - T_{\text{sat}})} \right]^{1/4} \tag{9.30}$$

The preceding expressions are based on heat transfer by convection through the vapor film, alone. However, when film boiling occurs much beyond $q_{\min}$ in water, the heater glows dull cherry-red to white-hot. Radiation in such cases can be enormous. One's first temptation might be simply to add a radiation heat transfer coefficient, $\overline{h}_{\text{rad}}$ to $\overline{h}_{\text{boiling}}$ as obtained from eqn. (9.28) or (9.30), where

$$\overline{h}_{\text{rad}} = \frac{q_{\text{rad}}}{T_w - T_{\text{sat}}} = \frac{\varepsilon\sigma\left(T_w^4 - T_{\text{sat}}^4\right)}{T_w - T_{\text{sat}}}$$

and where ε is a surface radiation property of the heater called the emittance (see Section 10.1).

Unfortunately, such addition is not correct, because the additional radiative heat transfer will increase the vapor blanket thickness, reducing the convective contribution. Bromley [9.24] suggested for cylinders the approximate relation

$$\overline{h}_{\text{total}} = \overline{h}_{\text{boiling}} + \tfrac{3}{4}\,\overline{h}_{\text{rad}}, \qquad \overline{h}_{\text{rad}} < \overline{h}_{\text{boiling}} \tag{9.31}$$

More accurate corrections that have subsequently been offered are considerably more complex than this [9.10]. One of the most comprehensive is that of Pitschmann and Grigull [9.27]. Their correlation, which is fairly intricate, brings together an enormous range of heat transfer data for cylinders, within 20%. It is worth noting that radiation is seldom important when the heater temperature is less than 300°C.

The use of the analogy between film condensation and film boiling is somewhat questionable during film boiling on a vertical surface. In this case, the liquid–vapor interface becomes Helmholtz-unstable at a short distance from the leading edge. However, Leonard, Sun, and Dix [9.28] have shown that by using $\lambda_{d_1}/\sqrt{3}$ in place of D in eqn. (9.28), one obtains a very satisfactory prediction of $\overline{h}$ for rather tall vertical plates.

The analogy between film condensation and film boiling also deteriorates when it is applied to small curved bodies. The reason is that the thickness of the vapor film in boiling is far greater than the liquid film during condensation. Consequently, a curvature correction, which could be ignored in film condensation, must be included during film boiling from small cylinders, spheres, and other curved bodies. The first curvature correction to be made was an empirical one given by Westwater and Breen [9.29] in 1962. They showed that the equation

$$\boxed{\overline{\text{Nu}}_D = \left[\left(0.715 + \frac{0.263}{R'}\right)\left(R'\right)^{1/4}\right]\overline{\text{Nu}}_{D_{\text{Bromley}}}} \tag{9.32}$$

applies when $R' < 1.86$. Otherwise, Bromley's equation should be used directly.

9.5 Minimum heat flux

Zuber [9.17] also provided a prediction of the minimum heat flux, q_{min}, along with his prediction of q_{max}. He assumed that as $T_w - T_{\text{sat}}$ is reduced in the film boiling regime, the rate of vapor generation eventually becomes too small to sustain the Taylor wave action that characterizes film boiling. Zuber's q_{min} prediction, based on this assumption, has to include an arbitrary constant. The result for flat horizontal heaters is

$$q_{\text{min}} = C\,\rho_g h_{fg} \sqrt[4]{\frac{\sigma g(\rho_f - \rho_g)}{(\rho_f + \rho_g)^2}} \tag{9.33}$$

Zuber guessed a value of C which Berenson [9.30] subsequently corrected on the basis of experimental data. Berenson used measured values of q_{min} on horizontal heaters to get

$$\boxed{q_{\text{min}_{\text{Berenson}}} = 0.09\,\rho_g h_{fg} \sqrt[4]{\frac{\sigma g(\rho_f - \rho_g)}{(\rho_f + \rho_g)^2}}} \tag{9.34}$$

Lienhard and Wong [9.31] did the parallel prediction for horizontal wires and found that

$$\boxed{q_{\text{min}} = 0.515\left[\frac{18}{R'^2(2R'^2 + 1)}\right]^{1/4} q_{\text{min}_{\text{Berenson}}}} \tag{9.35}$$

The problem with all of these expressions is that some contact frequently occurs between the liquid and the heater wall at film boiling heat fluxes higher than the minimum. When this happens, the boiling curve deviates above the film boiling curve and finds a higher minimum than those reported above. The values of the constants shown above should therefore be viewed as practical lower limits of q_{min}. We return to this matter subsequently.

Example 9.8

Check the value of q_{min} shown in Fig. 9.2.

Solution. The heater is a flat surface, so we use eqn. (9.34) and the physical properties given in Example 9.5.

$$q_{\text{min}} = 0.09(0.597)(2,257,000)\sqrt[4]{\frac{9.8(0.0589)(958)}{(959)^2}}$$

or

$$q_{\min} = 18,990 \text{ W/m}^2$$

From Fig. 9.2 we read 20,000 W/m^2, which is the same, within the accuracy of the graph. ■

9.6 Transition boiling and system influences

Many system features influence the pool boiling behavior we have discussed thus far. These include forced convection, subcooling, gravity, surface roughness and surface chemistry, and the heater configuration, among others. To understand one of the most serious of these—the influence of surface roughness and surface chemistry—we begin by thinking about transition boiling, which is extremely sensitive to both.

Surface condition and transition boiling

Less is known about transition boiling than about any other mode of boiling. Data are limited, and there is no comprehensive body of theory. The first systematic sets of accurate measurements of transition boiling were reported by Berenson [9.30] in 1960. Figure 9.14 shows two sets of his data.

The upper set of curves shows the typical influence of surface chemistry on transition boiling. It makes it clear that a change in the surface chemistry has little effect on the boiling curve except in the transition boiling region and the low heat flux film boiling region. The oxidation of the surface has the effect of changing the *contact angle* dramatically—making it far easier for the liquid to wet the surface when it touches it. Transition boiling is more susceptible than any other mode to such a change.

The bottom set of curves shows the influence of surface roughness on boiling. In this case, nucleate boiling is far more susceptible to roughness than any other mode of boiling except, perhaps, the very lowest end of the film boiling range. That is because as roughness increases the number of active nucleation sites, the heat transfer rises in accordance with the Yamagata relation, eqn. (9.3).

It is important to recognize that neither roughness nor surface chemistry affects film boiling, because the liquid does not touch the heater.

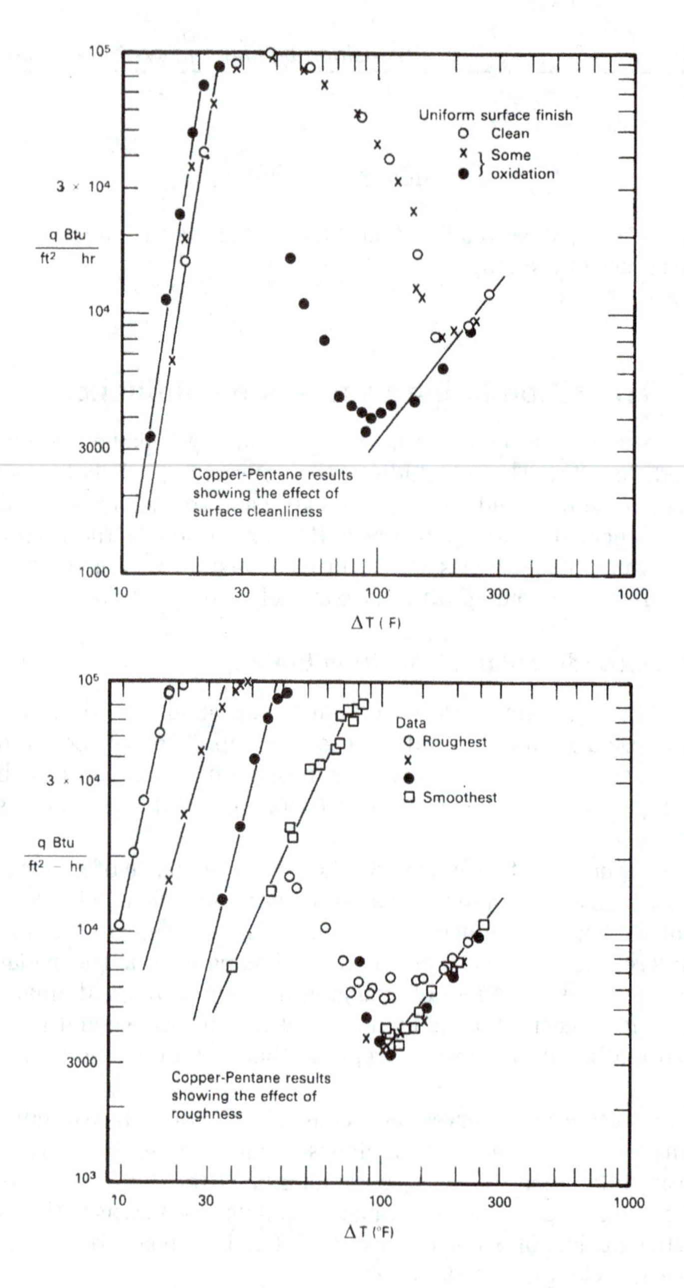

Figure 9.14 Typical data from Berenson's [9.30] study of the influence of surface condition on the boiling curve.

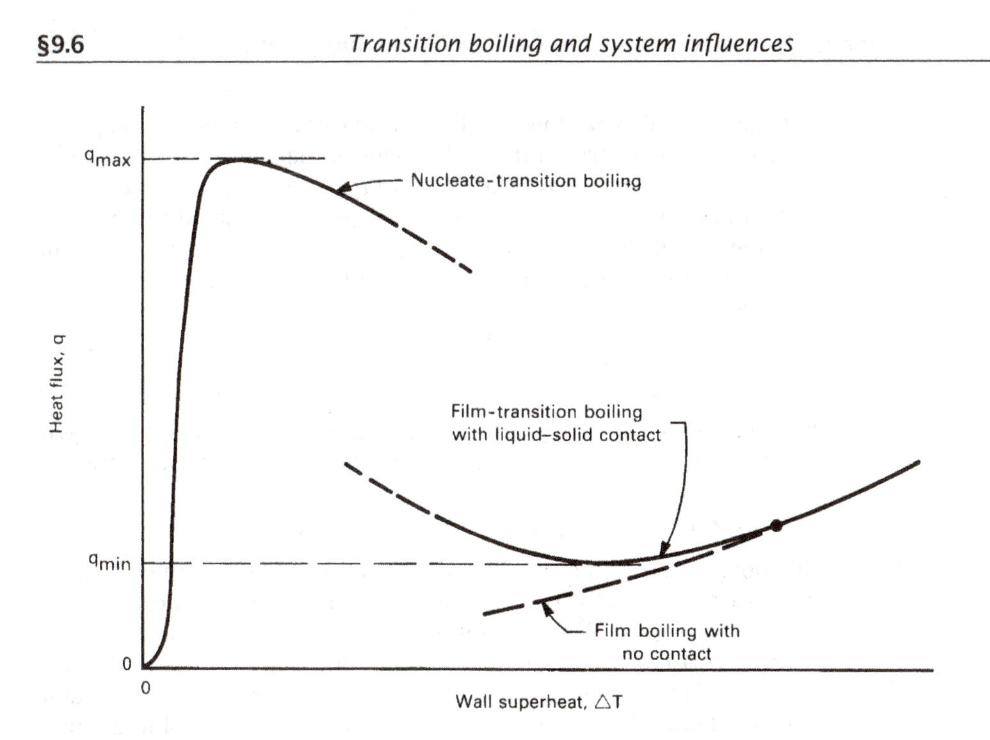

Figure 9.15 The transition boiling regime.

The fact that both effects appear to influence the lower film boiling range means that they actually cause film boiling to break down by initiating liquid–solid contact at low heat fluxes.

Figure 9.15 shows what an actual boiling curve looks like under the influence of a wetting (or even slightly wetting) contact angle. This figure is based on the work of Witte and Lienhard ([9.32] and [9.33]). On it are identified a *nucleate-transition* and a *film-transition* boiling region. These are continuations of nucleate boiling behavior with decreasing liquid-solid contact (as shown in Fig. 9.3c) and of film boiling behavior with increasing liquid–solid contact, respectively.

These two regions of transition boiling are often connected by abrupt jumps. However, no one has yet seen how to predict where such jumps take place. Reference [9.33] is a full discussion of the hydrodynamic theory of boiling, which includes an extended discussion of the transition boiling problem and a correlation for the transition-film boiling heat flux by Ramilison and Lienhard [9.34].

Figure 9.14 also indicates fairly accurately the influence of roughness and surface chemistry on q_{max}. It suggests that these influences normally can cause significant variations in q_{max} that are not predicted in the hydrodynamic theory. Ramilison et al. [9.35] correlated these effects for large flat-plate heaters using the rms surface roughness, r in μm, and the receding contact angle for the liquid on the heater material, β_r in radians:

$$\frac{q_{max}}{q_{max_Z}} = 0.0336\,(\pi - \beta_r)^{3.0}\, r^{0.0125} \tag{9.36}$$

This correlation collapses the data to ±6%. Uncorrected, variations from the predictions of hydrodynamic theory reached 40% as a result of roughness and finish. Equivalent results are needed for other geometries.

Subcooling

A stationary pool will normally not remain below its saturation temperature over an extended period of time. When heat is transferred to the pool, the liquid soon becomes saturated—as it does in a teakettle (recall Experiment 9.1). However, before a liquid comes up to temperature, or if a very small rate of forced convection continuously replaces warm liquid with cool liquid, we can justly ask what the effect of a cool liquid bulk might be.

Figure 9.16 shows how a typical boiling curve might be changed if $T_{bulk} < T_{sat}$: We know, for example, that in *laminar natural convection*, q will increase as $(T_w - T_{bulk})^{5/4}$ or as $[(T_w - T_{sat}) + \Delta T_{sub}]^{5/4}$, where $\Delta T_{sub} \equiv T_{sat} - T_{bulk}$. During *nucleate boiling*, the influence of subcooling on q is known to be small. The *peak and minimum heat fluxes* are known to increase linearly with ΔT_{sub}. These increases are quite significant. The *film boiling* heat flux increases rather strongly, especially at lower heat fluxes. The influence of ΔT_{sub} on transitional boiling is not well documented.

Gravity

The influence of gravity (or any other such body force) is of concern because boiling processes frequently take place in rotating or accelerating systems. The reduction of gravity has a significant impact on boiling processes aboard space vehicles. Since g appears explicitly in the equations for q_{max}, q_{min}, and $q_{\text{film boiling}}$, we know what its influence is. Both q_{max} and q_{min} increase directly as $g^{1/4}$ in finite bodies, and there is an additional gravitational influence through the parameter L'. However, when gravity is small enough to reduce R' below about 0.15, the hydrodynamic transitions deteriorate and eventually vanish altogether. Although

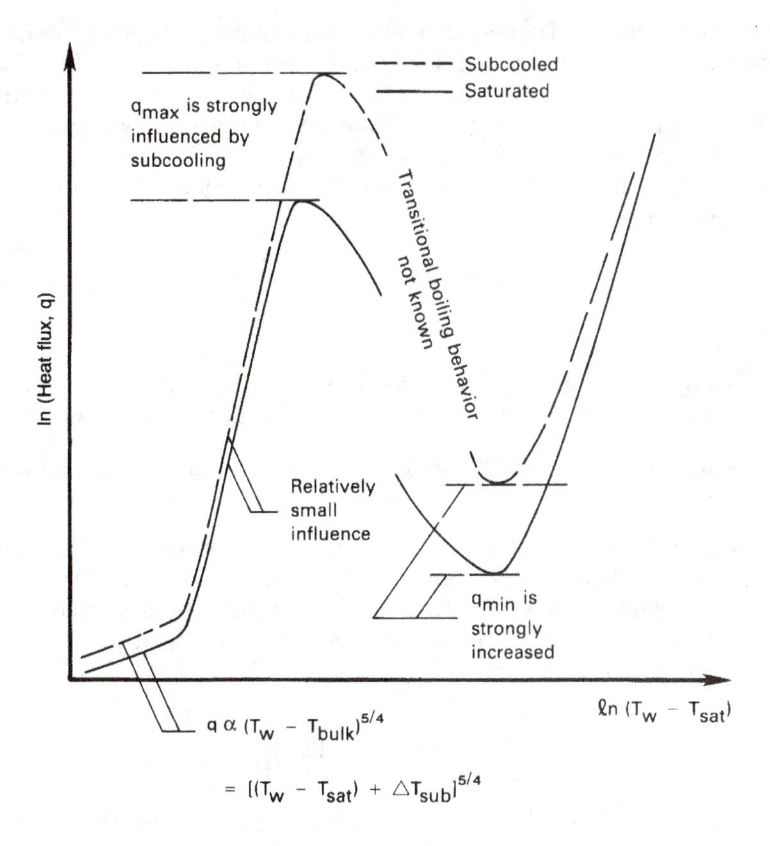

Figure 9.16 The influence of subcooling on the boiling curve.

Rohsenow's equation suggests that q is proportional to $g^{1/2}$ in the nucleate boiling regime, other evidence suggests that the influence of gravity on the nucleate boiling curve is very slight, apart from an indirect effect on the onset of boiling.

Forced convection

A superposed flow over a given heater (e.g., Fig. 9.2) generally improves heat transfer in all regimes of the boiling curve. But flow is particularly effective in raising q_{max}. Let us look at the influence of flow on the different regimes of boiling.

Influences of forced convection on nucleate boiling. Figure 9.17 shows nucleate boiling during the forced convection of water over a flat plate. Bergles and Rohsenow [9.36] offer an empirical strategy for predicting the heat flux during nucleate flow boiling when the net vapor generation is still relatively small. (The photograph in Fig. 9.17 shows how a substantial buildup of vapor can radically alter flow boiling behavior.) They suggest that

$$q = q_{\text{FC}}\sqrt{1 + \left[\frac{q_B}{q_{\text{FC}}}\left(1 - \frac{q_i}{q_B}\right)\right]^2} \tag{9.37}$$

where

- q_{FC} is the single-phase forced convection heat transfer for the heater, as one might calculate using the methods of Chapters 6 and 7.
- q_B is the *pool* boiling heat flux for that liquid and that heater from eqn. (9.4).
- q_i is the heat flux from the pool boiling curve evaluated at the value of $(T_w - T_{\text{sat}})$ where boiling begins during flow boiling (see Fig. 9.17). An estimate of $(T_w - T_{\text{sat}})_{\text{onset}}$ can be made by intersecting the forced convection equation $q = h_{\text{FC}}(T_w - T_b)$ with the following equation [9.37]:

$$(T_w - T_{\text{sat}})_{\text{onset}} = \left(\frac{8\sigma T_{\text{sat}} q}{\rho_g h_{fg} k_f}\right)^{1/2} \tag{9.38}$$

Equation (9.37) will provide a first approximation in most boiling configurations, but it is restricted to subcooled flows or other situations in which vapor generation is not too great.

Peak heat flux in external flows. The peak heat flux on a submerged body is strongly augmented by an external flow around it. Although knowledge of this area is still evolving, we do know from dimensional analysis that

$$\frac{q_{\text{max}}}{\rho_g h_{fg} u_\infty} = \text{fn}\left(\text{We}_D, \rho_f/\rho_g\right) \tag{9.39}$$

where the Weber number, We, is

$$\text{We}_L \equiv \frac{\rho_g u_\infty^2 L}{\sigma} = \frac{\text{inertia force}/L}{\text{surface force}/L}$$

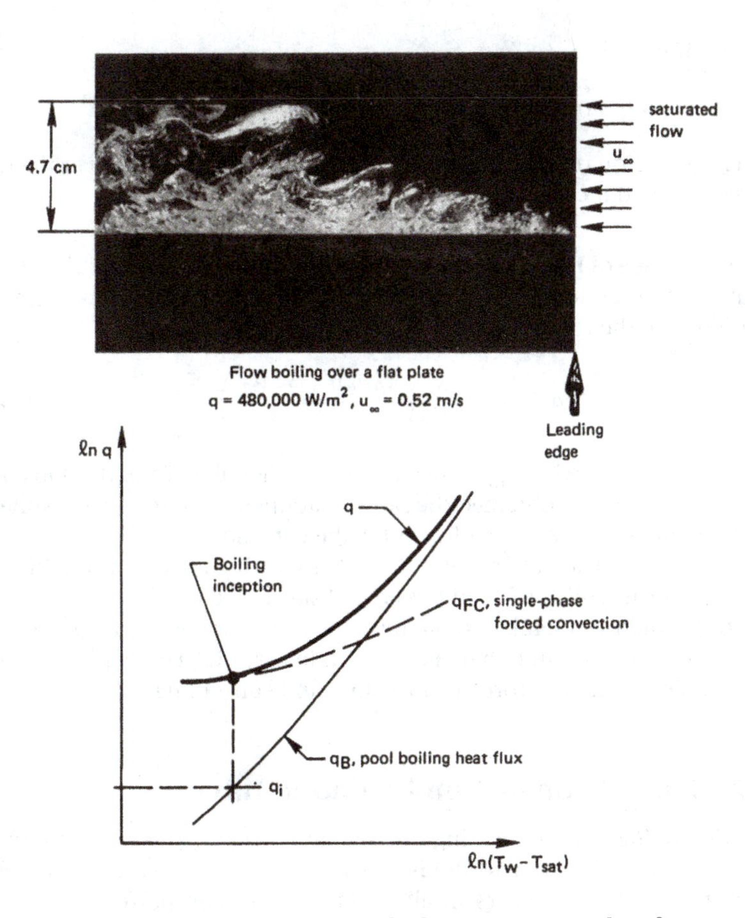

Figure 9.17 Forced convection boiling on an external surface.

and where L is any characteristic length.

Kheyrandish and Lienhard [9.38] suggest fairly complex expressions of this form for q_{max} on horizontal cylinders in cross flows. For a cylindrical liquid jet impinging on a heated disk of diameter D, Sharan and Lienhard [9.39] obtained

$$\frac{q_{\text{max}}}{\rho_g h_{fg} u_{\text{jet}}} = \left(0.21 + 0.0017\rho_f/\rho_g\right)\left(\frac{d_{\text{jet}}}{D}\right)^{1/3}\left(\frac{1000\rho_g/\rho_f}{\text{We}_D}\right)^{A} \qquad (9.40)$$

where, if we call $\rho_f/\rho_g \equiv r$,

$$A = 0.486 + 0.06052 \ln r - 0.0378\,(\ln r)^2 + 0.00362\,(\ln r)^3 \qquad (9.41)$$

This correlation represents all the existing data within ±20% over the full range of the data.

The influence of fluid flow on film boiling. Bromley et al. [9.40] showed that the film boiling heat flux during forced flow normal to a cylinder should take the form

$$q = \text{constant}\left(\frac{k_g \rho_g h'_{fg} \Delta T u_\infty}{D}\right)^{1/2} \qquad (9.42)$$

for $u_\infty^2/(gD) \geq 4$ with h'_{fg} from eqn. (9.29). Their data fixed the constant at 2.70. Witte [9.41] obtained the same relationship for flow over a sphere and recommended a value of 2.98 for the constant.

Additional work in the literature deals with forced film boiling on plane surfaces and combined forced and subcooled film boiling in a variety of geometries [9.42]. Although these studies are beyond our present scope, it is worth noting that one may attain very high cooling rates using film boiling with both forced convection and subcooling.

9.7 Forced convection boiling in tubes

Flowing fluids undergo boiling or condensation in many of the cases in which we transfer heat to fluids moving through tubes. For example, such phase change occurs in all vapor-compression power cycles and refrigerators. When we use the terms *boiler, condenser, steam generator,* or *evaporator* we usually refer to equipment that involves heat transfer within tubes. The prediction of heat transfer coefficients in these systems is often essential to determining U and sizing the equipment. So let us consider the problem of predicting boiling heat transfer to liquids flowing through tubes.

Relationship between heat transfer and temperature difference

Forced convection boiling in a tube or duct is a process that becomes very hard to delineate because it takes so many forms. In addition to the usual system variables that must be considered in pool boiling, the formation

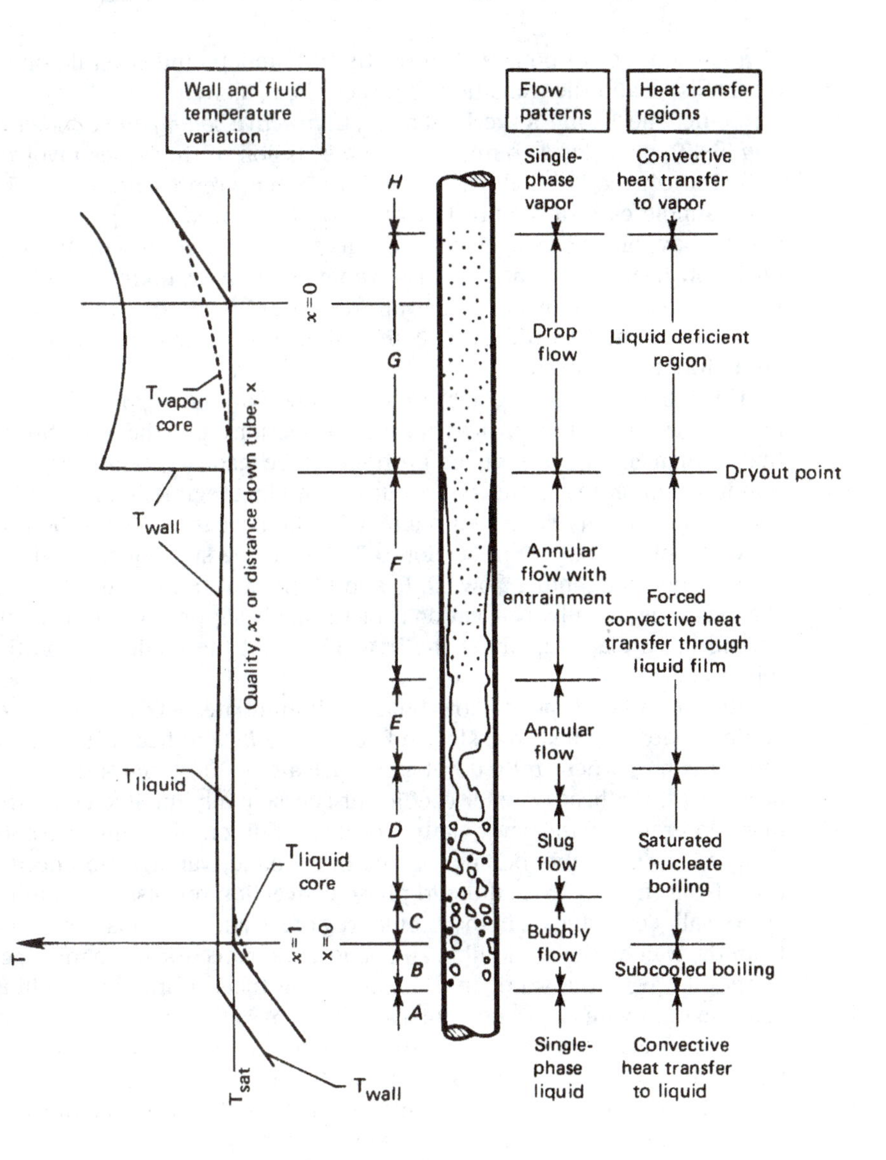

Figure 9.18 The development of a two-phase flow in a vertical tube with a uniform wall heat flux (not to scale).

of many regimes of boiling requires that we understand several boiling mechanisms and the transitions between them, as well.

Collier and Thome's excellent book, *Convective Boiling and Condensation* [9.43], provides a comprehensive discussion of the issues involved in forced convection boiling. Figure 9.18 is their representation of the fairly simple case of flow of liquid in a *uniform wall heat flux* tube in which body forces can be neglected. This situation is representative of a fairly low heat flux at the wall. The vapor fraction, or *quality*, x, of the flow increases steadily until the wall "dries out." Then the wall temperature rises rapidly. With a very high wall heat flux, the pipe could burn out before dryout occurs.

Figure 9.19, also provided by Collier, shows how the regimes shown in Fig. 9.18 are distributed in heat flux and in position along the tube. Notice that, at high enough heat fluxes, burnout can be made to occur at any station in the pipe. In the subcooled nucleate boiling regime (B in Fig. 9.18) and the low quality saturated regime (C), the heat transfer can be predicted using eqn. (9.37) in Section 9.6. But in the subsequent regimes of slug flow and annular flow (D, E, and F) the heat transfer mechanism changes substantially. Nucleation is increasingly suppressed, and vaporization takes place mainly at the free surface of the liquid film on the tube wall.

Most efforts to model flow boiling differentiate between nucleate-boiling-controlled heat transfer and *convective boiling* heat transfer. In those regimes where fully developed nucleate boiling occurs (the later parts of C), the heat transfer coefficient is essentially unaffected by the mass flow rate and the flow quality. Locally, conditions are similar to pool boiling. In convective boiling, on the other hand, vaporization occurs away from the wall, with a liquid-phase convection process dominating at the wall. For example, in the annular regions E and F, heat is convected from the wall by the liquid film, and vaporization occurs at the interface of the film with the vapor in the core of the tube. Convective boiling can also dominate at low heat fluxes or high mass flow rates, where wall nucleate is again suppressed. Vaporization then occurs mainly on entrained bubbles in the core of the tube. In convective boiling, the heat transfer coefficient is essentially independent of the heat flux, but it is strongly affected by the mass flow rate and quality.

Building a model to capture these complicated and competing trends has presented a challenge to researchers for several decades. One early effort by Chen [9.44] used a weighted sum of a nucleate boiling heat transfer coefficient and a convective boiling coefficient, where the weighting

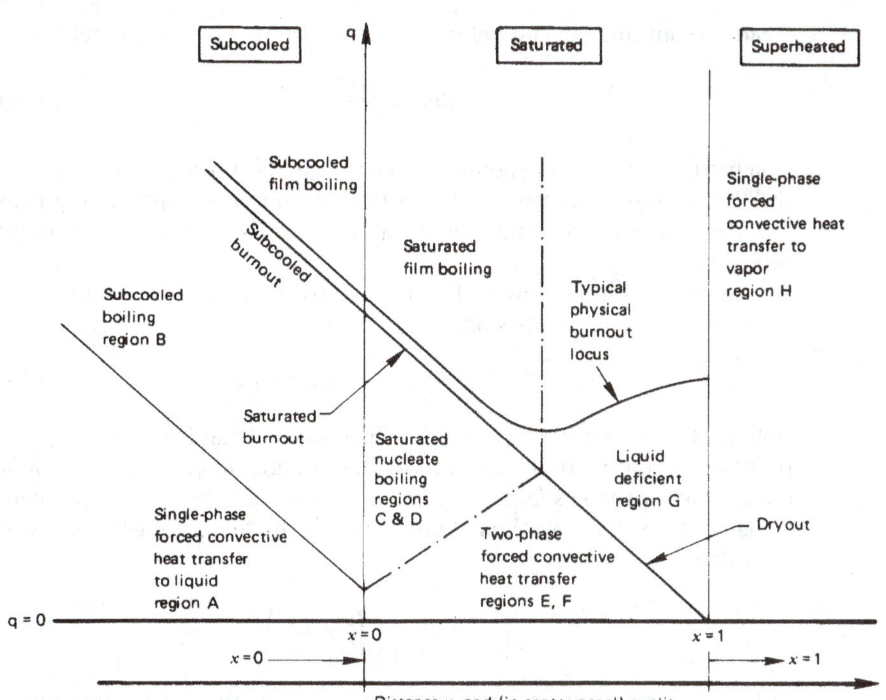

Figure 9.19 The influence of heat flux on two-phase flow behavior.

depended on local flow conditions. This model represents water data to an accuracy of about ±30% [9.45], but does not work well with most other fluids. Steiner and Taborek [9.46] substantially improved Chen's mechanistic prediction. Many other investigators have instead pursued correlations built from dimensional analysis and physical reasoning.

To do a dimensional analysis, we first note that the liquid and vapor phases may have different velocities. Thus, we avoid introducing a flow speed and instead rely on the the superficial mass flux, G, through the pipe:

$$G \equiv \frac{\dot{m}}{A_{\text{pipe}}} \quad (\text{kg/m}^2\text{s}) \tag{9.43}$$

This mass flow per unit area is constant along the duct if the flow is steady. From this, we can define a "liquid only" Reynolds number

$$\mathrm{Re}_{\mathrm{lo}} \equiv \frac{GD}{\mu_f} \tag{9.44}$$

which would be the Reynolds number if all the flowing mass were in the liquid state. Then we may use $\mathrm{Re}_{\mathrm{lo}}$ to compute a liquid-only heat transfer coefficient, h_{lo} from Gnielinski's equation, eqn. (7.43), using liquid properties at T_{sat}.

We then write the flow boiling heat transfer coefficient, h_{fb} for saturated flow in *vertical* tubes as:

$$h_{\mathrm{fb}} = \mathrm{fn}\left(h_{\mathrm{lo}}, G, x, h_{fg}, q_w, \rho_f, \rho_g, D\right) \tag{9.45}$$

Note that other liquid properties, such as viscosity and conductivity, are represented indirectly through h_{lo}. This functional equation has eight dimensional variables (and one dimensionless variable, x) in five dimensions (m, kg, s, J, K). We thus obtain three more dimensionless groups to go with x, specifically

$$\frac{h_{\mathrm{fb}}}{h_{\mathrm{lo}}} = \mathrm{fn}\left(x, \frac{q_w}{Gh_{fg}}, \frac{\rho_g}{\rho_f}\right) \tag{9.46}$$

In fact, the situation is even a bit simpler than this, since arguments related to the pressure gradient show that the quality and the density ratio can be combined into a single group, called the *convection number*:

$$\mathrm{Co} \equiv \left(\frac{1-x}{x}\right)^{0.8} \left(\frac{\rho_g}{\rho_f}\right)^{0.5} \tag{9.47}$$

The other dimensionless group in eqn. (9.46) is called the *boiling number*:

$$\mathrm{Bo} \equiv \frac{q_w}{Gh_{fg}} \tag{9.48}$$

so that

$$\frac{h_{\mathrm{fb}}}{h_{\mathrm{lo}}} = \mathrm{fn}\,(\mathrm{Bo}, \mathrm{Co}) \tag{9.49}$$

When the convection number is large ($\mathrm{Co} \gtrsim 1$), as for low quality, nucleate boiling dominates. In this range, $h_{\mathrm{fb}}/h_{\mathrm{lo}}$ rises with increasing Bo and is approximately independent of Co. When the convection number is smaller, as at higher quality, the effect of the boiling number declines and $h_{\mathrm{fb}}/h_{\mathrm{lo}}$ increases with decreasing Co.

Table 9.4 Fluid-dependent parameter F in the Kandlikar correlation for copper tubing. Additional values are given in [9.47].

Fluid	F	*Fluid*	F
Water	1.0	R-124	1.90
Propane	2.15	R-125	1.10
R-12	1.50	R-134a	1.63
R-22	2.20	R-152a	1.10
R-32	1.20	R-410a	1.72

Correlations having the general form of eqn. (9.49) were developed by Schrock and Grossman [9.48], Shah [9.49], and Gungor and Winterton [9.50]. Kandlikar [9.45, 9.47, 9.51] refined this approach further, obtaining good accuracy and better capturing the parametric trends. His method is to calculate $h_{\text{fb}}/h_{\text{lo}}$ from each of the following two correlations and to choose the larger value:

$$\left.\frac{h_{\text{fb}}}{h_{\text{lo}}}\right|_{\text{nbd}} = (1-x)^{0.8}\left[0.6683\,\text{Co}^{-0.2} f_o + 1058\,\text{Bo}^{0.7} F\right] \tag{9.50a}$$

$$\left.\frac{h_{\text{fb}}}{h_{\text{lo}}}\right|_{\text{cbd}} = (1-x)^{0.8}\left[1.136\,\text{Co}^{-0.9} f_o + 667.2\,\text{Bo}^{0.7} F\right] \tag{9.50b}$$

where "nbd" means "nucleate boiling dominant" and "cbd" means "convective boiling dominant".

In these equations, the orientation factor, f_o, is set to one for vertical tubes[4] and F is a fluid-dependent parameter whose value is given in Table 9.4. The parameter F arises here for the same reason that fluid-dependent parameters appear in nucleate boiling correlations: surface tension, contact angles, and other fluid-dependent variables influence nucleation and bubble growth. The values in Table 9.4 are for commercial grades of copper tubing. For stainless steel tubing, Kandlikar recommends $F = 1$ for all fluids. Equations (9.50) are applicable for the saturated boiling regimes (C through F) with quality in the range $0 < x \leq 0.8$. For subcooled conditions, see Problem 9.21.

[4]The value for horizontal tubes is given in eqn. (9.52).

Example 9.9

0.6 kg/s of saturated H_2O at $T_b = 207°C$ flows in a 5 cm diameter vertical tube heated at a rate of 184,000 W/m^2. Find the wall temperature at a point where the quality x is 20%.

Solution. Data for water are taken from Tables A.3–A.5. We first compute h_{lo}.

$$G = \frac{\dot{m}}{A_{\text{pipe}}} = \frac{0.6}{0.001964} = 305.6 \text{ kg/m}^2\text{s}$$

and

$$\text{Re}_{lo} = \frac{GD}{\mu_f} = \frac{(305.6)(0.05)}{1.297 \times 10^{-4}} = 1.178 \times 10^5$$

From eqns. (7.42) and (7.43):

$$f = \frac{1}{\left[1.82 \log_{10}(1.178 \times 10^5) - 1.64\right]^2} = 0.01736$$

$$\text{Nu}_D = \frac{(0.01736/8)\,(1.178 \times 10^5 - 1000)\;(0.892)}{1 + 12.7\sqrt{0.01736/8}\,[(0.892)^{2/3} - 1]} = 236.3$$

Hence,

$$h_{lo} = \frac{k_f}{D}\text{Nu}_D = \frac{0.6590}{0.05}236.3 = 3,115 \text{ W/m}^2\text{K}$$

Next, we find the parameters for eqns. (9.50). From Table 9.4, $F = 1$ for water, and for a vertical tube, $f_o = 1$. Also,

$$\text{Co} = \left(\frac{1-x}{x}\right)^{0.8}\left(\frac{\rho_g}{\rho_f}\right)^{0.5} = \left(\frac{1-0.20}{0.2}\right)^{0.8}\left(\frac{9.014}{856.5}\right)^{0.5} = 0.3110$$

$$\text{Bo} = \frac{q_w}{Gh_{fg}} = \frac{184,000}{(305.6)(1,913,000)} = 3.147 \times 10^{-4}$$

Substituting into eqns. (9.50):

$$h_{\text{fb}}\Big|_{\text{nbd}} = (3,115)(1-0.2)^{0.8}\Big[0.6683\,(0.3110)^{-0.2}(1) + 1058\,(3.147 \times 10^{-4})^{0.7}(1)\Big] = 11,950 \text{ W/m}^2\text{K}$$

$$h_{\text{fb}}\Big|_{\text{cbd}} = (3,115)(1-0.2)^{0.8}\Big[1.136\,(0.3110)^{-0.9}(1) + 667.2\,(3.147 \times 10^{-4})^{0.7}(1)\Big] = 14,620 \text{ W/m}^2\text{K}$$

Since the second value is larger, we use it: $h_{\text{fb}} = 14{,}620$ W/m^{2}K. Then,

$$T_w = T_b + \frac{q_w}{h_{\text{fb}}} = 207 + \frac{184{,}000}{14{,}620} = 220°\text{C} \qquad \blacksquare$$

The Kandlikar correlation leads to mean deviations of 16% for water and 19% for the various refrigerants. The Gungor and Winterton correlation [9.50], which is popular for its simplicity, does not contain fluid-specific coefficients, but it is somewhat less accurate than either the Kandlikar equations or the more complex Steiner and Taborek method [9.45, 9.46]. These three approaches, however, are among the best available.

Two-phase flow and heat transfer in horizontal tubes

The preceding discussion of flow boiling in tubes is largely restricted to vertical tubes. Several of the flow regimes in Fig. 9.18 will be altered as shown in Fig. 9.20 if the tube is oriented horizontally. The reason is that, especially at low quality, liquid will tend to flow along the bottom of the pipe and vapor along the top. The patterns shown in Fig. 9.20, by the way, will also be observed during the reverse process—condensation—or during adiabatic two-phase flow.

Which flow pattern actually occurs depends on several parameters in a fairly complex way. While many methods have been suggested to

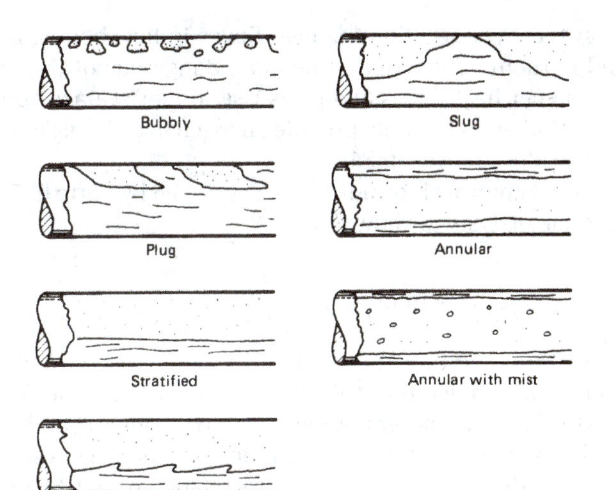

Figure 9.20 The discernible flow regimes during boiling, condensation, or adiabatic flow from left to right in horizontal tubes.

predict what flow pattern will result for a given set of conditions in the pipe, one of the best is that developed by Dukler, Taitel, and their co-workers. Their two-phase flow-regime maps are summarized in [9.52] and [9.53].

For the prediction of heat transfer, the most important additional parameter is the Froude number, $\mathrm{Fr_{lo}}$, which characterizes the strength of the flow's inertia (or momentum) relative to the gravitational forces that drive the separation of the liquid and vapor phases:

$$\mathrm{Fr_{lo}} \equiv \frac{G^2}{\rho_f^2 g D} \tag{9.51}$$

When $\mathrm{Fr_{lo}} < 0.04$, the top of the tube becomes relatively dry and $h_{\mathrm{fb}}/h_{\mathrm{lo}}$ begins to decline as the Froude number decreases further.

Kandlikar found that he could modify his correlation to account for gravitational effects in horizontal tubes by changing the value of f_o in eqns. (9.50):

$$f_o = \begin{cases} 1 & \text{for } \mathrm{Fr_{lo}} \geq 0.04 \\ (25\,\mathrm{Fr_{lo}})^{0.3} & \text{for } \mathrm{Fr_{lo}} < 0.04 \end{cases} \tag{9.52}$$

Peak heat flux

We have seen that there are two limiting heat fluxes in flow boiling in a tube: dryout and burnout. The latter is the more dangerous of the two since it occurs at higher heat fluxes and gives rise to more catastrophic temperature rises. Collier and Thome provide an extensive discussion of the subject [9.43], as does Hewitt [9.54].

One effective set of empirical formulas was developed by Katto [9.55]. He used dimensional analysis to show that

$$\frac{q_{\max}}{G h_{fg}} = \mathrm{fn}\left(\frac{\rho_g}{\rho_f}, \frac{\sigma \rho_f}{G^2 L}, \frac{L}{D}\right)$$

where L is the length of the tube and D its diameter. Since $G^2 L/\sigma \rho_f$ is a Weber number, we can see that this equation is of the same form as eqn. (9.39). Katto identifies several regimes of flow boiling with both saturated and subcooled liquid entering the pipe. For each of these regions, he and Ohne [9.56] later fit a successful correlation of this form to existing data.

Pressure gradients in flow boiling

Pressure gradients in flow boiling interact with the flow pattern and the void fraction, and they can change the local saturation temperature of the fluid. Gravity, flow acceleration, and friction all contribute to pressure change, and friction can be particularly hard to predict. In particular, the frictional pressure gradient can increase greatly as the flow quality rises from the pure liquid state to the pure vapor state; the change can amount to more than two orders of magnitude at low pressures. Data correlations are usually used to estimate the frictional pressure loss, but they are, at best, accurate to within about ±30%. Whalley [9.57] provides a nice introduction such methods. Certain complex models, designed for use in computer codes, can be used to make more accurate predictions [9.58].

9.8 Forced convective condensation heat transfer

When vapor is blown or forced past a cool wall, it exerts a shear stress on the condensate film. If the direction of forced flow is downward, it will drag the condensate film along, thinning it out and enhancing heat transfer. It is not hard to show (see Problem 9.22) that

$$\frac{4\mu k(T_{\text{sat}} - T_w)x}{gh'_{fg}\rho_f(\rho_f - \rho_g)} = \delta^4 + \frac{4}{3}\left[\frac{\tau_\delta \delta^3}{(\rho_f - \rho_g)g}\right] \tag{9.53}$$

where τ_δ is the shear stress exerted by the vapor flow on the condensate film.

Equation (9.53) is the starting point for any analysis of forced convection condensation on an external surface. Notice that if τ_δ is negative—if the shear opposes the direction of gravity—then it will have the effect of thickening δ and reducing heat transfer. Indeed, if for any value of δ,

$$\tau_\delta = -\frac{3g(\rho_f - \rho_g)}{4}\,\delta, \tag{9.54}$$

the shear stress will have the effect of halting the flow of condensate completely for a moment until δ grows to a larger value.

Heat transfer solutions based on eqn. (9.53) are complex because they require that one solve the boundary layer problem in the vapor in order to evaluate τ_δ; and this solution must be matched with the velocity at the outside surface of the condensate film. Collier and Thome [9.43, §10.5] discuss such solutions in some detail. One explicit result has been obtained in this way for condensation on the outside of a horizontal cylinder by Shekriladze and Gomelauri [9.59]:

$$\overline{\text{Nu}}_D = 0.64 \left\{ \frac{\rho_f u_\infty D}{\mu_f} \left[1 + \left(1 + 1.69 \frac{g h'_{fg} \mu_f D}{u_\infty^2 k_f (T_{\text{sat}} - T_w)} \right)^{1/2} \right] \right\}^{1/2} \tag{9.55}$$

where u_∞ is the free stream velocity and $\overline{\text{Nu}}_D$ is based on the liquid conductivity. Equation (9.55) is valid up to $\text{Re}_D \equiv \rho_f u_\infty D/\mu_f = 10^6$. Notice, too, that under appropriate flow conditions (large values of u_∞, for example), gravity becomes unimportant and

$$\overline{\text{Nu}}_D \longrightarrow 0.64 \sqrt{2\text{Re}_D} \tag{9.56}$$

The prediction of heat transfer during forced convective condensation in tubes becomes a different problem for each of the many possible flow regimes. The reader is referred to [9.43, §10.5] or [9.60] for details.

9.9 Dropwise condensation

An automobile windshield normally is covered with droplets during a light rainfall. They are hard to see through, and one must keep the windshield wiper moving constantly to achieve any kind of visibility. A glass windshield is normally quite clean and is free of any natural oxides, so the water forms a contact angle on it and any film will be unstable. The water tends to pull into droplets, which intersect the surface at the contact angle. Visibility can be improved by mixing a surfactant chemical into the window-washing water to reduce surface tension. It can also be improved by preparing the surface with a "wetting agent" to reduce the contact angle.[5]

Such behavior can also occur on a metallic condensing surface, but there is an important difference: Such surfaces are generally wetting. Wetting can be temporarily suppressed, and dropwise condensation can be encouraged, by treating an otherwise clean surface (or the vapor) with oil, kerosene, or a fatty acid. But these contaminants wash away fairly quickly. More permanent solutions have proven very elusive, with the result the liquid condensed in heat exchangers almost always forms a film.

[5]A way in which one can accomplish these ends is by wiping the wet window with a cigarette. It is hard to tell which of the two effects the many nasty chemicals in the cigarette achieve.

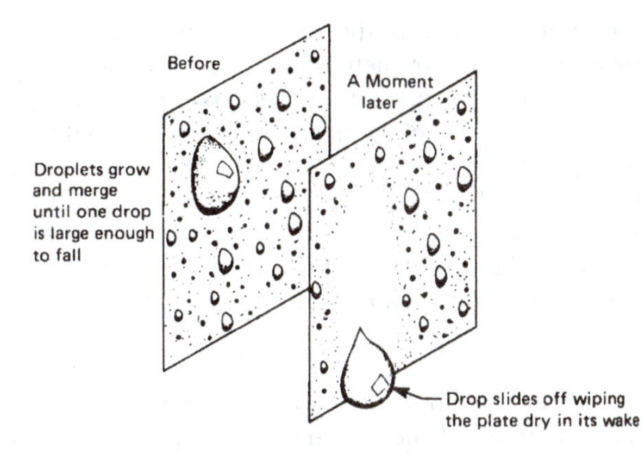

a. The process of liquid removal during dropwise condensation.

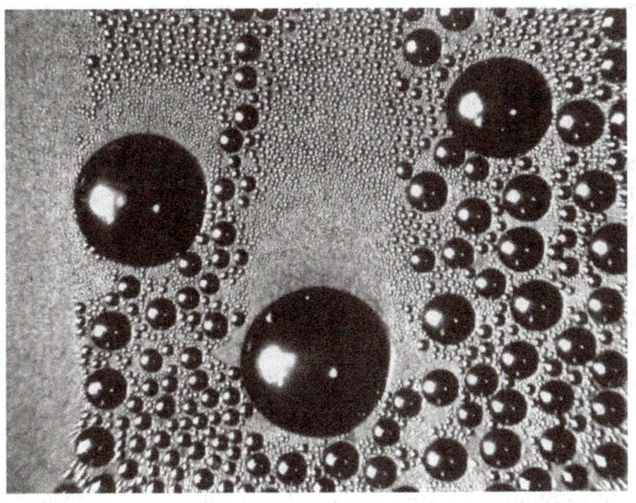

b. Typical photograph of dropwise condensation provided by Professor Borivoje B. Mikić. Notice the dry paths on the left and in the wake of the middle droplet.

Figure 9.21 Dropwise condensation.

It is regrettable that this is the case, because what is called *dropwise* condensation is an extremely effective heat removal mechanism. Figure 9.21 shows how it works. Droplets grow from active nucleation sites on the surface, and in this sense there is a great similarity between nucleate boiling and dropwise condensation. The similarity persists as the droplets grow, touch, and merge with one another until one is large enough to be pulled away from its position by gravity. It then slides off, wiping away the smaller droplets in its path and leaving a dry swathe in its wake. New droplets immediately begin to grow at the nucleation sites in the path.

The repeated re-creation of the early droplet growth cycle creates a very efficient heat removal mechanism. It is typically ten times more effective than film condensation under the same temperature difference. Indeed, condensing heat transfer coefficients as high as 200,000 W/m^2K can be obtained with water at 1 atm. Were it possible to sustain dropwise condensation, we would certainly design equipment in such a way as to make use of it.

Unfortunately, laboratory experiments on dropwise condensation are almost always done on surfaces that have been prepared with oleic, stearic, or other fatty acids, or, more recently, with dioctadecyl disulphide. These nonwetting agents, or *promoters* as they are called, are discussed in [9.60, 9.61]. While promoters are normally impractical for industrial use, since they either wash away or oxidize, experienced plant engineers have sometimes added rancid butter through the cup valves of commercial condensers to get at least temporary dropwise condensation.

Finally, we note that the obvious tactic of coating the surface with a thin, nonwetting, polymer film (such as PTFE, or Teflon) adds just enough conduction resistance to reduce the overall heat transfer coefficient to a value similar to film condensation, fully defeating its purpose! (Sufficiently thin polymer layers have not been found to be durable.) Noble metals, such as gold, platinum, and palladium, can also be used as nonwetting coating, and they have sufficiently high thermal conductivity to avoid the problem encountered with polymeric coatings. For gold, however, the minimum effective coating thickness is about 0.2 μm, or about 1/8 Troy ounce per square meter [9.62]. Such coatings are far too expensive for the vast majority of technical applications.

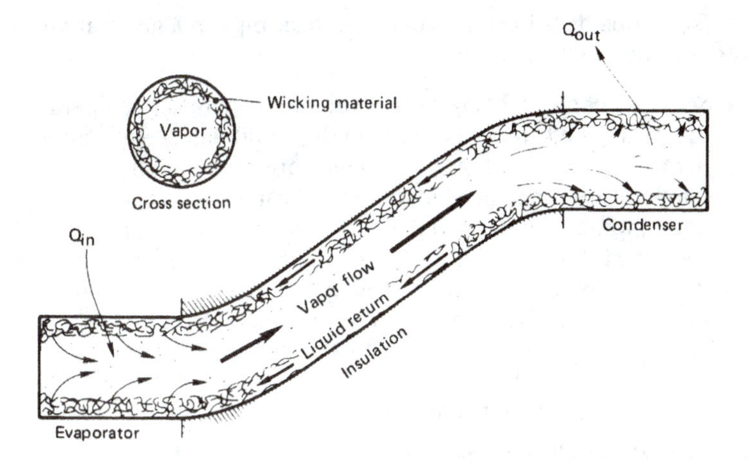

Figure 9.22 A typical heat pipe configuration.

9.10 The heat pipe

A *heat pipe* is a device that combines the high efficiencies of boiling and condensation. It is aptly named because it literally pipes heat from a hot region to a cold one.

The operation of a heat pipe is shown in Fig. 9.22. The pipe is a tube that can be bent or turned in any way that is convenient. The inside of the tube is lined with a layer of wicking material. The wick is wetted with an appropriate liquid. One end of the tube is exposed to a heat source that evaporates the liquid from the wick. The vapor then flows from the hot end of the tube to the cold end, where it is condensed. Capillary action moves the condensed liquid axially along the wick, back to the evaporator where it is again vaporized.

Placing a heat pipe between a hot region and a cold one is thus similar to connecting the regions with a material of extremely high thermal conductivity—potentially orders of magnitude higher than any solid material. Such devices are used not only for achieving high heat transfer rates between a source and a sink but for a variety of less obvious purposes. They are used, for example, to level out temperatures in systems, since they function almost isothermally and offer very little thermal resistance.

Design considerations in matching a heat pipe to a given application center on the following issues.

- *Selection of the right liquid.* The intended operating temperature of the heat pipe can be met only with a fluid whose saturation temperatures cover the design temperature range. Depending on the temperature range needed, the liquid can be a cryogen, an organic liquid, water, a liquid metal, or, in principle, almost any fluid. However, the following characteristics will serve to limit the vapor mass flow per watt, provide good capillary action in the wick, and control the temperature rise between the wall and the wick:

  i) High latent heat
  ii) High surface tension
  iii) Low liquid viscosities
  iv) High thermal conductivity

  Two liquids that meet these four criteria admirably are water and mercury, although toxicity and wetting problems discourage the use of the latter. Ammonia is useful at temperatures that are a bit too low for water. At high temperatures, sodium and lithium have good characteristics, while nitrogen is good for cryogenic temperatures. Fluids can be compared using the *merit number*, $M = h_{fg}\sigma/\nu_f$ (see Problem 9.36).

- *Selection of the tube material.* The tube material must be compatible with the working fluid. Gas generation and corrosion are particular considerations. Copper tubes are widely used with water, methanol, and acetone, but they cannot be used with ammonia. Stainless steel tubes can be used with ammonia and many liquid metals, but are not suitable for long term service with water. In some aerospace applications, aluminum is used for its low weight; however, it is compatible with working fluids other than ammonia.

- *Selection and installation of the wick.* Like the tube material, the wick material must be compatible with the working fluid. In addition, the working fluid must be able to wet the wick. Wicks can be fabricated from a metallic mesh, from a layer of sintered beads, or simply by scoring grooves along the inside surface of the tube. Many ingenious schemes have been created for bonding the wick to the inside of the pipe and keeping it at optimum porosity.

- *Operating limits of the heat pipe.* The heat transfer through a heat pipe is restricted by

  i) Viscous drag in the wick at low temperature
  ii) The sonic, or choking, speed of the vapor
  iii) Drag of the vapor on the counterflowing liquid in the wick
  iv) Ability of capillary forces in the wick to pump the liquid through the pressure rise between evaporator and condenser
  v) The boiling burnout heat flux in the evaporator section.

  These items much each be dealt with in detail during the design of a new heat pipe [9.63].

- *Control of the pipe performance.* Often a given heat pipe will be called upon to function over a range of conditions—under varying evaporator heat loads, for example. One way to vary its performance is through the introduction of a non-condensible gas in the pipe. This gas will collect at the condenser, limiting the area of the condenser that vapor can reach. By varying the amount of gas, the thermal resistance of the heat pipe can be controlled. In the absence of active control of the gas, an increase in the heat load at the evaporator will raise the pressure in the pipe, compressing the noncondensible gas and lowering the thermal resistance of the pipe. The result is that the temperature at the evaporator remains essentially constant even as the heat load rises as falls.

Heat pipes have proven useful in cooling high power-density electronic devices. The evaporator is located on a small electronic component to be cooled, perhaps a microprocessor, and the condenser is finned and cooled by a forced air flow (in a desktop or mainframe computer) or is unfinned and cooled by conduction into the exterior casing or structural frame (in a laptop computer). These applications rely on having a heat pipe with much larger condenser area than evaporator area. Thus, the heat fluxes on the condenser are kept relatively low. This facilitates such uncomplicated means for the ultimate heat disposal as using a small fan to blow air over the condenser. Typical heat pipe cooling systems for personal computer equipment are shown in Fig. 9.23.

The reader interested in designing or selecting a heat pipe will find a broad discussion of such devices in the book by Dunn and Reay [9.63].

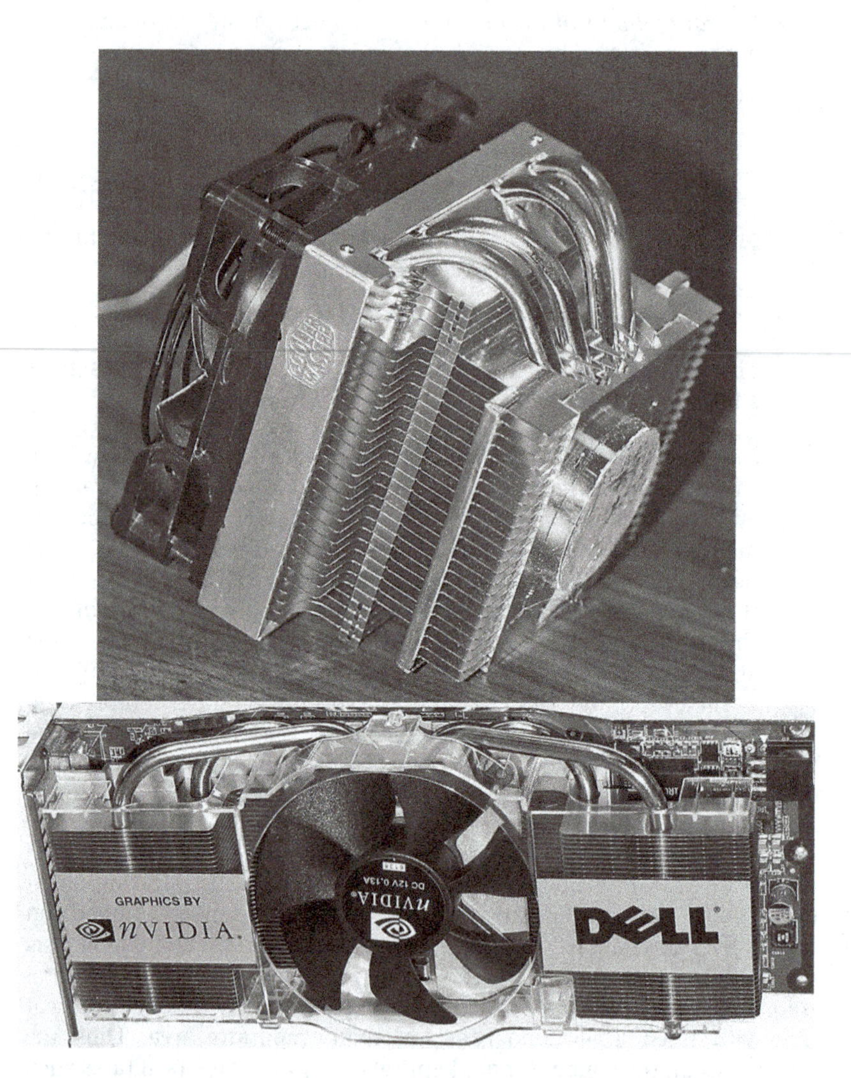

Figure 9.23 Two heat-pipe-cooled computer heat sinks. Top: A Cooler Master unit for cooling a CPU. The copper disk on the right affixes to the CPU. The visible heat pipes carry heat to the fin array which is cooled by the fan on the left. Bottom: A Dell *n*Vidia graphics card cooled by a copper block beneath the fan in the center. It feeds four heat pipes which carry heat to the fin arrays on either side of the fan in the center. Courtesy of Gene Quach, PC&C Computers, Houston, TX.

Problems

9.1 A large square tank with insulated sides has a copper base 1.27 cm thick. The base is heated to 650°C and saturated water is suddenly poured in the tank. Plot the temperature of the base as a function of time on the basis of Fig. 9.2 if the bottom of the base is insulated. In your graph, indicate the regimes of boiling and note the temperature at which cooling is most rapid.

9.2 Predict q_{max} for the two heaters in Fig. 9.3b. At what percentage of q_{max} is each one operating?

9.3 A very clean glass container of water at 70°C is depressurized until it is subcooled 30°C. Then it suddenly and explosively "flashes" (or boils). What is the pressure at which this happens? Approximately what diameter of gas bubble, or other disturbance in the liquid, caused it to flash?

9.4 Plot the unstable bubble radius as a function of liquid superheat for water at 1 atm. Comment on the significance of your curve.

9.5 In chemistry class you have probably witnessed the phenomenon of "bumping" in a test tube (the explosive boiling that blows the contents of the tube all over the ceiling). Yet you have never seen this happen in a kitchen pot. Explain why not.

9.6 Use van der Waal's equation of state to approximate the highest reduced temperature to which water can be superheated at low pressure. How many degrees of superheat does this suggest that water can sustain at the low pressure of 1 atm? (It turns out that this calculation is accurate within about 10%.) What would R_b be at this superheat?

9.7 Use Yamagata's equation, (9.3), to determine how nucleation site density increases with ΔT for Berenson's curves in Fig. 9.14. (That is, find c in the relation n = constant ΔT^c.)

9.8 Suppose that C_{sf} for a given surface is high by 50%. What will be the percentage error in q calculated for a given value of ΔT? [Low by 70%.]

9.9 Water at 100 atm boils on a nickel heater whose temperature is 6°C above T_{sat}. Find h and q.

9.10 Water boils on a large flat plate at 1 atm. Calculate q_{max} if the plate is operated on the surface of the moon (at $\frac{1}{6}$ of $g_{earth-normal}$). What would q_{max} be in a space vehicle experiencing 10^{-4} of $g_{earth-normal}$?

9.11 Water boils on a 0.002 m diameter horizontal copper wire. Plot, to scale, as much of the boiling curve on $\log q$ vs. $\log \Delta T$ coordinates as you can. The system is at 1 atm.

9.12 Redo Problem 9.11 for a 0.03 m diameter sphere in water at 10 atm.

9.13 Verify eqn. (9.17).

9.14 Make a sketch of the q vs. $(T_w - T_{sat})$ relation for a pool boiling process, and invent a graphical method for locating the points where h is maximum and minimum. What numerical values do you get for these two h's from Fig. 9.2?

9.15 A 2 mm diameter jet of methanol is directed normal to the center of a 1.5 cm diameter disk heater at 1 m/s. How many watts can safely be supplied by the heater?

9.16 Saturated water at 1 atm boils on a ½ cm diameter platinum rod. Estimate the temperature of the rod at burnout.

9.17 Plot $(T_w - T_{sat})$ and the quality x as a function of position x for the conditions in Example 9.9. Set $x = 0$ where $x = 0$ and end the plot where the quality reaches 80%.

9.18 Plot $(T_w - T_{sat})$ and the quality x as a function of position in an 8 cm I.D. pipe if 0.3 kg/s of water at 100°C passes through it and $q_w = 200,000$ W/m^2.

9.19 Use dimensional analysis to verify the form of eqn. (9.8).

9.20 Compare the peak heat flux calculated from the data given in Problem 5.6 with the appropriate prediction. [The prediction is within 11%.]

9.21 The Kandlikar correlation, eqn. (9.50a), can be adapted subcooled flow boiling, with $x = 0$ (region B in Fig. 9.19). Noting that $q_w = h_{\text{fb}}(T_w - T_{\text{sat}})$, show that

$$q_w = \left[1058\, h_{\text{lo}} F (G h_{fg})^{-0.7} (T_w - T_{\text{sat}})\right]^{1/0.3}$$

in subcooled flow boiling [9.47].

9.22 Verify eqn. (9.53) by repeating the analysis following eqn. (8.49) but using the b.c. $(\partial u/\partial y)_{y=\delta} = \tau_\delta/\mu$ in place of $(\partial u/\partial y)_{y=\delta} = 0$. Verify the statement involving eqn. (9.54).

9.23 A cool-water-carrying pipe 7 cm in outside diameter has an outside temperature of 40°C. Saturated steam at 80°C flows across it. Plot $\overline{h}_{\text{condensation}}$ over the range of Reynolds numbers $0 \leqslant \text{Re}_D \leqslant 10^6$. Do you get the value at $\text{Re}_D = 0$ that you would anticipate from Chapter 8?

9.24 (a) Suppose that you have pits of roughly 0.002 mm diameter in a metallic heater surface. At about what temperature might you expect water to boil on that surface if the pressure is 20 atm. (b) Measurements have shown that water at atmospheric pressure can be superheated about 200°C above its normal boiling point. Roughly how large an embryonic bubble would be needed to trigger nucleation in water in such a state.

9.25 Obtain the dimensionless functional form of the pool boiling q_{max} equation and the q_{max} equation for flow boiling on external surfaces, using dimensional analysis.

9.26 A chemist produces a nondegradable additive that will increase σ by a factor of ten for water at 1 atm. By what factor will the additive improve q_{max} during pool boiling on (a) infinite flat plates and (b) small horizontal cylinders? By what factor will it improve burnout in the flow of jet on a disk?

9.27 Steam at 1 atm is blown at 26 m/s over a 1 cm O.D. cylinder at 90°C. What is $\overline{h}$? Can you suggest any physical process within the cylinder that could sustain this temperature in this flow?

9.28 The water shown in Fig. 9.17 is at 1 atm, and the Nichrome heater can be approximated as nickel. What is $T_w - T_{\text{sat}}$?

9.29 For film boiling on horizontal cylinders, eqn. (9.6) is modified to

$$\lambda_d = 2\pi\sqrt{3}\left[\frac{g(\rho_f - \rho_g)}{\sigma} + \frac{2}{(\text{diam.})^2}\right]^{-1/2}.$$

If ρ_f is 748 kg/m^3 for saturated acetone, compare this λ_d, and the flat plate value, with Fig. 9.3d.

9.30 Water at 47°C flows through a 13 cm diameter thin-walled tube at 8 m/s. Saturated water vapor, at 1 atm, flows across the tube at 50 m/s. Evaluate T_{tube}, U, and q.

9.31 A 1 cm diameter thin-walled tube carries liquid metal through saturated water at 1 atm. The throughflow of metal is increased until burnout occurs. At that point the metal temperature is 250°C and h inside the tube is 9600 W/m^{2}K. What is the wall temperature at burnout?

9.32 At about what velocity of liquid metal flow does burnout occur in Problem 9.31 if the metal is mercury?

9.33 Explain, in physical terms, why eqns. (9.23) and (9.24), instead of differing by a factor of two, are almost equal. How do these equations change when H' is large?

9.34 A liquid enters the heated section of a pipe at a location $z = 0$ with a specific enthalpy $\hat{h}_{\text{in}}$. If the wall heat flux is q_w and the pipe diameter is D, show that the enthalpy a distance $z = L$ downstream is

$$\hat{h} = \hat{h}_{\text{in}} + \frac{\pi D}{\dot{m}}\int_0^L q_w\, dz.$$

Since the quality may be defined as $x \equiv (\hat{h} - \hat{h}_{f,\text{sat}})/h_{fg}$, show that for constant q_w

$$x = \frac{\hat{h}_{\text{in}} - \hat{h}_{f,\text{sat}}}{h_{fg}} + \frac{4q_w L}{GD}$$

9.35 Consider again the x-ray monochrometer described in Problem 7.44. Suppose now that the mass flow rate of liquid nitrogen is 0.023 kg/s, that the nitrogen is saturated at 110 K when it enters the heated section, and that the passage horizontal. Estimate the quality and the wall temperature at end of the heated section if $F = 4.70$ for nitrogen in eqns. (9.50). As before, assume the silicon to conduct well enough that the heat load is distributed uniformly over the surface of the passage.

9.36 Use data from Appendix A and Sect. 9.1 to calculate the merit number, M, for the following potential heat-pipe working fluids over the range 200 K to 600 K in 100 K increments: water, mercury, methanol, ammonia, and HCFC-22. If data are unavailable for a fluid in some range, indicate so. What fluids are best suited for particular temperature ranges?

References

[9.1] S. Nukiyama. The maximum and minimum values of the heat q transmitted from metal to boiling water under atmospheric pressure. *J. Jap. Soc. Mech. Eng.*, 37:367–374, 1934. (transl.: *Int. J. Heat Mass Transfer*, vol. 9, 1966, pp. 1419–1433).

[9.2] T. B. Drew and C. Mueller. Boiling. *Trans. AIChE*, 33:449, 1937.

[9.3] International Association for the Properties of Water and Steam. Release on surface tension of ordinary water substance. Technical report, September 1994. Available from the Executive Secretary of IAPWS or on the internet: http://www.iapws.org/.

[9.4] J. J. Jasper. The surface tension of pure liquid compounds. *J. Phys. Chem. Ref. Data*, 1(4):841–1010, 1972.

[9.5] M. Okado and K. Watanabe. Surface tension correlations for several fluorocarbon refrigerants. *Heat Transfer: Japanese Research*, 17 (1):35–52, 1988.

[9.6] A. P. Fröba, S. Will, and A. Leipertz. Saturated liquid viscosity and surface tension of alternative refrigerants. *Intl. J. Thermophys.*, 21 (6):1225–1253, 2000.

[9.7] V.G. Baidakov and I.I. Sulla. Surface tension of propane and isobutane at near-critical temperatures. *Russ. J. Phys. Chem.*, 59(4):551–554, 1985.

[9.8] P.O. Binney, W.-G. Dong, and J. H. Lienhard. Use of a cubic equation to predict surface tension and spinodal limits. *J. Heat Transfer*, 108(2):405–410, 1986.

[9.9] Y. Y. Hsu. On the size range of active nucleation cavities on a heating surface. *J. Heat Transfer, Trans. ASME, Ser. C*, 84:207–216, 1962.

[9.10] G. F. Hewitt. Boiling. In W. M. Rohsenow, J. P. Hartnett, and Y. I. Cho, editors, *Handbook of Heat Transfer*, chapter 15. McGraw-Hill, New York, 3rd edition, 1998.

[9.11] K. Yamagata, F. Hirano, K. Nishiwaka, and H. Matsuoka. Nucleate boiling of water on the horizontal heating surface. *Mem. Fac. Eng. Kyushu*, 15:98, 1955.

[9.12] W. M. Rohsenow. A method of correlating heat transfer data for surface boiling of liquids. *Trans. ASME*, 74:969, 1952.

[9.13] I. L. Pioro. Experimental evaluation of constants for the Rohsenow pool boiling correlation. *Int. J. Heat. Mass Transfer*, 42:2003–2013, 1999.

[9.14] R. Bellman and R. H. Pennington. Effects of surface tension and viscosity on Taylor instability. *Quart. Appl. Math.*, 12:151, 1954.

[9.15] V. Sernas. Minimum heat flux in film boiling—a three dimensional model. In *Proc. 2nd Can. Cong. Appl. Mech.*, pages 425–426, Canada, 1969.

[9.16] H. Lamb. *Hydrodynamics.* Dover Publications, Inc., New York, 6th edition, 1945.

[9.17] N. Zuber. Hydrodynamic aspects of boiling heat transfer. AEC Report AECU-4439, Physics and Mathematics, 1959.

[9.18] J. H. Lienhard and V. K. Dhir. Extended hydrodynamic theory of the peak and minimum pool boiling heat fluxes. NASA CR-2270, July 1973.

[9.19] J. H. Lienhard, V. K. Dhir, and D. M. Riherd. Peak pool boiling heat-flux measurements on finite horizontal flat plates. *J. Heat Transfer, Trans. ASME, Ser. C*, 95:477–482, 1973.

[9.20] J. H. Lienhard and V. K. Dhir. Hydrodynamic prediction of peak pool-boiling heat fluxes from finite bodies. *J. Heat Transfer, Trans. ASME, Ser. C*, 95:152–158, 1973.

[9.21] S. S. Kutateladze. On the transition to film boiling under natural convection. *Kotloturbostroenie*, (3):10, 1948.

[9.22] K. H. Sun and J. H. Lienhard. The peak pool boiling heat flux on horizontal cylinders. *Int. J. Heat Mass Transfer*, 13:1425–1439, 1970.

[9.23] J. S. Ded and J. H. Lienhard. The peak pool boiling heat flux from a sphere. *AIChE J.*, 18(2):337–342, 1972.

[9.24] A. L. Bromley. Heat transfer in stable film boiling. *Chem. Eng. Progr.*, 46:221–227, 1950.

[9.25] P. Sadasivan and J. H. Lienhard. Sensible heat correction in laminar film boiling and condensation. *J. Heat Transfer, Trans. ASME*, 109: 545–547, 1987.

[9.26] V. K. Dhir and J. H. Lienhard. Laminar film condensation on plane and axi-symmetric bodies in non-uniform gravity. *J. Heat Transfer, Trans. ASME, Ser. C*, 93(1):97–100, 1971.

[9.27] P. Pitschmann and U. Grigull. Filmverdampfung an waagerechten zylindern. *Wärme- und Stoffübertragung*, 3:75–84, 1970.

[9.28] J. E. Leonard, K. H. Sun, and G. E. Dix. Low flow film boiling heat transfer on vertical surfaces: Part II: Empirical formulations and application to BWR-LOCA analysis. In *Proc. ASME-AIChE Natl. Heat Transfer Conf.* St. Louis, August 1976.

[9.29] J. W. Westwater and B. P. Breen. Effect of diameter of horizontal tubes on film boiling heat transfer. *Chem. Eng. Progr.*, 58:67–72, 1962.

[9.30] P. J. Berenson. Transition boiling heat transfer from a horizontal surface. M.I.T. Heat Transfer Lab. Tech. Rep. 17, 1960.

[9.31] J. H. Lienhard and P. T. Y. Wong. The dominant unstable wavelength and minimum heat flux during film boiling on a horizontal cylinder. *J. Heat Transfer, Trans. ASME, Ser. C*, 86:220–226, 1964.

[9.32] L. C. Witte and J. H. Lienhard. On the existence of two transition boiling curves. *Int. J. Heat Mass Transfer*, 25:771–779, 1982.

[9.33] J. H. Lienhard and L. C. Witte. An historical review of the hydrodynamic theory of boiling. *Revs. in Chem. Engr.*, 3(3):187–280, 1985.

[9.34] J. R. Ramilison and J. H. Lienhard. Transition boiling heat transfer and the film transition region. *J. Heat Transfer*, 109, 1987.

[9.35] J. M. Ramilison, P. Sadasivan, and J. H. Lienhard. Surface factors influencing burnout on flat heaters. *J. Heat Transfer*, 114(1):287–290, 1992.

[9.36] A. E. Bergles and W. M. Rohsenow. The determination of forced-convection surface-boiling heat transfer. *J. Heat Transfer, Trans. ASME, Series C*, 86(3):365–372, 1964.

[9.37] E. J. Davis and G. H. Anderson. The incipience of nucleate boiling in forced convection flow. *AIChE J.*, 12:774–780, 1966.

[9.38] K. Kheyrandish and J. H. Lienhard. Mechanisms of burnout in saturated and subcooled flow boiling over a horizontal cylinder. In *Proc. ASME–AIChE Nat. Heat Transfer Conf.* Denver, Aug. 4–7 1985.

[9.39] A. Sharan and J. H. Lienhard. On predicting burnout in the jet-disk configuration. *J. Heat Transfer*, 107:398–401, 1985.

[9.40] A. L. Bromley, N. R. LeRoy, and J. A. Robbers. Heat transfer in forced convection film boiling. *Ind. Eng. Chem.*, 45(12):2639–2646, 1953.

[9.41] L. C. Witte. Film boiling from a sphere. *Ind. Eng. Chem. Fundamentals*, 7(3):517–518, 1968.

[9.42] L. C. Witte. External flow film boiling. In S. G. Kandlikar, M. Shoji, and V. K. Dhir, editors, *Handbook of Phase Change: Boiling and Condensation*, chapter 13, pages 311-330. Taylor & Francis, Philadelphia, 1999.

[9.43] J. G. Collier and J. R. Thome. *Convective Boiling and Condensation.* Oxford University Press, Oxford, 3rd edition, 1994.

[9.44] J. C. Chen. A correlation for boiling heat transfer to saturated fluids in convective flow. ASME Prepr. 63-HT-34, 5th ASME-AIChE Heat Transfer Conf. Boston, August 1963.

[9.45] S. G. Kandlikar. A general correlation for saturated two-phase flow boiling heat transfer inside horizontal and vertical tubes. *J. Heat Transfer*, 112(1):219–228, 1990.

[9.46] D. Steiner and J. Taborek. Flow boiling heat transfer in vertical tubes correlated by an asymptotic model. *Heat Transfer Engr.*, 13 (2):43–69, 1992.

[9.47] S. G. Kandlikar and H. Nariai. Flow boiling in circular tubes. In S. G. Kandlikar, M. Shoji, and V. K. Dhir, editors, *Handbook of Phase Change: Boiling and Condensation*, chapter 15, pages 367–402. Taylor & Francis, Philadelphia, 1999.

[9.48] V. E. Schrock and L. M. Grossman. Forced convection boiling in tubes. *Nucl. Sci. Engr.*, 12:474–481, 1962.

[9.49] M. M. Shah. Chart correlation for saturated boiling heat transfer: equations and further study. *ASHRAE Trans.*, 88:182–196, 1982.

[9.50] A. E. Gungor and R. S. H. Winterton. Simplified general correlation for flow boiling heat transfer inside horizontal and vertical tubes. *Chem. Engr. Res. Des.*, 65:148–156, 1987.

[9.51] S. G. Kandlikar, S. T. Tian, J. Yu, and S. Koyama. Further assessment of pool and flow boiling heat transfer with binary mixtures. In G. P. Celata, P. Di Marco, and R. K. Shah, editors, *Two-Phase Flow Modeling and Experimentation*. Edizioni ETS, Pisa, 1999.

[9.52] Y. Taitel and A. E. Dukler. A model for predicting flow regime transitions in horizontal and near horizontal gas-liquid flows. *AIChE J.*, 22(1):47–55, 1976.

[9.53] A. E. Dukler and Y. Taitel. Flow pattern transitions in gas–liquid systems measurement and modelling. In J. M. Delhaye, N. Zuber, and G. F. Hewitt, editors, *Advances in Multi-Phase Flow*, volume II. Hemisphere/McGraw-Hill, New York, 1985.

[9.54] G. F. Hewitt. Burnout. In G. Hetsroni, editor, *Handbook of Multiphase Systems*, chapter 6, pages 66–141. McGraw-Hill, New York, 1982.

[9.55] Y. Katto. A generalized correlation of critical heat flux for the forced convection boiling in vertical uniformly heated round tubes. *Int. J. Heat Mass Transfer*, 21:1527–1542, 1978.

[9.56] Y. Katto and H. Ohne. An improved version of the generalized correlation of critical heat flux for convective boiling in uniformly heated vertical tubes. *Int. J. Heat. Mass Transfer*, 27(9):1641–1648, 1984.

[9.57] P. B. Whalley. *Boiling, Condensation, and Gas-Liquid Flow.* Oxford University Press, Oxford, 1987.

[9.58] B. Chexal, J. Horowitz, G. McCarthy, M. Merilo, J.-P. Sursock, J. Harrison, C. Peterson, J. Shatford, D. Hughes, M. Ghiaasiaan, V.K. Dhir, W. Kastner, and W. Köhler. Two-phase pressure drop technology for design and analysis. Tech. Rept. 113189, Electric Power Research Institute, Palo Alto, CA, August 1999.

[9.59] I. G. Shekriladze and V. I. Gomelauri. Theoretical study of laminar film condensation of flowing vapour. *Int. J. Heat. Mass Transfer*, 9:581–591, 1966.

[9.60] P. J. Marto. Condensation. In W. M. Rohsenow, J. P. Hartnett, and Y. I. Cho, editors, *Handbook of Heat Transfer*, chapter 14. McGraw-Hill, New York, 3rd edition, 1998.

[9.61] J. Rose, Y. Utaka, and I. Tanasawa. Dropwise condensation. In S. G. Kandlikar, M. Shoji, and V. K. Dhir, editors, *Handbook of Phase Change: Boiling and Condensation*, chapter 20. Taylor & Francis, Philadelphia, 1999.

[9.62] D. W. Woodruff and J. W. Westwater. Steam condensation on electroplated gold: effect of plating thickness. *Int. J. Heat. Mass Transfer*, 22:629–632, 1979.

[9.63] P. D. Dunn and D. A. Reay. *Heat Pipes.* Pergamon Press Ltd., Oxford, UK, 4th edition, 1994.

Part IV

Thermal Radiation Heat Transfer

10. Radiative heat transfer

> *The sun that shines from Heaven shines but warm,*
> *And, lo, I lie between that sun and thee:*
> *The heat I have from thence doth little harm,*
> *Thine eye darts forth the fire that burneth me:*
> *And were I not immortal, life were done*
> *Between this heavenly and earthly sun.*
> ***Venus and Adonis,*** **Wm. Shakespeare, 1593**

10.1 The problem of radiative exchange

Chapter 1 included the elementary mechanisms of heat radiation. Before we proceed, you should reflect upon what you remember about the following key ideas from Chapter 1:

- Electromagnetic wave spectrum
- Heat radiation & infrared radiation
- Black body
- Absorptance, α
- Reflectance, ρ
- Transmittance, τ
- $\alpha + \rho + \tau = 1$
- $e(T)$ and $e_\lambda(T)$ for black bodies
- The Stefan-Boltzmann law
- Wien's law & Planck's law
- Radiant heat exchange
- Configuration factor, $F_{1\text{-}2}$
- Emittance, ε
- Transfer factor, $\mathcal{F}_{1\text{-}2}$
- Radiation shielding

The additional concept of a radiation heat transfer coefficient was developed in Section 2.3. We take these concepts for granted in what follows.

The heat exchange problem

Figure 10.1 shows two arbitrary surfaces radiating energy to one another. The net heat exchange, Q_{net}, from the hotter surface (1) to the cooler

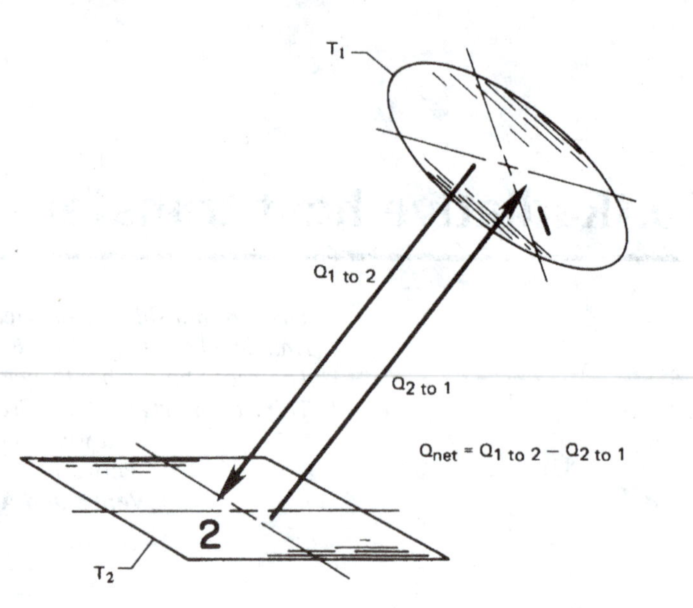

Figure 10.1 Thermal radiation between two arbitrary surfaces.

surface (2) depends on the following influences:

- T_1 and T_2.
- The areas of (1) and (2), A_1 and A_2.
- The shape, orientation, and spacing of (1) and (2).
- The radiative properties of the surfaces.
- Additional surfaces in the environment, which may reflect radiation from surface (1) to surface (2)
- The medium between (1) and (2) if it absorbs, emits, or "reflects" radiation. (When the medium is air, we can usually neglect these effects.)

If surfaces (1) and (2) are black, if they are surrounded by air, and if no heat flows between them by conduction or convection, then only the first three considerations are involved in determining Q_{net}. We saw some elementary examples of how this could be done in Chapter 1, leading to

$$Q_{\text{net}} = A_1 F_{1\text{-}2}\, \sigma\left(T_1^4 - T_2^4\right) \tag{10.1}$$

The last three considerations complicate the problem considerably. In Chapter 1, we saw that these nonideal factors are sometimes included in a *transfer factor* $\mathcal{F}_{1\text{-}2}$, such that

$$Q_{\text{net}} = A_1 \mathcal{F}_{1\text{-}2}\, \sigma\left(T_1^4 - T_2^4\right) \tag{10.2}$$

Before we undertake the problem of evaluating heat exchange among real bodies, we need several definitions.

Some definitions

Emittance. A real body at temperature T does not emit with the black body emissive power $e_b = \sigma T^4$ but rather with some fraction, ε, of e_b. The same is true of the monochromatic emissive power, $e_\lambda(T)$, which is always lower for a real body than the black body value given by Planck's law, eqn. (1.30). Thus, we define either the monochromatic emittance, ε_λ:

$$\varepsilon_\lambda \equiv \frac{e_\lambda(\lambda, T)}{e_{\lambda_b}(\lambda, T)} \tag{10.3}$$

or the total emittance, ε:

$$\varepsilon \equiv \frac{e(T)}{e_b(T)} = \frac{\displaystyle\int_0^\infty e_\lambda(\lambda, T)\, d\lambda}{\sigma T^4} = \frac{\displaystyle\int_0^\infty \varepsilon_\lambda\, e_{\lambda_b}(\lambda, T)\, d\lambda}{\sigma T^4} \tag{10.4}$$

For real bodies, both ε and ε_λ are greater than zero and less than one; for black bodies, $\varepsilon = \varepsilon_\lambda = 1$. The emittance is determined entirely by the properties of the surface of the particular body and its temperature. It is independent of the environment of the body.

Table 10.1 lists typical values of the total emittance for a variety of substances. Notice that most metals have quite low emittances, unless they are oxidized. Most nonmetals have emittances that are quite high—approaching the black body limit of one.

One particular kind of surface behavior is that for which ε_λ is independent of λ. We call such a surface a *gray body.* The monochromatic emissive power, $e_\lambda(T)$, for a gray body is a constant fraction, ε, of $e_{b_\lambda}(T)$, as indicated in the inset of Fig. 10.2. In other words, for a gray body, $\varepsilon_\lambda = \varepsilon$.

Table 10.1 Total emittances for a variety of surfaces [10.1]

Metals			*Nonmetals*		
Surface	*Temp.* (°C)	ε	*Surface*	*Temp.* (°C)	ε
Aluminum			Asbestos	40	0.93-0.97
Polished, 98% pure	200–600	0.04-0.06	Brick		
Commercial sheet	90	0.09	Red, rough	40	0.93
Heavily oxidized	90–540	0.20-0.33	Silica	980	0.80-0.85
Brass			Fireclay	980	0.75
Highly polished	260	0.03	Ordinary refractory	1090	0.59
Dull plate	40–260	0.22	Magnesite refractory	980	0.38
Oxidized	40–260	0.46-0.56	White refractory	1090	0.29
Copper			Carbon		
Highly polished electrolytic	90	0.02	Filament	1040–1430	0.53
Slightly polished to dull	40	0.12-0.15	Lampsoot	40	0.95
Black oxidized	40	0.76	Concrete, rough	40	0.94
Gold: pure, polished	90–600	0.02-0.035	Glass		
Iron and steel			Smooth	40	0.94
Mild steel, polished	150–480	0.14-0.32	Quartz glass (2 mm)	260–540	0.96-0.66
Steel, polished	40–260	0.07-0.10	Pyrex	260–540	0.94-0.74
Sheet steel, rolled	40	0.66	Gypsum	40	0.80-0.90
Sheet steel, strong rough oxide	40	0.80	Ice	0	0.97-0.98
Cast iron, oxidized	40–260	0.57-0.66	Limestone	400–260	0.95-0.83
Iron, rusted	40	0.61-0.85	Marble	40	0.93-0.95
Wrought iron, smooth	40	0.35	Mica	40	0.75
Wrought iron, dull oxidized	20–360	0.94	Paints		
Stainless, polished	40	0.07-0.17	Black gloss	40	0.90
Stainless, after repeated heating	230–900	0.50-0.70	White paint	40	0.89-0.97
			Lacquer	40	0.80-0.95
Lead			Various oil paints	40	0.92-0.96
Polished	40–260	0.05-0.08	Red lead	90	0.93
Oxidized	40–200	0.63	Paper		
Mercury: pure, clean	40–90	0.10-0.12	White	40	0.95-0.98
Platinum			Other colors	40	0.92-0.94
Pure, polished plate	200–590	0.05-0.10	Roofing	40	0.91
Oxidized at 590°C	260–590	0.07-0.11	Plaster, rough lime	40–260	0.92
Drawn wire and strips	40–1370	0.04-0.19	Quartz	100–1000	0.89-0.58
Silver	200	0.01-0.04	Rubber	40	0.86-0.94
Tin	40–90	0.05	Snow	10–20	0.82
Tungsten			Water, thickness ≥0.1 mm	40	0.96
Filament	540–1090	0.11-0.16	Wood	40	0.80-0.90
Filament	2760	0.39	Oak, planed	20	0.90

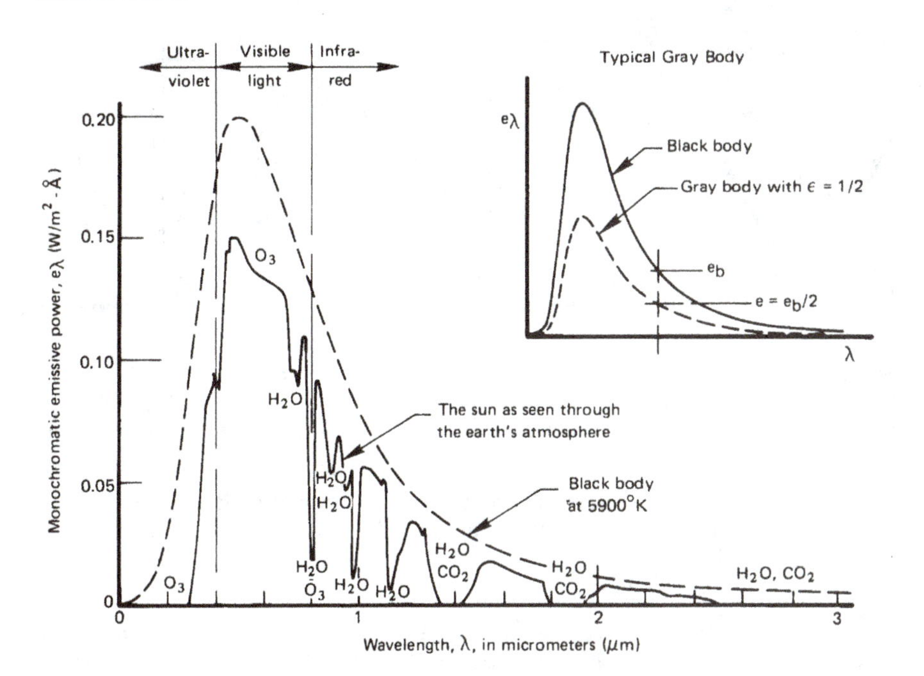

Figure 10.2 Comparison of the sun's energy as typically seen through the earth's atmosphere with that of a black body having the same mean temperature, size, and distance from the earth. (Notice that e_λ, just outside the earth's atmosphere, is far less than on the surface of the sun because the radiation has spread out over a much greater area.)

No real body is gray, but many exhibit approximately gray behavior. We see in Fig. 10.2, for example, that the sun appears to us on earth as an approximately gray body with an emittance of approximately 0.6. Some materials—for example, copper, aluminum oxide, and certain paints—are actually pretty close to being gray surfaces at normal temperatures.

Yet the emittance of most common materials and coatings varies with wavelength in the thermal range. The total emittance accounts for this behavior at a particular temperature. By using it, we can write the emissive power as if the body were gray, without integrating over wavelength:

$$e(T) = \varepsilon \sigma T^4 \tag{10.5}$$

We shall use this type of "gray body approximation" often in this chapter.

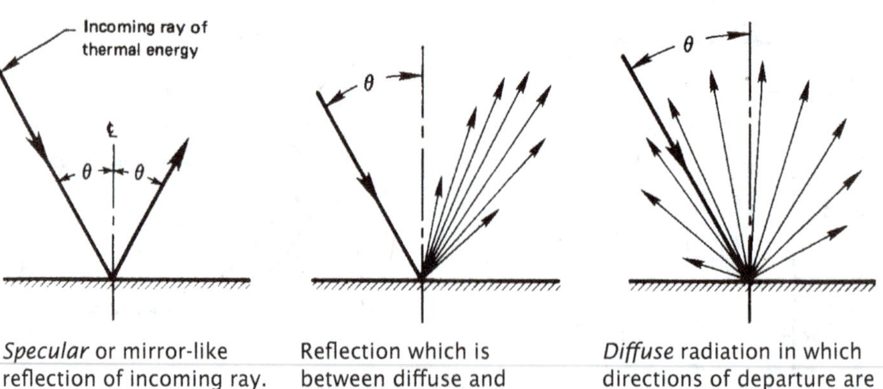

Figure 10.3 Specular and diffuse reflection of radiation. (Arrows indicate magnitude of the heat flux in the directions indicated.)

In situations where surfaces at very different temperatures are involved, the wavelength dependence of ε_λ must be dealt with explicitly. This occurs, for example, when sunlight heats objects here on earth. Solar radiation (from a high temperature source) is on visible wavelengths, whereas radiation from low temperature objects on earth is mainly in the infrared range. We look at this issue further in the next section.

Diffuse and specular emittance and reflection. The energy emitted by a non-black surface, together with that portion of an incoming ray of energy that is reflected by the surface, may leave the body *diffusely* or *specularly*, as shown in Fig. 10.3. That energy may also be emitted or reflected in a way that lies between these limits. A mirror reflects visible radiation in an almost perfectly *specular* fashion. (The "reflection" of a billiard ball as it rebounds from the side of a pool table is also specular.) When reflection or emission is diffuse, there is no preferred direction for outgoing rays. Black body emission is always diffuse.

The character of the emittance or reflectance of a surface will normally change with the wavelength of the radiation. If we take account of both directional and spectral characteristics, then properties like emittance and reflectance depend on wavelength, temperature, and angles of incidence and/or departure. In this chapter, we shall assume diffuse behavior for most surfaces. This approximation works well for many problems in engineering, in part because most tabulated spectral and total emittances have been averaged over all angles (in which case they are properly called *hemispherical* properties).

Experiment 10.1

Obtain a flashlight with as narrow a spot focus as you can find. Direct it at an angle onto a mirror, onto the surface of a bowl filled with sugar, and onto a variety of other surfaces, all in a darkened room. In each case, move the palm of your hand around the surface of an imaginary hemisphere centered on the point where the spot touches the surface. Notice how your palm is illuminated, and categorize the kind of reflectance of each surface—at least in the range of visible wavelengths.

Intensity of radiation. To account for the effects of geometry on radiant exchange, we must think about how angles of orientation affect the radiation between surfaces. Consider radiation from a circular surface element, dA, as shown at the top of Fig. 10.4. If the element is black, the radiation that it emits is indistinguishable from that which would be emitted from a black cavity at the same temperature, and that radiation is diffuse — the same in all directions. If it were non-black but diffuse, the heat flux leaving the surface would again be independent of direction. Thus, the rate at which energy is emitted in any direction from this diffuse element is proportional to the projected area of dA normal to the direction of view, as shown in the upper right side of Fig. 10.4.

If an aperture of area dA_a is placed at a radius r and angle θ from dA and is normal to the radius, it will see dA as having an area $\cos\theta\, dA$. The energy dA_a receives will depend on the solid angle,[1] $d\omega$, it subtends. Radiation that leaves dA within the solid angle $d\omega$ stays within $d\omega$ as it travels to dA_a. Hence, we define a quantity called the *intensity of radiation*, i (W/m$^2\cdot$steradian) using an energy conservation statement:

$$dQ_{\text{outgoing}} = (i\,d\omega)(\cos\theta\,dA) = \begin{cases}\text{radiant energy from } dA\\ \text{that is intercepted by } dA_a\end{cases} \tag{10.6}$$

[1]The unit of solid angle is the *steradian*. One steradian is the solid angle subtended by a spherical segment whose area equals the square of its radius. A full sphere therefore subtends $4\pi r^2/r^2 = 4\pi$ steradians. The aperture dA_a subtends $d\omega = dA_a/r^2$.

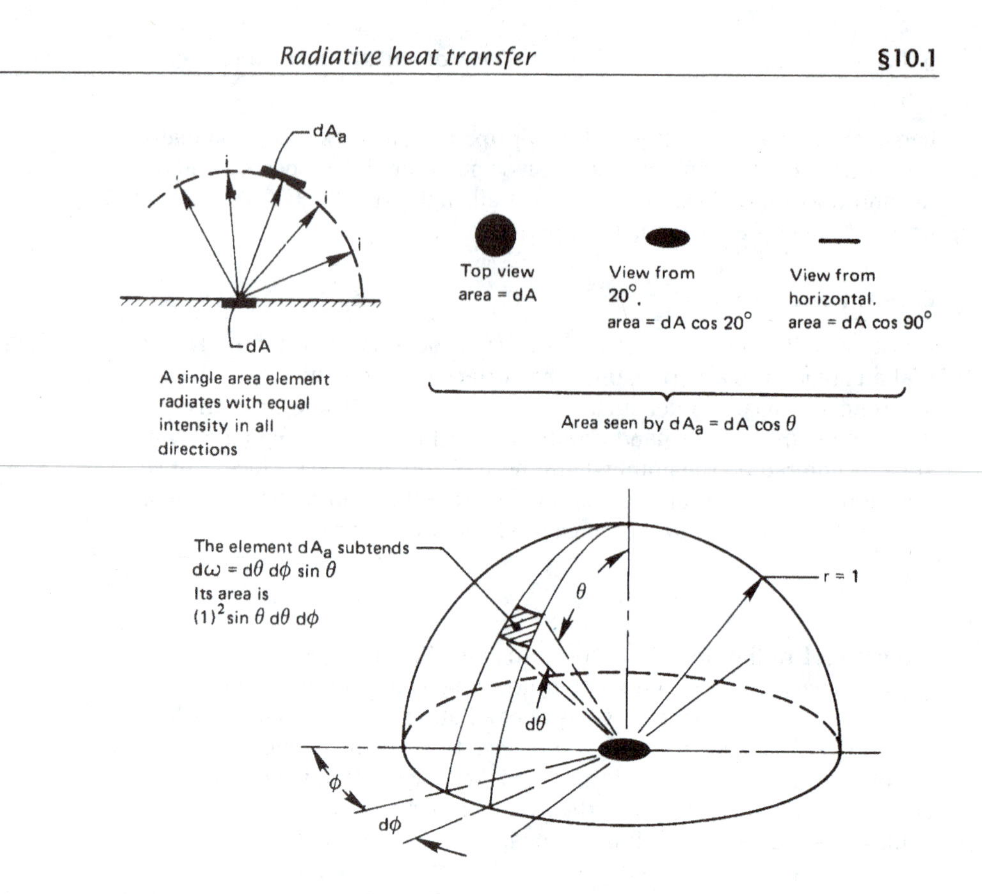

Figure 10.4 Radiation intensity through a unit sphere.

Notice that while the heat flux from dA decreases with θ (as indicated on the right side of Fig. 10.4), the intensity of radiation from a diffuse surface is uniform in all directions.

Finally, we we use our definition of i to express it in terms of the heat flux. from dA. We do this by dividing eqn. (10.6) by dA and integrating over the entire hemisphere. For convenience we set $r = 1$, and we note (see Fig. 10.4) that $d\omega = \sin\theta\, d\theta d\phi$.

$$q_{\text{outgoing}} = \int_{\phi=0}^{2\pi} \int_{\theta=0}^{\pi/2} i \cos\theta\, (\sin\theta\, d\theta d\phi) = \pi i \tag{10.7a}$$

In the particular case of a black body,

$$i_b = \frac{e_b}{\pi} = \frac{\sigma T^4}{\pi} = \text{fn}\,(T \text{ only}) \tag{10.7b}$$

For a given wavelength, we likewise define the monochromatic intensity

$$i_\lambda = \frac{e_\lambda}{\pi} = \text{fn}\,(T, \lambda) \tag{10.7c}$$

10.2 Kirchhoff's law

The problem of predicting α

The total emittance, ε, of a surface is determined only by the physical properties and temperature of that surface, as can be seen from eqn. (10.4). The total absorptance, α, on the other hand, depends on the *source* from which the surface absorbs radiation, as well as the surface's own characteristics. This happens because the surface may absorb some wavelengths better than others. Thus, the total absorptance will depend on the way that incoming radiation is distributed in wavelength. And that distribution, in turn, depends on the temperature and physical properties of the surface or surfaces from which radiation is absorbed.

The total absorptance α thus depends on the physical properties and temperatures of *all* bodies involved in the heat exchange process. Kirchhoff's law[2] is an expression that allows us to determine α under certain restrictions.

Kirchhoff's law

Kirchhoff's law is a relationship between the monochromatic, directional emittance and the monochromatic, directional absorptance for a surface that is in thermodynamic equilibrium with its surroundings

$$\boxed{\varepsilon_\lambda\,(T, \theta, \phi) = \alpha_\lambda\,(T, \theta, \phi)} \quad \text{exact form of Kirchhoff's law} \tag{10.8a}$$

Kirchhoff's law states that a body in thermodynamic equilibrium emits as much energy as it absorbs in each direction and at each wavelength. If this were not so, for example, a body might absorb more energy than it emits in one direction, θ_1, and might also emit more than it absorbs in another direction, θ_2. The body would thus pump heat out of its surroundings from the first direction, θ_1, and into its surroundings in the second

[2]Gustav Robert Kirchhoff (1824–1887) developed important new ideas in electrical circuit theory, thermal physics, spectroscopy, and astronomy. He formulated this particular "Kirchhoff's Law" when he was only 25. He and Robert Bunsen (inventor of the Bunsen burner) subsequently went on to do significant work on radiation from gases.

direction, θ_2. Since whatever matter lies in the first direction would be refrigerated without any work input, the Second Law of Thermodynamics would be violated. Similar arguments can be built for the wavelength dependence. In essence, then, Kirchhoff's law is a consequence of the laws of thermodynamics.

For a diffuse body, the emittance and absorptance do not depend on the angles, and Kirchhoff's law becomes

$$\boxed{\varepsilon_\lambda(T) = \alpha_\lambda(T)} \qquad \text{diffuse form of Kirchhoff's law} \tag{10.8b}$$

If, in addition, the body is gray, Kirchhoff's law is further simplified

$$\boxed{\varepsilon(T) = \alpha(T)} \qquad \text{diffuse, gray form of Kirchhoff's law} \tag{10.8c}$$

Equation (10.8c) is the most widely used form of Kirchhoff's law. Yet, it is a somewhat dangerous result, since many surfaces are not even approximately gray. If radiation is emitted on wavelengths much different from those that are absorbed, then a non-gray surface's variation of ε_λ and α_λ with wavelength will matter, as we discuss next.

Total absorptance during radiant exchange

Let us restrict our attention to diffuse surfaces, so that eqn. (10.8b) is the appropriate form of Kirchhoff's law. Consider two plates as shown in Fig. 10.5. Let the plate at T_1 be non-black and that at T_2 be black. Then net heat transfer from plate 1 to plate 2 is the difference between what plate 1 emits and what it absorbs. Since all the radiation reaching plate 1 comes from a black source at T_2, we may write

$$q_{\text{net}} = \underbrace{\int_0^\infty \varepsilon_{\lambda_1}(T_1)\, e_{\lambda_b}(T_1)\, d\lambda}_{\text{emitted by plate 1}} - \underbrace{\int_0^\infty \alpha_{\lambda_1}(T_1)\, e_{\lambda_b}(T_2)\, d\lambda}_{\text{radiation from plate 2 absorbed by plate 1}} \tag{10.9}$$

From eqn. (10.4), we may write the first integral in terms of total emittance, as $\varepsilon_1 \sigma T_1^4$. We *define* the total absorptance, $\alpha_1(T_1, T_2)$, as the second integral divided by σT_2^4. Hence,

$$q_{\text{net}} = \underbrace{\varepsilon_1(T_1)\sigma T_1^4}_{\text{emitted by plate 1}} - \underbrace{\alpha_1(T_1, T_2)\sigma T_2^4}_{\text{absorbed by plate 1}} \tag{10.10}$$

We see that the total absorptance depends on T_2, as well as T_1.

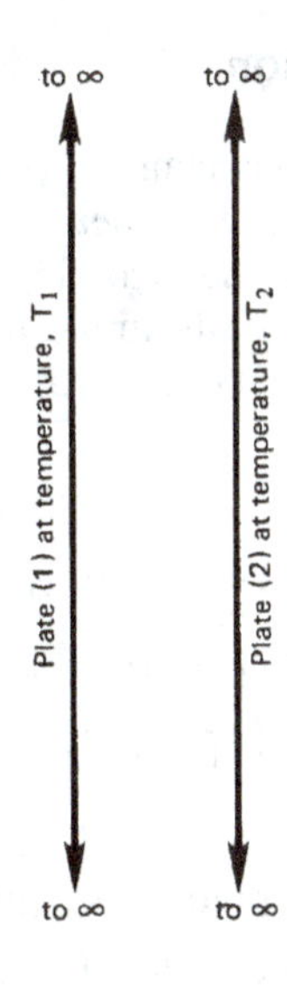

Figure 10.5 Heat transfer between two infinite parallel plates.

Why does total absorptance depend on both temperatures? It depends on T_1 simply because α_{λ_1} is a property of plate 1 that may be temperature dependent. It depends on T_2 because the spectrum of radiation from plate 2 depends on the temperature of plate 2 according to Planck's law, as we saw in Fig. 1.15.

As a typical example, consider solar radiation incident on a warm roof, painted black. From Table 10.1, we see that ε is on the order of 0.94. It turns out that α is just about the same. If we repaint the roof white, ε will not change noticeably. However, much of the energy arriving from the sun is carried in visible wavelengths, owing to the sun's very high temperature (about 5800 K).[3] Our eyes tell us that white paint reflects sunlight very strongly in these wavelengths, and indeed this is the case — 80 to 90% of the sunlight is reflected. The absorptance of white paint to energy from the sun is only 0.1 to 0.2 — much less than ε for the energy it emits, which is mainly at infrared wavelengths. For both paints, eqn. (10.8b) applies. However, in this situation, eqn. (10.8c) is only accurate for the black paint.

[3]Ninety percent of the sun's energy is on wavelengths between 0.33 and 2.2 μm (see Figure 10.2). For a black object at 300 K, 90% of the radiant energy is between 6.3 and 42 μm, in the infrared. This fact is at the heart of the "greenhouse effect."

The gray body approximation

Let us consider our facing plates again. If plate 1 is painted with white paint, and plate 2 is at a temperature near plate 1 (say $T_1 = 400$ K and $T_2 = 300$ K, to be specific), then the incoming radiation from plate 2 has a wavelength distribution not too dissimilar to plate 1. We might be very safe in approximating $\varepsilon_1 \cong \alpha_1$. The net heat flux between the plates can be expressed very simply

$$\begin{aligned} q_{\text{net}} &= \varepsilon_1 \sigma T_1^4 - \alpha_1(T_1, T_2)\sigma T_2^4 \\ &\cong \varepsilon_1 \sigma T_1^4 - \varepsilon_1 \sigma T_2^4 \\ &= \varepsilon_1 \sigma \left(T_1^4 - T_2^4\right) \end{aligned} \tag{10.11}$$

In effect, we are approximating plate 1 as a gray body.

In general, the simplest first estimate for total absorptance is the diffuse, gray body approximation, eqn. (10.8c). It is accurate either if the monochromatic emittance does not vary strongly with wavelength *or* if the bodies exchanging radiation are at similar absolute temperatures. More advanced texts describe techniques for calculating total absorptance (by integration) in other situations [10.2, 10.3].

We should always mistrust eqn. (10.8c) when solar radiation is absorbed by a low temperature object — a space vehicle or something on earth's surface, say. In this case, the best first approximation is to set total absorptance to a value for visible wavelengths of radiation (near 0.5 μm). Total emittance may be taken at the object's actual temperature, typically for infrared wavelengths. We return to solar absorptance in Section 10.6.

10.3 Radiant heat exchange between two finite black bodies

Let us now return to the purely geometric problem of evaluating the view factor, $F_{1\text{-}2}$. Although the evaluation of $F_{1\text{-}2}$ is also used in the calculation of heat exchange among diffuse, nonblack bodies, it is the *only* correction of the Stefan-Boltzmann law that we need for black bodies.

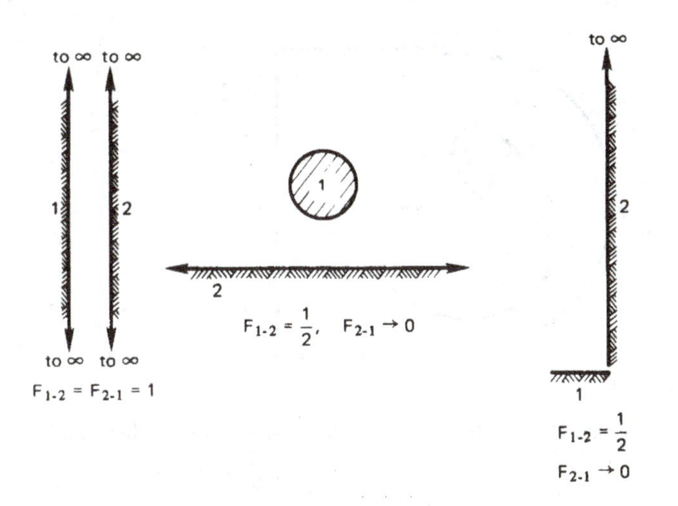

Figure 10.6 Some configurations for which the value of the view factor is immediately apparent.

Some evident results. Figure 10.6 shows three elementary situations in which the value of $F_{1\text{-}2}$ is evident using just the definition:

$$F_{1\text{-}2} \equiv \text{fraction of field of view of (1) occupied by (2).}$$

When the surfaces are each isothermal and diffuse, this corresponds to

$$F_{1\text{-}2} = \text{fraction of energy leaving (1) that reaches (2)}$$

A second apparent result in regard to the view factor is that all the energy leaving a body (1) reaches something else. Thus, conservation of energy requires

$$\boxed{1 = F_{1\text{-}1} + F_{1\text{-}2} + F_{1\text{-}3} + \cdots + F_{1\text{-}n}} \tag{10.12}$$

where (2), (3),...,(n) are all of the bodies in the neighborhood of (1). Figure 10.7 shows a representative situation in which a body (1) is surrounded by three other bodies. It sees all three bodies, but it also views itself, in part. This accounts for the inclusion of the view factor, $F_{1\text{-}1}$ in eqn. (10.12).

By the same token, it should also be apparent from Fig. 10.7 that the kind of sum expressed by eqn. (10.12) would also be true for any subset

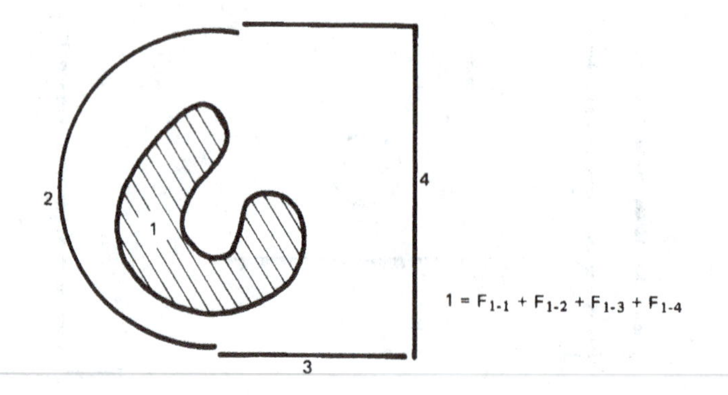

Figure 10.7 A body (1) that views three other bodies and itself as well.

of the bodies seen by surface 1. Thus,

$$F_{1\text{-}(2+3)} = F_{1\text{-}2} + F_{1\text{-}3}$$

Of course, such a sum makes sense only when all the view factors are based on the same viewing surface (surface 1 in this case). One might be tempted to write this sort of sum in the opposite direction, but it would clearly be untrue,

$$F_{(2+3)\text{-}1} \neq F_{2\text{-}1} + F_{3\text{-}1},$$

since each view factor is for a different viewing surface—(2 + 3), 2, and 3, in this case.

View factor reciprocity. So far, we have referred to the net radiation from black surface (1) to black surface (2) as Q_{net}. Let us refine our notation a bit, and call this $Q_{\text{net}_{1\text{-}2}}$:

$$Q_{\text{net}_{1\text{-}2}} = A_1 F_{1\text{-}2}\, \sigma \left(T_1^4 - T_2^4 \right) \tag{10.13}$$

Likewise, the net radiation from (2) to (1) is

$$Q_{\text{net}_{2\text{-}1}} = A_2 F_{2\text{-}1}\, \sigma \left(T_2^4 - T_1^4 \right) \tag{10.14}$$

Of course, $Q_{\text{net}_{1\text{-}2}} = -Q_{\text{net}_{2\text{-}1}}$. It follows that

$$A_1 F_{1\text{-}2}\, \sigma \left(T_1^4 - T_2^4 \right) = -A_2 F_{2\text{-}1}\, \sigma \left(T_2^4 - T_1^4 \right)$$

or

$$\boxed{A_1F_{1\text{-}2} = A_2F_{2\text{-}1}} \tag{10.15}$$

This result, called *view factor reciprocity*, is very useful in calculations.

Example 10.1

A jet of liquid metal at 2000°C pours from a crucible. It is 3 mm in diameter. A long cylindrical radiation shield, 5 cm diameter, surrounds the jet through an angle of 330°, but there is a 30° slit in it. The jet and the shield radiate as black bodies. They sit in a room at 30°C, and the shield has a temperature of 700°C. Calculate the net heat transfer: from the jet to the room through the slit; from the jet to the shield; and from the inside of the shield to the room.

Solution. By inspection, we see that $F_{\text{jet-room}} = 30/360 = 0.08333$ and $F_{\text{jet-shield}} = 330/360 = 0.9167$. Thus,

$$\begin{aligned} Q_{\text{net}_{\text{jet-room}}} &= A_{\text{jet}}F_{\text{jet-room}}\,\sigma\left(T_{\text{jet}}^4 - T_{\text{room}}^4\right) \\ &= \left[\frac{\pi(0.003)\ \text{m}^2}{\text{m length}}\right](0.08333)(5.67\times10^{-8})\left(2273^4 - 303^4\right) \\ &= 1{,}188\ \text{W/m} \end{aligned}$$

Likewise,

$$\begin{aligned} Q_{\text{net}_{\text{jet-shield}}} &= A_{\text{jet}}F_{\text{jet-shield}}\,\sigma\left(T_{\text{jet}}^4 - T_{\text{shield}}^4\right) \\ &= \left[\frac{\pi(0.003)\ \text{m}^2}{\text{m length}}\right](0.9167)(5.67\times10^{-8})\left(2273^4 - 973^4\right) \\ &= 12{,}637\ \text{W/m} \end{aligned}$$

The heat absorbed by the shield leaves it by radiation and convection to the room. (A balance of these effects can be used to *calculate* the shield temperature given here.)

To find the radiation from the *inside* of the shield to the room, we need $F_{\text{shield-room}}$. Since any radiation passing out of the slit goes to the room, we can find this view factor equating view factors to the room with view factors to the slit. The slit's area is $A_{\text{slit}} = \pi(0.05)30/360 = 0.01309\ \text{m}^2/\text{m}$ length. Hence, using our reciprocity and summation rules, eqns. (10.12) and (10.15),

$$F_{\text{slit-jet}} = \frac{A_{\text{jet}}}{A_{\text{slit}}} F_{\text{jet-room}} = \frac{\pi(0.003)}{0.01309}(0.0833) = 0.0600$$

$$F_{\text{slit-shield}} = 1 - F_{\text{slit-jet}} - \underbrace{F_{\text{slit-slit}}}_{\cong 0} = 1 - 0.0600 - 0 = 0.940$$

$$\begin{aligned} F_{\text{shield-room}} &= \frac{A_{\text{slit}}}{A_{\text{shield}}} F_{\text{slit-shield}} \\ &= \frac{0.01309}{\pi(0.05)(330)/(360)}(0.940) = 0.08545 \end{aligned}$$

Hence, for heat transfer from the inside of the shield only,

$$\begin{aligned} Q_{\text{net}_{\text{shield-room}}} &= A_{\text{shield}} F_{\text{shield-room}}\, \sigma\left(T^4_{\text{shield}} - T^4_{\text{room}}\right) \\ &= \left[\frac{\pi(0.05)330}{360}\right](0.08545)(5.67 \times 10^{-8})\left(973^4 - 303^4\right) \\ &= 619 \text{ W/m} \end{aligned}$$

Both the jet and the inside of the shield have relatively small view factors to the room, so that comparatively little heat is lost through the slit. ■

Calculation of the black-body view factor, $F_{1\text{-}2}$. When a view factor is not obvious as those in Fig. 10.6 were or when it cannot be obtained from other view factors using such equations as (10.12) or (10.15), one must resort to direct integration. Let us see how to do that.

Consider two elements, dA_1 and dA_2, of larger black bodies (1) and (2), as shown in Fig. 10.8. Body (1) and body (2) are each isothermal. Since element dA_2 subtends a solid angle $d\omega_1$, we use eqn. (10.6) to write

$$dQ_{1 \text{ to } 2} = (i_1 d\omega_1)(\cos\beta_1\, dA_1)$$

But from eqn. (10.7b),

$$i_1 = \frac{\sigma T_1^4}{\pi}$$

Note that because black bodies radiate diffusely, i_1 does not vary with angle; and because these bodies are isothermal, it does not vary with position. The element of solid angle is given by

$$d\omega_1 = \frac{\cos\beta_2\, dA_2}{s^2}$$

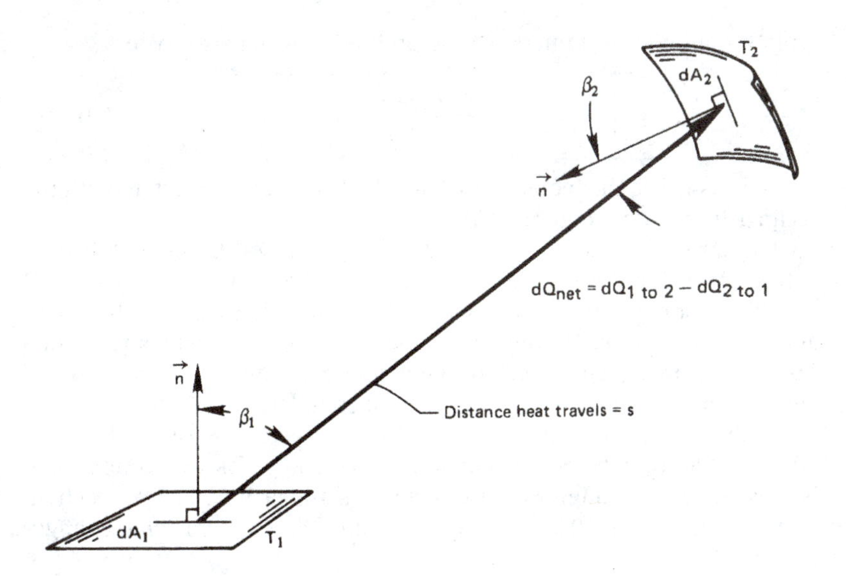

Figure 10.8 Radiant exchange between two black elements that are part of the bodies (1) and (2).

where s is the distance from (1) to (2) and $\cos\beta_2$ enters because dA_2 is not necessarily normal to s. Thus,

$$dQ_{1\text{ to }2} = \frac{\sigma T_1^4}{\pi}\left(\frac{\cos\beta_1\cos\beta_2\, dA_1 dA_2}{s^2}\right)$$

By the same token,

$$dQ_{2\text{ to }1} = \frac{\sigma T_2^4}{\pi}\left(\frac{\cos\beta_2\cos\beta_1\, dA_2 dA_1}{s^2}\right)$$

Then

$$Q_{\text{net}_{1\text{-}2}} = \sigma\left(T_1^4 - T_2^4\right)\int_{A_1}\int_{A_2}\frac{\cos\beta_1\cos\beta_2}{\pi s^2}\, dA_1 dA_2 \tag{10.16}$$

The view factors $F_{1\text{-}2}$ and $F_{2\text{-}1}$ are immediately obtainable from eqn. (10.16). If we compare this result with $Q_{\text{net}_{1\text{-}2}} = A_1 F_{1\text{-}2}\sigma(T_1^4 - T_2^4)$, we get

$$\boxed{F_{1\text{-}2} = \frac{1}{A_1}\int_{A_1}\int_{A_2}\frac{\cos\beta_1\cos\beta_2}{\pi s^2}\, dA_1 dA_2} \tag{10.17a}$$

From the inherent symmetry of the problem, we can also write

$$F_{2\text{-}1} = \frac{1}{A_2}\int_{A_2}\int_{A_1} \frac{\cos\beta_2 \cos\beta_1}{\pi s^2}\, dA_2 dA_1 \tag{10.17b}$$

We can easily see that eqns. (10.17a) and (10.17b) are consistent with the reciprocity relation, eqn. (10.15).

The direct evaluation of $F_{1\text{-}2}$ from eqn. (10.17a) becomes fairly involved, even for the simplest configurations. Siegel and Howell [10.4] provide a comprehensive discussion of such calculations and a large catalog of their results. Howell [10.5] gives an even more extensive tabulation of view factor equations, which is now available on the World Wide Web. At present, no other reference is as complete.

We list some typical expressions for view factors in Tables 10.2 and 10.3. Table 10.2 gives calculated values of $F_{1\text{-}2}$ for two-dimensional bodies—various configurations of cylinders and strips that approach infinite length. Table 10.3 gives $F_{1\text{-}2}$ for some three-dimensional configurations.

Many view factors have been evaluated numerically and presented in graphical form for easy reference. Figure 10.9, for example, includes graphs for configurations 1, 2, and 3 from Table 10.3. The reader should study these results and be sure that the trends they show make sense. Is it clear, for example, that $F_{1\text{-}2} \longrightarrow$ constant, which is < 1 in each case, as the abscissa becomes large? Can you locate the configuration on the right-hand side of Fig. 10.6 in Fig. 10.9? And so forth.

Figure 10.10 shows view factors for another kind of configuration—one in which one area is very small in comparison with the other one. Many solutions like this exist because they are a bit less difficult to calculate, and they can often be very useful in practice.

Example 10.2

A heater (h) as shown in Fig. 10.11 radiates to the partially conical shield (s) that surrounds it. If the heater and shield are black, calculate the net heat transfer from the heater to the shield.

Solution. First imagine a plane (i) laid across the open top of the shield:

$$F_{h-s} + F_{h-i} = 1$$

But F_{h-i} can be obtained from Fig. 10.9 or case 3 of Table 10.3,

Table 10.2 View factors for a variety of two-dimensional configurations (infinite in extent normal to the paper)

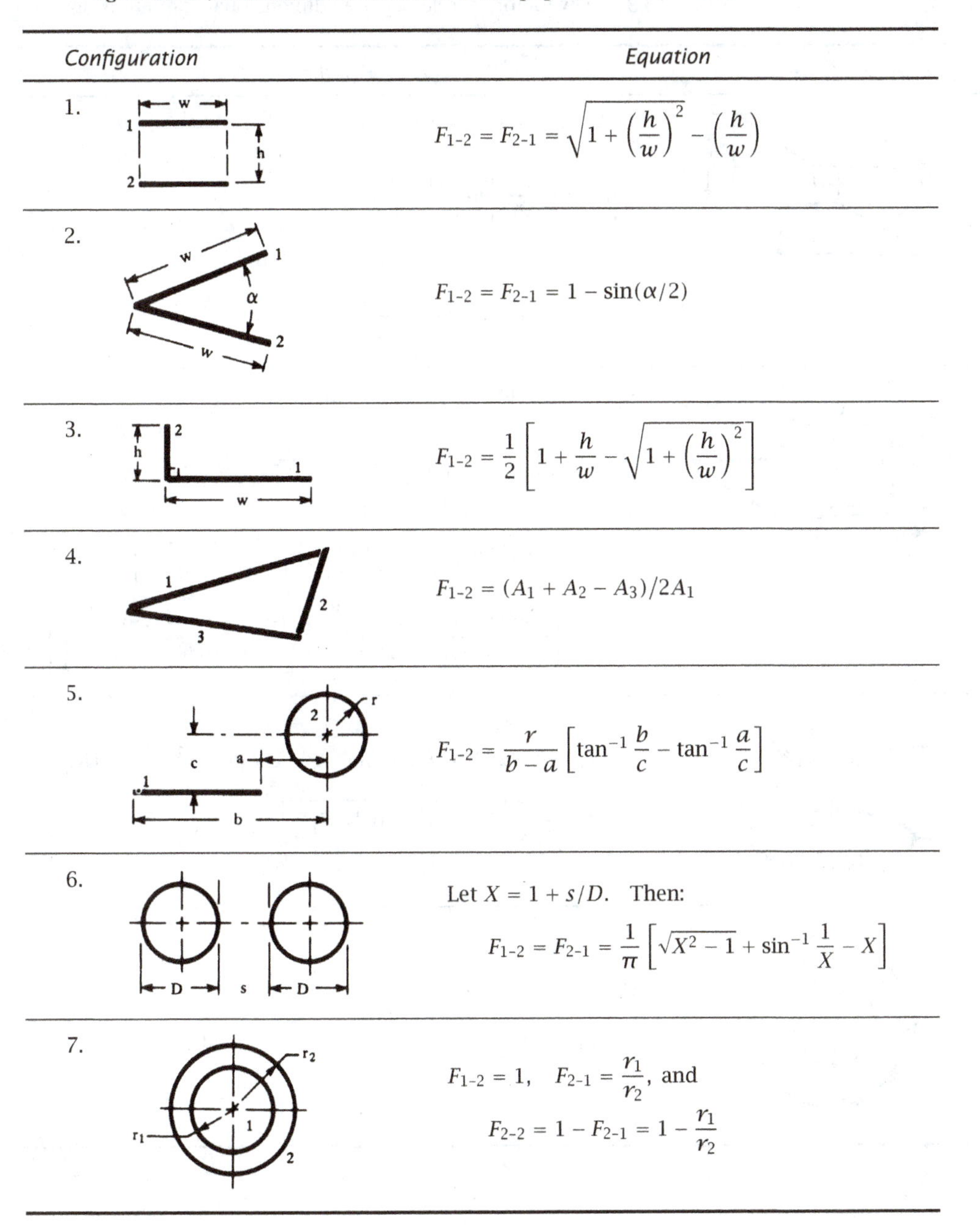

Configuration	Equation
1.	$F_{1\text{-}2} = F_{2\text{-}1} = \sqrt{1 + \left(\frac{h}{w}\right)^2} - \left(\frac{h}{w}\right)$
2.	$F_{1\text{-}2} = F_{2\text{-}1} = 1 - \sin(\alpha/2)$
3.	$F_{1\text{-}2} = \frac{1}{2}\left[1 + \frac{h}{w} - \sqrt{1 + \left(\frac{h}{w}\right)^2}\right]$
4.	$F_{1\text{-}2} = (A_1 + A_2 - A_3)/2A_1$
5.	$F_{1\text{-}2} = \frac{r}{b - a}\left[\tan^{-1}\frac{b}{c} - \tan^{-1}\frac{a}{c}\right]$
6.	Let $X = 1 + s/D$. Then: $F_{1\text{-}2} = F_{2\text{-}1} = \frac{1}{\pi}\left[\sqrt{X^2 - 1} + \sin^{-1}\frac{1}{X} - X\right]$
7.	$F_{1\text{-}2} = 1, \quad F_{2\text{-}1} = \frac{r_1}{r_2}$, and $F_{2\text{-}2} = 1 - F_{2\text{-}1} = 1 - \frac{r_1}{r_2}$

Table 10.3 View factors for some three-dimensional configurations

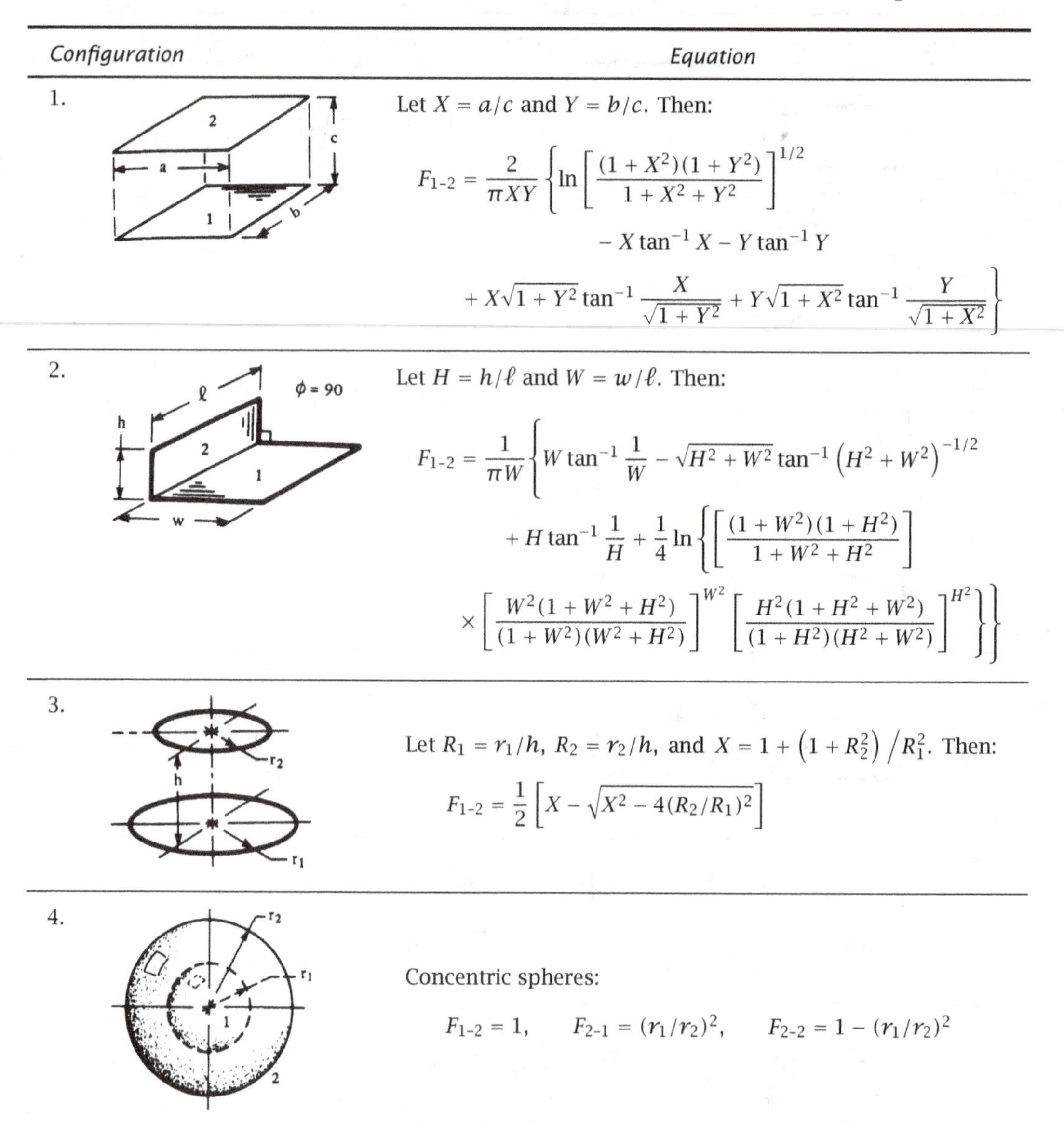

Configuration	Equation
1.	Let $X = a/c$ and $Y = b/c$. Then: $$F_{1-2} = \frac{2}{\pi XY}\left\{\ln\left[\frac{(1+X^2)(1+Y^2)}{1+X^2+Y^2}\right]^{1/2} - X\tan^{-1}X - Y\tan^{-1}Y + X\sqrt{1+Y^2}\tan^{-1}\frac{X}{\sqrt{1+Y^2}} + Y\sqrt{1+X^2}\tan^{-1}\frac{Y}{\sqrt{1+X^2}}\right\}$$
2.	Let $H = h/\ell$ and $W = w/\ell$. Then: $$F_{1-2} = \frac{1}{\pi W}\left\{W\tan^{-1}\frac{1}{W} - \sqrt{H^2+W^2}\tan^{-1}\left(H^2+W^2\right)^{-1/2} + H\tan^{-1}\frac{1}{H} + \frac{1}{4}\ln\left\{\left[\frac{(1+W^2)(1+H^2)}{1+W^2+H^2}\right] \times \left[\frac{W^2(1+W^2+H^2)}{(1+W^2)(W^2+H^2)}\right]^{W^2}\left[\frac{H^2(1+H^2+W^2)}{(1+H^2)(H^2+W^2)}\right]^{H^2}\right\}\right\}$$
3.	Let $R_1 = r_1/h$, $R_2 = r_2/h$, and $X = 1 + \left(1 + R_2^2\right)/R_1^2$. Then: $$F_{1-2} = \frac{1}{2}\left[X - \sqrt{X^2 - 4(R_2/R_1)^2}\right]$$
4.	Concentric spheres: $$F_{1-2} = 1, \qquad F_{2-1} = (r_1/r_2)^2, \qquad F_{2-2} = 1 - (r_1/r_2)^2$$

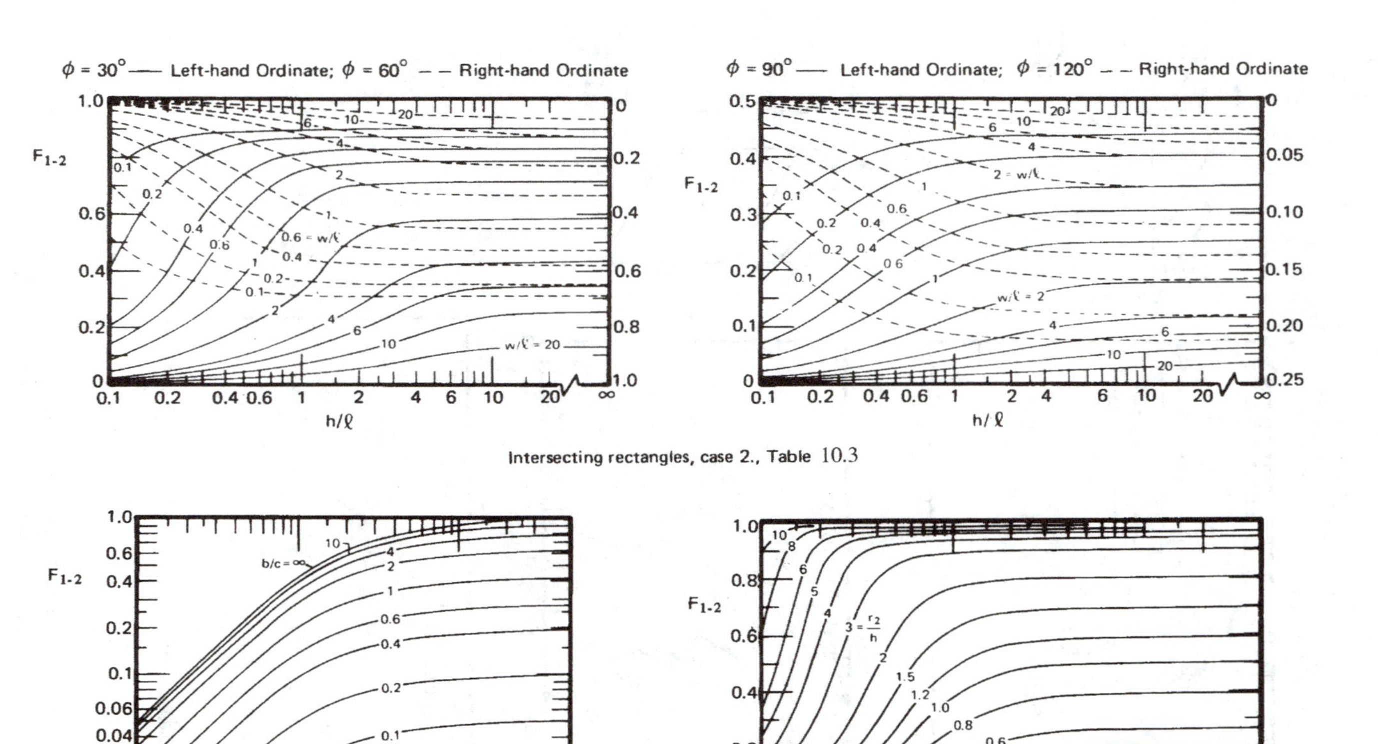

Intersecting rectangles, case 2., Table 10.3

Facing rectangles, case 1., Table 10.3

Facing concentric discs, case 3., Table 10.3

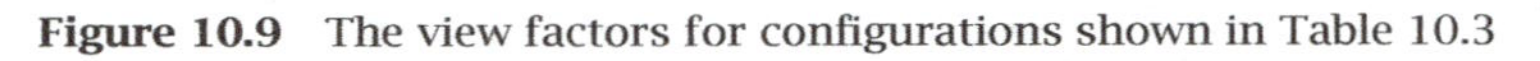

Figure 10.9 The view factors for configurations shown in Table 10.3

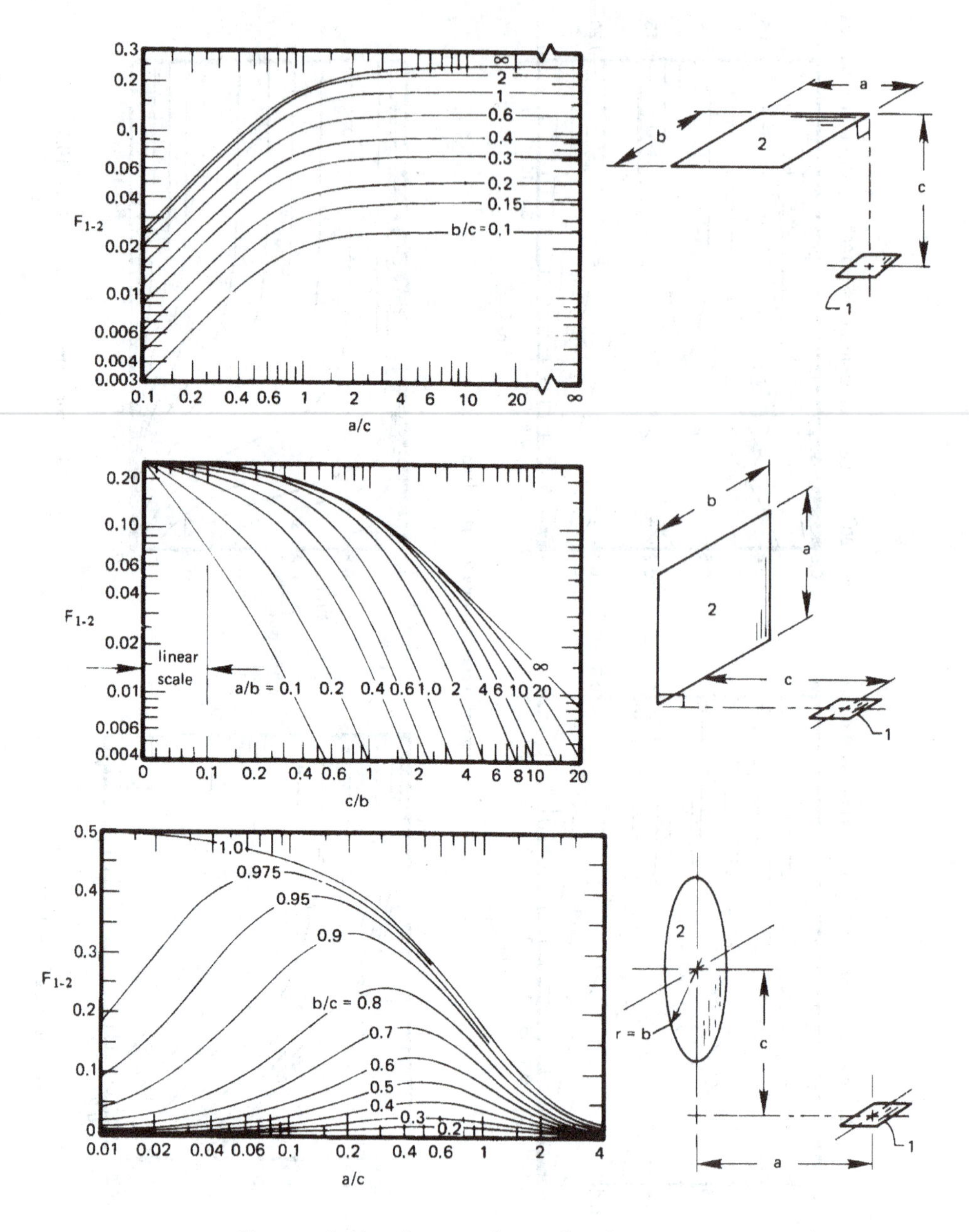

Figure 10.10 The view factor for three very small surfaces "looking at" three large surfaces ($A_1 \ll A_2$).

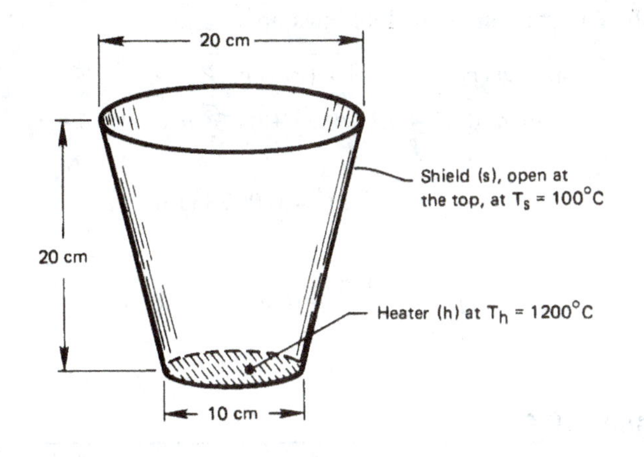

Figure 10.11 Heat transfer from a disc heater to its radiation shield.

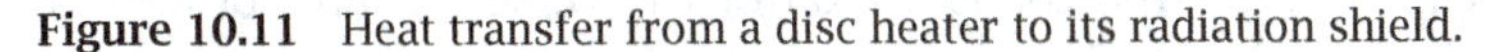

for $R_1 = r_1/h = 5/20 = 0.25$ and $R_2 = r_2/h = 10/20 = 0.5$. The result is $F_{h-i} = 0.192$. Then

$$F_{h-s} = 1 - 0.192 = 0.808$$

Thus,

$$\begin{aligned} Q_{\text{net}_{h-s}} &= A_h F_{h-s}\,\sigma\left(T_h^4 - T_s^4\right) \\ &= \frac{\pi}{4}(0.1)^2(0.808)(5.67\times 10^{-8})\left[(1200+273)^4 - 373^4\right] \\ &= 1687\ \text{W} \end{aligned}$$

■

Example 10.3

Suppose that the shield in Example 10.2 were heating the region where the heater is presently located. What would F_{s-h} be?

Solution. From eqn. (10.15) we have

$$A_s F_{s-h} = A_h F_{h-s}$$

But the frustrum-shaped shield has an area of

$$A_s = \pi(r_1 + r_2)\sqrt{h^2 + (r_2 - r_1)^2}$$
$$= \pi(0.05 + 0.1)\sqrt{0.2^2 + 0.05^2} = 0.09715 \text{ m}^2$$

and

$$A_h = \frac{\pi}{4}(0.1)^2 = 0.007854 \text{ m}^2$$

so

$$F_{s-h} = \frac{0.007854}{0.09715}(0.808) = 0.0653$$ ■

Example 10.4

Find $F_{1\text{-}2}$ for the configuration of two offset squares of area A, as shown in Fig. 10.12.

Solution. In this case we see how to obtain a view factor by the creative use of the various equations relating view factors to one another. Consider two fictitious areas 3 and 4 as indicated by the dotted lines. The view factor between the combined areas, (1+3) and (2+4), can be obtained from Fig. 10.9. In addition, we can write that view factor in terms of the unknown $F_{1\text{-}2}$ and other known view factors:

$$(2A)F_{(1+3)\text{-}(4+2)} = AF_{1\text{-}4} + AF_{1\text{-}2} + AF_{3\text{-}4} + AF_{3\text{-}2}$$
$$2F_{(1+3)\text{-}(4+2)} = 2F_{1\text{-}4} + 2F_{1\text{-}2}$$
$$F_{1\text{-}2} = F_{(1+3)\text{-}(4+2)} - F_{1\text{-}4}$$

And $F_{(1+3)\text{-}(4+2)}$ can be read from Fig. 10.9 (at $\phi = 90$, $w/\ell = 1/2$, and $h/\ell = 1/2$) as 0.245 and $F_{1\text{-}4}$ as 0.20. Thus,

$$F_{1\text{-}2} = (0.245 - 0.20) = 0.045$$ ■

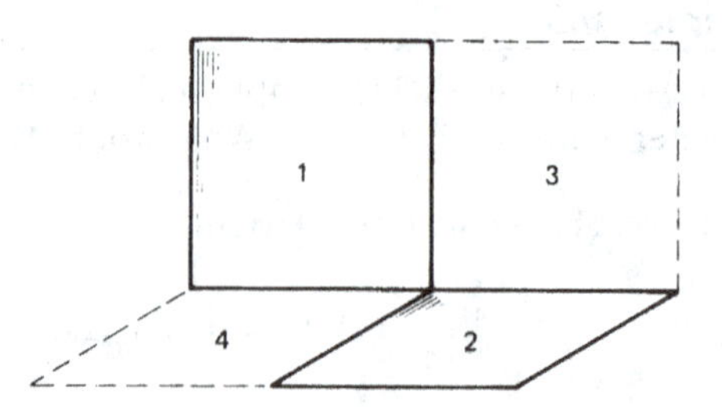

Figure 10.12 Radiation between two offset perpendicular squares.

10.4 Heat transfer among gray bodies

Electrical analogy for gray body heat exchange

An electric circuit analogy for heat exchange among diffuse gray bodies was developed by Oppenheim [10.6] in 1956. It begins with the definition of two new quantities:

$$H \ \ (\text{W/m}^2) \equiv \textit{irradiance} = \begin{cases}\text{flux of energy that irradiates the}\\ \text{surface}\end{cases}$$

and

$$B \ \ (\text{W/m}^2) \equiv \textit{radiosity} = \begin{cases}\text{total flux of radiative energy}\\ \text{away from the surface}\end{cases}$$

The radiosity can be expressed as the sum of the irradiated energy that is reflected by the surface and the radiation emitted by it. Thus,

$$B = \rho H + \varepsilon e_b \tag{10.18}$$

We can immediately write the net heat flux leaving any particular surface as the difference between B and H for that surface. Then, with the help of eqn. (10.18), we get

$$q_{\text{net}} = B - H = B - \frac{B - \varepsilon e_b}{\rho} \tag{10.19}$$

This can be rearranged as

$$q_{\text{net}} = \frac{\varepsilon}{\rho} e_b - \frac{1-\rho}{\rho} B \tag{10.20}$$

If the surface is opaque ($\tau = 0$), $1 - \rho = \alpha$, and if it is gray, $\alpha = \varepsilon$. Then, eqn. (10.20) gives

$$q_{\text{net}} A = Q_{\text{net}} = \frac{e_b - B}{\rho/\varepsilon A} = \frac{e_b - B}{(1-\varepsilon)/\varepsilon A} \tag{10.21}$$

Equation (10.21) may be viewed as a form of Ohm's law. It tells us that $(e_b - B)$ can be seen as a driving potential for transferring heat away from a surface through an effective surface resistance, $(1 - \varepsilon)/\varepsilon A$.

Now consider heat transfer from one infinite gray plate to another parallel to it. Radiant energy flows past an imaginary surface, parallel to the first infinite plate and quite close to it, as shown as a dotted line

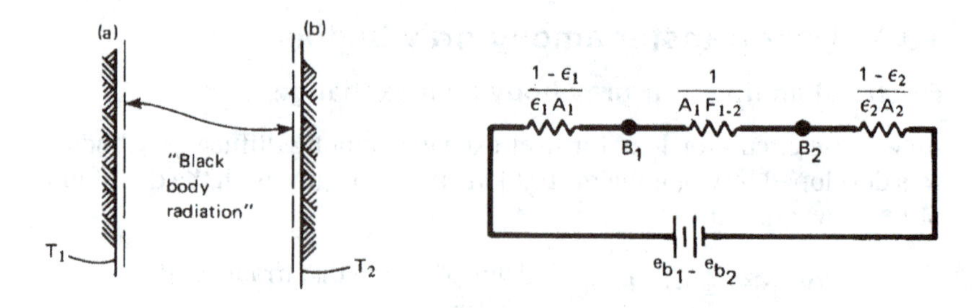

Figure 10.13 The electrical circuit analogy for radiation between two gray infinite plates.

in Fig. 10.13. If the gray plate is diffuse, its radiation has the same geometrical distribution as that from a black body, and it will travel to other objects in the same way that black body radiation would. Therefore, we can treat the radiation leaving the imaginary surface — the radiosity, that is — as though it were black body radiation travelling to an imaginary surface above the other plate. Thus, by analogy to eqn. (10.13),

$$Q_{\text{net}_{1\text{-}2}} = A_1 F_{1\text{-}2} \left(B_1 - B_2\right) = \frac{B_1 - B_2}{\left(\dfrac{1}{A_1 F_{1\text{-}2}}\right)} \tag{10.22}$$

where the final fraction shows that this is also a form of Ohm's law: the radiosity difference $(B_1 - B_2)$, can be said to drive heat through the geometrical resistance, $1/A_1 F_{1\text{-}2}$, that describes the field of view between the two surfaces.

When two gray surfaces exchange radiation only with each other, the net radiation flows through a surface resistance for each surface and a geometric resistance for the configuration. The electrical circuit shown in Fig. 10.13 expresses the analogy and gives us means for calculating $Q_{\text{net}_{1\text{-}2}}$ from Ohm's law. Recalling that $e_b = \sigma T^4$, we obtain

$$Q_{\text{net}_{1\text{-}2}} = \frac{e_{b_1} - e_{b_2}}{\sum \text{resistances}} = \frac{\sigma\left(T_1^4 - T_2^4\right)}{\left(\dfrac{1-\varepsilon}{\varepsilon A}\right)_1 + \dfrac{1}{A_1 F_{1\text{-}2}} + \left(\dfrac{1-\varepsilon}{\varepsilon A}\right)_2} \tag{10.23}$$

For the particular case of infinite parallel plates, $F_{1\text{-}2} = 1$ and $A_1 = A_2$

(Fig. 10.6), and, with $q_{\text{net}_{1\text{-}2}} = Q_{\text{net}_{1\text{-}2}}/A_1$, we find

$$q_{\text{net}_{1\text{-}2}} = \frac{1}{\left(\dfrac{1}{\varepsilon_1} + \dfrac{1}{\varepsilon_2} - 1\right)} \sigma\left(T_1^4 - T_2^4\right) \tag{10.24}$$

Comparing eqn. (10.24) with eqn. (10.2), we may identify

$$\mathcal{F}_{1\text{-}2} = \frac{1}{\left(\dfrac{1}{\varepsilon_1} + \dfrac{1}{\varepsilon_2} - 1\right)} \tag{10.25}$$

for infinite parallel plates. Notice, too, that if the plates are both black ($\varepsilon_1 = \varepsilon_2 = 1$), then both surface resistances are zero and

$$\mathcal{F}_{1\text{-}2} = 1 = F_{1\text{-}2}$$

which, of course, is what we would have expected.

Example 10.5 One gray body enclosed by another

Evaluate the heat transfer and the transfer factor for one gray body enclosed by another, as shown in Fig. 10.14.

Solution. The electrical circuit analogy is exactly the same as that shown in Fig. 10.13, and $F_{1\text{-}2}$ is still one. Therefore, with eqn. (10.23),

$$Q_{\text{net}_{1\text{-}2}} = A_1 q_{\text{net}_{1\text{-}2}} = \frac{\sigma(T_1^4 - T_2^4)}{\left(\dfrac{1-\varepsilon_1}{\varepsilon_1 A_1} + \dfrac{1}{A_1} + \dfrac{1-\varepsilon_2}{\varepsilon_2 A_2}\right)} \tag{10.26}$$

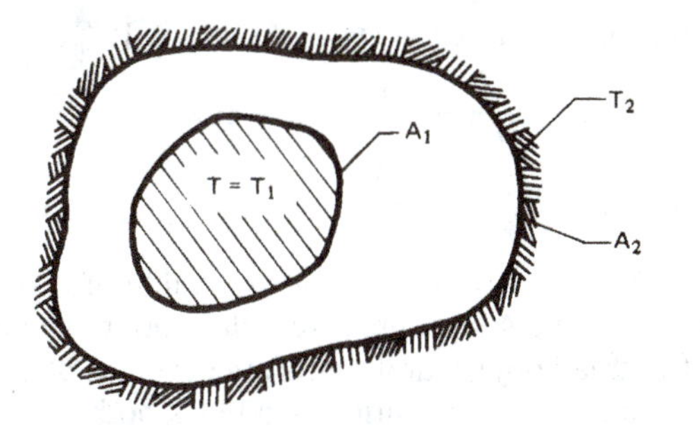

Figure 10.14 Heat transfer between an enclosed body and the body surrounding it.

The transfer factor may again be identified by comparison to eqn. (10.2):

$$Q_{\text{net}_{1\text{-}2}} = A_1 \underbrace{\frac{1}{\dfrac{1}{\varepsilon_1} + \dfrac{A_1}{A_2}\left(\dfrac{1}{\varepsilon_2} - 1\right)}}_{=\mathcal{F}_{1\text{-}2}} \sigma\left(T_1^4 - T_2^4\right) \tag{10.27}$$

This calculation is valid only when body (1) does not view itself. ■

Example 10.6 Transfer factor reciprocity

Derive $\mathcal{F}_{2\text{-}1}$ for the enclosed bodies shown in Fig. 10.14.

Solution.

$$Q_{\text{net}_{1\text{-}2}} = -Q_{\text{net}_{2\text{-}1}}$$

$$A_1\mathcal{F}_{1\text{-}2}\,\sigma\left(T_1^4 - T_2^4\right) = -A_2\mathcal{F}_{2\text{-}1}\,\sigma\left(T_2^4 - T_1^4\right)$$

from which we obtain the reciprocity relationship for transfer factors:

$$\boxed{A_1\mathcal{F}_{1\text{-}2} = A_2\mathcal{F}_{2\text{-}1}} \tag{10.28}$$

Hence, with the result of Example 10.5, we have

$$\mathcal{F}_{2\text{-}1} = \frac{A_1}{A_2}\mathcal{F}_{1\text{-}2} = \frac{1}{\dfrac{1}{\varepsilon_1}\dfrac{A_2}{A_1} + \left(\dfrac{1}{\varepsilon_2} - 1\right)} \tag{10.29}$$

■

Example 10.7 Small gray object in a large environment

Derive $\mathcal{F}_{1\text{-}2}$ for a small gray object (1) in a large isothermal environment (2), the result that was given as eqn. (1.35).

Solution. We may use eqn. (10.27) with $A_1/A_2 \ll 1$:

$$\mathcal{F}_{1\text{-}2} = \frac{1}{\dfrac{1}{\varepsilon_1} + \underbrace{\dfrac{A_1}{A_2}}_{\ll 1}\left(\dfrac{1}{\varepsilon_2} - 1\right)} \cong \varepsilon_1 \tag{10.30}$$

Note that the same result is obtained for *any* value of A_1/A_2 if the enclosure is black ($\varepsilon_2 = 1$). A large enclosure does not reflect much radiation back to the small object, and therefore becomes like a perfect absorber of the small object's radiation — a black body. ■

Additional two-body exchange problems

Radiation shields. A radiation shield is a surface, usually of high reflectance, that is placed between a high-temperature source and its cooler environment. Earlier examples in this chapter and in Chapter 1 show how such a surface can reduce heat exchange. Let us now examine the role of reflectance (or emittance: $\varepsilon = 1 - \rho$) in the performance of a radiation shield.

Consider a gray body (1) surrounded by another gray body (2), as discussed in Example 10.5. Suppose now that a thin sheet of reflective material is placed between bodies (1) and (2) as a radiation shield. The sheet will reflect radiation arriving from body (1) back toward body (1); likewise, owing to its low emittance, it will radiate little energy to body (2). The radiation from body (1) to the inside of the shield and from the outside of the shield to body (2) are each two-body exchange problems, coupled by the shield temperature. We may put the various radiation resistances in series to find (see Problem 10.46)

$$Q_{\text{net}_{1\text{-}2}} = \frac{\sigma(T_1^4 - T_2^4)}{\left(\dfrac{1-\varepsilon_1}{\varepsilon_1 A_1} + \dfrac{1}{A_1} + \dfrac{1-\varepsilon_2}{\varepsilon_2 A_2}\right) + \underbrace{2\left(\dfrac{1-\varepsilon_s}{\varepsilon_s A_s}\right) + \dfrac{1}{A_s}}_{\text{added by shield}}} \tag{10.31}$$

assuming $F_{1\text{-s}} = F_{\text{s-}2} = 1$. Note that the radiation shield reduces $Q_{\text{net}_{1\text{-}2}}$ more if its emittance is smaller, i.e., if it is highly reflective.

Specular surfaces. The electrical circuit analogy that we have developed is for diffuse surfaces. If the surface reflection or emission has directional characteristics, different methods of analysis must be used [10.3].

One important special case deserves to be mentioned. If the two gray surfaces in Fig. 10.14 are diffuse emitters but are perfectly specular reflectors — that is, if they each have only mirror-like reflections — then the transfer factor becomes

$$\mathcal{F}_{1\text{-}2} = \frac{1}{\left(\dfrac{1}{\varepsilon_1} + \dfrac{1}{\varepsilon_2} - 1\right)} \qquad \text{for specularly reflecting bodies} \tag{10.32}$$

This result is interestingly identical to eqn. (10.25) for parallel plates. Since parallel plates are a special case of the situation in Fig. 10.14, it follows that eqn. (10.25) is true for either specular or diffuse reflection.

Example 10.8

A physics experiment uses liquid nitrogen as a coolant. Saturated liquid nitrogen at 80 K flows through 6.35 mm O.D. stainless steel line ($\varepsilon_l = 0.2$) inside a vacuum chamber. The chamber walls are at $T_c = 230$ K and are at some distance from the line. Determine the heat gain of the line per unit length. If a second stainless steel tube, 12.7 mm in diameter, is placed around the line to act as radiation shield, to what rate is the heat gain reduced? Find the temperature of the shield.

Solution. The nitrogen coolant will hold the surface of the line at essentially 80 K, since the thermal resistances of the tube wall and the internal convection or boiling process are small. Without the shield, we can model the line as a small object in a large enclosure, as in Example 10.7:

$$\begin{aligned} Q_{\text{gain}} &= (\pi D_l)\varepsilon_l \sigma(T_c^4 - T_l^4) \\ &= \pi(0.00635)(0.2)(5.67\times10^{-8})(230^4 - 80^4) = 0.624 \text{ W/m} \end{aligned}$$

With the shield, eqn. (10.31) applies. Assuming that the chamber area is large compared to the shielded line ($A_c \gg A_l$),

$$\begin{aligned} Q_{\text{gain}} &= \frac{\sigma(T_c^4 - T_l^4)}{\left(\dfrac{1-\varepsilon_l}{\varepsilon_l A_l} + \dfrac{1}{A_l} + \underbrace{\dfrac{1-\varepsilon_c}{\varepsilon_2 A_c}}_{\text{neglect}}\right) + 2\left(\dfrac{1-\varepsilon_s}{\varepsilon_s A_s}\right) + \dfrac{1}{A_s}} \\ &= \frac{\pi(0.00635)(5.67\times10^{-8})(230^4 - 80^4)}{\left(\dfrac{1-0.2}{0.2} + 1\right) + \dfrac{0.00635}{0.0127}\left[2\left(\dfrac{1-0.2}{0.2}\right) + 1\right]} \\ &= 0.328 \text{ W/m} \end{aligned}$$

The radiation shield would cut the heat gain by 47%.

The temperature of the shield, T_s, may be found using the heat loss and considering the heat flow from the chamber to the shield, with the shield now acting as a small object in a large enclosure:

$$\begin{aligned} Q_{\text{gain}} &= (\pi D_s)\varepsilon_s \sigma(T_c^4 - T_s^4) \\ 0.328 \text{ W/m} &= \pi(0.0127)(0.2)(5.67\times10^{-8})(230^4 - T_s^4) \end{aligned}$$

Solving, we find $T_s = 213$ K. ■

The electrical circuit analogy when more than two gray bodies are involved in heat exchange

Let us first consider a three-body transaction, as pictured in at the bottom and left-hand sides of Fig. 10.15. The triangular circuit for three bodies is not so easy to analyze as the in-line circuits obtained in two-body problems. The basic approach is to apply energy conservation at each radiosity node in the circuit, setting the net heat transfer from any one of the surfaces (which we designate as i)

$$Q_{\text{net}_i} = \frac{e_{b_i} - B_i}{\dfrac{1-\varepsilon_i}{\varepsilon_i A_i}} \tag{10.33a}$$

equal to the sum of the net radiation to each of the other surfaces (call them j)

$$Q_{\text{net}_i} = \sum_j \left(\frac{B_i - B_j}{1 \big/ A_i F_{i-j}} \right) \tag{10.33b}$$

For the three body situation shown in Fig. 10.15, this leads to three equations

$$Q_{\text{net}_1}, \text{ at node } B_1: \quad \frac{e_{b_1} - B_1}{\dfrac{1-\varepsilon_1}{\varepsilon_1 A_1}} = \frac{B_1 - B_2}{\dfrac{1}{A_1 F_{1\text{-}2}}} + \frac{B_1 - B_3}{\dfrac{1}{A_1 F_{1\text{-}3}}} \tag{10.34a}$$

$$Q_{\text{net}_2}, \text{ at node } B_2: \quad \frac{e_{b_2} - B_2}{\dfrac{1-\varepsilon_2}{\varepsilon_2 A_2}} = \frac{B_2 - B_1}{\dfrac{1}{A_1 F_{1\text{-}2}}} + \frac{B_2 - B_3}{\dfrac{1}{A_2 F_{2\text{-}3}}} \tag{10.34b}$$

$$Q_{\text{net}_3}, \text{ at node } B_3: \quad \frac{e_{b_3} - B_3}{\dfrac{1-\varepsilon_3}{\varepsilon_3 A_3}} = \frac{B_3 - B_1}{\dfrac{1}{A_1 F_{1\text{-}3}}} + \frac{B_3 - B_2}{\dfrac{1}{A_2 F_{2\text{-}3}}} \tag{10.34c}$$

If the temperatures T_1, T_2, and T_3 are known (so that e_{b_1}, e_{b_2}, e_{b_3} are known), these equations can be solved simultaneously for the three unknowns, B_1, B_2, and B_3. After they are solved, one can compute the net heat transfer to or from any body (i) from either of eqns. (10.33).

Thus far, we have considered only cases in which the surface temperature is known for each body involved in the heat exchange process. Let us consider two other possibilities.

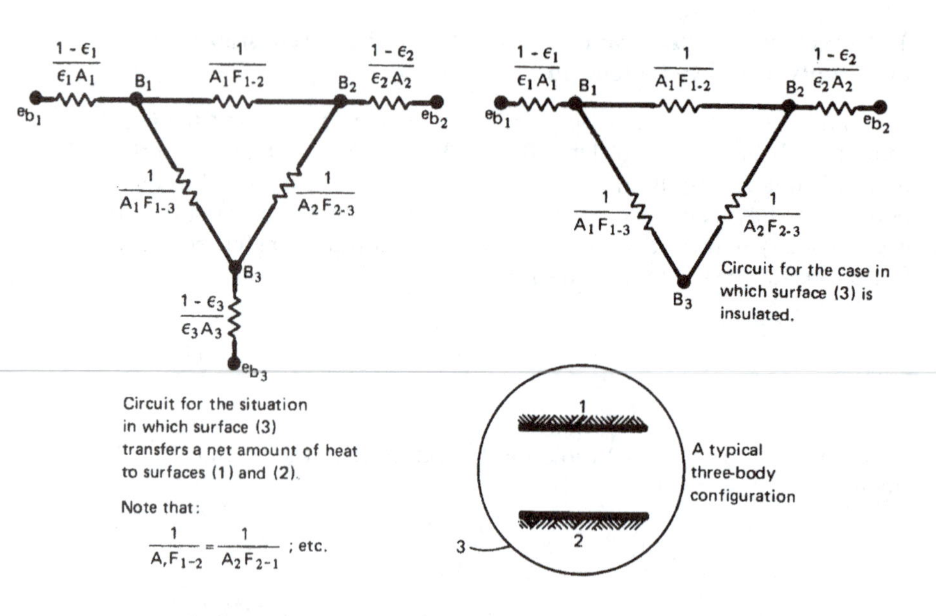

Figure 10.15 The electrical circuit analogy for radiation among three gray surfaces.

An insulated wall. If a wall is adiabatic, $Q_{\text{net}} = 0$ at that wall. For example, if wall (3) in Fig. 10.15 is insulated, then eqn. (10.33b) shows that $e_{b_3} = B_3$. We can eliminate one leg of the circuit, as shown on the right-hand side of Fig. 10.15; likewise, the left-hand side of eqn. (10.34c) equals zero. This means that all radiation absorbed by an adiabatic wall is immediately reemitted. Such walls are sometimes called "refractory surfaces" in discussing thermal radiation.

The circuit for an insulated wall can be treated as a series-parallel circuit, since all the heat from body (1) flows to body (2), even if it does so by travelling first to body (3). Then

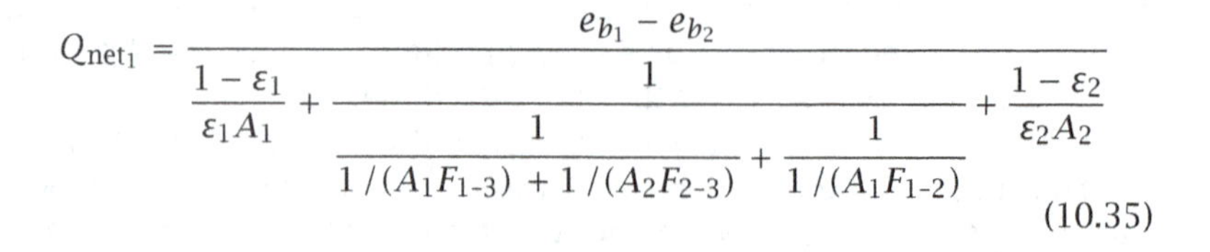

$$Q_{\text{net}_1} = \frac{e_{b_1} - e_{b_2}}{\dfrac{1-\varepsilon_1}{\varepsilon_1 A_1} + \dfrac{1}{\dfrac{1}{1/(A_1F_{1\text{-}3}) + 1/(A_2F_{2\text{-}3})} + \dfrac{1}{1/(A_1F_{1\text{-}2})}} + \dfrac{1-\varepsilon_2}{\varepsilon_2 A_2}} \tag{10.35}$$

A specified wall heat flux. The heat flux leaving a surface may be known, if, say, it is an electrically powered radiant heater. In this case, the left-hand side of one of eqns. (10.34) can be replaced with the surface's known Q_{net}, via eqn. (10.33b).

For the adiabatic wall case just considered, if surface (1) had a specified heat flux, then eqn. (10.35) could be solved for e_{b_1} and the unknown temperature T_1.

Example 10.9

Two very long strips 1 m wide and 2.40 m apart face each other, as shown in Fig. 10.16. (a) Find $Q_{net_{1\text{-}2}}$ (W/m) if the surroundings are black and at 250 K. (b) Find $Q_{net_{1\text{-}2}}$ (W/m) if they are connected by an insulated diffuse reflector between the edges on both sides. Also evaluate the temperature of the reflector in part (b).

Solution. From Table 10.2, case 1, we find $F_{1\text{-}2} = 0.2 = F_{2\text{-}1}$. In addition, $F_{2\text{-}3} = 1 - F_{2\text{-}1} = 0.8$, irrespective of whether surface (3) represents the surroundings or the insulated shield.

In case (a), the two nodal equations (10.34a) and (10.34b) become

$$\frac{1451 - B_1}{2.333} = \frac{B_1 - B_2}{1/0.2} + \frac{B_1 - B_3}{1/0.8}$$

$$\frac{459.3 - B_2}{1} = \frac{B_2 - B_1}{1/0.2} + \frac{B_2 - B_3}{1/0.8}$$

Equation (10.34c) cannot be used directly for black surroundings, since $\varepsilon_3 = 1$ and the surface resistance in the left-hand side denominator would be zero. But the numerator is also zero in this case, since $e_{b_3} = B_3$ for black surroundings. And since we now know $B_3 = \sigma T_3^4 = 221.5$ W/m²K, we can use it directly in the two equations above.

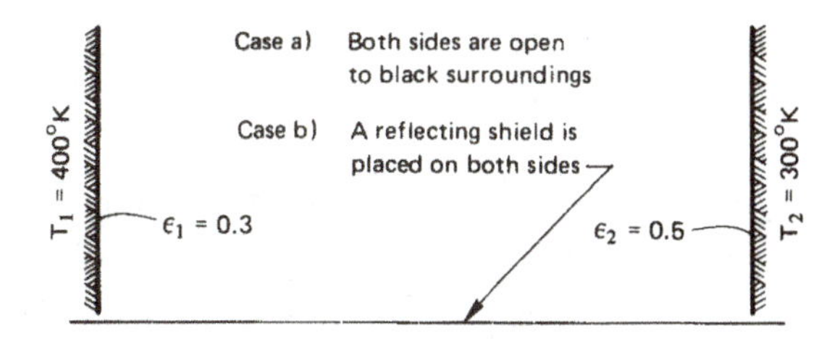

Figure 10.16 Illustration for Example 10.9.

Thus,

$$\begin{aligned} B_1 - \ 0.14\,B_2 - 0.56(221.5) &= \ 435.6 \\ -B_1 + 10.00\,B_2 - 4.00(221.5) &= 2296.5 \end{aligned}$$

or

$$\left.\begin{aligned} B_1 - \ 0.14\,B_2 &= \ 559.6 \\ -B_1 + 10.00\,B_2 &= 3182.5 \end{aligned}\right\} \quad \text{so} \quad \left\{\begin{aligned} B_1 &= 612.1 \text{ W/m}^2 \\ B_2 &= 379.5 \text{ W/m}^2 \end{aligned}\right.$$

Thus, the net flow from (1) to (2) is quite small:

$$Q_{\text{net}_{1\text{-}2}} = \frac{B_1 - B_2}{1/(A_1 F_{1\text{-}2})} = 46.53 \text{ W/m}$$

Since each strip also loses heat to the surroundings, $Q_{\text{net}_1} \neq Q_{\text{net}_2} \neq Q_{\text{net}_{1\text{-}2}}$.

For case (b), with the adiabatic shield in place, eqn. (10.34c) can be combined with the other two nodal equations:

$$0 = \frac{B_3 - B_1}{1/0.8} + \frac{B_3 - B_2}{1/0.8}$$

The three equations can be solved manually, by the use of determinants, or with a computerized matrix algebra package. The result is

$$B_1 = 987.7 \text{ W/m}^2 \qquad B_2 = 657.4 \text{ W/m}^2 \qquad B_3 = 822.6 \text{ W/m}^2$$

In this case, because surface (3) is adiabatic, all net heat transfer from surface (1) is to surface (2): $Q_{\text{net}_1} = Q_{\text{net}_{1\text{-}2}}$. Then, from eqn. (10.33a), we get

$$Q_{\text{net}_{1\text{-}2}} = \left[\frac{987.7 - 657.4}{1/(1)(0.2)} + \frac{987.7 - 822.6}{1/(1)(0.8)}\right] = 198 \text{ W/m}$$

Of course, because node (3) is insulated, it is much easier to use eqn. (10.35) to get $Q_{\text{net}_{1\text{-}2}}$:

$$Q_{\text{net}_{1\text{-}2}} = \frac{5.67 \times 10^{-8}(400^4 - 300^4)}{\dfrac{0.7}{0.3} + \dfrac{1}{\dfrac{1}{1/0.8 + 1/0.8} + 0.2} + \dfrac{0.5}{0.5}} = 198 \text{ W/m}$$

The result, of course, is the same. We note that the presence of the reflector increases the net heat flow from (1) to (2).

The temperature of the reflector (3) is obtained from eqn. (10.33b) with $Q_{\text{net}_3} = 0$:

$$0 = e_{b_3} - B_3 = 5.67 \times 10^{-8} T_3^4 - 822.6$$

so

$$T_3 = 347\text{ K}$$ ■

Algebraic solution of multisurface enclosure problems

An *enclosure* can consist of any number of surfaces that exchange radiation with one another. The evaluation of radiant heat transfer among these surfaces proceeds in essentially the same way as for three surfaces. For multisurface problems, however, the electrical circuit approach is less convenient than a formulation based on matrices. The matrix equations are usually solved on a computer.

An enclosure formed by n surfaces is shown in Fig. 10.17. As before, we will assume that:

- Each surface is diffuse, gray, and opaque, so that $\varepsilon = \alpha$ and $\rho = 1 - \varepsilon$.
- The temperature and net heat flux are uniform over each surface (more precisely, the radiosity must be uniform and the other properties are averages for each surface). Either temperature or flux must be specified on every surface.
- The view factor, F_{i-j}, between any two surfaces i and j is known.
- Conduction and convection within the enclosure can be neglected, and any fluid in the enclosure is transparent and nonradiating.

We are interested in determining the heat fluxes at the surfaces where temperatures are specified, and vice versa.

The rate of heat loss from the ith surface of the enclosure can conveniently be written in terms of the radiosity, B_i, and the irradiation, H_i, from eqns. (10.19) and (10.21)

$$q_{\text{net}_i} = B_i - H_i = \frac{\varepsilon_i}{1 - \varepsilon_i}\left(\sigma T_i^4 - B_i\right) \tag{10.36}$$

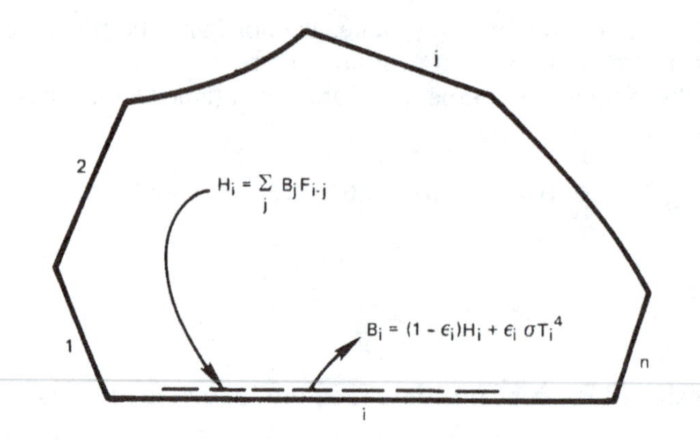

Figure 10.17 An enclosure composed of n diffuse, gray surfaces.

where

$$B_i = \rho_i H_i + \varepsilon_i e_{b_i} = (1-\varepsilon_i)\, H_i + \varepsilon_i\, \sigma T_i^4 \tag{10.37}$$

However, $A_i H_i$, the irradiating heat transfer incident on surface i, is the sum of energies reaching i from all other surfaces, including itself

$$A_i H_i = \sum_{j=1}^{n} A_j B_j F_{j-i} = \sum_{j=1}^{n} B_j A_i F_{i-j}$$

where we have used the reciprocity rule, $A_j F_{j-i} = A_i F_{i-j}$. Thus

$$H_i = \sum_{j=1}^{n} B_j F_{i-j} \tag{10.38}$$

It follows from eqns. (10.37) and (10.38) that

$$B_i = (1-\varepsilon_i) \sum_{j=1}^{n} B_j F_{i-j} + \varepsilon_i\, \sigma T_i^4 \tag{10.39}$$

This equation applies to every surface, $i = 1, \ldots, n$. When all the surface temperatures are specified, the result is a set of n linear equations for the n unknown radiosities. For numerical purposes, it is sometimes convenient to introduce the Kronecker delta,

$$\delta_{ij} = \begin{cases} 1 & \text{for } i = j \\ 0 & \text{for } i \neq j \end{cases} \tag{10.40}$$

and to rearrange eqn. (10.39) as

$$\sum_{j=1}^{n} \underbrace{\left[\delta_{ij} - (1-\varepsilon_i)F_{i-j}\right]}_{\equiv C_{ij}} B_j = \varepsilon_i \sigma T_i^4 \quad \text{for } i = 1, \ldots, n \tag{10.41}$$

The radiosities are then found by inverting the matrix C_{ij}. The rate of heat loss from the ith surface, $Q_{\text{net}_i} = A_i q_{\text{net}_i}$, can be obtained from eqn. (10.36).

For those surfaces where heat fluxes are prescribed, we can eliminate the $\varepsilon_i \sigma T_i^4$ term in eqn. (10.39) or (10.41) using eqn. (10.36). We again obtain a matrix equation that can be solved for the B_i's. Finally, eqn. (10.36) is solved for the unknown temperature of surface in question.

In many cases, the radiosities themselves are of no particular interest. The heat flows are what is really desired. With a bit more algebra (see Problem 10.45), one can formulate a matrix equation for the n unknown values of Q_{net_i}:

$$\sum_{j=1}^{n} \left[\frac{\delta_{ij}}{\varepsilon_i} - \frac{(1-\varepsilon_j)}{\varepsilon_j A_j} A_i F_{i-j}\right] Q_{\text{net}_j} = \sum_{j=1}^{n} A_i F_{i-j}\left(\sigma T_i^4 - \sigma T_j^4\right) \tag{10.42}$$

Example 10.10

Two sides of a long triangular duct, as shown in Fig. 10.18, are made of stainless steel ($\varepsilon = 0.5$) and are maintained at 500°C. The third side is of copper ($\varepsilon = 0.15$) and has a uniform temperature of 100°C. Calculate the rate of heat transfer to the copper base per meter of length of the duct.

Solution. Assume the duct walls to be gray and diffuse and that convection is negligible. The view factors can be calculated from configuration 4 of Table 10.2:

$$F_{1\text{-}2} = \frac{A_1 + A_2 - A_3}{2A_1} = \frac{0.5 + 0.3 - 0.4}{1.0} = 0.4$$

Similarly, $F_{2\text{-}1} = 0.67$, $F_{1\text{-}3} = 0.6$, $F_{3\text{-}1} = 0.75$, $F_{2\text{-}3} = 0.33$, and $F_{3\text{-}2} = 0.25$. The surfaces cannot "see" themselves, so $F_{1\text{-}1} = F_{2\text{-}2} = F_{3\text{-}3} = 0$. Equation (10.39) leads to three algebraic equations for the three

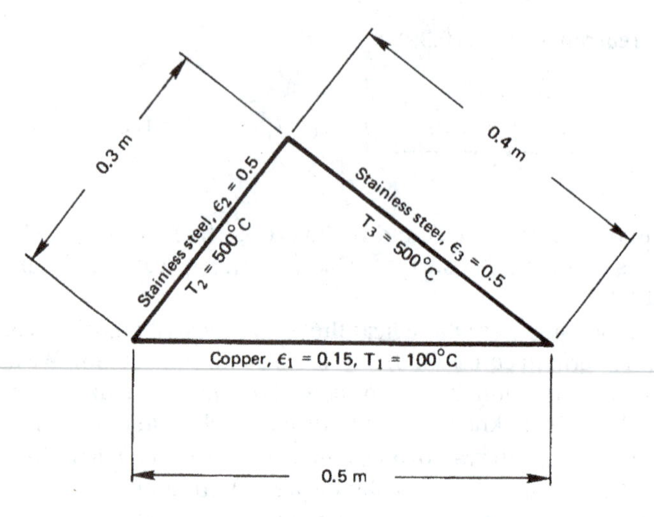

Figure 10.18 Illustration for Example 10.10.

unknowns, B_1, B_2, and B_3.

$$B_1 = \underbrace{(1-\varepsilon_1)}_{0.85}\Big(\underbrace{F_{1\text{-}1}}_{0} B_1 + \underbrace{F_{1\text{-}2}}_{0.4} B_2 + \underbrace{F_{1\text{-}3}}_{0.6} B_3\Big) + \underbrace{\varepsilon_1}_{0.15} \sigma T_1^4$$

$$B_2 = \underbrace{(1-\varepsilon_2)}_{0.5}\Big(\underbrace{F_{2\text{-}1}}_{0.67} B_1 + \underbrace{F_{2\text{-}2}}_{0} B_2 + \underbrace{F_{2\text{-}3}}_{0.33} B_3\Big) + \underbrace{\varepsilon_2}_{0.5} \sigma T_2^4$$

$$B_3 = \underbrace{(1-\varepsilon_3)}_{0.5}\Big(\underbrace{F_{3\text{-}1}}_{0.75} B_1 + \underbrace{F_{3\text{-}2}}_{0.25} B_2 + \underbrace{F_{3\text{-}3}}_{0} B_3\Big) + \underbrace{\varepsilon_3}_{0.5} \sigma T_3^4$$

It would be easy to solve this system numerically using matrix methods. Alternatively, we can substitute the third equation into the first two to eliminate B_3, and then use the second equation to eliminate B_2 from the first. The result is

$$B_1 = 0.232\,\sigma T_1^4 + 0.319\,\sigma T_2^4 + 0.447\,\sigma T_3^4$$

Equation (10.36) gives the rate of heat loss by surface (1) as

$$\begin{aligned} Q_{\text{net}_1} &= A_1 \frac{\varepsilon_1}{1-\varepsilon_1}\left(\sigma T_1^4 - B_1\right) \\ &= A_1 \frac{\varepsilon_1}{1-\varepsilon_1}\,\sigma\left(T_1^4 - 0.232\,T_1^4 - 0.319\,T_2^4 - 0.447\,T_3^4\right) \end{aligned}$$

$$= (0.5)\left(\frac{0.15}{0.85}\right)(5.67\times10^{-8})$$
$$\times\left[(373)^4 - 0.232(373)^4 - 0.319(773)^4 - 0.447(773)^4\right]\ \text{W/m}$$
$$= -1294\ \text{W/m} \qquad \blacksquare$$

The negative sign indicates that the copper base is gaining heat.

Enclosures with nonisothermal, nongray, or nondiffuse surfaces

The representation of enclosure heat exchange by eqn. (10.41) or (10.42) is actually quite powerful. For example, if the primary surfaces in an enclosure are not isothermal, they may be subdivided into a larger number of smaller surfaces, each of which is approximately isothermal. Then either equation may be used to calculate the heat exchange among the set of smaller surfaces.

For those cases in which the gray surface approximation, eqn. (10.8c), cannot be applied (owing to very different temperatures or strong wavelength dependence in ε_λ), eqns. (10.41) and (10.42) may be applied on a monochromatic basis, since the monochromatic form of Kirchhoff's law, eqn. (10.8b), remains valid. The results must, of course, be integrated over wavelength to get the heat exchange. The calculation is usually simplified by breaking the wavelength spectrum into a few discrete bands within which radiative properties are approximately constant [10.3, Chpt. 7].

When the surfaces are not diffuse — when emission or reflection vary with angle — a variety of other methods can be applied. Among them, the Monte Carlo technique is probably the most widely used. The Monte Carlo technique tracks emissions and reflections through various angles among the surfaces and estimates the probability of absorption or re-reflection [10.4, 10.7]. This method allows complex situations to be numerically computed with relative ease, provided that one is careful to obtain statistical convergence.

10.5 Gaseous radiation

We have treated every radiation problem thus far as though radiant heat flow in the space separating the surfaces of interest were completely unobstructed by any fluid in between. However, all gases interact with photons to some extent, by absorbing or deflecting them, and they can even emit additional photons. The result is that fluids can play a role in the thermal radiation to the the surfaces that surround them.

We have ignored this effect so far because it is generally very small, especially in air and if the distance between the surfaces is on the order of meters or less. When other gases are involved, especially at high temperatures, as in furnaces, or when long distances are involved, as in the atmosphere, gas radiation can become an important part of the heat exchange process.

How gases interact with photons

The photons of radiant energy passing through a gaseous region can be impeded in two ways. Some can be "scattered," or deflected, in various directions, and some can be absorbed into the molecules. Scattering is a fairly minor influence in most gases unless they contain foreign particles, such as dust or fog. In cloudless air, for example, we are aware of the scattering of sunlight only when it passes through many miles of the atmosphere. Then the shorter wavelengths of sunlight are scattered (short wavelengths, as it happens, are far more susceptible to scattering by gas molecules than longer wavelengths, through a process known as *Rayleigh scattering*). That scattered light gives the sky its blue hues.

At sunset, sunlight passes through the atmosphere at a shallow angle for hundreds of miles. Radiation in the blue wavelengths has all been scattered out before it can be seen. Thus, we see only the unscattered red hues, just before dark.

When particles suspended in a gas have diameters near the wavelength of light, a more complex type of scattering can occur, known as *Mie scattering*. Such scattering occurs from the water droplets in clouds (often making them a brilliant white color). It also occurs in gases that contain soot or in pulverized coal combustion. Mie scattering has a strong angular variation that changes with wavelength and particle size [10.8].

The absorption or emission of radiation by molecules, rather than particles, will be our principal focus. The interaction of molecules with radiation — photons, that is — is governed by quantum mechanics. It's helpful at this point to recall a few facts from molecular physics. Each photon has an energy hc_o/λ, where h is Planck's constant, c_o is the speed of light, and λ is the wavelength of light. Thus, photons of shorter wavelengths have higher energies: ultraviolet photons are more energetic than visible photons, which are in turn more energetic than infrared photons.

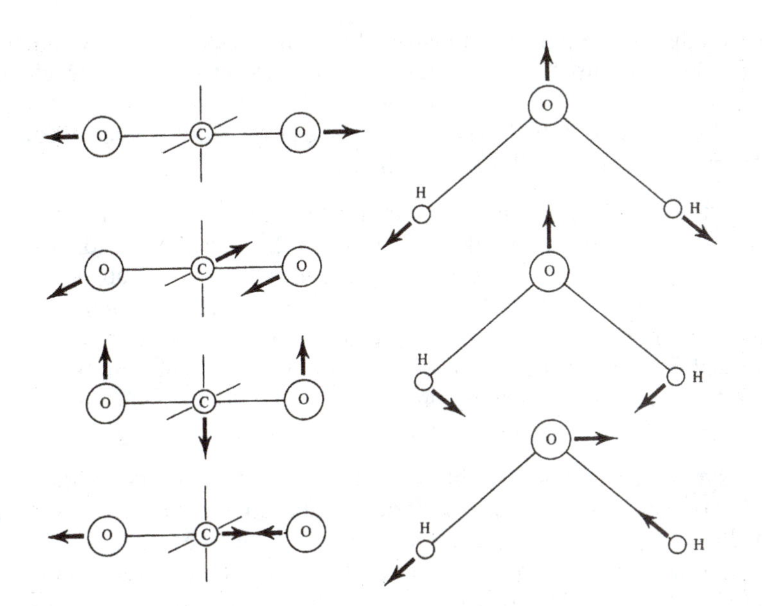

Figure 10.19 Vibrational modes of carbon dioxide and water.

It is not surprising that hotter objects emit more visible photons.

Molecules can store energy by rotation, by vibration (Fig. 10.19), or in their electrons. Whereas the possible energy of a photon varies smoothly with wavelength, the energies of molecules are constrained by quantum mechanics to change only in discrete steps between the molecule's allowable "energy levels." The available energy levels depend on the molecule's chemical structure.

When a molecule emits a photon, its energy drops in a discrete step from a higher energy level to a lower one. The energy given up is carried away by the photon. As a result, the wavelength of that photon is determined by the specific change in molecular energy level that caused it to be emitted. Just the opposite happens when a photon is absorbed: the photon's wavelength must match a specific energy level change available to that particular molecule. As a result, each molecular species can absorb only photons at, or very close to, particular wavelengths! Often, these wavelengths are tightly grouped into so-called *absorption bands*, outside of which the gas is essentially transparent to photons.

The fact that a molecule's structure determines how it absorbs and emits light has been used extensively by chemists as a tool for deducing molecular structure. A knowledge of the energy levels in a molecule, in conjunction with quantum theory, allows specific atoms and bonds to be identified. This is called *spectroscopy* (see [10.9, Chpt. 18 & 19] for an introduction; see [10.10] to go overboard).

At the wavelengths that correspond to thermal radiation at typical temperatures, it happens that transitions in the vibrational and rotation modes of molecules have the greatest influence on radiative absorptance. Such transitions can be driven by photons only when the molecule has some asymmetry.[4] Thus, for all practical purposes, monatomic and symmetrical diatomic molecules are transparent to thermal radiation. The major components of air—N_2 and O_2—are therefore nonabsorbing; so, too, are H_2 and such monatomic gases as argon.

Asymmetrical molecules like CO_2, H_2O, CH_4, O_3, NH_3, N_2O, and SO_2, on the other hand, each absorb thermal radiation of certain wavelengths. The first two of these, CO_2 and H_2O, are always present in air. To understand how the interaction works, consider the possible vibrations of CO_2 and H_2O shown in Fig. 10.19. For CO_2, the topmost mode of vibration is symmetrical and has no interaction with thermal radiation at normal pressures. The other three modes produce asymmetries in the molecule when they occur; each is important to thermal radiation.

The primary absorption wavelength for the two middle modes of CO_2 is 15 μm, which lies in the thermal infrared. The wavelength for the bottommost mode is 4.3 μm. For H_2O, middle mode of vibration interacts strongly with thermal radiation at 6.3 μm. The other two both affect 2.7 μm radiation, although the bottom one does so more strongly. In addition, H_2O has a rotational mode that absorbs thermal radiation having wavelengths of 14 μm or more. Both of these molecules show additional absorption lines at shorter wavelengths, which result from the superposition of two or more vibrations and their harmonics (e.g., at 2.7 μm for CO_2 and at 1.9 and 1.4 μm for H_2O, as seen in Fig. 10.2). Additional absorption bands can appear at high temperature or high pressure.

---

[4]The asymmetry required is in the distribution of electric charge — the dipole moment. A vibration of the molecule must create a fluctuating dipole moment in order to interact with photons. A rotation interacts with photons only if the molecule has a permanent dipole moment.

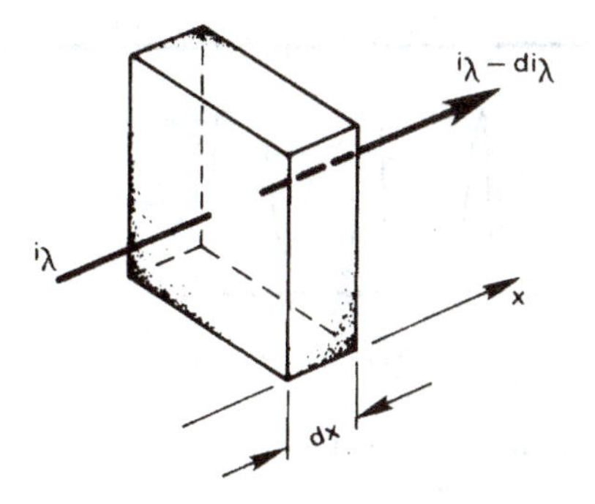

Figure 10.20 The attenuation of radiation through an absorbing (and/or scattering) gas.

Absorptance, transmittance, and emittance

Figure 10.20 shows radiant energy passing through an absorbing gas with a monochromatic intensity i_λ. As it passes through an element of thickness dx, the intensity will be reduced by an amount di_λ:

$$di_\lambda = -\rho\kappa_\lambda i_\lambda \, dx \tag{10.43}$$

where ρ is the gas density and κ_λ is called the *monochromatic absorption coefficient.* If the gas scatters radiation, we replace κ_λ with γ_λ, the *monochromatic scattering coefficient.* If it both absorbs and scatters radiation, we replace κ_λ with $\beta_\lambda \equiv \kappa_\lambda + \gamma_\lambda$, the *monochromatic extinction coefficient.*[5] The dimensions of κ_λ, β_λ, and γ_λ are all m^2/kg.

If $\rho\kappa_\lambda$ is constant through the gas, eqn. (10.43) can be integrated from an initial intensity i_{λ_0} at $x = 0$ to obtain

$$i_\lambda(x) = i_{\lambda_0} e^{-\rho\kappa_\lambda x} \tag{10.44}$$

This result is called *Beer's law* (pronounced "Bayr's" law). For a gas layer of a given depth $x = L$, the ratio of final to initial intensity defines that layer's monochromatic transmittance, τ_λ:

$$\tau_\lambda \equiv \frac{i_\lambda(L)}{i_{\lambda_0}} = e^{-\rho\kappa_\lambda L} \tag{10.45}$$

Further, since gases do not reflect radiant energy, $\tau_\lambda + \alpha_\lambda = 1$. Thus, the monochromatic absorptance, α_λ, is

$$\alpha_\lambda = 1 - e^{-\rho\kappa_\lambda L} \tag{10.46}$$

[5]All three coefficients, κ_λ, γ_λ, and β_λ, are expressed on a mass basis. They could, alternatively, have been expressed on a volumetric basis.

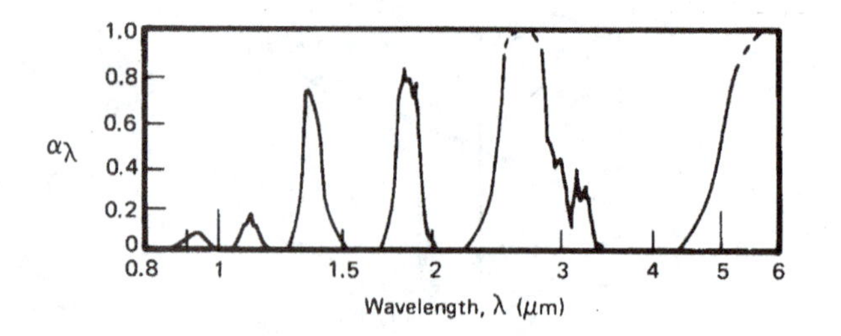

Figure 10.21 The monochromatic absorptance of a 1.09 m thick layer of steam at 127°C.

Both τ_λ and α_λ depend on the density and thickness of the gas layer. The product $\rho\kappa_\lambda L$ is sometimes called the *optical depth* of the gas. For very small values of $\rho\kappa_\lambda L$, the gas is transparent to the wavelength λ.

The dependence of α_λ on λ is normally very strong. As we have seen, a given molecule will absorb radiation in certain wavelength bands, while allowing radiation with somewhat higher or lower wavelengths to pass almost unhindered. Figure 10.21 shows the absorptance of water vapor as a function of wavelength for a fixed depth. We can see the absorption bands at wavelengths of 6.3, 2.7, 1.9, and 1.4 μm that were mentioned before.

A comparison of Fig. 10.21 with Fig. 10.2 readily shows why radiation from the sun, as viewed from the earth's surface, shows a number of spikey indentations at certain wavelengths. Several of those indentations occur in bands where atmospheric water vapor absorbs incoming solar radiation, in accordance with Fig. 10.21. The other indentations in Fig. 10.2 occur where ozone and CO_2 absorb radiation. The sun itself does not have these regions of low emittance; it is just that much of the radiation in these bands is absorbed by gases in the atmosphere before it can reach the ground.

Just as α_λ and ε_λ are equal to one another for a diffuse solid surface, they are equal for a gas. We may demonstrate this by considering an isothermal gas that is in thermal equilibrium with a black enclosure that contains it. The radiant intensity within the enclosure is that of a black body, i_{λ_b}, at the temperature of the gas and enclosure. Equation (10.43) shows that a small section of gas absorbs radiation, reducing the intensity by an amount $\rho\kappa_\lambda i_{\lambda_b}\,dx$. To maintain equilibrium, the gas must therefore emit an equal amount of radiation:

$$di_\lambda = \rho\kappa_\lambda i_{\lambda_b}\, dx \tag{10.47}$$

Now, if radiation from some other source is transmitted through a nonscattering isothermal gas, we can combine the absorption from eqn. (10.43) with the emission from eqn. (10.47) to form an energy balance called the *equation of transfer*

$$\frac{di_\lambda}{dx} = -\rho\kappa_\lambda i_\lambda + \rho\kappa_\lambda i_{\lambda_b} \tag{10.48}$$

Integration of this equation yields a result similar to eqn. (10.44):

$$i_\lambda(L) = i_{\lambda_0}\underbrace{e^{-\rho\kappa_\lambda L}}_{=\tau_\lambda} + i_{\lambda_b}\underbrace{\left(1 - e^{-\rho\kappa_\lambda L}\right)}_{\equiv\varepsilon_\lambda} \tag{10.49}$$

The first righthand term represents the transmission of the incoming intensity, as in eqn. (10.44), and the second is the radiation emitted by the gas itself. The coefficient of the second righthand term defines the monochromatic emittance, ε_λ, of the gas layer. Finally, comparison to eqn. (10.46) shows that

$$\varepsilon_\lambda = \alpha_\lambda = 1 - e^{-\rho\kappa_\lambda L} \tag{10.50}$$

Again, we see that for very small $\rho\kappa_\lambda L$ the gas will neither absorb nor emit radiation of wavelength λ.

Heat transfer from gases to walls

We now see that predicting the total emittance, ε_g, of a gas layer will be complex. We have to take account of the gases' absorption bands as well as the layer's thickness and density. Such predictions can be done [10.11], but they are laborious. For making simpler (but less accurate) estimates, correlations of ε_g have been developed.

Such correlations are based on the following model: An isothermal gas of temperature T_g and thickness L, is bounded by walls at the single temperature T_w. The gas consists of a small fraction of an absorbing species (say CO_2) mixed into a nonabsorbing species (say N_2). If the absorbing gas has a partial pressure p_a and the mixture has a total pressure p, the correlation takes this form:

$$\varepsilon_g = \text{fn}\left(p_a L, p, T_g\right) \tag{10.51}$$

The parameter p_aL is a measure of the layer's optical depth; p and T_g account for changes in the absorption bands with pressure and temperature.

Hottel and Sarofim [10.12] provide such correlations for CO_2 and H_2O, built from research by Hottel and others before 1960. The correlations take the form

$$\varepsilon_g\left(p_aL, p, T_g\right) = f_1\left(p_aL, T_g\right) \times f_2\left(p, p_a, p_aL\right) \tag{10.52}$$

where the experimental functions f_1 and f_2 are plotted in Figs. 10.22 and 10.23 for CO_2 and H_2O, respectively. The first function, f_1, is a correlation for a total pressure of $p = 1$ atm with a very small partial pressure of the absorbing species. The second function, f_2, is a correction factor to account for other values of p_a or p. Additional corrections must be applied if both CO_2 and H_2O are present in the same mixture.

To find the net heat transfer between the gas and the walls, we must also find the total absorptance, α_g, of the gas. Despite the equality of the monochromatic emittance and absorptance, ε_λ and α_λ, the total values, ε_g and α_g, will not generally be equal. This is because the absorbed radiation may come from, say, a wall having a much different temperature than the gas with a correspondingly different wavelength distribution. Hottel and Sarofim show that α_g may be estimated from the correlation for ε_g as follows:[6]

$$\alpha_g = \left(\frac{T_g}{T_w}\right)^{1/2} \cdot \varepsilon\left(p_aL\frac{T_w}{T_g}, p, T_w\right) \tag{10.53}$$

Finally, we need to determine an appropriate value of L for a given enclosure. The correlations just given for ε_g and α_g are based on L as a one-dimensional path through the gas. Even for a pair of flat plates a distance L apart, this won't be appropriate since radiation can travel much farther if it follows a path that is not perpendicular to the plates.

For enclosures that have black walls at a uniform temperature, we can use an effective path length, L_0, called the *geometrical mean beam length*, to represent both the size and the configuration of a gaseous region. The geometrical mean beam length is defined as

$$L_0 \equiv \frac{4\,(\text{volume of gas})}{\text{boundary area that is irradiated}} \tag{10.54}$$

[6] Hottel originally recommended replacing the exponent 1/2 by 0.65 for CO_2 and 0.45 for H_2O. Theory, and more recent work, both suggest using the value 1/2 [10.13].

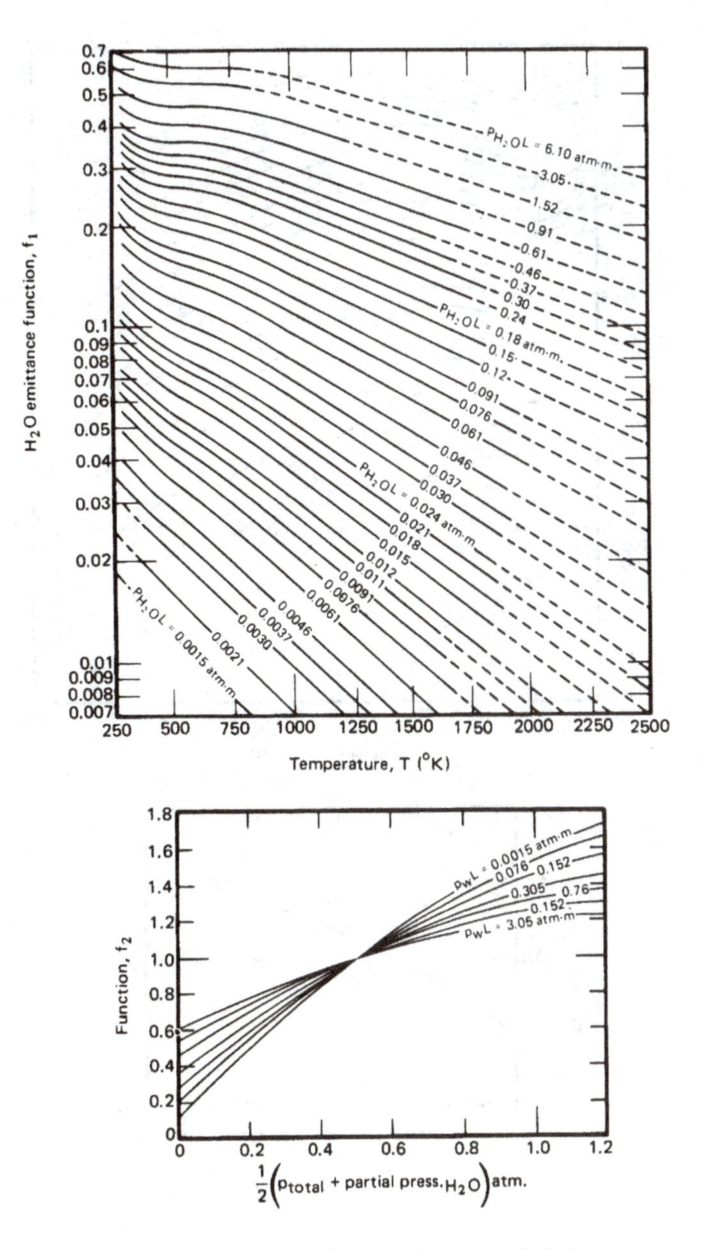

Figure 10.22 Functions used to predict $\varepsilon_g = f_1 f_2$ for water vapor in air.

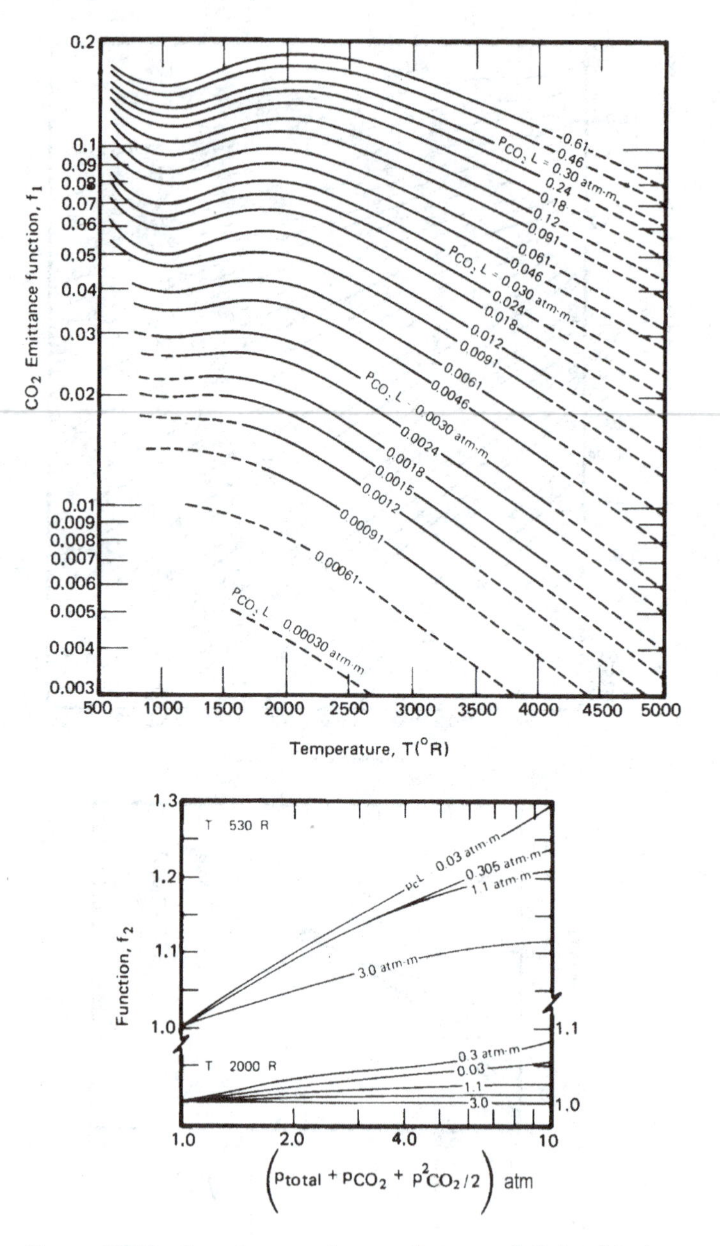

Figure 10.23 Functions used to predict $\varepsilon_g = f_1 f_2$ for CO_2 in air. All pressures in atmospheres.

Thus, for two infinite parallel plates a distance ℓ apart, $L_0 = 4A\ell/2A = 2\ell$. Some other values of L_0 for gas volumes exchanging heat with all points on their boundaries are as follow:

- For a sphere of diameter D, $L_0 = 2D/3$
- For an infinite cylinder of diameter D, $L_0 = D$
- For a cube of side L, $L_0 = 2L/3$
- For a cylinder with height $= D$, $L_0 = 2D/3$

For cases where the gas is strongly absorbing, better accuracy can be obtained by replacing the constant 4 in eqn. (10.54) by 3.5, lowering the mean beam length about 12%.

We are now in position to treat a problem in which hot gases (say the products of combustion) radiate to a black container. Consider an example:

Example 10.11

A long cylindrical combustor 40 cm in diameter contains a gas at 1200°C consisting of 0.8 atm N_2 and 0.2 atm CO_2. What is the net heat radiated to the walls if they are at 300°C?

Solution. Let us first obtain ε_g. We have $L_0 = D = 0.40$ m, a total pressure of 1.0 atm, $p_{CO_2} = 0.2$ atm, and $T = 1200°C = 2651°R$. Then Fig. 10.23a gives f_1 as 0.098 and Fig. 10.23b gives $f_2 \cong 1$, so $\varepsilon_g = 0.098$. Next, we use eqn. (10.53) to obtain α_g, with $T_w = 1031°R$, $p_{H_2O} L T_w / T_g = 0.031$:

$$\alpha_g = \left(\frac{1200 + 273}{300 + 273}\right)^{0.5} (0.074) = 0.12$$

Now we can calculate $Q_{net_{g\text{-}w}}$. For these problems with one wall surrounding one gas, the use of the mean beam length in finding ε_g and α_g accounts for all geometrical effects, and no view factor is required. The net heat transfer is calculated using the surface area of the wall:

$$\begin{aligned} Q_{net_{g\text{-}w}} &= A_w\left(\varepsilon_g \sigma T_g^4 - \alpha_g \sigma T_w^4\right) \\ &= \pi(0.4)(5.67\times10^{-8})\left[(0.098)(1473)^4 - (0.12)(573)^4\right] \\ &= 32 \text{ kW/m} \end{aligned}$$

■

Total emittance charts and the mean beam length provide a simple, but crude, tool for dealing with gas radiation. Since the introduction of these ideas in the mid-twentieth century, major advances have been made in our knowledge of the radiative properties of gases and in the tools available for solving gas radiation problems. In particular, band models of gas radiation, and better measurements, have led to better procedures for dealing with the total radiative properties of gases (see, in particular, References [10.11] and [10.13]). Tools for dealing with radiation in complex enclosures have also improved. The most versatile of these is the previously-mentioned Monte Carlo method [10.4, 10.7], which can deal with nongray, nondiffuse, and nonisothermal walls with nongray, scattering, and nonisothermal gases. An extensive literature also deals with approximate analytical techniques, many of which are based on the idea of a "gray gas" — one for which ε_λ and α_λ are independent of wavelength. However, as we have pointed out, the gray gas model is not even a *qualitative* approximation to the properties of real gases.[7]

Finally, it is worth noting that gaseous radiation is frequently less important than one might imagine. Consider, for example, two flames: a bright orange candle flame and a "cold-blue" hydrogen flame. Both have a great deal of water vapor in them, as a result of oxidizing H_2. But the candle will warm your hands if you place them near it and the hydrogen flame will not. Yet the temperature in the hydrogen flame is *higher.* It turns out that what is radiating both heat and light from the candle is soot — small solid particles of almost thermally black carbon. The CO_2 and H_2O in the candle flame actually contribute relatively little to radiation.

10.6 Solar energy

The sun

The sun continually irradiates the earth at a rate of about 1.74×10^{14} kW. If we imagine this energy to be distributed over a circular disk with the earth's diameter, the solar irradiation is about 1367 W/m^2, as measured

[7]Edwards [10.11] describes the gray gas as a "myth." He notes, however, that spectral variations may be overlooked for a gas containing spray droplets or particles [in a range of sizes] or for some gases that have wide, weak absorption bands within the spectral range of interest [10.2]. Some accommodation of molecular properties can be achieved using the *weighted sum of gray gases* concept [10.12], which treats a real gas as superposition of gray gases having different properties.

by satellites above the atmosphere. Much of this energy reaches the ground, where it sustains the processes of life.

The temperature of the sun varies from tens of millions of kelvin in its core to between 4000 and 6000 K at its surface, where most of the sun's thermal radiation originates. The wavelength distribution of the sun's energy is not quite that of a black body, but it may be approximated as such. A straightforward calculation (see Problem 10.49) shows that a black body of the sun's size and distance from the earth would produce the same irradiation as the sun if its temperature were 5777 K.

The solar radiation reaching the earth's surface is always less than that above the atmosphere owing to atmospheric absorption and the earth's curvature and rotation. Solar radiation usually arrives at an angle of less than 90° to the surface because the sun is rarely directly overhead. We have seen that a radiant heat flux arriving at an angle less than 90° is reduced by the cosine of that angle (Fig. 10.4). The sun's angle varies with latitude, time of day, and day of year. Trigonometry and data for the earth's rotation can be used to find the appropriate angle.

Figure 10.2 shows the reduction of solar radiation by atmospheric absorption for one particular set of atmospheric conditions. In fact, when the sun passes through the atmosphere at a low angle (near the horizon), the path of radiation through the atmosphere is longer, providing relatively more opportunity for atmospheric absorption and scattering. Additional moisture in the air can increase the absorption by H_2O, and, of course, clouds can dramatically reduce the solar radiation reaching the ground. The consequence of these various effects is that the solar radiation received on the ground is almost never more than 1200 W/m^2 and is often only a few hundred W/m^2. Extensive data are available for estimating the ground level solar irradiation at a given location, time, and date [10.14, 10.15].

The distribution of the Sun's energy and atmospheric irradiation

Figure 10.24 shows what becomes of the solar energy that impinges on the earth if we average it over the year and the globe, taking account of all kinds of weather. Only 45% of the sun's energy actually reaches the earth's surface. The mean energy received is about 235 W/m^2 if averaged over the surface and the year. The lower left-hand portion of the figure shows how this energy is, in turn, all returned to the atmosphere and to space.

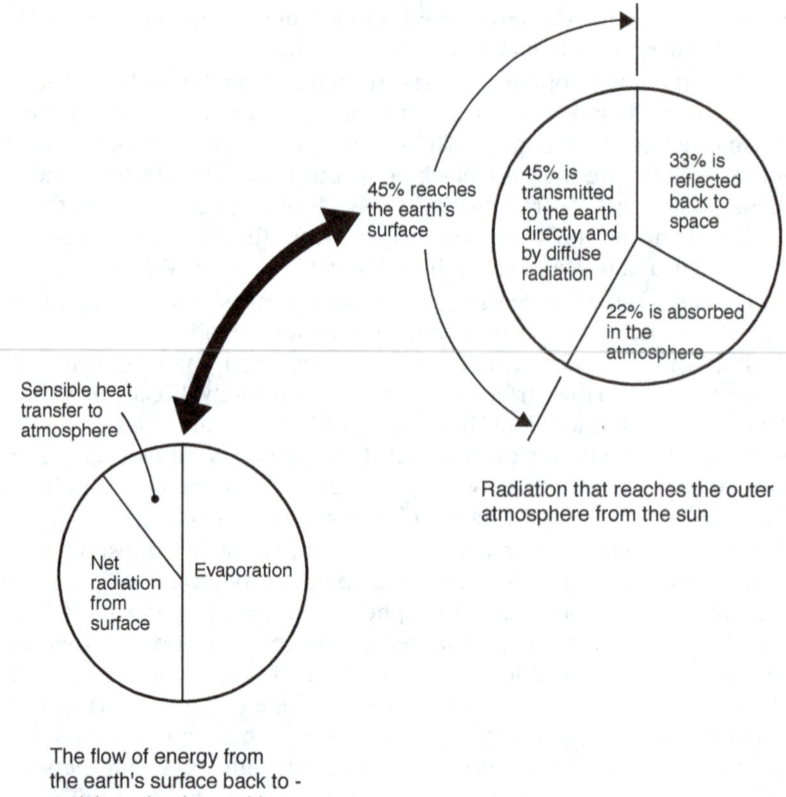

Figure 10.24 The approximate distribution of the flow of the sun's energy to and from the earth's surface [10.16].

The solar radiation reaching the earth's surface includes direct radiation that has passed through the atmosphere and diffuse radiation that has been scattered, but not absorbed, by the atmosphere. Atmospheric gases also irradiate the surface. This irradiation is quite important to maintaining the temperature of objects on the surface.

In Section 10.5, saw that the energy radiated by a gas depends upon the depth of the gas, its temperature, and the molecules present in it. The emittance of the atmosphere has been characterized in detail [10.16,

10.17, 10.18]. For practical calculations, however, it is often convenient to treat the sky as a black radiator having some appropriate temperature. This effective *sky temperature* usually lies between 5 and 30 K below the ground level air temperature. The sky temperature decreases as the amount of water vapor in the air goes down. For cloudless skies, the sky temperature may be estimated using the dew-point temperature, T_{dp}, and the hour past midnight, t:

$$T_{sky} = T_{air}\Big[0.711 + 0.0056\,T_{dp} + 7.3\times 10^{-5}\,T_{dp}^2 + 0.013\cos(2\pi t/24)\Big]^{1/4} \qquad (10.55)$$

where T_{sky} and T_{air} are in kelvin and T_{dp} is in °C. This equation applies for dew points from −20°C to 30°C [10.19].

It is fortunate that sky temperatures are relatively warm. In the absence of an atmosphere, not only would more of the sun's radiation reach the ground during the day, but at night heat would radiate directly into the bitter cold of outer space. Such conditions prevail on the Moon, where average daytime surface temperatures are about 110°C while average nighttime temperatures plunge to about −150°C.

Selective emitters, absorbers, and transmitters

We have noted that most of the sun's energy lies at wavelengths near the visible region of the electromagnetic spectrum and that most of the radiation from objects at temperatures typical of the earth's surface is on much longer, infrared wavelengths (see pg. 537). One result is that materials may be chosen or designed to be selectively good emitters or reflectors of both solar and infrared radiation.

Table 10.4 shows the infrared emittance and solar absorptance for several materials. Among these, we identify several particularly selective solar absorbers and solar reflectors. The selective absorbers have a high absorptance for solar radiation and a low emittance for infrared radiation. Consequently, they do not strongly reradiate the solar energy that they absorb. The selective solar reflectors, on the other hand, reflect solar energy strongly and also radiate heat efficiently in the infrared. Solar reflectors stay much cooler than solar absorbers in bright sunlight.

A wide range of selective coatings have been developed for solar absorbers operating in various temperature ranges. Coatings with solar absorptance above 90% and infrared emittance below 10% are commercially available. A comprehensive review of selective absorber materials is given in [10.20].

Table 10.4 Solar absorptance and infrared emittance for several surfaces near 300 K [10.4, 10.14].

Surface	α_{solar}	ε_{IR}
Aluminum, pure	0.09	0.1
Carbon black in acrylic binder	0.94	0.83
Copper, polished	0.3	0.04
Selective Solar absorbers		
Black Cr on Ni plate	0.95	0.09
CuO on Cu (Ebanol C)	0.90	0.16
Nickel black on steel	0.81	0.17
Sputtered cermet on steel	0.96	0.16
Selective Solar Reflectors		
Magnesium oxide	0.14	0.7
Snow	0.2–0.35	0.82
White paint		
Acrylic	0.26	0.90
Zinc Oxide	0.12–0.18	0.93

Example 10.12

In Section 10.2, we discussed white paint on a roof as a selective solar absorber. Consider now a barn roof under a sunlit sky. The solar radiation on the plane of the roof is 600 W/m^2, the air temperature is 35°C, and a light breeze produces a convective heat transfer coefficient of $\overline{h}$ = 8 W/m^{2}K. The sky temperature is 18°C. Find the temperature of the roof if it is painted with white acrylic paint, and find it again if painted with a non-selective black paint having $\varepsilon = 0.9$.

Solution. Heat loss from the roof to the inside of the barn will lower the roof temperature. Since we don't have enough information to evaluate that loss, we can make an upper bound on the roof temperature by assuming that no heat is transferred to the interior. Then, an energy balance on the roof must account for radiation absorbed from the sun and the sky and for heat lost by convection and reradiation:

$$\alpha_{\text{solar}} q_{\text{solar}} + \varepsilon_{\text{IR}} \sigma T_{\text{sky}}^4 = \overline{h}\,(T_{\text{roof}} - T_{\text{air}}) + \varepsilon_{\text{IR}} \sigma T_{\text{roof}}^4$$

Rearranging and substituting the given numbers,

$$8\left[T_{\text{roof}} - (273+35)\right] + \varepsilon_{\text{IR}}(5.67\times10^{-8})\left[T_{\text{roof}}^4 - (273+18)^4\right] = \alpha_{\text{solar}}(600)$$

For the non-selective black paint, $\alpha_{\text{solar}} = \varepsilon_{\text{IR}} = 0.90$. Solving by iteration, we find

$$T_{\text{roof}} = 338\text{ K} = 65^\circ\text{C}$$

For white acrylic paint, from Table 10.4, $\alpha_{\text{solar}} = 0.26$ and $\varepsilon_{\text{IR}} = 0.90$. We find

$$T_{\text{roof}} = 312\text{ K} = 39^\circ\text{C}$$

The white painted roof is only a few degrees warmer than the air. ■

Ordinary window glass is a very selective transmitter of solar radiation. Glass is nearly transparent to wavelengths below 2.7 μm or so, passing more than 90% of the incident solar energy. At longer wavelengths, in the infrared, glass is virtually opaque to radiation. A consequence of this fact is that solar energy passing through a window cannot pass back out as infrared reradiation. This is precisely why we make greenhouses out of glass. A *greenhouse* is a structure in which we use glass capture solar energy in the interior of a lower temperature space. The glass allows sunlight to enter the space, it stops air from flowing into the space, and it absorbs infrared reradiation from the interior rather than letting it pass directly back to the sky. All these factors help make the interior warm relative to the outside.

The atmospheric greenhouse effect and global warming

The atmosphere creates a *greenhouse effect* on the earth's surface that is very similar to that caused by a pane of glass. Solar energy passes through the atmosphere, arriving mainly on wavelengths between about 0.3 and 3 μm. The earth's surface, having a mean temperature of 15°C or so, radiates mainly on infrared wavelengths longer than 5 μm. Certain atmospheric gases have strong absorption bands at these longer wavelengths. Those gases absorb energy radiated from the surface, and then reemit it toward both the surface and outer space, reducing the net rate of radiative heat loss from the surface to outer space. The result is that the surface remains some 30 K warmer than the atmosphere. In effect,

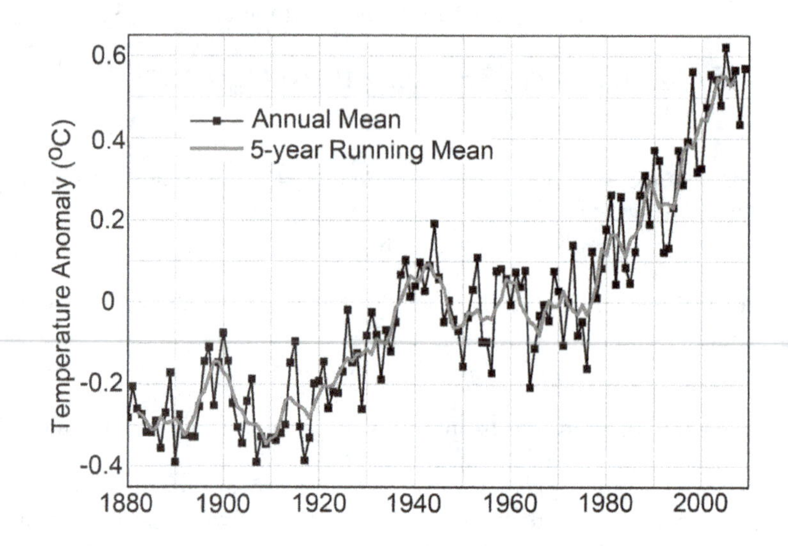

Figure 10.25 Global surface temperature change relative to the mean temperature from 1950–1980 (Courtesy of the NASA Goddard Institute for Space Studies [10.22, 10.23]).

the atmosphere functions as a radiation shield against infrared heat loss to space.

The gases mainly responsible for the the atmospheric greenhouse effect are CO_2, H_2O, CH_4, N_2O, O_3, and some chlorofluorcarbons [10.21]. If the concentration of these gases rises or falls, the strength of the greenhouse effect will change and the surface temperature will also rise or fall. With the exception of the chlorofluorocarbons, each of these gases is created, in part, by natural processes: H_2O by evaporation, CO_2 by animal respiration, CH_4 through plant decay and digestion by livestock, and so on. Human activities, however, have significantly increased the concentrations of all of the gases. Fossil fuel combustion increased the CO_2 concentration by more than 30% during the twentieth century. Methane concentrations have risen through the transportation and leakage of hydrocarbon fuels. Ground level ozone concentrations have risen as a result of photochemical interactions of other pollutants. Chlorofluorocarbons are human-made chemicals.

In parallel to the rising concentrations of these gases, the surface temperature of the earth has risen significantly. Over the course of the twentieth century, a rise of 0.6–0.7 K occurred, with 0.4–0.5 K of that rise coming after 1950 (see Fig. 10.25). The data showing this rise are extensive, are derived from multiple sources, and have been the subject of detailed scrutiny: there is relatively little doubt that surface temperatures have increased [10.22, 10.23, 10.24]. The question of how much of the rise should be attributed to anthropogenic greenhouse gases, however, remains a subject of intense debate.

Many factors must be considered in examining the causes of global warming. Carbon dioxide, for example, is present in such high concentrations that adding more of it increases absorption less rapidly than might be expected. Other gases that are present in smaller concentrations, such as methane, have far stronger effects per additional kilogram. The concentration of water vapor in the atmosphere rises with increasing surface temperature, amplifying any warming trend. Increased cloud cover has both warming and cooling effects. The melting of polar ice caps as temperatures rise reduces the planet's reflectance, or *albedo*, allowing more solar energy to be absorbed. Small temperature rises that have been observed in the oceans represent enormous amounts of stored energy that must taken into account. Atmospheric aerosols (two-thirds of which are produced by sulfate and carbon pollution from fossil fuels) also tend to reduce the greenhouse effect. All of these factors must be built into an accurate climate model (see, for example, [10.25]).

The current consensus among mainstream researchers is that the global warming seen during the last half of the twentieth century is mainly attributable to human activity, principally through the combustion of fossil fuels [10.24]. Numerical models have been used to project a continuing temperature rise in the twenty-first century. These are based on various scenarios of future fossil fuel use and future government policies for reducing greenhouse gas emissions. Regrettably, the outlook is not very positive, with best estimates of twenty-first century warming ranging from roughly 1.8–4.0 K.

The potential for solar power

One alternative to the continuing use of fossil fuels is solar energy. With so much solar energy falling upon all parts of the world, and with the apparent safety, reliability, and cleanliness of most schemes for utilizing solar energy, one might ask why we do not generally use solar power already. The reason is that solar power involves many serious heat transfer and thermodynamics design problems and may pose environmental threats of its own. We shall discuss the problems qualitatively and refer the reader to [10.14], [10.26], or [10.27] for detailed discussions of the design of solar energy systems.

Solar energy reaches the earth with very low intensity. We began this discussion in Chapter 1 by noting that human beings can interface with only a few hundred watts of energy. We could not live on earth if the sun were not relatively gentle. It follows that any large solar power source must concentrate the energy that falls on a huge area. By way of illustration, suppose that we sought to photovoltaically convert 615 W/m^2 of solar energy into electric power with a 15% efficiency (which is not pessimistic) during 8 hr of each day. This would correspond to a daily average of 31 W/m^2, and we would need almost 26 square kilometers (10 square miles) of collector area to match the steady output of an 800 MW power plant.

Other forms of solar energy conversion require similarly large areas. Hydroelectric power — the result of evaporation under the sun's warming influence — requires a large reservoir, and watershed, behind the dam. The burning of organic matter, as wood or grain-based ethanol, requires a large cornfield or forest to be fed by the sun, and so forth. Any energy supply that is served by the sun must draw from a large area of the earth's surface. Thus, they introduce their own kinds of environmental complications.

A second problem stems from the intermittent nature of solar devices. To provide steady power—day and night, rain or shine—requires thermal storage systems, which add both complication and cost.

These problems are minimal when one uses solar energy merely to heat air or water to moderate temperatures (50 to 90°C). In this case the efficiency will improve from just a few percent to as high as 70%. Such heating can be used for industrial processes (crop drying, for example), or it can be used on a small scale for domestic heating of air or water.

Figure 10.26 shows a typical configuration of a domestic solar collector of the flat-plate type. Solar radiation passes through one or more glass plates and impinges on a plate that absorbs the solar wavelengths. The absorber plate would be a selective solar absorber, perhaps blackened copper or nickel. The glass plates might be treated with anti-reflective coatings, raising their solar transmissivity to 98% or more. Once the energy is absorbed, it is reemitted as long-wavelength infrared radiation. Glass is almost opaque in this range, and energy is retained in the collector by a greenhouse effect. Multiple layers of glass serve to reduce both reradiative and convective losses from the absorber plate.

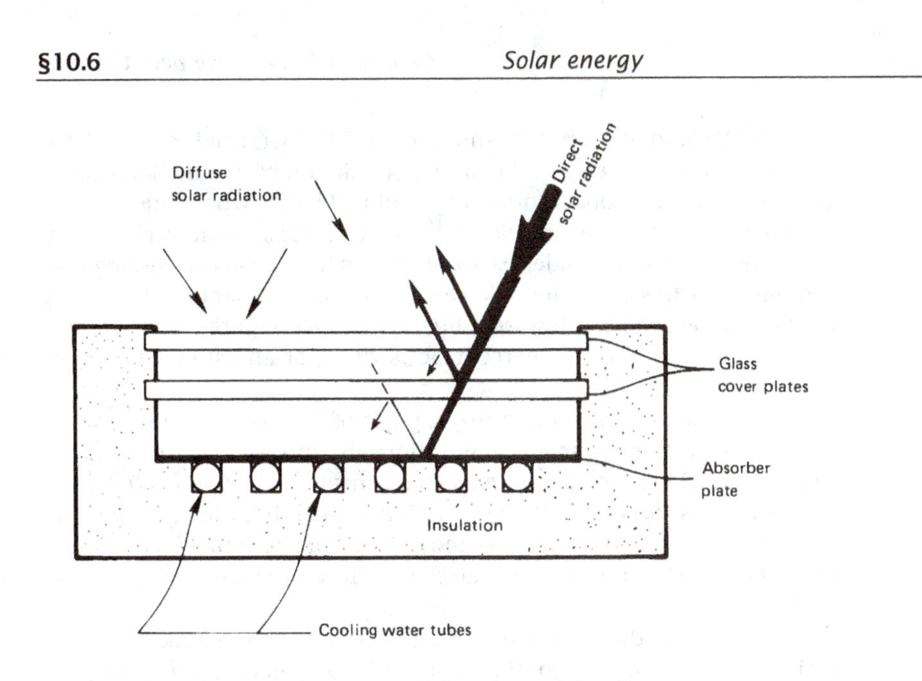

Figure 10.26 A typical flat-plate solar collector.

Water flowing through tubes, which may be brazed to the absorber plate, carries the energy away for use. The flow rate is adjusted to give an appropriate temperature rise.

If the working fluid is to be brought to a fairly high temperature, the direct radiation from the sun must be focused from a large area down to a very small region, using reflecting mirrors. Collectors equipped with a small parabolic reflector, focused on a water or air pipe, can raise the fluid to between 100 and 200°C. In any scheme intended to produce electrical power with a conventional thermal cycle, energy must be focused in an area ratio on the order of 1000 : 1 to achieve a practical cycle efficiency.

A question of over-riding concern as we enter the 21st century is "How much of the renewable energy that reaches Earth, can we hope to utilize?" Of the 1.74×10^{14} kW arriving from the sun, 33% is simply reflected back into outer space. If we were able to collect and use the remainder, 1.16×10^{14} kW, before it too was reradiated to space, each of the 6 billion or so people on the planet would have 19 MW at his or her disposal. Of course, the vast majority of that power must be used to sustain natural processes in the world around us.

In the USA, total energy consumption in 2002 averaged roughly 3.2×10^9 kW, and, dividing this value into a population of 280 million people gives a per capita consumption of roughly 11 kW. Worldwide, energy was consumed at a rate just over 10^{10} kW. That means that world energy consumption was just under 0.01% of the renewable energy passing into and out of Earth's ecosystem. Since many countries that once used very little energy are moving toward a life-style which requires much greater energy consumption, this percentage is rising at an estimated rate of 2%/year.

We must also bear in mind two aspects of this 0.01% figure. First, it is low enough that we might aim, ultimately, to take all of our energy from renewable sources, and thus avoid consuming irreplaceable terrestrial resources. Second, although 0.01% is a small fraction, the absolute amount of power it represents is enormous. It is, therefore, unclear just how much renewable energy we can claim before we create new ecological problems.

There is little doubt that our short-term needs can be met by fossil fuel reserves. However, continued use of those fuels can only amplify the now-well-documented global warming trend. Our long-term hope for a sustainable energy supply may be partially met (but only partially) with solar power, wind power, and other renewable sources. Nuclear fission remains an important option if we are willing to accept one or more of the means for nuclear waste disposal that are presently available to us. Nuclear fusion—the process by which we might manage to create mini-suns upon the earth—may also be a hope for the future. Under any scenario, however, we will serve our best term interests by seeking to bridle the continuing growth of energy consumption.

Problems

10.1 What will ε_λ of the sun appear to be to an observer on the earth's surface at $\lambda = 0.2$ µm and 0.65 µm? How do these emittances compare with the real emittances of the sun? [At 0.65 µm, $\varepsilon_\lambda \simeq 0.77$.]

10.2 Plot e_{λ_b} against λ for $T = 300$ K and 10,000 K with the help of eqn. (1.30). About what fraction of energy from each black body is visible?

10.3 A 0.6 mm diameter wire is drawn out through a mandril at 950°C. Its emittance is 0.85. It then passes through a long cylindrical shield of commercial aluminum sheet, 7 cm in diameter. The shield is horizontal in still air at 25°C. What is the temperature of the shield? Is it reasonable to neglect natural convection inside and radiation outside? [$T_{\text{shield}} = 153$°C.]

10.4 A 1 ft^2 shallow pan with adiabatic sides is filled to the brim with water at 32°F. It radiates to a night sky whose temperature is −18°F, while a 50°F breeze blows over it at 1.5 ft/s. Will the water freeze or warm up?

10.5 A thermometer is held vertically in a room with air at 10°C and walls at 27°C. What temperature will the thermometer read if everything can be considered black? State your assumptions.

10.6 Rework Problem 10.5, taking the room to be wall-papered and considering the thermometer to be nonblack.

10.7 Two thin aluminum plates, the first polished and the second painted black, are placed horizontally outdoors, where they are cooled by air at 10°C. The heat transfer coefficient is 5 W/m^{2}K on both the top and the bottom. The top is irradiated with 750 W/m^2 and it radiates to the sky at 250 K. The earth below the plates is black at 10°C. Find the equilibrium temperature of each plate.

10.8 A sample holder of 99% pure aluminum, 1 cm in diameter and 16 cm in length, protrudes from a small housing on an orbital space vehicle. The holder "sees" almost nothing but outer space at an effective temperature of 30 K. The base of the holders is 0°C and you must find the temperature of the sample at its tip. It will help if you note that aluminum is used, so that the temperature of the tip stays quite close to that of the root. [$T_{\text{end}} = -0.7$°C.]

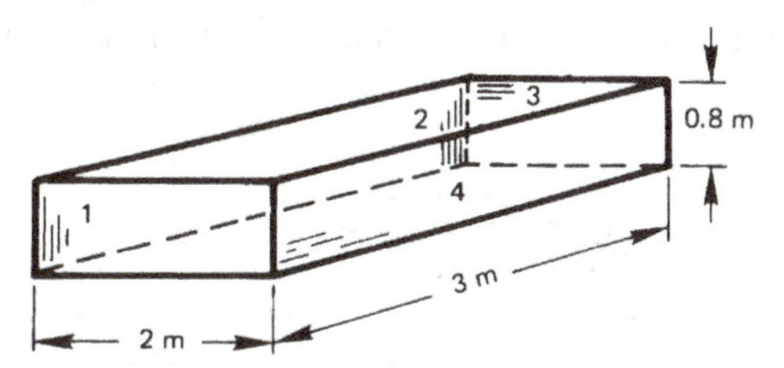

Figure 10.27 Configuration for Prob. 10.9.

10.9 The bottom of the box shown in Fig. 10.27 is a radiant heater. What percentage of the heat goes out the top? What fraction

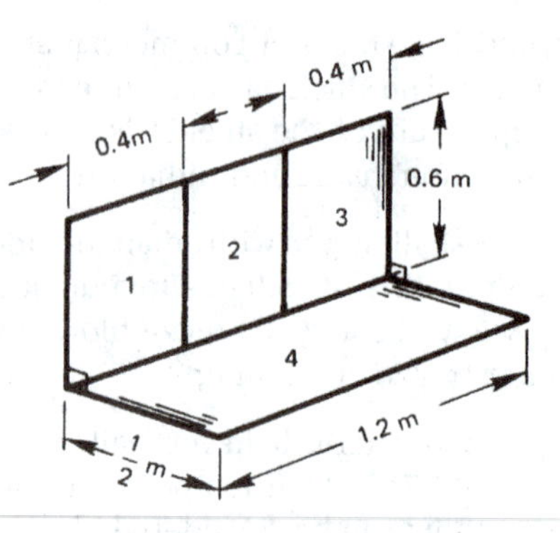

Figure 10.28 Configuration for Prob. 10.11.

impinges on each of the four sides? (Remember that the percentages must add up to 100.)

10.10 With reference to Fig. 10.12, find $F_{1\text{-}(2+4)}$ and $F_{(2+4)\text{-}1}$.

10.11 Find $F_{2\text{-}4}$ for the surfaces shown in Fig. 10.28. [0.315.]

10.12 What is $F_{1\text{-}2}$ for the squares shown in Fig. 10.29?

10.13 A particular internal combustion engine has an exhaust manifold at 600°C running parallel to a water cooling line at 20°C. If both the manifold and the cooling line are 4 cm in diameter, their centers are 7 cm apart, and both are approximately black, how much heat will be transferred to the cooling line by radiation? [383 W/m.]

10.14 Prove that $F_{1\text{-}2}$ for any pair of two-dimensional plane surfaces, as shown in Fig. 10.30, is equal to $[(a + b) - (c + d)]/2L_1$. This is called the *string rule* because we can imagine that the numerator equals the difference between the lengths of a set of crossed strings (a and b) and a set of uncrossed strings (c and d).

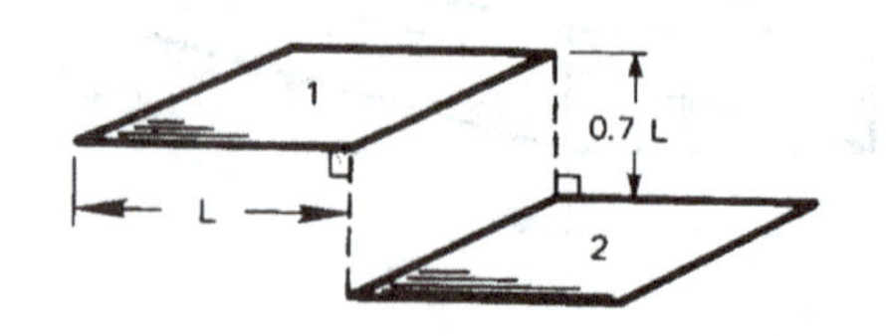

Figure 10.29 Configuration for Prob. 10.12.

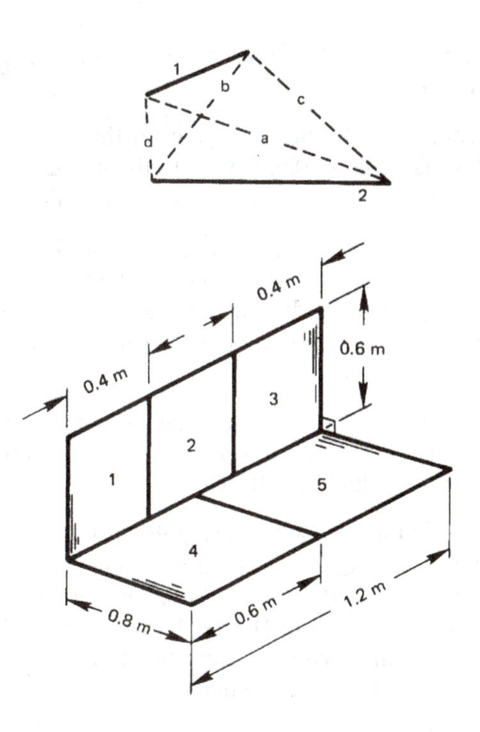

Figure 10.30 Configuration for Prob. 10.14.

Figure 10.31 Configuration for Prob. 10.15.

10.15 Find $F_{1\text{-}5}$ for the surfaces shown in Fig. 10.31.

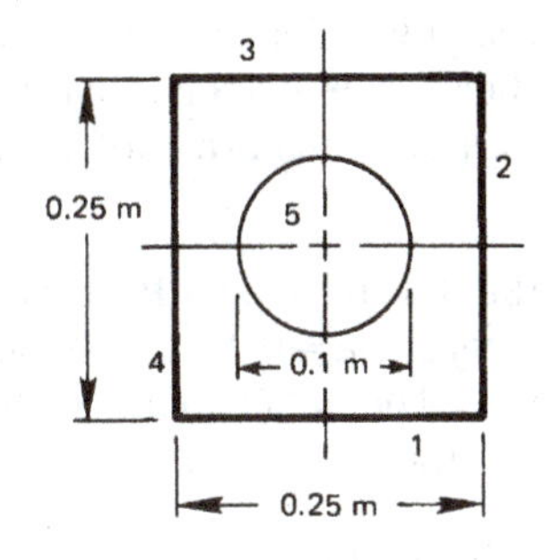

Figure 10.32 Configuration for Prob. 10.16.

10.16 Find $F_{1\text{-}(2+3+4)}$ for the surfaces shown in Fig. 10.32.

10.17 A cubic box 1 m on the side is black except for one side, which has an emittance of 0.2 and is kept at 300°C. An adjacent side is kept at 500°C. The other sides are insulated. Find Q_{net} inside the box. [2494 W.]

10.18 Rework Problem 10.17, but this time set the emittance of the insulated walls equal to 0.6. Compare the insulated wall temperature with the value you would get if the walls were black.

10.19 An insulated black cylinder, 10 cm in length and with an inside diameter of 5 cm, has a black cap on one end and a cap with an emittance of 0.1 on the other. The black end is kept at 100°C and the reflecting end is kept at 0°C. Find Q_{net} inside the cylinder and $T_{cylinder}$.

10.20 Rework Example 10.2 if the shield has an inside emittance of 0.34 and the room is at 20°C. How much cooling must be provided to keep the shield at 100°C?

10.21 A 0.8 m long cylindrical burning chamber is 0.2 m in diameter. The hot gases within it are at a temperature of 1500°C and a pressure of 1 atm, and the absorbing components consist of 12% by volume of CO_2 and 18% H_2O. Neglect end effects and determine how much cooling must be provided the walls to hold them at 750°C if they are black.

10.22 A 30 ft by 40 ft house has a conventional 30° sloping roof with a peak running in the 40 ft direction. Calculate the temperature of the roof in 20°C still air when the sun is overhead (a) if the roofing is of wooden shingles and (b) if it is commercial aluminum sheet. The incident solar energy is 670 W/m^2, Kirchhoff's law applies for both roofs, and the effective sky temperature is 22°C.

10.23 Calculate the radiant heat transfer from a 0.2 m diameter stainless steel hemisphere ($\varepsilon_{ss} = 0.4$) to a copper floor ($\varepsilon_{Cu} = 0.15$) that forms its base. The hemisphere is kept at 300°C and the base at 100°C. Use the algebraic method. [21.24 W.]

10.24 A hemispherical indentation in a smooth wrought-iron plate has an 0.008 m radius. How much heat radiates from the 40°C dent to the −20°C surroundings?

10.25 A conical hole in a block of metal for which $\varepsilon = 0.5$ is 5 cm in diameter at the surface and 5 cm deep. By what factor will the radiation from the area of the hole be changed by the presence of the hole? (This problem can be done to a close approximation using the methods in this chapter if the cone does not become very deep and slender. If it does, then the fact that the apex is receiving far less radiation makes it incorrect to use the network analogy.)

10.26 A single-pane window in a large room is 4 ft wide and 6 ft high. The room is kept at 70°F, but the pane is at 67°F owing to heat loss to the colder outdoor air. Find (a) the heat transfer by radiation to the window; (b) the heat transfer by natural convection to the window; and (c) the fraction of heat transferred to the window by radiation.

10.27 Suppose that the windowpane temperature is unknown in Problem 10.26. The outdoor air is at 40°F and $\overline{h}$ is 62 W/m²K on the outside of the window. It is nighttime and the effective temperature of the sky is 15°F. Assume $F_{\text{window-sky}} = 0.5$. Take the rest of the surroundings to be at 40°F. Find T_{window} and draw the analogous electrical circuit, giving numerical values for all thermal resistances. Discuss the circuit. (It will simplify your calculation to note that the window is opaque to infrared radiation but that it offers very little resistance to conduction. Thus, the window temperature is almost uniform.)

10.28 A very effective low-temperature insulation is made by evacuating the space between parallel metal sheets. Convection is eliminated, conduction occurs only at spacers, and radiation is responsible for what little heat transfer occurs. Calculate q between 150 K and 100 K for three cases: (a) two sheets of *highly* polished aluminum, (b) three sheets of highly polished aluminum, and (c) three sheets of rolled sheet steel.

10.29 Three parallel black walls, 1 m wide, form an equilateral triangle. One wall is held at 400 K, one is at 300 K, and the third is insulated. Find Q W/m and the temperature of the third wall.

10.30 Two 1 cm diameter rods run parallel, with centers 4 cm apart. One is at 1500 K and black. The other is unheated, and $\varepsilon = 0.66$. They are both encircled by a cylindrical black radiation shield at 400 K. Evaluate Q W/m and the temperature of the unheated rod.

10.31 A small-diameter heater is centered in a large cylindrical radiation shield. Discuss the relative importance of the emittance of the shield during specular and diffuse radiation.

10.32 Two 1 m wide commercial aluminum sheets are joined at a 120° angle along one edge. The back (or 240° angle) side is insulated. The plates are both held at 120°C. The 20°C surroundings are distant. What is the net radiant heat transfer from the left-hand plate: to the right-hand side, and to the surroundings?

10.33 Two parallel discs of 0.5 m diameter are separated by an infinite parallel plate, midway between them, with a 0.2 m diameter hole in it. The discs are centered on the hole. What is the view factor between the two discs if they are 0.6 m apart?

10.34 An evacuated spherical cavity, 0.3 m in diameter in a zero-gravity environment, is kept at 300°C. Saturated steam at 1 atm is then placed in the cavity. (a) What is the initial flux of radiant heat transfer to the steam? (b) Determine how long it will take for $q_{\text{conduction}}$ to become less than $q_{\text{radiation}}$. (Correct for the rising steam temperature if it is necessary to do so.)

10.35 Verify cases (1), (2), and (3) in Table 10.2 using the string method described in Problem 10.14.

10.36 Two long parallel heaters consist of 120° segments of 10 cm diameter parallel cylinders whose centers are 20 cm apart. The segments are those nearest each other, symmetrically placed on the plane connecting their centers. Find $F_{1\text{-}2}$ using the string method described in Problem 10.14.)

10.37 Two long parallel strips of rolled sheet steel lie along sides of an imaginary 1 m equilateral triangular cylinder. One piece is 1 m wide and kept at 20°C. The other is $\frac{1}{2}$ m wide, centered in an adjacent leg, and kept at 400°C. The surroundings are distant and they are insulated. Find Q_{net}. (You will need a shape factor; it can be found using the method described in Problem 10.14.)

10.38 Find the shape factor from the hot to the cold strip in Problem 10.37 using Table 10.2, not the string method. If your instructor asks you to do so, complete Problem 10.37 when you have $F_{1\text{-}2}$.

10.39 Prove that, as the figure becomes very long, the view factor for the second case in Table 10.3 reduces to that given for the third case in Table 10.2.

10.40 Show that $F_{1\text{-}2}$ for the first case in Table 10.3 reduces to the expected result when plates 1 and 2 are extended to infinity.

10.41 In Problem 2.26 you were asked to neglect radiation in showing that q was equal to 8227 W/m^2 as the result of conduction alone. Discuss the validity of the assumption quantitatively.

10.42 A 100°C sphere with $\varepsilon = 0.86$ is centered within a second sphere at 300°C with $\varepsilon = 0.47$. The outer diameter is 0.3 m and the inner diameter is 0.1 m. What is the radiant heat flux?

10.43 Verify $F_{1\text{-}2}$ for case 4 in Table 10.2. (*Hint:* This can be done without integration.)

10.44 Consider the approximation made in eqn. (10.30) for a small gray object in a large isothermal enclosure. How small must A_1/A_2 be in order to introduce less than 10% error in $\mathcal{F}_{1\text{-}2}$ if the small object has an emittance of $\varepsilon_1 = 0.5$ and the enclosure is: a) commerical aluminum sheet; b) rolled sheet steel; c) rough red brick; d) oxidized cast iron; or e) polished electrolytic copper. Assume both the object and its environment have temperatures of 40 to 90°C.

10.45 Derive eqn. (10.42), starting with eqns. (10.36–10.38).

10.46 (a) Derive eqn. (10.31), which is for a single radiation shield between two bodies. Include a sketch of the radiation network. (b) Repeat the calculation in the case when two radiation shields lie between body (1) and body (2), with the second shield just outside the first.

10.47 Use eqn. (10.32) to find the net heat transfer from between two specularly reflecting bodies that are separated by a specularly reflecting radiation shield. Compare the result to eqn. (10.31). Does specular reflection reduce the heat transfer?

10.48 Some values of the monochromatic absorption coefficient for liquid water, as $\rho\kappa_\lambda$ (cm^{-1}), are listed below [10.4]. For each wavelength, find the thickness of a layer of water for which the transmittance is 10%. On this basis, discuss the colors one might see underwater and water's infrared emittance.

λ (μm)	$\rho\kappa_\lambda$ (cm^{-1})	*Color*
0.3	0.0067	
0.4	0.00058	violet
0.5	0.00025	green
0.6	0.0023	orange
0.8	0.0196	
1.0	0.363	
2.0	69.1	
2.6-10.0	$> 100.$	

10.49 The sun has a diameter of 1.391×10^6 km. The earth has a diameter of 12,740 km and lies at a mean distance of 1.496×10^8 km from the center of the sun. (a) If the earth is treated as a flat disk normal to the radius from sun to earth, determine the view factor $F_{\text{sun-earth}}$. (b) Use this view factor and the measured solar irradiation of 1367 W/m^2 to show that the effective black body temperature of the sun is 5777 K.

References

[10.1] E. M. Sparrow and R. D. Cess. *Radiation Heat Transfer.* Hemisphere Publishing Corp./McGraw-Hill Book Company, Washington, D.C., 1978.

[10.2] D. K. Edwards. *Radiation Heat Transfer Notes.* Hemisphere Publishing Corp., Washington, D.C., 1981.

[10.3] M. F. Modest. *Radiative Heat Transfer.* McGraw-Hill, New York, 1993.

[10.4] R. Siegel and J. R. Howell. *Thermal Radiation Heat Transfer.* Taylor and Francis-Hemisphere, Washington, D.C., 4th edition, 2001.

[10.5] J. R. Howell. *A Catalog of Radiation Heat Transfer Configuration Factors.* University of Texas, Austin, 2nd edition, 2001. Available online at http://www.me.utexas.edu/~howell/.

[10.6] A. K. Oppenheim. Radiation analysis by the network method. *Trans. ASME,* 78:725–735, 1956.

[10.7] W.-J. Yang, H. Taniguchi, and K. Kudo. Radiative heat transfer by the Monte Carlo method. In T.F. Irvine, Jr., J. P. Hartnett, Y. I. Cho, and G. A. Greene, editors, *Advances in Heat Transfer,* volume 27. Academic Press, Inc., San Diego, 1995.

[10.8] H. C. van de Hulst. *Light Scattering by Small Particles.* Dover Publications Inc., New York, 1981.

[10.9] P. W. Atkins. *Physical Chemistry.* W. H. Freeman and Co., New York, 3rd edition, 1986.

[10.10] G. Herzberg. *Molecular Spectra and Molecular Structure.* Kreiger Publishing, Malabar, Florida, 1989. In three volumes.

[10.11] D. K. Edwards. Molecular gas band radiation. In T. F. Irvine, Jr. and J. P. Hartnett, editors, *Advances in Heat Transfer,* volume 12, pages 119–193. Academic Press, Inc., New York, 1976.

[10.12] H. C. Hottel and A. F. Sarofim. *Radiative Transfer.* McGraw-Hill Book Company, New York, 1967.

[10.13] D. K. Edwards and R. Matavosian. Scaling rules for total absorptivity and emissivity of gases. *J. Heat Transfer,* 106(4):684–689, 1984.

[10.14] J. A. Duffie and W. A. Beckman. *Solar Engineering of Thermal Processes.* John Wiley & Sons, Inc., New York, 2nd edition, 1991.

[10.15] M. Iqbal. *An Introduction to Solar Radiation.* Academic Press, Inc., New York, 1983.

[10.16] H. G. Houghton. *Physical Meteorology.* MIT Press, Cambridge, MA, 1985.

[10.17] P. Berdahl and R. Fromberg. The thermal radiance of clear skies. *Solar Energy,* 29:299–314, 1982.

[10.18] A. Skartveit, J. A. Olseth, G. Czeplak, and M. Rommel. On the estimation of atmospheric radiation from surface meteorological data. *Solar Energy*, 56:349–359, 1996.

[10.19] P. Berdahl and M. Martin. The emissivity of clear skies. *Solar Energy*, 32:663–664, 1984.

[10.20] C. E. Kennedy. Review of mid- to high-temperature solar selective materials. Technical Report TP-520-31267, National Renewable Energy Laboratory, Golden, Colorado, July 2002.

[10.21] J. A. Fay and D. S. Gollub. *Energy and Environment*. Oxford University Press, New York, 2002.

[10.22] J. Hansen, R. Ruedy, M. Sato, M. Imhoff, W. Lawrence, D. Easterling, T. Peterson, and T. Karl. A closer look at United States and global surface temperature change. *J. Geophysical Research*, 106: 23947, 2001.

[10.23] J. Hansen, R. Ruedy, M. Sato, and K. Lo. Global surface temperature change. *In preparation*, 2010. Additional data and updates are at http://data.giss.nasa.gov/gistemp/.

[10.24] Core Writing Team, R. K. Pachauri, and A. Reisinger, editors. *Climate Change 2007: Synthesis Report. Contribution of Working Groups I, II and III to the Fourth Assessment Report of the Intergovernmental Panel on Climate Change*. IPCC, Geneva, 2007. Also available at http://www.ipcc.ch.

[10.25] P. A. Stott, S. F. B. Tett, G. S. Jones, M. R. Allen, J. F. B. Mitchell, and G. J. Jenkins. External control of 20th century temperature by natural and anthropogenic forcings. *Science*, 290:2133–2137, 2000.

[10.26] F. Kreith and J. F. Kreider. *Principles of Solar Engineering*. Hemisphere Publishing Corp./McGraw-Hill Book Company, Washington, D.C., 1978.

[10.27] U.S. Department of Commerce. *Solar Heating and Cooling of Residential Buildings*, volume 1 and 2. Washington, D.C., October 1977.

Part V

Mass Transfer

11. An introduction to mass transfer

> *The edge of a colossal jungle, so dark-green as to be almost black, fringed with white surf, ran straight, like a ruled line, far, far away along a blue sea whose glitter was blurred by a creeping mist. The sun was fierce, the land seemed to glisten and drip with steam.*
>
> ***Heart of Darkness,*** **Joseph Conrad, 1902**

11.1 Introduction

We have, so far, dealt with heat transfer by convection, radiation, and diffusion. The word diffusion refers to transport by random molecular action. Heat diffuses as hotter molecules mix or agitate colder ones. Mass diffuses as molecules of one kind randomly penetrate regions occupied by molecules of another kind. We have largely limited our considerations to single component media in which mass diffusion is meaningless and which experience only heat diffuses. Many heat transfer processes, however, occur in mixtures of more than one substance. They are often coupled with mass diffusion—or *mass transfer* as we refer to it here.

A wall exposed to a hot air stream may be cooled evaporatively by bleeding water through its surface. Water vapor may condense out of damp air onto cool surfaces. Heat will flow through an air-water mixture in these situations, but water vapor will diffuse or convect through air as well. In this chapter, we study mass transfer phenomena with an eye toward predicting heat and mass transfer rates such situations.

During mass transfer processes, an individual chemical species travels from regions where it has a high concentration to regions where it has a low concentration. When liquid water is exposed to a dry air stream, its vapor pressure may produce a comparatively high concentration of water vapor in the air near the water surface. The concentration difference between the water vapor near the surface and that in the air stream will

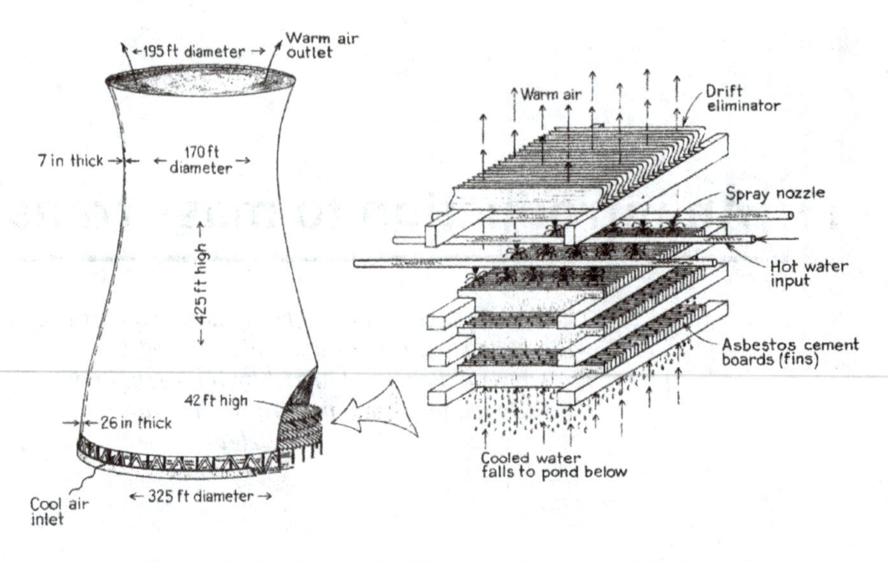

Figure 11.1 Schematic diagram of a natural-draft cooling tower at the Rancho Seco nuclear power plant. (From [11.1], courtesy of W. C. Reynolds.)

drive the diffusion of vapor into the air stream. We call this evaporation.

In this and other respects, mass transfer is analogous to heat transfer. Just as thermal energy diffuses from regions of high temperature to regions of low temperature (following the temperature gradient), the mass of one species diffuses from regions high concentration to regions of low concentration (following its concentration gradient.) Just as the diffusional (or conductive) heat flux is directly proportional to a temperature gradient, so the diffusional mass flux of a species is often directly proportional to its concentration gradient; this is called Fick's law of diffusion. Just as conservation of energy and Fourier's law lead to equations for the convection and diffusion of heat, conservation of mass and Fick's law lead to equations for the convection and diffusion of species in a mixture.

The great similarity of the equations of heat convection and diffusion to those of mass convection and diffusion extends to the use of convective mass transfer coefficients, which, like heat transfer coefficients, relate convective fluxes to concentration differences. In fact, with simple modifications, the heat transfer coefficients of previous chapters may

Figure 11.2 A mechanical-draft cooling tower. The fans are located within the cylindrical housings at the top. Air is drawn in through the louvres on the side.

often be applied to mass transfer calculations.

Mass transfer, by its very nature, is intimately concerned with mixtures of chemical species. We begin by learning how to quantify the concentration of chemical species and by defining rates of movement of species. We make frequent reference to an arbitrary "species i," the ith component of a mixture of N different species. These definitions are reminiscent of a first course in chemistry. We also spend some time, in Section 11.4, calculating such transport properties of mixtures as diffusion coefficients and viscosities.

Consider a typical technology that is dominated by mass transfer processes. Figure 11.1 shows a huge cooling tower used to cool the water leaving power plant condensers or other large heat exchangers. It is essentially an empty shell, at the bottom of which are arrays of cement boards or plastic louvres over which is sprayed the hot water to be cooled. The hot water runs down this packing, and a small portion of it evaporates into cool air that enters the tower from below. The remaining water, having been cooled by the evaporation, falls to the bottom, where it is collected and recirculated.

The temperature of the air rises as it absorbs the warm vapor and, in the *natural-draft* form of cooling tower shown, the upper portion of the tower acts as an enormous chimney through which the warm, moist air buoys, pulling in cool air at the base. In a *mechanical-draft* cooling tower (Fig. 11.2), fans are used to pull air through the packing. Mechanical-draft towers are much shorter and can sometimes be seen on the roofs of buildings.

The working mass transfer process in a cooling tower is the evaporation of water into air. The rate of evaporation depends on the temperature and humidity of the incoming air, the feed-water temperature, and the air-flow characteristics of the tower and the packing. When the air flow is buoyancy-driven, the flow rates are directly coupled. Thus, mass transfer lies at the core of the complex design of a cooling tower.

11.2 Mixture compositions and species fluxes

The composition of mixtures

A mixture of various chemical species displays its own density, molecular weight, and other overall thermodynamic properties. These properties depend on the types and relative amounts of the components, which might vary from point to point in the mixture. To determine the local properties of a mixture, we must identify the local proportion of each species composing the mixture.

One way to describe the amount of a particular species in a mixture is by the mass of that species per unit volume, known as the *partial density*. The mass of species i in a small volume of mixture, in kg, divided by that volume, in m^3, is the partial density, ρ_i, for that species, in kg of i per m^3. The composition of the mixture may be describe by stating the partial density of each of its components. The mass density of the mixture itself, ρ, is the total mass of all species per unit volume; therefore,

$$\rho = \sum_i \rho_i \tag{11.1}$$

The relative amount of species i in the mixture can be described by the mass of i per unit mass of the mixture, which is simply ρ_i/ρ. This ratio is called the *mass fraction*, m_i:

$$m_i \equiv \frac{\rho_i}{\rho} = \frac{\text{mass of species } i}{\text{mass of mixture}} \tag{11.2}$$

This definition leads to the following two results:

$$\sum_i m_i = \sum_i \rho_i/\rho = 1 \qquad \text{and} \qquad 0 \leqslant m_i \leqslant 1 \tag{11.3}$$

The *molar concentration* of species i in kmol/m^3, c_i, expresses concentration in terms of moles rather than mass. If M_i is the molecular weight of species i in kg/kmol, then

$$c_i \equiv \frac{\rho_i}{M_i} = \frac{\text{moles of } i}{\text{volume}} \tag{11.4}$$

The molar concentration of the mixture, c, is the total number of moles for all species per unit volume; thus,

$$c = \sum_i c_i. \tag{11.5}$$

The *mole fraction* of species i, x_i, is the number of moles of i per mole of mixture:

$$x_i \equiv \frac{c_i}{c} = \frac{\text{moles of } i}{\text{mole of mixture}} \tag{11.6}$$

Just as for the mass fraction, it follows for mole fraction that

$$\sum_i x_i = \sum_i c_i/c = 1 \qquad \text{and} \qquad 0 \leqslant x_i \leqslant 1 \tag{11.7}$$

The molecular weight of the mixture is the number of kg of mixture per kmol of mixture: $M \equiv \rho/c$. Using eqns. (11.1), (11.4), and (11.6) and (11.5), (11.4), and (11.2), respectively, we write M in terms of either mole or mass fraction

$$M = \sum_i x_i M_i \qquad \text{or} \qquad \frac{1}{M} = \sum_i \frac{m_i}{M_i} \tag{11.8}$$

We can easily derive (Problem 11.1) the following relations to convert mole fraction to mass fraction:

$$m_i = \frac{x_i M_i}{M} = \frac{x_i M_i}{\sum_k x_k M_k} \quad \text{and} \quad x_i = \frac{M m_i}{M_i} = \frac{m_i/M_i}{\sum_k m_k/M_k} \tag{11.9}$$

In some circumstances, such as kinetic theory calculations, one works directly with the number of molecules of i per unit volume. This *number density*, $\mathcal{N}_i$, is given by

$$\mathcal{N}_i = N_A c_i \tag{11.10}$$

where N_A is Avogadro's number, 6.02214×10^{26} molecules/kmol.

Ideal gases

The relations we have developed so far involve densities and concentrations that vary in as yet unknown ways with temperature or pressure. To get a useful, though more restrictive, set of results, we now combine the preceding relations with the ideal gas law. For any individual component, i, we may write the *partial pressure*, p_i, exerted by i as:

$$p_i = \rho_i R_i T \tag{11.11}$$

In eqn. (11.11), R_i is the ideal gas constant for species i:

$$R_i \equiv \frac{R^\circ}{M_i} \tag{11.12}$$

where R° is the universal gas constant, 8314.472 J/kmol·K. Equation (11.11) can alternatively be written in terms of c_i:

$$\begin{aligned} p_i = \rho_i R_i T &= (M_i c_i)\left(\frac{R^\circ}{M_i}\right) T \\ &= c_i R^\circ T \end{aligned} \tag{11.13}$$

Equations (11.5) and (11.13) can be used to relate c to p and T

$$c = \sum_i c_i = \sum_i \frac{p_i}{R^\circ T} = \frac{p}{R^\circ T} \tag{11.14}$$

Multiplying the last two parts of eqn. (11.14) by $R^\circ T$ yields Dalton's law of partial pressures,[1]

$$p = \sum_i p_i \tag{11.15}$$

Finally, we combine eqns. (11.6), (11.13), and (11.15) to obtain a very useful relationship between x_i and p_i:

$$x_i = \frac{c_i}{c} = \frac{p_i}{c\, R^\circ T} = \frac{p_i}{p} \tag{11.16}$$

in which the last two equalities are restricted to ideal gases.

[1]John Dalton offered his "law" as an empirical principle in 1801. But we can deduce it for ideal gases using molecular principles. We obtain it here from eqn. (11.11) which is true for ideal gas molecules because they occupy mixtures without influencing one another. While Dalton's law is strictly true only for ideal gases, it happens to be quite accurate even when gases deviate greatly from ideality. This fact allows us to greatly simplify many calculations.

Example 11.1

The most common mixture that we deal with is air. It has the following composition:

Species	*Mass Fraction*
N_2	0.7556
O_2	0.2315
Ar	0.01289
trace gases	< 0.01

Determine x_{O_2}, p_{O_2}, c_{O_2}, and ρ_{O_2} for air at 1 atm.

Solution. To make these calcuations, we need the molecular weights, which are given in Table 11.2 on page 618. We can start by checking the value of M_{air}, using the second of eqns. (11.8):

$$\begin{aligned} M_{\text{air}} &= \left(\frac{m_{N_2}}{M_{N_2}} + \frac{m_{O_2}}{M_{O_2}} + \frac{m_{\text{Ar}}}{M_{\text{Ar}}} \right)^{-1} \\ &= \left(\frac{0.7556}{28.02 \text{ kg/kmol}} + \frac{0.2315}{32.00 \text{ kg/kmol}} + \frac{0.01289}{39.95 \text{ kg/kmol}} \right)^{-1} \\ &= 28.97 \text{ kg/kmol} \end{aligned}$$

We may calculate the mole fraction using the second of eqns. (11.9)

$$x_{O_2} = \frac{m_{O_2} M}{M_{O_2}} = \frac{(0.2315)(28.97 \text{ kg/kmol})}{32.00 \text{ kg/kmol}} = 0.2095$$

The partial pressure of oxygen in air at 1 atm is [eqn. (11.16)]

$$p_{O_2} = x_{O_2}\, p = (0.2095)(101{,}325 \text{ Pa}) = 2.123 \times 10^4 \text{ Pa}$$

We may now obtain c_{O_2} from eqn. (11.13):

$$\begin{aligned} c_{O_2} &= \frac{p_{O_2}}{R^\circ T} \\ &= (2.123 \times 10^4 \text{ Pa})/(300 \text{ K})(8314.5 \text{ J/kmol}\cdot\text{K}) \\ &= 0.008510 \text{ kmol/m}^3 \end{aligned}$$

Finally, eqn. (11.4) gives the partial density

$$\begin{aligned} \rho_{O_2} &= c_{O_2} M_{O_2} = (0.008510 \text{ kmol/m}^3)(32.00 \text{ kg/kmol}) \\ &= 0.2723 \text{ kg/m}^3 \end{aligned}$$

■

Velocities and fluxes

Each species in a mixture undergoing a mass transfer process has an *species-average velocity*, $\vec{v}_i$, which can be different for each species in the mixture, as suggested by Fig. 11.3. We may obtain the *mass-average velocity*,[2] $\vec{v}$, for the entire mixture from the species average velocities using the formula

$$\rho\vec{v} = \sum_i \rho_i \vec{v}_i. \tag{11.17}$$

This equation is essentially a local calculation of the mixture's net momentum per unit volume. We refer to $\rho\vec{v}$ as the mixture's *mass flux*, $\vec{n}$, and we call its scalar magnitude $\dot{m}''$; each has units of kg/m^2·s. Likewise, the mass flux of species i is

$$\vec{n}_i = \rho_i \vec{v}_i \tag{11.18}$$

and, from eqn. (11.17), we see that the mixture's mass flux equals the sum of all species' mass fluxes

$$\vec{n} = \sum_i \vec{n}_i = \rho\vec{v} \tag{11.19}$$

Since each species diffusing through a mixture has some velocity relative to the mixture's mass-average velocity, the *diffusional mass flux*, $\vec{j}_i$, of a species relative to the mixture's mean flow may be identified:

$$\vec{j}_i = \rho_i \left(\vec{v}_i - \vec{v}\right). \tag{11.20}$$

The total mass flux of the ith species, $\vec{n}_i$, includes both this diffusional mass flux and bulk convection by the mean flow, as is easily shown:

$$\boxed{\begin{aligned} \vec{n}_i &= \rho_i \vec{v}_i = \rho_i \vec{v} + \rho_i \left(\vec{v}_i - \vec{v}\right) \\ &= \rho_i \vec{v} + \vec{j}_i \\ &= \underbrace{m_i \vec{n}}_{\text{convection}} + \underbrace{\vec{j}_i}_{\text{diffusion}} \end{aligned}} \tag{11.21}$$

[2] The mass average velocity, $\vec{v}$, given by eqn. (11.17) is identical to the fluid velocity, $\vec{u}$, used in previous chapters. This is apparent if one applies eqn. (11.17) to a "mixture" composed of only one species. We use the symbol $\vec{v}$ here because $\vec{v}$ is the more common notation in the mass transfer literature.

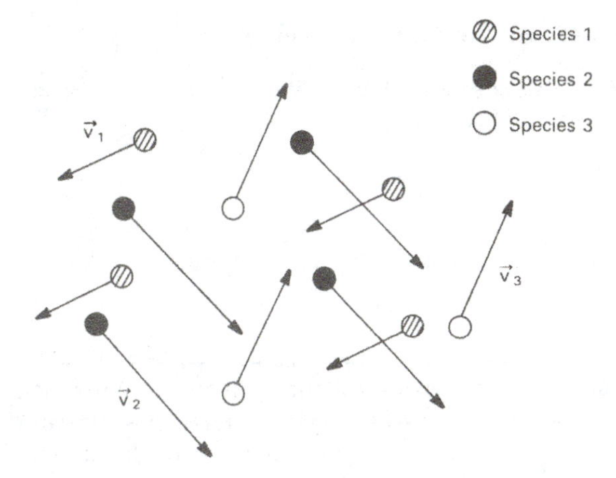

Figure 11.3 Molecules of different species in a mixture moving with different average velocities. The velocity $\vec{v}_i$ is the average over all molecules of species i.

Although the convective transport contribution is fully determined as soon as we know the velocity field and partial densities, the causes of diffusion need further discussion, which we defer to Section 11.3.

Combining eqns. (11.19) and (11.21), we find that

$$\vec{n} = \sum_i \vec{n}_i = \sum_i \rho_i \vec{v} + \sum_i \vec{j}_i = \rho \vec{v} + \sum_i \vec{j}_i = \vec{n} + \sum_i \vec{j}_i$$

Hence

$$\sum_i \vec{j}_i = 0 \tag{11.22}$$

Diffusional mass fluxes must sum to zero because they are each defined relative to the mean mass flux.

Velocities may also be stated in molar terms. The *mole flux* of the ith species, $\vec{N}_i$, is $c_i \vec{v}_i$, in kmol/m$^2 \cdot$ s. The *mixture's mole flux*, $\vec{N}$, is obtained by summing over all species

$$\vec{N} = \sum_i \vec{N}_i = \sum_i c_i \vec{v}_i = c\vec{v}^* \tag{11.23}$$

where we define the *mole-average velocity*, $\vec{v}^*$, as shown. The last flux we define is the *diffusional mole flux*, $\vec{J}_i^*$:

$$\vec{J}_i^* = c_i \left(\vec{v}_i - \vec{v}^*\right) \tag{11.24}$$

It may be shown, using these definitions, that

$$\boxed{\vec{N}_i = x_i \vec{N} + \vec{J}_i^*} \tag{11.25}$$

Substitution of eqn. (11.25) into eqn. (11.23) gives

$$\vec{N} = \sum_i \vec{N}_i = \vec{N} \sum_i x_i + \sum_i \vec{J}_i^* = \vec{N} + \sum_i \vec{J}_i^*$$

so that

$$\sum_i \vec{J}_i^* = 0. \tag{11.26}$$

Thus, *both* the $\vec{J}_i^*$'s and the $\vec{j}_i$'s sum to zero.

Example 11.2

At low temperatures, carbon oxidizes (burns) in air through reaction at the carbon surface: $C + O_2 \longrightarrow CO_2$. Figure 11.4 shows the carbon-air interface in a coordinate system that moves into the stationary carbon at the same speed that the carbon burns away—as though the observer were seated on the moving interface. Oxygen flows toward the carbon surface and carbon dioxide flows away, with a net flow of carbon through the interface. If the system is at steady state and, if a separate analysis shows that carbon is consumed at the rate of 0.00241 kg/m^2·s, find the mass and mole fluxes through an imaginary surface, s, that stays close to the gas side of the interface. For this case, concentrations at the s-surface turn out to be $m_{O_2,s} = 0.20$, $m_{CO_2,s} = 0.052$, and $\rho_s = 0.29$ kg/m^3.

Solution. The mass balance for the reaction is

$$12.0 \text{ kg C} + 32.0 \text{ kg O}_2 \longrightarrow 44.0 \text{ kg CO}_2$$

Since carbon flows through a second imaginary surface, u, moving through the stationary carbon just below the interface, the mass fluxes are related by

$$n_{C,u} = -\frac{12}{32} n_{O_2,s} = \frac{12}{44} n_{CO_2,s}$$

The minus sign arises because the O_2 flow is opposite the C and CO_2 flows, as shown in Figure 11.4. In steady state, if we apply mass conservation to the control volume between the u and s surfaces, we find that the total mass flux entering the u-surface equals that leaving the s-surface

$$n_{C,u} = n_{CO_2,s} + n_{O_2,s} = 0.00241 \text{ kg/m}^2\text{·s}$$

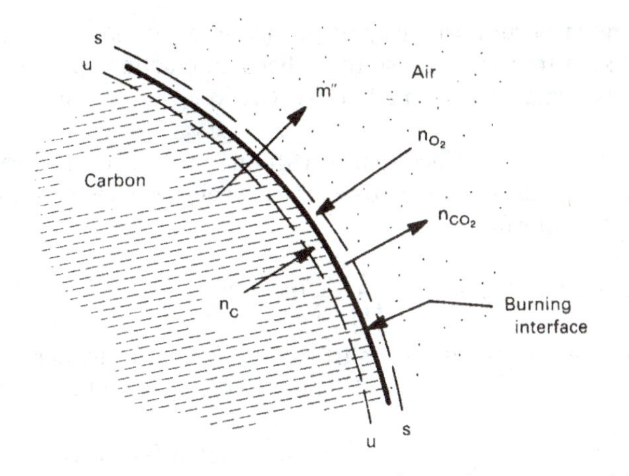

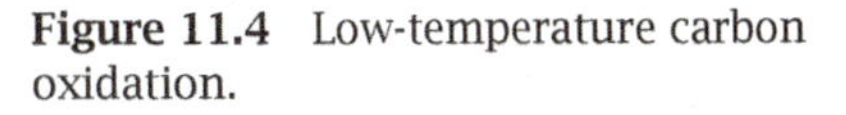

Figure 11.4 Low-temperature carbon oxidation.

Hence,

$$n_{O_2,s} = -\frac{32}{12}(0.00241\ \text{kg/m}^2{\cdot}\text{s}) = -0.00643\ \text{kg/m}^2{\cdot}\text{s}$$

$$n_{CO_2,s} = \frac{44}{12}(0.00241\ \text{kg/m}^2{\cdot}\text{s}) = 0.00884\ \text{kg/m}^2{\cdot}\text{s}$$

To get the diffusional mass flux, we need species and mass average speeds from eqns. (11.18) and (11.19):

$$v_{O_2,s} = \frac{n_{O_2,s}}{\rho_{O_2,s}} = \frac{-0.00643\ \text{kg/m}^2{\cdot}\text{s}}{0.2\,(0.29\ \text{kg/m}^3)} = -0.111\ \text{m/s}$$

$$v_{CO_2,s} = \frac{n_{CO_2,s}}{\rho_{CO_2,s}} = \frac{0.00884\ \text{kg/m}^2{\cdot}\text{s}}{0.052\,(0.29\ \text{kg/m}^3)} = 0.586\ \text{m/s}$$

$$v_s = \frac{1}{\rho_s}\sum_i n_i = \frac{(0.00884 - 0.00643)\ \text{kg/m}^2{\cdot}\text{s}}{0.29\ \text{kg/m}^3} = 0.00831\ \text{m/s}$$

Thus, from eqn. (11.20),

$$j_{i,s} = \rho_{i,s}\,(v_{i,s} - v_s) = \begin{cases} -0.00691\ \text{kg/m}^2{\cdot}\text{s for } O_2 \\ 0.00871\ \text{kg/m}^2{\cdot}\text{s for } CO_2 \end{cases}$$

The diffusional mass fluxes, $j_{i,s}$, are very nearly equal to the species mass fluxes, $n_{i,s}$. That is because the mass-average speed, v_s, is much less than the species speeds, $v_{i,s}$, in this case. Thus, the convective contribution to $n_{i,s}$ is much smaller than the diffusive contribution, and mass transfer occurs primarily by diffusion. Note that $j_{O_2,s}$ and $j_{CO_2,s}$ do *not* sum to zero because the other, nonreacting species in air must diffuse against the small convective velocity, v_s (see Section 11.7).

One mole of carbon surface reacts with one mole of O_2 to form one mole of CO_2. Thus, the mole fluxes of each species have the same magnitude at the interface:

$$N_{CO_2,s} = -N_{O_2,s} = N_{C,u} = \frac{n_{C,u}}{M_C} = 0.000201 \text{ kmol/m}^2\cdot\text{s}$$

The mole average velocity at the s-surface, v_s^*, is identically zero by eqn. (11.23), since $N_{CO_2,s} + N_{O_2,s} = 0$. The diffusional mole fluxes are

$$J_{i,s}^* = c_{i,s}(v_{i,s} - \underbrace{v_s^*}_{=0}) = N_{i,s} = \begin{cases} -0.000201 \text{ kmol/m}^2\cdot\text{s for } O_2 \\ 0.000201 \text{ kmol/m}^2\cdot\text{s for } CO_2 \end{cases}$$

These two diffusional mole fluxes sum to zero themselves because there is *no* convective mole flux for other species to diffuse against (i.e., for the other species $J_{i,s}^* = 0$).

The reader may calculate the velocity of the interface from $n_{c,u}$. That calculation would show the interface to be receding so slowly that the velocities we calculate are almost equal to those that would be seen by a stationary observer. ■

11.3 Diffusion fluxes and Fick's law

When the composition of a mixture is nonuniform, the concentration gradient in any species, i, of the mixture provides a driving potential for the diffusion of that species. The diffusing species flows from a region where it is highly concentrated to one where its concentration is low—like the diffusion of heat from a region of high temperature to one of low temperature. We have already noted in Section 2.1 that mass diffusion obeys Fick's law

$$\boxed{\vec{j}_i = -\rho \mathcal{D}_{im} \nabla m_i} \tag{11.27}$$

which is analogous to Fourier's law.

The constant of proportionality, $\rho\mathcal{D}_{im}$, between the local diffusive mass flux of species i and the local concentration gradient of i involves a physical property called the *diffusion coefficient*, $\mathcal{D}_{im}$, for species i diffusing in the mixture m. Like the thermal diffusivity, α, or the kinematic viscosity (a momentum diffusivity), ν, the mass diffusivity $\mathcal{D}_{im}$ has the units of m^2/s. These three diffusivities can form three dimensionless groups, among which is the Prandtl number:

$$\begin{aligned} \text{The Prandtl number,}\quad & \Pr \equiv \nu/\alpha \\ \text{The Schmidt number,}^{3}\quad & \text{Sc} \equiv \nu/\mathcal{D}_{im} \\ \text{The Lewis number,}^{4}\quad & \text{Le} \equiv \alpha/\mathcal{D}_{im} = \text{Sc}/\Pr \end{aligned} \tag{11.28}$$

Each of these groups compares the relative strength of two different diffusive processes. We make considerable use of the Schmidt number later in this chapter.

When diffusion occurs in mixtures of only two species—so-called *binary mixtures*—$\mathcal{D}_{im}$ reduces to the *binary diffusion coefficient*, $\mathcal{D}_{12}$[5]. In binary diffusion, species 1 has the same diffusivity through species 2 as does species 2 through species 1 (see Problem 11.5); in other words,

$$\mathcal{D}_{12} = \mathcal{D}_{21} \tag{11.29}$$

A kinetic model of diffusion

Diffusion coefficients depend upon composition, temperature, and pressure. Equations that predict $\mathcal{D}_{12}$ and $\mathcal{D}_{im}$ are given in Section 11.4. For now, let us see how Fick's law arises from the same sort of elementary molecular kinetics that gave Fourier's and Newton's laws in Section 6.4.

[3]Ernst Schmidt (1892–1975) served successively as the professor of thermodynamics at the Technical Universities of Danzig, Braunschweig, and Munich (Chapter 6, footnote 3). His many contributions to heat and mass transfer include the introduction of aluminum foil as radiation shielding, the first measurements of velocity and temperature fields in a natural convection boundary layer, and a once widely-used graphical procedure for solving unsteady heat conduction problems. He was among the first to develop the analogy between heat and mass transfer.

[4]Warren K. Lewis (1882–1975) was a professor of chemical engineering at M.I.T. from 1910 to 1975 and headed the department throughout the 1920s. He defined the original paradigm of chemical engineering, that of "unit operations", and, through his textbook with Walker and McAdams, *Principles of Chemical Engineering*, he laid the foundations of the discipline. He was a prolific inventor in the area of industrial chemistry, holding more than 80 patents. He also did important early work on simultaneous heat and mass transfer in connection with evaporation problems.

[5]Actually, Fick's Law is strictly valid only for binary mixtures. It can, however, often be applied to multicomponent mixtures with an appropriate choice of $\mathcal{D}_{im}$ (see Section 11.4).

We consider a two-component *dilute* gas (one with a low density) in which the molecules A of one species are very similar to the molecules A' of a second species (as though some of the molecules of a pure gas had merely been labeled without changing their properties.) The resulting process is called *self-diffusion*.

If we have a one-dimensional concentration distribution, as shown in Fig. 11.5, molecules of A diffuse down their concentration gradient in the x-direction. This process is entirely analogous to the transport of energy and momentum shown in Fig. 6.13. We take the temperature and pressure of the mixture (and thus its number density) to be uniform and the mass-average velocity to be zero.

Individual molecules move at a speed C, which varies randomly from molecule to molecule and is called the *thermal* or *peculiar* speed. The average speed of the molecules is $\overline{C}$. The average rate at which molecules cross the plane $x = x_0$ in either direction is proportional to $\mathcal{N}\overline{C}$, where $\mathcal{N}$ is the number density (molecules/m^3). Prior to crossing the x_0-plane, the molecules travel a distance close to one mean free path, ℓ—call it $a\ell$, where a is a number on the order of one.

The molecular flux travelling rightward across x_0, from its plane of origin at $x_0 - a\ell$, then has a fraction of molecules of A equal to the value of $\mathcal{N}_A/\mathcal{N}$ at $x_0 - a\ell$. The leftward flux, from $x_0 + a\ell$, has a fraction equal to the value of $\mathcal{N}_A/\mathcal{N}$ at $x_0 + a\ell$. Since the mass of a molecule of A is M_A/N_A (where N_A is Avogadro's number), the net mass flux in the x-direction is then

$$j_A\Big|_{x_0} = \eta\left(\mathcal{N}\overline{C}\right)\left(\frac{M_A}{N_A}\right)\left(\frac{\mathcal{N}_A}{\mathcal{N}}\bigg|_{x_0-a\ell} - \frac{\mathcal{N}_A}{\mathcal{N}}\bigg|_{x_0+a\ell}\right) \tag{11.30}$$

where η is a constant of proportionality. Since $\mathcal{N}_A/\mathcal{N}$ changes little in a distance of two mean free paths (in most real situations), we can expand the right side of eqn. (11.30) in a two-term Taylor series expansion about x_0 and obtain Fick's law:

$$\begin{aligned} j_A\Big|_{x_0} &= \eta\left(\mathcal{N}\overline{C}\right)\left(\frac{M_A}{N_A}\right)\left(-2a\ell\,\frac{d(\mathcal{N}_A/\mathcal{N})}{dx}\bigg|_{x_0}\right) \\ &= -2\eta a(\overline{C}\ell)\rho\frac{dm_A}{dx}\bigg|_{x_0} \end{aligned} \tag{11.31}$$

(for details, see Problem 11.6). Thus, we identify

$$\mathcal{D}_{AA'} = (2\eta a)\overline{C}\ell \tag{11.32}$$

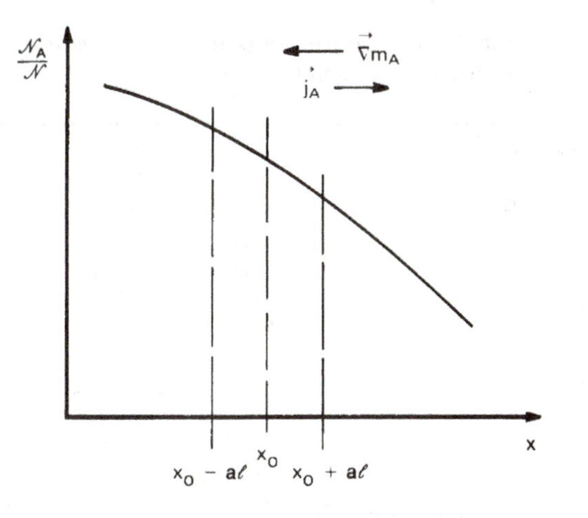

Figure 11.5 One-dimensional diffusion.

and Fick's law takes the form

$$j_A = -\rho \mathcal{D}_{AA'} \frac{dm_A}{dx} \tag{11.33}$$

The constant, ηa, in eqn. (11.32) can be fixed only with the help of a more detailed kinetic theory calculation [11.2], the result of which is given in Section 11.4.

The choice of j_i and m_i for the description of diffusion is really somewhat arbitrary. The molar diffusion flux, J_i^*, and the mole fraction, x_i, are often used instead, in which case Fick's law reads

$$\boxed{\vec{J}_i^{\,*} = -c\mathcal{D}_{im} \nabla x_i} \tag{11.34}$$

Obtaining eqn. (11.34) from eqn. (11.27) for a binary mixture is left as an exercise (Problem 11.4).

Typical values of the diffusion coefficient

Fick's law works well in low density gases and in dilute liquid and solid solutions, but for concentrated liquid and solid solutions the diffusion coefficient is found to vary with the concentration of the diffusing species. In part, the concentration dependence of those diffusion coefficients reflects the inadequacy of the concentration gradient in representing the

Table 11.1 Typical diffusion coefficients for binary gas mixtures at 1 atm and dilute liquid solutions [11.4].

Gas mixture	T (K)	$\mathcal{D}_{12}$ (m²/s)
air-carbon dioxide	276	1.42×10^{-5}
air-ethanol	313	1.45
air-helium	276	6.24
air-napthalene	303	0.86
air-water	313	2.88
argon-helium	295	8.3
	628	32.1
	1068	81.0
(dilute solute, 1)-(liquid solvent, 2)	T (K)	$\mathcal{D}_{12}$ (m²/s)
ethanol-benzene	288	2.25×10^{-9}
benzene-ethanol	298	1.81
water-ethanol	298	1.24
carbon dioxide-water	298	2.00
ethanol-water	288	1.00
methane-water	275	0.85
	333	3.55
pyridene-water	288	0.58

driving force for diffusion in nondilute solutions. Gradients in the chemical potential actually drive diffusion. In concentrated liquid or solid solutions, chemical potential gradients are not always equivalent to concentration gradients [11.3, 11.4, 11.5].

Table 11.1 lists some experimental values of the diffusion coefficient in binary gas mixtures and dilute liquid solutions. For gases, the diffusion coefficient is typically on the order of 10^{-5} m^2/s near room temperature. For liquids, the diffusion coefficient is much smaller, on the order of 10^{-9} m^2/s near room temperature. For both liquids and gases, the diffusion coefficient rises with increasing temperature. Typical diffusion coefficients in solids (not listed) may range from about 10^{-20} to about 10^{-9} m^2/s, depending upon what substances are involved and the temperature [11.6].

The diffusion of water vapor through air is of particular technical importance, and it is therefore useful to have an empirical correlation specifically for that mixture:

$$\mathcal{D}_{\mathrm{H_2O,air}} = 1.87 \times 10^{-10} \left(\frac{T^{2.072}}{p} \right) \quad \text{for} \quad 282\ \mathrm{K} \le T \le 450\ \mathrm{K} \qquad (11.35)$$

where $\mathcal{D}_{\mathrm{H_2O,air}}$ is in $\mathrm{m^2/s}$, T is in kelvin, and p is in atm [11.7]. The scatter of the available data around this equation is about 10%.

Coupled diffusion phenomena

Mass diffusion can be driven by factors other than concentration gradients, although the latter are of primary importance. For example, temperature gradients can induce mass diffusion in a process known as *thermal diffusion* or the *Soret effect.* The diffusional mass flux resulting from both temperature and concentration gradients in a binary gas mixture is then [11.2]

$$\vec{j}_i = -\rho \mathcal{D}_{12} \left[\nabla m_1 + \frac{M_1 M_2}{M^2} k_T \nabla \ln(T) \right] \qquad (11.36)$$

where k_T is called the *thermal diffusion ratio* and is generally quite small. Thermal diffusion is occasionally used in chemical separation processes. Pressure gradients and body forces acting unequally on the different species can also cause diffusion. Again, these effects are normally small. A related phenomenon is the generation of a heat flux by a concentration gradient (as distinct from heat convected by diffusing mass), called the *diffusion-thermo* or *Dufour effect.*

In this chapter, we deal only with mass transfer produced by concentration gradients.

11.4 Transport properties of mixtures[6]

Direct measurements of mixture transport properties are not always available for the temperature, pressure, or composition of interest. Thus, we must often rely upon theoretical predictions or experimental correlations for estimating mixture properties. In this section, we discuss methods

[6]This section may be regarded as reference material for the rest of the chapter, and it may be omitted without loss of continuity. The property predictions of this section are used only in Examples 11.11, 11.14, and 11.16, and in some of the end-of-chapter problems.

for computing $\mathcal{D}_{im}$, k, and μ in gas mixtures using equations from kinetic theory—particularly the Chapman-Enskog theory [11.2, 11.8, 11.9]. We also consider some methods for computing $\mathcal{D}_{12}$ in dilute liquid solutions.

The diffusion coefficient for binary gas mixtures

As a starting point, we return to our simple model for the self-diffusion coefficient of a dilute gas, eqn. (11.32). We can approximate the average molecular speed, $\overline{C}$, by Maxwell's equilibrium formula (see, e.g., [11.9]):

$$\overline{C} = \left(\frac{8k_B N_A T}{\pi M}\right)^{1/2} \tag{11.37}$$

where $k_B = R^\circ / N_A$ is Boltzmann's constant. If we assume the molecules to be rigid and spherical, then the mean free path turns out to be

$$\ell = \frac{1}{\pi\sqrt{2}\mathcal{N}d^2} = \frac{k_B T}{\pi\sqrt{2}d^2 p} \tag{11.38}$$

where d is the effective molecular diameter. Substituting these values of $\overline{C}$ and ℓ in eqn. (11.32) and applying a kinetic theory calculation that shows $2\eta a = 1/2$, we find

$$\begin{aligned} \mathcal{D}_{AA'} &= (2\eta a)\overline{C}\ell \\ &= \frac{(k_B/\pi)^{3/2}}{d^2}\left(\frac{N_A}{M}\right)^{1/2}\frac{T^{3/2}}{p} \end{aligned} \tag{11.39}$$

The diffusion coefficient varies as p^{-1} and $T^{3/2}$, based on the simple model for self-diffusion.

Of course, molecules are not really hard spheres. We should also allow for differences in the molecular sizes of different species and for nonuniformities in the bulk properties of the gas. The Chapman-Enskog kinetic theory takes all these factors into account [11.8]. It yields the following formula for $\mathcal{D}_{AB}$:

$$\mathcal{D}_{AB} = \frac{(1.8583\times 10^{-7})T^{3/2}}{p\Omega_D^{AB}(T)}\sqrt{\frac{1}{M_A}+\frac{1}{M_B}}$$

where the units of p, T, and $\mathcal{D}_{AB}$ are atm, K, and m^2/s, respectively. The function $\Omega_D^{AB}(T)$, called a collision integral, reflects the collisions between molecules of A and B. It depends, in general, on the specific type of molecules involved and the temperature.

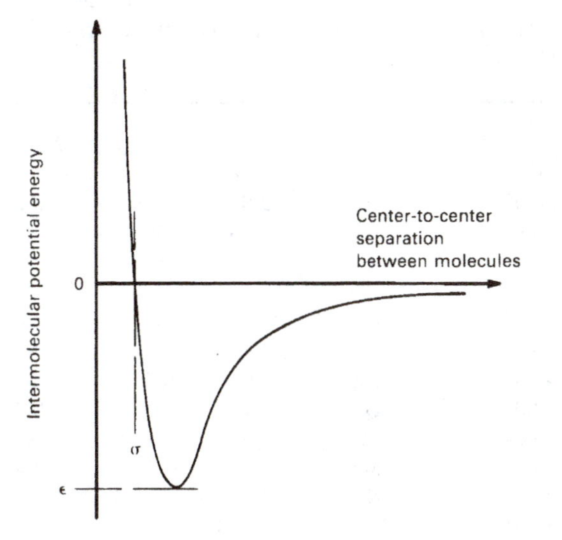

Figure 11.6 The Lennard-Jones potential.

The type of molecule matters because of the intermolecular forces of attraction and repulsion that arise when molecules collide. A good approximation to those forces is given by the Lennard-Jones intermolecular potential (see Fig. 11.6)[7]. This potential is based on two parameters, a molecular diameter, σ, and a *potential well depth*, ε. The potential well depth is the energy required to separate two molecules from one another. Both constants can be inferred from physical property data. Some values are given in Table 11.2 together with the associated molecular weights (from [11.10], with values for calculating the diffusion coefficients of water from [11.11]).

An accurate approximation to $\Omega_D^{AB}(T)$ can be obtained using the Lennard-Jones potential function. The result is

$$\Omega_D^{AB}(T) = \sigma_{AB}^2\, \Omega_D(k_B T/\varepsilon_{AB})$$

where, the *collision diameter*, σ_{AB}, may be viewed as an effective molecular diameter for collisions of A and B. If σ_A and σ_B are the cross-sectional

[7]John Edward Jones was born in 1894 and educated in theoretical physics at Bristol University. He served as a flyer in the First World War, then proposed his attraction formula in his doctorate at Manchester. The next year he married Katherine Lennard and attached her maiden name to his own. The formula thereupon became the Lennard-Jones potential.

Table 11.2 Lennard-Jones constants and molecular weights of selected species.

Species	σ(Å)	ε/k_B(K)	$M\left(\frac{\text{kg}}{\text{kmol}}\right)$	*Species*	σ(Å)	ε/k_B(K)	$M\left(\frac{\text{kg}}{\text{kmol}}\right)$
Al	2.655	2750	26.98	H_2	2.827	59.7	2.016
Air	3.711	78.6	28.96	H_2O	2.655[a]	363[a]	18.02
Ar	3.542	93.3	39.95	H_2O	2.641[b]	809.1[b]	
Br_2	4.296	507.9	159.8	H_2O_2	4.196	289.3	34.01
C	3.385	30.6	12.01	H_2S	3.623	301.1	34.08
CCl_2F_2	5.25	253	120.9	He	2.551	10.22	4.003
CCl_4	5.947	322.7	153.8	Hg	2.969	750	200.6
CH_3OH	3.626	481.8	32.04	I_2	5.160	474.2	253.8
CH_4	3.758	148.6	16.04	Kr	3.655	178.9	83.80
CN	3.856	75.0	26.02	Mg	2.926	1614	24.31
CO	3.690	91.7	28.01	NH_3	2.900	558.3	17.03
CO_2	3.941	195.2	44.01	N_2	3.798	71.4	28.01
C_2H_6	4.443	215.7	30.07	N_2O	3.828	232.4	44.01
C_2H_5OH	4.530	362.6	46.07	Ne	2.820	32.8	20.18
CH_3COCH_3	4.600	560.2	58.08	O_2	3.467	106.7	32.00
C_6H_6	5.349	412.3	78.11	SO_2	4.112	335.4	64.06
Cl_2	4.217	316.0	70.91	Xe	4.047	231.0	131.3
F_2	3.357	112.6	38.00				

[a] Based on mass diffusion data.
[b] Based on viscosity and thermal conductivity data.

diameters of A and B, in Å,[8] then

$$\sigma_{AB} = (\sigma_A + \sigma_B)/2 \tag{11.40}$$

The *collision integral*, Ω_D is a result of kinetic theory calculations calculations based on the Lennard-Jones potential. Table 11.3 gives values of Ω_D from [11.12]. The effective potential well depth for collisions of A and B is

$$\varepsilon_{AB} = \sqrt{\varepsilon_A \varepsilon_B} \tag{11.41}$$

[8] One Ångström (1 Å) is equal to 0.1 nm.

Hence, we may calculate the binary diffusion coefficient from

$$\boxed{\mathcal{D}_{AB} = \frac{(1.8583 \times 10^{-7})T^{3/2}}{p\sigma_{AB}^2\Omega_D}\sqrt{\frac{1}{M_A} + \frac{1}{M_B}}} \tag{11.42}$$

where, again, the units of p, T, and $\mathcal{D}_{AB}$ are atm, K, and m^2/s, respectively, and σ_{AB} is in Å.

Equation (11.42) indicates that the diffusivity varies as p^{-1} and is independent of mixture concentrations, just as the simple model indicated that it should. The temperature dependence of Ω_D, however, increases the overall temperature dependence of $\mathcal{D}_{AB}$ from $T^{3/2}$, as suggested by eqn. (11.39), to approximately $T^{7/4}$.

Air, by the way, can be treated as a single substance in Table 11.2 owing to the similarity of its two main constituents, N_2 and O_2.

Example 11.3

Compute $\mathcal{D}_{AB}$ for the diffusion of hydrogen in air at 276 K and 1 atm.

Solution. Let air be species A and H_2 be species B. Then we read from Table 11.2

$$\sigma_A = 3.711\text{ Å}, \quad \sigma_B = 2.827\text{ Å}, \quad \frac{\varepsilon_A}{k_B} = 78.6\text{ K}, \quad \frac{\varepsilon_B}{k_B} = 59.7\text{ K}$$

and calculate these values

$$\sigma_{AB} = (3.711 + 2.827)/2 = 3.269\text{ Å}$$
$$\varepsilon_{AB}/k_B = \sqrt{(78.6)(59.7)} = 68.5\text{ K}$$

Hence, $k_B T/\varepsilon_{AB} = 4.029$, and $\Omega_D = 0.8822$ from Table 11.3. Then

$$\begin{aligned}\mathcal{D}_{AB} &= \frac{(1.8583 \times 10^{-7})(276)^{3/2}}{(1)(3.269)^2(0.8822)}\sqrt{\frac{1}{2.016} + \frac{1}{28.97}}\ \text{m}^2/\text{s} \\ &= 6.58 \times 10^{-5}\ \text{m}^2/\text{s}\end{aligned}$$

An experimental value from Table 11.1 is 6.34×10^{-5} m^2/s, so the prediction is high by 5%. ■

Table 11.3 Collision integrals for diffusivity, viscosity, and thermal conductivity based on the Lennard-Jones potential.

$k_B T/\varepsilon$	Ω_D	$\Omega_\mu = \Omega_k$	$k_B T/\varepsilon$	Ω_D	$\Omega_\mu = \Omega_k$
0.30	2.662	2.785	2.70	0.9770	1.069
0.35	2.476	2.628	2.80	0.9672	1.058
0.40	2.318	2.492	2.90	0.9576	1.048
0.45	2.184	2.368	3.00	0.9490	1.039
0.50	2.066	2.257	3.10	0.9406	1.030
0.55	1.966	2.156	3.20	0.9328	1.022
0.60	1.877	2.065	3.30	0.9256	1.014
0.65	1.798	1.982	3.40	0.9186	1.007
0.70	1.729	1.908	3.50	0.9120	0.9999
0.75	1.667	1.841	3.60	0.9058	0.9932
0.80	1.612	1.780	3.70	0.8998	0.9870
0.85	1.562	1.725	3.80	0.8942	0.9811
0.90	1.517	1.675	3.90	0.8888	0.9755
0.95	1.476	1.629	4.00	0.8836	0.9700
1.00	1.439	1.587	4.10	0.8788	0.9649
1.05	1.406	1.549	4.20	0.8740	0.9600
1.10	1.375	1.514	4.30	0.8694	0.9553
1.15	1.346	1.482	4.40	0.8652	0.9507
1.20	1.320	1.452	4.50	0.8610	0.9464
1.25	1.296	1.424	4.60	0.8568	0.9422
1.30	1.273	1.399	4.70	0.8530	0.9382
1.35	1.253	1.375	4.80	0.8492	0.9343
1.40	1.233	1.353	4.90	0.8456	0.9305
1.45	1.215	1.333	5.00	0.8422	0.9269
1.50	1.198	1.314	6.00	0.8124	0.8963
1.55	1.182	1.296	7.0	0.7896	0.8727
1.60	1.167	1.279	8.0	0.7712	0.8538
1.65	1.153	1.264	9.0	0.7556	0.8379
1.70	1.140	1.248	10.0	0.7424	0.8242
1.75	1.128	1.234	20.0	0.6640	0.7432
1.80	1.116	1.221	30.0	0.6232	0.7005
1.85	1.105	1.209	40.0	0.5960	0.6718
1.90	1.094	1.197	50.0	0.5756	0.6504
1.95	1.084	1.186	60.0	0.5596	0.6335
2.00	1.075	1.175	70.0	0.5464	0.6194
2.10	1.057	1.156	80.0	0.5352	0.6076
2.20	1.041	1.138	90.0	0.5256	0.5973
2.30	1.026	1.122	100.0	0.5170	0.5882
2.40	1.012	1.107	200.0	0.4644	0.5320
2.50	0.9996	1.093	300.0	0.4360	0.5016
2.60	0.9878	1.081	400.0	0.4172	0.4811

Limitations of the diffusion coefficient prediction. Equation (11.42) is not valid for all gas mixtures. We have already noted that concentration gradients cannot be too steep; thus, it cannot be applied, say, across a shock wave when the Mach number is significantly greater than one. Furthermore, the gas must be dilute, and its molecules should be nonpolar and approximately spherically symmetric.

Reid et al. [11.4] compared values of $\mathcal{D}_{12}$ calculated using eqn. (11.42) with data for binary mixtures of monatomic, polyatomic, nonpolar, and polar gases of the sort appearing in Table 11.2. They reported an average absolute error of 7.3 percent. Better results can be obtained by using values of σ_{AB} and ε_{AB} that have been fit specifically to the pair of gases involved, rather than using eqns. (11.40) and (11.41), or by constructing a mixture-specific equation for $\Omega_D^{AB}(T)$ [11.13, Chap. 11].

The density of the gas also affects the accuracy of kinetic theory predictions, which is based on the gas being dilute in the sense that its molecules interact with one another only during brief two-molecule collisions. Childs and Hanley [11.14] have suggested that the transport properties of gases are within 1% of the dilute values if the gas densities do not exceed the following limiting value

$$\rho_{\text{max}} = 22.93M/(\sigma^3\Omega_\mu) \tag{11.43}$$

Here, σ (the collision diameter of the gas) and ρ are expressed in Å and kg/m^3, and Ω_μ—a second collision integral for viscosity—is included in Table 11.3. Equation (11.43) normally gives ρ_{max} values that correspond to pressures substantially above 1 atm.

At higher gas densities, transport properties can be estimated by a variety of techniques, such as corresponding states theories, absolute reaction-rate theories, or modified Enskog theories [11.13, Chap. 6] (also see [11.4, 11.8]). Conversely, if the gas density is so very *low* that the mean free path is on the order of the dimensions of the system, we have what is called *free molecule flow*, and the present kinetic models are again invalid (see, e.g., [11.15]).

Diffusion coefficients for multicomponent gases

We have already noted that an effective binary diffusivity, $\mathcal{D}_{im}$, can be used to represent the diffusion of species i into a mixture m. The preceding equations for the diffusion coefficient, however, are strictly applicable only when one pure substance diffuses through another. Different equations are needed when there are three or more species present.

If a low concentration of species i diffuses into a homogeneous mixture of n species, then $\vec{J}_j^{\,*} \cong 0$ for $j \neq i$, and one may show (Problem 11.14) that

$$\mathcal{D}_{im}^{-1} = \sum_{\substack{j=1 \\ j\neq i}}^{n} \frac{x_j}{\mathcal{D}_{ij}} \tag{11.44}$$

where $\mathcal{D}_{ij}$ is the binary diffusion coefficient for species i and j alone. This rule is sometimes called *Blanc's law* [11.4].

If a mixture is dominantly composed of one species, A, and includes only small traces of several other species, then the diffusion coefficient of each trace gas is approximately the same as it would be if the other trace gases were not present. In other words, for any particular trace species i,

$$\mathcal{D}_{im} \cong \mathcal{D}_{iA} \tag{11.45}$$

Finally, if the binary diffusion coefficient has the same value for each pair of species in a mixture, then one may show (Problem 11.14) that $\mathcal{D}_{im} = \mathcal{D}_{ij}$, as one might expect.

Diffusion coefficients for binary liquid mixtures

Each molecule in a liquid is always in contact with several neighboring molecules, and a kinetic theory like that used in gases, which relies on detailed descriptions of two-molecule collisions, is no longer feasible. Instead, a less precise theory can be developed and used to correlate experimental measurements.

For a dilute solution of substance A in liquid B, the so-called *hydrodynamic model* has met some success. Suppose that, when a force per molecule of F_A is applied to molecules of A, they reach an average steady speed of v_A relative to the liquid B. The ratio v_A/F_A is called the *mobility* of A. If there is no applied force, then the molecules of A diffuse as a result of random molecular motions—called *Brownian motion*[9]. Kinetic and thermodynamic arguments, such as those given by Einstein [11.16] and Sutherland [11.17], lead to an expression for the diffusion coefficient

[9]Robert Brown was a Scottish botanist, born in 1773. In 1827, he was examining pollen grains suspended in water, and noted that they continuously jerked about. Were they alive? No, he found the same behavior in dust motes. Without a full atomic theory, he could not tell that the particles were small enough to be knocked about by molecules. Mathematical descriptions of the process had to wait for late 19th and early 20th century.

of A in B as a result of Brownian motion:

$$\mathcal{D}_{AB} = k_B T \left(v_A / F_A \right) \tag{11.46}$$

Equation (11.46) is usually called the *Nernst-Einstein equation.*

To evaluate the mobility of a molecular (or particulate) solute, we may make the rather bold approximation that *Stokes' law* [11.18] applies, even though it is really a drag law for spheres at low Reynolds number ($\text{Re}_D < 1$)

$$F_A = 6\pi \mu_B v_A R_A \left(\frac{1 + 2\mu_B/\beta R_A}{1 + 3\mu_B/\beta R_A} \right) \tag{11.47}$$

Here, R_A is the radius of sphere A and β is a coefficient of "sliding" friction, for a friction force proportional to the velocity. Substituting eqn. (11.47) in eqn. (11.46), we get

$$\frac{\mathcal{D}_{AB}\mu_B}{T} = \frac{k_B}{6\pi R_A} \left(\frac{1 + 3\mu_B/\beta R_A}{1 + 2\mu_B/\beta R_A} \right) \tag{11.48}$$

This model is valid if the concentration of solute A is so low that the molecules of A do not interact with one another.

For viscous liquids one usually assumes that no slip occurs between the liquid and a solid surface that it touches; but, for particles whose size is on the order of the molecular spacing of the solvent molecules, some slip may very well occur. This is the reason for the unfamiliar factor in parentheses on the right side of eqn. (11.47). For large solute particles, there should be no slip, so $\beta \longrightarrow \infty$ and the factor in parentheses tends to one, as expected. Equation (11.48) then reduces to[10]

$$\frac{\mathcal{D}_{AB}\mu_B}{T} = \frac{k_B}{6\pi R_A} \tag{11.49a}$$

For smaller molecules—close in size to those of the solvent—we expect that $\beta \longrightarrow 0$, leading to [11.19]

$$\frac{\mathcal{D}_{AB}\mu_B}{T} = \frac{k_B}{4\pi R_A} \tag{11.49b}$$

[10]While earlier investigators attempted mathematical descriptions of Brownian motion, eqn. (11.49a) was first presented by Einstein in May 1905. The more general form, eqn. (11.48), was presented independently by Sutherland in June 1905. Equations (11.48) and (11.49a) are commonly called the *Stokes-Einstein* equation, although Stokes had no hand in applying eqn. (11.47) to diffusion. Equation (11.48) might better be called the *Sutherland-Einstein* equation.

Table 11.4 Molal specific volumes and latent heats of vaporization for selected liquids at their normal boiling points.

Substance	V_m (m^3/kmol)	h_{fg} (MJ/kmol)
Methanol	0.042	35.53
Ethanol	0.064	39.33
n-Propanol	0.081	41.97
Isopropanol	0.072	40.71
n-Butanol	0.103	43.76
tert-Butanol	0.103	40.63
n-Pentane	0.118	25.61
Cyclopentane	0.100	27.32
Isopentane	0.118	24.73
Neopentane	0.118	22.72
n-Hexane	0.141	28.85
Cyclohexane	0.117	33.03
n-Heptane	0.163	31.69
n-Octane	0.185	34.14
n-Nonane	0.207	36.53
n-Decane	0.229	39.33
Acetone	0.074	28.90
Benzene	0.096	30.76
Carbon tetrachloride	0.102	29.93
Ethyl bromide	0.075	27.41
Nitromethane	0.056	25.44
Water	0.0187	40.62

The most important feature of eqns. (11.48), (11.49a), and (11.49b) is that, so long as the solute is dilute, the primary determinant of the group $\mathcal{D}\mu/T$ is the size of the diffusing species, with a secondary dependence on intermolecular forces (i.e., on β). More complex theories, such as the absolute reaction-rate theory of Eyring [11.20], lead to the same dependence. Moreover, experimental studies of dilute solutions verify that the group $\mathcal{D}\mu/T$ is essentially temperature-independent for a given solute-solvent pair, with the only exception occuring in very high viscosity solutions. Thus, most correlations of experimental data have used some form of eqn. (11.48) as a starting point.

Many such correlations have been developed. One fairly successful correlation is by King et al. [11.21]. They expressed the molecular size in terms of molal volumes at the normal boiling point, $V_{m,A}$ and $V_{m,B}$, and accounted for intermolecular association forces using the latent heats of vaporization at the normal boiling point, $h_{fg,A}$ and $h_{fg,B}$. They obtained

$$\frac{\mathcal{D}_{AB}\mu_B}{T} = (4.4 \times 10^{-15}) \left(\frac{V_{m,B}}{V_{m,A}}\right)^{1/6} \left(\frac{h_{fg,B}}{h_{fg,A}}\right)^{1/2} \tag{11.50}$$

which has an rms error of 19.5% and for which the units of $\mathcal{D}_{AB}\mu_B/T$ are kg·m/K·s^2. Values of h_{fg} and V_m are given for various substances in Table 11.4. Equation (11.50) is valid for nonelectrolytes at high dilution, and it appears to be satisfactory for both polar and nonpolar substances. The difficulties with polar solvents of high viscosity led the authors to limit eqn. (11.50) to values of $\mathcal{D}\mu/T < 1.5\times10^{-14}$ kg·m/K·s^2. The predictions of eqn. (11.50) are compared with experimental data in Fig. 11.7. Reid et al. [11.4] review several other liquid-phase correlations and provide an assessment of their accuracies.

The thermal conductivity and viscosity of dilute gases

In any convective mass transfer problem, we must know the viscosity of the fluid and, if heat is also being transferred, we must also know its thermal conductivity. Accordingly, we now consider the calculation of μ and k for mixtures of gases.

Two of the most important results of the kinetic theory of gases are the predictions of μ and k for a pure, monatomic gas of species A:

$$\mu_A = \left(2.6693 \times 10^{-6}\right) \frac{\sqrt{M_A T}}{\sigma_A^2 \Omega_\mu} \tag{11.51}$$

and

$$k_A = \frac{0.083228}{\sigma_A^2 \Omega_k} \sqrt{\frac{T}{M_A}} \tag{11.52}$$

where Ω_μ and Ω_k are collision integrals for the viscosity and thermal conductivity. In fact, Ω_μ and Ω_k are equal to one another, but they are different from Ω_D. In these equations μ is in kg/m·s, k is in W/m·K, T is in kelvin, and σ_A again has units of Å.

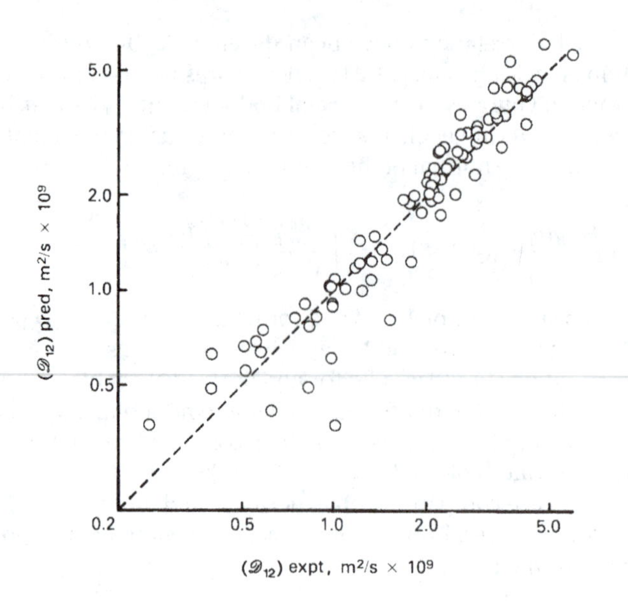

Figure 11.7 Comparison of liquid diffusion coefficients predicted by eqn. (11.50) with experimental values for assorted substances from [11.4].

The equation for μ_A applies equally well to polyatomic gases, but k_A must be corrected to account for internal modes of energy storage—chiefly molecular rotation and vibration. Eucken (see, e.g., [11.9]) gave a simple analysis showing that this correction was

$$k = \left(\frac{9\gamma - 5}{4\gamma}\right) \mu c_p \tag{11.53}$$

for an ideal gas, where $\gamma \equiv c_p/c_v$. You may recall from your thermodynamics courses that γ is 5/3 for monatomic gases, 7/5 for diatomic gases at modest temperatures, and approaches one for very complex molecules. Equation (11.53) should be used with tabulated data for c_p; on average, it will underpredict k by perhaps 10 to 20% [11.4].

An approximate formula for μ for multicomponent gas mixtures was developed by Wilke [11.22], based on the kinetic theory of gases. He introduced certain simplifying assumptions and obtained, for the mixture viscosity,

$$\mu_m = \sum_{i=1}^{n} \frac{x_i \mu_i}{\sum_{j=1}^{n} x_j \phi_{ij}} \tag{11.54}$$

where

$$\phi_{ij} = \frac{\left[1 + (\mu_i/\mu_j)^{1/2}(M_j/M_i)^{1/4}\right]^2}{2\sqrt{2}\left[1 + (M_i/M_j)\right]^{1/2}}$$

The analogous equation for the thermal conductivity of mixtures was developed by Mason and Saxena [11.23]:

$$k_m = \sum_{i=1}^{n} \frac{x_i k_i}{\sum_{j=1}^{n} x_j \phi_{ij}} \tag{11.55}$$

(We have followed [11.4] in omitting a minor empirical correction factor proposed by Mason and Saxena.)

Equation (11.54) is accurate to about 2 % and eqn. (11.55) to about 4% for mixtures of nonpolar gases. For higher accuracy or for mixtures with polar components, refer to [11.4] and [11.13].

Example 11.4

Compute the transport properties of normal air at 300 K.

Solution. The mass composition of air was given in Example 11.1. Using the methods of Example 11.1, we obtain the mole fractions as $x_{N_2} = 0.7808$, $x_{O_2} = 0.2095$, and $x_{Ar} = 0.0093$.

We first compute μ and k for the three species to illustrate the use of eqns. (11.51) to (11.53), although we could simply use tabled data in eqns. (11.54) and (11.55). From Tables 11.2 and 11.3, we obtain

Species	σ(Å)	ε/k_B(K)	M	Ω_μ
N_2	3.798	71.4	28.02	0.9599
O_2	3.467	106.7	32.00	1.057
Ar	3.542	93.3	39.95	1.021

Substitution of these values into eqn. (11.51) yields

Species	μ_{calc} (kg/m·s)	μ_{data} (kg/m·s)
N_2	1.767×10^{-5}	1.80×10^{-5}
O_2	2.059×10^{-5}	2.07×10^{-5}
Ar	2.281×10^{-5}	2.29×10^{-5}

where we show data from Appendix A (Table A.6) for comparison. We then read c_p from Appendix A and use eqn. (11.52) and (11.53) to get the thermal conductivities of the components:

Species	c_p (J/kg·K)	k_{calc} (W/m·K)	k_{data} (W/m·K)
N_2	1041.	0.02500	0.0260
O_2	919.9	0.02569	0.02615
Ar	521.5	0.01782	0.01787

The predictions thus agree with the data to within about 2% for μ and within about 4% for k.

To compute μ_m and k_m, we use eqns. (11.54) and (11.55) and the tabulated values of μ and k. Identifying N_2, O_2, and Ar as species 1, 2, and 3, we get

$$\begin{aligned} \phi_{12} &= 0.9894, & \phi_{21} &= 1.010 \\ \phi_{13} &= 1.043, & \phi_{31} &= 0.9445 \\ \phi_{23} &= 1.058, & \phi_{32} &= 0.9391 \end{aligned}$$

and $\phi_{ii} = 1$. The sums appearing in the denominators are

$$\sum x_j \phi_{ij} = \begin{cases} 0.9978 & \text{for} \quad i = 1 \\ 1.008 & \text{for} \quad i = 2 \\ 0.9435 & \text{for} \quad i = 3 \end{cases}$$

When they are substituted in eqns. (11.54) and (11.55), these values give

$$\begin{aligned} \mu_{m,\text{calc}} &= 1.861 \times 10^{-5} \text{ kg/m·s}, & \mu_{m,\text{data}} &= 1.857 \times 10^{-5} \text{ kg/m·s} \\ k_{m,\text{calc}} &= 0.02596 \text{ W/m·K}, & k_{m,\text{data}} &= 0.02623 \text{ W/m·K} \end{aligned}$$

so the mixture values are also predicted within 0.3 and 1.0%, respectively.

Finally, we need c_{p_m} to compute the Prandtl number of the mixture. This is merely the mass weighted average of c_p, or $\sum_i m_i c_{p_i}$, and it is equal to 1006 J/kg·K. Then

$$\text{Pr} = (\mu c_p/k)_m = (1.861 \times 10^{-5})(1006)/0.02596 = 0.721.$$

This is 1% above the tabled value of 0.713. The reader may wish to compare these values with those obtained directly using the values for air in Table 11.2 or to explore the effects of neglecting argon in the preceding calculations. ■

11.5 The equation of species conservation

Conservation of species

Just as we formed an equation of energy conservation in Chapter 6, we now form an equation of *species conservation*—one that applies to each substance in a mixture. This equation should account not only for the convection and diffusion of each species; it should also allow the possibility that a species may be created or destroyed by chemical reactions occurring in the bulk medium (so-called *homogeneous reactions*). Reactions on surfaces surrounding the medium (*heterogeneous reactions*) will not enter the equation itself, but will appear in the boundary conditions.

We again begin with an arbitrary control volume, R, with a boundary, S, as shown in Fig. 11.8. The control volume is fixed in space, and fluid might move through it. Species i may accumulate in R, it may travel in and out of R by bulk convection or by diffusion, and it may be created within R by homogeneous reactions. We denote the rate of creation of species i as $\dot{r}_i$(kg/m^3·s); and, because the overall mass cannot change during chemical reactions, the net mass creation is $\dot{r} = \sum \dot{r}_i = 0$. The rate of change of the mass of each species i in R is then described by the

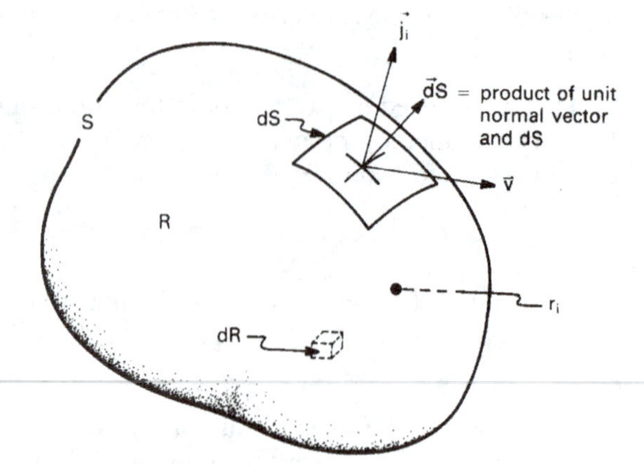

Figure 11.8 Control volume in a fluid-flow and mass-diffusion field.

following balance:

$$\underbrace{\frac{d}{dt}\int_R \rho_i \, dR}_{\text{rate of increase of } i \text{ in } R} = -\int_S \vec{n}_i \cdot d\vec{S} + \int_R \dot{r}_i \, dR$$

$$= -\underbrace{\int_S \rho_i \vec{v} \cdot d\vec{S}}_{\text{rate of convection of } i \text{ out of } R} - \underbrace{\int_S \vec{j}_i \cdot d\vec{S}}_{\text{diffusion of } i \text{ out of } R} + \underbrace{\int_R \dot{r}_i \, dR}_{\text{rate of creation of } i \text{ in } R} \tag{11.56}$$

This species conservation statement is identical to our energy conservation statement, eqn. (6.36) on page 295, except that mass of species i has taken the place of energy and heat.

We next convert the surface integrals to volume integrals using Gauss's theorem [eqn. (2.8)] and rearrange the result:

$$\int_R \left[\frac{\partial \rho_i}{\partial t} + \nabla \cdot (\rho_i \vec{v}) + \nabla \cdot \vec{j}_i - \dot{r}_i \right] dR = 0 \tag{11.57}$$

Since the control volume is selected arbitrarily, the integrand must be identically zero. Thus, we obtain the general form of the species conservation equation:

$$\boxed{\frac{\partial \rho_i}{\partial t} + \nabla \cdot (\rho_i \vec{v}) = -\nabla \cdot \vec{j}_i + \dot{r}_i} \tag{11.58}$$

We can obtain a mass conservation equation for the entire mixture by summing eqn. (11.58) over all species, applying eqns. (11.1), (11.17), and (11.22) and requiring that there be no net creation of mass:

$$\sum_i \left[\frac{\partial \rho_i}{\partial t} + \nabla \cdot (\rho_i \vec{v}) \right] = \sum_i (-\nabla \cdot \vec{j}_i + \dot{r}_i) = 0$$

so that

$$\frac{\partial \rho}{\partial t} + \nabla \cdot (\rho \vec{v}) = 0 \tag{11.59}$$

This equation applies to any mixture, including those with varying density (see Problem 6.36).

Incompressible mixtures. For an incompressible mixture, $\nabla \cdot \vec{v} = 0$ (see Sect. 6.2 or Problem 11.22), and the second term in eqn. (11.58) may therefore be rewritten as

$$\nabla \cdot (\rho_i \vec{v}) = \vec{v} \cdot \nabla \rho_i + \rho_i \underbrace{\nabla \cdot \vec{v}}_{=0} = \vec{v} \cdot \nabla \rho_i \tag{11.60}$$

Compare the resulting, incompressible species equation to the incompressible energy equation, eqn. (6.37)

$$\frac{D\rho_i}{Dt} = \quad \frac{\partial \rho_i}{\partial t} + \vec{v} \cdot \nabla \rho_i \quad = -\nabla \cdot \vec{j}_i + \dot{r}_i \tag{11.61}$$

$$\rho c_p \frac{DT}{Dt} = \rho c_p \left(\frac{\partial T}{\partial t} + \vec{v} \cdot \nabla T \right) = -\nabla \cdot \vec{q} + \dot{q} \tag{6.37}$$

and we see that the reaction term, $\dot{r}_i$, is analogous to the heat generation term, $\dot{q}$; the diffusional mass flux, $\vec{j}_i$, is analogous to the heat flux, $\vec{q}$; and $d\rho_i$ is analogous to $\rho c_p dT$.

We can use Fick's law to eliminate $\vec{j}_i$ in eqn. (11.61). The resulting equation may be written in various ways. If the product $\rho \mathcal{D}_{im}$ is independent of (x, y, z)—if it is spatially uniform—then eqn. (11.61) becomes

$$\boxed{\frac{D}{Dt} m_i = \mathcal{D}_{im} \nabla^2 m_i + \dot{r}_i / \rho} \tag{11.62}$$

where the material derivative, D/Dt, is defined in eqn. (6.38). If ρ is also spatially uniform, then

$$\frac{D\rho_i}{Dt} = \mathcal{D}_{im} \nabla^2 \rho_i + \dot{r}_i \tag{11.63}$$

The equation of species conservation and its particular forms may also be stated in molar variables, using c_i or x_i, N_i, and J_i^* (see Problem 11.24.) Molar analysis sometimes has advantages over mass-based analysis, as we discover in Section 11.7.

Interfacial boundary conditions

We are already familiar with the general issue of boundary conditions from our study of the heat equation. To find a temperature distribution, we specified temperatures or heat fluxes at the boundaries of the domain of interest. Likewise, to find a concentration distribution, we must specify the concentration or flux of species i at the boundaries of the medium of interest.

Temperature and concentration behave differently at interfaces. An interfacial temperature obviously has to be the same in both media. Concentration, on the other hand, need *not* be continuous across an interface, even in a state of thermodynamic equilibrium. Water in a drinking glass, for example, shows discontinous changes in the concentration of water at both the glass-water interface on the sides and the air-water interface above. The concentration of oxygen in the air is likewise 21 percent while its concentration in the adjacent water is far less.

That situation also arises when gaseous ammonia is absorbed into water in some types of refrigeration cycles. A gas mixture containing some mass fraction of ammonia will produce a different mass fraction of ammonia just inside an adjacent body of water, as shown in Fig. 11.9. To characterize conditions at an interface, we imagine surfaces, s and u, very close to either side of the interface. In the ammonia absorption process, then, we have a mass fraction $m_{\mathrm{NH_3},s}$ on the gas side of the interface and a *different* mass fraction $m_{\mathrm{NH_3},u}$ on the liquid side.

We must often find the concentration distribution of a species in one medium, given only its concentration at the interface in the *adjacent* medium. We might wish to find the distribution of ammonia in the body of water knowing only the concentration of ammonia on the gas side of the interface. We would need to find $m_{\mathrm{NH_3},u}$ from $m_{\mathrm{NH_3},s}$ and the interfacial temperature and pressure, since $m_{\mathrm{NH_3},u}$ is the appropriate boundary condition for the species conservation equation in the water.

Thus, for the general mass transfer boundary condition, we must specify not only the concentration of species i in the medium adjacent to the medium of interest but also the *solubility* of species i from one medium to the other. Although a detailed study of solubility and phase

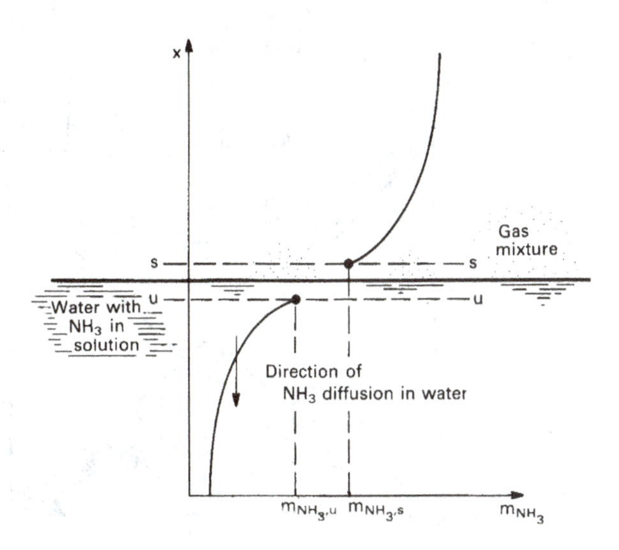

Figure 11.9 Absorption of ammonia into water.

equilibria is far beyond our scope (see, for example, [11.5, 11.24]), we illustrate these concepts with the following simple solubility relations.

Gas-liquid interfaces. For a gas mixture in contact with a liquid mixture, two simplified rules dictate the vapor composition. When the liquid is *rich* in species i, the partial pressure of species i in the gas phase, p_i, can be characterized approximately with *Raoult's law*, which says that

$$p_i = p_{\mathrm{sat},i}\, x_i \qquad \text{for } x_i \approx 1 \tag{11.64}$$

where $p_{\mathrm{sat},i}$ is the saturation pressure of pure i at the interface temperature and x_i is the mole fraction of i in the liquid. When the species i is *dilute* in the liquid, *Henry's law* applies. It says that

$$p_i = H\, x_i \qquad \text{for } x_i \ll 1 \tag{11.65}$$

where H is a temperature-dependent empirical constant which may be found in tables online or in handbooks. Figure 11.10 shows how the vapor pressure varies over a liquid mixture of species i and another species, and it indicates the regions of validity of Raoult's and Henry's laws. For

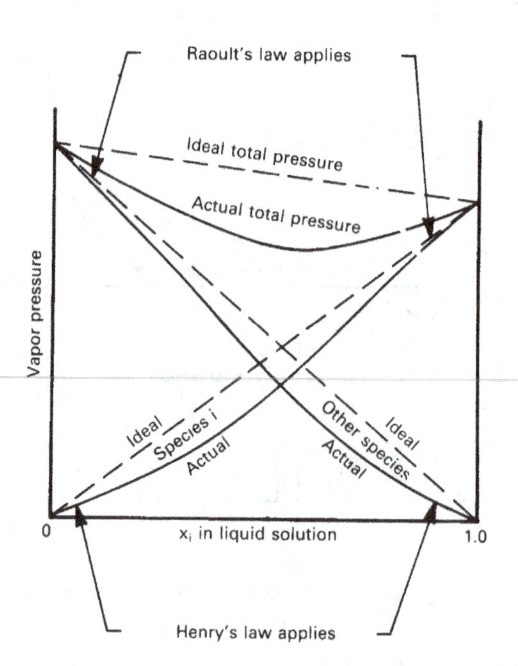

Figure 11.10 Typical partial and total vapor-pressure plot for the vapor in contact with a liquid solution, illustrating the regions of validity of Raoult's and Henry's laws.

example, when x_i is near one, Raoult's law applies to species i; when x_i is near zero, Raoult's law applies to the other species.

If the vapor pressure were to obey Raoult's law over the entire range of liquid composition, we would have what is called an *ideal solution* (see Fig. 11.10). When x_i is much less than one, the ideal solution approximation is usually very poor.

Example 11.5

A cup of tea sits in air at 1 atm total pressure. It starts at 100°C and cools toward room temperature. What is the mass fraction of water vapor above the surface of the tea as a function of the surface temperature?

Solution. We'll approximate tea as having the properties of pure water. Raoult's law applies almost exactly in this situation, since it happens that the concentration of air in water is very small. Thus, by eqn. (11.64), $p_{H_2O,s} = p_{sat,H_2O}(T)$. We can read the saturation pressure of water for several temperatures from a steam table or from

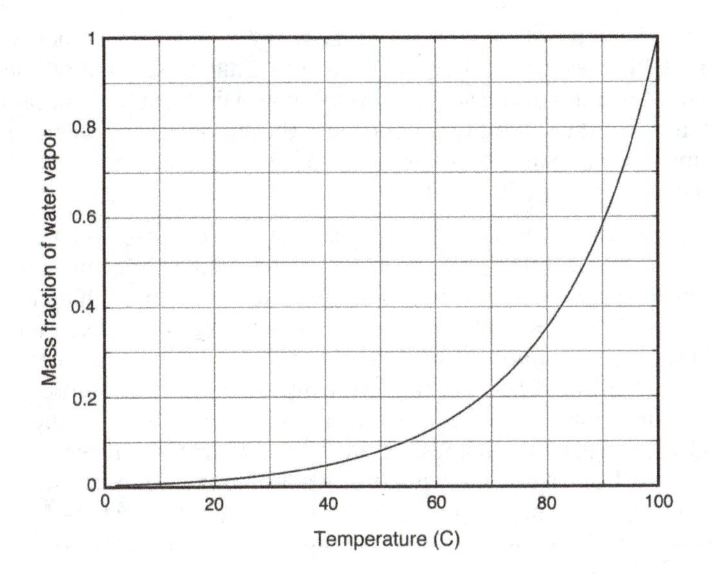

Figure 11.11 Mass fraction of water vapor in air adjacent to a liquid water (or tea) surface as a function of surface temperature (1 atm total pressure).

Table A.5 on pg. 717. From the vapor pressure, $p_{H_2O,s}$, we can compute the mole fraction with eqn. (11.16),

$$x_{H_2O,s} = p_{H_2O,s}/p_{atm} = p_{sat,H_2O}(T)/(101,325 \text{ Pa}) \tag{11.66}$$

The mass fraction can be calculated from eqn. (11.9), noting that $x_{air} = 1 - x_{H_2O}$ and substituting $M_{H_2O} = 18.02$ kg/kmol and $M_{air} =$ 28.96 kg/kmol

$$m_{H_2O,s} = \frac{(x_{H_2O,s})(18.02)}{[(x_{H_2O,s})(18.02) + (1 - x_{H_2O,s})(28.96)]} \tag{11.67}$$

The result is plotted in Fig. 11.11. Note that the mass fraction is less than 10% if the surface temperature is below about 54°C. ■

Gas-solid interfaces. When a solid is exposed to a gas, some of it will vaporize. This process is quite apparent, for example, when dry ice (solid CO_2) is placed in air. For other materials and temperatures, the vaporization rate may be indetectably tiny. We call a direct solid-to-vapor phase transition *sublimation*.

The solubility of most gases in most solids is so small that we can ignore their presence in the solid. The only issue is then that of knowing the concentration of the sublimed solid material within the adjacent gas. Most data for the solubility of solids into the gas phase are given in terms of vapor pressure as a function of surface temperature (see, e.g., [11.25] for a compilation of such data).

Despite the small amounts of gas that can be absorbed into most inorganic solids, their consequences can be quite significant. Material properties may be altered by absorbed gases, and, through absorption and diffusion, gases might leak through metal pressure-vessel walls. The process of absorption may include dissociation of the gas on the solid surface prior to its absorption into the bulk material. For example, when molecular hydrogen gas, H_2, is absorbed into iron, it first dissociates into two hydrogen atoms, 2H. At low temperatures, the dissociation reaction may be so slow that equilibrium conditions cannot be established between the bulk metal and the gas. Solubility relationships for gases entering solids are thus somewhat complex, and they will not be covered here (see [11.26]).

One important application of gas absorption into solids is the case-hardening of low-carbon steel by a process called *carburization*. The steel is exposed to a hot carbon-rich gas, such as CO or CO_2. That causes carbon to be absorbed on the surface of the metal. The elevated concentration of carbon within the surface causes carbon to diffuse inward. A typical goal is to raise the carbon mass fraction to 0.8% over a depth of about 2 mm (see Problem 11.27).

Example 11.6

Ice at $-10°$C is exposed to 1 atm air. What is the mass fraction of water vapor above the surface of the ice?

Solution. To begin, we need the vapor pressure, p_v, of water above ice. A typical local curve-fit is

$$\ln p_v(\text{kPa}) = 21.99 - 6141/(T\ \text{K}) \qquad \text{for } 243\ \text{K} \le T \le 273\ \text{K}$$

At $T = -10°\text{C} = 263.15$ K this yields $p_v = 0.260$ kPa. The remainder of the calculation follows exactly the approach of Example 11.5.

$$x_{\mathrm{H_2O},s} = 0.260/101.325 = 0.00257$$

$$m_{\mathrm{H_2O},s} = \frac{(0.00257)(18.02)}{[(0.00257)(18.02) + (1 - 0.00257)(28.96)]}$$

$$= 0.00160$$

■

11.6 Mass transfer at low rates

We have seen how mass flows within a mixture as the result of concentration gradients. When those flow rates are sufficiently low, the velocity of mass transfer is negligible. Thus, a stationary medium will remain at rest and a flowing fluid will have the same velocity field as if there were no mass transfer. More generally, when the diffusing species is dilute, its total mass flux is principally carried by diffusion.

Such diffusion has a direct correspondence to the heat transfer problems we considered Chapters 1 through 8. We refer to this correspondence as the *analogy between heat and mass transfer.* We will focus our attention on nonreacting systems, for which $\dot{r}_i = 0$ in the species conservation equation.

Steady mass diffusion in stationary media

Equations (11.58) and (11.21) show that steady mass transfer without reactions is described by the equation

$$\boxed{\nabla \cdot (\rho_i \vec{v}) + \nabla \cdot \vec{j}_i = \nabla \cdot \vec{n}_i = 0} \tag{11.68}$$

or, in one dimension,

$$\frac{dn_i}{dx} = \frac{d}{dx}(\rho_i v + j_i) = \frac{d}{dx}(m_i n + j_i) = 0 \tag{11.69}$$

that is, the mass flux of species i, n_i, is independent of x.

When the convective mass flux of i, $\rho_i v = m_i n$, is small, the transport of i is mainly by the diffusional flux, j_i. The following pair of examples show how this situation might arise.

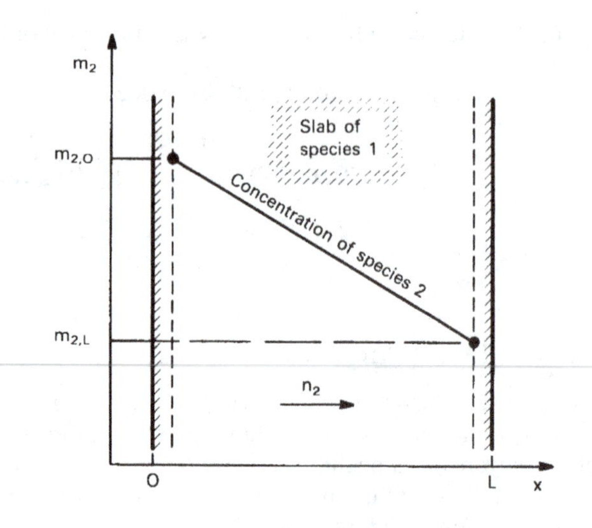

Figure 11.12 One-dimensional, steady diffusion in a slab.

Example 11.7

A thin slab, made of species 1, separates two volumes of gas. On one side, the pressure of species 2 is high, and on the other it is low. Species 2 two is soluble in the slab material and thus has different concentrations at each inside face of the slab, as shown in Fig. 11.12. What is the mass transfer rate of species 2 through the slab if the concentration of species 2 is low?

Solution. The mass transfer rate through the slab satisfies eqn. (11.69)

$$\frac{dn_2}{dx} = 0$$

If species 2 is dilute, with $m_2 \ll 1$, the convective transport will be small

$$n_2 = m_2 n + j_2 \cong j_2$$

With Fick's law, we have

$$\frac{dn_2}{dx} \cong \frac{dj_2}{dx} = \frac{d}{dx}\left(-\rho \mathcal{D}_{21}\frac{dm_2}{dx}\right) = 0$$

If $\rho\mathcal{D}_{21} \cong$ constant, the mass fraction satisfies

$$\frac{d^2 m_2}{dx^2} = 0$$

Integrating and applying the boundary conditions, $m_2(x = 0) = m_{2,0}$ and $m_2(x = L) = m_{2,L}$, we obtain the concentration distribution:

$$m_2(x) = m_{2,0} + (m_{2,L} - m_{2,0})\left(\frac{x}{L}\right)$$

The mass flux is then

$$n_2 \cong j_2 = -\rho\mathcal{D}_{21}\frac{dm_2}{dx} = -\frac{\rho\mathcal{D}_{21}}{L}(m_{2,L} - m_{2,0}) \tag{11.70}$$

This, in essence, is the same calculation we made in Example 2.2 in Chapter 2. ■

Example 11.8

Suppose that the concentration of species 2 in the slab were not small in the preceding example. How would the total mass flux of species 1 differ from the diffusional flux?

Solution. As before, the total mass flux each species would be constant in the steady state, and if the slab material is not transferred into the gas its mass flux is zero

$$n_1 = 0 = \rho_1 v + j_1$$

Therefore, the mass-average velocity in the slab is

$$v = -\frac{j_1}{\rho_1} = \frac{j_2}{\rho_1}$$

since $j_1 + j_2 = 0$. The mass flux for species 2 is

$$\begin{aligned} n_2 &= \rho_2 v + j_2 \\ &= j_2\left(\frac{\rho_2}{\rho_1} + 1\right) \\ &= j_2\left(\frac{m_2}{m_1} + 1\right) = j_2\left(\frac{1}{1 - m_2}\right) \end{aligned}$$

since $m_1 + m_2 = 1$.

When $m_2 \ll 1$, the diffusional flux will approximate n_2. On the other hand, if, say, $m_2 = 0.5$, then $n_2 = 2j_2$! In that case, the convective transport $\rho_2 v$ is *equal* to the diffusive transport j_2. ■

In the second example, we see that the stationary material of the slab had a diffusion velocity, j_1. For the slab to remain at rest, the opposing velocity v must be present. For this reason, an induced velocity of this sort is sometimes called a *counterdiffusion velocity*.

From these two examples, we see that steady mass diffusion is directly analogous to heat conduction only if the convective transport is negligible. That can generally be ensured if the transferred species is dilute. When the transferred species has a high concentration, nonnegligible convective transport can occur, even in a solid medium.

Unsteady mass diffusion in stationary media

Similar conclusions apply to unsteady mass diffusion. Consider a medium at rest through which a dilute species i diffuses. From eqn. (11.58) with $r_i = 0$,

$$\begin{aligned} \frac{\partial \rho_i}{\partial t} &= -\nabla \cdot \left(\rho_i \vec{v} + \vec{j}_i\right) \\ &= -\nabla \cdot \left(m_i \vec{n} + \vec{j}_i\right) \end{aligned} \tag{11.71}$$

If $m_i \ll 1$, only diffusion contributes significantly to the mass flux of i, and we may neglect $m_i n$

$$\frac{\partial \rho_i}{\partial t} \approx -\nabla \cdot \vec{j}_i = \nabla \cdot (\rho \mathcal{D}_{im} \nabla m_i)$$

With small m_i, the density ρ and the diffusion coefficient $\mathcal{D}_{im}$ will not vary much, and we can factor ρ through the equation

$$\frac{\partial m_i}{\partial t} = \mathcal{D}_{im} \nabla^2 m_i \tag{11.72}$$

This is called the *mass diffusion equation.* It has the same form as the equation of heat conduction. Solutions for the unsteady diffusion of a dilute species in a stationary medium are thus entirely analogous to those for heat conduction when the boundary conditions are the same.

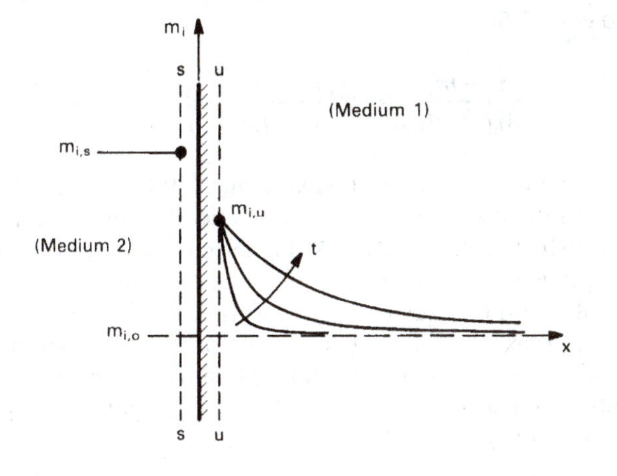

Figure 11.13 Mass diffusion into a semi-infinite stationary medium.

Example 11.9

A semi-infinite stationary medium (medium 1) has an initially uniform concentration, $m_{i,0}$ of species i. From time $t = 0$ onward, we place the end plane at $x = 0$ in contact with a second medium (medium 2) with a concentration $m_{i,s}$. What is the resulting distribution of species in medium 1 if species 1 remains dilute?

Solution. Once $m_{i,s}$ and the solubility data are known, the mass fraction just inside the solid surface, $m_{i,u}$, can be determined (see Fig. 11.13). This concentration provides the boundary condition at $x = 0$ for $t > 0$. Our mathematical problem then becomes

$$\frac{\partial m_i}{\partial t} = \mathcal{D}_{im_1} \frac{\partial^2 m_i}{\partial x^2} \tag{11.73}$$

with

$$\begin{aligned} m_i &= m_{i,0} && \text{for} \quad t = 0 \quad (\text{all } x) \\ m_i &= m_{i,u} && \text{for} \quad x = 0 \quad (t > 0) \\ m_i &\to m_{i,0} && \text{for} \quad x \to \infty \quad (t > 0) \end{aligned}$$

This math problem is identical to that for transient heat conduction into a semi-infinite region (Section 5.6), and its solution is completely analogous to eqn. (5.50):

$$\frac{m_i - m_{i,u}}{m_{i,0} - m_{i,u}} = \text{erf}\left(\frac{x}{2\sqrt{\mathcal{D}_{im_1} t}}\right) \qquad \blacksquare$$

The reader can solve all sorts of unsteady mass diffusion problems by direct analogy to the methods of Chapters 4 and 5 when the concentration of the diffusing species is low. At higher concentrations of the diffusing species, however, counterdiffusion velocities can be induced, as in Example 11.8. Counterdiffusion may be significant in concentrated metallic alloys, as, for example, during annealing of a butt-welded junction between two dissimilar metals. In those situations, eqn. (11.72) is sometimes modified to use a concentration-dependent, spatially varying *interdiffusion coefficient* (see [11.6]).

Convective mass transfer at low rates

Convective mass transfer is analogous to convective heat transfer when two conditions apply:

1. The mass flux normal to the surface, $n_{i,s}$, must be essentially equal to the diffusional mass flux, $j_{i,s}$ from the surface. In general, this requires that the concentration of the diffusing species, m_i, be low.[11]

2. The diffusional mass flux must be low enough that it does not affect the imposed velocity field.

The first condition ensures that mass flow from the wall is diffusional, as is the heat flow in a convective heat transfer problem. The second condition ensures that the flow field will be the same as for the heat transfer problem.

As a concrete example, consider a laminar flat-plate boundary layer in which species i is transferred from the wall to the free stream, as shown in Fig. 11.14. Free stream values, at the edge of the b.l., are labeled with the subscript e, and values at the wall (the s-surface) are labeled with the subscript s. The mass fraction of species i varies from $m_{i,s}$ to $m_{i,e}$ across a *concentration boundary layer* on the wall. If the mass fraction of species i at the wall, $m_{i,s}$, is small, then $n_{i,s} \approx j_{i,s}$, as we saw earlier in this section. Mass transfer from the wall will be essentially diffusional. (This is the first condition.)

[11]In a few situations, such as catalysis, no mass flows through the wall, and no convective transport occurs irrespective of the concentration (see Problems 11.9 and 11.44).

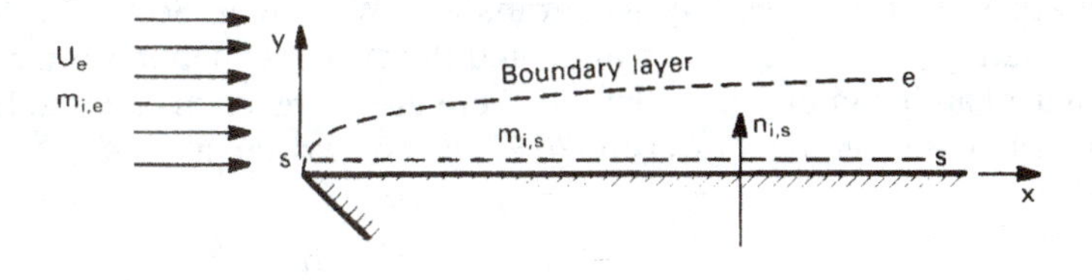

Figure 11.14 Concentration boundary layer on a flat plate.

In regard to the second condition, when the concentration difference, $m_{i,s} - m_{i,e}$, is small, then the diffusional mass flux of species i through the wall, $j_{i,s}$, will be small compared to the bulk mass flow in the streamwise direction, and it will have little influence on the velocity field. Hence, we would expect that $\vec{v}$ is essentially that for the Blasius boundary layer.

These two conditions can be combined into a single requirement for low-rate mass transfer, as will be described in Section 11.8. Specifically, low-rate mass transfer can be assumed if

$$B_{m,i} \equiv \left(\frac{m_{i,s} - m_{i,e}}{1 - m_{i,s}} \right) \lesssim 0.2 \quad \text{condition for low-rate mass transfer} \tag{11.74}$$

The quantity $B_{m,i}$ is called the *mass transfer driving force.* It is written here in the form that applies when only one species is transferred through the s-surface. The evaporation of water into air is typical example of single-species transfer: only water vapor crosses the s-surface.

The mass transfer coefficient. In convective *heat* transfer problems, we have found it useful to express the heat flux from a surface, q, as the product of a heat transfer coefficient, h, and a driving force for heat transfer, ΔT. Thus, in the notation of Fig. 11.14,

$$q_s = h\,(T_s - T_e) \tag{1.17}$$

In convective mass transfer problems, we would therefore like to express the diffusional mass flux from a surface, $j_{i,s}$, as the product of a mass transfer coefficient and the concentration difference between the s-surface and the free stream. Hence, we define the *mass transfer coefficient* for species i, $g_{m,i}$ (kg/m^2·s), as follows:

$$j_{i,s} \equiv g_{m,i}\,(m_{i,s} - m_{i,e}) \tag{11.75}$$

We expect $g_{m,i}$, like h, to be determined mainly by the flow field, fluid, and geometry of the problem.

The analogy to convective heat transfer. We saw in Sect. 11.5 that the equation of species conservation and the energy equation were quite similar in an incompressible flow. If there are no reactions and no heat generation, then eqns. (11.61) and (6.37) can be written as

$$\frac{\partial \rho_i}{\partial t} + \vec{v} \cdot \nabla \rho_i = -\nabla \cdot \vec{j}_i \tag{11.61}$$

$$\rho c_p \left(\frac{\partial T}{\partial t} + \vec{v} \cdot \nabla T \right) = -\nabla \cdot \vec{q} \tag{6.37}$$

These conservation equations describe changes in, respectively, the amount of mass or energy per unit volume that results from convection by a given velocity field and from diffusion under either Fick's or Fourier's law.

We may identify the analogous quantities in these equations. For the capacity of mass or energy per unit volume, we see that

$$d\rho_i \quad \text{is analogous to} \quad \rho c_p dT \tag{11.76a}$$

or, in terms of the mass fraction,

$$\rho\, dm_i \quad \text{is analogous to} \quad \rho c_p dT \tag{11.76b}$$

The flux laws may be rewritten to show the capacities explicitly

$$\vec{j}_i = -\rho \mathcal{D}_{im} \nabla m_i = -\mathcal{D}_{im} (\rho \nabla m_i)$$

$$\vec{q} = -k \nabla T \qquad = -\frac{k}{\rho c_p} \left(\rho c_p \nabla T \right)$$

Hence, we find the analogy of the diffusivities:

$$\mathcal{D}_{im} \quad \text{is analogous to} \quad \frac{k}{\rho c_p} = \alpha \tag{11.76c}$$

It follows that the Schmidt number and the Prandtl number are directly analogous:

$$\text{Sc} = \frac{\nu}{\mathcal{D}_{im}} \quad \text{is analogous to} \quad \text{Pr} = \frac{\nu}{\alpha} = \frac{\mu c_p}{k} \tag{11.76d}$$

Thus, a high Schmidt number signals a thin concentration boundary layer, just as a high Prandtl number signals a thin thermal boundary layer. Finally, we may write the transfer coefficients in terms of the capacities

$$j_{i,s} = g_{m,i}\,(m_{i,s} - m_{i,e}) = \left(\frac{g_{m,i}}{\rho}\right)\rho\,(m_{i,s} - m_{i,e})$$

$$q_s = h\,(T_s - T_e) = \left(\frac{h}{\rho c_p}\right)\rho c_p\,(T_s - T_e)$$

from which we see that

$$g_{m,i} \text{ is analogous to } \frac{h}{c_p} \qquad (11.76e)$$

From these comparisons, we conclude that the solution of a heat convection problem becomes the solution of a low-rate mass convection problem when we replace the variables in the heat transfer problem with the analogous mass transfer variables given by eqns. (11.76).

Convective heat transfer coefficients are usually expressed in terms of the Nusselt number as a function of Reynolds and Prandtl number

$$\mathrm{Nu}_x = \frac{hx}{k} = \frac{(h/c_p)x}{\rho(k/\rho c_p)} = \mathrm{fn}\,(\mathrm{Re}_x, \mathrm{Pr}) \qquad (11.77)$$

For convective mass transfer problems, we expect the same functional dependence after we make the substitutions indicated above. Specifically, if we replace h/c_p by $g_{m,i}$, $k/\rho c_p$ by $\mathcal{D}_{i,m}$, and Pr by Sc, we obtain

$$\boxed{\mathrm{Nu}_{m,x} \equiv \frac{g_{m,i}x}{\rho \mathcal{D}_{im}} = \mathrm{fn}\,(\mathrm{Re}_x, \mathrm{Sc})} \qquad (11.78)$$

where $\mathrm{Nu}_{m,x}$, the Nusselt number for mass transfer, is defined as indicated. Nu_m is sometimes called the *Sherwood number*[12], Sh.

Example 11.10

A napthalene model of a printed circuit board (PCB) is placed in a wind tunnel. The napthalene sublimes slowly as a result of forced

[12]Thomas K. Sherwood (1903–1976) obtained his doctoral degree at M.I.T. under Warren K. Lewis in 1929 and was a professor of Chemical Engineering there from 1930 to 1969. He served as Dean of Engineering from 1946 to 1952. His research dealt with mass transfer and related industrial processes. Sherwood was also the author of very influential textbooks on mass transfer.

convective mass transfer. If the first 5 cm of the napthalene model is a flat plate, calculate the average rate of loss of napthalene from that part of the model. Assume that conditions are isothermal at 303 K and that the air speed is 5 m/s. Also, explain how napthalene sublimation might be used to determine heat transfer coefficients .

Solution. Let us first find the mass fraction of napthalene just above the model surface. A relationship for the vapor pressure of napthalene (in mmHg) is $\log_{10} p_v = 11.450 - 3729.3/(T\,\text{K})$. At 303 K, this gives $p_v = 0.1387$ mmHg $= 18.49$ Pa. The mole fraction of napthalene is thus $x_{\text{nap},s} = 18.49/101325 = 1.825 \times 10^{-4}$, and with eqn. (11.9), the mass fraction is, with $M_{\text{nap}} = 128.2$ kg/kmol,

$$m_{\text{nap},s} = \frac{(1.825 \times 10^{-4})(128.2)}{(1.825 \times 10^{-4})(128.2) + (1 - 1.825 \times 10^{-4})(28.96)}$$
$$= 8.074 \times 10^{-4}$$

The mass fraction of napthalene in the free stream, $m_{\text{nap},s}$, is zero. With these numbers, we can check to see if the mass transfer rate is low enough to use the analogy of heat and mass transfer, with eqn. (11.74):

$$B_{m,\text{nap}} = \left(\frac{8.074 \times 10^{-4} - 0}{1 - 8.074 \times 10^{-4}}\right) = 8.081 \times 10^{-4} \ll 0.2$$

The analogy therefore applies.

The convective heat transfer coefficient for this situation is that for a flat plate boundary layer. The Reynolds number is

$$\text{Re}_L = \frac{u_\infty L}{\nu} = \frac{(5)(0.05)}{1.867 \times 10^{-5}} = 1.339 \times 10^4$$

where we have used the viscosity of pure air, since the concentration of napthalene is very low. The flow is laminar, so the applicable heat transfer relationship is eqn. (6.68)

$$\overline{\text{Nu}}_L = \frac{\overline{h}L}{k} = 0.664\,\text{Re}_L^{1/2}\,\text{Pr}^{1/3} \tag{6.68}$$

Under the analogy, the Nusselt number for mass transfer is

$$\overline{\text{Nu}}_{m,L} = \frac{\overline{g_{m,i}}\,L}{\rho \mathcal{D}_{im}} = 0.664\,\text{Re}_L^{1/2}\,\text{Sc}^{1/3}$$

The diffusion coefficient for napthalene in air, from Table 11.1, is $\mathcal{D}_{\text{nap,air}} = 0.86 \times 10^{-5}$ m/s, and thus $\text{Sc} = 1.867 \times 10^{-5}/0.86 \times 10^{-5} = 2.17$. Hence,

$$\overline{\text{Nu}}_{m,L} = 0.664\ (1.339 \times 10^4)^{1/2}\ (2.17)^{1/3} = 99.5$$

and, using the density of pure air,

$$\begin{aligned}\overline{g_{m,\text{nap}}} &= \frac{\rho \mathcal{D}_{\text{nap,air}}}{L}\, \overline{\text{Nu}}_{m,L} \\ &= \frac{(1.166)(0.86 \times 10^{-5})}{0.05}(99.5) = 0.0200\ \text{kg/m}^2\text{s}\end{aligned}$$

The average mass flux from this part of the model is then

$$\begin{aligned}\overline{n_{\text{nap},s}} &= \overline{g_{m,\text{nap}}}\left(m_{\text{nap},s} - m_{\text{nap},e}\right) \\ &= (0.0200)(8.074 \times 10^{-4} - 0) \\ &= 1.61 \times 10^{-5}\ \text{kg/m}^2\text{s} = 58.0\ \text{g/m}^2\text{h}\end{aligned}$$

Napthalene sublimation can be used to infer heat transfer coefficients by measuring the loss of napthalene from a model over some length of time. Experiments are run at several Reynolds numbers. The lost mass fixes the subliming rate and the mass transfer coefficient. The mass transfer coefficient is then substituted in the analogy to heat transfer to determine a heat transfer Nusselt number at each Reynolds number. Since the Schmidt number of napthalene is not generally equal to the Prandtl number under the conditions of interest, some assumption about the dependence of the Nusselt number on the Prandtl number must usually be introduced. ■

Boundary conditions. When we apply the analogy between heat transfer and mass transfer to calculate $g_{m,i}$, we must consider the boundary condition at the wall. We have dealt with two common types of wall condition in the study of heat transfer: uniform temperature and uniform heat flux. The analogous mass transfer wall conditions are uniform concentration and uniform mass flux. We used the mass transfer analog of the uniform wall temperature solution in the preceding example, since the *mass fraction* of napthalene was uniform over the entire model. Had the *mass flux* been uniform at the wall, we would have used the analog of a uniform heat flux solution.

Natural convection in mass transfer. In Chapter 8, we saw that the density differences produced by temperature variations can lead to flow and convection in a fluid. Variations in fluid composition can also produce density variations that result in natural convection mass transfer. This type of natural convection flow is still governed by eqn. (8.3),

$$u\frac{\partial u}{\partial x} + v\frac{\partial u}{\partial y} = (1 - \rho_\infty/\rho)g + \nu\frac{\partial^2 u}{\partial y^2} \tag{8.3}$$

but the species equation is now used in place of the energy equation in determining the variation of density. Rather than solving eqn. (8.3) and the species equation for specific mass transfer problems, we again turn to the analogy between heat and mass transfer.

In analyzing natural convection heat transfer, we eliminated ρ from eqn. (8.3) using $(1 - \rho_\infty/\rho) = \beta(T - T_\infty)$, and the resulting Grashof and Rayleigh numbers came out in terms of an appropriate $\beta\Delta T$ instead of $\Delta\rho/\rho$. These groups could just as well have been written for the heat transfer problem as

$$\text{Gr}_L = \frac{g\Delta\rho L^3}{\rho\nu^2} \quad \text{and} \quad \text{Ra}_L = \frac{g\Delta\rho L^3}{\rho\alpha\nu} = \frac{g\Delta\rho L^3}{\mu\alpha} \tag{11.79}$$

although $\Delta\rho$ would still have had to have been evaluated from ΔT.

With Gr and Pr expressed in terms of density differences instead of temperature differences, the analogy between heat transfer and low-rate mass transfer may be used directly to adapt natural convection heat transfer predictions to natural convection mass transfer. As before, we replace Nu by Nu_m and Pr by Sc. But this time we also write

$$\text{Ra}_L = \text{Gr}_L\text{Sc} = \frac{g\Delta\rho L^3}{\mu\mathcal{D}_{12}} \tag{11.80}$$

or calculate Gr_L as in eqn. (11.79). The densities must now be calculated from the concentrations.

Example 11.11

Helium is bled through a porous vertical wall, 40 cm high, into surrounding air at a rate of 87.0 mg/m^2·s. Both the helium and the air are at 300 K, and the environment is at 1 atm. What is the average concentration of helium at the wall, $\overline{m_{\text{He},s}}$?

Solution. This is a uniform flux natural convection problem. Here $\overline{g_{m,\text{He}}}$ and $\overline{\Delta\rho}$ depend on $\overline{m_{\text{He},s}}$, so the calculation is not as straightforward as it was for thermally driven natural convection.

To begin, let us assume that the concentration of helium at the wall will be small enough that the mass transfer rate is low. Since $m_{\text{He},e} = 0$, if $m_{\text{He},s} \ll 1$, then $\overline{m_{\text{He},s} - m_{\text{He},e}} \ll 1$ as well. Both conditions for the analogy to heat transfer will be met.

The mass flux of helium at the wall, $n_{\text{He},s}$, is known, and because low rates prevail,

$$n_{\text{He},s} \approx j_{\text{He},s} = \overline{g_{m,\text{He}}}\,(\overline{m_{\text{He},s} - m_{\text{He},e}})$$

Hence,

$$\overline{\text{Nu}}_{m,L} = \frac{\overline{g_{m,\text{He}}}\,L}{\rho \mathcal{D}_{\text{He,air}}} = \frac{n_{\text{He},s}\,L}{\rho \mathcal{D}_{\text{He,air}}\,(\overline{m_{\text{He},s} - m_{\text{He},e}})}$$

The appropriate Nusselt number is obtained from the mass transfer analog of eqn. (8.44b):

$$\overline{\text{Nu}}_{m,L} = \frac{6}{5}\left(\frac{\text{Ra}_L^*\,\text{Sc}}{4 + 9\sqrt{\text{Sc}} + 10\,\text{Sc}}\right)^{1/5}$$

with

$$\text{Ra}_L^* = \text{Ra}_L \overline{\text{Nu}}_{m,L} = \frac{g\overline{\Delta\rho}\,n_{\text{He},s}\,L^4}{\mu\rho\mathcal{D}_{\text{He,air}}^2\,(\overline{m_{\text{He},s} - m_{\text{He},e}})}$$

The Rayleigh number cannot easily be evaluated without assuming a value of the mass fraction of helium at the wall. As a first guess, we pick $\overline{m_{\text{He},s}} = 0.010$. Then the *film composition* is

$$m_{\text{He},f} = (0.010 + 0)/2 = 0.005$$

From eqn. (11.8) and the ideal gas law, we obtain estimates for the film density (at the film composition) and the wall density

$$\rho_f = 1.141 \text{ kg/m}^3 \quad \text{and} \quad \overline{\rho_s} = 1.107 \text{ kg/m}^3$$

From eqn. (11.42) the diffusion coefficient is

$$\mathcal{D}_{\text{He,air}} = 7.119 \times 10^{-5} \text{ m}^2/\text{s}.$$

At this low concentration of helium, we expect the film viscosity to be close to that of pure air. From Appendix A, for air at 300 K

$$\mu_f \cong \mu_{\text{air}} = 1.857 \times 10^{-5} \text{ kg/m·s}.$$

The corresponding Schmidt number is $\mathrm{Sc} = (\mu_f/\rho_f)/\mathcal{D}_{\mathrm{He,air}} = 0.2286$. Furthermore,

$$\rho_e = \rho_{\mathrm{air}} = 1.177\ \mathrm{kg/m^3}$$

From these values,

$$\begin{aligned}\mathrm{Ra}_L^* &= \frac{9.806(1.177-1.107)(87.0\times 10^{-6})(0.40)^4}{(1.857\times 10^{-5})(1.141)(7.119\times 10^{-5})^2(0.010)}\\ &= 1.424\times 10^9\end{aligned}$$

We may now evaluate the mass transfer Nusselt number

$$\overline{\mathrm{Nu}}_{m,L} = \frac{6\left[(1.424\times 10^9)(0.2286)\right]^{1/5}}{5\left[4+9\sqrt{0.2286}+10(0.2286)\right]^{1/5}} = 37.73$$

From this we calculate

$$\begin{aligned}(\overline{m_{\mathrm{He},s}} - m_{\mathrm{He},e}) &= \frac{n_{\mathrm{He},s}\,L}{\rho\mathcal{D}_{\mathrm{He,air}}\overline{\mathrm{Nu}}_{m,L}}\\ &= \frac{(87.0\times 10^{-6})(0.40)}{(1.141)(7.119\times 10^{-5})(37.73)}\\ &= 0.01136\end{aligned}$$

We have already noted that $(\overline{m_{\mathrm{He},s}} - m_{\mathrm{He},e}) = \overline{m_{\mathrm{He},s}}$, so we have obtained an average wall concentration 14% higher than our initial guess of 0.010.

Using $\overline{m_{\mathrm{He},s}} = 0.01136$ as our second guess, we repeat the preceding calculations with revised values of the densities to obtain

$$\overline{m_{\mathrm{He},s}} = 0.01142$$

Since this result is within 0.5% of our second guess, a third iteration is unnecesary. ■

Concentration variations alone gave rise to buoyancy in the preceding example. If *both* temperature and density vary in a natural convection problem, the appropriate Gr or Ra may be calculated using density differences based on both the local m_i and the local T, *provided* that the Prandtl and Schmidt numbers are approximately equal (that is, if the Lewis number $\cong 1$). This is usually true in gases.

If the Lewis number is far from one, the analogy between heat and mass transfer breaks down in natural convection problems that involve both heat and mass transfer. That's the case because the concentration and thermal boundary layers may take on very different thicknesses, complicating the density distributions that drive the velocity field.

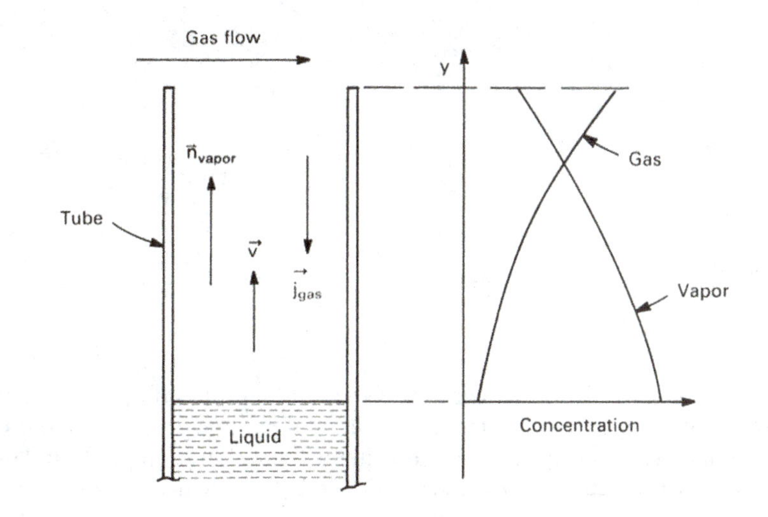

Figure 11.15 The Stefan tube.

11.7 Steady mass transfer with counterdiffusion

So far, we have studied mass transfer for situations in which mass diffusion does not affect the velocity of the medium through which mass diffuses. In this section and the one that follows, we develop models for cases in which mass transfer does affect the velocity. We begin with a simple configuration, the Stefan tube.

In 1874, Josef Stefan presented his solution for evaporation from a liquid pool at the bottom of a vertical tube over which a gas flows (Fig. 11.15). This configuration, often called a Stefan tube, has often been used to measure diffusion coefficients. Vapor leaving the liquid surface diffuses through the gas in the tube and is carried away by the gas flow across top of the tube. If the gas stream itself has a low vapor concentration, then diffusion is driven by the higher concentration of vapor over the liquid pool.

A typical Stefan tube is 5 to 10 mm in diameter and 10 to 20 cm long. If the air flow at the top is not too vigorous, and if density variations in the tube do not give rise to natural convection, then the transport of vapor from the liquid pool to the top of the tube will be a one-dimensional upflow.

The other gas in the tube is stationary if it is not being absorbed by the liquid (e.g., if it is insoluble in the liquid or if the liquid is saturated with

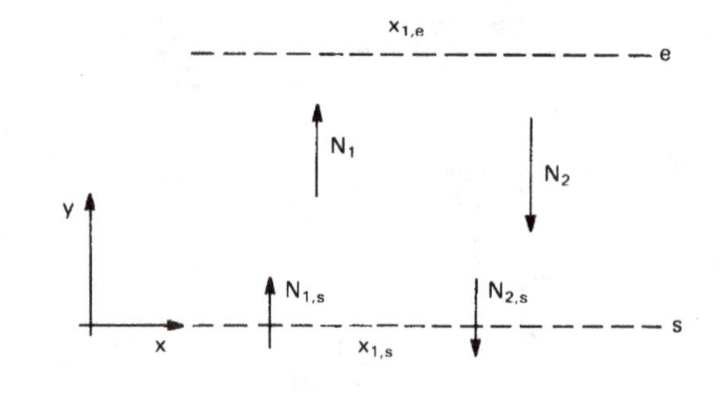

Figure 11.16 Mass flow across a one-dimensional layer.

it). Yet, because there is a concentration gradient of vapor, there must also be an opposing concentration gradient of gas and an associated diffusional mass flux of gas, similar to what we found in Example 11.8. For the gas in the tube to have a net diffusion flux when it is stationary, there must be an induced upward convective velocity — a counterdiffusion velocity — against which the gas diffuses. As in Example 11.8, the counterdiffusion velocity can be found in terms of the diffusional mass fluxes:

$$v = -j_{\text{gas}}/\rho_{\text{gas}} = j_{\text{vapor}}/\rho_{\text{gas}}$$

In this section, we determine the mass transfer rate and concentration profiles in the tube, treating it as the one-dimensional layer shown in Fig. 11.16. The s-surface lies above the liquid and the e-surface lies at the top end of the tube. We allow for the possibility that the counterdiffusion velocity may not be negligible, so that both diffusion and vertical convection may occur. We also allow for the possibility that the gas passes through the liquid surface ($N_{2,s} \neq 0$). The results obtained here form an important prototype for our subsequent analyses of convective mass transfer at high rates.

The solution of the mass transfer problem begins with an appropriate form of the equation of species conservation. Since the mixture composition varies along the length of the tube, the density may vary as well. If the temperature and pressure are constant, however, the molar concentration of the mixture does not change through the tube [cf. (11.14)]. The system is therefore most accurately analyzed using the molar form of species conservation.

For one-dimensional steady mass transfer, the mole fluxes N_1 and N_2 have only vertical components and depend only on the vertical coordinate, y. Using eqn. (11.69), we get, with $n_i = M_i N_i$,

$$\frac{dN_1}{dy} = \frac{dN_2}{dy} = 0$$

so that N_1 and N_2 are constant throughout the layer. They have s-surface values, $N_{1,s}$ and $N_{2,s}$, everywhere. These constants will be positive for upward mass flow. (For the orientations in Fig. 11.16, $N_{1,s} > 0$ and $N_{2,s} < 0$.) These results are a straightforward consequence of steady-state species conservation.

Recalling the general expression for N_i, eqn. (11.25), and introducing Fick's law, eqn. (11.34), we write

$$N_1 = x_1 N - c\mathcal{D}_{12}\frac{dx_1}{dy} = N_{1,s} \tag{11.81}$$

The term xN_1 represents vertical convective transport induced by mass transfer. The total mole flux, N, must also be constant at its s-surface value; by eqn. (11.23), this is

$$N = N_{1,s} + N_{2,s} = N_s \tag{11.82}$$

Substituting this result into eqn. (11.81), we obtain a differential equation for x_1:

$$c\mathcal{D}_{12}\frac{dx_1}{dy} = N_s x_1 - N_{1,s} \tag{11.83}$$

In this equation, x_1 is a function of y, the N's are constants, and $c\mathcal{D}_{12}$ depends on temperature and pressure. If the temperature and pressure are constant, so too is $c\mathcal{D}_{12}$. Integration then yields

$$\frac{N_s y}{c\mathcal{D}_{12}} = \ln\left(N_s x_1 - N_{1,s}\right) + \text{ constant} \tag{11.84}$$

We need to fix the constant and the two mole fluxes, $N_{1,s}$ and N_s. To do this, we apply the boundary conditions at either end of the layer. The first boundary condition is the mole fraction of species 1 at the bottom of the layer

$$x_1 = x_{1,s} \quad \text{at} \quad y = 0$$

and it requires that

$$\text{constant} \;=\; -\ln(N_s x_{1,s} - N_{1,s}) \tag{11.85}$$

so

$$\frac{N_s y}{c\mathcal{D}_{12}} = \ln\left(\frac{N_s x_1 - N_{1,s}}{N_s x_{1,s} - N_{1,s}}\right) \tag{11.86}$$

The second boundary condition is the mole fraction at the top of the layer

$$x_1 = x_{1,e} \quad \text{at} \quad y = L$$

which yields

$$\frac{N_s L}{c\mathcal{D}_{12}} = \ln\left(\frac{x_{1,e} - N_{1,s}/N_s}{x_{1,s} - N_{1,s}/N_s}\right) \tag{11.87}$$

or

$$\boxed{N_s = \frac{c\mathcal{D}_{12}}{L} \ln\left(1 + \frac{x_{1,e} - x_{1,s}}{x_{1,s} - N_{1,s}/N_s}\right)} \tag{11.88}$$

The last boundary condition is the value of $N_{1,s}/N_s$. Since we have allowed for the possiblity that species 2 passes through the bottom of the layer, $N_{1,s}/N_s$ may not equal one. The ratio depends on the specific problem at hand, as shown in the two following examples.

Example 11.12

Find an equation for the evaporation rate of the liquid in the Stefan tube described at the beginning of this section.

Solution. Species 1 is the evaporating vapor, and species 2 be the stationary gas. Only vapor is transferred through the s-surface, since the gas is not significantly absorbed into the already gas-saturated liquid. Thus, $N_{2,s} = 0$, and $N_s = N_{1,s} = N_{\text{vapor},s}$ is simply the evaporation rate of the liquid. The s-surface is just above the surface of the liquid. The mole fraction of the evaporating liquid can be determined from solubility data; for example, if the gas is more-or-less insoluble in the liquid, Raoult's law, eqn. (11.64), may be used. The e-surface is at the mouth of the tube. The gas flow over the top may contain some concentration of the vapor, although it should generally be near zero. The ratio $N_{1,s}/N_s$ is one, and the rate of evaporation is

$$N_s = N_{\text{vapor},s} = \frac{c\mathcal{D}_{12}}{L} \ln\left(1 + \frac{x_{1,e} - x_{1,s}}{x_{1,s} - 1}\right) \tag{11.89}$$

■

Example 11.13

What is the evaporation rate in the Stefan tube if the gas is bubbled up to the liquid surface at some fixed rate, N_{gas}?

Solution. Again, $N_{1,s} = N_{\text{vapor},s}$ is the evaporation rate. However, the total mole flux is

$$N_s = N_{\text{gas}} + N_{1,s}$$

Thus,

$$N_{\text{gas}} + N_{1,s} = \frac{c\mathcal{D}_{12}}{L} \ln\left[1 + \frac{x_{1,e} - x_{1,s}}{x_{1,s} - N_{1,s}/(N_{1,s} + N_{\text{gas}})}\right] \tag{11.90}$$

This equation fixes $N_{1,s}$, but it must be solved iteratively. ■

Once we have found the mole fluxes, we may compute the concentration distribution, $x_1(y)$, using eqn. (11.86):

$$x_1(y) = \frac{N_{1,s}}{N_s} + (x_{1,s} - N_{1,s}/N_s)\exp(N_s y/c\mathcal{D}_{12}) \tag{11.91}$$

Alternatively, we may eliminate N_s between eqns. (11.86) and (11.87) to obtain the concentration distribution in a form that depends only on the ratio $N_{1,s}/N_s$:

$$\frac{x_1 - N_{1,s}/N_s}{x_{1,s} - N_{1,s}/N_s} = \left(\frac{x_{1,e} - N_{1,s}/N_s}{x_{1,s} - N_{1,s}/N_s}\right)^{y/L} \tag{11.92}$$

Example 11.14

Find the concentration distribution of water vapor in a helium-water Stefan tube at 325 K and 1 atm. The tube is 20 cm in length. Assume the helium stream at the top of the tube to have a mole fraction of water of 0.01.

Solution. Let water be species 1 and helium be species 2. The vapor pressure of the liquid water is approximately the saturation pressure at the water temperature. Using the steam tables, we get $p_v = 1.341 \times 10^4$ Pa and, from eqn. (11.16),

$$x_{1,s} = \frac{1.341 \times 10^4 \text{ Pa}}{101{,}325 \text{ Pa}} = 0.1323$$

We use eqn. (11.14) to evaluate the mole concentration in the tube:

$$c = \frac{101{,}325}{8314.5(325)} = 0.03750 \text{ kmol/m}^3$$

From eqn. (11.42) we obtain $\mathcal{D}_{12}(325\text{ K}, 1\text{ atm}) = 1.067 \times 10^{-4}\text{ m}^2/\text{s}$. Then eqn. (11.89) gives the molar evaporation rate:

$$N_{1,s} = \frac{0.03750(1.067 \times 10^{-4})}{0.20} \ln\left(1 + \frac{0.01 - 0.1323}{0.1323 - 1}\right)$$
$$= 2.638 \times 10^{-6} \text{ kmol/m}^2{\cdot}\text{s}$$

This corresponds to a mass evaporation rate:

$$n_{1,s} = 4.754 \times 10^{-5} \text{ kg/m}^2{\cdot}\text{s}$$

The concentration distribution of water vapor [eqn. (11.91)] is

$$x_1(y) = 1 - 0.8677 \exp(0.6593y)$$

where y is expressed in meters. ■

Stefan tubes have been widely used to measure mass transfer coefficients, by observing the change in liquid level over a long period of time and solving eqn. (11.89) for $\mathcal{D}_{12}$. These measurements are subject to a variety of experimental errors, however. For example, the latent heat of vaporization may tend to cool the gas mixture near the interface, causing a temperature gradient in the tube. Vortices near the top of the tube, where it meets the gas stream, may cause additional mixing, and density gradients may cause buoyant circulation. Additional sources of error and alternative measurement techniques are described by Marrero and Mason [11.7].

The problem dealt with in this section can alternatively be solved on a mass basis, assuming a constant value of $\rho\mathcal{D}_{12}$ (see Problem 11.33 and Problem 11.34). The mass-based solution of this problem provides an important approximation in our analysis of high-rate convective mass transfer in the next section.

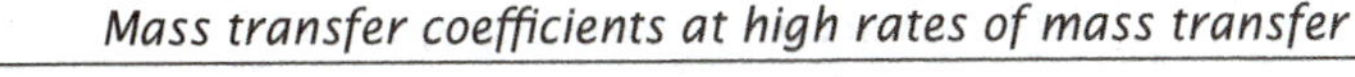

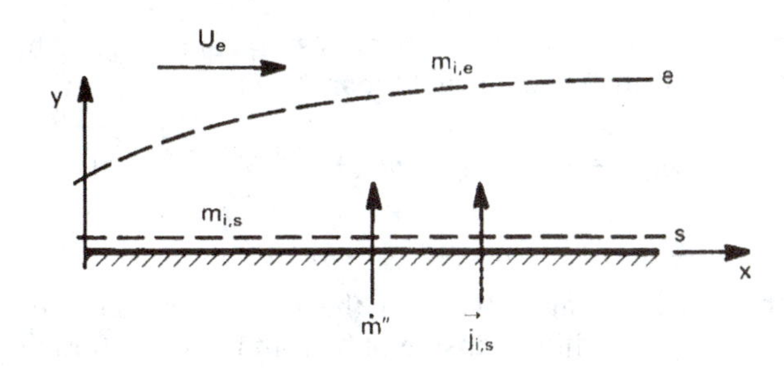

Figure 11.17 The mass concentration boundary layer.

11.8 Mass transfer coefficients at high rates of mass transfer

We developed an analogy between heat and mass transfer in Section 11.6, and it let us calculate mass transfer coefficients when the rate of mass transfer was low. This analogy required that the velocity field be unaffected by mass transfer and that the transferred species be dilute.

When those conditions are not met, the mass transfer coefficient will be different than the value given by the analogy. It can be either greater or less, and the difference can range from a few percent to an order of magnitude or more, depending upon the concentrations of the diffusing species. In addition to the diffusive transport represented by a mass transfer coefficient, the convective transport can contribute substantially to the total mass flux.

In this section, we model mass convection when the transferred species affects the velocity field and is not necessarily dilute. First, we define the *mass transfer driving force*, which governs the total mass flux from the wall. Then, we relate the mass transfer coefficient at high mass transfer rates to that at low mass transfer rates.

The mass transfer driving force

Figure 11.17 shows a boundary layer over a wall through which there is a net mass transfer, $n_s \equiv \dot{m}''$, of the various species in the direction normal to the wall.[13] In particular, we will focus on species i. In the free stream, i has a concentration $m_{i,e}$; at the wall, it has a concentration $m_{i,s}$.

The mass flux of i leaving the wall is obtained from eqn. (11.21):

$$n_{i,s} = m_{i,s}\dot{m}'' + j_{i,s} \tag{11.93}$$

[13]In this context, we denote the total mass flux through the wall as $\dot{m}''$, rather than n_s, so as to be consistent with other literature on the subject.

We seek to obtain $\dot{m}''$ in terms of the concentrations $m_{i,s}$ and $m_{i,e}$. As before, we define the mass transfer coefficient for species i, $g_{m,i}$ (kg/m^2·s), as

$$g_{m,i} = j_{i,s} / (m_{i,s} - m_{i,e}) \tag{11.94}$$

Thus,

$$n_{i,s} = m_{i,s}\dot{m}'' + g_{m,i}(m_{i,s} - m_{i,e}) \tag{11.95}$$

The mass transfer coefficient is again based on the *diffusive* transfer from the wall; however, it may now differ considerably from the value for low-rate transport.

Equation (11.95) may be rearranged as

$$\dot{m}'' = g_{m,i}\left(\frac{m_{i,e} - m_{i,s}}{m_{i,s} - n_{i,s}/\dot{m}''}\right) \tag{11.96}$$

which express the total mass flux of all species through the wall, $\dot{m}''$, as the product of the mass transfer coefficient and a ratio of concentrations. This ratio is called the *mass transfer driving force* for species i:

$$B_{m,i} \equiv \left(\frac{m_{i,e} - m_{i,s}}{m_{i,s} - n_{i,s}/\dot{m}''}\right) \tag{11.97}$$

The ratio of mass fluxes in the denominator is called the *mass fraction in the transferred state*, denoted as $m_{i,t}$:

$$m_{i,t} \equiv n_{i,s}/\dot{m}'' \tag{11.98}$$

The mass fraction in the transferred state is simply the fraction of the total mass flux, $\dot{m}''$, which is made up of species i. It is not really a mass fraction in the sense of Section 11.2 because it can have any value from $-\infty$ to $+\infty$, depending on the relative magnitudes of $\dot{m}''$ and $n_{i,s}$. If, for example, $n_{1,s} \cong -n_{2,s}$ in a binary mixture, then $\dot{m}''$ is very small and both $m_{1,t}$ and $m_{2,t}$ are very large.

Equations (11.96), (11.97), and (11.98) provide a formulation of mass transfer problems in terms of the mass transfer coefficient, $g_{m,i}$, and the driving force for mass transfer, $B_{m,i}$:

$$\boxed{\dot{m}'' = g_{m,i} B_{m,i}} \tag{11.99}$$

where

$$\boxed{B_{m,i} = \left(\frac{m_{i,e} - m_{i,s}}{m_{i,s} - m_{i,t}}\right), \qquad m_{i,t} = n_{i,s}/\dot{m}''} \tag{11.100}$$

These relations are based on an arbitrary species, i. The mass transfer rate may equally well be calculated using any species in a mixture; one obtains the same result for each. This is well illustrated in a binary mixture for which one may show that (Problem 11.36)

$$g_{m,1} = g_{m,2} \quad \text{and} \quad B_{m,1} = B_{m,2}$$

In many situations, only one species is transferred through the wall. If species i is the only one passing through the s-surface, then $n_{i,s} = \dot{m}''$, so that $m_{t,i} = 1$. The mass transfer driving force is simply

$$\boxed{B_{m,i} = \left(\frac{m_{i,e} - m_{i,s}}{m_{i,s} - 1}\right) \quad \text{one species transferred}} \tag{11.101}$$

In all the cases described in Section 11.6, only one species is transferred.

Example 11.15

A pan of hot water with a surface temperature of 75°C is placed in an air stream that has a mass fraction of water equal to 0.05. If the average mass transfer coefficient for water over the pan is $\overline{g_{m,H_2O}} = 0.0169$ kg/m^2·s and the pan has a surface area of 0.04 m^2, what is the evaporation rate?

Solution. Only water vapor passes through the liquid surface, since air is not strongly absorbed into water under normal conditions. Thus, we use eqn. (11.101) for the mass transfer driving force. Reference to a steam table shows the saturation pressure of water to be 38.58 kPa at 75°C, so

$$x_{H_2O,s} = 38.58/101.325 = 0.381$$

Putting this value into eqn. (11.67), we obtain

$$m_{H_2O,s} = 0.277$$

so that

$$B_{m,H_2O} = \frac{0.05 - 0.277}{0.277 - 1.0} = 0.314$$

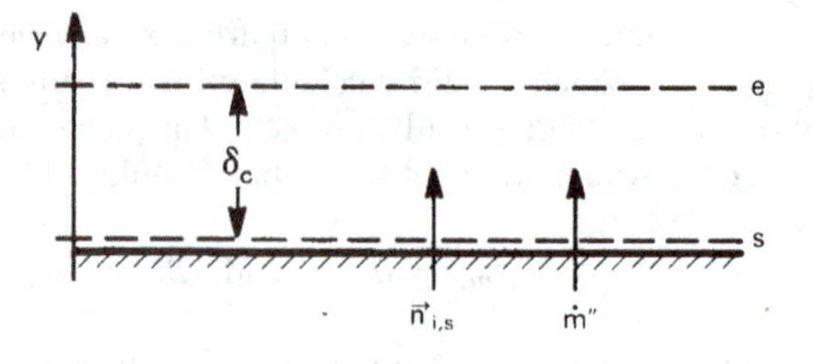

Figure 11.18 A stagnant film.

Thus,

$$\begin{aligned}\dot{m}_{H_2O} &= \overline{g_{m,H_2O}}\, \mathcal{B}_{m,H_2O}(0.04\ \text{m}^2)\\ &= (0.0169\ \text{kg/m}^2\!\cdot\!\text{s})(0.314)(0.04\ \text{m}^2)\\ &= 0.000212\ \text{kg/s} = 764\ \text{g/hr}\end{aligned}$$

■

The effect of mass transfer rates on the mass transfer coefficient

We still face the task of finding the mass transfer coefficient, $g_{m,i}$. The most obvious way to do this would be to apply the same methods we used to find the heat transfer coefficient in Chapters 6 through 8—solution of the momentum and species equations or through correlation of mass transfer data. These approaches are often used, but they are more complicated than the analogous heat transfer problems, owing to the coupling of the flow field and the mass transfer rate. Simple solutions are not so readily available for mass transfer problems. We instead employ a widely used approximate method that allows us to calculate $g_{m,i}$ from the low-rate mass transfer coefficient by applying a correction for the effect of finite mass transfer rates.

To isolate the effect of $\dot{m}''$ on the mass transfer coefficient, we first define the mass transfer coefficient at zero net mass transfer, $g^*_{m,i}$:

$$g^*_{m,i} \equiv \lim_{\dot{m}'' \to 0} g_{m,i}$$

The value $g^*_{m,i}$ is simply the mass transfer coefficient for low rates that would be obtained from the analogy between heat and mass transfer, as described in Section 11.6. Although $g_{m,i}$ depends directly on the rate of mass transfer, $g^*_{m,i}$ does not: it is determined only by flow configuration and physical properties.

In a boundary layer, the fluid near the wall is slowed by the no-slip condition. One way of modeling high-rate mass transfer effects on $g_{m,i}$ is to approximate the boundary layer as a *stagnant film*—a stationary layer of fluid with no horizontal gradients in it, as shown in Fig. 11.18. The film thickness, δ_c, is an effective local concentration boundary layer thickness.

The presence of a finite mass transfer rate across the film means that vertical convection—counterdiffusion effects—will be present. In fact, the stagnant film shown in Fig. 11.18 is identical to the configuration dealt with in the previous section (i.e., Fig. 11.16). Thus, the solution obtained in the previous section—eqn. (11.88)—also gives the rate of mass transfer across the stagnant film, taking account of vertical convective transport.

In the present mass-based analysis, it is convenient to use the mass-based analog of the mole-based eqn. (11.88). This analog can be shown to be (Problem 11.33)

$$\dot{m}'' = \frac{\rho \mathcal{D}_{im}}{\delta_c} \ln\left(1 + \frac{m_{i,e} - m_{i,s}}{m_{i,s} - n_{i,s}/\dot{m}''}\right)$$

which we may recast in the following, more suggestive form

$$\dot{m}'' = \frac{\rho \mathcal{D}_{im}}{\delta_c} \left[\frac{\ln(1 + B_{m,i})}{B_{m,i}}\right] B_{m,i} \tag{11.102}$$

Comparing this equation with eqn. (11.99), we see that

$$g_{m,i} = \frac{\rho \mathcal{D}_{im}}{\delta_c} \left[\frac{\ln(1 + B_{m,i})}{B_{m,i}}\right]$$

and when $\dot{m}''$ approaches zero,

$$g^*_{m,i} = \lim_{\dot{m}'' \to 0} g_{m,i} = \lim_{B_{m,i} \to 0} g_{m,i} = \frac{\rho \mathcal{D}_{im}}{\delta_c} \tag{11.103}$$

Hence,

$$\boxed{g_{m,i} = g^*_{m,i} \left[\frac{\ln(1 + B_{m,i})}{B_{m,i}}\right]} \tag{11.104}$$

The appropriate value $g^*_{m,i}$ (or δ_c) may be found from the solution of corresponding low-rate mass transfer problem, using the analogy of heat and mass transfer. (The value of $g^*_{m,i}$, in turn, defines the effective concentration b.l. thickness, δ_c.)

The group $[\ln(1 + B_{m,i})]/B_{m,i}$ is called the *blowing factor.* It accounts the effect of mass transfer on the velocity field. When $B_{m,i} > 0$, we have mass flow away from the wall (or *blowing*.) In this case, the blowing factor is always a positive number less than one, so blowing *reduces* $g_{m,i}$. When $B_{m,i} < 0$, we have mass flow toward the wall (or *suction*), and the blowing factor is always a positive number greater than one. Thus, $g_{m,i}$ is *increased* by suction. These trends may be better understood if we note that wall suction removes the slow fluid at the wall and thins the boundary layer. The thinner b.l. offers less resistance to mass transfer. Likewise, blowing tends to thicken the b.l., increasing the resistance to mass transfer.

The stagnant film b.l. model ignores details of the flow in the b.l. and focuses on the balance of mass fluxes across it. It is equally valid for both laminar and turbulent flows. Analogous stagnant film analyses of heat and momentum transport may also be made, as discussed in Problem 11.37.

Example 11.16

Calculate the mass transfer coefficient for Example 11.15 if the air speed is 5 m/s, the length of the pan in the flow direction is 20 cm, and the air temperature is 45°C. Assume that the air flow does not generate waves on the water surface.

Solution. The water surface is essentially a flat plate, as shown in Fig. 11.19. To find the appropriate equation for the Nusselt number, we must first compute Re_L.

The properties are evaluated at the film temperature, $T_f = (75 + 45)/2 = 60°\mathrm{C}$, and the film composition,

$$m_{f,\mathrm{H_2O}} = (0.050 + 0.277)/2 = 0.164$$

For these conditions, we find the mixture molecular weight from eqn. (11.8) as $M_f = 26.34$ kg/kmol. Thus, from the ideal gas law,

$$\rho_f = (101,325)(26.34)/(8314.5)(331.15) = 0.969 \text{ kg/m}^3$$

From Appendix A, we get $\mu_{\text{air}} = 1.986 \times 10^{-5}$ kg/m·s and $\mu_{\text{water vapor}} = 1.088 \times 10^{-5}$ kg/m·s. Then eqn. (11.54), with $x_{\mathrm{H_2O},f} = 0.240$ and $x_{\text{air},f} = 0.760$, yields

$$\mu_f = 1.78 \times 10^{-5} \text{ kg/m·s} \quad \text{so} \quad \nu_f = (\mu/\rho)_f = 1.84 \times 10^{-5} \text{ m}^2\text{/s}$$

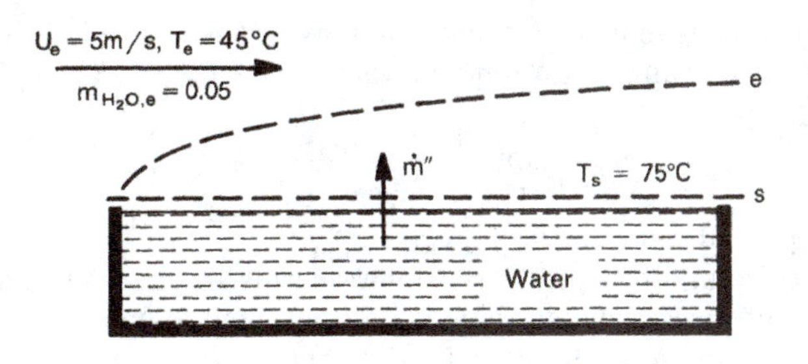

Figure 11.19 Evaporation from a tray of water.

We compute $\text{Re}_L = 5(0.2)/(1.84 \times 10^{-5}) = 54,300$, so the flow is laminar.

The appropriate Nusselt number is obtained from the mass transfer version of eqn. (6.68):

$$\overline{\text{Nu}}_{m,L} = 0.664\ \text{Re}_L^{1/2}\text{Sc}^{1/3}$$

Equation (11.35) yields $\mathcal{D}_{H_2O,\text{air}} = 3.11 \times 10^{-5}\ \text{m}^2/\text{s}$, so

$$\text{Sc} = 1.78/3.11 = 0.572$$

and

$$\overline{\text{Nu}}_{m,L} = 128$$

Hence,

$$\overline{g^*_{m,H_2O}} = \overline{\text{Nu}}_{m,L}(\rho\mathcal{D}_{H_2O,\text{air}}/L) = 0.0194\ \text{kg/m}^2{\cdot}\text{s}$$

Finally,

$$\begin{aligned}\overline{g_{m,H_2O}} &= g^*_{m,H_2O}\left[\ln(1 + B_{m,H_2O})/B_{m,H_2O}\right]\\ &= 0.0194\ \ln(1.314)/0.314 = 0.0169\ \text{kg/m}^2{\cdot}\text{s}\end{aligned}$$

In this case, the blowing factor is 0.870. Thus, mild blowing has reduced the mass transfer coefficient by about 13%. ■

Conditions for low-rate mass transfer. When the mass transfer driving force is small enough, the low-rate mass transfer coefficient itself is an adequate approximation to the actual mass transfer coefficient. This is because the blowing factor tends toward one as $B_{m,i} \longrightarrow 0$:

$$\lim_{B_{m,i}\longrightarrow 0} \frac{\ln(1 + B_{m,i})}{B_{m,i}} = 1$$

Thus, for small values of $B_{m,i}$, $g_{m,i} \cong g^*_{m,i}$.

The calculation of mass transfer proceeds in one of two ways for low rates of mass transfer, depending upon how the limit of small $\dot{m}''$ is reached. The first situation is when the ratio $n_{i,s}/\dot{m}''$ is fixed at a nonzero value while $\dot{m}'' \longrightarrow 0$. This would be the case when only one species is transferred, since $n_{i,s}/\dot{m}'' = 1$. Then the mass flux at low rates is

$$\dot{m}'' \cong g^*_{m,i} B_{m,i} \tag{11.105}$$

In this case, convective and diffusive contributions to $n_{i,s}$ are of the same order of magnitude, in general. To reach conditions for which the analogy of heat and mass transfer applies, it is also necessary that $m_{i,s} \ll 1$, so that convective effects will be negligible, as discussed in Section 11.6. When that condition also applies, and if only one species is transferred, we have

$$\begin{aligned}\dot{m}'' = n_{i,s} &\cong g^*_{m,i} B_{m,i}\\ &= g^*_{m,i}\left(\frac{m_{i,e} - m_{i,s}}{m_{i,s} - 1}\right)\\ &\cong g^*_{m,i}(m_{i,s} - m_{i,e})\end{aligned}$$

In the second situation, $n_{i,s}$ remains finite while $\dot{m}'' \longrightarrow 0$. Then, from eqn. (11.93),

$$n_{i,s} \cong j_{i,s} \cong g^*_{m,i}(m_{i,s} - m_{i,e}) \tag{11.106}$$

The transport in this case is purely diffusive, irrespective of the size of $m_{i,s}$. This situation arises is catalysis, where two species flow to a wall and react, creating a third species that flows away from the wall. Since the reaction conserves mass, the net mass flow through the s-surface is zero, even though $n_{i,s}$ is not (see Problem 11.44).

An estimate of the blowing factor can be used to determine whether $B_{m,i}$ is small enough to justify using the simpler low-rate theory. If, for example, $B_{m,i} = 0.20$, then $[\ln(1+B_{m,i})]/B_{m,i} = 0.91$ and an error of only 9 percent is introduced by assuming low rates. This level of accuracy is adequate for many engineering calculations.

11.9 Simultaneous heat and mass transfer

Many important engineering mass transfer processes occur simultaneously with heat transfer. Cooling towers, dryers, and combustors are just a few examples of equipment that intimately couple heat and mass transfer.

Coupling can arise when temperature-dependent mass transfer processes cause heat to be released or absorbed at a surface. For example, during evaporation, latent heat is absorbed at a liquid surface when vapor is created. This tends to cool the surface, lowering the vapor pressure and reducing the evaporation rate. Similarly, in the carbon oxidation problem discussed in Example 11.2, heat is released when carbon is oxidized, and the rate of oxidation is a function of temperature. The balance between convective cooling and the rate of reaction determines the surface temperature of the burning carbon.

Simultaneous heat and mass transfer processes may be classified as low-rate or high-rate. At low rates of mass transfer, mass transfer has only a negligible influence on the velocity field, and heat transfer rates may be calculated as if mass transfer were not occurring. At high rates of mass transfer, the heat transfer coefficient must be corrected for the effect of counterdiffusion. In this section, we consider these two possibilities in turn.

Heat transfer at low rates of mass transfer

One very common case of low-rate heat and mass transfer is the evaporation of water into air at low or moderate temperatures. An archetypical example of such a process is provided by a *sling psychrometer*, which is a device used to measure the humidity of air.

In a sling psychrometer, a wet cloth is wrapped about the bulb of a thermometer, as shown in Fig. 11.20. This so-called *wet-bulb* thermometer is mounted, along with a second *dry-bulb* thermometer, on a swivel handle, and the pair are "slung" in a rotary motion until they reach steady state.

The wet-bulb thermometer is cooled, as the latent heat of the vaporized water is given up, until it reaches the temperature at which the rate of cooling by evaporation just balances the rate of convective heating by the warmer air. This temperature, called the *wet-bulb temperature*, is directly related to the concentration of water in the surrounding air.[14]

[14]The wet-bulb temperature for air-water systems is very nearly the *adiabatic satu-*

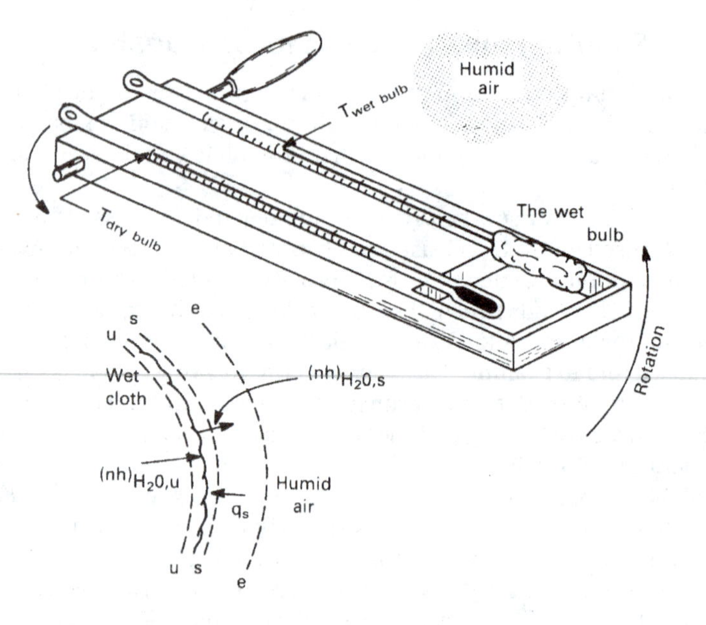

Figure 11.20 The wet bulb of a sling psychrometer.

The highest ambient air temperatures we normally encounter are low enough that the rate of mass transfer remains modest. We can test this suggestion by computing an upper bound on B_{m,H_2O}, under conditions that should maximize the evaporation rate: using the highest likely air temperature and the lowest humidity. Let us set those values, say, at 120°F (49°C) and zero humidity ($m_{H_2O,e} = 0$).

We know that the vapor pressure on the wet bulb will be less than the saturation pressure at 120°F, since evaporation will keep the bulb at a lower temperature:

$$x_{H_2O,s} \lesssim p_{sat}(120°F)/p_{atm} = (11,671 \text{ Pa})/(101,325 \text{ Pa}) = 0.115$$

so, with eqn. (11.67),

$$m_{H_2O,s} \lesssim 0.0750$$

ration temperature of the air-water mixture — the temperature reached by a mixture if it is brought to saturation with water by adding water vapor without adding heat. The adiabatic saturation temperature is a thermodynamic property of an air-water combination.

Thus, our criterion for low-rate mass transfer, eqn. (11.74), is met:

$$B_{m,\mathrm{H_2O}} = \left(\frac{m_{\mathrm{H_2O},s} - m_{\mathrm{H_2O},e}}{1 - m_{\mathrm{H_2O},s}} \right) \lesssim 0.0811$$

Alternatively, in terms of the blowing factor, eqn. (11.104),

$$\frac{\ln(1 + B_{m,\mathrm{H_2O}})}{B_{m,\mathrm{H_2O}}} \lesssim 0.962$$

This means that under the worst normal circumstances, the low-rate theory should deviate by only 4 percent from the actual rate of evaporation.

We may form an energy balance on the wick by considering the u, s, and e surfaces shown in Fig. 11.20. At the steady temperature, no heat is conducted past the u-surface (into the wet bulb), but liquid water flows through it to the surface of the wick where it evaporates. An energy balance on the region between the u and s surfaces gives

$$\underbrace{n_{\mathrm{H_2O},s}\hat{h}_{\mathrm{H_2O},s}}_{\substack{\text{enthalpy of water}\\ \text{vapor leaving}}} - \underbrace{q_s}_{\substack{\text{heat convected}\\ \text{to the wet bulb}}} = \underbrace{n_{\mathrm{H_2O},u}\hat{h}_{\mathrm{H_2O},u}}_{\substack{\text{enthalpy of liquid}\\ \text{water arriving}}}$$

Since mass is conserved, $n_{\mathrm{H_2O},s} = n_{\mathrm{H_2O},u}$, and because the enthalpy change results from vaporization, $\hat{h}_{\mathrm{H_2O},s} - \hat{h}_{\mathrm{H_2O},u} = h_{fg}$. Hence,

$$n_{\mathrm{H_2O},s}\, h_{fg}\big|_{T_{\text{wet-bulb}}} = h(T_e - T_{\text{wet-bulb}})$$

For low-rate mass transfer, $n_{\mathrm{H_2O},s} \cong j_{\mathrm{H_2O},s}$, and this equation can be written in terms of the mass transfer coefficient

$$g_{m,\mathrm{H_2O}}\left(m_{\mathrm{H_2O},s} - m_{\mathrm{H_2O},e}\right) h_{fg}\big|_{T_{\text{wet-bulb}}} = h(T_e - T_{\text{wet-bulb}}) \qquad (11.107)$$

The heat and mass transfer coefficients depend on the geometry and flow rates of the psychrometer, so it would appear that $T_{\text{wet-bulb}}$ should depend on the device used to measure it. The two coefficients are not independent, however, owing to the analogy between heat and mass transfer. For forced convection in cross flow, we saw in Chapter 7 that the heat transfer coefficient had the general form

$$\frac{hD}{k} = C\,\mathrm{Re}^a\mathrm{Pr}^b$$

where C is a constant, and typical values of a and b are $a \cong 1/2$ and $b \cong 1/3$. From the analogy,

$$\frac{g_m D}{\rho \mathcal{D}_{12}} = C\,\mathrm{Re}^a\mathrm{Sc}^b$$

Dividing the second expression into the first, we find

$$\frac{h}{g_m c_p}\frac{\mathcal{D}_{12}}{\alpha} = \left(\frac{\text{Pr}}{\text{Sc}}\right)^b$$

Both $\alpha/\mathcal{D}_{12}$ and Sc/Pr are equal to the Lewis number, Le. Hence,

$$\frac{h}{g_m c_p} = \text{Le}^{1-b} \cong \text{Le}^{2/3} \tag{11.108}$$

Equation (11.108) shows that the ratio of h to g_m depends primarily on the physical properties of the gas mixture, Le and c_p, rather than the geometry or flow rate. The Lewis number for air–water systems is about 0.847; and, because the concentration of water vapor is generally low, c_p can often be approximated by $c_{p_{\text{air}}}$.

This type of relationship between h and g_m was first developed by W. K. Lewis in 1922 for the case in which Le = 1 [11.27]. (Lewis's primary interest was in air–water systems, so the approximation was not too bad.) The more general form, eqn. (11.108), is another Reynolds-Colburn type of analogy, similar to eqn. (6.76). It was given by Chilton and Colburn [11.28] in 1934.

We can now write eqn. (11.107) as

$$T_e - T_{\text{wet-bulb}} = \left(\frac{h_{fg}|_{T_{\text{wet-bulb}}}}{c_{p_{\text{air}}}\,\text{Le}^{2/3}}\right)(m_{\text{H}_2\text{O},s} - m_{\text{H}_2\text{O},e}) \tag{11.109}$$

This expression can be solved iteratively with steam table data to obtain the wet-bulb temperature as a function of the dry-bulb temperature, T_e, and the humidity of the ambient air, $m_{\text{H}_2\text{O},e}$. The psychrometric charts found in engineering handbooks, thermodynamics texts, and online can be generated in this way. We ask the reader to make such calculations in Problem 11.49.

The wet-bulb temperature is a helpful concept in many phase-change processes. When a small body without internal heat sources evaporates or sublimes, it cools to a steady "wet-bulb" temperature at which convective heating is balanced by latent heat removal. The body will stay at that temperature until the phase-change process is complete. Thus, the wet-bulb temperature appears in the evaporation of water droplets, the sublimation of dry ice, the combustion of fuel sprays, and so on. If the body is massive, however, it could take a long time to reach steady state.

Stagnant film model of heat transfer at high mass transfer rates

The multicomponent energy equation. Each species in a mixture carries its own enthalpy, $\hat{h}_i$. In a flow with mass transfer, different species move with different velocities, so that enthalpy transport by individual species must enter the energy equation along with heat conduction through the fluid mixture. For steady, low-speed flow without internal heat generation or chemical reactions, we may rewrite the energy balance, eqn. (6.36), as

$$-\int_S (-k\nabla T) \cdot d\vec{S} - \int_S \left(\sum_i \rho_i \hat{h}_i \vec{v}_i \right) \cdot d\vec{S} = 0$$

where the second term accounts for the enthalpy transport by each species in the mixture. The usual procedure of applying Gauss's theorem and requiring the integrand to vanish identically gives

$$\nabla \cdot \left(-k\nabla T + \sum_i \rho_i \hat{h}_i \vec{v}_i \right) = 0 \tag{11.110}$$

This equation shows that the total energy flux—the sum of heat conduction and enthalpy transport—is conserved in steady flow.[15]

The stagnant film model. Let us restrict attention to the transport of a single species, i, across a boundary layer. We again use the stagnant film model for the thermal boundary layer and consider the one-dimensional flow of energy through it (see Fig. 11.21). Equation (11.110) simplifies to

$$\frac{d}{dy}\left(-k\frac{dT}{dy} + \rho_i \hat{h}_i v_i \right) = 0 \tag{11.111}$$

From eqn. (11.69) for steady, one-dimensional mass conservation

$$n_i = \text{constant in } y = n_{i,s}$$

[15]The multicomponent energy equation becomes substantially more complex when kinetic energy, body forces, and thermal or pressure diffusion are taken into account. The complexities are such that most published derivations of the multicomponent energy equation are incorrect, as shown by Mills in 1998 [11.29]. The main source of error has been the assignment of an independent kinetic energy to the ordinary diffusion velocity. This leads to such inconsistencies as a mechanical work term in the thermal energy equation.

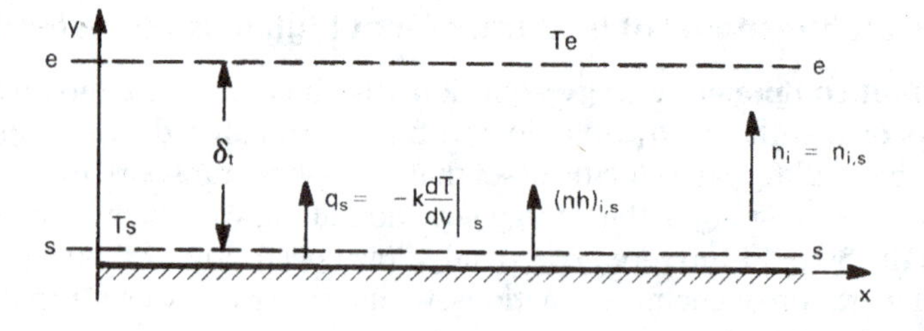

Figure 11.21 Energy transport in a stagnant film.

If we neglect pressure variations and assume a constant specific heat capacity (as in Sect. 6.3), the enthalpy may be written as $\hat{h}_i = c_{p,i}(T - T_{\text{ref}})$, and eqn. (11.111) becomes

$$\frac{d}{dy}\left(-k\frac{dT}{dy} + n_{i,s}c_{p,i}T\right) = 0$$

Integrating twice and applying the boundary conditions

$$T(y = 0) = T_s \quad \text{and} \quad T(y = \delta_t) = T_e$$

we obtain the temperature profile of the stagnant film:

$$\frac{T - T_s}{T_e - T_s} = \frac{\exp\left(\dfrac{n_{i,s}c_{p,i}}{k}y\right) - 1}{\exp\left(\dfrac{n_{i,s}c_{p,i}}{k}\delta_t\right) - 1} \tag{11.112}$$

The temperature distribution may be used to find the heat transfer coefficient according to its definition [eqn. (6.5)]:

$$h \equiv \frac{-k\dfrac{dT}{dy}\Big|_s}{T_s - T_e} = \frac{n_{i,s}c_{p,i}}{\exp\left(\dfrac{n_{i,s}c_{p,i}}{k}\delta_t\right) - 1} \tag{11.113}$$

We define the heat transfer coefficient in the limit of zero mass transfer, h^*, as

$$h^* \equiv \lim_{n_{i,s}\to 0} h = \frac{k}{\delta_t} \tag{11.114}$$

Substitution of eqn. (11.114) into eqn. (11.113) yields

$$\boxed{h = \frac{n_{i,s} c_{p,i}}{\exp(n_{i,s} c_{p,i}/h^*) - 1}} \tag{11.115}$$

To use this result, one first calculates the heat transfer coefficient as if there were *no* mass transfer, using the methods of Chapters 6 through 8. The value obtained is h^*, which is then placed in eqn. (11.115) to determine h in the presence of mass transfer. Note that h^* defines the effective film thickness δ_t through eqn. (11.114).

Equation (11.115) shows the primary effects of mass transfer on h. When $n_{i,s}$ is large and positive—the blowing case—h becomes smaller than h^*. Thus, blowing decreases the heat transfer coefficient, just as it decreases the mass transfer coefficient. Likewise, when $n_{i,s}$ is large and negative—the suction case—h becomes very large relative to h^*: suction increases the heat transfer coefficient just as it increases the mass transfer coefficient.

Condition for the low-rate approximation. When the rate of mass transfer is small, we may approximate h by h^*, just as we approximated g_m by g_m^* at low mass transfer rates. The approximation $h = h^*$ may be tested by considering the ratio $n_{i,s} c_{p,i}/h^*$ in eqn. (11.115). For example, if $n_{i,s} c_{p,i}/h^* = 0.2$, then $h/h^* = 0.90$, and $h = h^*$ within an error of only 10 percent. This is within the uncertainty to which h^* can be predicted in most flows. In gases, if $B_{m,i}$ is small, $n_{i,s} c_{p,i}/h^*$ will usually be small as well.

Property reference state. In Section 11.8, we calculated $g_{m,i}^*$ (and thus $g_{m,i}$) at the film temperature *and* film composition, as though mass transfer were occurring at the mean mixture composition and temperature. We may evaluate h^* and $g_{m,i}^*$ in the same way when heat and mass transfer occur simultaneously. If composition variations are not large, as in many low-rate problems, it may be adequate to use the freestream composition and film temperature. When large properties variations are present, other schemes may be required [11.30].

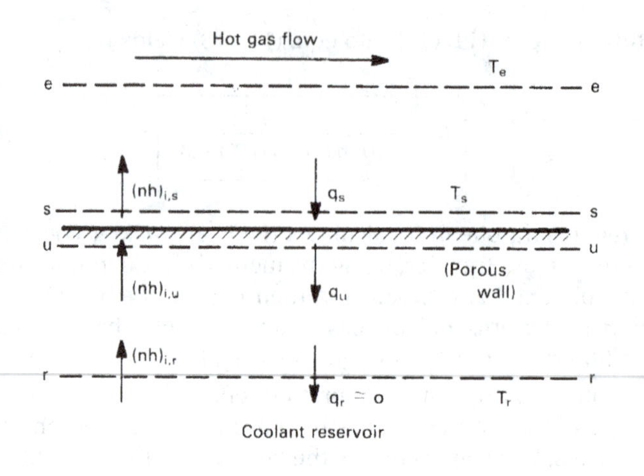

Figure 11.22 Transpiration cooling.

Energy balances in simultaneous heat and mass transfer

Transpiration cooling. To calculate simultaneous heat and mass transfer rates, one must generally look at the energy balance below the wall as well as those at the surface and across the boundary layer. Consider, for example, the process of *transpiration cooling*, shown in Fig. 11.22. Here a wall exposed to high temperature gases is protected by injecting a cooler gas into the flow through a porous section of the surface. A portion of the heat transfer to the wall is taken up in raising the temperature of the transpired gas. Blowing serves to thicken the boundary layer and reduce h, as well. This process is frequently used to cool turbine blades and combustion chamber walls.

Let us construct an energy balance for a steady state in which the wall has reached a temperature T_s. The enthalpy and heat fluxes are as shown in Fig. 11.22. We take the coolant reservoir to be far enough back from the surface that temperature gradients at the r-surface are negligible and the conductive heat flux, q_r, is zero. An energy balance between the r- and u-surfaces gives

$$n_{i,r}\hat{h}_{i.r} = n_{i,u}\hat{h}_{i,u} - q_u \tag{11.116}$$

and between the u- and s-surfaces,

$$n_{i,u}\hat{h}_{i,u} - q_u = n_{i,s}\hat{h}_{i,s} - q_s \tag{11.117}$$

Since the enthalpy of the transpired species does not change when it passes out of the wall,

$$\hat{h}_{i,u} = \hat{h}_{i,s} \tag{11.118}$$

and, because the process is steady, conservation of mass gives

$$n_{i,r} = n_{i,u} = n_{i,s} \tag{11.119}$$

Thus, eqn. (11.117) reduces to

$$q_s = q_u \tag{11.120}$$

The flux q_u is the conductive heat flux into the wall, while q_s is the convective heat transfer from the gas stream,

$$q_s = h(T_e - T_s) \tag{11.121}$$

Combining eqns. (11.116) through (11.121), we find

$$n_{i,s}\left(\hat{h}_{i,s} - \hat{h}_{i,r}\right) = h(T_e - T_s) \tag{11.122}$$

This equation shows that, at steady state, the heat convection to the wall is absorbed by the enthalpy rise of the transpired gas. Writing the enthalpy as $\hat{h}_i = c_{p,i}(T_s - T_{\text{ref}})$, we obtain

$$n_{i,s}c_{p,i}(T_s - T_r) = h(T_e - T_s) \tag{11.123}$$

or

$$T_s = \frac{hT_e + n_{i,s}c_{p,i}T_r}{h + n_{i,s}c_{p,i}} \tag{11.124}$$

It is left as an exercise (Problem 11.47) to show that

$$T_s = T_r + (T_e - T_r)\exp(-n_{i,s}c_{p,i}/h^*) \tag{11.125}$$

The wall temperature thus decreases exponentially to T_r as the mass flux of the transpired gas increases. That means we can enhance transpiration cooling by injecting a gas with a high specific heat.

Sweat Cooling. A common variation on transpiration cooling is *sweat cooling,* in which a *liquid* is bled through a porous wall. The liquid is vaporized by convective heat flow to the wall, and the latent heat of vaporization acts as a sink. Figure 11.22 also represents this process. The balances, eqns. (11.116) and (11.117), as well as mass conservation, eqn. (11.119), still apply, but the enthalpies at the interface now differ by the latent heat of vaporization:

$$\hat{h}_{i,u} + h_{fg} = \hat{h}_{i,s} \tag{11.126}$$

Thus, eqn. (11.120) becomes

$$q_s = q_u + h_{fg} n_{i,s}$$

and eqn. (11.122) takes the form

$$n_{i,s}\left[h_{fg} + c_{p,i_f}(T_s - T_r)\right] = h(T_e - T_s) \tag{11.127}$$

where c_{p,i_f} is the specific heat of *liquid* i. Since the latent heat is generally much larger than the sensible heat, a comparison of eqn. (11.127) to eqn. (11.123) exposes the greater efficiency per unit mass flow of sweat cooling relative to transpiration cooling.

Thermal radiation. When thermal radiation falls on the surface through which mass is transferred, the additional heat flux must enter the energy balances. For example, suppose that thermal radiation were present during transpiration cooling. Radiant heat flux, $q_{\text{rad},e}$, originating above the e-surface would be absorbed *below* the u-surface.[16] Thus, eqn. (11.116) becomes

$$n_{i,r}\hat{h}_{i,r} = n_{i,u}\hat{h}_{i,u} - q_u - \alpha q_{\text{rad},e} \tag{11.128}$$

where α is the radiation absorptance. Equation (11.117) is unchanged. Similarly, thermal radiation emitted by the wall is taken to originate below the u-surface, so eqn. (11.128) is now

$$n_{i,r}\hat{h}_{i,r} = n_{i,u}\hat{h}_{i,u} - q_u - \alpha q_{\text{rad},e} + q_{\text{rad},u} \tag{11.129}$$

or, in terms of radiosity and irradiation (see Section 10.4)

$$n_{i,r}\hat{h}_{i,r} = n_{i,u}\hat{h}_{i,u} - q_u - (H - B) \tag{11.130}$$

for an opaque surface.

[16]Remember that the s- and u-surfaces are fictitious elements of the enthalpy balances at the phase interface. The apparent space between them need be only a few molecules thick. Thermal radiation therefore passes through the u-surface and is absorbed below it.

Chemical Reactions. The heat and mass transfer analyses in this section and Section 11.8 assume that the transferred species undergo no homogeneous reactions. If reactions do occur, the mass balances of Section 11.8 are invalid, because the mass flux of a reacting species will vary across the region of reaction. Likewise, the energy balance of this section will fail because it does not include the heat of reaction.

For heterogeneous reactions, the complications are not so severe. Reactions at the boundaries release the heat of reaction released between the s- and u-surfaces, altering the boundary conditions. The proper stoichiometry of the mole fluxes to and from the surface must be taken into account, and the heat transfer coefficient [eqn. (11.115)] must be modified to account for the transfer of more than one species [11.30].

Problems

11.1 Derive: (a) eqns. (11.8); (b) eqns. (11.9).

11.2 A 1000 liter cylinder at 300 K contains a gaseous mixture composed of 0.10 kmol of NH_3, 0.04 kmol of CO_2, and 0.06 kmol of He. (a) Find the mass fraction for each species and the pressure in the cylinder. (b) After the cylinder is heated to 600 K, what are the new mole fractions, mass fractions, and molar concentrations? (c) The cylinder is now compressed isothermally to a volume of 600 liters. What are the molar concentrations, mass fractions, and partial densities? (d) If 0.40 kg of gaseous N_2 is injected into the cylinder while the temperature remains at 600 K, find the mole fractions, mass fractions, and molar concentrations. [(a) $m_{CO_2} = 0.475$; (c) $c_{CO_2} = 0.0667\ \text{kmol/m}^3$; (d) $x_{CO_2} = 0.187$.]

11.3 Planetary atmospheres show significant variations of temperature and pressure in the vertical direction. Observations suggest that the atmosphere of Jupiter has the following composition at the tropopause level:

$$\begin{aligned}
\text{number density of } H_2 &= 5.7 \times 10^{21}\ (\text{molecules/m}^3)\\
\text{number density of He} &= 7.2 \times 10^{20}\ (\text{molecules/m}^3)\\
\text{number density of } CH_4 &= 6.5 \times 10^{18}\ (\text{molecules/m}^3)\\
\text{number density of } NH_3 &= 1.3 \times 10^{18}\ (\text{molecules/m}^3)
\end{aligned}$$

Find the mole fraction and partial density of each species at this level if $p = 0.1$ atm and $T = 113$ K. Estimate the number densities at the level where $p = 10$ atm and $T = 400$ K, deeper within the Jovian troposphere. (Deeper in the Jupiter's atmosphere, the pressure may exceed 10^5 atm.)

11.4 Using the definitions of the fluxes, velocities, and concentrations, derive eqn. (11.34) from eqn. (11.27) for *binary* diffusion.

11.5 Show that $\mathcal{D}_{12} = \mathcal{D}_{21}$ in a binary mixture.

11.6 Fill in the details involved in obtaining eqn. (11.31) from eqn. (11.30).

11.7 Batteries commonly contain an aqueous solution of sulfuric acid with lead plates as electrodes. Current is generated by the reaction of the electrolyte with the electrode material. At the negative electrode, the reaction is

$$\text{Pb}(s) + \text{SO}_4^{2-} \rightleftharpoons \text{PbSO}_4(s) + 2e^-$$

where the (s) denotes a solid phase component and the charge of an electron is -1.609×10^{-19} coulombs. If the current density at such an electrode is $J = 5$ milliamperes/cm^2, what is the mole flux of SO_4^{2-} to the electrode? (1 amp =1 coulomb/s.) What is the mass flux of SO_4^{2-}? At what mass rate is $PbSO_4$ produced? If the electrolyte is to remain electrically neutral, at what rate does H^+ flow toward the electrode? Hydrogen does not react at the negative electrode. [$\dot{m}''_{\text{PbSO}_4} = 7.83 \times 10^{-5}$ kg/m^2·s.]

11.8 The salt concentration in the ocean increases with increasing depth, z. A model for the concentration distribution in the upper ocean is

$$S = 33.25 + 0.75 \tanh(0.026z - 3.7)$$

where S is the salinity in grams of salt per kilogram of ocean water and z is the distance below the surface in meters. (a) Plot the mass fraction of salt as a function of z. (The region of rapid transition of $m_{\text{salt}}(z)$ is called the *halocline*.) (b) Ignoring the effects of waves or currents, compute $j_{\text{salt}}(z)$. Use a value of

$\mathcal{D}_{\text{salt,water}} = 1.5 \times 10^{-5}$ cm^2/s. Indicate the position of maximum diffusion on your plot of the salt concentration. (c) The upper region of the ocean is well mixed by wind-driven waves and turbulence, while the lower region and halocline tend to be calmer. Using $j_{\text{salt}}(z)$ from part (b), make a simple estimate of the amount of salt carried upward in one week in a 5 km^2 horizontal area of the sea.

11.9 In *catalysis*, one gaseous species reacts with another on a passive surface (the catalyst) to form a gaseous product. For example, butane reacts with hydrogen on the surface of a nickel catalyst to form methane and propane. This heterogeneous reaction, referred to as *hydrogenolysis*, is

$$C_4H_{10} + H_2 \xrightarrow{\text{Ni}} C_3H_8 + CH_4$$

The molar rate of consumption of C_4H_{10} per unit area in the reaction is $\dot{R}_{C_4H_{10}} = A(e^{-\Delta E/R^\circ T}) p_{C_4H_{10}} p_{H_2}^{-2.4}$, where $A = 6.3 \times 10^{10}$ kmol/m^2·s, $\Delta E = 1.9 \times 10^8$ J/kmol, and p is in atm. (a) If $p_{C_4H_{10},s} = p_{C_3H_8,s} = 0.2$ atm, $p_{CH_4,s} = 0.17$ atm, and $p_{H_2,s} = 0.3$ atm at a nickel surface with conditions of 440°C and 0.87 atm total pressure, what is the rate of consumption of butane? (b) What are the mole fluxes of butane and hydrogen to the surface? What are the mass fluxes of propane and ethane away from the surface? (c) What is $\dot{m}''$? What are v, v^*, and $v_{C_4H_{10}}$? (d) What is the diffusional mole flux of butane? What is the diffusional mass flux of propane? What is the flux of Ni? [(b) $n_{CH_4,s} = 0.0441$ kg/m^2·s; (d) $j_{C_3H_8} = 0.121$ kg/m^2·s.]

11.10 Consider two chambers held at temperatures T_1 and T_2, respectively, and joined by a small insulated tube. The chambers are filled with a binary gas mixture, with the tube open, and allowed to come to steady state. If the Soret effect is taken into account, what is the concentration difference between the two chambers? Assume that an effective mean value of the thermal diffusion ratio is known.

11.11 Compute $\mathcal{D}_{12}$ for oxygen gas diffusing through nitrogen gas at $p = 1$ atm, using eqns. (11.39) and (11.42), for $T = 200$ K, 500 K, and 1000 K. Observe that eqn. (11.39) shows large deviations from eqn. (11.42), even for such simple and similar molecules.

11.12 (a) Compute the binary diffusivity of each of the noble gases when they are individually mixed with nitrogen gas at 1 atm and 300 K. Plot the results as a function of the molecular weight of the noble gas. What do you conclude? (b) Consider the addition of a small amount of helium ($x_{\text{He}} = 0.04$) to a mixture of nitrogen ($x_{\text{N}_2} = 0.48$) and argon ($x_{\text{Ar}} = 0.48$). Compute $\mathcal{D}_{\text{He},m}$ and compare it with $\mathcal{D}_{\text{Ar},m}$. Note that the higher concentration of argon does not improve its ability to diffuse through the mixture.

11.13 (a) One particular correlation shows that gas phase diffusion coefficients vary as $T^{1.81}$ and p^{-1}. If an experimental value of $\mathcal{D}_{12}$ is known at T_1 and p_1, develop an equation to predict $\mathcal{D}_{12}$ at T_2 and p_2. (b) The diffusivity of water vapor (1) in air (2) was measured to be 2.39×10^{-5} m^2/s at 8°C and 1 atm. Provide a formula for $\mathcal{D}_{12}(T, p)$.

11.14 Kinetic arguments lead to the *Stefan-Maxwell equation* for a dilute-gas mixture:

$$\nabla x_i = \sum_{j=1}^{n} \frac{c_i c_j}{c^2 \mathcal{D}_{ij}} \left(\frac{\vec{J}_j^*}{c_j} - \frac{\vec{J}_i^*}{c_i} \right)$$

(a) Show that eqn. (11.44) is a special case of this, under the appropriate assumptions. (b) Show that if $\mathcal{D}_{ij}$ has the same value for each pair of species, then $\mathcal{D}_{im} = \mathcal{D}_{ij}$.

11.15 Compute the diffusivity of methane in air using (a) eqn. (11.42) and (b) Blanc's law. For part (b), treat air as a mixture of oxygen and nitrogen, ignoring argon. Let $x_{\text{methane}} = 0.05$, $T = 420$°F, and $p = 10$ psia. [(a) $\mathcal{D}_{\text{CH}_4,\text{air}} = 7.66 \times 10^{-5}$ m^2/s; (b) $\mathcal{D}_{\text{CH}_4,\text{air}} = 8.13 \times 10^{-5}$ m^2/s.]

11.16 Diffusion of solutes in liquids is driven by the chemical potential, μ. Work is required to move a mole of solute A from a region of low chemical potential to a region of high chemical potential; that is,

$$dW = d\mu_A = \frac{d\mu_A}{dx}\, dx$$

under isothermal, isobaric conditions. For an ideal (very dilute) solute, μ_A is given by

$$\mu_A = \mu_0 + R^\circ T \ln(c_A)$$

where μ_0 is a constant. Using an elementary principle of mechanics, derive the Nernst-Einstein equation. Note that the solution must be assumed to be very dilute.

11.17 A dilute aqueous solution at 300 K contains potassium ions, K^+. If the velocity of aqueous K^+ ions is 6.61×10^{-4} cm^2/s·V per unit electric field (1 V/cm), estimate the effective radius of K^+ ions in an aqueous solution. Criticize this estimate. (The charge of an electron is -1.609×10^{-19} coulomb and a volt = 1J/coulomb.)

11.18 (a) Obtain diffusion coefficients for: (1) dilute CCl_4 diffusing through liquid methanol at 340 K; (2) dilute benzene diffusing through water at 290 K; (3) dilute ethyl alcohol diffusing through water at 350 K; and (4) dilute acetone diffusing through methanol at 370 K. (b) Estimate the effective radius of a methanol molecule in a dilute aqueous solution.
[(a) $\mathcal{D}_{\text{acetone,methanol}} = 6.8 \times 10^{-9}$ m^2/s.]

11.19 Calculate values of the viscosity, μ, for methane, hydrogen sulfide, and nitrous oxide, under the following conditions: 250 K and 1 atm, 500 K and 1 atm, 250 K and 2 atm, 250 K and 12 atm, 500 K and 12 atm. Is the calculation possible in every case?

11.20 (a) Show that $k = (5/2)\mu c_v$ for a monatomic gas. (b) Obtain Eucken's formula for the Prandtl number of a dilute gas:

$$\text{Pr} = 4\gamma/(9\gamma - 5)$$

(c) Recall that for an ideal gas, $\gamma \cong (D + 2)/D$, where D is the number of modes of energy storage of its molecules. Obtain an expression for Pr as a function of D and describe what it means. (d) Use Eucken's formula to compute Pr for gaseous Ar, N_2, and H_2O. Compare the result to data in Appendix A over the range of temperatures. Explain the results obtained for steam as opposed to Ar and N_2. (Note that for each mode of vibration, there are two modes of energy storage but that vibration is normally inactive until T is very high.)

11.21 A student is studying the combustion of a premixed gaseous fuel with the following molar composition: 10.3% methane,

15.4% ethane, and 74.3% oxygen. She passes 0.006 ft^3/s of the mixture (at 70°F and 18 psia) through a smooth 3/8 inch I.D. tube, 47 inches long. (a) What is the pressure drop? (b) The student's advisor recommends preheating the fuel mixture, using a Nichrome strip heater wrapped around the last 5 inches of the duct. If the heater produces 0.8 W/inch, what is the wall temperature at the outlet of the duct? Let $c_{p,CH_4} = 2280$ J/kg·K, $\gamma_{CH_4} = 1.3$, $c_{p,C_2H_6} = 1730$ J/kg·K, and $\gamma_{C_2H_6} = 1.2$, and evaluate the properties at the inlet conditions.

11.22 (a) Work Problem 6.36. (b) A fluid is said to be incompressible if the density of a fluid particle does not change as it moves about in the flow (i.e., if $D\rho/Dt = 0$). Show that an incompressible flow satisfies $\nabla \cdot \vec{u} = 0$. (c) How does the condition of incompressibility differ from that of "constant density"? Describe a flow that is incompressible but that does not have "constant density."

11.23 Carefully derive eqns. (11.62) and (11.63). Note that ρ is not assumed constant in eqn. (11.62).

11.24 Derive the equation of species conservation on a molar basis, using c_i rather than ρ_i. Also obtain an equation in c_i alone, similar to eqn. (11.63) but without the assumption of incompressibility. What assumptions must be made to obtain the latter result?

11.25 Find the following concentrations: (a) the mole fraction of air in solution with water at 5°C and 1 atm, exposed to air at the same conditions, $H = 4.88 \times 10^4$ atm; (b) the mole fraction of ammonia in air above an aqueous solution, with $x_{NH_3} = 0.05$ at 0.9 atm and 40°C and $H = 1522$ mm Hg; (c) the mole fraction of SO_2 in an aqueous solution at 15°C and 1 atm, if $p_{SO_2} = 28.0$ mm Hg and $H = 1.42 \times 10^4$ mm Hg; and (d) the partial pressure of ethylene over an aqueous solution at 25°C and 1 atm, with $x_{C_2H_4} = 1.75 \times 10^{-5}$ and $H = 11.4 \times 10^3$ atm.

11.26 Use steam table data to estimate (a) the mass fraction of water vapor in air over water at 1 atm and 20°C, 50°C, 70°C, and 90°C; (b) the partial pressure of water over a 3 percent-by-weight aqueous solution of HCl at 50°C; (c) the boiling point at 1 atm of salt water with a mass fraction $m_{\text{NaCl}} = 0.18$. [(c) $T_{B.P.} = 101.8°\text{C}$.]

11.27 Suppose that a steel fitting with a carbon mass fraction of 0.2% is put into contact with carburizing gases at 940°C, and that these gases produce a steady mass fraction, $m_{C,u}$, of 1.0% carbon just within the surface of the metal. The diffusion coefficient of carbon in this steel is

$$\mathcal{D}_{\text{C,Fe}} = (1.50 \times 10^{-5}\ \text{m}^2/\text{s}) \exp\left[-(1.42 \times 10^8\ \text{J/kmol})/(R^\circ T)\right]$$

for T in kelvin. How long does it take to produce a carbon concentration of 0.6% by mass at a depth of 0.5 mm? How much less time would it take if the temperature were 980°C?

11.28 (a) Write eqn. (11.62) in its boundary layer form. (b) Write this concentration boundary layer equation and its b.c.'s in terms of a nondimensional mass fraction, ψ, analogous to the dimensionless temperature in eqn. (6.42). (c) For $\nu = \mathcal{D}_{im}$, relate ψ to the Blasius function, f, for flow over a flat plate. (d) Note the similar roles of Pr and Sc in the two boundary layer transport processes. Infer the mass concentration analog of eqn. (6.55) and sketch the concentration and momentum b.l. profiles for $\text{Sc} = 1$ and $\text{Sc} \gg 1$.

11.29 When Sc is large, momentum diffuses more easily than mass, and the concentration b.l. thickness, δ_c, is much less than the momentum b.l. thickness, δ. On a flat plate, the small part of the velocity profile within the concentration b.l. is approximately $u/U_e = 3y/2\delta$. Compute $\text{Nu}_{m,x}$ based on this velocity profile, assuming a constant wall concentration. (*Hint*: Use the mass transfer analogs of eqn. (6.47) and (6.50) and note that $q_w/\rho c_p$ becomes $j_{i,s}/\rho$.).

11.30 Consider a one-dimensional, binary gaseous diffusion process in which species 1 and 2 travel in opposite directions along the z-axis at equal molar rates. (The gas mixture will be at rest, with $v = 0$ if the species have identical molecular weights). This process is known as *equimolar counter-diffusion.* (a) What are the relations between N_1, N_2, J_1^*, and J_2^*? (b) If steady state prevails and conditions are isothermal and isobaric, what is the concentration of species 1 as a function of z? (c) Write the mole flux in terms of the difference in partial pressure of species 1 between locations z_1 and z_2.

11.31 Consider steady mass diffusion from a small sphere. When convection is negligible, the mass flux in the radial direction is $n_{r,i} = j_{r,i} = -\rho \mathcal{D}_{im} dm_i/dr$. If the concentration is $m_{i,\infty}$ far from the sphere and $m_{i,s}$ at its surface, use a mass balance to obtain the surface mass flux in terms of the overall concentration difference (assuming that $\rho \mathcal{D}_{im}$ is constant). Then apply the definition eqns. (11.94) and (11.78) to show that $\text{Nu}_{m,D} = 2$ for this situation.

11.32 An experimental Stefan tube is 1 cm in diameter and 10 cm from the liquid surface to the top. It is held at 10°C and 8.0×10^4 Pa. Pure argon flows over the top and liquid CCl_4 is at the bottom. The pool level is maintained while 0.086 ml of liquid CCl_4 evaporates during a period of 12 hours. What is the diffusivity of carbon tetrachloride in argon measured under these conditions? The specific gravity of liquid CCl_4 is 1.59 and its vapor pressure is $\log_{10} p_v = 8.004 - 1771/T$, where p_v is expressed in mm Hg and T in K.

11.33 Repeat the analysis given in Section 11.7 on the basis of *mass* fluxes, assuming that $\rho \mathcal{D}_{im}$ is constant and neglecting any buoyancy-driven convection. Obtain the analog of eqn. (11.88).

11.34 In Sections 11.5 and 11.7, it was assumed at points that $c\mathcal{D}_{12}$ or $\rho \mathcal{D}_{12}$ was independent of position. (a) If the mixture composition (e.g., x_1) varies in space, this assumption may be poor. Using eqn. (11.42) and the definitions from Section 11.2, examine the composition dependence of these two groups. For what type of mixture is $\rho \mathcal{D}_{12}$ most sensitive to composition? What does this indicate about molar versus mass-based analysis? (b) How do each of these groups depend on pressure and temperature? Is the analysis of Section 11.7 really limited to isobaric conditions? (c) Do the Prandtl and Schmidt numbers depend on composition, temperature, or pressure?

11.35 A Stefan tube contains liquid bromine at 320 K and 1.2 atm. Carbon dioxide flows over the top and is also bubbled up through the liquid at the rate of 4.4 ml/hr. If the distance from the liquid surface to the top is 16 cm and the diameter is 1 cm, what is the evaporation rate of Br_2? ($p_{sat,Br_2} = 0.680$ bar at 320 K.) [$N_{Br_2,s} = 1.90 \times 10^{-6}$ kmol/m$^2\cdot$s.]

11.36 Show that $g_{m,1} = g_{m,2}$ and $B_{m,1} = B_{m,2}$ in a binary mixture.

11.37 Demonstrate that stagnant film models of the momentum and thermal boundary layers reproduce the proper dependence of $C_{f,x}$ and Nu_x on Re_x and Pr. Using eqns. (6.31b) and (6.55) to obtain the dependence of δ and δ_t on Re_x and Pr, show that stagnant film models gives eqns. (6.33) and (6.58) within a constant on the order of one. [The constants in these results will differ from the exact results because the effective b.l. thicknesses of the stagnant film model are not the same as the exact values—see eqn. (6.57).]

11.38 (a) What is the largest value of the mass transfer driving force when one species is transferred? What is the smallest value? (b) Plot the blowing factor as a function of $B_{m,i}$ for one species transferred. Indicate on your graph the regions of blowing, suction, and low-rate mass transfer. (c) Verify the two limits used to show that $g^*_{m,i} = \rho \mathcal{D}_{im}/\delta_c$.

11.39 Nitrous oxide is bled through the surface of a porous 3/8 in. O.D. tube at 0.025 liter/s per meter of tube length. Air flows over the tube at 25 ft/s. Both the air and the tube are at 18°C, and the ambient pressure is 1 atm. Estimate the mean concentration of N_2O at the tube surface. (*Hint*: First estimate the concentration using properties of pure air; then correct the properties if necessary.)

11.40 *Film absorbtion* is a process whereby gases are absorbed into a falling liquid film. Typically, a thin film of liquid runs down the inside of a vertical tube through which the gas flows. Analyze this process under the following assumptions: The film flow is laminar and of constant thickness, δ_0, with a velocity profile given by eqn. (8.50); the gas is only slightly soluble in the liquid, so that it does not penetrate far beyond the liquid surface and so that liquid properties are unaffected by it; and, the gas concentration at the s- and u-surfaces (above and below the liquid-vapor interface, respectively) does not vary along the length of the tube. The inlet concentration of gas in the liquid is $m_{1,0}$. Show that the mass transfer is given by

$$\mathrm{Nu}_{m,x} = \left(\frac{u_0 x}{\pi \mathcal{D}_{12}}\right)^{1/2} \quad \text{with} \quad u_0 = \frac{(\rho_f - \rho_g) g \delta_0^2}{2\mu_f}$$

The mass transfer coefficient here is based on the concentration difference between the u-surface and the bulk liquid at $m_{1,0}$. (*Hint*: The small penetration assumption can be used to reduce the species equation for the film to the diffusion equation, eqn. 11.72.)

11.41 Benzene vapor flows through a 3 cm I.D. vertical tube. A thin film of initially pure water runs down the inside wall of the tube at a flow rate of 0.3 liter/s. If the tube is 0.5 m long and 40°C, estimate the rate (in kg/s) at which benzene is absorbed into water over the entire length of the tube. The mass fraction of benzene at the u-surface is 0.206. (*Hint*: Use the result stated in Problem 11.40. Obtain δ_0 from the results in Chapter 8.)

11.42 A mothball consists of a 2.5 cm diameter sphere of naphthalene ($C_{10}H_8$) that is hung by a wire in a closet. The solid naphthalene slowly sublimes to vapor, which drives off the moths. The latent heat of sublimation and evaporation rate are low enough that the wet-bulb temperature is essentially the ambient temperature. Estimate the lifetime of this mothball in a closet with a mean temperature of 20°C. Use the following data:

$$\sigma = 6.18\ \text{Å}, \qquad \varepsilon/k_B = 561.5\ \text{K} \quad \text{for} \quad C_{10}H_8,$$

and, for solid naphthalene,

$$\rho_{C_{10}H_8} = 1145\ \text{kg/m}^3 \text{ at } 20°\text{C}$$

The vapor pressure (in mmHg) of solid naphthalene near room temperature is given approximately by $\log_{10} p_v = 11.450 - 3729.3/(T\ \text{K})$. The integral you obtain can be evaluated numerically.

11.43 In contrast to the napthalene mothball described in Prob. 11.42, other mothballs are made from paradichlorobenzene (PDB). Estimate the lifetime of a 2.5 cm diameter PDB mothball using the following room temperature property data:

$$\sigma = 5.76\ \text{Å} \qquad \varepsilon/k_B = 578.9\ \text{K} \qquad M_{\text{PDB}} = 147.0\ \text{kg/kmol}$$

$$\log_{10}(p_v \text{ mmHg}) = 11.985 - 3570/(T \text{ K})$$

$$\rho_{\text{PDB}} = 1248 \text{ kg/m}^3$$

11.44 Consider the process of catalysis as described in Problem 11.9. The mass transfer process involved is the diffusion of the reactants to the surface and diffusion of products away from it. (a) What is $\dot{m}''$ in catalysis? (b) Reaction rates in catalysis are of the form:

$$\dot{R}_{\text{reactant}} = A\, e^{-\Delta E/R^\circ T} (p_{\text{reactant}})^n (p_{\text{product}})^m \text{ kmol/m}^2\cdot\text{s}$$

for the rate of consumption of a reactant per unit surface area. The p's are partial pressures and A, ΔE, n, and m are constants. Suppose that $n = 1$ and $m = 0$ for the reaction $B + C \longrightarrow D$. Approximate the reaction rate, in terms of mass, as

$$\dot{r}_B = A' e^{-\Delta E/R^\circ T} \rho_{B,s} \text{ kg/m}^2\cdot\text{s}$$

and find the rate of consumption of B in terms of $m_{B,e}$ and the mass transfer coefficient for the geometry in question. (c) The ratio $\text{Da} \equiv \rho A' e^{-\Delta E/R^\circ T}/g_m^*$ is called the *Damkohler number.* Explain its significance in catalysis. What features dominate the process when Da approaches 0 or ∞? What temperature range characterizes each?

11.45 One typical kind of mass exchanger is a fixed-bed catalytic reactor. A flow chamber of length L is packed with a catalyst bed. A gas mixture containing some species i to be consumed by the catalytic reaction flows through the bed at a rate $\dot{m}$. The effectiveness of such a exchanger (cf. Chapter 3) is

$$\varepsilon = 1 - e^{-\text{NTU}}, \quad \text{where NTU} = g_{m,\text{oa}} PL/\dot{m}$$

where $g_{m,\text{oa}}$ is the overall mass transfer coefficient for the catalytic packing, P is the surface area per unit length, and ε is defined in terms of mass fractions. In testing a 0.5 m catalytic reactor for the removal of ethane, it is found that the ethane concentration drops from a mass fraction of 0.36 to 0.05 at a flow rate of 0.05 kg/s. The packing is known to have a surface area of 11 m^2. What is the exchanger effectiveness? What is the overall mass transfer coefficient in this bed?

11.46 (a) Perform the integration to obtain eqn. (11.112). Then take the derivative and the limit needed to get eqns. (11.113) and (11.114). (b) What is the general form of eqn. (11.115) when more than one species is transferred?

11.47 (a) Derive eqn. (11.125) from eqn. (11.124). (b) Suppose that 1.5 m^2 of the wing of a spacecraft re-entering the earth's atmosphere is to be cooled by transpiration; 900 kg of the vehicle's weight is allocated for this purpose. The low-rate heat transfer coefficient is 1800 $W/m^2{\cdot}K$, and the hottest period of re-entry is expected to last 3 minutes. If the air behind the shock wave ahead of the wing is at 2500°C and the reservior is at 5°C , which of these gases—H_2, He, and N_2—keeps the surface coolest? (Of course, the result for H_2 is invalidated by the fact that H_2 would burn under these conditions.)

11.48 Dry ice (solid CO_2) is used to cool medical supplies transported by a small plane to a remote village in Alaska. A roughly spherical chunk of dry ice, 5 cm in diameter, falls from the plane through air at 5°C with a terminal velocity of 15 m/s. If steady state is reached quickly, what are the temperature and sublimation rate of the dry ice? The latent heat of sublimation is 574 kJ/kg and $\log_{10}(p_v \text{ mmHg}) = 9.9082 - 1367.3/(T \text{ K})$. The temperature will be well below the "sublimation point" of CO_2 (solid-to-vapor saturation temperature), which is −78.6°C at 1 atm. Use the heat transfer relation for spheres in a laminar flow, $\overline{\text{Nu}}_D = 2 + 0.3\,\text{Re}_D^{0.6}\,\text{Pr}^{1/3}$. (*Hint*: First estimate the surface temperature using properties for pure air; then correct the properties if necessary.)

11.49 The following data were taken at a weather station over a period of several months:

Date	$T_{\text{dry-bulb}}$	$T_{\text{wet-bulb}}$
3/15	15.5°C	11.0°C
4/21	22.0	16.8
5/13	27.3	25.8
5/31	32.7	20.0
7/4	39.0	31.2

Use eqn. (11.109) to find the mass fraction of water in the air at each date. Compare to values from a psychrometric chart.

11.50 Biff Harwell has taken Deb sailing. Deb, and Biff's towel, fall into the harbor. Biff rescues them both from a passing dolphin and then spreads his wet towel out to dry on the fiberglas foredeck of the boat. The incident solar radiation is 1050 W/m^2; the ambient air is at 31°C, with $m_{H_2O} = 0.017$; the wind speed is 8 knots relative to the boat (1 knot = 1.151 mph); $\varepsilon_{\text{towel}} \cong \alpha_{\text{towel}} \cong 1$; and the sky has the properties of a black body at 280 K. The towel is 3 ft long in the windward direction and 2 ft wide. Help Biff figure out how rapidly (in kg/s) water evaporates from the towel.

11.51 Steam condenses on a 25 cm high, cold vertical wall in a low-pressure condenser unit. The wall is isothermal at 25°C, and the ambient pressure is 8000 Pa. Air has leaked into the unit and has reached a mass fraction of 0.04. The steam–air mixture is at 45°C and is blown downward past the wall at 8 m/s. (a) Estimate the rate of condensation on the wall. (*Hint*: The surface of the condensate film is *not* at the mixture or wall temperature.) (b) Compare the result of part (a) to condensation without air in the steam. What do you conclude?

11.52 During a coating process, a thin film of ethanol is wiped onto a thick flat plate, 0.1 m by 0.1 m. The initial thickness of the liquid film is 0.1 mm, and the initial temperature of both the plate and the film is 303 K. The air above the film is at 303 K and moves at 10 m/s. (a) Assume that the plate is a poor conductor, so that heat transfer from it is negligible. After a short initial transient, the liquid film reaches a steady temperature. Find this temperature and calculate the time required for the film to evaporate. (b) Discuss what happens when the plate is a very good conductor of heat, and estimate the shortest time required for evaporation. Properties of ethanol are as follow: $\log_{10}(p_v \text{ mmHg}) = 9.4432 - 2287.8/(T \text{ K})$; $h_{fg} = 9.3 \times 10^5$ J/kg; liquid density, $\rho_{\text{eth}} = 789$ kg/m^3; Sc = 1.30 for ethanol vapor in air; vapor specific heat capacity, $c_{p_{\text{eth}}} = 1420$ J/kg·K.

11.53 Ice cubes left in a freezer will slowly sublime into the air. Suppose that a tray of ice cubes is left in a freezer with air at −10°C and a relative humidity of 50%. The air in the freezer is circulated by a small fan, creating a heat transfer coefficient from the top of the ice cube tray of 5 W/m^{2}K. If a 20 g ice cube is rectangular and has a top surface area of 8 cm^2, find the temperature of the ice cube and estimate the time required for it to sublime completely. Assume that no heat is transferred through the ice cube tray. For ice, take $h_{sg} = 2.837 \times 10^6$ J/kg, and for water vapor in air, take Sc = 0.63. The vapor pressure of ice is given in Example 11.6.

11.54 Bikram yoga is a strenuous yoga done in a room at 38 to 41°C with relative humidity from 20 to 50%. People doing this yoga will generate body heat $\dot{Q}_b$ of 300 to 600 W, which must be removed to avoid heat stroke. Calculate the rate at which one's body can cool under these conditions and compare it to the rate of heat generation.

The body sweats more as its need to cool increases, but the amount of sweat evaporated on the skin depends on air temperature and humidity. Sweating cannot exceed about 2 L/h, of which only about half evaporates (the rest will simply drip).

Assume that sweating skin has a temperature of 36°C and an emittance of 0.95, and that an average body surface area is $A_b = 1.8$ m^2. Assume that the walls in the yoga studio are at the air temperature. Ignore the thermal effects of clothing. Convection to a person active in still air can be estimated from the following equation [11.31]:

$$\overline{h} = (5.7\ \text{W/m}^2\text{K}) \left(\frac{\dot{Q}_b}{(58.1\ \text{W/m}^2)\, A_b} - 0.8 \right)^{0.39}$$

Note that at high humidity and temperature, some people become overheated and must stop exercising.

11.55 We're off on a drive across West Texas. It's going to be hot today—40°C—but we're unsure of the humidity. We attach a desert water bag to the outside of our pickup truck. It's made of canvas, and it holds a liter and a half of drinking water. When we fill it, we make sure to saturate the canvas inside and out. Water will continue to permeate the fabric, but the weave is tight enough that no water drips from it. Plot the temperature of the water inside the bag as a function of the outdoor humidity. (*Hint:* These bags were once widely used in the Western US, but they never found much use along the US Gulf coast.)

References

[11.1] W. C. Reynolds. *Energy, from Nature to Man.* McGraw-Hill Book Company, New York, 1974.

[11.2] S. Chapman and T. G. Cowling. *The Mathematical Theory of Nonuniform Gases.* Cambridge University Press, New York, 2nd edition, 1964.

[11.3] R. K. Ghai, H. Ertl, and F. A. L. Dullien. Liquid diffusion of non-electrolytes: Part 1. *AIChE J.*, 19(5):881–900, 1973.

[11.4] R. C. Reid, J. M. Prausnitz, and B. E. Poling. *The Properties of Gases and Liquids.* McGraw-Hill Book Company, New York, 4th edition, 1987.

[11.5] P. W. Atkins. *Physical Chemistry.* W. H. Freeman and Co., New York, 3rd edition, 1986.

[11.6] D. R. Poirier and G. H. Geiger. *Transport Phenomena in Materials Processing.* The Minerals, Metals & Materials Society, Warrendale, Pennsylvania, 1994.

[11.7] T. R. Marrero and E. A. Mason. Gaseous diffusion coefficients. *J. Phys. Chem. Ref. Data*, 1:3–118, 1972.

[11.8] J. O. Hirschfelder, C. F. Curtiss, and R. B. Bird. *Molecular Theory of Gases and Liquids.* John Wiley & Sons, Inc., New York, 1964.

[11.9] C. L. Tien and J. H. Lienhard. *Statistical Thermodynamics.* Hemisphere Publishing Corp., Washington, D.C., rev. edition, 1978.

[11.10] R. A. Svehla. Estimated viscosities and thermal conductivities of gases at high temperatures. NASA TR R-132, 1962. (Nat. Tech. Inf. Svcs. N63-22862).

[11.11] C. R. Wilke and C. Y. Lee. Estimation of diffusion coefficients for gases and vapors. *Ind. Engr. Chem.*, 47:1253, 1955.

[11.12] J. O. Hirschfelder, R. B. Bird, and E. L. Spotz. The transport properties for non-polar gases. *J. Chem. Phys.*, 16(10):968–981, 1948.

[11.13] J. Millat, J. H. Dymond, and C. A. Nieto de Castro. *Transport Properties of Fluids: Their Correlation, Prediction and Estimation.* Cambridge University Press, Cambridge, UK, 1996.

[11.14] G. E. Childs and H. J. M. Hanley. Applicability of dilute gas transport property tables to real gases. *Cryogenics*, 8:94–97, 1968.

[11.15] C. Cercignani. *Rarefied Gas Dynamics.* Cambridge University Press, Cambridge, UK, 2000.

[11.16] A. Einstein. *Investigations of the Theory Brownian Movement.* Dover Publications, Inc., New York, 1956. (This book is a collection of Einstein's original papers on the subject, which were published between 1905 and 1908.).

[11.17] W. Sutherland. A dynamical theory of diffusion for non-electrolytes and the molecular mass of albumin. *Phil. Mag., Ser. 6*, 9(54):781–785, 1905.

[11.18] H. Lamb. *Hydrodynamics.* Dover Publications, Inc., New York, 6th edition, 1945.

[11.19] J. C. M. Li and P. Chang. Self-diffusion coefficient and viscosity in liquids. *J. Chem. Phys.*, 23(3):518–520, 1955.

[11.20] S. Glasstone, K. J. Laidler, and H. Eyring. *The Theory of Rate Processes.* McGraw-Hill Book Company, New York, 1941.

[11.21] C. J. King, L. Hsueh, and K-W. Mao. Liquid phase diffusion of nonelectrolytes at high dilution. *J. Chem. Engr. Data*, 10(4):348–350, 1965.

[11.22] C. R. Wilke. A viscosity equation for gas mixtures. *J. Chem. Phys.*, 18(4):517–519, 1950.

[11.23] E. A. Mason and S. C. Saxena. Approximate formula for the thermal conductivity of gas mixtures. *Phys. Fluids*, 1(5):361–369, 1958.

[11.24] J. M. Prausnitz, R. N. Lichtenthaler, and E. G. de Azevedo. *Molecular Thermodynamics of Fluid-Phase Equilibria.* Prentice-Hall, Englewood Cliffs, N.J., 2nd edition, 1986.

[11.25] R. C. Weast, editor. *Handbook of Chemistry and Physics.* Chemical Rubber Co., Cleveland, Ohio, 56th edition, 1976.

[11.26] D. S. Wilkinson. *Mass Transfer in Solids and Fluids.* Cambridge University Press, Cambridge, 2000.

[11.27] W. K. Lewis. The evaporation of a liquid into a gas. *Mech. Engr.*, 44(7):445–446, 1922.

[11.28] T. H. Chilton and A. P. Colburn. Mass transfer (absorption) coefficients: Prediction from data on heat transfer and fluid friction. *Ind. Eng. Chem.*, 26:1183–1187, 1934.

[11.29] A. F. Mills. The use of the diffusion velocity in conservation equations for multicomponent gas mixtures. *Int. J. Heat. Mass Transfer*, 41(13):1955–1968, 1998.

[11.30] A. F. Mills. *Mass Transfer.* Prentice-Hall, Inc., Upper Saddle River, 2001.

[11.31] American Society of Heating, Refrigerating, and Air-Conditioning Engineers, Inc. *2001 ASHRAE Handbook—Fundamentals.* Altanta, 2001.

PART VI

APPENDICES

A. Some thermophysical properties of selected materials

A property is that which, once disjoined
And severed from a thing, undoes its nature:
As Weight to a rock,
Heat to a fire,
Flow to the wide waters,
Touch to corporeal things,
Intangibility to the viewless void.
de Rerum Natura, **Lucretius, 50 BCE**

A *primary source* of thermophysical properties is a document in which the experimentalist who obtained the data reports the details and results of his or her measurements. The term *secondary source* generally refers to a document, based on primary sources, that presents other peoples' data and does so critically. This appendix is neither a primary nor a secondary source, since it has been assembled from a variety of secondary and even tertiary sources.

We attempted to cross-check these data against different sources, and this often led to contradictory values. Such contradictions are usually the result of differences among the experimental samples that are reported or of differences in the accuracy of experiments themselves. We resolved such differences by judging the source, by reducing the number of significant figures to accommodate the conflict, or by omitting the substance from the table. We attempt to report no more decimal places than are accurate. The resulting numbers will suffice for most calculations. However, the reader who needs high accuracy should be sure of the physical constitution of the material and should then consult the latest secondary or primary sources to see if better information has been obtained.

The format of these tables is quite close to that established by R. M. Drake, Jr., in his excellent appendix on thermophysical data [A.1]. However, although we use a few of Drake's numbers directly in Table A.6, many of his other values have been superseded by more recent measurements. One secondary source from which many of the data here were obtained was the Purdue University series *Thermophysical Properties of Matter* [A.2]. The Purdue series is the result of an enormous property-gathering effort carried out under the direction of Y. S. Touloukian and several coworkers. The various volumes in the series are dated since 1970, and addenda were issued throughout the following decade. In more recent years, IUPAC, NIST, and other agencies have been developing critically reviewed, standard reference data for various substances, some of which are contained in [A.3, A.4, A.5, A.6, A.7, A.8, A.9, A.10, A.11]. We have taken many data for fluids from those publications. A third secondary source that we have used is the G. E. *Heat Transfer Data Book* [A.12].

Numbers that did not come directly from [A.1], [A.2], [A.12] or the sources of standard reference data were obtained from a variety of manufacturers' tables, handbooks, and other technical literature. While we have not documented all these diverse sources and the various compromises that were made in quoting them, specific citations are given below for the bulk of the data in these tables.

Table A.1 gives the density, specific heat, thermal conductivity, and thermal diffusivity for various metallic solids. These values were obtained from volumes 1 and 4 of [A.2] or from [A.3] whenever it was possible to find them there. Most thermal conductivity values in the table have been rounded off to two significant figures. The reason is that k is sensitive to very minor variations in physical structure that cannot be detailed fully here. Notice, for example, the significant differences between pure silver and 99.9% pure silver, or between pure aluminum and 99% pure aluminum. Additional information on the characteristics and use of these metals can be found in the ASM *Metals Handbook* [A.13].

The effect of temperature on thermal conductivity is shown for most of the metals in Table A.1. The specific heat capacity is shown only at 20°C. For most materials, the heat capacity is much lower at cryogenic temperatures. For example, c_p for aluminum, iron, molybdenum, and titanium decreases by two orders of magnitude as temperature decreases from 200 K to 20 K. On the other hand, for most of these metals, c_p changes more gradually for temperatures between 300 K and 800 K, varying by tens of percent to a factor of two. At still higher temperatures, some of these metals (iron and titanium) show substantial spikes in c_p. These are associated with solid-to-solid phase transitions.

Table A.2 gives the same properties as Table A.1 (where they are available) but for nonmetallic substances. Volumes 2 and 5 of [A.2] and also

[A.3] provided many of the data here, and they revealed even greater variations in k than the metallic data did. For the various sands reported, k varied by a factor of 500, and for the various graphites by a factor of 50, for example. The sensitivity of k to small variations in the packing of fibrous materials or to the water content of hygroscopic materials forced us to restrict many of the k values to a single significant figure. The effect of water content is illustrated for soils. Additional data for many building materials can be found in [A.14].

The data for polymers come mainly from their manufacturers' data and are substantially less reliable than, say, those given in Table A.1 for metals. The values quoted are mainly those for room temperature. In processing operations, however, most of these materials are taken to temperatures of several hundred degrees Celsius, where they flow more easily. The specific heat capacity may double from room temperature to such temperatures. These polymers are also produced in a variety of modified forms; and in many applications they may be loaded with significant portions of reinforcing fillers (e.g., 10 to 40% by weight glass fiber). The fillers, in particular, can have a significant effect on thermal properties.

Table A.3 gives ρ, c_p, k, α, ν, Pr, and β for several liquids. Data for water are from [A.4] and [A.15]; they are in agreement with IAPWS recommendations through 1998. Data for ammonia are from [A.5, A.16, A.17], data for carbon dioxide are from [A.6, A.7, A.8], and data for oxygen are from [A.9, A.10]. Data for HFC-134a, HCFC-22, and nitrogen are from [A.11] and [A.18]. For these liquids, ρ has uncertainties less than 0.2%, c_p has uncertainties of 1–2%, while μ and k have typical uncertainties of 2–5%. Uncertainties may be higher near the critical point. Thermodynamic data for methanol follow [A.19], while most viscosity data follow [A.20]. Data for mercury follow [A.3] and [A.21]. Sources of olive oil data include [A.20, A.22, A.23]. Data for Freon 12 are from [A.14]. Volumes 3, 6, 10, and 11 of [A.2] gave many of the other values of c_p, k, and $\mu = \rho\nu$, and occasional independently measured values of α. Additional values came from [A.24]. Values of α that disagreed only slightly with $k/\rho c_p$ were allowed to stand. Densities for other substances came from [A.24] and a variety of other sources. A few values of ρ and c_p were taken from [A.25].

Table A.5 provides thermophysical data for saturated vapors. The sources and the uncertainties are as described for gases in the next paragraph.

Table A.6 gives thermophysical properties for gases at 1 atmosphere pressure. The values were drawn from a variety of sources: air data are from [A.26, A.27], except for ρ and c_p above 850 K which came from [A.28]; argon data are from [A.29, A.30, A.31]; ammonia data were taken from [A.5, A.16, A.17]; carbon dioxide properties are from [A.6, A.7, A.8]; carbon monoxide properties are from [A.18]; helium data are from [A.32, A.33, A.34]; nitrogen data came from [A.35]; oxygen data are from [A.9, A.10]; water data were taken from [A.4] and [A.15] (in agreement with IAPWS recommendations through 1998); and a few high-temperature hydrogen data are from [A.24] with the remaining hydrogen data drawn from [A.1]. Uncertainties in these data vary among the gases; typically, ρ has uncertainties of 0.02–0.2%, c_p has uncertainties of 0.2–2%, μ has uncertainties of 0.3–3%, and k has uncertainties of 2–5%. The uncertainties are generally lower in the dilute gas region and higher near the saturation line or the critical point. The values for hydrogen and for low temperature helium have somewhat larger uncertainties.

Table A.7 lists values for some fundamental physical constants, as given in [A.36] and its successors. Table A.8 points out physical data that are listed in other parts of this book.

References

[A.1] E. R. G. Eckert and R. M. Drake, Jr. *Analysis of Heat and Mass Transfer.* McGraw-Hill Book Company, New York, 1972.

[A.2] Y. S. Touloukian. *Thermophysical Properties of Matter.* vols. 1–6, 10, and 11. Purdue University, West Lafayette, IN, 1970 to 1975.

[A.3] C. Y. Ho, R. W. Powell, and P. E. Liley. Thermal conductivity of the elements: A comprehensive review. *J. Phys. Chem. Ref. Data*, 3, 1974. Published in book format as Supplement No. 1 to the cited volume.

[A.4] C.A. Meyer, R. B. McClintock, G. J. Silvestri, and R.C. Spencer. *ASME Steam Tables.* American Society of Mechanical Engineers, New York, NY, 6th edition, 1993.

[A.5] A. Fenghour, W. A. Wakeham, V. Vesovic, J. T. R. Watson, J. Millat, and E. Vogel. The viscosity of ammonia. *J. Phys. Chem. Ref. Data*, 24(5):1649–1667, 1995.

[A.6] A. Fenghour, W. A. Wakeham, and V. Vesovic. The viscosity of carbon dioxide. *J. Phys. Chem. Ref. Data*, 27(1):31–44, 1998.

[A.7] V. Vesovic, W. A. Wakeham, G. A. Olchowy, J. V. Sengers, J. T. R. Watson, and J. Millat. The transport properties of carbon dioxide. *J. Phys. Chem. Ref. Data*, 19(3):763–808, 1990.

[A.8] R. Span and W. Wagner. A new equation of state for carbon dioxide covering the fluid region from the triple-point temperature to 1100 K at pressures up to 800 MPa. *J. Phys. Chem. Ref. Data*, 25 (6):1509–1596, 1996.

[A.9] A. Laesecke, R. Krauss, K. Stephan, and W. Wagner. Transport properties of fluid oxygen. *J. Phys. Chem. Ref. Data*, 19(5):1089–1122, 1990.

[A.10] R. B. Stewart, R. T. Jacobsen, and W. Wagner. Thermodynamic properties of oxygen from the triple point to 300 K with pressures to 80 MPa. *J. Phys. Chem. Ref. Data*, 20(5):917–1021, 1991.

[A.11] R. Tillner-Roth and H. D. Baehr. An international standard formulation of the thermodynamic properties of 1,1,1,2-tetrafluoroethane (HFC-134a) covering temperatures from 170 K to 455 K at pressures up to 70 MPa. *J. Phys. Chem. Ref. Data*, 23: 657–729, 1994.

[A.12] R. H. Norris, F. F. Buckland, N. D. Fitzroy, R. H. Roecker, and D. A. Kaminski, editors. *Heat Transfer Data Book.* General Electric Co., Schenectady, NY, 1977.

[A.13] ASM Handbook Committee. *Metals Handbook.* ASM, International, Materials Park, OH, 10th edition, 1990.

[A.14] American Society of Heating, Refrigerating, and Air-Conditioning Engineers, Inc. *2001 ASHRAE Handbook—Fundamentals.* Altanta, 2001.

[A.15] A. H. Harvey, A. P. Peskin, and S. A. Klein. *NIST/ASME Steam Properties.* National Institute of Standards and Technology, Gaithersburg, MD, March 2000. NIST Standard Reference Database 10, Version 2.2.

[A.16] R. Tufeu, D. Y. Ivanov, Y. Garrabos, and B. Le Neindre. Thermal conductivity of ammonia in a large temperature and pressure range including the critical region. *Ber. Bunsenges. Phys. Chem.*, 88:422–427, 1984.

[A.17] R. Tillner-Roth, F. Harms-Watzenberg, and H. D. Baehr. Eine neue Fundamentalgleichung fuer Ammoniak. *DKV-Tagungsbericht*, 20: 167–181, 1993.

[A.18] E. W. Lemmon, A. P. Peskin, M. O. McLinden, and D. G. Friend. *Thermodynamic and Transport Properties of Pure Fluids — NIST Pure Fluids.* National Institute of Standards and Technology, Gaithersburg, MD, September 2000. NIST Standard Reference Database Number 12, Version 5. Property values are based upon the most accurate standard reference formulations then available.

[A.19] K. M. deReuck and R. J. B. Craven. *Methanol: International Thermodynamic Tables of the Fluid State-12.* Blackwell Scientific Publications, Oxford, 1993. Developed under the sponsorship of the International Union of Pure and Applied Chemistry (IUPAC).

[A.20] D. S. Viswanath and G. Natarajan. *Data Book on the Viscosity of Liquids.* Hemisphere Publishing Corp., New York, 1989.

[A.21] N. B. Vargaftik, Y. K. Vinogradov, and V. S. Yargin. *Handbook of Physical Properties of Liquids and Gases.* Begell House, Inc., New York, 3rd edition, 1996.

[A.22] D. Dadarlat, J. Gibkes, D. Bicanic, and A. Pasca. Photopyroelectric (PPE) measurement of thermal parameters in food products. *J. Food Engr.*, 30:155–162, 1996.

[A.23] H. Abramovic and C. Klofutar. The temperature dependence of dynamic viscosity for some vegetable oils. *Acta Chim. Slov.*, 45(1): 69–77, 1998.

[A.24] N. B. Vargaftik. *Tables on the Thermophysical Properties of Liquids and Gases.* Hemisphere Publishing Corp., Washington, D.C., 2nd edition, 1975.

[A.25] E. W. Lemmon, M. O. McLinden, and D. G. Friend. Thermophysical properties of fluid systems. In W. G. Mallard and P. J. Linstrom, editors, *NIST Chemistry WebBook, NIST Standard Reference Database Number 69.* National Institute of Standards and Technology, Gaithersburg, MD, 2000. http://webbook.nist.gov.

[A.26] K. Kadoya, N. Matsunaga, and A. Nagashima. Viscosity and thermal conductivity of dry air in the gaseous phase. *J. Phys. Chem. Ref. Data*, 14(4):947–970, 1985.

[A.27] R.T. Jacobsen, S.G. Penoncello, S.W. Breyerlein, W.P. Clark, and E.W. Lemmon. A thermodynamic property formulation for air. *Fluid Phase Equilibria*, 79:113–124, 1992.

[A.28] E.W. Lemmon, R.T. Jacobsen, S.G. Penoncello, and D. G. Friend. Thermodynamic properties of air and mixtures of nitrogen, argon, and oxygen from 60 to 2000 K at pressures to 2000 MPa. *J. Phys. Chem. Ref. Data*, 29(3):331–385, 2000.

[A.29] Ch. Tegeler, R. Span, and W. Wagner. A new equation of state for argon covering the fluid region for temperatures from the melting line to 700 K at pressures up to 1000 MPa. *J. Phys. Chem. Ref. Data*, 28(3):779–850, 1999.

[A.30] B. A. Younglove and H. J. M. Hanley. The viscosity and thermal conductivity coefficients of gaseous and liquid argon. *J. Phys. Chem. Ref. Data*, 15(4):1323–1337, 1986.

[A.31] R. A. Perkins, D. G. Friend, H. M. Roder, and C. A. Nieto de Castro. Thermal conductivity surface of argon: A fresh analysis. *Intl. J. Thermophys.*, 12(6):965–984, 1991.

[A.32] R. D. McCarty and V. D. Arp. A new wide range equation of state for helium. *Adv. Cryo. Eng.*, 35:1465–1475, 1990.

[A.33] E. Bich, J. Millat, and E. Vogel. The viscosity and thermal conductivity of pure monatomic gases from their normal boiling point up to 5000 K in the limit of zero density and at 0.101325 MPa. *J. Phys. Chem. Ref. Data*, 19(6):1289–1305, 1990.

[A.34] V. D. Arp, R. D. McCarty, and D. G. Friend. Thermophysical properties of helium-4 from 0.8 to 1500 K with pressures to 2000 MPa. Technical Note 1334, National Institute of Standards and Technology, Boulder, CO, 1998.

[A.35] B. A. Younglove. Thermophysical properties of fluids: Argon, ethylene, parahydrogen, nitrogen, nitrogen trifluoride, and oxygen. *J. Phys. Chem. Ref. Data*, 11, 1982. Published in book format as Supplement No. 1 to the cited volume.

[A.36] P. J. Mohr and B. N. Taylor. CODATA recommended values of the fundamental physical constants: 2002. *Rev. Mod. Phys.*, 77(1):1–107, 2005.

Table A.1 Properties of metallic solids

Metal	Properties at 20°C				Thermal Conductivity, k (W/m·K)									
	ρ (kg/m^3)	c_p (J/kg·K)	k (W/m·K)	α (10^{-5} m^2/s)	−170°C	−100°C	0°C	100°C	200°C	300°C	400°C	600°C	800°C	1000°C
Aluminums														
Pure	2,707	905	237	9.61	302	242	236	240	238	234	228	215	≈95 (liq.)	
99% pure			211		220	206	209							
Duralumin (≈4% Cu, 0.5% Mg)	2,787	883	164	6.66		126	164	182	194					
Alloy 6061-T6	2,700	896	167	6.90			166	172	177	180				
Alloy 7075-T6	2,800	841	130	5.52	76	100	121	137	172	177				
Chromium	7,190	453	90	2.77	158	120	95	88	85	82	77	69	64	62
Cupreous metals														
Pure Copper	8,954	384	398	11.57	483	420	401	391	389	384	378	366	352	336
DS-C15715*	8,900	≈384	365	≈10.7			367	355	345	335	320			
Beryllium copper (2.2% Be)	8,250	420	103	2.97				117						
Brass (30% Zn)	8,522	385	109	3.32	73	89	106	133	143	146	147			
Bronze (25% Sn)§	8,666	343	26	0.86										
Constantan (40% Ni)	8,922	410	22	0.61	17	19	22	26	35					
German silver (15% Ni, 22% Zn)	8,618	394	25	0.73	18	19	24	31	40	45	48			
Gold	19,320	129	318	12.76	327	324	319	313	306	299	293	279	264	249
Ferrous metals														
Pure iron	7,897	447	80	2.26	132	98	84	72	63	56	50	39	30	29.5
Cast iron (4% C)	7,272	420	52	1.70										
Steels ($C \le 1.5\%$)‖														
AISI 1010††	7,830	434	64	1.88		70	65	61	55	50	45	36	29	
0.5% carbon	7,833	465	54	1.47			55	52	48	45	42	35	31	29
1.0% carbon	7,801	473	43	1.17			43	43	42	40	36	33	29	28
1.5% carbon	7,753	486	36	0.97			36	36	36	35	33	31	28	28

* Dispersion-strengthened copper (0.3% Al_2O_3 by weight); strength comparable to stainless steel.
§ Conductivity data for this and other bronzes vary by a factor of about two.
‖ k and α for carbon steels can vary greatly, owing to trace elements.
†† 0.1% C, 0.42% Mn, 0.28% Si; hot-rolled.

Table A.1 Properties of metallic solids...*continued.*

Metal	*Properties at 20°C*				*Thermal Conductivity, k* (W/m·K)									
	ρ (kg/m^3)	c_p (J/kg·K)	k (W/m·K)	α (10^{-5} m^2/s)	−170°C	−100°C	0°C	100°C	200°C	300°C	400°C	600°C	800°C	1000°C
Stainless steels:														
AISI 304	8,000	400	13.8	0.4				15	17^+	19^-	21	25		
AISI 316	8,000	460	13.5	0.37		12		15	16	17^+	19^-	21^+	24	26^+
AISI 347	8,000	420	15	0.44		13		16^+	18^-	19	20	23	26	28
AISI 410	7,700	460	25	0.7				25^+	26	27	27^+	28^+		
AISI 446	7,500	460						18	19^-	19	20	21	22	
Lead	11,373	130	35	2.34	40	37	36	34	33	32	17 (liq.)	20 (liq.)		
Magnesium	1,746	1023	156	8.76	169	160	157	154	152	150	148	145	89 (liq.)	
Mercury†					32	30	7.8 (liq.)							
Molybdenum	10,220	251	138	5.38	175	146	139	135	131	127	123	116	109	103
Nickels														
Pure	8,906	445	91	2.30	156	114	94	83	74	67	64	69	73	78
Alumel§§	8,600	532						30	32	35	38			
Chromel P (10% Cr)	8,730	428						19	21	23	25			
Inconel X-750¶	8,510	442	11.6	0.23	8.8	10.6	11.3	13.0	14.7	16.0	18.3	21.8	25.3	29
Nichrome♭	8,250	448		0.34				13	15	16	18^-			
Nichrome V**	8,410	466	10	0.26				11	13	15	17	20	24	
Platinum	21,450	133	71	2.50	78	73	72	72	72	73	74	77	80	84
Silicon‡	2,330	705.5	153	9.31	856	342	168	112	82	66	54	38	29	25
Silver														
99.99^+% pure	10,524	236	427	17.19	449	431	428	422	417	409	401	386	370	176 (liq.)
99.9% pure	10,524	236	411	16.55		422	405		373	367	364			
Tin†	7,304	228	67	4.17	85	76	68	63	60	32 (liq.)	34 (liq.)	38 (liq.)		
Titanium														
Pure†	4,540	523	22	0.93	31	26	22	21	20	20	19	21	21	22
Ti-6%Al-4%V	4,430	580	7.1	0.28				7.8	8.8	10	12^-			
Tungsten	19,350	133	178	6.92	235	223	182	166	153	141	134	125	122	114
Uranium	18,700	116	28	1.29	22	24	27	29	31	33	36	41	46	
Zinc	7,144	388	121	4.37	124	122	122	117	110	106	100	60 (liq.)		

† Polycrystalline form. §§ 2% Al, 2% Mn, 1% Si ¶ 73% Ni, 15% Cr, 6.75% Fe, 2.5% Ti, 0.85% Nb, 0.8% Al, 0.7% Mn, 0.3% Si. ♭ 23% Fe, 16% Cr ** 20% Cr, 1.4% Si
‡ Single crystal form.

Table A.2 Properties of nonmetallic solids

Material	Temperature Range (°C)	Density ρ (kg/m^3)	Specific Heat c_p (J/kg·K)	Thermal Conductivity k (W/m·K)	Thermal Diffusivity α (m^2/s)
Aluminum oxide (Al_2O_3)					
plasma sprayed coating	20			≈ 4	
HVOF sprayed coating	20			≈ 14	
polycrystalline (98% dense)	0		725	40	
	27	3900	779	36	1.19×10^{-5}
	127		940	26	
	577		1200	10	
	1077		1270	6.1	
	1577		1350	5.6	
single crystal (sapphire)	0		725	52	
	27	3980	779	46	1.48×10^{-5}
	127		940	32	
	577		1180	13	
Asbestos					
Cement board	20	1920		0.6	
Fiber, densely packed	20	1930		0.8	
Fiber, loosely packed	20	980		0.14	
Asphalt	20–25			0.75	
Beef (lean, fresh)	25	1070	3400	0.48	1.35×10^{-7}
Brick					
B & W, K-28 insulating	300			0.3	
	1000			0.4	
Cement	10	720		0.34	
Common	0–1000			0.7	
Chrome	100			1.9	
Facing	20			1.3	
Firebrick, insulating	300	2000	960	0.1	5.4×10^{-8}
	1000			0.2	
Butter	20	920	2520	0.22	9.5×10^{-6}
Carbon					
Diamond (type IIb)	20	≈ 3250	510	1350.0	8.1×10^{-4}
Graphites	20	≈ 1730	≈ 710	k varies with structure	
AGOT graphite					
$\perp$ to extrusion axis	0			141	
	27	1700	800	138	
	500		1600	59.1	
$\parallel$ to extrusion axis	0			230	
	27	1700	800	220	
	500		1600	93.6	

Table A.2...*continued.*

Material	*Temperature Range* (°C)	*Density* ρ (kg/m^3)	*Specific Heat* c_p (J/kg·K)	*Thermal Conductivity* k (W/m·K)	*Thermal Diffusivity* α (m^2/s)
Pyrolitic graphite					
$\perp$ to layer planes	0			10.6	
	27	2200	710	9.5	
	227			5.4	
	1027			1.9	
$\parallel$ to layer planes	0			2230	
	27	2200	710	2000	
	227			1130	
	1027			400	
Cardboard	0–20	790		0.14	
Cement, Portland	34	2010		0.7	
Clay					
Fireclay	500–750			1.0	
Sandy clay	20	1780		0.9	
Coal					
Anthracite	900	$\approx$1500		$\approx$0.2	
Brown coal	900			$\approx$0.1	
Bituminous in situ		$\approx$1300		0.5–0.7	3 to 4×10^{-7}
Concrete					
Limestone gravel	20	1850		0.6	
Sand : cement (3 : 1)	230			0.1	
Sand and gravel	24	2400		1.4–2.9	
	24	2240	900	1.3–2.6	
	24	2080		1.0–1.9	
Corkboard (medium ρ)	30	170		0.04	
Egg white	20		3400	0.56	1.37×10^{-7}
Glass					
Lead	44	3040		1.2	
Pyrex (borosilicate)	60–100	2210	753	1.3	7.8×10^{-7}
Soda-lime	−73		610	0.9	
	20	2480	750	1.1	
	93		866	1.3	
Glass wool	20	64–160		0.04	
Ice	0	917	2100	2.215	1.15×10^{-6}
Ivory	80			0.5	
Kapok	30			0.035	
Lunar surface dust (high vacuum)	250	1500±300	$\approx$600	$\approx$0.0006	$\approx 7 \times 10^{-10}$

Table A.2...*continued.*

Material	*Temperature Range* (°C)	*Density* ρ (kg/m^3)	*Specific Heat* c_p (J/kg·K)	*Thermal Conductivity* k (W/m·K)	*Thermal Diffusivity* α (m^2/s)
Magnesia, 85% (insulation)	38	≈200		0.067	
	93			0.071	
	150			0.074	
	204			0.08	
Magnesium oxide					
polycrystalline (98% dense)	27	3500	900	48	1.5×10^{-5}
single crystal	27	3580	900	60	1.9×10^{-5}
Polymers					
acetyl (POM, Delrin)	−18–100	1420	1470	0.30–0.37	
acrylic (PMMA, Plexiglas)	25	1180		0.17	
acrylonitrile butadiene styrene (ABS)		1060		0.14–0.31	
epoxy, bisphenol A (EP), cast	24–55	1200		≈ 0.22	
epoxy/glass-cloth laminate (G-10, FR4)		1800	≈1600	0.29	$\approx 1.0 \times 10^{-7}$
polyamide (PA)					
nylon 6,6	0–49	1120	1470	0.25	1.5×10^{-7}
nylon 6,12	0–49	1060	1680	0.22	1.2×10^{-7}
polycarbonate (PC, Lexan)	23	1200	1250	0.29	1.9×10^{-7}
polyethylene (PE)					
HDPE		960	2260	0.33	1.5×10^{-7}
LDPE		920	≈2100	0.33	$\approx 1.7 \times 10^{-7}$
polyimide (PI)		1430	1130	0.35	2.2×10^{-7}
polypropylene (PP)		905	1900	0.17–0.20	
polystyrene (PS)		1040	≈ 1350	0.10–0.16	
expanded (EPS)	4–55	13–30		0.035	
polytetrafluoroethylene (PTFE, Teflon)	20	2200	1050	0.25	$\approx 1.1 \times 10^{-7}$
polyvinylchloride (PVC)	25	1600		0.16	
Rock wool	−5	≈130		0.03	
	93			0.05	
Rubber (hard)	0	1200	2010	0.15	6.2×10^{-8}
Silica aerogel	0	140		0.024	
	120	136		0.022	
Silo-cel (diatomaceous earth)	0	320		0.061	
Silicon dioxide					
Fused silica glass	0		703	1.33	
	27	2200	745	1.38	8.4×10^{-7}
	227		988	1.62	

Table A.2...*continued.*

Material	*Temperature Range* (°C)	*Density* ρ (kg/m^3)	*Specific Heat* c_p (J/kg·K)	*Thermal Conductivity* k (W/m·K)	*Thermal Diffusivity* α (m^2/s)
Single crystal (quartz)					
⊥ to c-axis	0		709	6.84	
	27	2640	743	6.21	
	227		989	3.88	
‖ to c-axis	0		709	11.6	
	27	2640	743	10.8	
	227		989	6.00	
Soil (mineral)					
Dry	15	1500	1840	1.	4×10^{-7}
Wet	15	1930		2.	
Soil (k dry to wet, by type)					
Clays				1.1–1.6	
Loams				0.95–2.2	
Sands				0.78–2.2	
Silts				1.6–2.2	
Stone					
Granite (NTS)	20	≈2640	≈820	1.6	$\approx 7.4 \times 10^{-7}$
Limestone (Indiana)	100	2300	≈900	1.1	$\approx 5.3 \times 10^{-7}$
Sandstone (Berea)	25			≈3	
Slate	100			1.5	
Wood (perpendicular to grain)					
Ash	15	740		0.15–0.3	
Balsa	15	100		0.05	
Cedar	15	480		0.11	
Fir	15	600	2720	0.12	7.4×10^{-8}
Mahogany	20	700		0.16	
Oak	20	600	2390	0.1–0.4	
Particle board (medium ρ)	24	800	1300	0.14	1.3×10^{-7}
Pitch pine	20	450		0.14	
Plywood, Douglas fir	24	550	1200	0.12	1.8×10^{-7}
Sawdust (dry)	17	128		0.05	
Sawdust (dry)	17	224		0.07	
Spruce	20	410		0.11	
Wool (sheep)	20	145		0.05	

Table A.3 Thermophysical properties of saturated liquids

Temperature K	°C	ρ (kg/m^3)	c_p (J/kg·K)	k (W/m·K)	α (m^2/s)	ν (m^2/s)	Pr	β (K^{-1})
				Ammonia				
200	−73	728	4227	0.803	2.61×10^{-7}	6.967×10^{-7}	2.67	0.00147
220	−53	706	4342	0.733	2.39	4.912	2.05	0.00165
240	−33	682	4488	0.665	2.19	3.738	1.70	0.00182
260	−13	656	4548	0.600	2.01	3.007	1.50	0.00201
280	7	629	4656	0.539	1.84	2.514	1.37	0.00225
300	27	600	4800	0.480	1.67	2.156	1.29	0.00258
320	47	568	5018	0.425	1.49	1.882	1.26	0.00306
340	67	532	5385	0.372	1.30	1.663	1.28	0.00387
360	87	490	6082	0.319	1.07	1.485	1.39	0.00542
380	107	436	7818	0.267	0.782	1.337	1.71	0.00952
400	127	345	22728	0.216	0.276	1.214	4.40	0.04862
				Carbon dioxide				
220	−53	1166	1962	0.176	7.70×10^{-8}	2.075×10^{-7}	2.70	0.00318
230	−43	1129	1997	0.163	7.24	1.809	2.50	0.00350
240	−33	1089	2051	0.151	6.75	1.588	2.35	0.00392
250	−23	1046	2132	0.139	6.21	1.402	2.26	0.00451
260	−13	999	2255	0.127	5.61	1.245	2.22	0.00538
270	−3	946	2453	0.115	4.92	1.110	2.26	0.00677
280	7	884	2814	0.102	4.10	0.993	2.42	0.00934
290	17	805	3676	0.0895	3.03	0.887	2.93	0.0157
300	27	679	8698	0.0806	1.36	0.782	5.73	0.0570
302	29	634	15787	0.0845	0.844	0.756	8.96	0.119
				Freon 12 (dichlorodifluoromethane)				
180	−93	1661	823	0.113	8.27×10^{-8}	5.27×10^{-7}	6.37	
200	−73	1608	837	0.104	7.73	3.82	4.94	
220	−53	1553	858	0.0959	7.20	2.97	4.12	0.00263
240	−33	1496	882	0.0880	6.67	2.40	3.60	
260	−13	1437	912	0.0806	6.15	1.99	3.24	
280	7	1373	948	0.0734	5.63	1.68	2.99	
300	27	1304	994	0.0665	5.13	1.43	2.80	
320	47	1226	1059	0.0597	4.97	1.32	2.67	
340	67	1134	1170	0.0530	3.99	1.04	2.61	

Table A.3: saturated liquids...*continued*

Temperature K	°C	ρ (kg/m^3)	c_p (J/kg·K)	k (W/m·K)	α (m^2/s)	ν (m^2/s)	Pr	β (K^{-1})
				Glycerin (or glycerol)				
273	0	1276	2200	0.282	1.00×10^{-7}	0.0083	83,000	
293	20	1261	2350	0.285	0.962	0.001120	11,630	0.00048
303	30	1255	2400	0.285	0.946	0.000488	5,161	0.00049
313	40	1249	2460	0.285	0.928	0.000227	2,451	0.00049
323	50	1243	2520	0.285	0.910	0.000114	1,254	0.00050
				20% glycerin, 80% water				
293	20	1047	3860	0.519	1.28×10^{-7}	1.681×10^{-6}	13.1	0.00031
303	30	1043	3860	0.532	1.32	1.294	9.8	0.00036
313	40	1039	3915	0.540	1.33	1.030	7.7	0.00041
323	50	1035	3970	0.553	1.35	0.849	6.3	0.00046
				40% glycerin, 60% water				
293	20	1099	3480	0.448	1.20×10^{-7}	3.385×10^{-6}	28.9	0.00041
303	30	1095	3480	0.452	1.22	2.484	20.4	0.00045
313	40	1090	3570	0.461	1.18	1.900	16.1	0.00048
323	50	1085	3620	0.469	1.19	1.493	12.5	0.00051
				60% glycerin, 40% water				
293	20	1154	3180	0.381	1.04×10^{-7}	9.36×10^{-6}	90.0	0.00048
303	30	1148	3180	0.381	1.04	6.89	66.3	0.00050
313	40	1143	3240	0.385	1.04	4.44	42.7	0.00052
323	50	1137	3300	0.389	1.04	3.31	31.8	0.00053
				80% glycerin, 20% water				
293	20	1209	2730	0.327	0.99×10^{-7}	4.97×10^{-5}	502	0.00051
303	30	1203	2750	0.327	0.99	2.82	282	0.00052
313	40	1197	2800	0.327	0.98	1.74	178	0.00053
323	50	1191	2860	0.331	0.97	1.14	118	0.00053

Helium I and Helium II

- k for He I is about 0.020 W/m·K near the λ-transition ($\approx$ 2.17 K).
- k for He II below the λ-transition is hard to measure. It appears to be about 80,000 W/m·K between 1.4 and 1.75 K and it might go as high as 340,000 W/m·K at 1.92 K. These are the highest conductivities known (cf. copper, silver, and diamond).

Table A.3: saturated liquids...*continued*

Temperature K	°C	ρ (kg/m^3)	c_p (J/kg·K)	k (W/m·K)	α (m^2/s)	ν (m^2/s)	Pr	β (K^{-1})
				HCFC-22 (R22)				
160	−113	1605	1061	0.1504	8.82×10^{-8}	7.10×10^{-7}	8.05	0.00163
180	−93	1553	1061	0.1395	8.46	4.77	5.63	0.00170
200	−73	1499	1064	0.1291	8.09	3.55	4.38	0.00181
220	−53	1444	1076	0.1193	7.67	2.79	3.64	0.00196
240	−33	1386	1100	0.1099	7.21	2.28	3.16	0.00216
260	−13	1324	1136	0.1008	6.69	1.90	2.84	0.00245
280	7	1257	1189	0.0918	6.14	1.61	2.62	0.00286
300	27	1183	1265	0.0828	5.53	1.37	2.48	0.00351
320	47	1097	1390	0.0737	4.83	1.17	2.42	0.00469
340	67	990.1	1665	0.0644	3.91	0.981	2.51	0.00756
360	87	823.4	3001	0.0575	2.33	0.786	3.38	0.02388
				Heavy water (D_2O)				
589	316	740	2034	0.0509	0.978×10^{-7}	1.23×10^{-7}	1.257	
				HFC-134a (R134a)				
180	−93	1564	1187	0.1391	7.49×10^{-8}	9.45×10^{-7}	12.62	0.00170
200	−73	1510	1205	0.1277	7.01	5.74	8.18	0.00180
220	−53	1455	1233	0.1172	6.53	4.03	6.17	0.00193
240	−33	1397	1266	0.1073	6.06	3.05	5.03	0.00211
260	−13	1337	1308	0.0979	5.60	2.41	4.30	0.00236
280	7	1271	1360	0.0890	5.14	1.95	3.80	0.00273
300	27	1199	1432	0.0803	4.67	1.61	3.45	0.00330
320	47	1116	1542	0.0718	4.17	1.34	3.21	0.00433
340	67	1015	1750	0.0631	3.55	1.10	3.11	0.00657
360	87	870.1	2436	0.0541	2.55	0.883	3.46	0.0154
				Lead				
644	371	10,540	159	16.1	1.084×10^{-5}	2.276×10^{-7}	0.024	
755	482	10,442	155	15.6	1.223	1.85	0.017	
811	538	10,348	145	15.3	1.02	1.68	0.017	

Table A.3: saturated liquids...*continued*

Temperature K	°C	ρ (kg/m³)	c_p (J/kg·K)	k (W/m·K)	α (m²/s)	ν (m²/s)	Pr	β (K⁻¹)
				Mercury				
234	−39		141.5	6.97	3.62×10^{-6}	1.5×10^{-7}	0.041	
250	−23		140.5	7.32	3.83	1.4	0.037	
300	27	13,529	139.3	8.34	4.43	1.12	0.0253	0.000181
350	77	13,407	137.7	9.15	4.96	0.974	0.0196	0.000181
400	127	13,286	136.6	9.84	5.42	0.88	0.016	0.000181
500	227	13,048	135.3	11.0	6.23	0.73	0.012	0.000183
600	327	12,809	135.5	12.0	6.91	0.71	0.010	0.000187
700	427	12,567	136.9	12.7	7.38	0.67	0.0091	0.000195
800	527	12,318	139.8	12.8	7.43	0.64	0.0086	0.000207
				Methyl alcohol (methanol)				
260	−13	823	2336	0.2164	1.126×10^{-7}	1.21×10^{-6}	10.8	0.00113
280	7	804	2423	0.2078	1.021	0.883	8.65	0.00119
300	27	785	2534	0.2022	1.016	0.675	6.65	0.00120
320	47	767	2672	0.1965	0.959	0.537	5.60	0.00123
340	67	748	2856	0.1908	0.893	0.442	4.94	0.00135
360	87	729	3036	0.1851	0.836	0.36	4.3	0.00144
380	107	710	3265	0.1794	0.774	0.30	3.9	0.00164
				NaK (eutectic mixture of sodium and potassium)				
366	93	849	946	24.4	3.05×10^{-5}	5.8×10^{-7}	0.019	
672	399	775	879	26.7	3.92	2.67	0.0068	
811	538	743	872	27.7	4.27	2.24	0.0053	
1033	760	690	883			2.12		
				Nitrogen				
70	−203	838.5	2014	0.162	9.58×10^{-8}	2.62×10^{-7}	2.74	0.00513
77	−196	807.7	2040	0.147	8.90	2.02	2.27	0.00564
80	−193	793.9	2055	0.140	8.59	1.83	2.13	0.00591
90	−183	745.0	2140	0.120	7.52	1.38	1.83	0.00711
100	−173	689.4	2318	0.101	6.29	1.09	1.74	0.00927
110	−163	621.5	2743	0.0818	4.80	0.894	1.86	0.0142
120	−153	523.4	4507	0.0633	2.68	0.730	2.72	0.0359

Table A.3: saturated liquids...*continued*

Temperature K	°C	ρ (kg/m^3)	c_p (J/kg·K)	k (W/m·K)	α (m^2/s)	ν (m^2/s)	Pr	β (K^{-1})
				Oils (some approximate viscosities)				
273	0			MS-20		0.0076	100,000	
339	66			California crude (heavy)		0.00008		
289	16			California crude (light)		0.00005		
339	66			California crude (light)		0.000010		
289	16			Light machine oil (ρ = 907)		0.00016		
339	66			Light machine oil (ρ = 907)		0.000013		
289	16			SAE 30		0.00044	≈ 5,000	
339	66			SAE 30		0.00003		
289	16			SAE 30 (Eastern)		0.00011		
339	66			SAE 30 (Eastern)		0.00001		
289	16			Spindle oil (ρ = 885)		0.00005		
339	66			Spindle oil (ρ = 885)		0.000007		
				Olive Oil (1 atm, not saturated)				
283	10	920				14.9 $\times 10^{-5}$		
293	20	913	1800	0.24	1.46×10^{-7}	9.02	620	0.000728
303	30	906				5.76		
313	40	900				3.84		
323	50	893				2.67		
333	60	886				1.91		
343	70	880				1.41		
				Oxygen				
60	−213	1282	1673	0.195	9.09×10^{-8}	4.50×10^{-7}	4.94	0.00343
70	−203	1237	1678	0.181	8.72	2.84	3.26	0.00370
80	−193	1190	1682	0.167	8.33	2.08	2.49	0.00398
90	−183	1142	1699	0.153	7.88	1.63	2.07	0.00436
100	−173	1091	1738	0.139	7.33	1.34	1.83	0.00492
110	−163	1036	1807	0.125	6.67	1.13	1.70	0.00575
120	−153	973.9	1927	0.111	5.89	0.974	1.65	0.00708
130	−143	902.5	2153	0.0960	4.94	0.848	1.72	0.00953
140	−133	813.2	2691	0.0806	3.67	0.741	2.01	0.0155
150	−123	675.5	5464	0.0643	1.74	0.639	3.67	0.0495

Table A.3: saturated liquids...*continued*

Temperature K	°C	ρ (kg/m^3)	c_p (J/kg·K)	k (W/m·K)	α (m^2/s)	ν (m^2/s)	Pr	β (K^{-1})
				Water				
273.16	0.01	999.8	4220	0.5610	1.330×10^{-7}	17.91×10^{-7}	13.47	-6.80×10^{-5}
275	2	999.9	4214	0.5645	1.340	16.82	12.55	-3.55×10^{-5}
280	7	999.9	4201	0.5740	1.366	14.34	10.63	4.36×10^{-5}
285	12	999.5	4193	0.5835	1.392	12.40	8.91	0.000112
290	17	998.8	4187	0.5927	1.417	10.85	7.66	0.000172
295	22	997.8	4183	0.6017	1.442	9.600	6.66	0.000226
300	27	996.5	4181	0.6103	1.465	8.568	5.85	0.000275
305	32	995.0	4180	0.6184	1.487	7.708	5.18	0.000319
310	37	993.3	4179	0.6260	1.508	6.982	4.63	0.000361
320	47	989.3	4181	0.6396	1.546	5.832	3.77	0.000436
340	67	979.5	4189	0.6605	1.610	4.308	2.68	0.000565
360	87	967.4	4202	0.6737	1.657	3.371	2.03	0.000679
373.15	100.0	958.3	4216	0.6791	1.681	2.940	1.75	0.000751
400	127	937.5	4256	0.6836	1.713	2.332	1.36	0.000895
420	147	919.9	4299	0.6825	1.726	2.030	1.18	0.001008
440	167	900.5	4357	0.6780	1.728	1.808	1.05	0.001132
460	187	879.5	4433	0.6702	1.719	1.641	0.955	0.001273
480	207	856.5	4533	0.6590	1.697	1.514	0.892	0.001440
500	227	831.3	4664	0.6439	1.660	1.416	0.853	0.001645
520	247	803.6	4838	0.6246	1.607	1.339	0.833	0.001909
540	267	772.8	5077	0.6001	1.530	1.278	0.835	0.002266
560	287	738.0	5423	0.5701	1.425	1.231	0.864	0.002783
580	307	697.6	5969	0.5346	1.284	1.195	0.931	0.003607
600	327	649.4	6953	0.4953	1.097	1.166	1.06	0.005141
620	347	586.9	9354	0.4541	0.8272	1.146	1.39	0.009092
640	367	481.5	25,940	0.4149	0.3322	1.148	3.46	0.03971
642	369	463.7	34,930	0.4180	0.2581	1.151	4.46	0.05679
644	371	440.7	58,910	0.4357	0.1678	1.156	6.89	0.1030
646	373	403.0	204,600	0.5280	0.06404	1.192	18.6	0.3952
647.0	374	357.3	3,905,000	1.323	0.00948	1.313	138.	7.735

Table A.4 Some latent heats of vaporization, h_{fg} (kJ/kg), with temperatures at triple point, T_{tp} (K), and critical point, T_c (K).

T (K)	Water	Ammonia	CO_2	HCFC-22	HFC-134a	Mercury	Methanol	Nitrogen	Oxygen
60									238.4
70								208.1	230.5
80								195.7	222.3
90								180.5	213.2
100								161.0	202.6
110								134.3	189.7
120				300.4				92.0	173.7
130				294.0					153.1
140				287.9					125.2
150				281.8					79.2
160				275.9					
180				264.3	257.4				
200		1474		252.9	245.7		1310		
220		1424	344.9	241.3	233.9		1269		
230		1397	328.0	235.2	227.8		1258		
240		1369	309.6	228.9	221.5		1247		
250		1339	289.3	222.2	215.0		1235		
260		1307	266.5	215.1	208.2		1222		
270		1273	240.1	207.5	201.0		1209		
273	2501	1263	230.9	205.0	198.6	306.8	1205		
280	2485	1237	208.6	199.4	193.3	306.6	1196		
290	2462	1199	168.1	190.5	185.0	306.2	1181		
300	2438	1158	103.7	180.9	176.1	305.8	1166		
310	2414	1114		170.2	166.3	305.5	1168		
320	2390	1066		158.3	155.5	305.1	1150		
330	2365	1015		144.7	143.3	304.8	1116		
340	2341	957.9		128.7	129.3	304.4	1096		
350	2315	895.2		109.0	112.5	304.1	1078		
360	2290	824.8		81.8	91.0	303.8	1054		
373	2257	717.0				303.3	1022		
400	2183	346.9				302.4	945		
500	1828					299.2	391		
600	1173					295.9			
700						292.3			
T_{tp}	273.16	195.5	216.6	115.7	169.9	234.2	175.5	63.2	54.3
T_c	647.1	405.4	304.3	369.3	374.2		512.5	126.2	154.6

Table A.5 Thermophysical properties of saturated vapors ($p \neq 1$ atm).

T (K)	p (MPa)	ρ (kg/m^3)	c_p (J/kg·K)	k (W/m·K)	μ (kg/m·s)	Pr	β (K^{-1})
			Ammonia				
200	0.008651	0.08908	2076	0.0197	6.952×10^{-6}	0.733	0.005141
220	0.03379	0.3188	2160	0.0201	7.485	0.803	0.004847
240	0.1022	0.8969	2298	0.0210	8.059	0.883	0.004724
260	0.2553	2.115	2503	0.0223	8.656	0.973	0.004781
280	0.5509	4.382	2788	0.0240	9.266	1.08	0.005042
300	1.062	8.251	3177	0.0264	9.894	1.19	0.005560
320	1.873	14.51	3718	0.0296	10.56	1.33	0.006462
340	3.080	24.40	4530	0.0339	11.33	1.51	0.008053
360	4.793	40.19	5955	0.0408	12.35	1.80	0.01121
380	7.140	67.37	9395	0.0546	14.02	2.42	0.01957
400	10.30	131.1	34924	0.114	18.53	5.70	0.08664
			Carbon dioxide				
220	0.5991	15.82	930.3	0.0113	1.114×10^{-5}	0.917	0.006223
230	0.8929	23.27	1005.	0.0122	1.169	0.962	0.006615
240	1.283	33.30	1103.	0.0133	1.227	1.02	0.007223
250	1.785	46.64	1237.	0.0146	1.290	1.09	0.008154
260	2.419	64.42	1430.	0.0163	1.361	1.19	0.009611
270	3.203	88.37	1731.	0.0187	1.447	1.34	0.01203
280	4.161	121.7	2277.	0.0225	1.560	1.58	0.01662
290	5.318	172.0	3614.	0.0298	1.736	2.10	0.02811
300	6.713	268.6	11921.	0.0537	2.131	4.73	0.09949
302	7.027	308.2	23800.	0.0710	2.321	7.78	0.2010
			HCFC-22 (R22)				
160	0.0005236	0.03406	479.2	0.00398	6.69×10^{-6}	0.807	0.006266
180	0.003701	0.2145	507.1	0.00472	7.54	0.810	0.005622
200	0.01667	0.8752	539.1	0.00554	8.39	0.816	0.005185
220	0.05473	2.649	577.8	0.00644	9.23	0.828	0.004947
240	0.1432	6.501	626.2	0.00744	10.1	0.847	0.004919
260	0.3169	13.76	688.0	0.00858	10.9	0.877	0.005131
280	0.6186	26.23	769.8	0.00990	11.8	0.918	0.005661
300	1.097	46.54	885.1	0.0116	12.8	0.977	0.006704
320	1.806	79.19	1071.	0.0140	14.0	1.07	0.008801
340	2.808	133.9	1470.	0.0181	15.7	1.27	0.01402
360	4.184	246.7	3469.	0.0298	19.3	2.24	0.04233

Table A.5: saturated vapors ($p \neq 1$ atm)...*continued.*

T (K)	p (MPa)	ρ (kg/m^3)	c_p (J/kg·K)	k (W/m·K)	μ (kg/m·s)	Pr	β (K^{-1})
			HFC-134a (R134a)				
180	0.001128	0.07702	609.7	0.00389	6.90×10^{-6}	1.08	0.005617
200	0.006313	0.3898	658.6	0.00550	7.75	0.929	0.005150
220	0.02443	1.385	710.9	0.00711	8.59	0.859	0.004870
240	0.07248	3.837	770.5	0.00873	9.40	0.829	0.004796
260	0.1768	8.905	841.8	0.0104	10.2	0.826	0.004959
280	0.3727	18.23	929.6	0.0121	11.0	0.845	0.005421
300	0.7028	34.19	1044.	0.0140	11.9	0.886	0.006335
320	1.217	60.71	1211.	0.0163	12.9	0.961	0.008126
340	1.972	105.7	1524.	0.0197	14.4	1.11	0.01227
360	3.040	193.6	2606.	0.0274	17.0	1.62	0.02863
			Nitrogen				
70	0.03854	1.896	1082.	0.00680	4.88×10^{-6}	0.776	0.01525
77	0.09715	4.437	1121.	0.00747	5.41	0.812	0.01475
80	0.1369	6.089	1145.	0.00778	5.64	0.830	0.01472
90	0.3605	15.08	1266.	0.00902	6.46	0.906	0.01553
100	0.7783	31.96	1503.	0.0109	7.39	1.02	0.01842
110	1.466	62.58	2062.	0.0144	8.58	1.23	0.02646
120	2.511	125.1	4631.	0.0235	10.6	2.09	0.06454
			Oxygen				
60	0.0007258	0.04659	947.5	0.00486	3.89×10^{-6}	0.757	0.01688
70	0.006262	0.3457	978.0	0.00598	4.78	0.781	0.01471
80	0.03012	1.468	974.3	0.00711	5.66	0.776	0.01314
90	0.09935	4.387	970.5	0.00826	6.54	0.769	0.01223
100	0.2540	10.42	1006.	0.00949	7.44	0.789	0.01207
110	0.5434	21.28	1101.	0.0109	8.36	0.847	0.01277
120	1.022	39.31	1276.	0.0126	9.35	0.951	0.01462
130	1.749	68.37	1600.	0.0149	10.5	1.13	0.01868
140	2.788	116.8	2370.	0.0190	12.1	1.51	0.02919
150	4.219	214.9	6625.	0.0318	15.2	3.17	0.08865

Table A.5: saturated vapors ($p \neq 1$ atm)...*continued.*

T (K)	p (MPa)	ρ (kg/m^3)	c_p (J/kg·K)	k (W/m·K)	μ (kg/m·s)	Pr	β (K^{-1})
			Water vapor				
273.16	0.0006177	0.004855	1884	0.01707	0.9216×10^{-5}	1.02	0.003681
275.0	0.0006985	0.005507	1886	0.01717	0.9260	1.02	0.003657
280.0	0.0009918	0.007681	1891	0.01744	0.9382	1.02	0.003596
285.0	0.001389	0.01057	1897	0.01773	0.9509	1.02	0.003538
290.0	0.001920	0.01436	1902	0.01803	0.9641	1.02	0.003481
295.0	0.002621	0.01928	1908	0.01835	0.9778	1.02	0.003428
300.0	0.003537	0.02559	1914	0.01867	0.9920	1.02	0.003376
305.0	0.004719	0.03360	1920	0.01901	1.006	1.02	0.003328
310.0	0.006231	0.04366	1927	0.01937	1.021	1.02	0.003281
320.0	0.01055	0.07166	1942	0.02012	1.052	1.02	0.003195
340.0	0.02719	0.1744	1979	0.02178	1.116	1.01	0.003052
360.0	0.06219	0.3786	2033	0.02369	1.182	1.01	0.002948
373.15	0.1014	0.5982	2080	0.02510	1.227	1.02	0.002902
380.0	0.1289	0.7483	2110	0.02587	1.250	1.02	0.002887
400.0	0.2458	1.369	2218	0.02835	1.319	1.03	0.002874
420.0	0.4373	2.352	2367	0.03113	1.388	1.06	0.002914
440.0	0.7337	3.833	2560	0.03423	1.457	1.09	0.003014
460.0	1.171	5.983	2801	0.03766	1.526	1.13	0.003181
480.0	1.790	9.014	3098	0.04145	1.595	1.19	0.003428
500.0	2.639	13.20	3463	0.04567	1.665	1.26	0.003778
520.0	3.769	18.90	3926	0.05044	1.738	1.35	0.004274
540.0	5.237	26.63	4540	0.05610	1.815	1.47	0.004994
560.0	7.106	37.15	5410	0.06334	1.901	1.62	0.006091
580.0	9.448	51.74	6760	0.07372	2.002	1.84	0.007904
600.0	12.34	72.84	9181	0.09105	2.135	2.15	0.01135
620.0	15.90	106.3	14,940	0.1267	2.337	2.76	0.02000
640.0	20.27	177.1	52,590	0.2500	2.794	5.88	0.07995
642.0	20.76	191.5	737,900	0.2897	2.894	7.37	0.1144
644.0	21.26	211.0	1,253,000	0.3596	3.034	10.6	0.1988
646.0	21.77	243.5	3,852,000	0.5561	3.325	23.0	0.6329
647.0	22.04	286.5	53,340,000	1.573	3.972	135.	9.274

Table A.6 Thermophysical properties of gases at atmospheric pressure (101325 Pa)

T (K)	ρ (kg/m^3)	c_p (J/kg·K)	μ (kg/m·s)	ν (m^2/s)	k (W/m·K)	α (m^2/s)	Pr
				Air			
100	3.605	1039	0.711×10^{-5}	0.197×10^{-5}	0.00941	0.251×10^{-5}	0.784
150	2.368	1012	1.035	0.437	0.01406	0.587	0.745
200	1.769	1007	1.333	0.754	0.01836	1.031	0.731
250	1.412	1006	1.606	1.137	0.02241	1.578	0.721
260	1.358	1006	1.649	1.214	0.02329	1.705	0.712
270	1.308	1006	1.699	1.299	0.02400	1.824	0.712
280	1.261	1006	1.747	1.385	0.02473	1.879	0.711
290	1.217	1006	1.795	1.475	0.02544	2.078	0.710
300	1.177	1007	1.857	1.578	0.02623	2.213	0.713
310	1.139	1007	1.889	1.659	0.02684	2.340	0.709
320	1.103	1008	1.935	1.754	0.02753	2.476	0.708
330	1.070	1008	1.981	1.851	0.02821	2.616	0.708
340	1.038	1009	2.025	1.951	0.02888	2.821	0.707
350	1.008	1009	2.090	2.073	0.02984	2.931	0.707
400	0.8821	1014	2.310	2.619	0.03328	3.721	0.704
450	0.7840	1021	2.517	3.210	0.03656	4.567	0.703
500	0.7056	1030	2.713	3.845	0.03971	5.464	0.704
550	0.6414	1040	2.902	4.524	0.04277	6.412	0.706
600	0.5880	1051	3.082	5.242	0.04573	7.400	0.708
650	0.5427	1063	3.257	6.001	0.04863	8.430	0.712
700	0.5040	1075	3.425	6.796	0.05146	9.498	0.715
750	0.4704	1087	3.588	7.623	0.05425	10.61	0.719
800	0.4410	1099	3.747	8.497	0.05699	11.76	0.723
850	0.4150	1110	3.901	9.400	0.05969	12.96	0.725
900	0.3920	1121	4.052	10.34	0.06237	14.19	0.728
950	0.3716	1131	4.199	11.30	0.06501	15.47	0.731
1000	0.3528	1142	4.343	12.31	0.06763	16.79	0.733
1100	0.3207	1159	4.622	14.41	0.07281	19.59	0.736
1200	0.2940	1175	4.891	16.64	0.07792	22.56	0.738
1300	0.2714	1189	5.151	18.98	0.08297	25.71	0.738
1400	0.2520	1201	5.403	21.44	0.08798	29.05	0.738
1500	0.2352	1211	5.648	23.99	0.09296	32.64	0.735

Table A.6: gases at 1 atm...*continued.*

T(K)	ρ (kg/m^3)	c_p (J/kg·K)	μ (kg/m·s)	ν (m^2/s)	k (W/m·K)	α (m^2/s)	Pr
			Argon				
100	4.982	547.4	0.799×10^{-5}	0.160×10^{-5}	0.00632	0.232×10^{-5}	0.692
150	3.269	527.7	1.20	0.366	0.00939	0.544	0.673
200	2.441	523.7	1.59	0.652	0.01245	0.974	0.669
250	1.950	522.2	1.95	1.00	0.01527	1.50	0.668
300	1.624	521.5	2.29	1.41	0.01787	2.11	0.667
350	1.391	521.2	2.59	1.86	0.02029	2.80	0.666
400	1.217	520.9	2.88	2.37	0.02256	3.56	0.666
450	1.082	520.8	3.16	2.92	0.02470	4.39	0.666
500	0.9735	520.7	3.42	3.51	0.02675	5.28	0.666
550	0.8850	520.6	3.67	4.14	0.02870	6.23	0.665
600	0.8112	520.6	3.91	4.82	0.03057	7.24	0.665
650	0.7488	520.5	4.14	5.52	0.03238	8.31	0.665
700	0.6953	520.5	4.36	6.27	0.03412	9.43	0.665
			Ammonia				
240	0.8888	2296	8.06×10^{-6}	0.907×10^{-5}	0.0210	0.1028×10^{-4}	0.882
273	0.7719	2180	9.19	1.19	0.0229	0.1361	0.874
323	0.6475	2176	11.01	1.70	0.0274	0.1943	0.876
373	0.5589	2238	12.92	2.31	0.0334	0.2671	0.866
423	0.4920	2326	14.87	3.01	0.0407	0.3554	0.850
473	0.4396	2425	16.82	3.82	0.0487	0.4565	0.838
			Carbon dioxide				
220	2.4733	783	11.06×10^{-6}	4.472×10^{-6}	0.01090	0.05628×10^{-4}	0.795
250	2.1657	804	12.57	5.804	0.01295	0.07437	0.780
300	1.7973	853	15.02	8.357	0.01677	0.1094	0.764
350	1.5362	900	17.40	11.33	0.02092	0.1513	0.749
400	1.3424	942	19.70	14.68	0.02515	0.1989	0.738
450	1.1918	980	21.88	18.36	0.02938	0.2516	0.730
500	1.0732	1013	24.02	22.38	0.03354	0.3085	0.725
550	0.9739	1047	26.05	26.75	0.03761	0.3688	0.725
600	0.8938	1076	28.00	31.33	0.04159	0.4325	0.724

Table A.6: gases at 1 atm...*continued.*

T(K)	ρ (kg/m^3)	c_p (J/kg·K)	μ (kg/m·s)	ν (m^2/s)	k (W/m·K)	α (m^2/s)	Pr
			Carbon monoxide				
250	1.367	1042	1.54×10^{-5}	1.13×10^{-5}	0.02306	1.62×10^{-5}	0.697
300	1.138	1040	1.77	1.56	0.02656	2.24	0.694
350	0.975	1040	1.99	2.04	0.02981	2.94	0.693
400	0.853	1039	2.19	2.56	0.03285	3.70	0.692
450	0.758	1039	2.38	3.13	0.03571	4.53	0.691
500	0.682	1040	2.55	3.74	0.03844	5.42	0.691
600	0.5687	1041	2.89	5.08	0.04357	7.36	0.690
700	0.4874	1043	3.20	6.56	0.04838	9.52	0.689
800	0.4265	1046	3.49	8.18	0.05297	11.9	0.689
900	0.3791	1049	3.77	9.94	0.05738	14.4	0.689
1000	0.3412	1052	4.04	11.8	0.06164	17.2	0.689
			Helium				
50	0.9732	5201	0.607×10^{-5}	0.0624×10^{-4}	0.0476	0.0940×10^{-4}	0.663
100	0.4871	5194	0.953	0.196	0.0746	0.295	0.664
150	0.3249	5193	1.25	0.385	0.0976	0.578	0.665
200	0.2437	5193	1.51	0.621	0.118	0.932	0.667
250	0.1950	5193	1.76	0.903	0.138	1.36	0.665
300	0.1625	5193	1.99	1.23	0.156	1.85	0.664
350	0.1393	5193	2.22	1.59	0.174	2.40	0.663
400	0.1219	5193	2.43	1.99	0.190	3.01	0.663
450	0.1084	5193	2.64	2.43	0.207	3.67	0.663
500	0.09753	5193	2.84	2.91	0.222	4.39	0.663
600	0.08128	5193	3.22	3.96	0.252	5.98	0.663
700	0.06967	5193	3.59	5.15	0.281	7.77	0.663
800	0.06096	5193	3.94	6.47	0.309	9.75	0.664
900	0.05419	5193	4.28	7.91	0.335	11.9	0.664
1000	0.04877	5193	4.62	9.46	0.361	14.2	0.665
1100	0.04434	5193	4.95	11.2	0.387	16.8	0.664
1200	0.04065	5193	5.27	13.0	0.412	19.5	0.664
1300	0.03752	5193	5.59	14.9	0.437	22.4	0.664
1400	0.03484	5193	5.90	16.9	0.461	25.5	0.665
1500	0.03252	5193	6.21	19.1	0.485	28.7	0.665

Table A.6: gases at 1 atm...*continued.*

T(K)	ρ (kg/m^3)	c_p (J/kg·K)	μ (kg/m·s)	ν (m^2/s)	k (W/m·K)	α (m^2/s)	Pr
			Hydrogen				
30	0.8472	10840	1.606×10^{-6}	1.805×10^{-6}	0.0228	0.0249×10^{-4}	0.759
50	0.5096	10501	2.516	4.880	0.0362	0.0676	0.721
100	0.2457	11229	4.212	17.14	0.0665	0.2408	0.712
150	0.1637	12602	5.595	34.18	0.0981	0.475	0.718
200	0.1227	13540	6.813	55.53	0.1282	0.772	0.719
250	0.09819	14059	7.919	80.64	0.1561	1.130	0.713
300	0.08185	14314	8.963	109.5	0.182	1.554	0.706
350	0.07016	14436	9.954	141.9	0.206	2.031	0.697
400	0.06135	14491	10.86	177.1	0.228	2.568	0.690
450	0.05462	14499	11.78	215.6	0.251	3.164	0.682
500	0.04918	14507	12.64	257.0	0.272	3.817	0.675
600	0.04085	14537	14.29	349.7	0.315	5.306	0.664
700	0.03492	14574	15.89	455.1	0.351	6.903	0.659
800	0.03060	14675	17.40	569	0.384	8.563	0.664
900	0.02723	14821	18.78	690	0.412	10.21	0.675
1000	0.02451	14968	20.16	822	0.445	12.13	0.678
1100	0.02227	15165	21.46	965	0.488	14.45	0.668
1200	0.02050	15366	22.75	1107	0.528	16.76	0.661
1300	0.01890	15575	24.08	1273	0.568	19.3	0.660
			Nitrogen				
100	3.484	1072	6.80×10^{-6}	1.95×10^{-6}	0.00988	0.0265×10^{-4}	0.738
200	1.711	1043	12.9	7.54	0.0187	0.105	0.720
300	1.138	1041	18.0	15.8	0.0260	0.219	0.721
400	0.8533	1044	22.2	26.0	0.0326	0.366	0.711
500	0.6826	1055	26.1	38.2	0.0388	0.539	0.709
600	0.5688	1074	29.5	51.9	0.0448	0.733	0.708
700	0.4876	1096	32.8	67.3	0.0508	0.951	0.708
800	0.4266	1120	35.8	83.9	0.0567	1.19	0.707
900	0.3792	1143	38.7	102.	0.0624	1.44	0.709
1000	0.3413	1165	41.5	122.	0.0680	1.71	0.711
1100	0.3103	1184	44.2	142.	0.0735	2.00	0.712
1200	0.2844	1201	46.7	164.	0.0788	2.31	0.712
1400	0.2438	1229	51.7	212.	0.0889	2.97	0.715
1600	0.2133	1250	56.3	264.	0.0984	3.69	0.715

Table A.6: gases at 1 atm...*continued.*

T(K)	ρ (kg/m^3)	c_p (J/kg·K)	μ (kg/m·s)	ν (m^2/s)	k (W/m·K)	α (m^2/s)	Pr
			Oxygen				
100	3.995	935.6	0.738×10^{-5}	0.185×10^{-5}	0.00930	0.249×10^{-5}	0.743
150	2.619	919.8	1.13	0.431	0.01415	0.587	0.733
200	1.956	914.6	1.47	0.754	0.01848	1.03	0.730
250	1.562	915.0	1.79	1.145	0.02244	1.57	0.729
300	1.301	919.9	2.07	1.595	0.02615	2.19	0.730
350	1.114	929.1	2.34	2.101	0.02974	2.87	0.731
400	0.9749	941.7	2.59	2.657	0.03324	3.62	0.734
450	0.8665	956.4	2.83	3.261	0.03670	4.43	0.737
500	0.7798	972.2	3.05	3.911	0.04010	5.29	0.739
600	0.6498	1003	3.47	5.340	0.04673	7.17	0.745
700	0.5569	1031	3.86	6.930	0.05309	9.24	0.750
800	0.4873	1054	4.23	8.673	0.05915	11.5	0.753
900	0.4332	1073	4.57	10.56	0.06493	14.0	0.757
1000	0.3899	1089	4.91	12.59	0.07046	16.6	0.759
			Steam (H_2O vapor)				
373.15	0.5976	2080	12.28×10^{-6}	20.55×10^{-6}	0.02509	2.019×10^{-5}	1.018
393.15	0.5652	2021	13.04	23.07	0.02650	2.320	0.994
413.15	0.5365	1994	13.81	25.74	0.02805	2.622	0.982
433.15	0.5108	1980	14.59	28.56	0.02970	2.937	0.973
453.15	0.4875	1976	15.38	31.55	0.03145	3.265	0.966
473.15	0.4665	1976	16.18	34.68	0.03328	3.610	0.961
493.15	0.4472	1980	17.00	38.01	0.03519	3.974	0.956
513.15	0.4295	1986	17.81	41.47	0.03716	4.357	0.952
533.15	0.4131	1994	18.63	45.10	0.03919	4.758	0.948
553.15	0.3980	2003	19.46	48.89	0.04128	5.178	0.944
573.15	0.3840	2013	20.29	52.84	0.04341	5.616	0.941
593.15	0.3709	2023	21.12	56.94	0.04560	6.077	0.937
613.15	0.3587	2034	21.95	61.19	0.04784	6.554	0.934
673.15	0.3266	2070	24.45	74.86	0.05476	8.100	0.924
773.15	0.2842	2134	28.57	100.5	0.06698	11.04	0.910
873.15	0.2516	2203	32.62	129.7	0.07990	14.42	0.899
973.15	0.2257	2273	36.55	161.9	0.09338	18.20	0.890
1073.15	0.2046	2343	40.38	197.4	0.1073	22.38	0.882

Table A.7 Physical constants from 2002 CODATA. The one standard deviation uncertainty of the last two digits is stated in parentheses.

Avogadro's number, N_A	$6.0221415\ (10) \times 10^{26}$	molecules/kmol
Boltzmann's constant, k_B	$1.3806505\ (24) \times 10^{-23}$	J/K
Universal gas constant, R°	8314.472 (15)	J/kmol·K
Speed of light in vacuum, c	299,792,458 (0)	m/s
Standard acceleration of gravity, g	9.80665 (0)	m/s^2
Stefan-Boltzmann constant, σ	$5.670400\ (40) \times 10^{-8}$	W/m^2K^4

Table A.8 Additional physical property data in the text

Page no.	*Data*
28	Electromagnetic wave spectrum
52, 53	Additional thermal conductivities of metals, liquids, and gases
467, 468	Surface tension
530	Total emittances
618	Lennard-Jones constants and molecular weights
620	Collision integrals
624	Molal specific volumes and latent heats

B. Units and conversion factors

A'RTABA, a Persian measure of capacity, principally used as a corn-measure, which contained, according to Herodotus, 1 medimnus and 3 choenices, i.e. 51 choenices = 102 Roman sextarii = 12-3/4 gallons nearly; but, according to Suidas, Hesychius, Polyaenus (Strat. IV.3, 32), and Epiphanius (Pond. 24) only 1 Attic medimnus = 96 sextarii.

A Dictionary of Greek and Roman Antiquities **W. Smith, 1875**

The underlying standard for all our units is ultimately the *Système International d' Unités* (the "S.I. System"). But the need to deal with English units will remain with us for many years to come. We therefore list some conversion factors from English units to S.I. units in this appendix. Many more conversion factors and an extensive discussion of the S.I. system and may be found in [B.1]. The dimensions that are used consistently in the subject of heat transfer are length, mass, force, energy, temperature, and time. We generally avoid using both force and mass dimensions in the same equation, since force is always expressible in dimensions of mass, length, and time, and vice versa. We do not make a practice of eliminating energy in terms of force times length because work and heat must often be accounted separately in heat transfer problems. The text makes occasional reference to electrical units; however, these are conventional and do not have counterparts in the English system, so no electrical units are discussed here.

We present conversion factors in the form of multipliers that may be applied to English units so as to obtain S.I units. For example, the relationship between Btu and J is

$$1\ \text{Btu} = 1055.05\ \text{J}. \tag{B.1}$$

We may rearrange eqn. (B.1) to display a conversion factor whose numerical worth is one:

$$1 = 1055.05\ \frac{\text{J}}{\text{Btu}} \tag{B.2}$$

Table B.1 SI Multiplying Factors

Multiple	*Prefix*	*Symbol*	*Multiple*	*Prefix*	*Symbol*
10^{24}	yotta	Y	10^{-24}	yocto	y
10^{21}	zetta	Z	10^{-21}	zepto	z
10^{18}	exa	E	10^{-18}	atto	a
10^{15}	peta	P	10^{-15}	femto	f
10^{12}	tera	T	10^{-12}	pico	p
10^{9}	giga	G	10^{-9}	nano	n
10^{6}	mega	M	10^{-6}	micro	µ
10^{3}	kilo	k	10^{-3}	milli	m
10^{2}	hecto	h	10^{-2}	centi	c
10^{1}	deka	da	10^{-1}	deci	d

Thus, if we were to multiply a given number of Btus by this factor, we would obtain the corresponding number of joules. The latter form is quite useful in changing units within more complex equations. For example, the conversion factor

$$1 = 0.0001663 \, \frac{\text{m/s}}{\text{furlong/fortnight}}$$

could be multiplied by a velocity in furlongs per fortnight[1], on just one side of an equation, to convert it to meters per second.

Note that the S.I. units may have prefixes placed in front of them to indicate multiplication by various powers of ten. For example, the prefix "k" denotes multiplication by 1000 (e.g., 1 km = 1000 m). The complete set of S.I. prefixes is given in Table B.1.

Table B.2 provides multipliers for a selection of common units. As an example of their use, consider the first entry in the table which shows a conversion factor (in column "multiply no.") of 16.018 for changing lbm/ft^3 to kg/m^3. If we consider a liquid with a density of 62.40 lbm/ft^3, we may convert to density in kg/m^3 as follows:

$$62.40 \text{ lb/ft}^3 \times \left(16.018 \, \frac{\text{kg/m}^3}{\text{lbm/ft}^3}\right) = 999.5 \text{ kg/m}^3. \tag{B.3}$$

[1]Shortly after World War II, a group of staff physicists at Boeing Airplane Co. answered angry demands by engineers that calculations be presented in English units with a report translated entirely into such dimensions as these.

Table B.2 Selected Conversion Factors

Dimension	*To get SI*	=	*multiply no.*	×	*other unit*
Density	kg/m^3	=	16.018	×	lbm/ft^3
	kg/m^3	=	10^3	×	g/cm^3
Diffusivity (α, ν, $\mathcal{D}$)	m^2/s	=	0.092903	×	ft^2/s
	m^2/s	=	10^{-6}	×	centistokes
Energy	J	=	1055.05	×	Btu[a]
	J	=	4.1868	×	cal[b]
	J	=	10^{-7}	×	erg
	J	=	3.6×10^6	×	kW·hr
Energy per unit mass	J/kg	=	2326.0	×	Btu/lbm
	J/kg	=	4186.8	×	cal/g
Flow rate	m^3/s	=	6.3090×10^{-5}	×	gal/min (gpm)
	m^3/s	=	4.7195×10^{-4}	×	ft^3/min (cfm)
	m^3/s	=	10^{-3}	×	L/s
Force	N	=	10^{-5}	×	dyne
	N	=	4.4482	×	lbf
Heat flux	W/m^2	=	3.154	×	$Btu/hr{\cdot}ft^2$
	W/m^2	=	10^4	×	W/cm^2
Heat transfer coefficient	W/m^2K	=	5.6786	×	$Btu/hr{\cdot}ft^2{\circ}F$
Length	m	=	10^{-10}	×	ångströms (Å)
	m	=	0.0254	×	inches
	m	=	0.3048	×	feet
	m	=	201.168	×	furlongs
	m	=	1609.34	×	miles
	m	=	3.0857×10^{16}	×	parsecs
Mass	kg	=	0.45359	×	lbm
	kg	=	14.594	×	slug

Table B.2...*continued.*

Dimension	*To get SI*	=	*multiply no.*	×	*other unit*
Power	W	=	0.022597	×	ft·lbf/min
	W	=	0.29307	×	Btu/hr
	W	=	745.700	×	hp
Pressure	Pa	=	133.32	×	mmHg (@0°C)
	Pa	=	248.84	×	inH_2O (@60°F)
	Pa	=	3376.9	×	inHg (@60°F)
	Pa	=	6894.8	×	psi
	Pa	=	10^5	×	bar
	Pa	=	101325	×	atm
Specific heat capacity	J/kg·K	=	4186.8	×	Btu/lbm·°F
	J/kg·K	=	4186.8	×	cal/g·°C
Temperature	K	=	5/9	×	°R
	K	=	°C + 273.15		
	K	=	(°F + 459.67)/1.8		
Thermal conductivity	W/m·K	=	0.14413	×	Btu·in/hr·ft^2°F
	W/m·K	=	1.7307	×	Btu/hr·ft°F
	W/m·K	=	418.68	×	cal/s·cm°C
Viscosity (dynamic)	Pa·s	=	10^{-3}	×	centipoise
	Pa·s	=	1.4881	×	lbm/ft·s
	Pa·s	=	47.880	×	lbf·s/ft^2
Volume	m^3	=	10^{-3}	×	L
	m^3	=	3.7854×10^{-3}	×	gallons
	m^3	=	0.028317	×	ft^3

[a] The British thermal unit, originally defined as the heat that raises 1 lbm of water 1°F, has several values that depend mainly on the initial temperature of the water warmed. The above is the International Table (*i.e.*, steam table) Btu. A "mean" Btu of 1055.87 J is also common. Related quantities are: 1 therm = 10^5 Btu; 1 quad = 10^{15} Btu ≈ 1 EJ; 1 ton of refrigeration = 12,000 Btu/hr absorbed.

[b] The calorie represents the heat that raises 1 g of water 1°C. Like the Btu, the calorie has several values that depend on the initial temperature of the water warmed. The above is the International Table calorie, or IT calorie. A "thermochemical" calorie of 4.184 J has also been in common use. The dietitian's "Calorie" is actually 1 kilocalorie.

References

[B.1] B. N. Taylor. *Guide to the Use of the International System of Units (SI).* National Institute of Standards and Technology, Gaithersburg, MD, 1995. NIST Special Publication 811. May be downloaded from NIST's web pages.

C. Nomenclature

Count every day one letter of my name;
Before you reach the end, dear,
Will come to lead you to my palace halls
A guide whom I shall send, dear.

Abhijñana Sakuntala, Kalidasa, 5th C

Arbitrary constants, coefficients, and functions introduced in context are not included here; neither are most geometrical dimensions. Dimensions of symbols are given in S.I. units in parenthesis after the definition. Symbols without dimensions are noted with (-), where it is not obvious.

A, A_c, A_h, A_j — area (m^2) or function defined in eqn. (9.41); cross-sectional area (m^2); area of heater (m^2); jet cross-sectional area (m^2)

B — radiosity (W/m^2) or the function defined in Fig. 8.14.

$B_{m,i}$ — mass transfer driving force, eqn. (11.97) (-)

b.c. — boundary condition

b.l. — boundary layer

C, C_c, C_h — heat capacity rate (W/K) or electrical capacitance (s/ohm) or correction factor in Fig. 7.17; heat capacity rate for hot and cold fluids (W/K)

$\overline{C}$ — average thermal molecular speed

C_f — skin friction coefficient (-) [eqn. (6.33)]

C_{sf} — surface roughness factor (-). (see Table 9.2)

c, c_p, c_v — specific heat, specific heat at constant pressure, specific heat at constant volume (J/kg·K)

c — molar concentration of a mixture ($kmol/m^3$) or damping coefficient (N·s/m)

c — partial molar concentration of a species i ($kmol/m^3$)

c_o — speed of light, 2.99792458×10^8 m/s

D or d — diameter (m)

D_h — hydraulic diameter, $4A_c/P$ (m)

$\mathcal{D}_{12}, \mathcal{D}_{im}$ — binary diffusion coefficient for species 1 diffusing in species 2, effective binary diffusion coefficient for species i diffusing in mixture m (m^2/s)

E, E_0 voltage, initial voltage (V)

$e, e_b, e_\lambda, e_{\lambda b}$ emissive power (W/m^2) or energy equivalent of mass (J); black body emissive power(W/m^2); monochromatic emissive power (W/m^2·µm); black body monochromatic emissive power (W/m^2·µm)

F LMTD correction factor (-) or fluid parameter from Table 9.4 (-)

$F(t)$ time-dependent driving force (N)

$F_{1\text{-}2}$ radiation view factor for surface (1) seeing surface (2)

$\mathcal{F}_{1\text{-}2}$ gray-body transfer factor from surface (1) to surface (2)

f Darcy-Weisbach friction factor(-) [eqn. (3.25)] or Blasius function of η (-)

f_o orientation factor for eqns. (9.50)

f_v frequency of vibration (Hz)

G superficial mass flux $= \dot{m}/A_{\text{pipe}}$

g, g_{eff} gravitational body force (m/s^2), effective g defined in eqn. (8.62) (m/s^2)

$g_{m,i}$ mass transfer coefficient for species i, (kg/m^2·s)

H height of ribbon (m), head (m), irradiance (W/m^2), or Henry's law constant (N/m^2)

$h, \overline{h}, h_{\text{rad}}$ local heat transfer coefficient (W/m^{2}K), or enthalpy (J/kg), or height (m), or Planck's constant (6.6261×10^{-34} J·s); average heat transfer coefficient (W/m^{2}K); radiation heat transfer coefficient (W/m^{2}K)

$\hat{h}$ specific enthalpy (J/kg)

h_c interfacial conductance (W/m^{2}K)

h_{fg}, h_{sf}, h_{sg} latent heat of vaporization, latent heat of fusion, latent heat of sublimation (J/kg)

h'_{fg} latent heat corrected for sensible heat

$\hat{h}_i$ specific enthalpy of species i (J/kg)

h^* heat transfer coefficient at zero mass transfer, in Chpt. 11 only (W/m^{2}K)

I electric current (amperes) or number of isothermal increments (-)

$\vec{i}, \vec{j}, \vec{k}$ unit vectors in the x, y, z directions

i intensity of radiation (W/m^2·steradian)

$I_0(x)$ modified Bessel function of the first kind of order zero

i.c. initial condition

$J_0(x), J_1(x)$ Bessel function of the first kind of order zero, of order one

$\vec{j}_i$ diffusional mass flux of species i (kg/m^2·s)

$\vec{J}$ electric current density (amperes/m^2)

$\vec{J}_i^*$ diffusional mole flux of species i (kmol/m^2·s)

k thermal conductivity (W/m·K)

k_B Boltzmann's constant, $1.3806503 \times 10^{-23}$ J/K

k_T thermal diffusion ratio (-)

L any characteristic length (m)

L_0 — geometrical mean beam length (m)

LMTD — logarithmic mean temperature difference

ℓ — an axial length or length into the paper or mean free molecular path (m or Å) or mixing length (m)

M — molecular weight (of mixture if not subscripted) (kg/kmol) or merit number of heat pipe working fluid, $h_{fg}\sigma/\nu_f$.

m — fin parameter, $\sqrt{\overline{h}P/kA}$ ($\mathrm{m^{-1}}$)

m_0 — rest mass (kg)

$\dot{m}$ — mass flow rate (kg/s) or mass flux per unit width (kg/m · s)

m_i — mass fraction of species i (-)

$\dot{m}''$ — scalar mass flux of a mixture ($\mathrm{kg/m^2{\cdot}s}$)

N — number of adiabatic channels (-) or number of rows in a rod bundle (-)

$\vec{N}$ — mole flux (of mixture if not subscripted) ($\mathrm{kmol/m^2{\cdot}s}$)

N_A — Avogadro's number, $6.02214199 \times 10^{26}$ molecules/kmol

$\mathcal{N}$ — number density (of mixture if not subscripted) ($\mathrm{molecules/m^3}$)

$\vec{n}$ — mass flux (of mixture if not subscripted) ($\mathrm{kg/m^2{\cdot}s}$), unit normal vector

n — summation index (-) or nucleation site density ($\mathrm{sites/m^2}$)

P — factor (-) defined in eqn. (3.14) or pitch of a tube bundle (m) or perimeter (m)

p — pressure ($\mathrm{N/m^2}$)

p_i — partial pressure of species i ($\mathrm{N/m^2}$)

Q — rate of heat transfer (W)

$q, \vec{q}$ — heat flux ($\mathrm{W/m^2}$)

q_b, q_{FC}, q_i — defined in context of eqn. (9.37)

$q_{\max}$ or q_{burnout} — peak boiling heat flux ($\mathrm{W/m^2}$)

$q_{\min}$ — minimum boiling heat flux ($\mathrm{W/m^2}$)

$\dot{q}$ — volumetric heat generation ($\mathrm{W/m^3}$)

R — factor defined in eqn. (3.14) (-), radius (m), electrical resistance (ohm), or region ($\mathrm{m^3}$)

R — ideal gas constant per unit mass, R°/M (for mixture if not subscripted) (J/kg·K)

R° — ideal gas constant, 8314.472 (J/kmol·K)

R_t, R_f — thermal resistance (K/W or $\mathrm{m^2{\cdot}K/W}$), fouling resistance ($\mathrm{m^2{\cdot}K/W}$)

$r, \vec{r}$ — radial coordinate (m), position vector (m)

r_{crit} — critical radius of insulation (m)

$\dot{r}_i$ — volume rate of creation of mass of species i ($\mathrm{kg/m^3{\cdot}s}$)

S — entropy (J/K), or surface ($\mathrm{m^2}$), or shape factor (N/I)

S_L, S_T — rod bundle spacings (m). See Fig. 7.14

s — specific entropy (J/kg·K)

T, T_c, T_f, T_m — temperature (°C, K); thermodynamic critical temperature (K); film temperature (°C, K); mean

	temperature for radiation exchange (K)
T	time constant, $\rho cV/\overline{h}A$ (s)
T	a long time over which properties are averaged (s)
t	time (s)
U	overall heat transfer coefficient (W/m²K); internal thermodynamic energy (J); characteristic velocity (m/s)
$u, \vec{u}$	local x-direction fluid velocity (m/s) or specific energy (J/kg); vectorial velocity (m/s)
$u_{av}, \overline{u}, u_c, u_g$	average velocity over an area (m/s); local time-averaged velocity (m/s); characteristic velocity (m/s) [eqn. (8.18)]; Helmholtz-unstable velocity (m/s)
$\hat{u}$	specific internal energy (J/kg)
V	volume (m³); voltage (V)
V_m	molal specific volume (m³/kmol)
v	local y-direction fluid velocity (m/s)
$\vec{v}$	mass-average velocity, in Chapter 11 only (m/s)
$\vec{v}_i$	average velocity of species i (m/s)
$\vec{v}^*$	mole average velocity (m/s)
Wk	rate of doing work (W)
w	z-direction velocity (m/s) or width (m)
x, y, z	Cartesian coordinates (m); x is also used to denote any unknown quantity
x_i	mole fraction of species i (-)
x	quality of two-phase flow

Greek symbols

α	thermal diffusivity, $k/\rho c_p$ (m²/s), or helix angle (rad.)
α, α_g	absorptance (-); gaseous absorptance (-)
β	coefficient of thermal expansion (K^{-1}), or relaxation factor (-), or $h\sqrt{\alpha t}/k$, or coefficient of viscous friction (-)
β_λ	monochromatic extinction coefficient (m²/kg)
Γ, Γ_c	$\dot{g}L^2/k\Delta T$, mass flow rate in film (kg/m·s)
γ	c_p/c_v; electrical conductivity (V/ohm·m²)
γ_λ	monochromatic scattering coefficient (m²/kg)
ΔE	Activation energy of reaction (J/kmol)
Δp	pressure drop in any system (N/m²)
ΔT	any temperature difference; various values are defined in context.
$\delta, \delta_c, \delta_t, \delta_t'$	flow boundary layer thickness (m) or condensate film thickness (m); concentration boundary layer thickness (m); thermal boundary layer thickness (m); h/k (m).
ε	emittance (-); heat exchanger effectiveness (-); roughness (m)
$\varepsilon_A, \varepsilon_{AB}$	potential well depth for molecules of A, for collisions of A and B (J)
ε_f	fin effectiveness (-)
ε_g	gaseous emittance (-)
$\varepsilon_m, \varepsilon_h$	eddy diffusivity of mass (-), of heat (-)

η independent variable of Blasius function, $y\sqrt{u_\infty/\nu x}$ (-)

η_f fin efficiency

Θ a ratio of two temperature differences (-)

θ $(T - T_\infty)$ (K) or angular coordinate (rad)

ζ $x/\sqrt{\alpha t}$

κ_λ monochromatic absorption coefficient (m^2/kg)

$\lambda, \lambda_c, \lambda_H$ wavelength (m) or eigenvalue (m^{-1}); critical Taylor wavelength (m); Helmholtz-unstable wavelength (m)

$\lambda_d, \lambda_{d_1}, \lambda_{d_2}$ most dangerous Taylor-unstable wavelength (m); subscripts denote one- and two-dimensional values

$\hat{\lambda}$ dimensionless eigenvalue (-)

μ dynamic viscosity (kg/m·s) or chemical potential (J/mol)

ν kinematic viscosity, μ/ρ (m^2/s)

ξ x/L or $x\sqrt{\omega/2\alpha}$; also $(x/L + 1)$ or x/L (-)

ρ mass density (kg/m^3) or reflectance (-)

ρ_i partial density of ith species (kg/m^3)

σ surface tension (N/m) or Stefan-Boltzmann constant 5.670400×10^{-8} (W/m^2·K^4)

σ_A, σ_{AB} collision diameter of molecules of A, for collisions of A with B (Å)

τ transmittance (-) or dimensionless time $(T/\boldsymbol{T})$ or shear stress (N/m^2) or length of travel in b.l. (m)

τ_w, τ_{yx} shear stress on a wall (N/m^2), shear stress in the x-direction on the plane normal to the y-direction (N/m^2)

τ_δ shear stress exerted by liquid film (N/m^2)

Φ $\Delta T/(\dot{q}L^2/k)$ or fraction of total heat removed (see Fig. 5.10) (-)

ϕ angular coordinate (rad), or δ_t/δ (-), or factor defined in context of eqn. (6.120c) (-)

ϕ_{ij} weighting functions for mixture viscosity or thermal conductivity (-)

χ $d\Theta/d\zeta$

ψ $\omega L^2/\alpha$

Ω ωt

$\Omega_D, \Omega_k, \Omega_\mu$ collision integral for diffusivity, thermal conductivity, or dynamic viscosity (-)

ω frequency of a wave or of rotation (rad/s) or solid angle (sr)

General subscripts

av, avg denoting bulk or average values

b, body denoting any body

b denoting a black body

c denoting the critical state

cbd denoting a convective boiling dominated value

D denoting a value based on D

e, e_t denoting a dynamical entry length or a free stream variable; denoting a thermal entry length

f, g denoting saturated liquid and saturated vapor states

fb denoting a value for flow boiling

i denoting initial or inside value, or a value that changes with the index i, or values for the ith species in a mixture

in denoting a value at the inlet

L denoting a value based on L or at the left-hand side

lo denoting a value computed as if all fluid were in liquid state

m denoting values for mixtures

max, min denoting maximum or minimum values

n denoting a value that changes with the index n

nbd denoting a nucleate boiling dominated value

o denoting outside, in most cases

out denoting a value at the outlet

R denoting a value based on R or at the right-hand side

s denoting values above an interface

sfc denoting conditions at a surface

sup, sat, sub denoting superheated, saturated, or subcooled states

u denoting values below an interface

w denoting conditions at a wall

x denoting a local value at a given value of x

∞ denoting conditions in a fluid far from a surface

λ denoting radiative properties evaluated at a particular wavelength

General superscript

$*$ denoting values for zero net mass transfer (in Chpt. 11 only)

Dimensionless parameters

Bi Biot number, hL/k_{body}

Bo Bond number, $L^2 g(\rho_f - \rho_g)/\sigma$, or Boiling number, q_w/Gh_{fg}

Co Convection number, $[(1-x)/x]^{0.8}(\rho_g/\rho_f)^{0.5}$

Da Damkohler number, $\rho A' \exp(-\Delta E/R^\circ T)/g_m^*$

Ec Eckert number, $u^2/(c_p \Delta T)$

Fo Fourier number, $\alpha t/L^2$

Fr Froude number, $U^2/(gL)$

Gr_L Grashof number, $g\beta\Delta T L^3/\nu^2$ (for heat transfer), or $g(\Delta\rho/\rho)L^3/\nu^2$

Gz Graetz number, $\text{RePr}D/x$

H' L' based on $L \equiv H$

Ja Jakob number, $c_p\Delta T/h_{fg}$

j Colburn j-factor, $\text{St}\,\text{Pr}^{2/3}$

Ku Kutateladze number, $(\pi/24)(q_{\max}/q_{\max_z})$

L' $L\sqrt{g(\rho_f - \rho_g)/\sigma}$

Le Lewis number, $\text{Sc}/\text{Pr} = \alpha/\mathcal{D}_{im}$

Ma Mach number, u/(sound speed)

NTU number of transfer units, $UA/C_{\min}$

$\text{Nu}_x, \overline{\text{Nu}}_L$ local Nusselt number, hx/k_{fluid}; average Nusselt number, $\overline{h}L/k_{\text{fluid}}$

$Nu_{m,x}$, $\overline{Nu}_{m,L}$ — local Nusselt number for mass transfer (or Sherwood number) $g^*_{m,i}x/(\rho\mathcal{D}_{im})$; average Nusselt number for mass transfer, $\overline{g_{m,i}}^*L/(\rho\mathcal{D}_{im})$

Pe_L — Péclét number, $UL/\alpha = \mathrm{Re\,Pr}$

Pr, Pr_t — Prandtl number, $\mu c_p/k = \nu/\alpha$; turbulent Prandtl number, $\varepsilon_m/\varepsilon_h$

Ra_L, — Rayleigh number, $\mathrm{Gr\,Pr} = g\beta\Delta TL^3/(\nu\alpha)$ for heat transfer; $g(\Delta\rho/\rho)L^3/(\nu\mathcal{D}_{12})$ for mass transfer

Ra^*_L — $Ra_L Nu_L = g\beta q_w L^4/(k\nu\alpha)$

Re_L, Re_c, Re_f, Re_{lo} — Reynolds number, UL/ν; condensation Re equal to Γ_c/μ; Re for liquid; liquid-only Reynolds number, GD/μ_f

Sc — Schmidt number for species i in mixture m, $\nu/\mathcal{D}_{im}$

Sh_L — Sherwood number, $g^*_{m,i}L/(\rho\mathcal{D}_{im})$

St — Stanton number, $\mathrm{Nu}/(\mathrm{Re\,Pr}) = h/(\rho c_p u)$

Str — Strouhal number, $f_v D/u_\infty$

We_L — Weber number, $\rho_g U_\infty^2 L/\sigma$

Π — any dimensionless group

Citation Index

D

E

F

G

H

I

J

K

L

M

N

O

P

R

S

T

U

V

W

Y

Z

Ž

Subject Index

G

H

I

N

T

U

V

W

Y